Chemistry of Nucleosides and Nucleotides

Volume 2

Chemistry of Nucleosides and Nucleotides

Volume 2

Edited by

Leroy B. Townsend
University of Michigan
Ann Arbor, Michigan

Plenum Press • *New York and London*

Library of Congress Cataloging in Publication Data

(Revised for vol. 2)
Chemistry of nucleosides and nucleotides.

Includes bibliographical references and indexes.
1. Nucleotides. 2. Nucleosides. I. Townsend, Leroy, B.
QD436.N85C47   1988   547.7'9   88-22359
ISBN 0-306-43646-9

ISBN 0-306-43646-9

© 1991 Plenum Press, New York
A Division of Plenum Publishing Corporation
233 Spring Street, New York, N.Y. 10013

Printed in the United States of America

Contributors

Richard A. Glennon, Department of Medicinal Chemistry, School of Pharmacy, Medical College of Virginia, Virginia Commonwealth University, Richmond, Virginia 23298-0540

Alexander Hampton, Institute for Cancer Research, Fox Chase Cancer Center, Philadelphia, Pennsylvania 19111

D. W. Hutchinson, Department of Chemistry, Warwick University, Coventry CV4 7AL, United Kingdom

Leon M. Lerner, Department of Biochemistry, State University of New York Health Sciences Center, Brooklyn, New York 11203

Ganapathi R. Revankar, Triplex Pharmaceutical Corporation, The Woodlands, Texas 77380

Roland K. Robins, ICN Nucleic Acid Research Institute, Costa Mesa, California 92626

Shanaz Tejani-Butt, Department of Medicinal Chemistry, School of Pharmacy, Medical College of Virginia, Virginia Commonwealth University, Richmond, Virginia 23298-0540

Contents

Chapter 3

**The Synthesis, Reactions, and Properties of Nucleoside Mono-, Di-,
Tri-, and Tetraphosphates and Nucleotides with Changes in the
Phosphoryl Residue**

D. W. Hutchinson

Chapter 4

**The Synthesis and Chemistry of Heterocyclic Analogues
of Purine Nucleosides and Nucleotides**

Ganapathi R. Revankar and Roland K. Robins

Chapter 5

**Substrate Binding of Adenine Nucleosides and Nucleotides
to Certain Enzymes**

Alexander Hampton

Chapter 1

Mesoionic Nucleosides and Heterobases

Richard A. Glennon and Shanaz Tejani-Butt

1. Introduction

1.1. Modified Nucleosides

Before 1950, each of the known classes of nucleic acids was assumed to consist of only four basic nucleoside monomers. This concept has proven to be naive since many additional components have been discovered in both ribonucleic acids (RNA) and deoxyribonucleic acids (DNA). Modified nucleosides have been found in almost all major classes of nucleic acids, including messenger RNA and ribosomal RNA. The greatest number of structural variations per molecule have been found to occur in transfer RNA (tRNA).[1,2] The nucleic acid molecule can be visualized as a polymer constructed of several major nucleoside monomers: adenosine, guanosine, cytidine, and uridine for RNA, and deoxyadenosine, deoxyguanosine, deoxycytidine, and thymidine for DNA. Modifications in a specific manner at a specific location in the basic structure may be achieved by the addition or substitution of specific functional groups. Such types of structural modifications give rise to a class of nucleic acid components that have been referred to as minor, odd, or rare nucleosides.[3] The presence of modified nucleosides in nucleic acids was realized in 1948 when Hotchkiss[4] detected the first known modified component of a nucleic acid, 5-methyl-2′-deoxycytosine, in a sample of calf-thymus DNA. Since that time, at least 5 modified nucleosides in DNA and more than 30 modified nucleosides in RNA have been identified.[1,2] Much of the work on the identification and isolation of modified nucleosides has centered around tRNA since the tRNA molecule is relatively small and individual molecular species can be more readily isolated and identified. Transfer RNA is perhaps the most versatile of the known classes of nucleic acids in terms of variety and complexity of the chemical reactions in which it participates. Each

Richard A. Glennon and Shanaz Tejani-Butt • Department of Medicinal Chemistry, School of Pharmacy, Medical College of Virginia, Virginia Commonwealth University, Richmond, Virginia 23298-0540.

tRNA molecule carries a specificity for a particular amino acid, recognizes the corresponding aminoacyl synthetase, and maintains its reading fidelity for the codon. Thus, the role of modified nucleosides may be of significance.[3] Some modifications in the tRNA structure are relatively simple whereas others are more elaborate. In the latter case, the tRNA contains nucleosides that have been termed "hypermodified nucleosides".[3] However, the majority of modified nucleosides are of the "simple" type and are represented by major nucleosides in which, for example, the 5,6-double bond of a pyrimidine base has been reduced, a methyl group has been introduced, a carbonyl oxygen has been replaced by a sulfur atom, or the position of sugar attachment has been altered (e.g., C-nucleosides). In addition to the naturally occurring modified nucleosides, numerous synthetic modified nucleosides have been prepared in the laboratory. Such compounds have found use as tools for studying the structure and function of nucleic acids and have been exploited for their chemotherapeutic potential as, for example, antineoplastic and antiviral agents.

1.2. Mesoionic Nucleosides

One of the largest and simplest groups of modified nucleosides are those bearing a methyl (or some other alkyl) function. Various naturally occurring nucleosides have now been identified in which either the heterocyclic base or the sugar moiety has been methylated. One of these compounds is 7-methylguanosine. In 1956, Bredereck *et al.*[5,6] reported that treatment of guanosine with dimethyl sulfate at pH 4 affords 1-methylguanosine. Later, Lawley[7] noted that 7-methylguanine was a product when DNA was methylated with dimethyl sulfate in a neutral aqueous solution. Lawley and Wallick[8] suggested that guanosine reacts with dimethyl sulfate to yield a betaine, which was irreversibly destroyed by alkali. Haines *et al.*[9] alkylated 2',3',5'-tri-*O*-acetylguanosine with diazomethane and provided evidence for the 7-methylguanosine structure **1**. In 1963, Dunn[10] isolated 7-methylguanosine as the free base from acid hydrolysates of tRNA from pig liver, yeast cells, and the leaves of *Brassica chinensis*. Guanosine has also been alkylated by a wide variety of alkylating agents includ-

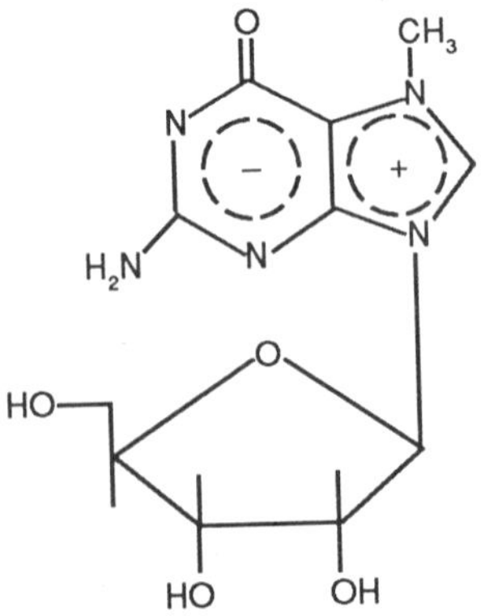

1

ing, for example, diethyl sulfate, sulfonate esters, and ethylene oxide.[11] At the time of its discovery, 7-methylguanosine was considered to be an oddity; it may now be looked upon as being the first reported example of a large class of compounds that we have termed mesoionic nucleosides.

Mesoionic nucleosides result from the combination of a mesoionic heterocycle and a sugar moiety. That is, mesoionic nucleosides are nucleosides in which a "normal" heterobase has been replaced by a mesoionic heterocycle. Baker, Ollis, and co-workers[12-14] introduced the concept of "mesoionic" in 1949—the term being derived from "mesomeric" and "ionic." A mesoionic compound was defined as a five-membered heterocycle that could not be satisfactorily represented by any single covalent or dipolar structure and that possessed a sextet of electrons in association with the five atoms comprising the ring. The ring bears a fractional positive charge which is balanced by a corresponding fractional negative charge located on a covalently attached atom or on a group of atoms. This definition has since been extended to include any heterocyclic ring system that shows extensive π-electron delocalization and for which no single dipolar or covalent structure can be drawn to indicate its true molecular structure.[15] Although some investigators prefer to describe six-membered mesoionic compounds as mesomeric betaines,[14,16,17] the exclusion of six-membered rings from the definition of "mesoionic" appears unwarranted.[18]

The present overview is not meant to be an exhaustive treatment of the subject; rather, it is intended to stir some interest in a large class of unexplored nucleoside analogues.

2. Mesoionic Purinones

Coburn *et al.*[15,19] reported the formulation and quantum-chemical study of a large class of mesoionic heterocycles that are isosteric and isoelectronic with the purinones; these mesoionic compounds were formally divided into two classes. Those mesoionic purinones that can be derived from a five-membered mesoionic ring structure have been termed Class I derivatives, whereas Class II derivatives are derived from six-membered ring mesoionic structures. The mesoionic purinones can be further divided into several subclasses: mesoionic hypoxanthines, mesoionic xanthines, and mesoionic 2-purinones (Fig. 1). The atoms X, Y, and Z can be varied among C/CH, N/NR, O, and S, so long as the resultant structures possess a π-electron system that is isoelectronic with that of their nonmesoionic counterpart. In this way, over 100 mesoionic ring systems that are isoelectronic with the purinones can be theoretically constructed.

Mesoionic compounds undoubtedly derive some of their stability from resonance stabilization. Choosing two examples for closer inspection, it can be seen that several possible resonance hybrids can be drawn for a Class I mesoionic hypoxanthine analogue (Fig. 2) and for a Class II mesoionic xanthine analogue (Fig. 3). Thus, it might be misleading to depict these structures as single dipolar entities. Several different methods have now been employed to describe the structures of these compounds. With respect to the Class I mesoionic hypoxanthines, the structure shown in Fig. 4a is perhaps the most common; however, this

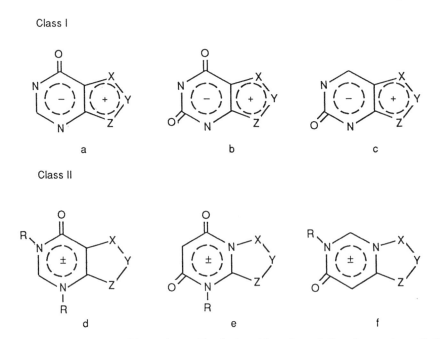

Figure 1. Mesoionic hypoxanthine (a,d), xanthine (b,e), and 2-purinone (c,f) analogues. Atoms X, Y, and Z can vary among carbon, oxygen, nitrogen, and sulfur so long as the resulting system is a six-π-electron heterocycle.

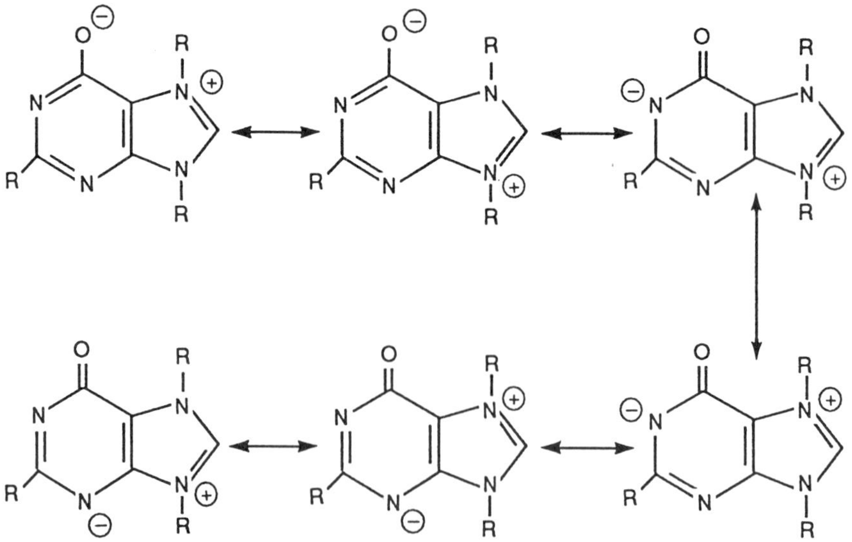

Figure 2. Resonance structures for an example of a Class I mesoionic hypoxanthine analogue.

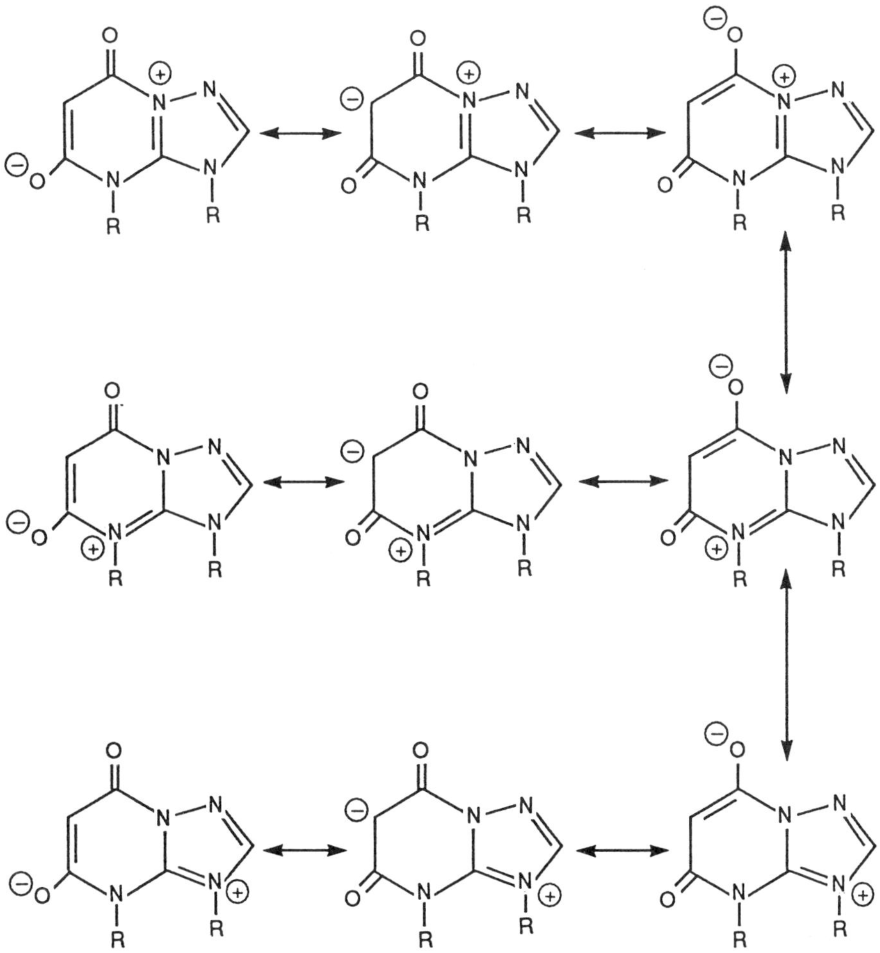

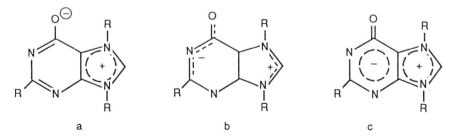

Figure 3. Resonance structures for an example of a Class II mesoionic xanthine analogue.

Figure 4. Notations that have been employed to represent Class I mesoionic hypoxanthine analogues.

structure fails to take into account the possible electron delocalization in the six-membered ring. Alternative representations are those shown in Fig. 4b and 4c. The Class II mesoionic xanthine analogues have been represented as shown in Fig. 5. Because Figs. 4c and 5c imply aromatic character (which these compounds do display), suggest electron delocalization, and readily identify and contrast Class I versus Class II analogues, such structural representations will be used herein.

These mesoionic compounds are mesoionic purinones, and, as such, they themselves belong to the larger class of five-membered or six-membered ring mesoionic heterocycles. Many more "nonpurinone" mesoionic heterocycles have been described and reviewed.[14,18] Before continuing, it should be made clear that these mesoionic derivatives are not usually salts; they simply suffer from our inability to satisfactorily depict their structures using standard "covalent" methods. However, some of these compounds do exist as their corresponding salts; for example, several Class II mesoionic 1,3-dialkylhypoxanthine derivatives have been prepared by the treatment of 1,3-dialkylhypoxanthinium salts by base,[20] and guanosine can be methylated with methyl iodide to afford 7-methylguanosinium iodide.[21] These will be discussed further under the appropriate headings.

Some of the mesoionic heterocycles to be described below possess numbering systems that differ from that of purine. *However, for the sake of simplicity and for ease of discussion, the purine numbering system will be used throughout.* It should also be emphasized that in addition to the Class I and Class II mesoionic nucleosides described below, it is entirely possible to formulate mesoionic nucleosides in which the mesoionic heterocycle is a five-membered ring, or a ring system not necessarily covered by Fig. 1; this will be further discussed in Section 3.

2.1. Class I Hypoxanthine Derivatives

Although a fairly large number of Class I hypoxanthine ring systems have been formulated and studied from a quantum-chemical standpoint,[19] relatively few examples have actually been synthesized. Ackermann and List[22] isolated herbipolin (**2**), a dimethylguanine, from Mediterranean sponges. This naturally occurring mesoionic 7,9-dimethylguanine derivative was later synthesized by Bredereck and co-workers.[23] The bis-hydroxyethyl derivative (i.e., **3**),[11]

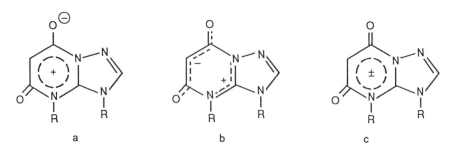

Figure 5. Notations that have been employed to represent Class II mesoionic xanthine analogues.

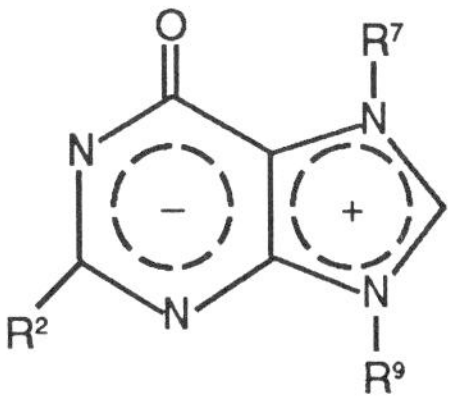

	R²	R⁷	R⁹
2	–NH₂	–CH₃	–CH₃
3	–NH₂	–CH₂CH₂OH	–CH₂CH₂OH
4	–NH₂	–CH₃	–C₂H₅
5	–H	–CH₃	–CH₃

7-methyl-9-ethylguanine (**4**),[24] and the 7,9-disubstituted hypoxanthine derivative **5**[25] are also known.

If the N-9 (or N-7) alkyl group of a mesoionic hypoxanthine is replaced by a sugar moiety, the result is a mesoionic nucleoside. Thus, 7-methylguanosine (**1**) may now be considered as being a Class I mesoionic hypoxanthine nucleoside. Other related derivatives include **6–14**. The 7-ethyl[24,26] and 7-benzyl[21] homologues of **1** (i.e., **6** and **7**, respectively) have been prepared. The 7-methyl-[7,27] and 7-ethyl-[24] 2-deoxyguanosine derivatives **8** and **9** and the mesoionic nucleotides 7-methylguanylic acid (**10**) and 7-methyl- and 7-ethyl-2′-deoxyguanylic acids **11** and **12** have also been reported[11,24] but have not been as thoroughly investigated as 7-methylguanosine (**1**) *per se*. Compounds **13** and **14** represent

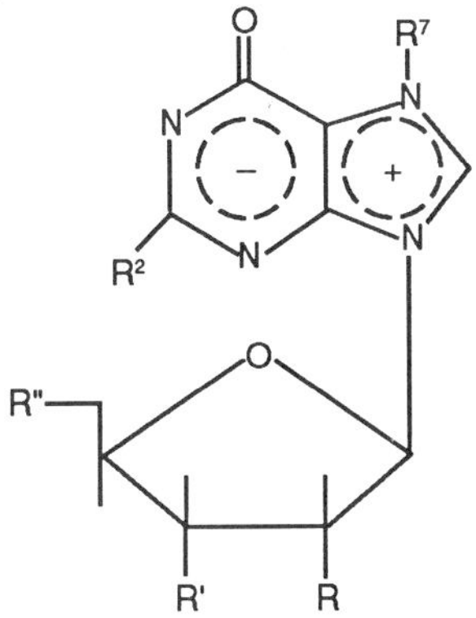

	R²	R⁷	R	R'	R"
6	–NH₂	–C₂H₅	–OH	–OH	–OH
7	–NH₂	–CH₂Ph	–OH	–OH	–OH
8	–NH₂	–CH₃	–H	–OH	–OH
9	–NH₂	–C₂H₅	–H	–OH	–OH
10	–NH₂	–CH₃	–OH	–OH	–OPO₃²⁻
11	–NH₂	–CH₃	–H	–OH	–OPO₃²⁻
12	–NH₂	–C₂H₅	–H	–OH	–OPO₃²⁻
13	–H	–CH₃	–OH	–OH	–OH
14	–H	–CH₃	–O	O–	–OH

mesoionic inosine derivatives.[28,29] Broom and Robins [30] synthesized the first example of a Class I mesoionic hypoxanthine nucleoside that contains a thiocarbonyl group, 7-methyl-6-thioguanosine (**15**).

Because of the potential mechanistic implications in the action of antineoplastic agents, methylation of nucleic acids has attracted a considerable amount of attention.[24,31,32] However, because chemical alkylation of nucleic acids can result in rather complex mixtures of products, much of the early work in this area employed the simpler ribonucleosides or deoxyribonucleosides as nucleic acid models.[9,24,31] This initial pioneering work provided information with regard to the principal sites of alkylation and the relative reactivities of the more common purine bases. In DNA, the major product of alkylation, 7-alkyl-2′-deoxyguanosine residues, can result in depurination and backbone breakage, whereas certain difunctional alkylating agents can result in cross-linking.[29,33] With respect to the nucleoside itself, it has been demonstrated that 7-methyl-2′-

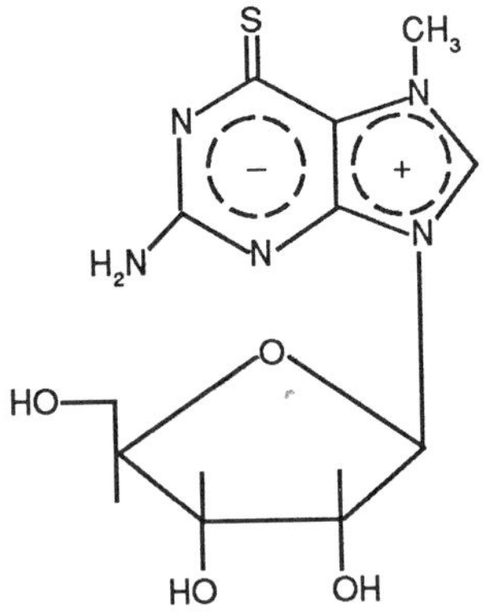

15

deoxyguanosine 5'-triphosphate can be substituted for 2'-deoxyguanosine 5'-triphosphate in DNA polymerase reactions utilizing calf-thymus DNA as the template[34] and that 7-methylguanosine has a higher turnover rate than other methylated derivatives in intact animals.[35] It was findings such as these that generated interest in the further study of these "unusual" nucleosides.

One of the first systematic studies, including product identification, on the alkylation of guanosine was that by Haines and co-workers.[9] Later, Jones and Robins[27] reported that methylation of guanosine with methyl iodide in *N,N*-dimethylacetamide at 30°C afforded high yields of 7-methylguanosine (**1**); the major site of alkylation in neutral or acidic media is the nitrogen atom at the 7-position, whereas in alkaline solution, the *N*-1-methyl derivative is the predominant product.[27] On occasion, it has been difficult to separate the mesoionic compound from its salt. For example, further methylation of 7-methylguanosine (**1**) with methyl iodide yields 1,7-dimethylguanosinium iodide (**16**). Brookes *et al.*[21] reacted guanosine with benzyl bromide to afford a mixture of mesoionic 7-benzylguanosine (**7**) and 7-benzylguanosinium bromide (**17**).

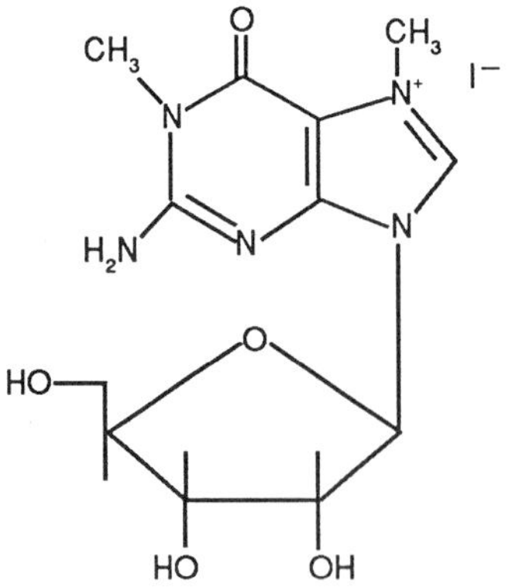

16 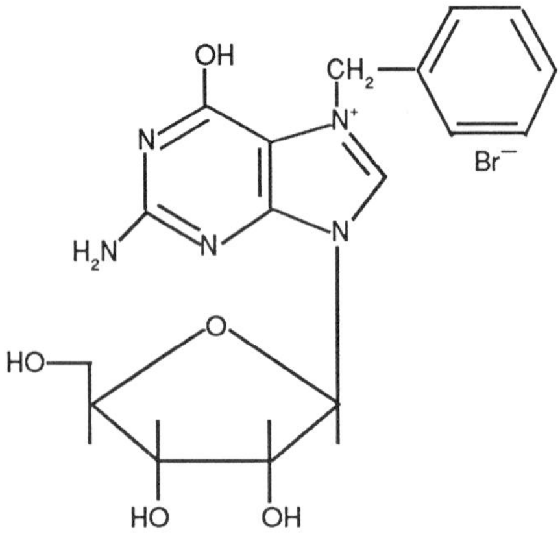   **17**

Although anhydrous 7-methylguanosine (**1**) is stable indefinitely, 7-methylinosine (**13**), when stored at room temperature, slowly decomposes to 7-methylhypoxanthine. 7-Methylguanosine and other Class I mesoionic hypoxanthine analogues exhibit properties that indicate a general electron deficiency in the imidazole ring. The somewhat greater stability of 7-methylguanosine (**1**) as compared to 7-methylinosine (**13**) has been interpreted as resulting from the stabilizing effect of the electron-donating amino group at the purine 2-position.[27] Hydrolysis studies also support the idea of an electron-deficient five-membered ring. Methylation of either 7- or 9-methylguanine affords the identical mesoionic 7,9-dimethylguanine derivative. Stability studies have shown that, in general, 7,9-dialkylguanines are relatively stable at room temperature below pH 8.5. However, alkaline (pH > 8.5) hydrolysis results in ring opening of the imidazole ring to give 5,6-diaminopyrimidine derivatives. On the other hand, heating 7-alkylpurine nucleosides in dilute acids results in a rapid (usually in less than 30 minutes) and quantitative conversion to the 7-alkylpurine.[36] Thus, the glycosidic bond is less stable than in the corresponding nonmesoionic nucleoside. Similarly, when the mesoionic 7-methyl-6-thioguanosine (**15**) is treated with acid, 2-amino-7-methyl-6-purinethiol can be isolated.[30] 7-Alkylpurine nucleosides are also rapidly decomposed by aqueous alkali, at neutral or alkaline pH. 7-Methyl-guanosine (**1**) undergoes ring opening to yield the 2-amino-4-hydroxy-5-(*N*-methyl-*N*-formylamino)pyrimidine derivative **18**, and this same compound has been isolated from an enzymatic hydrolysate of yeast tRNA.[36] Nucleoside **15** appears to undergo a similar reaction to afford **19**.[30] Under

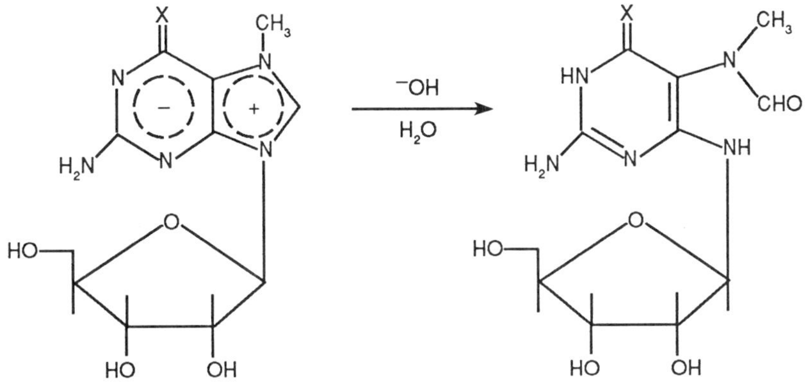

1,	X = O	**18,**	X = O
15,	X = S	**19,**	X = S

conditions of acid hydrolysis, there is little difference between the glycosidic bond stability of 7-ethylguanosine (**6**) and 7-methylguanosine (**1**)[21]; however, the glycosidic bond of 7-alkylguanosines is approximately 15 to 20 times more stable than that of 1,7-dialkylguanosines.[31] Studies on the glycosidic cleavage of the mesoionic 7-alkylpurine nucleosides may be useful inasmuch as these compounds might serve as models for studying the acid-catalyzed hydrolysis of natural nucleosides and suggest that acid hydrolysis of such natural nucleosides may involve an initial protonation at N-7.[27]

Methylation of the 2'-deoxyguanosine moieties of DNA has been reported. N-7 methylation labilizes the glycosidic bond, and the methylated base, 7-methylguanine, is readily split out of the nucleic acid. According to Lawley and Wallick,[8] it seems probable that 7-methyl-2'-deoxyguanosine may have only a temporary existence under these conditions. Jones and Robins[27] have methylated 2'-deoxyguanosine directly to afford mesoionic 7-methyl-2'-deoxyguanosine (8). Lawley and Brookes[24] have prepared both 7-methyl- and 7-ethyl-2'-deoxyguanosine 5'-monophosphates (11 and 12, respectively). The 5-methyl hydrogen phosphate ester of 8 has been prepared by Khorana.[33] Comparison studies of the stabilities of 7-methylguanosine (1) and 7-methyl-2'-deoxyguanosine (8) reveal that the mesoionic 2'-deoxy derivative is very unstable and decomposes almost spontaneously to give 7-methylguanine.[24] A comparison of the rates of hydrolysis of 7-methyl- and 7-ethyl-2'-deoxyguanosine (8 and 9, respectively) and 7-methyl and 7-ethyl-2'-deoxyguanosine 5'-monophosphate (11 and 12, respectively) at neutral pH, and over the range of pH 5–10, shows that the methylated nucleoside and nucleotide were less stable than the corresponding ethylated derivatives.[24] It has also been reported that, at pH 9, 7-methyl-2'-deoxyguanosine 5'-monophosphate (11) undergoes opening of the imidazole ring,[37] a reaction that apparently competes with hydrolysis of the glycosidic bond. However, at pH > 9, ring opening appears to be the predominant reaction.[24]

Michelson and Pochon[38] methylated polyguanylic acid and polyinosinic acid using techniques that avoided degradation of these polymers. They were successful in isolating the corresponding mesoionic polynucleotides. Acid hydrolysis of these products gave the corresponding methylated purine components, that is, 7-methylguanine and 7-methylhypoxanthine. Alkaline hydrolysis followed the same path as previously described for hydrolysis of the nucleotide.[8] Both polymers were resistant to the action of rattlesnake venom diesterase.[38]

As previously mentioned, some of these mesoionic nucleoside derivatives are naturally occurring. In 1963, Dunn isolated 7-methylguanosine (1) from acid hydrolysates of tRNA from pig liver, yeast cells, and the leaves of *Brassica chinensis*.[10] There is also evidence that 1 may be a component of calf- and rat-liver tRNA.[10] Igo-Kemenes and Zachau[39] have hypothesized that a modified nucleoside, such as 1, might stabilize certain regions of the tRNA structure. The observation that 1 is not reduced by sodium borohydride is pertinent to theoretical models of the three-dimensional structure of tRNA.[39] It may be reasonable to assume that these mesoionic nucleosides are involved in defining the molecular conformation of parts of the tRNA structure.

RNA from viral sources contains 7-methylguanosine (1) as part of the 5'-terminal cap structure. This is thought to play an important role in translation.[40-44] Antibodies specific for 1 have been described.[45-47] These antibodies have been employed to detect the presence of 1 in synthetic nucleic acid polymers and to fractionate oligonucleotides containing methylated bases. 7-Methylguanosine antibodies display specificity for methylated purines over nonmethylated purines and distinguish between the 1-, 3- and 7-monomethyl derivatives.[45] Although the antibodies can recognize 7-methylguanosine in the RNA cap, the degree of reactivity observed is less than with free 7-methylguanosine. This has been suggested as reflecting a constrained configuration for **1** as it exists in the cap. Proton NMR studies indicate that the 7-methylguanosine occurs in a stacked conformation with respect to the adjacent base, which may prevent maximum recognition by the antibody.[48,49]

Very few reactions involving these mesoionic nucleosides have been reported. For example, 7-methylguanosine (**1**), a Class I mesoionic hypoxanthine derivative, can be readily converted, by treatment with nitrous acid, to 7-methyl-xanthosine (**20**), a Class I mesoionic xanthine analogue.[27] Compound **1** has also been methylated to afford 1,7-dimethylguanosinium iodide.[21] Attempts to derivatize **1** by silylation resulted in oxygen insertion to give what is believed to be **21**.[50] Igo-Kemenes and Zachau[39,51] observed that tritium was not incorporated into 7-methylguanosine upon treatment with tritiated sodium borohydride. Pochon *et al.*[52] reported the reduction of **1** to 7-methyl-8-hydroxyguanosine, which was reoxidized almost instantaneously. These reactions, and the above-mentioned hydrolysis studies, are about all that is known about the chemistry of these compounds.

2.2. Class I Xanthine Derivatives

Bredereck *et al.*[53] reported the preparation of 7,9-dimethylxanthine (**22**), which now, by definition, is a Class I mesoionic xanthine analogue. Jones and Robins[25] established the structure of **22** by independent methylation of 7-methylxanthine and 9-methylxanthine. Compound **22** was also obtained upon treatment of the mesoionic 7,9-dimethylguanine (**2**) with nitrous acid.[27] Senga *et al.*[54] have reported the synthesis of the 1,2,3,-thiadiazolo[4,5-*d*]pyrimidine **23**, which is an example of a Class I mesoionic xanthine analogue.

7-Methylxanthosine (**20**), the first known example of a Class I mesoionic xanthine nucleoside, was obtained by treating 7-methylguanosine with nitrous

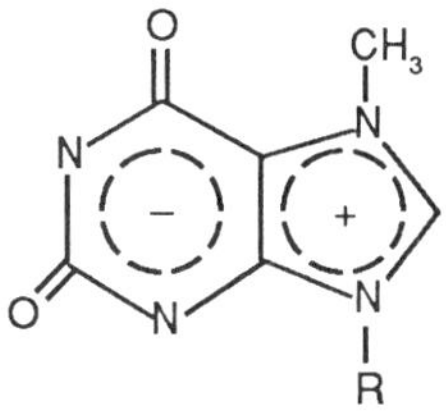

20, R = [ribose structure, HO with OH, OH hydroxyls]

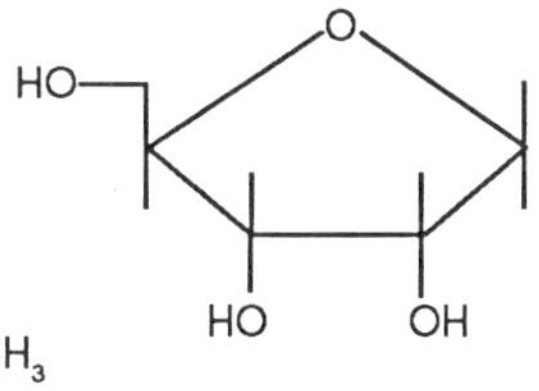

22, R = CH₃

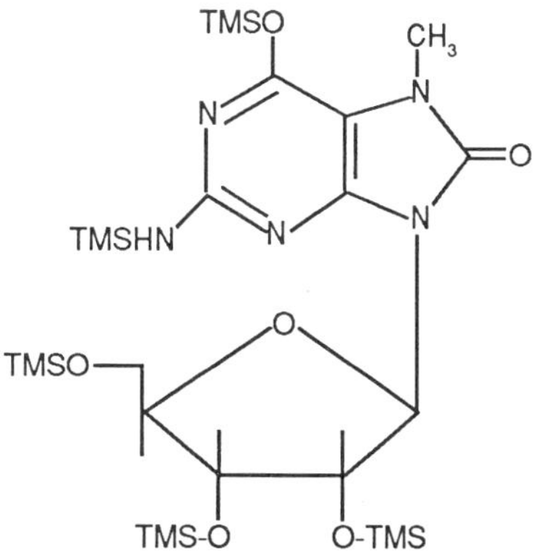

21

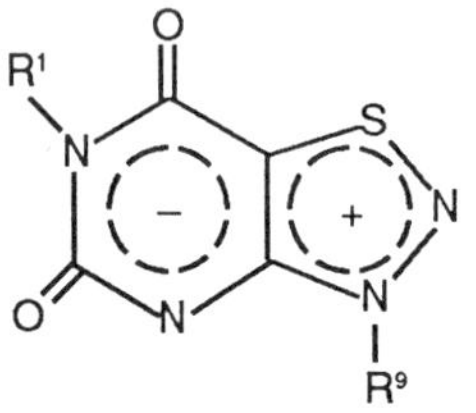

23,   R^1 = –CH$_2$—⟨benzene ring⟩ ; R^9 = –CH$_3$

acid.[27] At room temperature, **20** slowly decomposes to afford 7-methylxanthine whereas, in aqueous alkali, **20** rapidly decomposes to afford several ring-opened products similar to those observed for 7-methylguanosine.[27] Upon treatment with dilute acid, **20** is hydrolyzed to 7-methylxanthine.[27]

2.3. Class I 2-Purinone Derivatives

To date, no examples of Class I mesoionic 2-purinones or 2-purinone mesoionic nucleosides have been reported.

2.4. Class II Hypoxanthine Derivatives

In general, Class II mesoionic purinones may be prepared by the fusion of a five-membered ring with the hypothetical imino derivative of a 1,3-disubstituted mesoionic pyrimidine-4,6-dione (i.e., **24**) such that the imino nitrogen

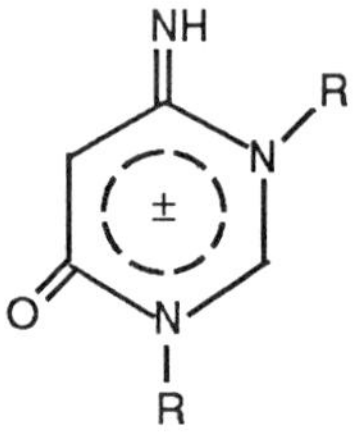

24

atom is part of the fused ring. Results of quantum-chemical studies suggest that there are both similarities and differences between these mesoionic ring systems and their nonmesoionic counterparts.[15]

In 1962, Townsend and Robins reported the synthesis of 3-methylguanine and speculated, based on observed chemical properties, that **25** and **26** must constitute significant resonance contributors.[55] Albert used a similar argument to explain some of the properties of the structurally related 3-methyl-8-azapurine-6-ones.[56] Structures such as **25** and **26** do not adhere to the definition of

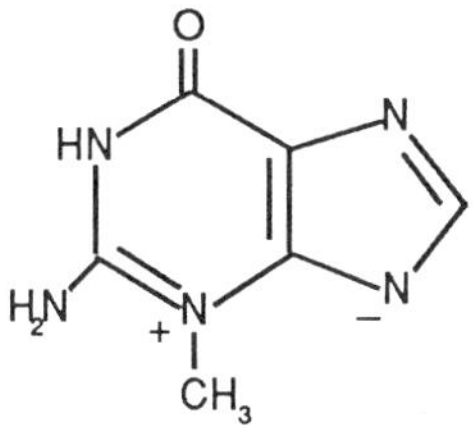

25 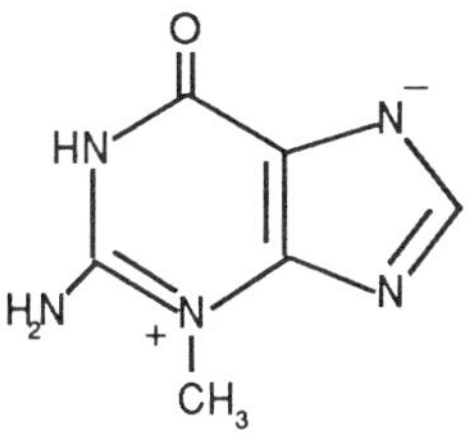   26 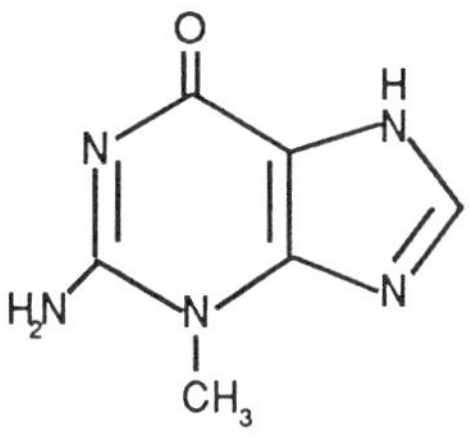   27

mesoionic in that the structure of 3-methylguanine can be represented by a single covalent structure (i.e., **27**). Thus, it was of interest, from a theoretical standpoint, to further investigate such structures. On the basis of quantum-chemical calculations, Pullman predicted that the "mesoionic" 3-methylguanine would be about 50 kcal/mol less stable than the more usual tautomers of guanine and concluded that the 7*H*-tautomer **27** would be the most stable.[57] Indeed, Abola *et al.*[58] provided crystallographic evidence that **27** is the preferred structure for 3-methylguanine.

Coburn and Carapellotti prepared the first examples of a true Class II mesoionic hypothanthine: **28**.[59] In contrast to hypoxanthinium salts, such as **29** (where X = Br or I), which darken rapidly upon exposure to air, the mesoionic compounds exist as white crystals that are stable to heat and light in air. The chemistry of these compounds was not studied. However, it was found that **28** (R = R' = benzyl) undergoes a slow conversion, upon treatment with 5% aqueous sodium bicarbonate, to the ring-opened product *N*-benzyl-5(4)-(*N*-benzyl-formamido)imidazole-4(5)-carboxamide.[59]

To date, no Class II mesoionic hypoxanthine nucleosides have been reported.

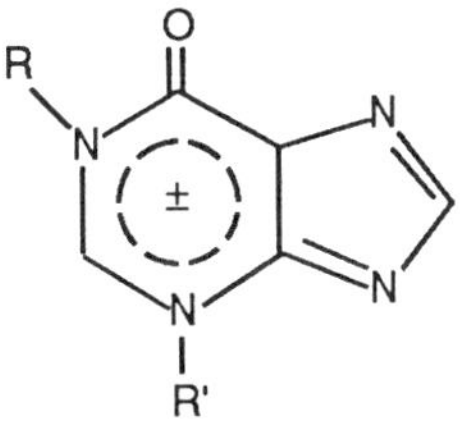

28 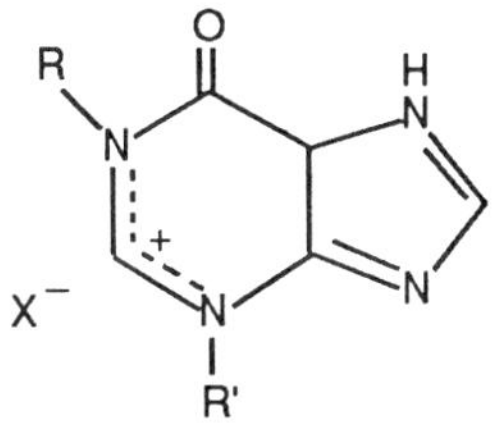   29

2.5. Class II Xanthine Derivatives

Perhaps the best studied of the three Class II subclasses is that of the xanthine derivatives, and yet of the 36 possible heterocyclic ring systems that are isoelectronic and isosteric with xanthine, examples of less than a dozen have been synthesized. Of these, only two or three ring systems have been studied in any detail. Glennon *et al.*[60] have prepared several examples of mesoionic imidazopyrimidines **30** and 1,2,4-triazolopyrimidines **31**, but their chemistry has

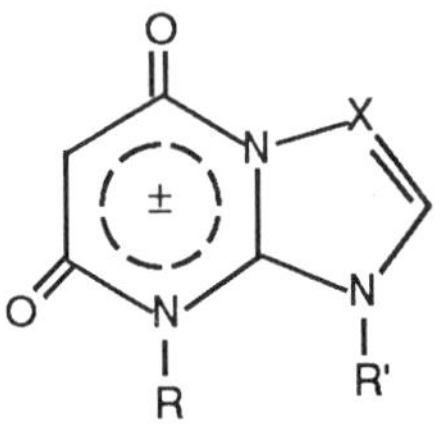

30, X = CH   **32**   **33**
31, X = N

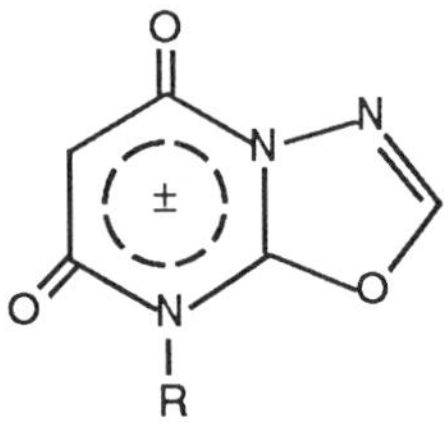

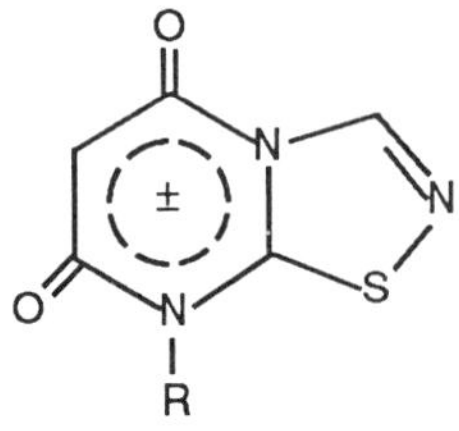

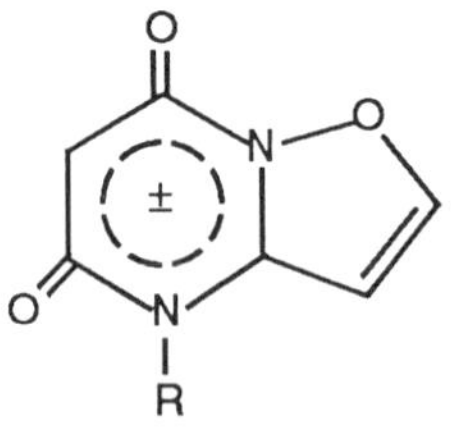

34 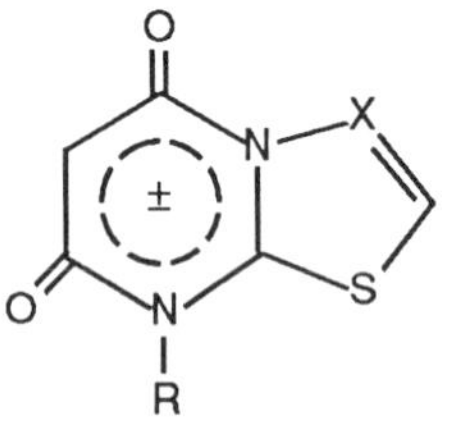   **35,** X = CH
                    36, X = N

not been explored. Examples of 1,3,4-oxadiazolo-, 1,2,4-thiadiazolo-, and isox-azolopyrimidines (**32–34**, respectively) have also been prepared.[61] The meso-ionic thiazolopyrimidines **35** and the 1,3,4-thiadiazolopyrimidines **36**, first syn-thesized in 1973,[62] are the two ring systems that have received the most attention. A large series of such compounds have been prepared and evaluated for their antimicrobial activity.[63,64] Because of the close structural and elec-tronic similarity between these mesoionic derivatives and xanthine, several such derivatives have been examined in biological assays in which nonmesoionic xanthines are typically active. Depending upon the particular mesoionic xan-thine ring system and pendant substituents, these mesoionic heterobases possess varying degrees of xanthine-like activity. For example, they have been examined as inhibitors of cyclic AMP phosphodiesterase[65–67] and for their ability to bind at adenosine A1 and A2 binding sites.[61,68] Several derivatives of **35** and **36** also produce a xanthine-like (i.e., a theophylline-like) hypotensive effect in rats.[67] Thus, there is some evidence that mesoionic xanthine analogues can mimic their nonmesoionic counterparts in various biological systems.

For the most part, Class II mesoionic xanthine analogues have been pre-pared by the cyclization of an appropriately substituted amino derivative of a five-membered heterocycle with what is essentially a reactive malonate synthon. Cyclization has been accomplished using chlorocarbonylphenylketene[70] or carbon suboxide[61,62] at 0°C, bis(2,3,4,5,6-pentachlorophenyl)malonate esters at room temperature,[61] or bis(2,4,6-trichlorophenyl)malonate esters at 160°C.[18,61–69,71] The latter method, developed by Kappe and Lube,[71] has been the most popu-lar; the other methods have generally been employed when either the product or the heterocyclic starting material is thermally unstable.

Derivatives of **35** and **36** are ordinarily stable to heat and light in air, are readily recrystallized from organic solvents, possess minimal aqueous solubility, and display typical heteroaromatic properties. Derivatives of **36** usually melt with decomposition whereas derivatives of **35** do not; melting points (depending upon substituents present) typically range from 100 to 250°C. The spectral properties of these compounds, including infrared,[62] proton NMR,[62] ^{13}C-NMR,[72] ultraviolet,[62] and mass spectra,[62,73] have been examined.

As with most mesoionic compounds, the chemistry of Class II mesoionic xanthine analogues has not been well studied. What little is known about the reactions of thiazolo- and 1,3,4-thiadiazolopyrimidines is shown in Fig. 6. When the purine 1-position of **35** is unsubstituted, electrophilic attack can occur at this position; **35** has been brominated and nitrated.[62] Compounds **35** and **36** are moderately sensitive to acid; however, certain derivatives of **35** have been re-crystallized from 5% HCl or from glacial acetic acid. Nevertheless, prolonged heating in aqueous acid results in a hydrolytic ring opening to afford **40**. Derivatives of **35** and **36** are also susceptible to nucleophilic attack. Attack by hydroxide or alkoxide presumably occurs at the purine 4-position to give complex mixtures of products.[62] Concomitant attack may also occur at one of the pseudo-carbonyl positions; however, this phenomenon has not been thoroughly investigated. Attack by amines occurs in a two-step process (Fig. 6). Initial attack occurs at the purine 6-position to afford **41**; under more vigorous conditions, **41** can be converted to **40**.[62,64,74]

It is entirely possible that the mesoionic xanthine analogues, by virtue of their susceptibility toward nucleophilic attack, might serve as latent acylating agents *in vivo* and be capable of interacting irreversibly with a receptor site or the active site of an enzyme (Fig. 7). For this reason, the kinetics of chemically induced ring opening of **35** have been studied in detail.[74] Steric bulk at the purine 1-position retards nucleophilic attack; the rate of ring opening is also influenced by the electronic nature of substituents at the purine 3-position, and the ring opening occurs more readily with derivatives of **36** than with the correspondingly substituted **35**.[64,74] There is some recent evidence that derivatives of **36** can act in an irreversible manner as inhibitors of platelet aggregation.[75]

Class II mesoionic xanthine analogues contain a "masked" 1,4-dipole and, as such, can undergo cycloaddition reactions with dipolarophiles.[61,62,76–78] For example, derivatives of **35** react with dimethylacetylene dicarboxylate to form the cycloadduct **42**, which, after loss of alkyl or aryl isocyanate, affords **43**.[62] Thus, the 1,4-cycloaddition reaction provides a versatile route to substituted 2-pyridones which might otherwise be difficult to prepare.

Several Class II mesoionic xanthine nucleosides have been reported. For example, Glennon and co-workers[79,80] have prepared the α- and β-anomers of **44** (R = H, Et) from their 2′,3′,5′-tri-O-acetyl derivatives. They have reported several 2′-deoxyribofuranosyl and glucopyranosyl derivatives of **35**.[80] The mass spectral characteristics of these compounds have been examined using low- and high-resolution mass spectrometry and metastable ion analysis.[73] The mass spectra of these compounds, although displaying some similarity to those of nonmesoionic nucleosides, differ significantly in that there is an absence of fragment ions associated with the base plus portions of the sugar. Compound **44** and related derivatives have also been studied using fast atom bombardment

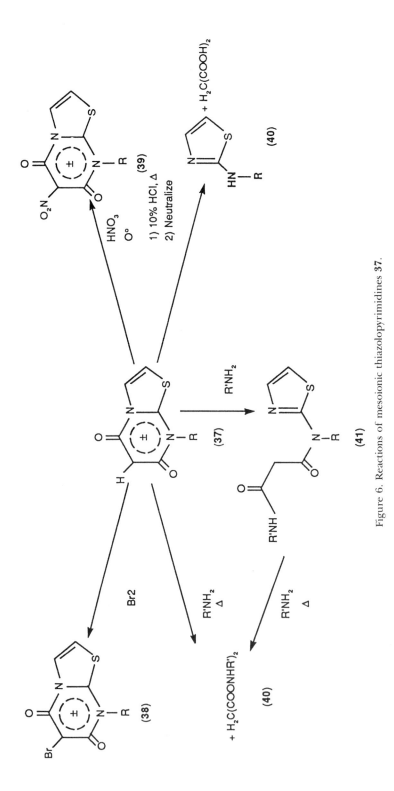

Figure 6. Reactions of mesoionic thiazolopyrimidines **37**.

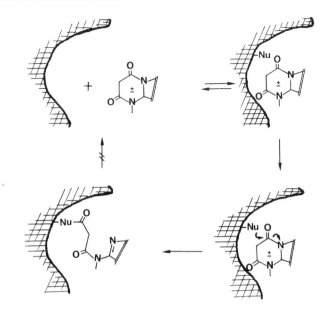

Figure 7. A possible mechanism for acylation of a biological substrate (e.g., receptor, enzyme) by a Class II mesoionic xanthine analogue.

(FAB) mass spectrometry.[73] In this latter study, comparisons were made with the FAB spectra of 7-methylguanosine (**1**) and several nonmesoionic nucleosides.

Several nucleoside derivatives of **36** have also been prepared (i.e., **45**; R = H, alkyl); however, all attempts to deprotect **45** resulted in hydrolytic ring opening of the six-membered ring.[81,82]

2.6. Class II 2-Purinone Derivatives

The first examples of a Class II mesoionic 2-purinone ring system, **46**, were recently reported[83]; no nucleoside derivatives have been prepared to date.

Compound **46** could exist as any one of several tautomers, and each tautomer may be represented by a number of dipolar canonical structures. On the basis of proton and ^{13}C-NMR studies, Coburn and Taylor[83] have concluded that these mesoionic compounds exist predominantly as the C_7-H tautomer **47**. Derivatives of **47** were found to be susceptible to hydrolytic ring opening reactions. Although **47** (R = CH_3) was stable in water at room temperature, heating at reflux or treatment with sodium hydroxide afforded compound **48**. The reaction of **47** (R = phenyl) with dimethylacetylene dicarboxylate (DMAD) gave a compound which was assigned structure **49**.[83]

3. Miscellaneous Mesoionic Derivatives

Up to this point, the discussion has been limited to mesoionic purinone derivatives. However, once the artificial constraint imposed by the definition of

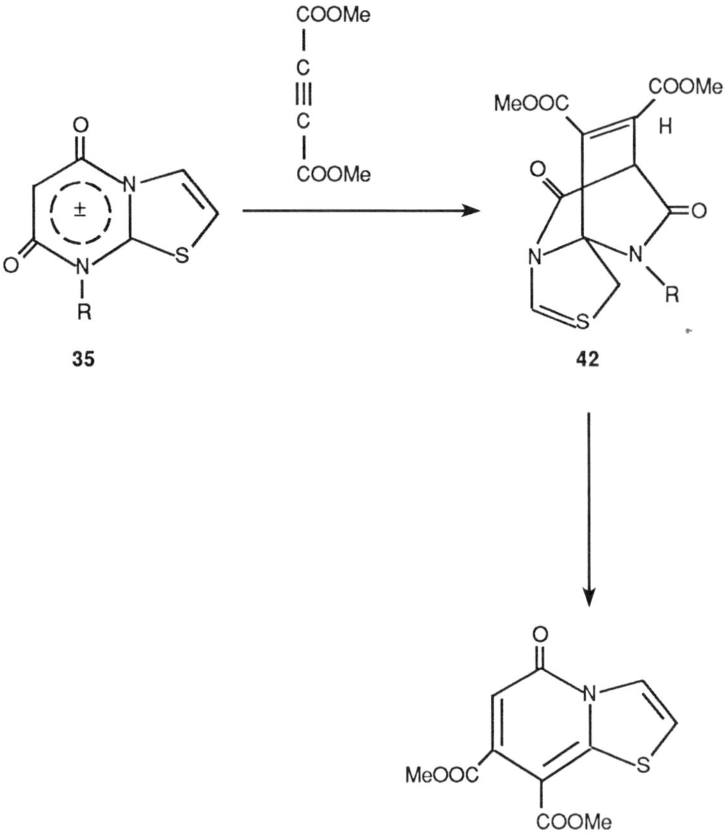

35

42

43

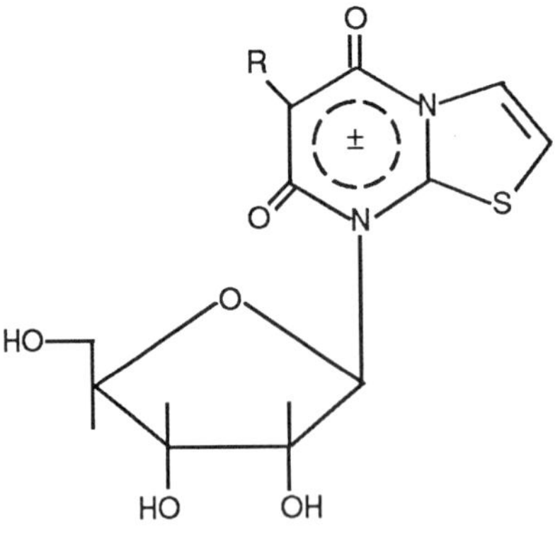

44a, R = H
44b, R = Et

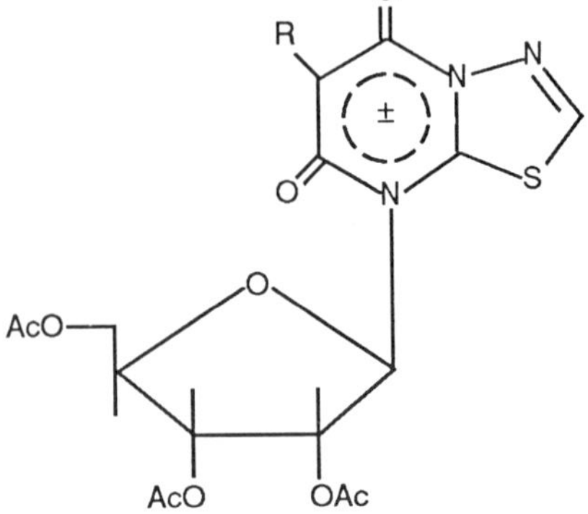

45

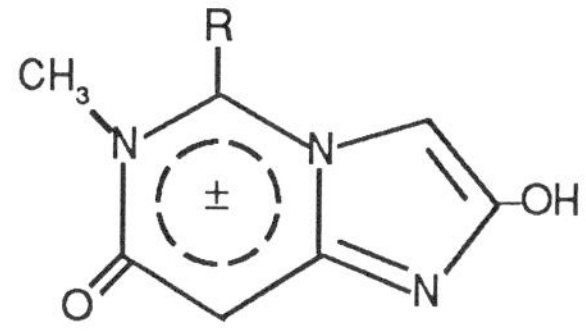

46

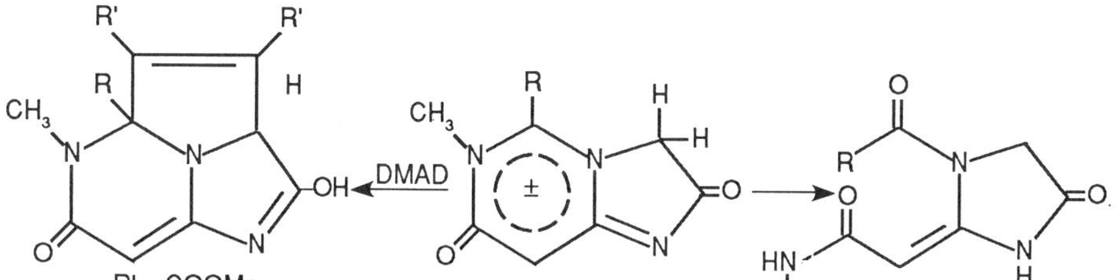

R' = COOMe
49   **47**   **48**

"mesoionic purinones" is removed, we are introduced to an even larger class of mesoionic heterocycles. In general, five-membered mesoionic heterocycles have received much more attention than six-membered mesoionic heterocycles, and several reviews are available; for an extensive review of the former, see Ollis and Ramsden.[14] Thus, it is conceptually possible to prepare mesoionic nucleosides such as **50**.

Monocyclic pyrimidine-type nucleosides such as, for example, **51** (X = N) are also possible; although the monocyclic mesoionic ring system is known,[71,84] nucleoside derivatives have not been reported. Recently, Bambury *et al.*[85] have described a related compound (i.e., **52**).

Monocyclic mesoionic 1,3,5-triazines and several ring-fused derivatives are known.[18] For example, **53** (X = CR or N) may be viewed as a 1-aza analogue of the Class II mesoionic xanthines **35** and **36**.[86,87] Compared to **35** and **36**, however, derivatives of **53** possess only limited solubility in organic solvents, are

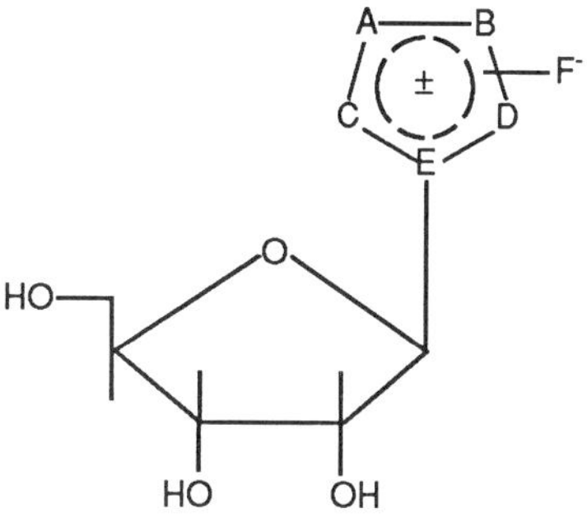

50 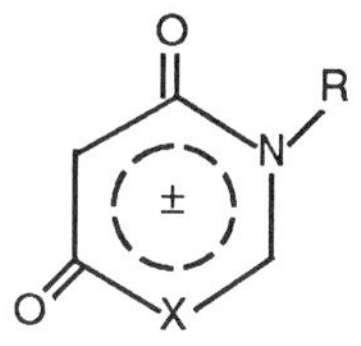   **51** 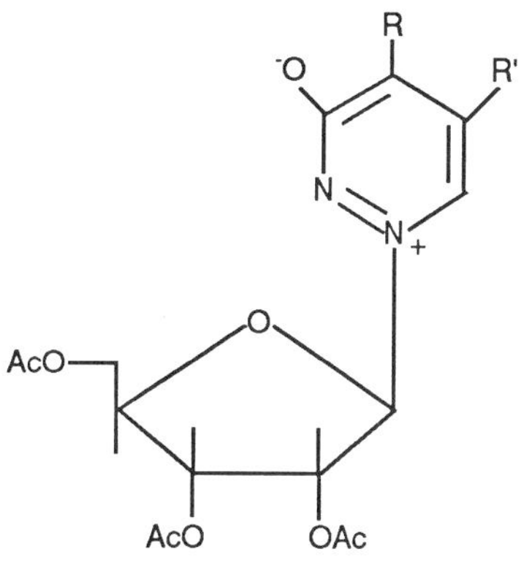   **52**

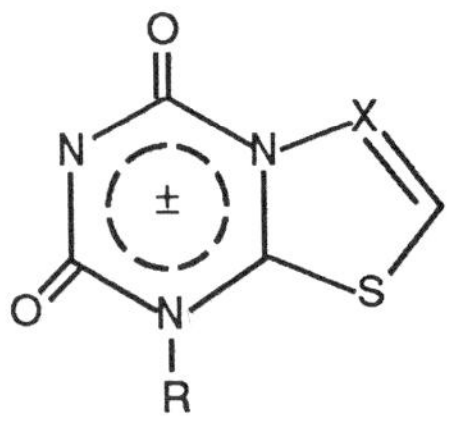

53

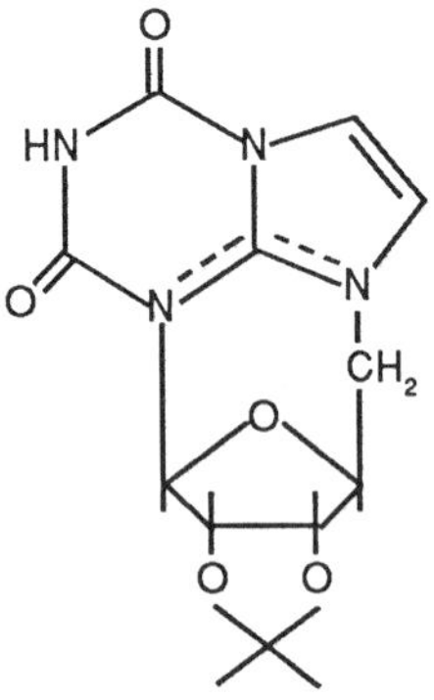

54 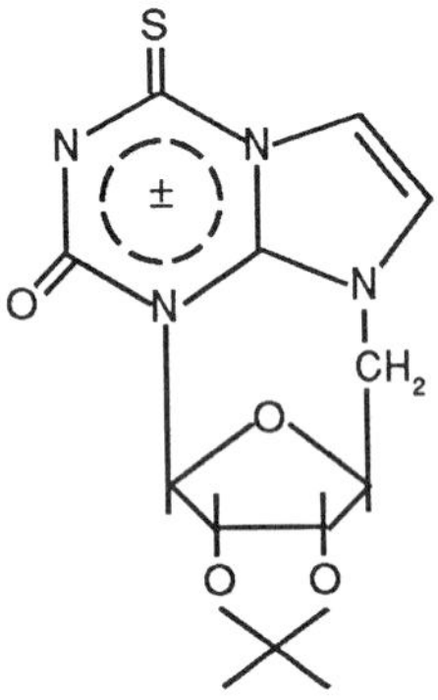   **55**

somewhat less stable, and are more prone to hydrolytic ring opening.[87] Prisbe *et al.*[88] reported that whereas 1-(2,3-*O*-isopropylidene-β-D-ribofuranosyl)imidazo[1,2-*a*]-1,3,5-triazine-2,4-(1*H*,3*H*)-dione is converted to **54** by treatment with methyltriphenylphosphonium iodide, the corresponding 6-thio derivative affords the mesoionic compound **55** when subjected to the same conditions.

Replacement of the ring nitrogen atom of **51** (X = N) with oxygen or sulfur affords the corresponding mesoionic 1,3-oxazines and 1,3-thiazines.[86,87,89–93] Whereas these compounds are fairly unstable and are even prone to hydrolytic ring opening upon exposure to moist air, ring fusion with a five-membered heterocycle (e.g., **56**; X = CH, N) appears to enhance their stability.[60,61,70,80,81,89,94,95] Glennon and Tejani[96] recently reported the synthesis of **57**; although spectral data support the assigned structure and although the compound is stable at 0°C, it apparently undergoes ring opening at room temperature to afford **58**. They also found that structures such as **56**, with X = CH, are much less stable when R = H than when R = alkyl[96]; thus, the stability of **57** might be enhanced by the presence of an alkyl group at the purine 1-position.

Numerous other six-membered monocyclic and fused mesoionic ring systems have been reported,[18] and it is conceptually possible to prepare nucleoside analogues of these.

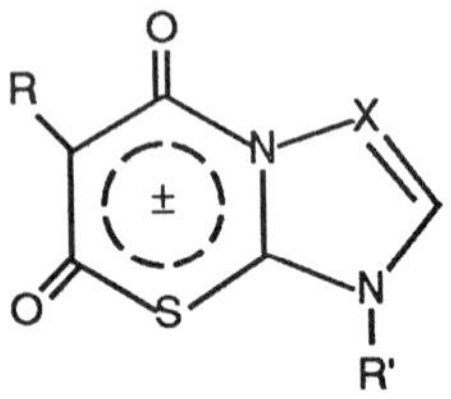

56

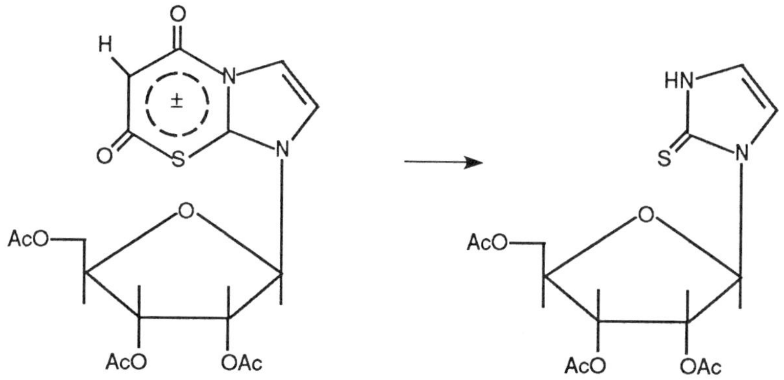

57   58

4. Conclusion

Modified nucleosides are important because of their structural similarity to the purine and pyrimidine nucleosides and because they have aided in studying complex processes such as RNA and DNA biosynthesis and exzymatic reaction mechanisms. Many more modified nucleosides have been prepared and evaluated for their chemotherapeutic potential (e.g., see reviews by Harmon *et al.*[97] and Walker *et al.*[98]). There is also some evidence that non-mesoionic nucleosides can be metabolized to mesoionic nucelosides.[99] Mesoionic nucleosides represent an entirely new class of modified nucleosides. Although the term "mesoionic nucleoside" was first introduced in 1981,[79] a review of the literature reveals that such compounds have been known for over 30 years. This review has been a first attempt to organize some of the available information on these compounds.

Many hundreds of mesoionic heterobases are possible, and examples of some of these have been presented herein. Among these, only a few nucleoside derivatives have been prepared. One mesoionic nucleoside, 7-methylguanosine (1), has generated considerable interest and has played a role in our current understanding of nucleic acid chemistry. Hopefully, research with this compound has only been a harbinger of things to come. Some of the mesoionic nucleosides (particularly, Class I purinones) appear susceptible to hydrolysis; this may be a limiting factor with respect to their utility in biological systems. On the other hand, it may be possible (a) to enhance stability by introduction of the appropriate substituents or (b) as with the Class II mesoionic xanthines, to take advantage of this reactivity in the development of irreversibly acting agents. Mesoionic nucleosides represent an enormous new class, indeed, a new field, of nucleosides that warrants further investigation.

5. References

1. R. P. Perry and D. E. Kelly, Existence of methylated messenger RNA in mouse L cells, *Cell* **1**, 37–42 (1974).
2. Y. Furuichi, Methylation-coupled transcription by virus-associated transcriptase of cytoplasmic polyhedrosis virus containing double-stranded RNA, *Nucleic Acids Res.* **1**, 809–822 (1974).

3. R. H. Hall, *The Modified Nucleosides in Nucleic Acids*, Columbia Press, New York (1971).

4. R. D. Hotchkiss, The quantitative separation of purines, pyrimidines and nucleosides by paper chromatography, *J. Biol. Chem.* **175**, 315–332 (1948).

5. H. Bredereck, H. Haas, and A. Martini, Uber Methylierte Nucleoside und Purine und ihre pharmakologischen Wirkungen, II. Mitteil.: Methylierung von Nucleosiden durch Dimethyl-sulfat, *Chem. Ber.* **81**, 307–313 (1948).

6. H. Bredereck, R. Sieber, L. Kamphenkel, and R. Bamberger, Uber Acylspal-tungen mit Diazo-methan, *Chem. Ber.* **89**, 1169–1176 (1956).

7. P. D. Lawley, Hydrolysis of methylated deoxyguanylic acid at pH 7.0 to yield 7-methylguanine, *Proc. Chem. Soc.* **1957**, 290–291.

8. P. D. Lawley and C. A. Wallick, Action of alkylating agents on deoxyribonucleic acid and guanylic acid, *Chem. Ind. (London)* **1957**, 633.

9. J. A. Haines, C. B. Reese, and A. R. Todd, The methylation of guanosine and related compounds with diazomethane, *J. Chem. Soc.* **1962**, 5281–5288.

10. D. B. Dunn, The isolation of 1-methyladenylic acid and 7-methylguanylic acid from ribonucleic acid, *Biochem. J.* **86**, 14–15 (1963).

11. P. Brookes and P. D. Lawley, The alkylation of guanosine and guanylic acid, *J. Chem. Soc.* **1961**, 3923–3928.

12. W. Baker, W. D. Ollis, and V. D. Poole, Cyclic mesoionic compounds. Part 1. The structure of the sydnones and related compounds, *J. Chem. Soc.* **1949**, 307–3314.

13. W. Baker and W. D. Ollis, Cyclic mesoionic compounds. Part III. Further properties of the sydnones and the mechanism of their formation, *J. Chem. Soc.* **1950**, 1542–1551.

14. W. D. Ollis and C. A. Ramsden, Mesoionic compounds, *Adv. Heterocycl. Chem.* **19**, 1–122 (1976).

15. R. A. Coburn, R. A. Carapellotti, and R. A. Glennon, Mesoionic purinone analogs. II. A PP π-SCF variable integral study of mesoionic analogs based upon six-membered ring mesoionic systems, *J. Heterocycl. Chem.* **10**, 479–485 (1973).

16. N. Demnis, A. R. Katritzsky, and Y. Takeuchi, Synthetic applications of heteroaromatic betaines with six-membered rings, *Angew. Chem. Int. Ed.* **15**, 1–9 (1976).

17. C. A. Ramsden, Structure and reactivity of dipolar heterocycles, *J. Chem. Soc., Chem. Commun.* **1977**, 109–110.

18. W. Friedrichsen, T. Kappe, and A. Bottcher, Six-membered mesoionic heterocycles of the *m*-quinodimethane dianion type, *Heterocycles* **19**, 1083–1148 (1982).

19. R. A. Coburn, Mesoionic purinone analogs I. Initial theoretical considerations, *J. Heterocycl. Chem.* **8**, 881–888 (1971).

20. A. Yamazaki, I. Kumashiro, and T. Takenshi, A simple method for the synthesis of inosine, 2-alkylinosine and xanthosine from 5-amino-1-β-D-ribofuranosyl-4-imidazole carboxamide, *J. Org. Chem.* **32**, 3258–3260 (1967).

21. P. Brookes, A. Dipple, and P. D. Lawley, The preparation and properties of some benzylated nucleosides, *J. Chem. Soc. (C)* **1968**, 2026–2028.

22. D. Ackermann and P. H. List, Constitution of herbipoline, a new purine base of animal origin, *Z. Physiol. Chem.* **309**, 286–290 (1958).

23. H. Bredereck, O. Christmann, and W. Koser, Struktur und Synthese des Herbipolins, *Chem. Ber.* **93**, 1206–1207 (1960).

24. P. D. Lawley and P. Brookes, Further studies on the alkylation of nucleic acids and their constituent nucleotides, *Biochem. J.* **89**, 127–128 (1963).

25. J. W. Jones and R. K. Robins, Potential purine antagonists. XXX. Purine betaines and related derivatives prepared by direct methylation of the simple purines, *J. Am. Chem. Soc.* **84**, 1914–1919 (1962).

26. B. Singer, Reaction of guanosine with ethylating agents, *Biochemistry* **11**, 3939–3947 (1972).

27. J. W. Jones and R. K. Robins, Purine nucleosides III. Methylation studies of certain naturally occurring purine nucleosides, *J. Am. Chem. Soc.* **85**, 193–201 (1963).

28. H. K. Scheit and A. Holy, Methylation of inosine and uridylyl-(3′-5′) inosine with dimethylsul-fate, *Biochim. Biophys. Acta* **149**, 344–354 (1967).

29. P. N. Magee and J. M. Barnes, Carcinogenic nitroso compounds, *Adv. Cancer Res.* **10**, 163–246 (1967).

30. A. D. Broom and R. K. Robins, Synthesis of a purine ribonucleoside thiobetaine, 7-methyl-5-thioguanosine, *J. Heterocycl. Chem.* **1**, 113–114 (1964).

31. A. D. Broom, L. B. Townsend, J. W. Jones, and R. K. Robins, Purine nucleosides. VI. Further methylation studies of naturally occurring purine nucleosides, *Biochemistry* **3**, 494–500 (1964).

32. P. D. Lawley, The action of alkylating agents on deoxyribonucleic acid (DNA), *J. Chim. Phys.* **58**, 1011–1120 (1961).

33. H. G. Khorana, Studies on polynucleotides. VII. Approaches to the marking of end groups in polynucleotide chains. The methylation of phosphomonoester groups, *J. Am. Chem. Soc.* **81**, 4657–4660 (1959).

34. S. Hendler, E. Furer, and P. R. Srinivasan, Synthesis and chemical properties of monomers and polymers containing 7-methylguanosine and an investigation of their substrate or template properties for bacterial deoxyribonucleic acid or ribonucleic acid polymerases, *Biochemistry* **9**, 4141–4153 (1970).

35. L. R. Mandel, P. R. Srinivasan, and E. Borek, Origin of urinary methylated purines, *Nature (London)* **209**, 586–588 (1966).

36. R. H. Hall, in: *Methods in Enzymology* (L. Grossman and K. Moldave, eds.), Vol. XII, Part A, pp. 305–315, Academic Press, New York (1967).

37. L. B. Townsend and R. K. Robins, Ring cleavage of purine nucleosides to yield possible biogenetic precursors of pteridines and riboflavin, *J. Am. Chem. Soc.* **85**, 242–243 (1963).

38. A. M. Michelson and F. Pochon, Polynucleotide analogues. VII. Methylation of polynucleotides, *Biochim. Biophys. Acta* **114**, 469–480 (1966).

39. T. Igo-Kemenes and H. G. Zachau, Involvement of 1-methyladenosine and 7-methylguanosine in the three-dimensional structure of (yeast) tRNA[Phe], *Eur. J. Biochem.* **18**, 292–298 (1971).

40. L. Rainen and D. B. Stollar, Antibodies distinguishing between intact and alkali-hydrolyzed 7-methylguanosine, *Nucleic Acid Res.* **5**, 4877–4889 (1978).

41. R. P. Perry and K. Scherrer, The methylated constituents of globin mRNA, *FEBS Lett.* **57**, 73–78 (1975).

42. Y. Uruichi, M. Morgan, S. Muthukrishnan, and A. J. Shatkin, Reovirus messenger RNA contains a methylated blocked 5'-terminal structure: m7G (5') Pppp (5') 9mpCp, *Proc. Natl. Acad. Sci. U.S.A.* **72**, 362–366 (1975).

43. G. W. Both, A. K. Banerjee, and A. J. Shatkin, Methylation-dependent translation of viral messenger RNAs *in vitro*, *Proc. Natl. Acad. Sci. U.S.A.* **72**, 1189–1193 (1975).

44. S. Muthukrishnan, G. W. Both, Y. Furuichi, and A. J. Shatkin, 5'-Terminal 7-methylguanosine in eukaryotic mRNA is required for translation, *Nature (London)* **255**, 33–37 (1975).

45. L. Levine, V. H. Vunakis, and R. C. Gallo, Serologic specificities of methylated base immune systems, *Biochemistry* **10**, 2009–2013 (1971).

46. D. L. Sawicki, P. R. Srinivasan, and S. M. Beiser, Antibodies specific for 7-methylguanosine, *Fed. Proc.* **31** (Abstract 3343), 808 (1972).

47. T. W. Munns, M. K. Liszewski, and H. F. Sims, Characterization of antibodies specific for N^6-methyladenosine and for 7-methylguanosine, *Biochemistry* **16**, 2163–2168 (1977).

48. R. Wagner and R. A. Garrett, Chemical evidence for a codon-induced allosteric change in tRNALys involving 7-methylguanosine residue 46, *Eur. J. Biochem.* **97**, 615–621 (1979).

49. R. E. Hurd and B. R. Reid, Nuclear magnetic resonance studies on the tertiary folding of transfer ribonucleic acid: Assignment of the 7-methylguanosine resonance, *Biochemistry* **18**, 4017–4025 (1979).

50. D. L. VonMilden, R. N. Stillwell, W. A. Koenig, K. J. Lyman, and J. A. McCloskey, Mass spectrometry of 7-methylpurine nucleosides: Studies of a characteristic oxygen incorporation reaction that occurs during trimethylsilylation, *Anal. Biochem.* **50**, 110–121 (1972).

51. T. Igo-Kemenes and H. G. Zachau, On the specificity of the reduction of transfer ribonucleic acids with sodium borohydride. *Eur. J. Biochem.* **10**, 549–556 (1969).

52. F. Pochon, Y. Pascal, P. Pitha, and A. M. Michelson, Photochimie des polynucléotides. Photochimie de quelques nucléosides puriques méthylés, *Biochim. Biophys. Acta* **213**, 273–281 (1970).

53. H. Bredereck, G. Kupsh, and H. Wieland, "Desoxyharnsauren," Ihre konstitution als xanthiniumbetaine und neue Synthesen, *Chem. Ber.* **92**, 566–582 (1959).

54. K. Senga, M. Ichiba, and S. Nishigaki, Synthesis and properties of [1,2,3]thiadiazolo[4,5-*d*]-

pyrimidine derivatives including their mesoionic compounds. A. New class of heterocycles, *J. Org. Chem.* **43**, 1677–1683 (1978).

55. L. B. Townsend and R. K. Robins, Potential purine antagonists. XXXIV. The synthesis of 3-methylguanine and a study of the structure and chemical reactivity of certain 3-methyl-purines, *J. Am. Chem. Soc.* **84**, 3008–3012 (1962).

56. A. J. Albert, *v*-Triazolo[4,5-*d*]pyrimidines (8-azapurines). Part 24. The 3-alkyl derivatives, *J. Chem. Soc., Perkin Trans. 1* **1981**, 2344–2351.

57. B. Pullman, in: *Quantum Aspects of Heterocyclic Compounds in Chemistry and Biochemistry* (E. D. Bergmann and B. Pullman, eds.), Vol II, pp. 292–397, Academic Press, New York (1970).

58. J. E. Abola, D. J. Abraham, and L. B. Townsend, The crystal and molecular structure of 3-methylguanine, a potentially miscoding base, *Tetrahedron Lett.* **1976**, 2483–3486.

59. R. A. Coburn and R. A. Carapellotti, Mesoionic purinone analogs. VI. Mesoionic 1,3-dialkyl-hypoxanthines, *Tetrahedron Lett.* **1974**, 663–666.

60. R. A. Glennon, M. E. Rogers, and M. K. El-Said, Imidazo[1,2-*a*]pyrimidines and 1,2,4-triazolo-[1,5-*a*]pyrimidines: Two new examples of mesoionic xanthine analogs, *J. Heterocycl. Chem.* **17**, 337–340 (1980).

61. I. Shehata and R. A. Glennon, Mesoionic isoxazolo[2,3-*a*]pyrimidinediones and 1,3,4-oxadiazolo [3,2-*a*]pyrimidinediones as potential adenosine antagonists, *J. Heterocycl. Chem.* **24**, 1291–1296 (1987) and unpublished data.

62. R. A. Coburn and R. A. Glennon, Mesoionic purinone analogs. III. The synthesis and properties of mesoionic thiazolo-[3,2-*a*]pyrimidine-5,7-diones, *J. Heterocycl. Chem.* **10**, 487–494 (1973).

63. R. A. Coburn and R. A. Glennon, Mesoionic purinone analogs. IV: Synthesis and *in vitro* antibacterial properties of mesoionic thiazolo[3,2-*a*]pyrimidin-5,7-diones, *J. Pharm. Sci.* **52**, 1785–1789 (1973).

64. R. A. Coburn, R. A. Glennon, and Z. F. Chmielewiez, Mesoionic purinone analogs. 7. *In vitro* antibacterial activity of mesoionic 1,3,4,-thiadiazolo[3,2-*a*]pyrimidine-5,7-diones. *J. Med. Chem.* **17**, 1025–1027 (1974).

65. R. A. Glennon, M. E. Rogers, R. G. Bass, and S. B. Ryan, Mesoionic xanthine analogs as inhibitors of cyclic AMP phosphodiesterase, *J. Pharm. Sci.* **67**, 1762–1765 (1978).

66. R. A. Glennon, J. J. Gaines, and M. E. Rogers, Benz-fused mesoionic xanthine analogues as inhibitors of cyclic-AMP phosphodiesterase, *J. Med. Chem.* **24**, 766–769 (1981).

67. R. A. Glennon, M. E. Rogers, J. D. Smith, M. K. El-Said, and J. L. Egle, Mesoionic xanthine analogues: Phosphodiesterase inhibitory and hypotensive activity, *J. Med. Chem.* **24**, 658–661 (1981).

68. R. A. Glennon, S. M. Tejani-Butt, W. Padgett, and J. W. Daly, Mesoionic xanthine analogues: Antagonists of adenosine receptors, *J. Med. Chem.* **27**, 1364–1367 (1984).

69. R. A. Glennon, R. G. Bass, and E. Schubert, Alkylation studies on 6-ethyl-2,3-dihydrothiazolo-[3,2-*a*]pyrimidine-5,7-diones, *J. Heterocycl. Chem.* **16**, 903–907 (1979).

70. K. T. Potts, R. Ehlinger, and S. Kanemasa, Reactions of (chlorocarbonyl)phenylketene formation of ring-fused 4*H*-1,3-thiazinium betaines from heterocyclic thiones, *J. Org. Chem.* **45**, 2474–2479 (1980).

71. T. Kappe and W. Lube, Zur Synthese Mesomerer Pyrimidinbetaine, *Monatsh. Chem.* **102**, 781–787 (1971).

72. G. O. Mbagwu, R. G. Bass, and R. A. Glennon, Carbon-13 nuclear magnetic resonance spectra of some mesoionic xanthine analogs, *Org. Magn. Reson.* **21**, 527–331 (1983).

73. E. M. Schubert, K. H. Schram, and R. A. Glennon, Mass spectrometry of modified nucleic acid bases and nucleosides: Class II mesoionic nucleosides and bases derived from thiazolo[3,2-*a*]-pyrimidine-5,7-diones, *J. Heterocycl. Chem.* **22**, 889–905 (1985).

74. G. O. Mbagwu, R. G. Bass, and R. A. Glennon, Studies of amine-induced ring opening of some mesoionic xanthines, *J. Heterocycl. Chem.* **22**, 465–474 (1985).

75. M. Hellberg, J. F. Stubbins and R. A. Glennon, Unpublished data.

76. K. T. Potts and R. K. C. Hsia, Mesoionic compounds. XXIII. The anhydro-2-hydroxy-4-oxo-1,6,8-trimethylpyrimido[1,2-*a*]pyrimidinium hydroxide system, *J. Org. Chem.* **38**, 3485–3486 (1973).

77. J. Honzl and M. Sorm, A new mesoionic system with a six-membered ring; reactions of 1,4-disubstituted 2,6-dioxo piperazines with benzene-sulfonyl and benzene sulfonyl chloride in pyridine, *Tetrahedron Lett.* **1969**, 3339–3342.

78. T. Kappe and E. Ziegler, Carbon suboxide in preparative organic chemistry, *Angew. Chem. Int. Ed. Engl.* **13**, 491–558 (1974).

79. R. A. Glennon, E. Schubert, and R. G. Bass, Synthesis of mesoionic xanthine nucleosides, *Tetrahedron Lett.* **22**, 2753–2756 (1981).

80. E. M. Schubert, R. G. Bass, and R. A. Glennon, Synthesis of mesoionic xanthine nucleosides, *Nucleosides Nucleotides* **2**, 127–146 (1983).

81. S. M. Tejani-Butt and R. A. Glennon, Unpublished data.

82. S. M. Tejani, Synthesis of mesoionic nucleosides as potential antineoplastic agents, *Diss. Abstr. Int.* **44B**, 3091-B (1984).

83. R. A. Coburn and M. D. Taylor, Mesoionic Purinone Analogs. VIII. Synthesis and properties of mesoionic 5-substituted 6-methylimidazo-[1,2-*c*]-pyrimidine-2,7-diones, *J. Heterocycl. Chem.* **19**, 567–572 (1982).

84. W. Friedrichsen, E. Kujath, G. Liebezeit, R. Schmidt, and I. Schwarz, Mesoionic six-membered heterocycles, I. Synthesis of 6-oxo-6*H*-1,3-oxa-zin-3-ium-4-olates, *Justus Liebigs Ann. Chem.* **1978**, 1655–1665.

85. R. E. Bambury, D. T. Fedey, G. C. Lowton, J. M. Weaver, and J. Wemple, Mesoionic pyridazine ribonucleosides: A novel biologically active nucleoside metabolite, *J. Med. Chem.* **27**, 1613–1621 (1984).

86. R. A. Coburn and B. Bhooshan, Mesoionic purinone analogs. V. Synthesis of mesoionic thiazolo[3,2-*a*]-*s*-triazine-5,7-diones, mesoionic 1,3,4-thiadiazolo[3,2-*a*]-*s*-triazine-5,7-diones, and their monothione derivatives, *J. Org. Chem.* **38**, 3868–3871 (1973).

87. R. A. Coburn, and B. Bhooshan, Mesoionic 1,3-disubstituted-*s*-triazin-4, 6-diones and their thione derivatives, *J. Heterocycl. Chem.* **12**, 187–189 (1975).

88. E. J. Prisbe, J. P. H. Verheyden, and J. G. Moffatt, 5-Aza-7-deazapurine nucleosides. I. Synthesis of some 1-(β-D-ribofuranosyl)imidazo[1,2-*a*]-1,3,5-triazines, *J. Org. Chem.* **43**, 4774–4783 (1978).

89. K. T. Potts and M. J. Sorm, Mesoionic compounds. XVI. 1,4-Dipolar type cycloaddition reactions utilizing pyrimidinium betaines, *J. Org. Chem.* **37**, 1422–1425 (1972).

90. T. Kappe, W. Golser, M. Hariri, and W. Stadlbauer, Synthese und Reaktionen messionischer 1,3-Oxazine, *Chem. Ber.* **112**, 1585–1594 (1979).

91. T. Debaerdemaeker and W. Friedrichsen, Mesoionische Sechsringheterocyclen; Struktur des 2-Ferrocenyl-3,5-diphenyl-6-oxo-6*H*-1,3-oxazin-3-ium-4-olats, *Z. Naturforsch. B* **37**, 217–221 (1982).

92. T. Kappe and W. Golser, Synthese Mesomerer 1,3-Thiazine-Betaine, *Synthesis* **1972**, 312–314.

93. K. T. Potts, R. Ehlinger, and W. M. Nichols, Mesoionic compounds. XXXIII. Thermal rearrangement of 4*H*-1,3-thiazinium betaines to 4-quinolones, *J. Org. Chem.* **40**, 2596–2600 (1975).

94. T. Kappe and W. Golser, Mesoionic six-membered heterocycles. VIII. 1,4-Dipolar cycloadditions to mesoionic 1,3-thiazines, *Chem. Ber.* **109**, 3668–3674 (1976).

95. R. A. Coburn, M. D. Taylor, and W. L. Wright, Preliminary evaluation of mesoionic 6-substituted 1-methylimidazo[2,1-*b*]-1,3-thiazine-5,7-diones as potential novel prodrugs of methimazole, *J. Pharm. Sci.* **70**, 1322–1324 (1981).

96. R. A. Glennon and S. M. Tejani-Butt, Mesoionic nucleosides, 3. Studies on mesoionic imidazo[1,1-*b*]-1,3-thiazine derivatives, *Nucleosides Nucleotides* **3**, 389–400 (1984).

97. R. E. Harmon, R. K. Robins, and L. B. Townsend, *Chemistry and Biology of Nucleosides and Nucleotides*, Academic Press, New York (1978).

98. R. T. Walker, E. DeClercq, and F. Eckstein, *Nucleoside Analogs: Chemistry, Biology and Medical Applications*, Plenum Press, New York (1979).

99. R. E. Bambury, D. T. Feeley, G. C. Lawton, J. M. Weaver, and J. Wemple, Mesoionic pyridazine and pyridine nucleosides. An unusual biologically active nucleoside metabolite, *J. Chem. Soc., Chem. Commun.* **1984**, 422–423.

Chapter 2

Synthesis and Properties of Various Disaccharide Nucleosides

Leon M. Lerner

1. Introduction

The discovery of amicetin and its close relatives plicacetin and bamicetin in the early 1950s helped to usher in a new era of interest in nucleoside antibiotics. It also aroused interest in the synthesis of fraudulent nucleosides having disaccharide components at a time when even the number of monosaccharide analogue substitutions for D-ribose and 2-deoxy-D-*erythro*-pentose were still relatively few. The preparation of disaccharide nucleosides was pursued either with the goal of studying the chemistry and biological properties of these novel compounds or with the intent of solving problems expected to arise with the synthesis of an antibiotic such as amicetin. Subsequently, other nucleoside antibiotics were discovered that contained more than one sugar residue. This chapter discusses only true nucleosides—those with a bond from the anomeric carbon of a disaccharide to the heterocyclic ring—and only those whose structures were completely known at the time of writing.

2. Disaccharide Nucleoside Antibiotics

2.1. Amicetin and Related Antibiotics

The isolation of amicetin occurred almost simultaneously in two laboratories. It was obtained from the culture broth of either *Streptomyces fasiculatus*[1] or *Streptomyces vinaceusdrappus*.[2,3] Shortly after, it was reported that an unidentified species of *Streptomyces* elaborated an antibiotic called sacromycin, which was shown to be identical to amicetin by its melting point behavior and IR spectrum.[4] Similarly, an antibiotic called allomycin was isolated from the cul-

Leon M. Lerner • Department of Biochemistry, State University of New York Health Sciences Center, Brooklyn, New York 11203.

ture broth of *Streptomyces sindenensis* and was found to be the same as amicetin only after considerable structural work had been accomplished.[5].

Amicetin exists in two isomorphic forms. Crystallization from an aqueous solution of pH 8.5 gave needles, mp 160-165°C, whereas crystals melting at 244–245°C could be obtained by heating the needles in methanol or by crystallization from water heated to 70°C.[2] The structure proof of amicetin required a number of years, and several wrong conclusions were reached before the correct structure was finally published. Early experiments[2] indicated the formula $C_{29}H_{44}N_6O_9$, pK_a values of 7 and 10.4, and two N-methyl groups, three C-methyl groups, and one primary amino group.[6]

When amicetin was hydrolyzed with a dilute solution of sodium hydroxide, an aminosugar-containing component named cytosamine was obtained.[6] Further hydrolysis of cytosamine with strong acid afforded cytosine. Mild acid hydrolysis of amicetin yielded another noncarbohydrate-containing fragment called cytimidine. Further hydrolysis of cytimidine produced cytosine, *p*-aminobenzoic acid, and an amino acid that was identified as (+)-α-methylserine by a comparison of its properties with those of a synthetic sample. Because cytimidine had no free carboxyl group, it was concluded that both ends of *p*-aminobenzoic acid were linked through amide bonds. Also, the pK_a of 10.4, which was indicative of an amide bond between cytosine and a carboxylic acid, disappeared upon alkaline hydrolysis. A model compound, 1-methyl-N^4-benzoylcytosine, exhibited the same behavior. Cytimidine rapidly consumed one mole of periodate, forming formaldehyde, ammonia, and crystalline pyruvyl-*p*-aminobenzoylcytosine. The structure of cytimidine was therefore proposed as **1**.[6]

It was recognized that the cytosine part of amicetin was common to both cytimidine and cytosamine. Acid hydrolysis broke the nucleoside C—N bond, yielding cytimidine; with alkaline hydrolysis, the amide bond at N^4 of cytosine was split to afford cytosamine. The sugar bond to N-1 of cytosine was demonstrated by first treating cytosamine with nitrous acid, followed by methylation with diazomethane, and finally hydrolysis. The product was 3-methyluracil, which excluded the possibility of N-3 substitution in cytosamine.[6] Acid hydrolysis of amicetin yielded a crystalline compound that had the properties of an aminosugar disaccharide.[7] This compound was named amicetamine and was recognized as part of the cytosamine fragment. Amicetamine was hydrolyzed to give an aminosugar called amosamine and a deoxysugar called amicetose (**2**). Both structures were incorrectly assigned at the time, although it was correctly shown that amosamine contained a dimethylamino group.[7]

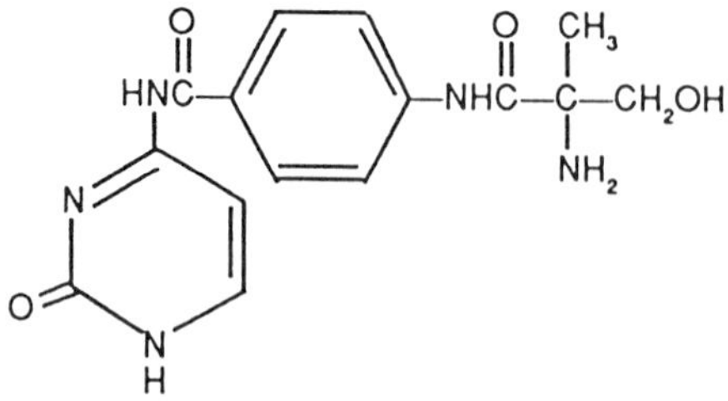

(1)

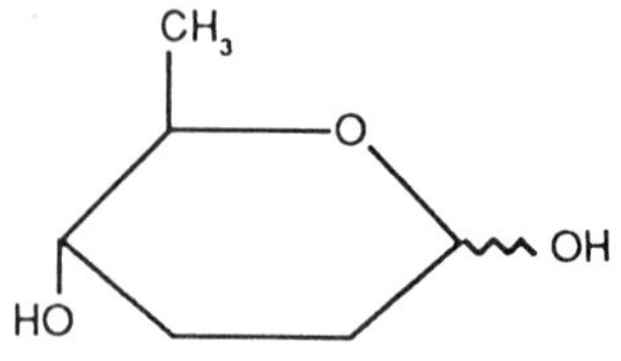

(2)

A few years later, Stevens *et al.*[8] reported the results of a meticulous reinvestigation of the structure of amicetin. Acid-catalyzed methanolysis of amicetin gave cytimidine hydrochloride together with the anomeric methyl glycosides of amicetose and amosamine, which were separated by column chromatography on a cation-exchange resin. Group analysis of the amicetose glycosides revealed one methyl group and one methoxyl group. There was no consumption of periodate unless the methoxyl group was removed by hydrolysis, whereupon one mole was taken up. The free sugar produced a positive iodoform test, which was interpreted to mean that a methylcarbinol was present. Amicetose was converted into a 2,4-dinitrophenylhydrazone, which consumed one mole of periodate and produced succindialdehyde and acetaldehyde. The configuration of the hydroxyl group at C-4 was determined by a comparison of the rate of periodate consumption of natural amicetose with that of the DL-*threo* and DL-*erythro* isomers, which were prepared synthetically.[9] The *erythro* isomer had the same rate of uptake as natural amicetose, whereas the *threo* isomer consumed periodate faster.

Group analysis of amosamine showed the presence of one C-methyl group and one methoxyl group in the glycoside mixture. Acid hydrolysis liberated amosamine, which gave a positive iodoform reaction. Borohydride reduction of amosamine afforded an alditol, amosaminol. The latter consumed four moles of periodate and liberated one mole of formaldehyde. These results supported the presence of a terminal methyl group in the structure of amosamine. In earlier work, Hochstein and Regna[10] compared the rate of deaminolysis of various aminosugars under alkaline conditions. Mycaminose and desosamine, C-3 aminosugars, had a half-life of 0.5 h, whereas the half-life of 2-amino-2-deoxy-D-glucose was 20 h. Applying this work to their studies on amosamine, Stevens *et al.*[8] found that the half-life of deaminolysis was 24 h. Therefore, the assignment of C-3 as the position of the dimethylamino group seemed incorrect, but they argued against C-2 on the basis of the ease of acid hydrolysis of the glycosides; 2-amino groups on glycosides have a protective effect due to the positive charge. An amino substituent at C-2 was also not considered because *N,N*-dimethylglucosaminol produced 0.5 mol of formaldehyde after 50 h of periodate oxidation, but amosaminol required only 1 h. The possibility of C-4 substitution was rejected based upon the assertion that such amines undergo alkaline deaminolysis very rapidly.[10] As a result, the dimethylamino group was placed at C-3 despite the kinetic evidence. This decision was reconsidered later. This reinvestigation was inspired by the low level of basicity observed for amosamine and amicetamine (pK_a 7) compared to mycaminose (3,6-dideoxy-3-dimethylamino-β-D-glucose) (pK_a 8.5) and desosamine (3-dimethylamino-3,4,6-trideoxy-β-D-*xylo*-hexose) (pK_a 8.6). The correct structure of amosamine was finally demonstrated to be 4,6-dideoxy-4-dimethylamino-D-glucose (**3**) by synthesis.[11]

The linkage of amicetose to cytosine was decided on the basis of the ease of hydrolysis of this bond. 2'-Deoxynucleosides are hydrolyzed in acid more easily than those containing a C-2' hydroxyl group. Since the only available hydroxyl group in amicetose to form a glycosidic bond to amosamine was at C-4, this had to be the position for the disaccharide linkage.[8]

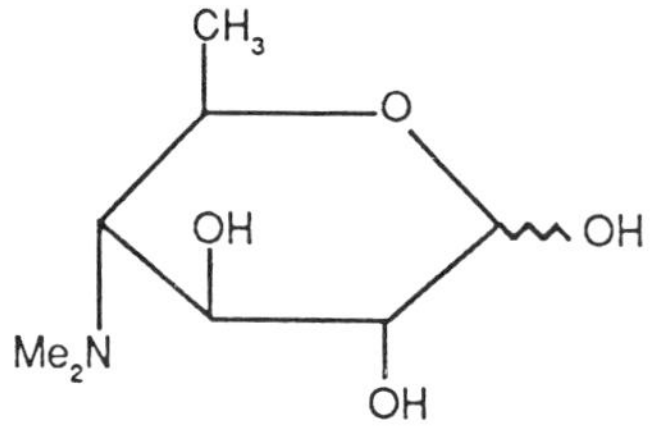

(3)

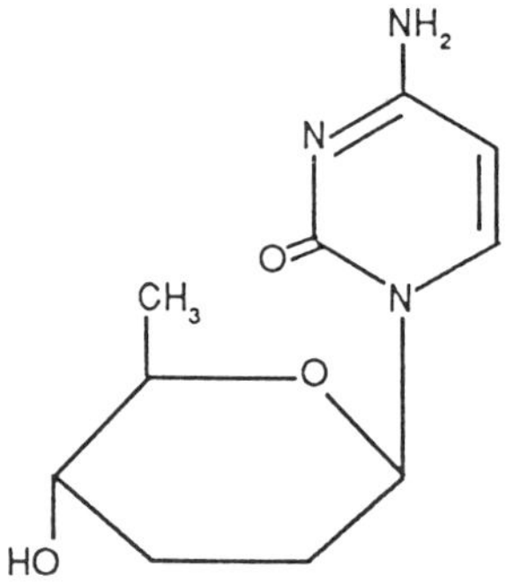

(4)

The complete structure of amicetin was finally solved with the determination of the configuration at the anomeric linkages using NMR spectroscopy.[12] Cytosamine was degraded to 1-(2,3,6-trideoxy-β-D-*erythro*-hexopyranosyl)cytosine (4) by a sequence of reactions consisting of periodate oxidation, borohydride reduction, and mild acid hydrolysis. The spectrum showed that the anomeric proton of 4 had an axial orientation, and the spectrum of the acetylated nucleoside supported the equatorial placement of the hydroxyl at C-4. In order to determine the configuration of the bond between amosamine and amicetose, amicetamine was reduced with sodium borohydride to amicetaminol to remove the contribution of the anomeric proton of amicetose. Since the structure of amosamine was known to have the D-*gluco* configuration, the small doublet, $J_{1,2} = 3.5$ Hz, for the anomeric proton suggested that the glycosidic bond to amicetose was α 1→4, giving amicetamine structure 5, cytosamine structure 6 with a β linkage at the nucleoside bond, and, thus, the structure of amicetin (7).

In 1957, when structural studies on amicetin were in progress, an antibacterial agent was isolated from a species of *Streptomyces* obtained from a soil sample from Chile.[13] This substance had the elemental formula $C_{25}H_{35}N_5O_7$ and resembled amicetin, but certain small details in the UV and IR spectra were different. It was accordingly named amicetin B. Shortly thereafter, three antibiotics were isolated from the culture filtrate of *Streptomyces plicatus*, and these were inhibitory against experimental mouse tuberculosis.[14] One of these compounds was amicetin; a second one, named plicacetin, was identical to amicetin B. The

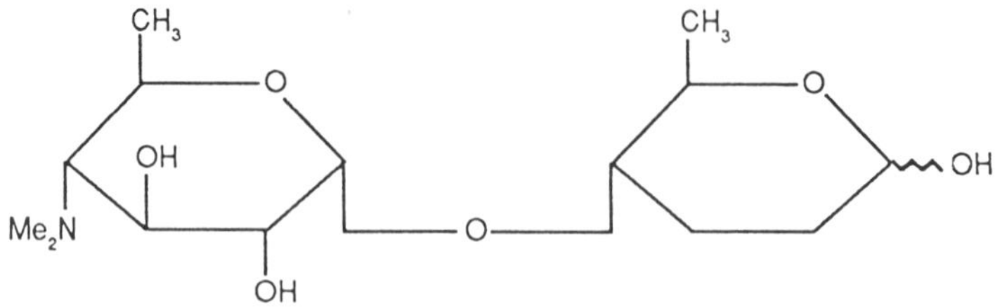

(5)

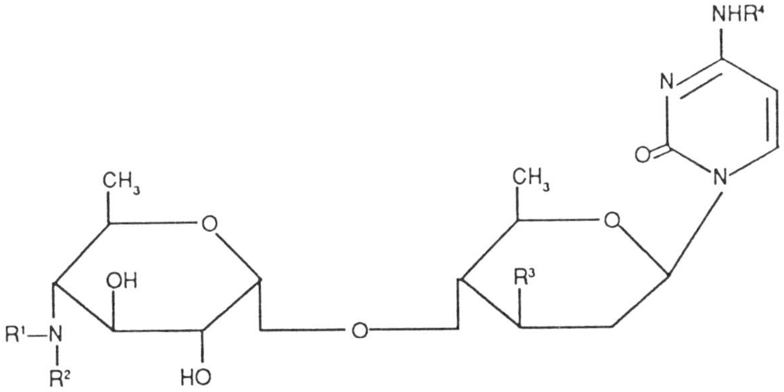

6, R^1 = R^2 = CH$_3$; R^3 = R^4 = H

7, R^1 = R^2 = CH$_3$; R^3 = H; R^4 =

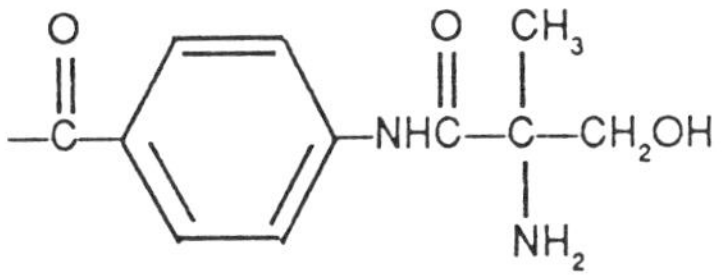

8, R^1 = R^2 = CH$_3$; R^3 = H; R^4 =

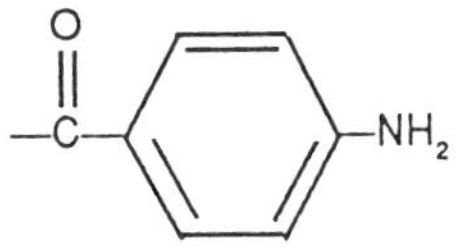

9, R^1 = CH$_3$; R^2 = R^3 = H; R^4 =

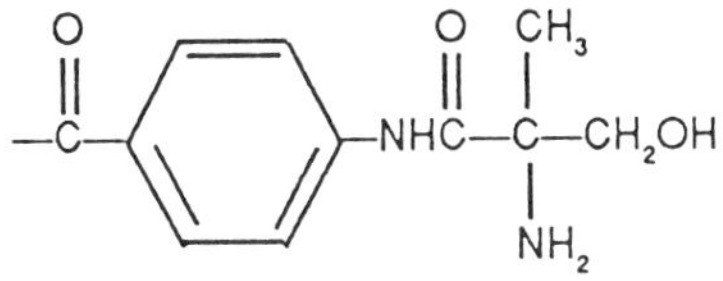

10, R^1 = R^2 = CH$_3$; R^3 = OH; R^4 =

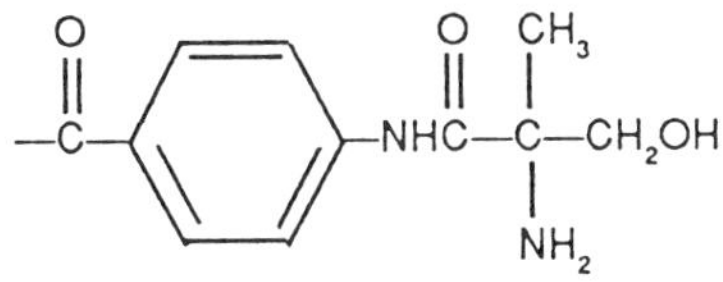

third compound was new and had one less carbon atom than amicetin. This compound was named bamicetin.

Plicacetin had a pK_a of 2.2, corresponding to an arylamine group, and the presence of this group was verified by a positive Ehrlich dimethylamino benzaldehyde test[14] and by diazotization and reaction with phenol.[13] It did not give a positive ninhydrin test for an amino acid. Upon alkaline hydrolysis, cytosamine was obtained, but acid hydrolysis gave only cytosine and *p*-aminobenzoic acid, with no (+)-α-methylserine.[13] Therefore, plicacetin (**8**) simply lacked the amino acid, and conclusive proof of its structure was obtained by the treatment of **6** with *p*-nitrobenzoyl chloride and reduction of the nitro group to an amino group.[15] Nearly 20 years later, a fourth amicetin family antibiotic was isolated from the same organism. Field desorption mass spectrometry and NMR spectroscopy revealed that the compound was related to plicacetin but had one less methyl group on the aminosugar nitrogen. Thus, the new antibiotic was called norplicacetin.[16]

Since acid hydrolysis of bamicetin gave cytimidine (**1**), it was clear that the structural difference with amicetin resided in the sugar.[14] The disaccharide from bamicetin was different, and although it consumed the same amount of periodate, one of the products of this oxidation was methylamine rather than dimethylamine. Bamicetin (**9**) simply contained a monomethylamino group in place of the dimethylamino group of amicetin.[15]

Another amicetin-like antibiotic, reported in 1973, was named oxamicetin in recognition that it contained one more oxygen atom than amicetin.[17] It was isolated from the fermentation broth of *Arthrobacter oxamicetus* by very nearly the same procedure as employed in the isolation of amicetin itself, except that the hydrochloride salt was purified by alumina chromatography. The free base was crystallized from aqueous methanol as a monohydrate and had the formula $C_{29}H_{42}N_6O_{10} \cdot H_2O$. UV, IR, and NMR spectroscopy revealed close structural similarities to amicetin.[17] Acid methanolysis gave cytimidine and the anomeric methyl glycosides of amosaminide.[18] This implied that the structural difference was in the amicetosyl moiety. A cytosine nucleoside was separated by chromatography. The NMR spectrum of this nucleoside indicated that it was 1-(2,6-dideoxy-β-D-*arabino*-hexopyranosyl)cytosine [1-(β-3-hydroxyamicetosyl)-cytosine]. Confirmation of the configuration of the hydroxyl group at C-3 was obtained by measuring the rotational shift in cuprammonium salt solution, which indicated an angle of 60° between the hydroxyl groups at C-3 and C-4, placing the hydroxyl at C-3 in an equatorial orientation. Alkaline hydrolysis yielded hydroxycytosamine. The glycosidic bond linking amosamine to 3-hydroxyamicetose was considered to have the α configuration based on a doublet, $J_{1,2} = 3.5$ Hz, at δ 5.05. Proof that the bond was to the hydroxyl group at C-4 rather than C-3 was also obtained by NMR spectroscopy. Acetylation of the nucleoside caused three sugar protons to shift downfield. Two of these were assigned to C-2 and C-3 of amosamine. The third one, at δ 5.14, was considered to belong to either C-3 or C-4 of 3-hydroxyamicetose. A double-resonance experiment established that this proton was vicinal to the proton at C-2′, which meant that it had to be the proton at C-3′, and the disaccharide bond had to be at the C-4′ hydroxyl. The mass spectrum also supported the structure proposed for the tetraacetate derivative. The structure of oxamicetin is shown as **10**.

2.2. Hikizimycin

In 1971, it was reported[19] that a new antibiotic had been isolated from the culture broth of a strain of *Streptomyces* obtained from a soil sample near the Hikizi River in Japan. The antibiotic, named hikizimycin after the location of its discovery, was found to be active against fungi, especially phytopathogenic fungi. Hikizimycin was not extractable from the culture broth with organic solvents but was isolable by a combination of chromatographic techniques, particularly ion-exchange and charcoal column chromatography. It was crystallized as a hydrobromide salt of indeterminate melting point.

The UV spectrum of hikizimycin suggested that it contained cytosine, and this was supported by the doublets at δ 6.30 and 7.95 in the NMR spectrum and the IR peak at 1675 cm^{-1}. Peracetyl hikizimycin had 13 acetyl groups, as determined by an integration of the acetyl peaks in the NMR spectrum. The mass spectrum had a molecular-ion peak at *m/e* 1129 and a peak at *m/e* 977, indicating loss of an *N*-acetylcytosine, which appeared as a peak at *m/e* 154. Methanolic sodium methoxide removed all but two of the *N*-acetyl groups. This material was treated with methyl iodide and sodium hydride in *N,N*-dimethylformamide to give permethyl-*N,N*-diacetylhikizimycin, which had 14 methyl groups. Since two of these were acetamido groups and two methyl groups were bonded to the amino group of cytosine, it was concluded that hikizimycin had 10 hydroxyl groups, with the empirical formula being $C_{21}H_{37}N_5O_{14}$.[20]

Acidic methanolysis of *N,N*-diacetylhikizimycin gave two fragments that were separated by ion-exchange chromatography. One of these fragments and its peracetyl derivative had properties that were identical to those of 3-amino-3-deoxy-D-glucose (**11**, kanosamine). The second fragment provided two more components by hydrogenation and acid hydrolysis. One of these components was identified as tetrahydropyrimidin-2-one by comparison with a sample prepared from cytidine in the same way.[20] Treatment of the reduced nucleoside fragment with methanolic hydrogen chloride instead of aqueous acid produced a mixture of α,β-glycosides of an aminosugar, aptly named hikosamine. The anomeric mixture was acetylated and separated by silica gel chromatography to afford the anomers as crystalline compounds in a 1:1 ratio. Base-catalyzed methanolysis of these anomers yielded crystalline methyl *N*-acetyl-β-hikosaminide and the α-anomer, which did not crystallize.[21]

Periodate oxidation of methyl *N*-acetyl-α-hikosaminide resulted in a rapid consumption of five moles and a slow consumption of one additional mole. This was interpreted to mean that the compound contained a long exocyclic polyhydroxyl chain and two *trans*-oriented vicinal hydroxyl groups in a pyranose ring. Termination of the periodate reaction after the first five moles had been consumed gave a product, which was reduced with sodium borohydride and acetylated. Methyl 4-acetamido-2,3,6-tri-*O*-acetyl-4-deoxy-α-D-glucopyranoside (**12**), a known compound, was identified as the product. It was proposed that hikosamine consisted of a pyranose ring with a 6-carbon alditol chain projecting

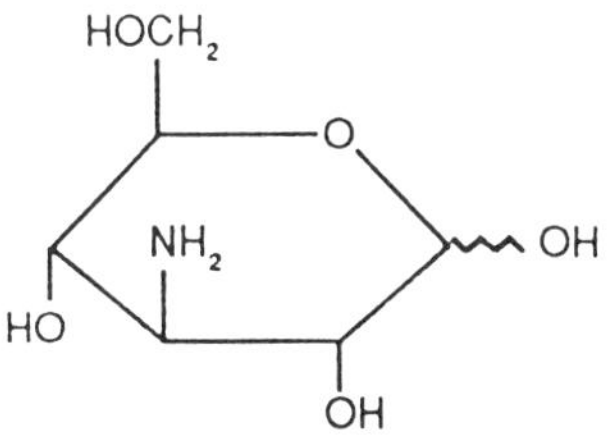

(11) 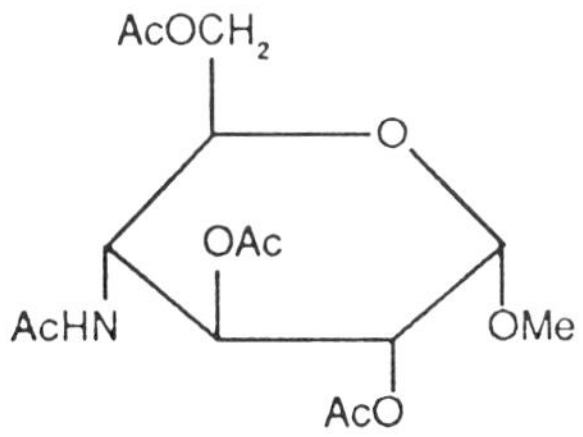   (12)

from C-5, and the mass spectrum supported this proposed structure. Therefore, it appeared that hikosamine was a rare 11-carbon aminosugar.[21]

In a rather clever experiment, the configurations of the hydroxyl groups in the alditol chain were elucidated.[22] Hikosamine was converted to the *N*-benzyloxycarbonyl derivative and then acetylated. The protected sugar was treated with piperidine in tetrahydrofuran, which liberated the anomeric hydroxyl group. Hydrogenolysis of the benzyloxycarbonyl group gave three products. This was explained by an acetyl migration to the amino group, which could occur from either side. The entire mixture was subjected to periodate oxidation. Deacetylation and preparative paper chromatography gave three sugars, which were identified as D-mannose, D-*glycero*-D-*galacto*-heptopyranose, and *N*-acetylhikosamine. These products were accounted for as those expected if the original mixture of acetates consisted of the three compounds **13**, **14**, and **15**. Compound **13** would be cleaved between C-4 and C-5 and give the acetylated aldoheptose. Oxidation of **14** would cause cleavage between C-5 and C-6, affording the acetylated aldohexose, and **15** would not be affected by periodate. The fact that D-mannose and D-*glycero*-D-*galacto*-heptopyranose were the products after deacetylation proved the absolute configurations of each of the hydroxyl groups in the entire exocyclic chain in one experiment. Therefore, the structure of hikosamine was 4-amino-4-deoxy-D-*glycero*-D-*galacto*-D-*gluco*-undecapyranose (**16**).[22]

The next problem to be solved was the position of the bond between 3-amino-3-deoxy-D-glucose and hikosamine. Since hikizimycin rapidly consumed five moles of periodate, it was considered unlikely that bonding was to the exocyclic chain. Hikizimycin consumed eight moles of periodate; therefore, substitution at C-3 was not possible. Bonding of **11** through C-4 of hikosamine would have been contrary to the known chemical properties of hikizimycin. Thus, the C-2' position of hikosaminylcytosine was concluded to be the point of attachment, and this was thought to be consistent with the data. The anomeric configurations were studied by NMR spectroscopy. The anomeric protons at δ

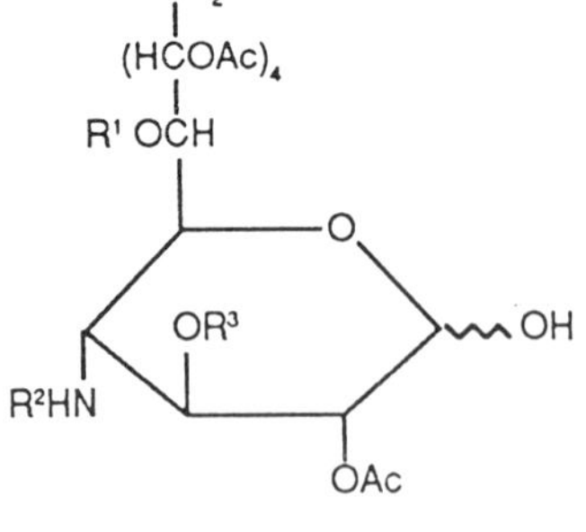

13, R¹ = Ac; R² = H; R³ = Ac

14, R¹ = H; R² = H; R³ = Ac

15, R¹ = R² = Ac; R³ = H

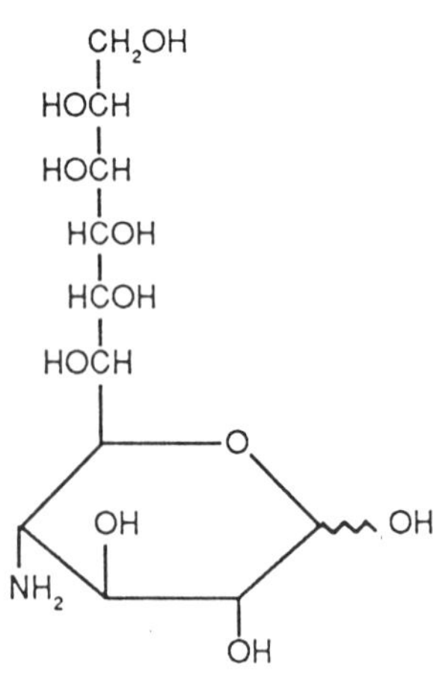

(16)

5.79 and 4.98 were easy to differentiate because the anomeric proton of hikosaminylcytosine itself appeared at δ 5.75. The large coupling constants, $J_{1',2'} = 8.8$ Hz and $J_{1'',2''} = 8.0$ Hz, indicated that the protons had diaxial orientations in both cases with C1 conformations. Thus, the anomeric configurations were β, and the complete structure of hikizimycin was proposed as **17**.[22] This structure was believed to be confirmed by a ^{13}C-NMR study.[23] However, another research group disputed the position of the linkage of 3-amino-3-deoxy-D-glucose to the C_{11} sugar.[24] They demonstrated by ^{13}C-NMR and X-ray powder diffraction patterns that hikizimycin and anthelmycin, the latter an antibiotic discovered in 1964,[25] were one and the same compound. They showed that the peak for the 2'-proton of the C_{11} sugar shifted downfield upon acetylation of hikizimycin. This was interpreted to mean that the 2'-hydroxyl group had been acetylated, and this view was supported by a ^{13}C-NMR spectrum. The position of the glycosidic linkage was proposed to be at C-6' from the results of two series of experiments. The pyrimidine ring of N,N'-diacetylhikizimycin was reduced by hydrogenation, and the product was permethylated and then hydrolyzed to give 3,4,5,6-tetrahydropyrimidin-2-one and the methylated disaccharide. The latter was reduced with sodium borohydride to the alditol disaccharide, the free hydroxyl groups were methylated, and the product was subjected to mass spectral analysis. The lack of peaks corresponding to either a 2'- or a 3'-linkage and the presence of peaks at m/e 268 and m/e 557 indicated bonding at C-6' and, therefore, structure **18** for hikizimycin. In a confirmatory experiment, N,N'-diacetylhikizimycin was treated with periodate and the products were reduced with sodium borohydride. The mixture of nucleoside disaccharides (**19** and **20**) was peracetylated and separated by column chromatography. Mass spectrometry of the acetates again supported C-6' as the position of the linkage. The purified nucleosides **19** and **20** were then obtained by deacetylation.[25] Further ^{13}C-NMR comparison studies using 1-(β-D-hexopyranosyl)-uracil models for hikizimycin led to the conclusion that the antibiotic favored the *anti* conformation in solution.[26]

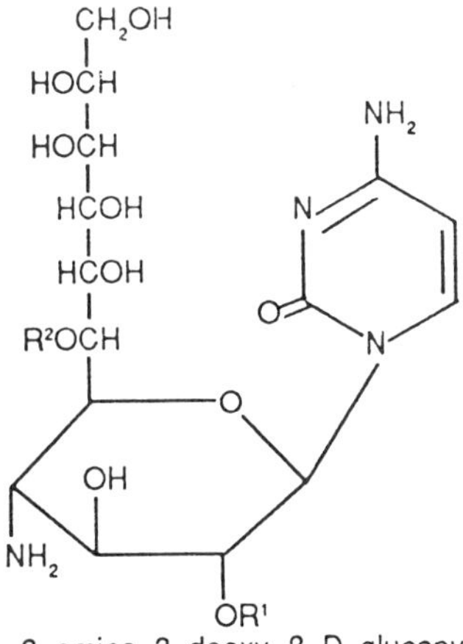

17, R^1 = 3–amino–3–deoxy–ß–D–glucopyranosyl; R^2 = H

18, R^1 = H; R^2 = 3–amino–3–deoxy–ß–D–glucopyranosyl

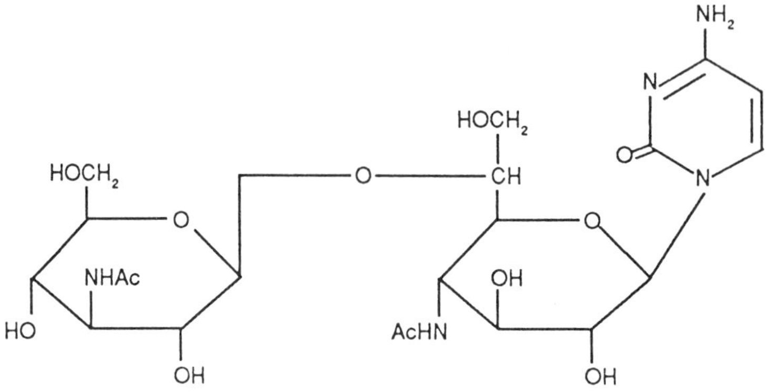

(19)

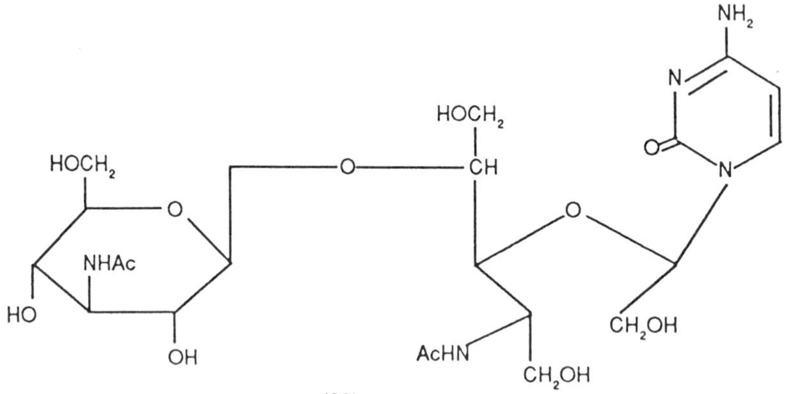

(20)

2.3. The Ezomycins

An antibiotic named ezomycin was isolated from the culture filtrate of a strain of *Streptomyces* and demonstrated activity against certain types of phytopathogenic fungi.[27] Sakata *et al.*[28] showed that this extract was really a mixture of components. These components were successfully separated by column chromatography. Initially, Sakata and co-workers assumed that there were only four components, designated A_1, A_2, B_1, and B_2, with the main components being A_1 and B_1. Later, they discovered that ezomycin B_1 was contaminated with a new ezomycin, C_1.[29] Further investigation of the original ezomycin complex revealed other compounds, which were likewise separated and designated C_2, D_1, and D_2.[29] Ezomycins B_2, C_2, and D_2 have been crystallized.

Acid hydrolysis of ezomycin A_1 yielded cytosine as one of the products. The UV spectra of ezomycins A_1 and A_2 were similar, and both were similar to that of cytidine. An amino acid was isolated from the hydrolysate after ion-exchange chromatography and was identified as L-cystathionine by degradation to L-alanine and L-α-amino-*n*-butyric acid after treatment with Raney nickel. Further identification was made by a comparison of the IR and ORD spectra. When ezomycin A_1 was treated with Raney nickel, L-alanine was liberated, which meant that the homocysteine moiety was linked to the nucleoside.[30] Careful hydrolysis of ezomycin A_1 with a cation-exchange resin, in the acid form, gave L-cystathionine and a nucleoside identical to ezomycin A_2. Thus, ezomycin A_2 was simply ezomycin A_1 minus the amino acid.

Vigorous acid hydrolysis of ezomycin A_2 and chromatography on a charcoal column afforded a new aminosugar. This aminosugar was named ezoaminuroic acid and structure-proofed as 3-amino-3,4-dideoxy-D-*xylo*-hexopyranuronic acid (**21**).[31] The aminosugar was converted to a mixture of anomeric methyl glycosides and benzoylated, and the anomers were separated by silica gel chromatography. The NMR spectra of the glycosides had signals for H_{4a} and H_{4e} at high field, suggesting a deoxy sugar. The large coupling constants, $J_{4a,3} = 11$ and $J_{4a,5} = 11–12$ Hz, indicated a C1 conformation. Equatorial orientations for the O-benzoyl and N-benzoyl groups were supported by the coupling constants, $J_{3,2} = 10.5$, $J_{3,4a} = 11$, and $J_{3,4e} = 4.5$ Hz. That this new sugar (**21**) belonged to the D series was argued on the basis of the greater dextrorotation of the α-anomer compared to the β-anomer. The "dibenzoate chirality rule"[32] was also applied to support this conclusion.

Alkaline hydrolysis of ezomycin A_1 afforded an anhydronucleoside whose UV spectrum and pK_a data indicated a conversion of the cytosine moiety to uracil. The anhydronucleoside gave a negative ninhydrin test, whereas ezomycin A_1 was ninhydrin positive. A test for a ureido group with *p*-dimethylaminobenzaldehyde was positive in both cases. The methyl ester of the anhydronucleoside was obtained by treatment with methanolic hydrogen chloride and subsequent acetylation to give a crystalline product. The NMR spectrum of the latter revealed that position C-2' was acetylated. Decoupling experiments demonstrated coupling of H-5' with an amide proton and gave proof that the ureido group was linked at C-5'. A double bond was confirmed by the maximum in the UV spectrum at 238 nm, which is typical of an α,β-unsaturated acid. Such an unsaturated group would arise due to β elimination under alkaline conditions. A deficiency of protons in the NMR spectrum suggested a bicyclic structure, which was proposed as **22**.

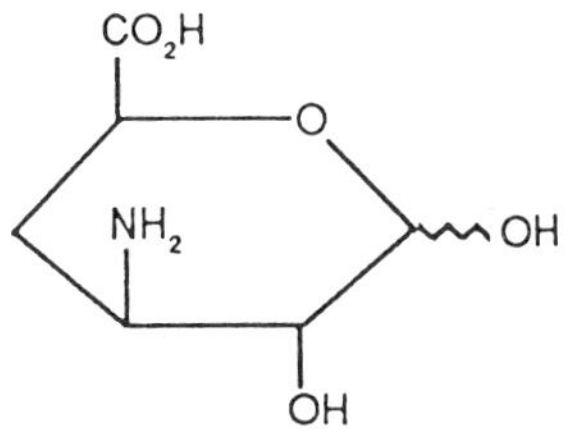

(21) 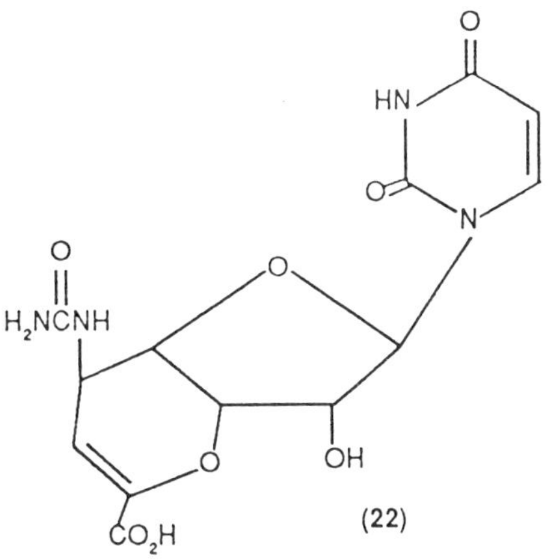   (22)

Periodate oxidation and acid hydrolysis of ezomycin A_1 gave a nucleoside, which was converted to the methyl ester and then acetylated. The protons at H-2' and H-6' were easily identified by NMR spectroscopy, because upon acetylation they moved 1.0 ppm downfield, which indicated that these were the positions of the hydroxyl groups. A broad multiplet assigned to H-5' became narrower upon the addition of deuterium oxide, and this was interpreted to mean that the ureido group was linked at C-5'. The anomeric proton appeared as a singlet, indicating a β configuration, and a *trans*-diaxial orientation of the protons at C-3' and C-4' was supported by the large coupling constant, $J_{3',4'} = 11$ Hz. The coupling constants $J_{5',6'} = 2.5-3$ and $J_{6',7'} = 1.8$ Hz also supported the proposed structure of the nucleoside as 1-(3,7-anhydro-5-deoxy-5-ureido-D-*threo*-β-D-allooctofuranosyluronic acid)cytosine (**23**).[33] The structure of the nucleoside was confirmed shortly thereafter by ^{13}C-NMR spectroscopy.[34] It is of interest that this structure is closely related to a group of naturally occurring nucleosides known as the octosyl acids, a group of bicyclic anhydrooctose uronic acid nucleosides (e.g., **24**), which have been isolated from culture broths of *Streptomyces cacaoi* var. *asoensis*.[35]

The periodate oxidation and hydrolysis of ezomycin A_1 served to provide further evidence concerning the configuration of ezoaminuroic acid at C-5. Two of the products were identified as glyoxal and a lactam-aminohemiacetal. The latter was reduced with sodium borohydride to give N-(2,4-dihydroxybutyryl)-L-cystathionine sulfoxide (**25**). Acid hydrolysis of **25** afforded L-cystathionine sulfoxide and α-hydroxybutyrolactone (**26**). The hydroxyl group of

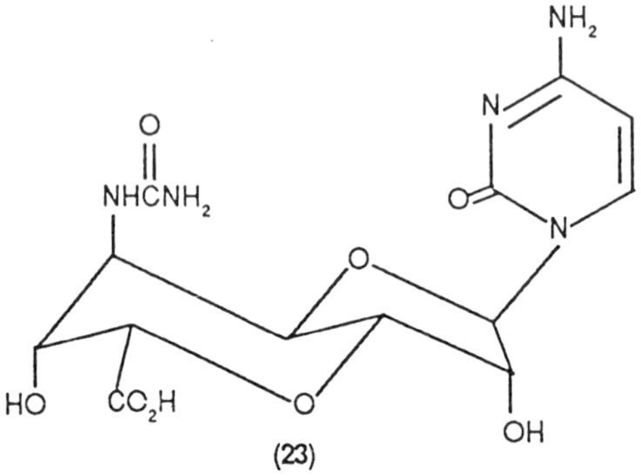

(23)

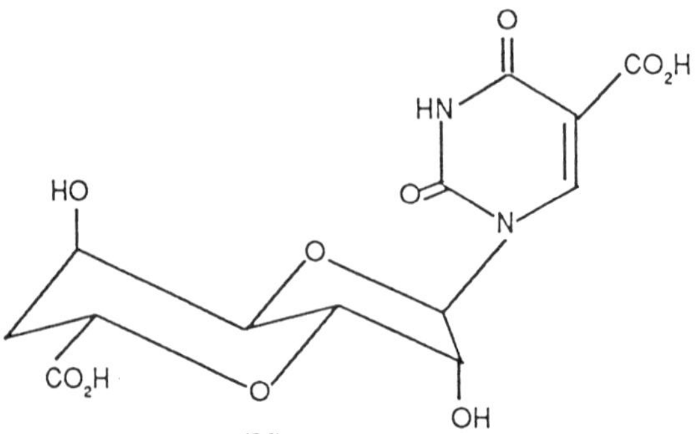

(24)

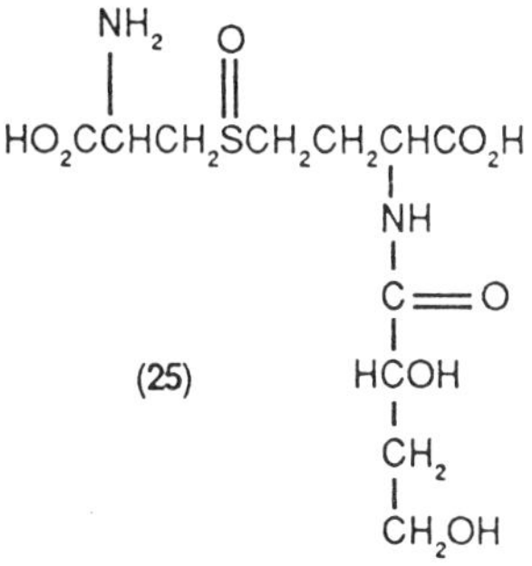

(25)

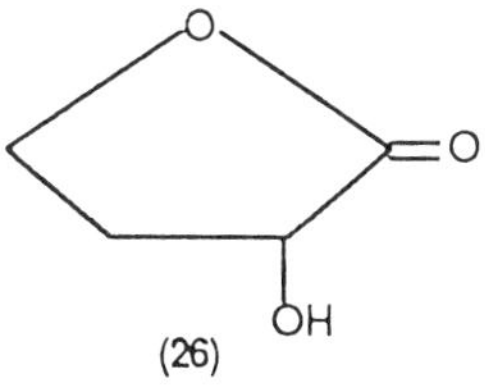

(26)

26 had the *S* configuration, and since this position was originally at C-5 of ezoaminuroic acid, the configuration was thus proved. The structure of the intermediate lactam-aminohemiacetal was proposed as **27**.[33]

The bonding of ezoaminuroic acid to **23** was considered to be through the C-6′ oxygen based upon a β elimination that occurred upon alkaline treatment. A peak at δ 4.74 with $J_{1',2'} = 6.5$ Hz favored a β glycoside configuration. The structure of ezomycin A_1 was proposed as **28**.[33] The NMR and CD spectra of **22, 23,** and **28** provided further evidence that the nucleoside C-N linkage was in the β configuration. The C-3′ position had the *endo* conformation.

The relationship between the members of the ezomycin pairs B_1 and B_2, C_1 and C_2, and D_1 and D_2 is the same as that between A_1 and A_2; namely, they differ

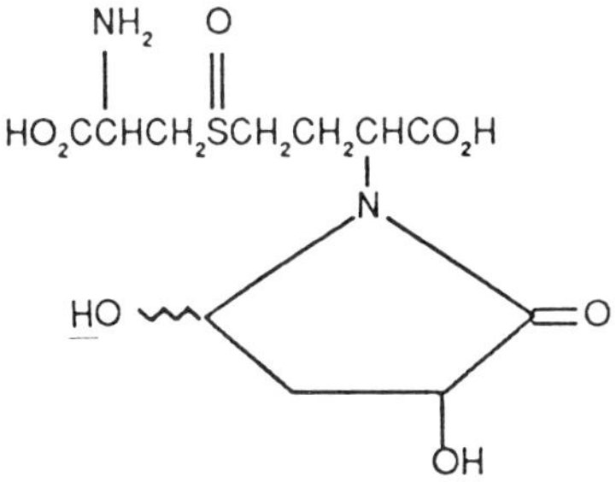

(27)

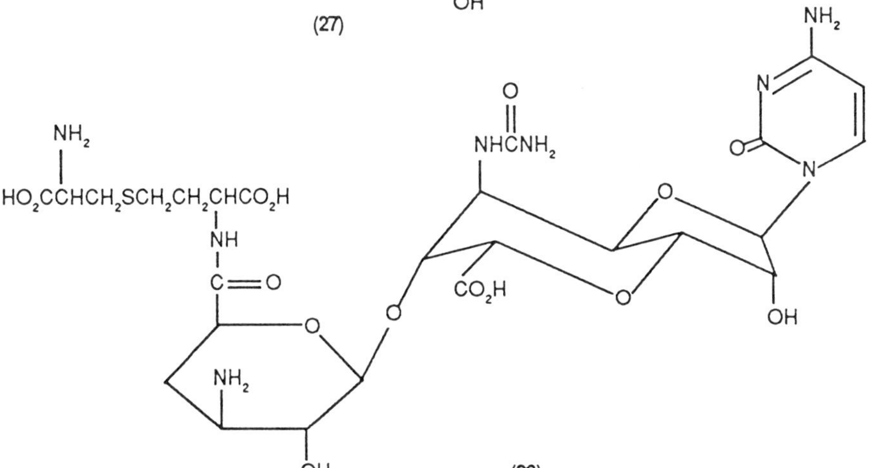

(28)

by the presence or absence of L-cystathionine.[29] The ureido group was present in all forms. The main structural difference, which was apparent from the UV, CD, and NMR spectra, was that these other ezomycins were 5-substituted uracil nucleosides rather than N-1-substituted cytosine nucleosides.[28,29] The same sequence of experiments as performed for ezomycin A_1 gave similar results except that the anhydrooctosyl uronic acid was linked at C-5. A β configuration for the nucleoside linkage of B_1 and B_2 and an α configuration for C_1 and C_2 was implied by the NMR and CD spectra. Because the interconversion of anomers occurred so easily under mild acid conditions, it was possible that ezomycins C_1 and C_2 were simply artifacts resulting from the method of isolation rather than natural products. Under acidic conditions, ezomycins B_1 and B_2 yielded D_1 and D_2 in addition to C_1 and C_2. Any of the ezomycins B, C, or D, when exposed to acid, produced a mixture of all three. The B form appeared to be the most stable. Periodate oxidation of ezomycin D_1 or D_2 revealed an extra vicinal hydroxyl pair, which could be accounted for by opening of the five-membered sugar ring. The part of the structure of these ezomycins that differs from that of A_1 is shown as **29–31**.[29]

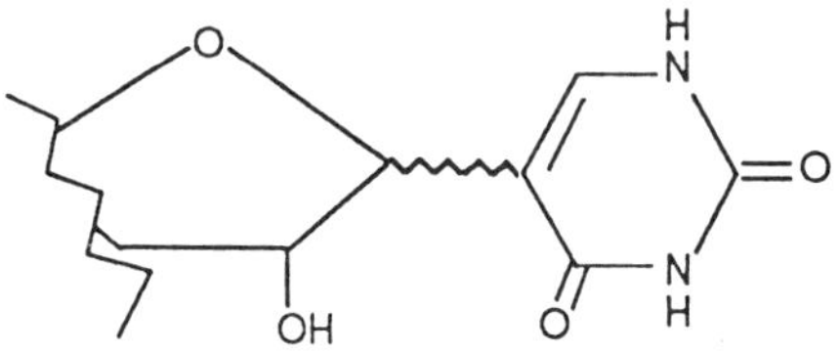

29, B_1 β
30, C_1 α

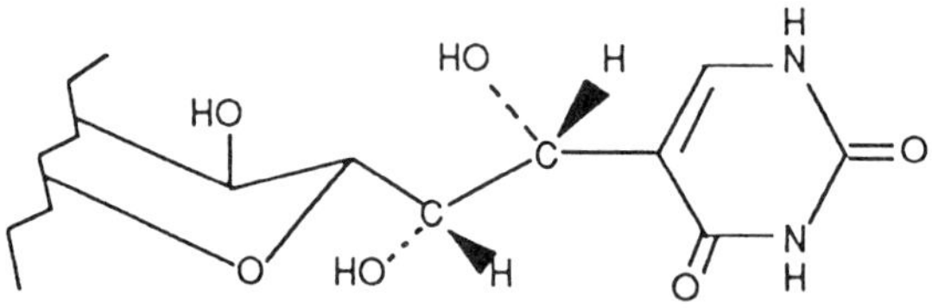

31, D_1

2.4. Thuringiensin

Thuringiensin is a specific exotoxin with insecticidal activity that forms during the growth of two variants of *Bacillus thuringiensis*. In most publications prior to 1977, it was simply called exotoxin. It was isolated from a culture medium by a combination of procedures and obtained as a lyophilized powder.[36–39]

Thuringiensin was isolated from *Bacillus thuringiensis* var. Berliner. It was determined that it either was a nucleotide or contained a nucleotide as part of its structure.[36–38] Adenine, D-ribose, and phosphate were identified in equivalent amounts, and a molecular weight of 800 was estimated based upon gel filtration studies and elemental analysis. Working with thuringiensin isolated from *Bacillus thuringiensis* var. *gelechiae*, Šebasta *et al.*[39] confirmed these results and determined pK_a values of 3.8 and 6.6. The phosphate group was readily removed with alkaline phophatase, which resulted in a complete loss of biological activity and the disappearance of the inflection point at pK_a 6.6. The UV spectrum indicated that the nucleoside component was adenosine.[37,39]

Bond and co-workers[40,41] refined the purification of thuringiensin, and their preliminary structural studies verified the above findings. In addition, enzymatically dephosphorylated thuringiensin migrated toward the anode upon electrophoresis, which was a good indication that a carboxyl group was present. The NMR spectrum supported N-9 substitution of adenine due to the 13-Hz separation between the signals of H-2 and H-8. However, the anomeric configuration was assigned on rather weak grounds, namely, the similarity of the anomeric signal at δ 6.13 to the signal for the anomeric proton of adenosine. The sign and amplitude of the Cotton effect in the ORD spectrum also supported the β configuration.[41] An important new contribution at this point was the isolation and identification of allaric acid (allomucic acid) from a hydrolysate of thuringiensin that had been fractionated on an anion-exchange column. Allaric acid was identified by a comparison to known samples prepared by nitric acid oxidation of either D-allono-1,4-lactone or 1,2:5,6-di-*O*-isopropylidene-α-D-allofuranose.[40]

A complete structure for thuringiensin was proposed by Šorm and co-workers.[42–46] Acid methanolysis and subsequent alkaline hydrolysis of thuringiensin yielded adenine, allaric acid, and a mixture of disaccharide methyl glycosides (**32**). The disaccharide mixture consumed two moles of periodate, the first mole being almost instantly consumed. The disaccharide mixture was converted to an isopropylidene derivative, and the structure of one of the glycosides was confirmed by mass spectrometry, and the structure of its derived acetate by NMR spectroscopy. Hydrolysis of the glycoside mixture with trifluoroacetic acid gave the free disaccharide, and extended hydrolysis liberated D-glucose. Partial hydrolysis removed only one of the methoxyl groups. Oxidation of this partial hydrolysate containing **33** with sodium hypoiodate gave an anomeric mixture of lactones (**34**). The IR spectrum showed a 1,4-lactone, and the ORD spectrum was similar to that of D-ribono-1,4-lactone. Conversion of **33** into the anomeric flavazoles **35** followed by acetylation caused a shift in the NMR signal for H-4 of the D-glucose moiety and supported the suggestion that the disaccharide had an unusual ether bond through the C-4 of D-glucose. These results were corroborated by acetylation and ^{1}H-NMR investigation of **34**. The next step was to ascertain that adenosine was really a component of thuringiensin, rather than 9-β-D-glucopyranosyladenine. The ORD spectrum and rate of acid hydrolysis of the dephosphorylated thuringiensin were compared with those of adenosine and 9-β-D-glucopyranosyladenine. These date closely approximated those for adenosine and ATP.

The bond to allaric acid was then studied in detail. On the basis of the chemical properties of dephosphorylated thuringiensin, particularly the ease of

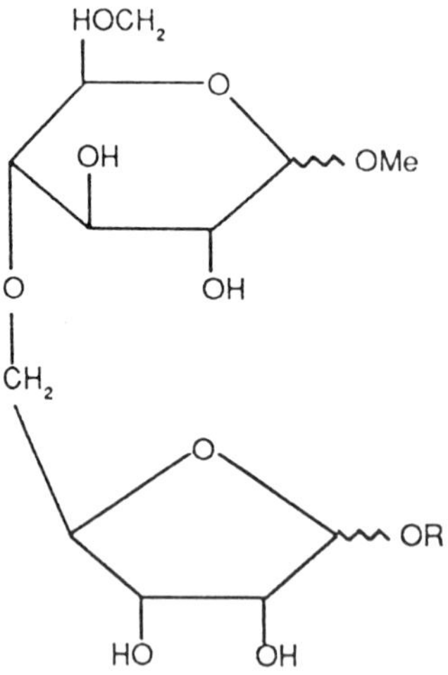

32, R = Me
33, R = H

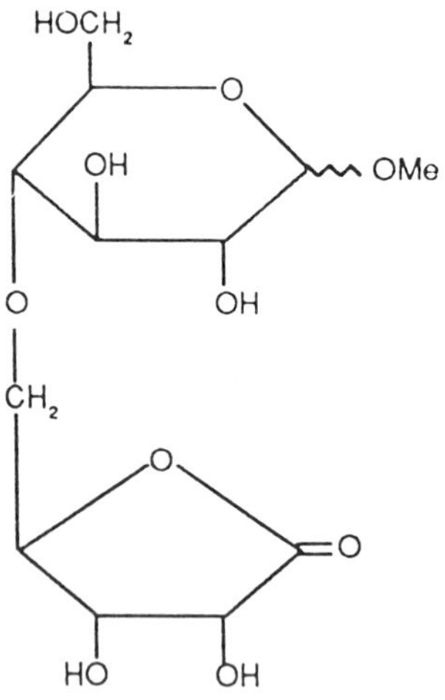

(34)

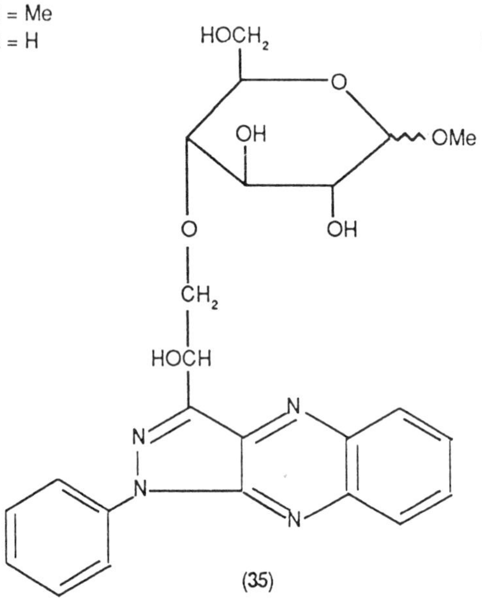

(35)

acid hydrolysis required to break off the allaric acid, and the typical glycoside spectrum obtained in the NMR investigation, it was concluded that D-glucose was bound to it. Acid-catalyzed cyclization of dephosphorylated thuringiensin gave a product having two five-membered lactones, which was evidence for substitution at C-2 of allaric acid. Dephosphorylated thuringiensin was oxidized with sodium periodate, reduced with sodium borohydride, and hydrolyzed under mild acid conditions. The hydroxyacid product was reacted with *o*-phenyl-enediamine, and the product that formed was identified as the benzimidazole derivative of D-glyceric acid. This proved that the bond was not at C-3 since the benzimidazole derivative of D-erythronic acid would have been the expected product, and C-5 was excluded because the configuration of the glyceric acid would have been L. Thus, it was concluded that D-glucose was linked at C-2 of allaric acid.

Bond and co-workers[41] had proposed that the anomeric configuration of the D-glucose linkage to allaric acid was α based upon the small coupling constant ($J_{1,2} = 3$ Hz) for the anomeric signal at δ 5.17. Šorm and co-workers[44,45] pointed out that the assignment was not unambiguous because of the number of signals in that region. Further ^{1}H-NMR studies were conducted in which modified thuringiensin was compared with model compounds. Dephosphorylated thuringiensin was converted to the dimethyl ester with diazomethane after first protecting the hydroxyl groups as the trimethylsilyl ethers. The ester groups were reduced with lithium aluminum hydride to give the alditol form of dephosphorylated thuringiensin. Permethylation gave a product whose ^{1}H-NMR spectrum was compared to those of permethylated adenosine (36) and permethylated (5S)-5-O-(α-D-glucopyranosyl)allitol (37). Because of the similarity of the chemical shift at δ 5.25 and the small coupling constant ($J_{1,2} = 3.3$ Hz) of 37 in comparison to the values for modified thuringiensin, the original designation of the configuration by Bond and co-workers was verified.[46]

Methanolysis of the modified permethylated thuringiensin gave a penta-O-methylallitol, the mass spectrum of which was identical to that of an authentic sample of (5R)-1,2,3,4,6-penta-O-methylallitol. However, the CD spectra of the 5-O-acetyl derivatives revealed that they were enantiomers. The allitol derivative prepared from thuringiensin was, therefore, (2S)-1,2,3,4,6-penta-O-methylallitol, and the original D-glucosylallaric acid portion was identified as (2R)-2-O-α-D-glucopyranosylallaric acid. The structure of dephosphorylated thuringiensin (38) was complete, and only the position of the phosphate remained to be established. This was done on the basis of periodate uptake studies,[41,42] and the phosphate was assigned to C-4 of the allaric acid moiety (39). This assignment was confirmed by ^{13}C-NMR spectroscopy.[47]

2.5. The Tunicamycins

An antibiotic noted for its activity against viruses, gram-positive bacteria, and fungi was isolated from a newly discovered soil organism, *Streptomyces*

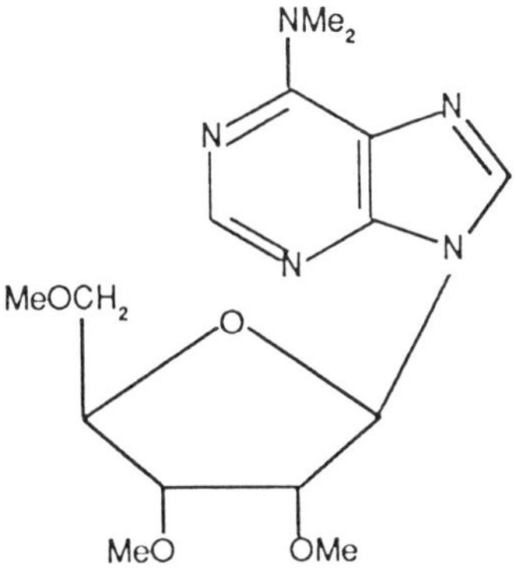

(36)

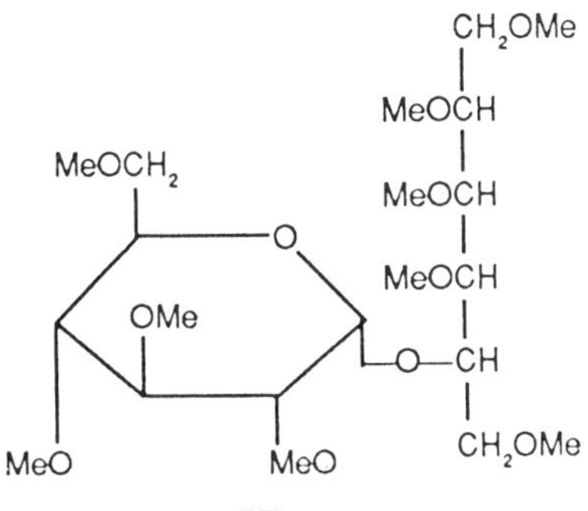

(37)

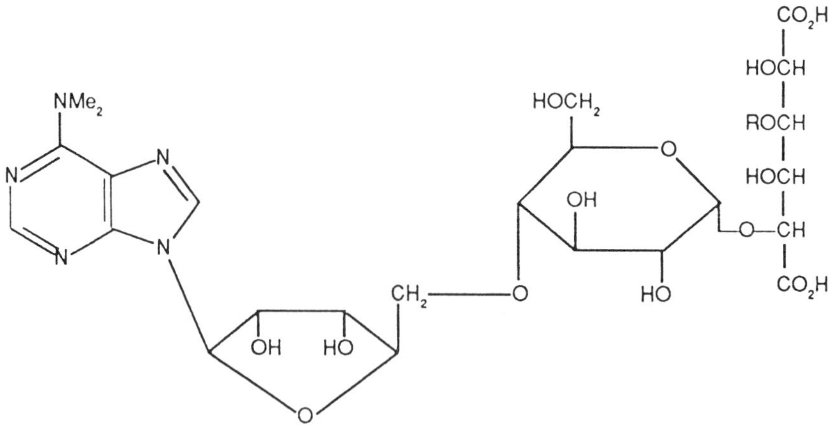

38, R = H
39, R = –PO₃H₂

lysosuperificus, and named tunicamycin because of its mode of action.[48] It was purified from the culture broth by extraction with organic solvents and silica gel chromatography. It was quickly recognized from spectroscopic data that the antibiotic was a nucleoside and that it contained glucosamine and another aminosugar based on hydrolytic studies and Elson–Morgan tests. Moreover, a careful examination of the hydrolysis products extracted into ether demonstrated that the antibiotic was not a single compound. It was found to be a homologous series of compounds differing from each other by the nature of the fatty acids that were linked to one of the aminosugars.[49] The major fatty acids were shown by combined gas chromatography–mass spectrometry and ¹H-NMR and IR spectroscopy to be *trans*-α,β-unsaturated iso acids, and originally four fatty acids were identified. However, the list of tunicamycins rapidly expanded to at least 10 compounds.[50] The four original tunicamycins were designated by different letters, each of which signified a different fatty acid: A, 13-methyl-2-tetradecenoic acid; B, 14-methyl-2-pentadecenoic acid; C, 12-methyl-2-tridecenoic acid; and D, 15-methyl-2-hexadecenoic acid.[49] The minor ones were α, β-unsaturated normal and iso acids.[50]

The hydrolysate remaining after the ether extraction contained D-glucosamine. The presence of an N-acetyl group in the glucosamine residue in the original tunicamycin was shown by NMR spectroscopy. If the tunicamycin mixture was hydrolyzed under mild conditions to just remove the glucosamine, it was found from IR data that the fatty acids were in an amide linkage to the nucleoside moiety. Further hydrolysis gave the nucleoside, called tunicaminyl-uracil. The nucleoside had an 11-carbon aminosugar, which was named tunicamine. Tunicaminyluracil had reducing sugar activity, whereas tunicamycin did not. This suggested that tunicamine was a dialdose and that glucosamine was linked through its anomeric carbon atom to the anomeric carbon atom of the tunicaminyluracil moiety. An α–α linkage was inferred from polarimetric data,[49] but this was later shown to be incorrect.

The structure of tunicaminyluracil was determined to be as shown in **40**.[51] The UV spectrum and ORD data were very similar to those of uridine. Extensive and vigorous acid hydrolysis liberated uracil. The ^{1}H-NMR spectrum in the region of the anomeric proton was also identical to that of uridine, indicating that the sugar moiety was a β-ribofuranosyl ring. The ^{13}C-NMR spectrum gave a pattern for the furanosyl moiety that was identical to that for uridine, with the exception of C-5'. This spectrum also showed that the remainder of the molecule was similar to D-galactosamine rather than to D-glucosamine. The position of the amino group was verified by a reduction with sodium borohydride, acetylation of the amino group to give **41**, and mass spectroscopy of the per-*O*-trimethylsilyl derivative. Because there was no cleavage between C-6' and C-7', it was concluded that C-6' was a methylene, which was supported by ^{1}H- and ^{13}C-NMR spectra. That the furanose ring of tunicaminyluracil had *cis*-oriented hydroxyl groups was supported by the formation of a dicyclohexylidene derivative, the second ketal being formed on the *galacto* ring.

A further proof of the structure of tunicamine was provided by the hydrolysis of **41**.[51] The aminosugar was isolated by ion-exchange chromatography, and the ^{1}H-NMR spectrum identified a β-pyranose ring identical to β-D-allopyranose. This demonstrated conclusively that the configuration at C-5' was R. Finally, the correct anomeric linkages between tunicamine and *N*-acetyl-D-glucosamine were determined by an extensive ^{1}H-NMR and IR study. It was concluded that the linkage at the D-glucosamine was α, and the bond at the tunicamine moiety was β, so that the structure of the tunicamycins could be written as **42**.[52]

Tunicamycins have recently been isolated from *Streptomyces clavuligerus*,[53] *Streptomyces griseoflavus*,[54] and *Corynebacterium rathayi*.[55] Many of these tunicamycins have different fatty acids than the previous group and include new saturated normal, iso, and anteiso acids, as well as some with β-hydroxyl groups. The great interest in these compounds on the part of biochemists is because tunicamycins inhibit the transfer of *N*-acetyl-D-glucosamine from UDP-*N*-acetylglucosamine to dolichol phosphate and thus are unique tools for inhibiting the biosynthesis of glycoproteins.[56]

3. Synthesis of Disaccharide Nucleosides

3.1. Synthetic Methods and Preparation of Nucleosides

The preparation of nucleosides derived from disaccharides follows the same general principles utilized in the synthesis of monosaccharide nucleosides. The coupling procedure by which the carbohydrate and the nitrogenous base become linked has generally involved the reaction of a glycosyl halide with an appropriate derivative of a purine or pyrimidine base in an inert solvent. Although there are a number of methods available for carrying this out, only relatively few have been used. A second procedure for the preparation of a disaccharide nucleoside is to form a new glycosidic bond between a preformed nucleoside and another monosaccharide. This is accomplished by applying the classical methods of glycoside synthesis, in which a glycosyl halide reacts with an

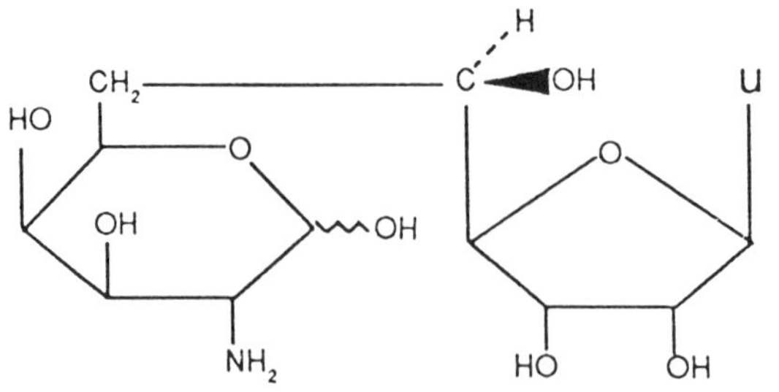
CH2
HO
O
OH
OH
NH2
H
C
OH
u
O
HO
OH

(40)

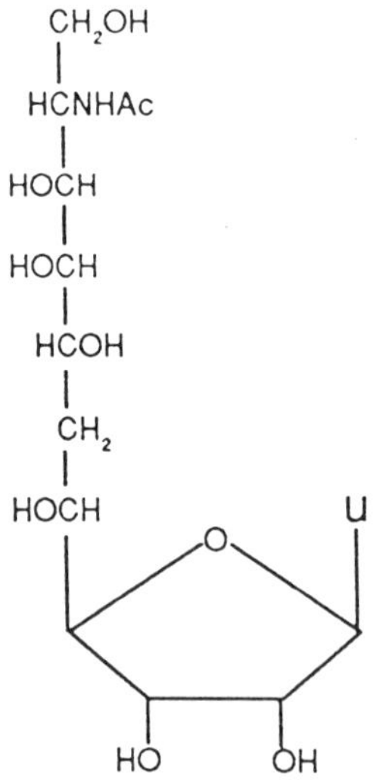
CH2OH
HCNHAc
HOCH
HOCH
HCOH
CH2
HOCH
u
O
HO
OH

(41)

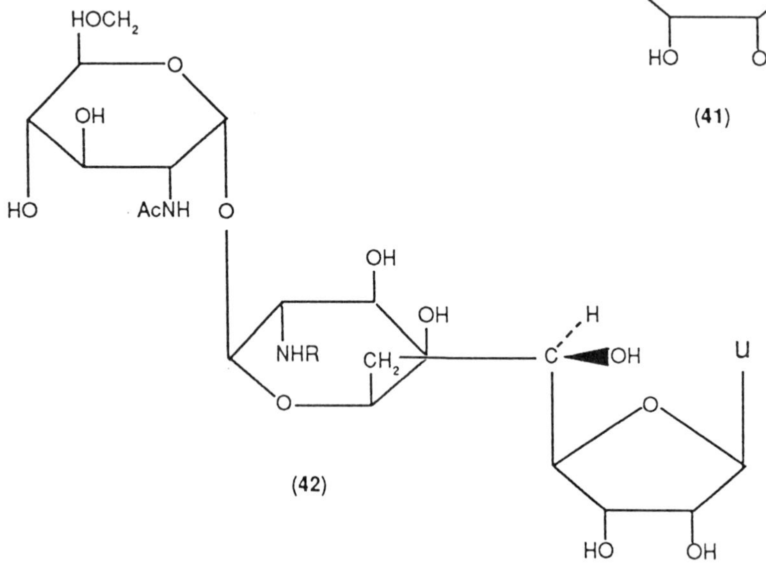
HOCH2
O
OH
HO
AcNH
O
OH
OH
NHR
CH2
O
H
C
OH
u
O
HO
OH

(42)

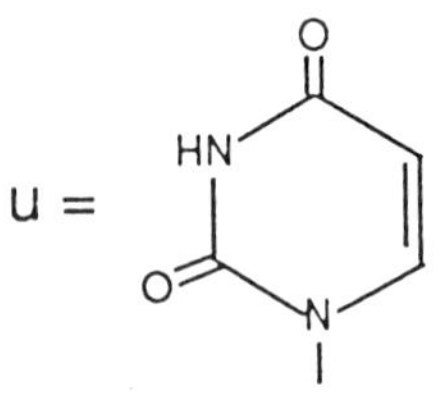
u =
O
HN
O
N

unprotected hydroxyl group on the nucleoside. Of course, a third route to a new disaccharide nucleoside is to transform a preformed nucleoside into a new one by altering either the sugar, the base, or both.

The first synthesis of disaccharide nucleosides was reported by Wolfrom and co-workers.[57] 6-Acetamidochloromercuripurine and 2,6-diacetamido-chloromercuripurine were each treated with hepta-O-acetyl-α-lactosyl bromide in boiling xylene, following, in general, the procedure of Davoll and Lowy.[58] Removal of the acetyl groups with methanolic ammonia afforded 9-β-lactosyl-adenine (**43**) and 2-acetamido-9-β-lactosyladenine (**44**), respectively. Complete deacetylation of 2,6-diacetamido-9-(hepta-O-acetyl-β-lactosyl)purine was effected in boiling methanolic sodium methoxide to give 2,6-diamino-9-β-lactosyl-purine (**45**) after purification via the picrate. Nucleosides derived from maltose and cellobiose were prepared using the same purine derivatives and coupling procedures.[59] 6-Acetamidochloromercuripurine was condensed with hepta-O-acetyl-α maltosyl bromide, and the blocking groups were removed in a solution of n-butylamine in methanol to give 9-β-maltosyladenine (**46**). The reaction of 6-acetamidochloromercuripurine with hepta-O-acetyl-α-cellobiosyl bromide gave 6-acetamido-9-(hepta-O-acetyl-β-cellobiosyl)purine (**47**), which was isolated by extrusion column chromatography. Deesterification with methanolic ammonia gave 9-β-cellobiosyladenine (**48**). Condensation of hepta-O-acetyl-α-cellobiosyl bromide with 2,6-diacetamidochloromercuripurine followed by a removal of the blocking groups afforded 2,6-diamino-9-β-cellobiosylpurine (**49**). No proof was offered for the designation of anomeric configuration of any of these nucleosides; in fact, the problem was not even mentioned. However, it is fairly safe to assume that the configurations were β based upon the knowledge that reactions of this type give, as the major products, nucleosides in which the bases are *trans* to the hydroxyl groups at C-2 of the sugars.[60]

Based upon this work, the preparation of 9-β-melibiosyladenine (**50**) was achieved.[61] Octa-O-benzoylmelibiose was chosen as the protected sugar instead of the octaacetate because the rather labile 1→6 glycosidic bond between D-galactose and D-glucose was more stable to the hydrogen bromide–acetic acid reagent used to prepare the glycosyl bromide. The latter was condensed with 6-benzamidochloromercuripurine in hot toluene, and debenzoylation with methanolic sodium methoxide gave **50**. An intermediate, 9-(hepta-O-benzoyl-β-melibiosyl)adenine (**51**) was obtained in crystalline form by a selective removal of the N-benzoyl group with ethanolic picric acid followed by a removal of picrate ion with an anion-exchange resin. Again, the anomeric configuration was assumed to be β. In related work,[62] hexa-O-acetylcellobial was treated with hydrogen chloride in benzene, and the resulting 2-deoxyglycosyl chloride was condensed with 6-benzamidochloromercuripurine in hot xylene. Partial purification of the protected nucleoside was achieved on a silicic acid column, and a removal of blocking groups with methanolic sodium methoxide afforded 9-(2-deoxycellobiosyl)adenine of unknown anomeric configuration.

In 1965, Yamaoka *et al.*[63] reported a coupling procedure in which a suitably protected purine or pyrimidine was reacted with a glycosyl halide in nitromethane in the presence of mercuric cyanide. As one of their examples, they prepared a disaccharide nucleoside. Hepta-O-acetylgentiobiosyl chloride was

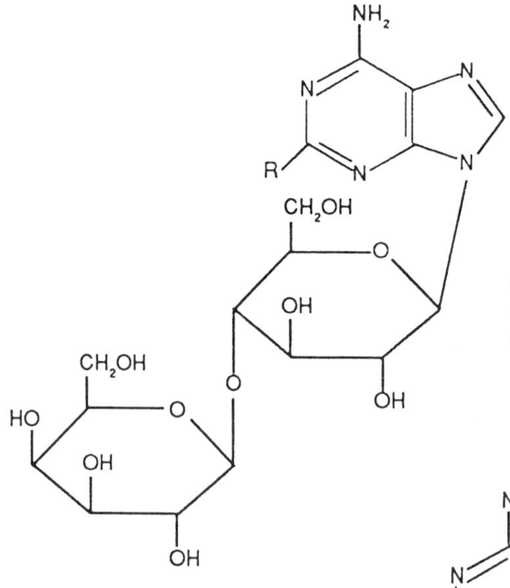

43, R = H
44, R = AcNH
45, R = NH_2

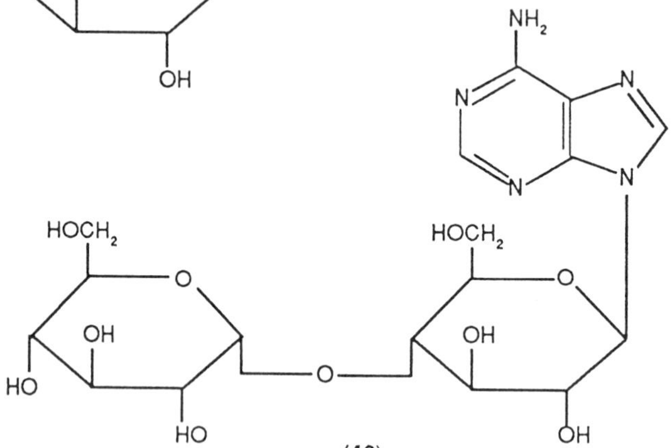

(**46**)

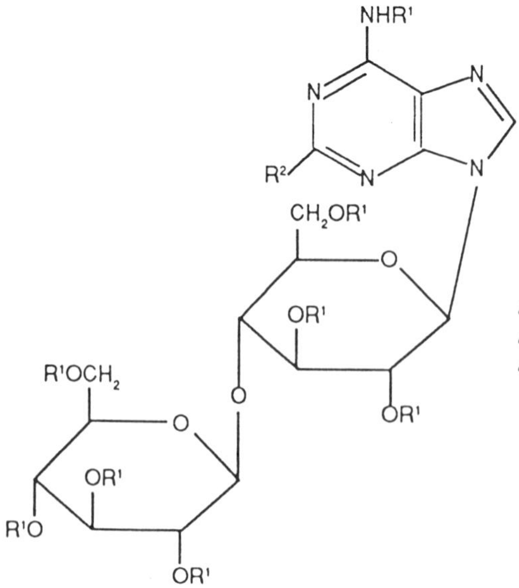

47, R^1 = Ac; R^2 = H
48, R^1 = R^2 = H
49, R^1 = H; R^2 = NH_2

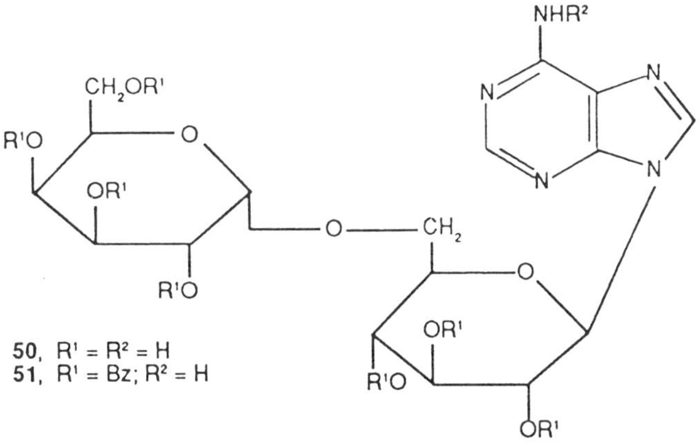

CH$_2$OR'

R'O

OR'

NHR^2

O——CH$_2$

OR'

R'O

50, R' = R^2 = H
51, R' = Bz; R^2 = H

R'O

OR'

condensed with 2,6,8-trichloropurine, and crystalline 2,6,8-trichloro-9-(hepta-*O*-acetyl-β-gentiobiosyl)purine (**52**) was obtained. 1-β-D-Glucopyranosylbenzimidazole was among the nucleoside analogues prepared by Yamaoka *et al.*[63] Using their reaction conditions, several disaccharide derivatives of benzimidazole were prepared. Hepta-*O*-acetyl-α-lactosyl bromide and hepta-*O*-acetyl-α-cellobiosyl bromide were each treated with benzimidazole in nitromethane containing mercuric cyanide. The acetyl groups were removed with methanolic sodium methoxide, and both nucleosides were isolated as picrates. After removal of the picrate ion with an anion-exchange resin, the free nucleosides were purified further by cellulose chromatography. 1-β-Cellobiosylbenzimidazole (**53**) crystallized, and 1-β-lactosylbenzimidazole (**54**) was obtained as a lyophilized powder.[64] The anomeric configuration was proved in each case by acid methanolysis of the disaccharide nucleoside to the known 1-β-D-glucopyranosylbenzimidazole, which was characterized as its picrate.

The first synthesis of disaccharide nucleosides of pyrimidines was reported by Stevens and Blumbergs.[65] Hepta-*O*-acetyl-α-lactosyl chloride and hepta-*O*-acetyl-α-cellobiosyl chloride were each coupled with 2,4-diethoxypyrimidine by the Hilbert–Johnson method[66] to afford 1-(hepta-*O*-acetyl-β-lactosyl)-4-ethoxypyimidin-2-one (**55**) and 1-(hepta-*O*-acetyl-β-cellobiosyl)-4-ethoxypyrimi-

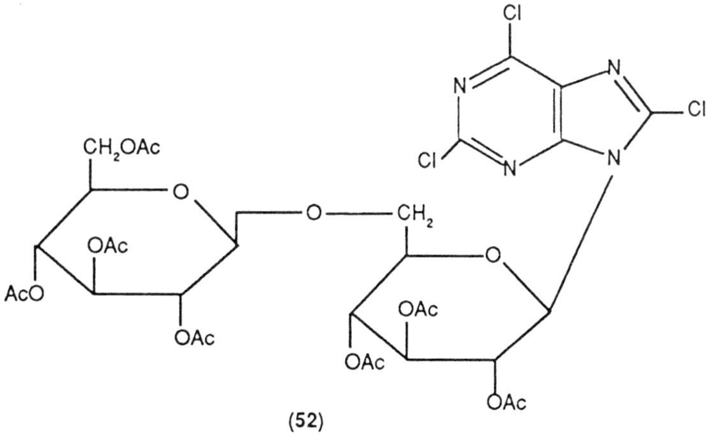

CH$_2$OAc

OAc

AcO

OAc

O——CH$_2$

Cl

Cl

Cl

OAc

OAc

OAc

(52)

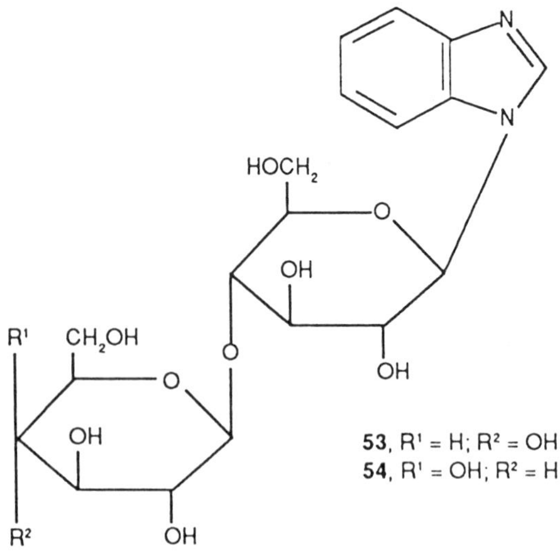

53, R¹ = H; R² = OH
54, R¹ = OH; R² = H

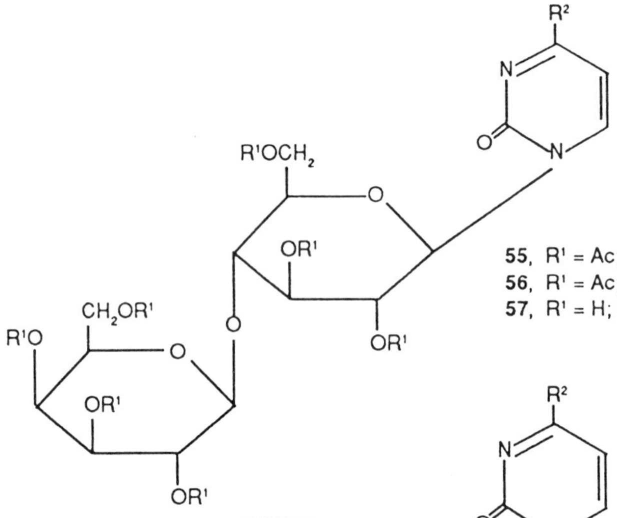

55, R¹ = Ac; R² = OEt
56, R¹ = Ac; R² = NHAc
57, R¹ = H; R² = NH₂

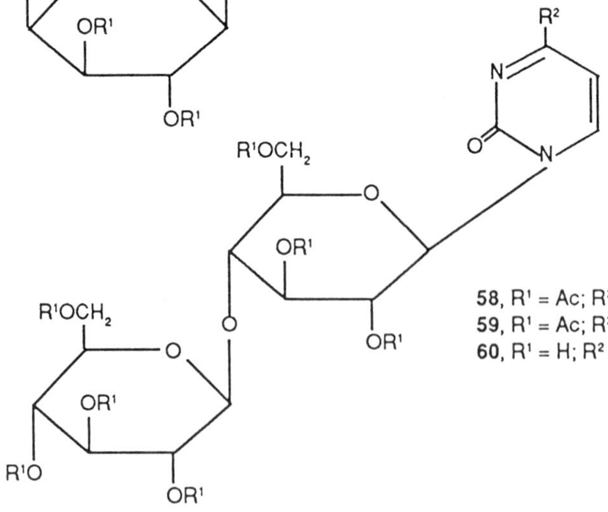

58, R¹ = Ac; R² = OEt
59, R¹ = Ac; R² = NHAc
60, R¹ = H; R² = NH₂

din-2-one (**58**), respectively. Treatment of **55** with ammonia gave 1-β-lactosyl-cytosine (**57**), and a similar reaction of **58** afforded 1-β-cellobiosylcytosine (**60**). These same nucleosides were obtained using the heavy metal salt procedure.[67] Hepta-*O*-acetyl-α-lactosyl bromide and hepta-*O*-acetyl-α-cellobiosyl bromide were each condensed with *N*-acetylcytosinemercury in hot benzene to give 1-(hepta-*O*-acetyl-β-lactosyl)-4-acetamidopyrimidin-2-one (**56**) and 1-(hepta-*O*-acetyl-β-cellobiosyl)-4-acetamidopyrimidin-2-one (**59**), respectively, which were deblocked to afford **57** and **60**. The anomeric configurations of **57** and **60** were demonstrated by acid methanolysis and acetylation to yield the known crystalline nucleoside derivative 1-(tetra-*O*-acetyl-β-D-glucopyranosyl)-4-acetamidopyrimidin-2-one.[65] In further work, the disaccharide derivative 1,2,3,4-tetra-*O*-acetyl-6-(2-deoxy-2-benzamido-3,4,6-tri-*O*-acetyl-β-D-glucopyranosyl)-β-D-glucopyranose (**61**) was treated with a mixture of hydrogen chloride and acetic anhydride and the chloride (**62**) was condensed with 2,4-diethoxypyrimidine to afford 1-[2,3,4-tri-*O*-acetyl-6-*O*-(2-deoxy-2-benzamido-3,4,6-tri-*O*-acetyl-β-D-glucopyranosyl)-β-D-glucopyranosyl]-4-ethoxypyrimidin-2-one (**63**).

Niedballa and Vorbrüggen[68] synthesized a number of disaccharide nucleosides of bases related to the pyrimidines. Various silylated pyrimidine analogues were reacted with the octaacetates of maltose, cellobiose, and lactose in 1,2-dichloroethane in the presence of stannic chloride, with yields of 60–80% in most cases and as high as 95% in a few instances. Maltosyl, cellobiosyl, and lactosyl nucleosides of the bases 6-azauracil, 5-methyl-6-azauracil, and 5-methyl-2-thio-6-azauracil were prepared. In addition, 1-β-maltosyluracil and 1-β-cellobiosyl-5-ethyluracil were obtained by the same procedure. In an interesting variation of this procedure, hexaacetyl anhydrolactose (**64**) was reacted with bis(trimethylsilyl)-5-fluorouracil to yield the hexaacetyl derivative of 1-(β-lactosyl)-5-fluorouracil (**65**).[69]

Octa-*O*-acetyl-β-cellobiose has been condensed with trimethylsilylthe-

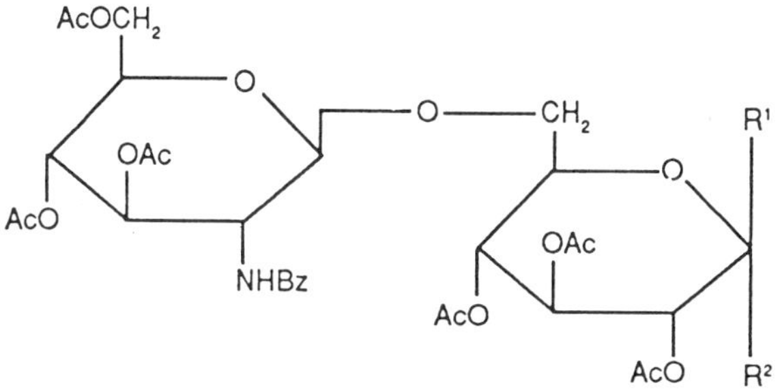

61, R¹ = OAc; R² = H
62, R¹ = H; R² = Cl

63, R¹ =

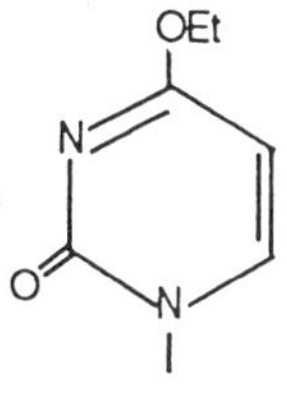

; R² = H

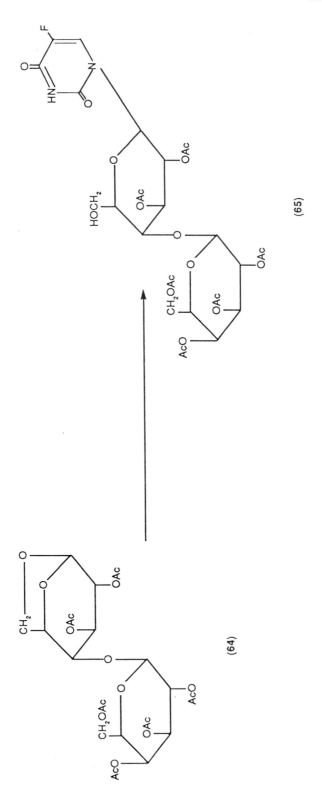

(64)

(65)

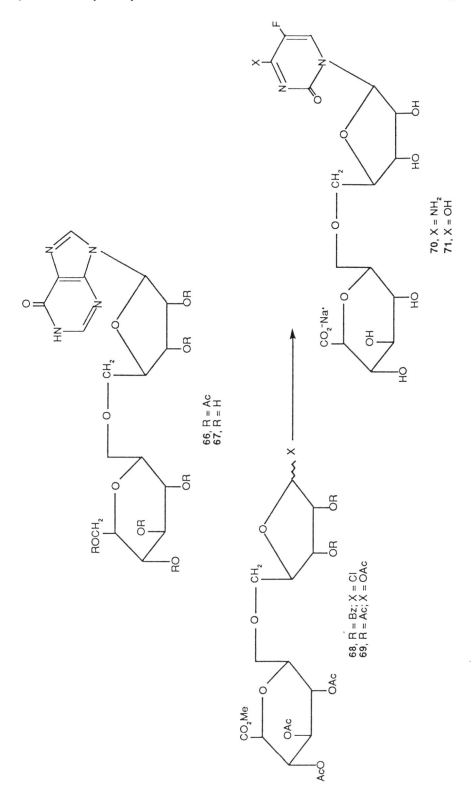

66, R = Ac
67, R = H

68, R = Bz; X = Cl
69, R = Ac; X = OAc

70, X = NH₂
71, X = OH

ophylline, with stannic chloride as the catalyst and acetonitrile as the solvent. A 90% yield of crystalline heptaacetylcellobiosyltheophylline was obtained, and deacetylation with methanolic ammonia afforded 7-β-cellobiosyltheophylline.[70] Coupling of 1,2,3-tri-O-acetyl-5-O-(2,3,4,6-tetra-O-acetyl-β-D-glucopyranosyl)-β-D-ribofuranose with N^9,O^6-bis(trimethylsilyl) hypoxanthine in dichloromethane with stannic chloride as catalyst gave the blocked nucleoside **66**. Removal of the acetyl groups with sodium methoxide gave 5'-O-(β-D-glucopyranosyl)inosine (**67**).[71]

5-O-(Methyl-2,3,4-tri-O-acetyl-β-D-glucopyranosyluronate)-2,3-di-O-benzoyl-D-ribofuranosyl chloride (**68**) was reacted with tris(trimethylsilyl)-5-fluorocytosine in acetonitrile. Saponification of the blocked product afforded 5'-O-(sodium β-D-glucopyranosyluronate)-5-fluorocytidine (**70**) in modest yield.[72] An improved procedure was to condense 5-O-(methyl 2,3,4,-tri-O-acetyl-β-D-glucopyranosyluronate)-1,2,3-tri-O-acetyl-D-ribofuranose (**69**) with tris(trimethylsilyl)-5-fluorocytosine in 1,2-dichloroethane containing stannic chloride. A 98% yield of the blocked nucleoside was obtained, from which crystalline **70** was obtained in very high yield after hydrolysis in sodium hydroxide solution. Similarly, **69** was coupled with bis(trimethylsilyl-5-fluorouracil to afford 5'-O-(sodium β-D-glucopyranosyluronate-5-fluorouridine (**71**) after saponification. In order to obtain some disaccharide nucleosides of 1,2,4-triazole, hepta-O-acetyl-β-maltosyl, -lactosyl, and -cellobiosyl isothiocyanates were condensed with acylhydrazines to give N-glycosyl-N'-acylhydrazinethiocarboxamides, which were then treated with acetic anhydride–phosphoric acid, to afford the disaccharide triazoles.[69,73]

The first synthesis of a disaccharide nucleoside by coupling of a preformed monosaccharide nucleoside with a glycosyl halide was reported by Stevens and Blumbergs.[65] 1,2,3,4-Tetra-O-acetyl-β-D-glucopyranose was converted to 1,2,3,4-tetra-O-acetyl-6-O-trichloroacetyl-β-D-glucopyranose (**72**) with trichloroacetyl chloride, and this was treated with hydrogen bromide in a mixture of acetic acid

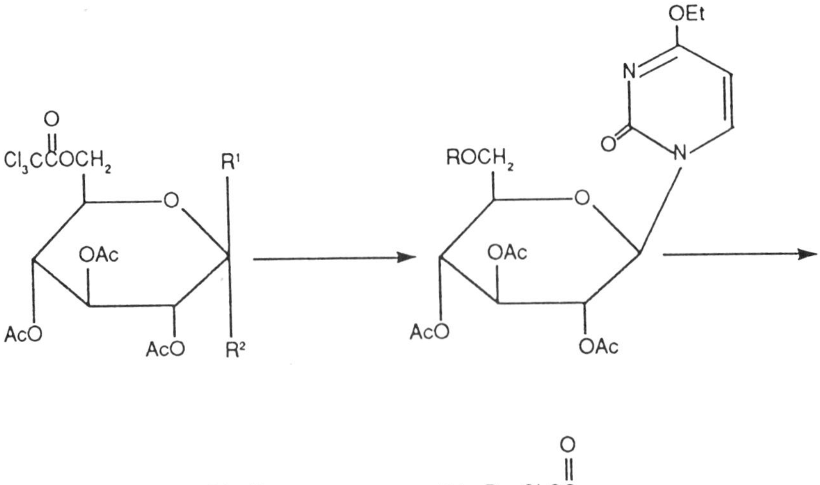

72, R¹ = OAc; R² = H
73, R¹ = H; R² = Br

74, R = Cl₃CC–
     ‖
     O
75, R = H

and acetic anhydride to obtain 2,3,4-tri-*O*-acetyl-6-*O*-trichloroacetyl-α-D-gluco-pyranosyl bromide (**73**). Coupling with 2,4-diethoxypyrimidine by the Hilbert–Johnson method afforded 1-(2,3,4-tri-*O*-acetyl-6-*O*-trichloroacetyl-β-D-glu-copyranosyl)-4-ethoxypyrimidin-2-one (**74**). Selective ammonolysis removed the trichloroacetyl group, giving **75**, which was reacted with 2-deoxy-2-benz-amido-3,4,6-tri-*O*-acetyl-α-D-glucopyranosyl bromide under Koenigs–Knorr reaction conditions to obtain the disaccharide nucleoside **63** in low yield. Acetylation of **75** gave the known nucleoside 1-(2,3,4,6-tetra-*O*-acetyl-β-D-glu-copyranosyl)-4-ethoxypyrimidin-2-one, thereby proving the anomeric configu-ration of both **75** and the disaccharide **63**.

A reaction of 3,4,6-tri-*O*-acetyl-2-*O*-benzyl-α-D-glucopyranosyl bromide with 2′,3′-*O*-isopropylideneinosine in chloroform with silver carbonate gave a 39% yield of the α,β-anomers.[74] After hydrolysis of the isopropylidene group, the anomers were separated on a silica gel column. The benzyl group was cleaved from each anomer by hydrogenolysis, and acetylation gave the respective crystal-line acetates. Deacetylation with methanolic ammonia afforded the free α- and β-linked glucosylinosines. The anomeric configurations were verified with spe-cific glucosidases. When N^6,N^6-bis(benzoyl)-2′,3′-*O*-isopropylideneadenosine was similarly glucosylated, the main product was the α-anomer in 58% yield after a removal of the isopropylidene group and the benzyl group, followed by column chromatography. Acetylation for purposes of further purification and deacetylation gave 5′-*O*-(α-D-glucopyranosyl)adenosine, the anomeric configu-ration of which was verified with a specific α-glucosidase.[74] Condensation of acetobromoglucose with 2′,3′-*O*-isopropylideneuridine with mercuric cyanide as the catalyst and nitromethane as the solvent, followed by hydrolysis of the isopropylidene group with trifluoroacetic acid, gave a crystalline tetraacetate. Deacetylation provided 5′-*O*-(β-D-glucopyranosyl)uridine. The same route was used to obtain 5′-*O*-(β-D-galactopyranosyl)uridine.[74] When acetobromogalac-tose was reacted with 5′-*O*-acetyluridine under the same conditions, the main product obtained after silica gel chromatography was the blocked 2′-*O*-galac-tosyl disaccharide nucleoside having the β configuration.[75] The position of substitution was verified by a separate synthesis in which 3′,5′-di-*O*-acetyl-uridine was used in place of the monoacetate. Further confidence in the assign-ment of the position of substitution was obtained from a ^{13}C-NMR study of the chemical shift patterns and decoupling behavior of 2′-, 3′-, and 5′-substi-tuted nucleosides.[76] When attempts were made to utilize nitromethane as the solvent in the reactions with 2′,3′-*O*-isopropylideneinosine, a large amount of the N^1-glucosylinosine was obtained in addition to the 5′-*O*-glucosyl product. Use of mercuric cyanide with nitromethane afforded a mixture of glucopyra-nosylhypoxanthines instead of the disaccharide nucleosides, due to sugar ex-change.[77]

Based upon the methods of disaccharide synthesis developed by Bredereck and co-workers,[78] several papers have appeared concerning the synthesis of disaccharide nucleosides by the reaction of a glycosyl halide with a trityl nucleo-side derivative. Thus, treatment of N^4,3′-*O*-diacetyl-5′-*O*-trityl-2′-deoxycytidine and 3′-*O*-acetyl-5′-*O*-tritylthymidine with acetobromoglucose in nitromethane in the presence of silver perchlorate afforded the respective 5′-*O*-glucosyl nucleosides.[79] Methyl (2,3,4-tri-*O*-acetyl-α-D-glucopyranosyl bromide)uronate was condensed with 2′,3′-di-*O*-acetyl-5′-*O*-trityl-6-azauridine to obtain, after hydrolysis of the blocking groups, 5′-*O*-(β-D-glucopyranosyluronate)-6-azauridine.[80] Treatment of 1-(2,3-di-*O*-acetyl-5-*O*-trityl-β-D-arabinofuranosyl)-*N*⁴-acetylcytosine with the same bromide and then deblocking afforded 5′-*O*-(β-D-gluco-pyranosyluronate)-1-(β-D-arabinofuranosyl)cytosine.[81]

Methyl (2,3,4-tri-*O*-acetyl-α-D-glucopyranosyl bromide)uronate was also reacted with 2′,3′-*O*-isopropylidene-5-fluorouridine, but in a solvent mixture of benzene and acetonitrile and in the presence of silver oxide, iodine, and Drierite.[72] A 65% yield of the blocked disaccharide nucleoside was obtained; however, the acid-catalyzed removal of the isopropylidene group also cleaved nearly half of the disaccharide glycoside bonds. Acetylation of the remaining disaccharide nucleoside followed by ammonolysis gave 5′-*O*-(β-D-glucopyranosyluronamide)-5-fluorouridine, whereas hydrolysis with sodium hydroxide and removal of the sodium ion with an ion-exchange resin afforded 5′-*O*-(β-D-glucopyranosyl-uronic acid)-5-fluorouridine. A poor yield of 5-*O*-(2,3,5-tri-*O*-benzoyl-α-L-arabinofuranosyl)-1-(2,3-di-*O*-acetyl-β-D-arabinofuranosyl)-*N*⁴-acetylcytosine was obtained when 2,3,5-tri-*O*-benzoyl-L-arabinofuranosyl chloride was condensed with 1-(2,3-di-*O*-acetyl-β-D-arabinofuranosyl)-*N*⁴-acetylcytosine in a mixture containing silver oxide, molecular sieve 4A, and dichloromethane.[82] Deesterification gave the free nucleoside as an oil.

The coupling of 7-(2,3-di-*O*-acetyl-6-deoxy-β-D-glucopyranosyl)theophylline with acetobromoglucose was not successful under standard Koenigs–Knorr reaction conditions; however, it worked satisfactorily when silver trifluoromethanesulfonate was the catalyst and dichloromethane the solvent.[70] The free nucleoside 7-[6-deoxy-4-*O*-(β-D-glucopyranosyl)-β-D-glucopyranosyl]theophylline (**76**) was obtained after ammonolysis. This product was the starting material for a transformation into another disaccharide nucleoside. Treatment of **76** with a benzaldehyde–zinc chloride mixture followed by benzoylation gave **77**. Hydrogenolysis followed by treatment of the product with *N*-bromosuccinimide–triphenylphosphine in *N*,*N*-dimethylformamide gave the 6″-bromo nucleoside derivative **78**, which was reduced to the 6′,6″-dideoxy nucleoside derivative **79** by a catalytic hydrogenation. Nucleoside **79** was also obtained, in lower yield, from 7-β-cellobiosyltheophylline by the same series of reactions, except that the 6′-position had to be blocked with a trityl group and the subsequent synthesis produced a 6′,6″-dibromo nucleoside, which was catalytically reduced to **79**. Treatment of **79** with dimethyl sulfoxide–acetic anhydride resulted in a simultaneous oxidation and elimination to give **80**.[70]

3.2. Synthesis of Nucleoside Antibiotics

3.2.1. Total Synthesis of Plicacetin

Amicetin, bamicetin, and oxamicetin have never been synthesized from their simplest structural components. A total synthesis of plicacetin (**8**) was reported by Stevens and co-workers.[83,84] A monosaccharide nucleoside useful for transformation to plicacetin was first synthesized. A suitable glycosyl halide

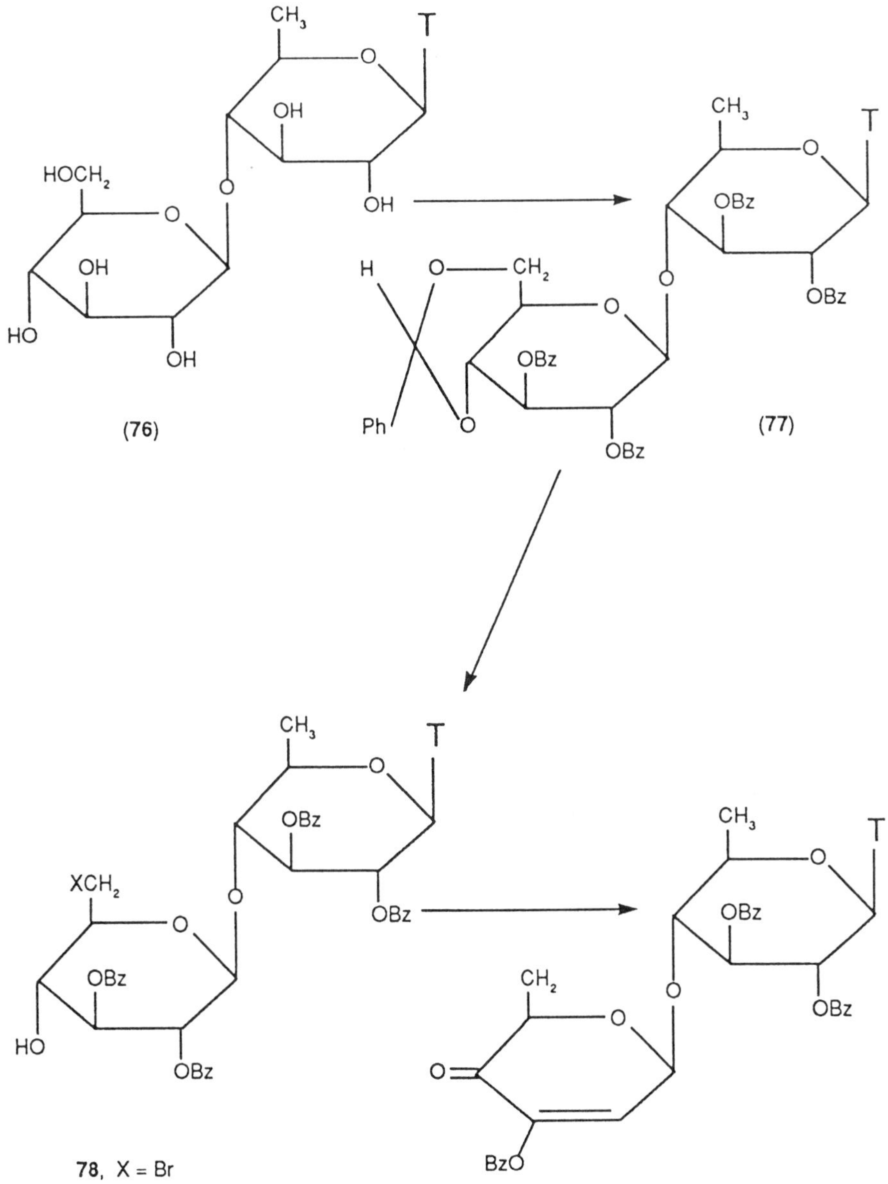

(76)

(77)

78, X = Br
79, X = H

(80)

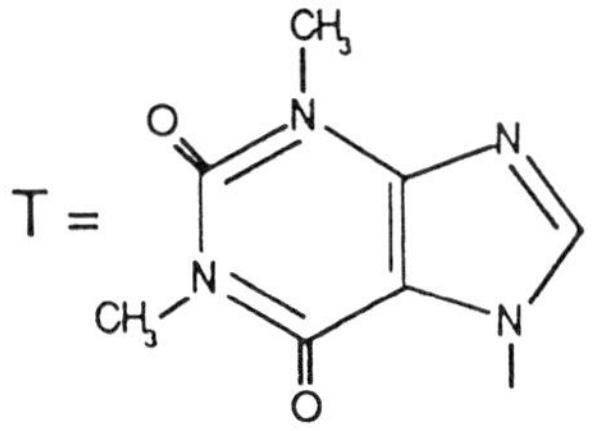

T =

was then condensed with it to form the disaccharide nucleoside, which was converted to **8**.

Ethyl 2,3-dideoxy-α-D-*erythro*-hexopyranoside (**81**)[85] was hydrolyzed with acid and treated with *p*-nitrobenzoyl chloride and then with hydrogen chloride in dichloromethane to give 2,3-dideoxy-4,6-di-*O*-*p*-nitrobenzoyl-α-D-*erythro*-hexopyranosyl chloride (**82**). Condensation of **82** with 2,4-diethoxypyrimidine and deesterification with ethoxide gave the nucleoside **83**. The anomeric configuration of the latter was proved by a conversion to **84**, a nucleoside derivative obtained from another nucleoside of known configuration. Treatment of **83** with

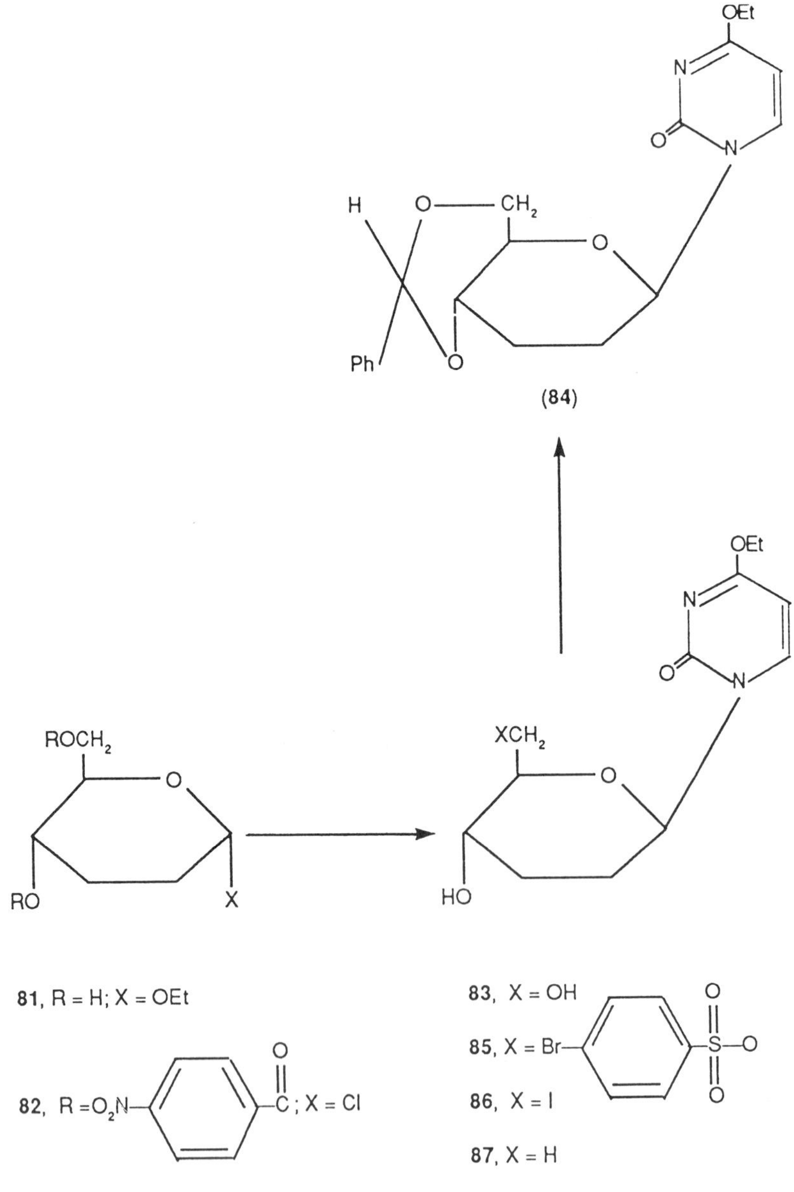

81, R = H; X = OEt

82, R =O$_2$N—⟨benzene⟩—C(=O)—; X = Cl

83, X = OH

85, X = Br—⟨benzene⟩—S(=O)$_2$—O

86, X = I

87, X = H

one equivalent of *p*-bromobenzenesulfonyl chloride in pyridine gave the 6'-brosylate **85**, which was converted to the 6'-iodo nucleoside **86** with sodium iodide in acetone. Hydrogenolysis over palladium on charcoal gave a good yield of 1-(2,3,6-trideoxy-β-D-*erythro*-hexopyranosyl)-4-ethoxypyrimidin-2-one (**87**).

The required glycosyl halide was synthesized from methyl α-D-glucopyranoside, which was converted to the 4,6-*O*-benzylidene derivative; the latter was treated with benzyl chloride to give **88**. Hydrolysis of the benzylidene group of **88** and treatment of the product with methylsulfonyl chloride afforded methyl 2,3-di-*O*-benzyl-4,6-di-*O*-methylsulfonyl-α-D-glucopyranoside (**89**). Reaction of **89** with sodium iodide in 2-butanone gave the 6-iodo derivative **90**, and reduction with lithium aluminum hydride in tetrahydrofuran resulted in an excellent yield of methyl 2,3-di-*O*-benzyl-6-deoxy-4-*O*-methylsulfonyl-α-D-glucopyranoside (**91**).[86] This was treated with sodium benzoate in *N,N*-dimethylformamide under reflux, and the benzoate group was removed with alkali to afford methyl 2,3-di-*O*-benzyl-6-deoxy-α-D-galactopyranoside (**92**). Mesylation of **92** provided **93**, and a reaction with sodium azide in *N,N*-dimethylformamide gave methyl 4-azido-2,3-di-*O*-benzyl-4,6-dideoxy-α-D-glucopyranoside (**94**). Acetolysis of **94** and deacetylation gave crystalline 4-azido-2,3-di-*O*-benzyl-4,6-dideoxy-D-glucopyranose (**95**), which was converted to an anomeric mixture of 1-*O*-*p*-nitrobenzoates (**96**). Treatment of **96** with hydrogen chloride in chloroform gave 4-azido-2,3-di-*O*-benzyl-4,6-dideoxy-β-D-glucopyranosyl chloride (**97**) in high yield and in crystalline form.[83] Coupling of a molten mixture of **97** and **87** in the presence of Dowex 1-X2 (OH⁻) ion-exchange resin gave a 64% yield of the protected disaccharide nucleoside **98**. The azido group was reduced with hydrogen over palladium on carbon catalyst to give **99**, and the amino group of **99** was reductively methylated with formaldehyde in a hydrogen atmosphere to afford an 84% yield of **100**. Treatment of **100** with liquid ammonia in ethanol gave a high yield of the cytosine derivative **101**, and a removal of the benzyl groups by hydrogenolysis afforded cytosamine (**6**).[84] The reaction of cytosamine with *p*-nitrobenzoyl chloride in chloroform gave a 65% yield of N^4-*p*-nitrobenzoyl-cytosamine. A catalytic reduction of the nitro group to an amino group afforded plicacetin.

3.2.2. Thuringiensin

The rather novel structure of thuringiensin (**39**) presented a significant challenge in synthetic chemistry. In particular, the main problems to be attacked were the unusual monosaccharide-to-monosaccharide ether linkage, the positions of the various bonds in the two sugar residues and the allaric acid, the α-glycoside bond of D-glucose, and the placement of the phosphate. These problems were admirably solved by Šorm and co-workers,[87–91] and two reaction sequences leading to thuringiensin have been published.

The problem of the D-ribose-to-D-glucose ether bond was approached first.[87] A route which worked well involved the condensation of methyl 2,3-*O*-isopropylidene-β-D-ribofuranoside (**102**) with 2-*O*-benzyl-1,6:3,4-dianhydro-β-D-galactopyranose (**103**) and boron trifluoride in benzene. A 59% yield of 2-*O*-benzyl-4-*O*-(methyl 5-deoxy-2,3-*O*-isopropylidene-β-D-ribofuranosid-5-yl)-1,6-

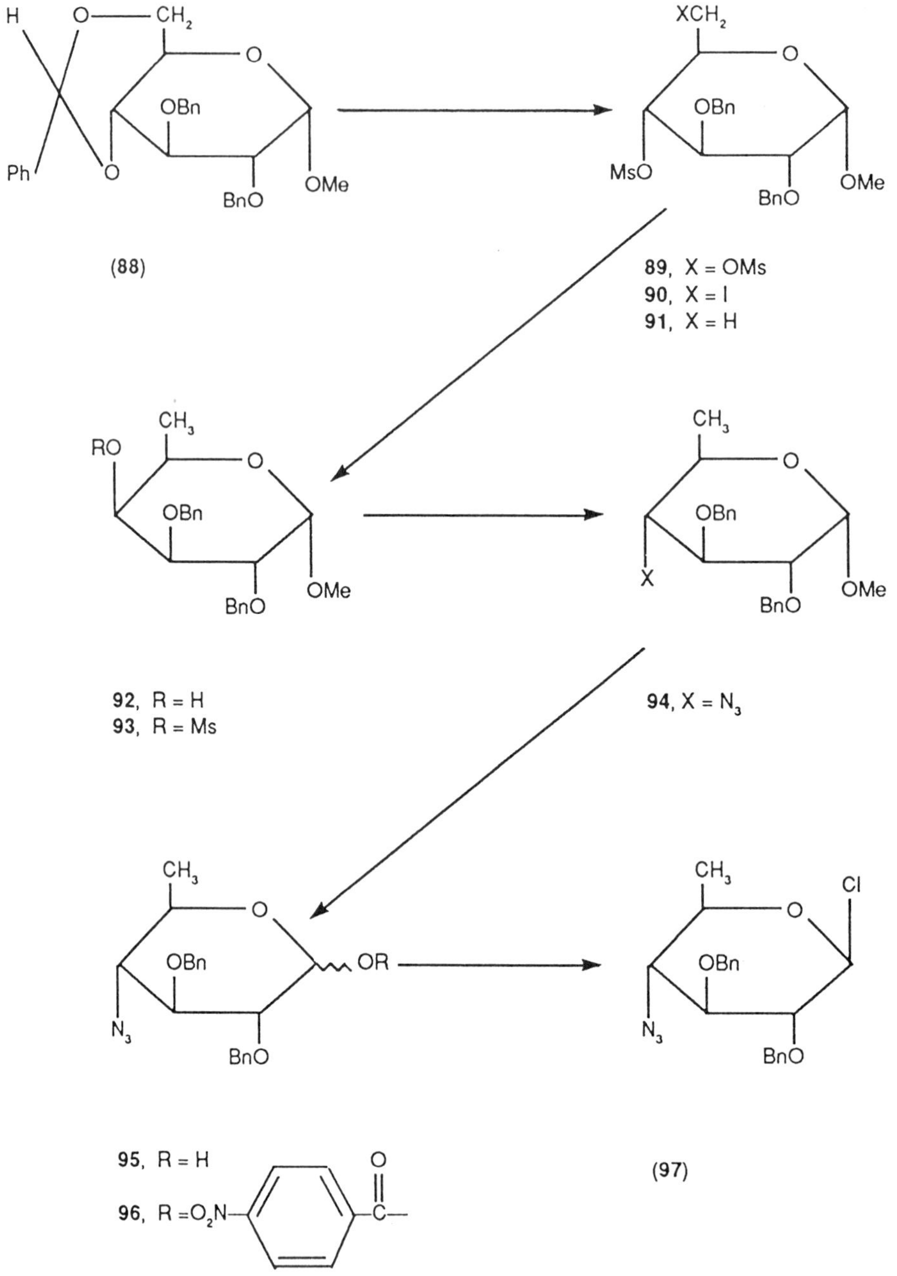

(88)

89, X = OMs
90, X = I
91, X = H

92, R = H
93, R = Ms

94, X = N$_3$

95, R = H

96, R =O$_2$N—⟨benzene ring⟩—C(=O)—

(97)

anhydro-β-D-glucopyranose (**104**) was obtained by a *trans*-diaxial ring opening. The isopropylidene group of **104** was removed, the unprotected hydroxyl groups were reprotected with benzoyl groups, the benzyl group was removed by hydrogenolysis, and the free hydroxyl group was benzoylated. Acetolysis gave the triacetate **105**, which was treated with hydrogen bromide in acetic acid to yield the 1,1'-dibromo derivative **106**. Treatment of **106** with methanol and silver

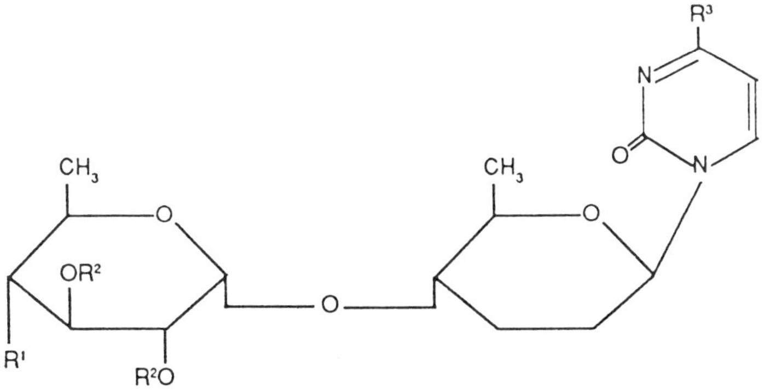

98, R¹ = N₃; R² = Bn; R³ = OEt

$$\textbf{98,} \quad R^1 = N_3;\ R^2 = Bn;\ R^3 = OEt$$
$$\textbf{99,} \quad R^1 = NH_2;\ R^2 = Bn;\ R^3 = OEt$$
$$\textbf{100,} \quad R^1 = NMe_2;\ R^2 = Bn;\ R^3 = OEt$$
$$\textbf{101,} \quad R^1 = NMe_2;\ R^2 = Bn;\ R^3 = NH_2$$
$$\textbf{6,} \quad R^1 = NMe_2;\ R^2 = H;\ R^3 = NH_2$$

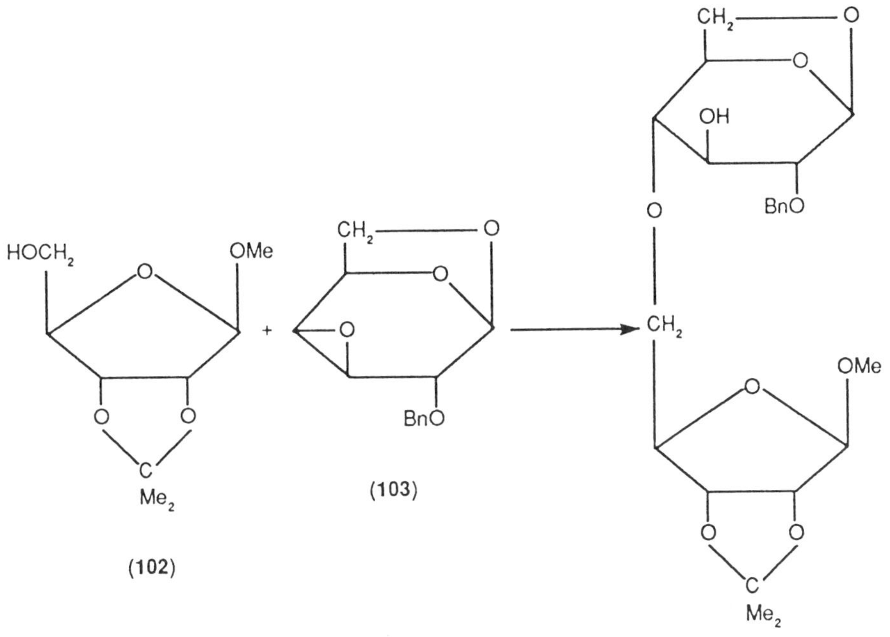

(102)   +   (103) $\longrightarrow$ (104)

carbonate gave the diglycoside. A removal of the ester groups with methanolic sodium methoxide and acetonation of the disaccharide followed by acetylation gave methyl 2,3,6-tri-O-acetyl-4-O-(methyl 2,3-O-isopropylidene-5-deoxy-β-D-ribofuranosid-5-yl)-β-D-glucopyranoside (**107**). Compound **107** had also been obtained after methanolysis of thuringiensin and identical derivatization.[42]

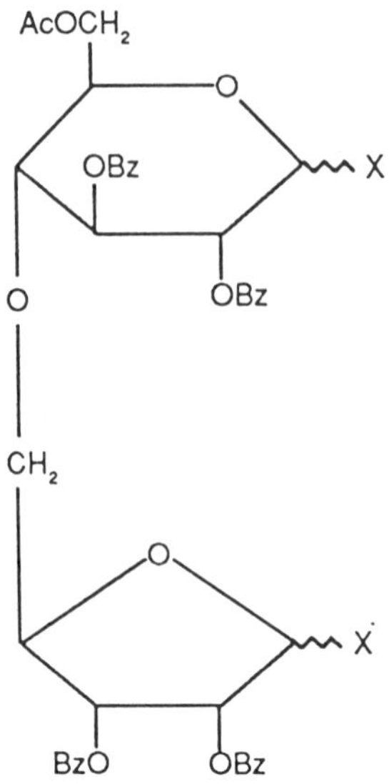

105, X = OAc
106, X = Br

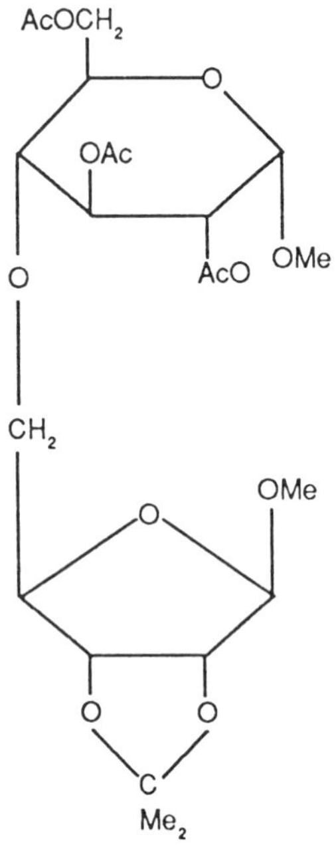

(107)

The next experiments were concerned with the synthesis of the disaccharide nucleoside.[88] In preliminary experiments, it was determined that the bromides of 2,3,5-tri-*O*-benzoyl-D-ribofuranose and 6-*O*-acetyl-2,3,4-tri-*O*-benzoyl-D-glucopyranose differed in their reactivity with 6-benzamidochloromercuripurine. Maximum reaction conditions were found, as related to the stoichiometry of the reactants and the nature of the solvent, which gave a 50% yield of the ribofuranosyl nucleoside but less than 1% of the glucopyranosyl nucleoside in a short reaction time. Thus, the 1,1'-dibromo derivative **106** was reacted with 0.4 equivalents of 6-benzamidochloromercuripurine for 4 min in boiling acetonitrile. The reaction mixture was decomposed with methanol and silver oxide to generate the glucoside. The protected nucleoside (**108**) was obtained in

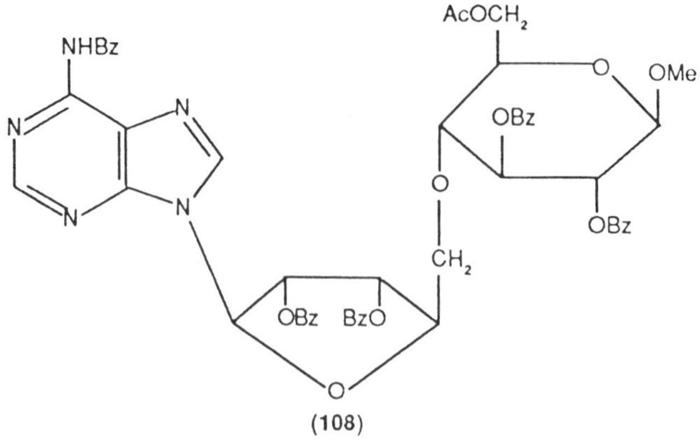

(108)

a 20% yield. It was claimed that when a fourfold excess of the mercuric chloride salt was used, a double-headed nucleoside having both anomeric positions in nucleoside linkages was formed in a 22% yield.

The preparation of thuringiensin proceeded in a manner somewhat different from the above. It was recognized that a reaction sequence that incorporated the α-D-glucoside bond to allaric acid before nucleoside synthesis would be highly advantageous and that the proper derivative of this nucleoside could then accept a phosphate in the appropriate position. It was also necessary to use a protected ether disaccharide glycoside that would have a D-ribofuranoside moiety of low reactivity compared to the D-glucose residue for the synthesis of the allaric acid glycoside. For this latter purpose, 2,2,2-trichloroethyl 2,3-*O*-cyclocarbonyl-β-D-ribofuranoside (**109**) was prepared by the treatment of tetra-*O*-acetyl-D-ribofuranose with 2,2,2-trichloroethanol and boron trifluoride, followed by esterolysis and reaction of the D-riboside with phosgene.[89] Condensation of **109** with **103** as previously described gave 2-*O*-benzyl-4-*O*-(2,2,2-trichloroethyl 2,3-*O*-cyclocarbonyl-β-D-ribofuranosid-5-yl)-1,6-anhydro-β-D-glucopyranose (**110**). Acetolysis of **110** gave the triacetate **111**, in which the 2,2,2-trichloroethoxy group remained intact.

The derivative of allaric acid used in the preparation of the D-glucoside linkage was synthesized from 1,2-*O*-isopropylidene-α-D-allofuranose (**112**).[90] Treatment of **112** with triphenylmethyl chloride gave the 6-*O*-trityl derivative (**113**), which was benzoylated and detritylated to yield 3,5-di-*O*-benzoyl-1,2-*O*-isopropylidene-α-D-allofuranose (**114**). Ruthenium tetraoxide oxidation afforded the uronic acid **115**, which was converted to methyl 3,5-di-*O*-benzoyl-1,2-*O*-isopropylidene-α-D-alluronate (**116**) with diazomethane. Acid hydrolysis of the isopropylidene group and bromine oxidation gave the desired methyl 3,5-di-*O*-benzoyl-(2*R*)-allaro-1,4-lactone-6-ate (**117**).

The triacetate **111** was coupled to **117** in chloroform containing boron trifluoride. The protected trisaccharide was obtained in low yield and was difficult to purify unless the benzyl group was first removed by hydrogenolysis, whereupon methyl 3,5-di-*O*-benzoyl-2-*O*-[3,6-di-*O*-acetyl-4-*O*-(2,2,2-trichloroethyl 2,3-*O*-cyclocarbonyl-β-D-ribofuranosid-5-yl)-α-D-glucopyranosyl]-(2*R*)-allaro-1,4-lactone-6-ate (**118**) could be separated by column chromatography on silicic acid. The free hydroxyl group was acetylated and the carbonate group removed by alkaline hydrolysis. The allaric acid residue had to be relactonized after this step and the remaining hydroxyl groups were acetylated. The trichloroethyl group was removed by a reductive cleavage with zinc, and the anomeric hydroxyl group was acetylated. The syrupy trisaccharide **119** was converted to the glycosyl bromide with hydrogen bromide and condensed with 6-benzamidochloromercuripurine in acetonitrile. The product **120** was treated with methanol and pyridine to obtain the dimethyl ester (**121**). The free hydroxyl group liberated in this step is the one required for phosphorylation and had been protected by the lactone ring throughout previous steps. The reaction of **121** with phosphorus oxychloride and alkaline hydrolysis of the blocking ester groups gave thuringiensin (**39**).[90]

In an alternative approach, published shortly after the above, thuringiensin was again prepared, with a number of improvements.[91] The use of benzoyl blocking groups in place of the cyclocarbonyl group on D-ribose removed the

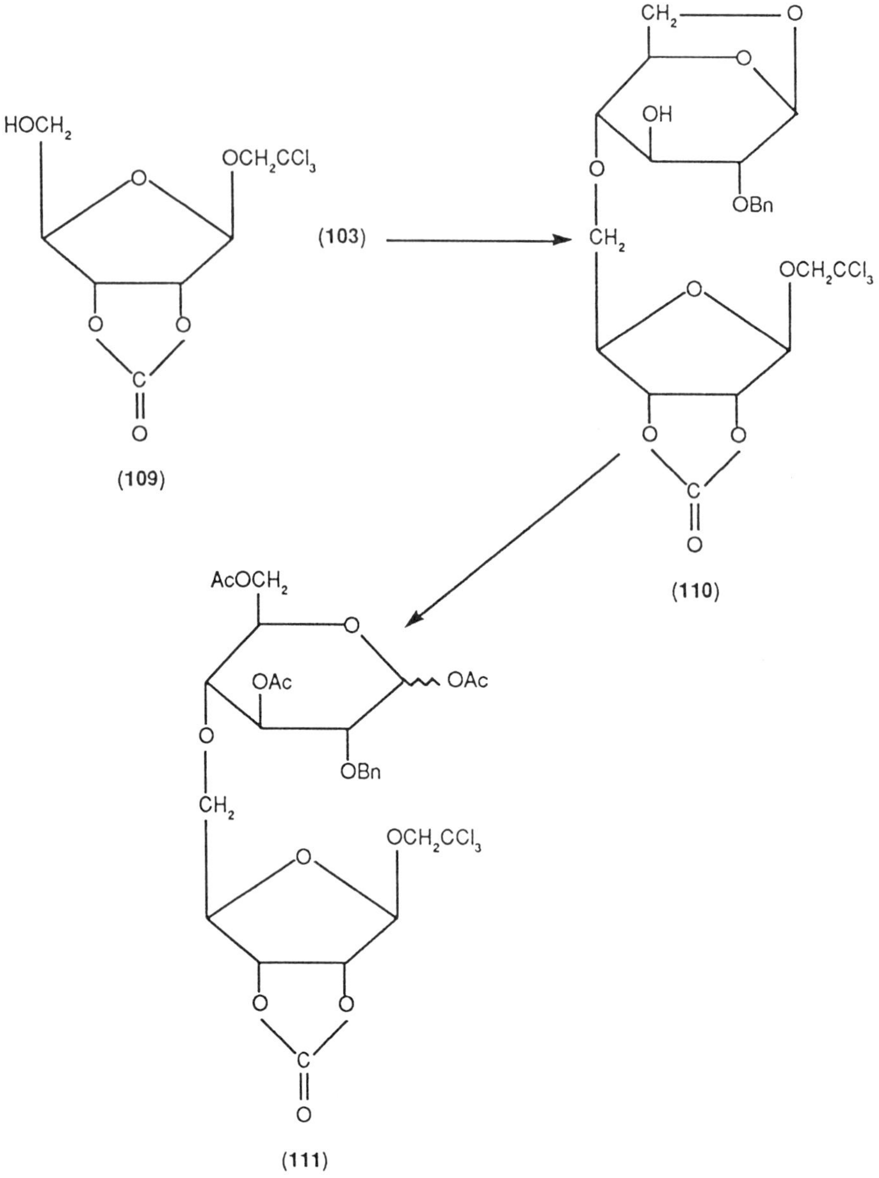

(109)

(103) ⟶

(110)

(111)

necessity of hydrolyzing this group later on to replace it with acetyl or benzoyl groups. This was also the step that caused the lactone to open up. Another innovation was the use of stannic chloride, in place of boron trifluoride, in the synthesis of the ether disaccharide bond. A third change was the use of a benzyl ether group at C-3 of the allaro-1,4-lactone derivative, because a benzoyl group in that position tended to migrate to C-2. 2,2,2-Trichloroethyl β-D-ribofurano-side was converted to the 5-O-triphenylmethyl derivative; the latter was benzoyl-ated, and then the trityl group was removed with acid to give the dibenzoate **122**. Treatment of the latter with **103** in benzene in the presence of stannic chloride

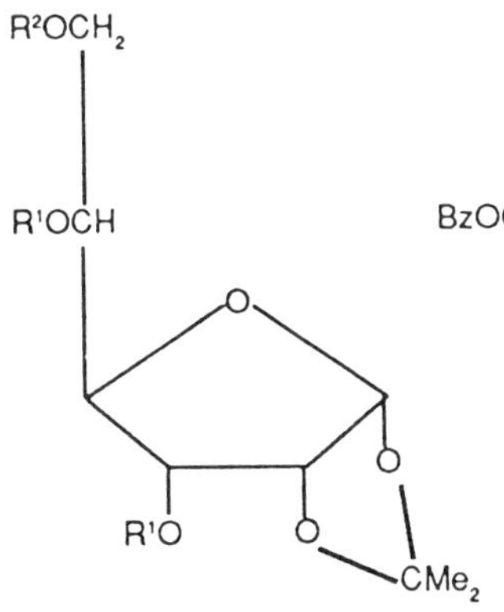

112, R¹ = R² = H
113, R¹ = H; R² = Tr
114, R¹ = Bz; R² = H

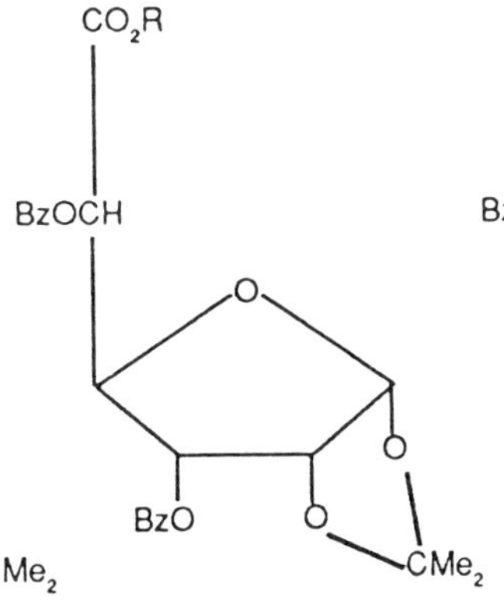

115, R = H
116, R = Me

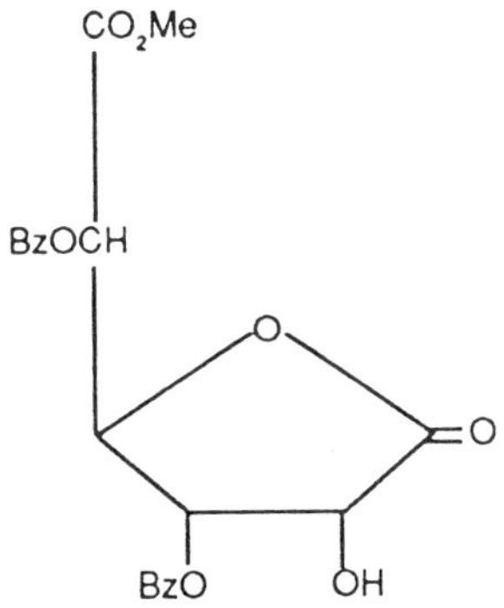

(117)

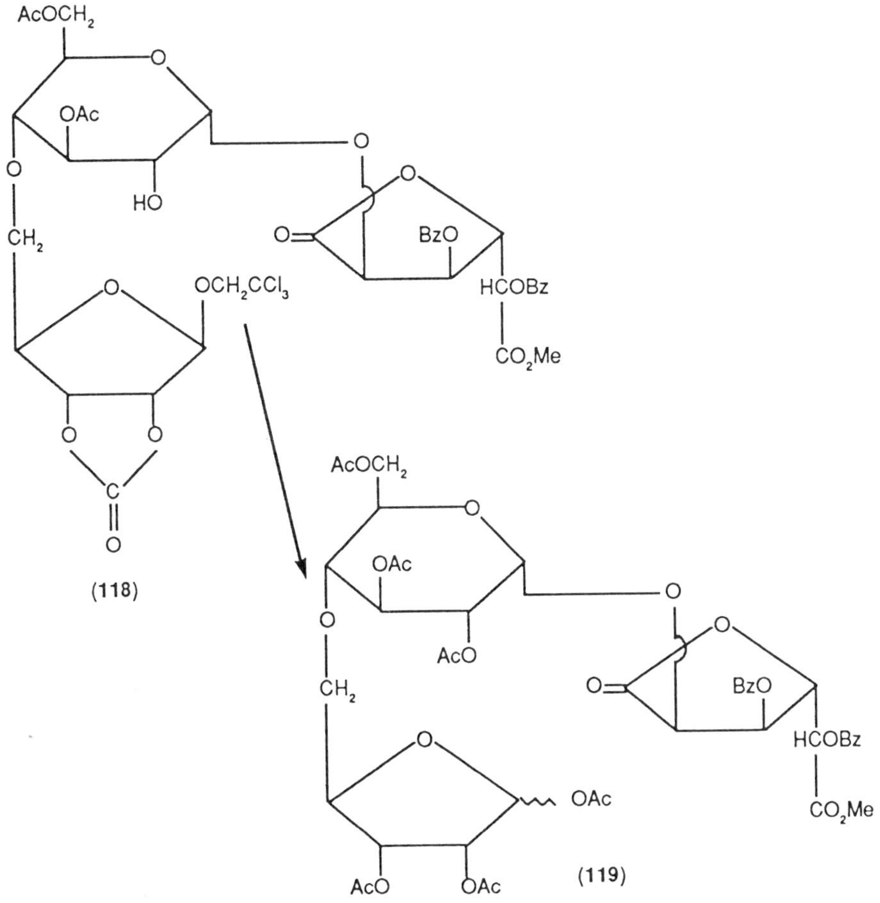

(118)

(119)

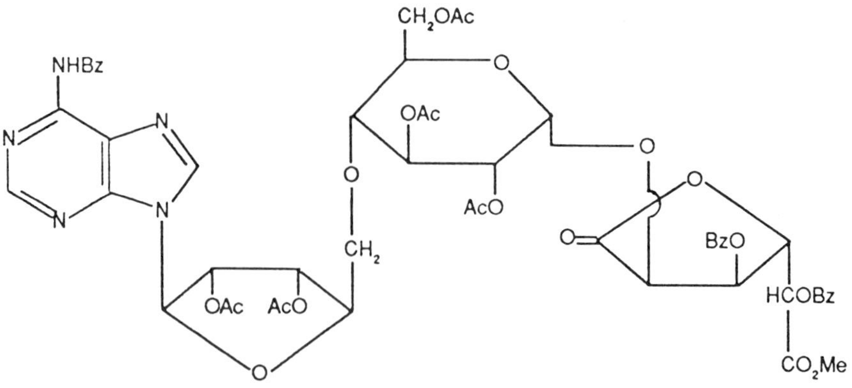

(120)

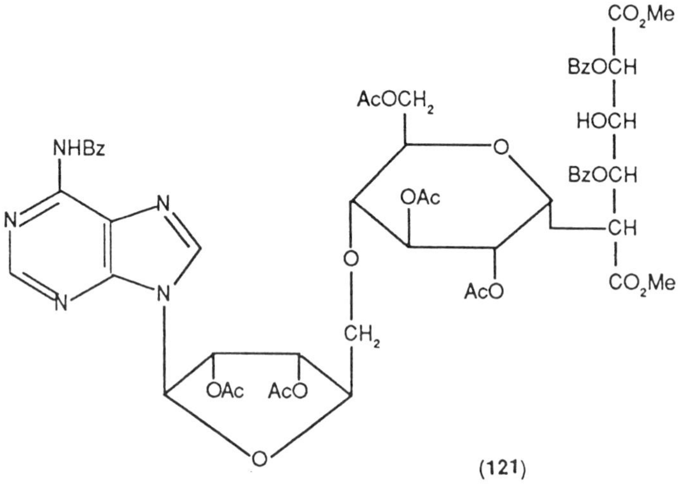

(121)

gave the desired ether, which was benzoylated and subjected to acetolysis to afford **123**.

The preparation of the allaro-1,4-lactone derivative, required for coupling with **123**, was achieved from 1,2-*O*-isopropylidene-α-D-glucofurano-3,6-lactone (**124**). Benzoylation of the free hydroxyl group and methanolysis with methanolic triethylamine gave methyl 5-*O*-benzoyl-1,2-*O*-isopropylidene-α-D-glucofuranuronate (**125**). Oxidation with ruthenium tetraoxide gave the 3-keto compound **126**, which was reduced with tri(*tert*-butyloxy) lithium aluminum hydride in tetrahydrofuran to give methyl 5-*O*-benzoyl-1,2-*O*-isopropylidene-α-D-allofuranuronate (**127**). The benzyl derivative **128** was prepared, the isopropylidene group was removed with aqueous formic acid, and bromine oxidation gave the desired lactone **129**.[86]

Reaction of **129** with **123** gave a 13% yield of the desired trisaccharide and a 2.3% yield of the β-anomer, as well as a mixture of other products, which were

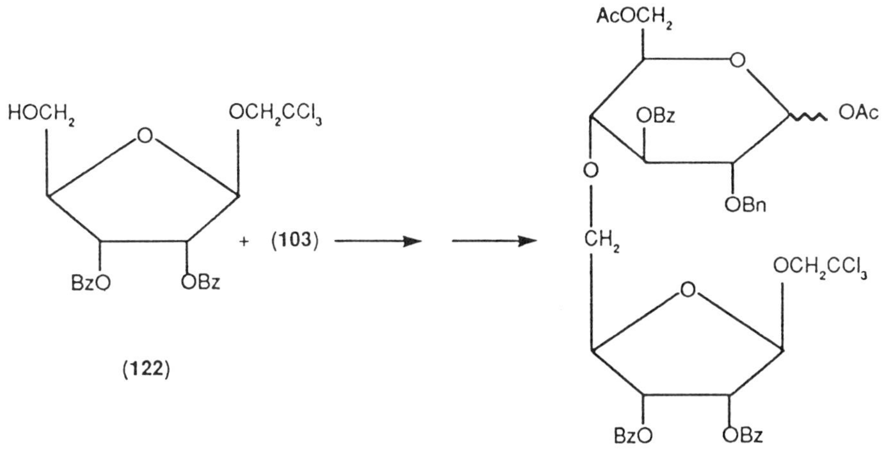

(122) + (103) ⟶ ⟶ (123)

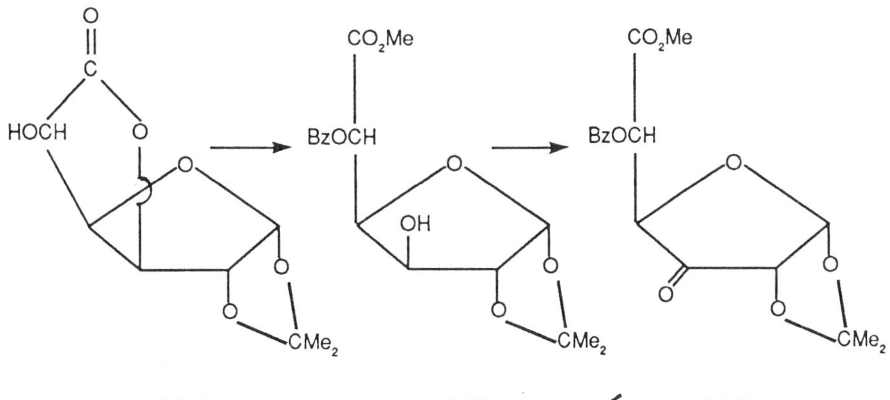

(124) ⟶ (125) ⟶ (126)

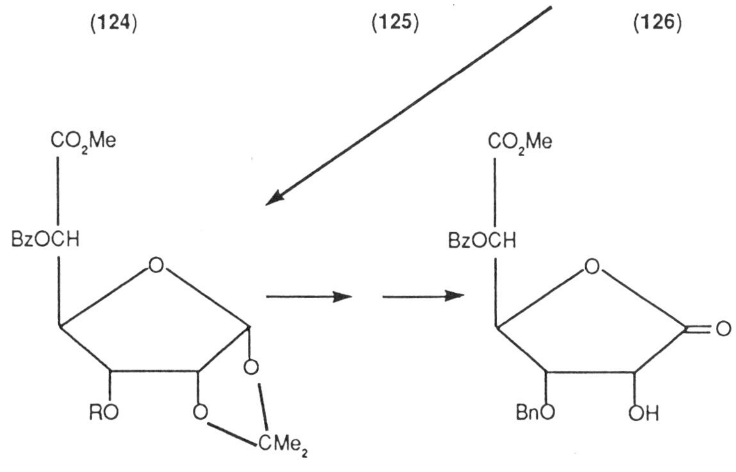

127, R = H
128, R = Bn

(129)

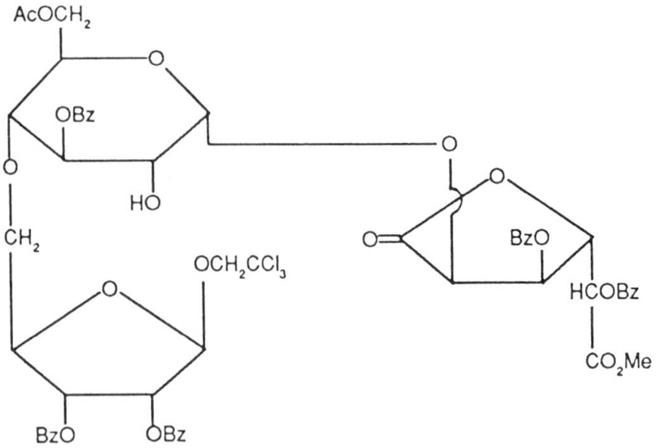

(130)

separated chromatographically. The benzyl groups were removed and a crystalline product (**130**) obtained. That this was the α-glycoside was verified by the
NMR spectrum of acetylated **130**, which has a chemical shift higher than that of
the β-anomer and $J_{1,2}$ = 3.5 Hz. Treatment of either **130** or its acetylated
derivative with zinc and acetic anhydride in the presence of trifluoroacetic acid
replaced the trichloroethoxyl group with an acetoxyl. The fully protected trisaccharide was converted to the bromide and then to the nucleoside in a manner
similar to that described above. The lactone residue of the nucleoside was
opened with sodium acetate in methanol, and the phosphorylation and alkaline
hydrolysis which followed again produced thuringiensin (**39**).

Using the same synthetic methods, three analogues of thuringiensin have
been prepared in which glucaric, ribaric, and allonic acids are substituted for
allaric acid, and their ability to inhibit DNA-dependent RNA polymerase has
been evaluated.[92] An analogue of thuringiensin containing uracil in place of
adenine has been reported.[93] The same hexaacetate **119** was converted to the
glycosyl bromide and condensed with 2,4-bis(trimethylsilyl)uracil in acetonitrile
in the presence of mercuric bromide. The same sequence of reactions as described
above followed, and the nucleoside product was purified by ion-exchange chromatography. The yield and properties were determined spectroscopically, and
the substance acted as an antagonist for UTP with bacterial DNA-dependent
RNA polymerase.

3.2.3 The Tunicamycins

The synthesis of tunicamycin V (earlier called tunicamycin A) has recently
been reported.[94,95] Potassium fluoride-catalyzed condensation of 3-*O*-acetyl-
5-deoxy-1,2-*O*-isopropylidene-5-nitro-α-D-ribofuranose (**131**) with methyl
2-(benzyloxycarbonyl)amino-2-deoxy-3,4-*O*-isopropylidene-α-D-galactodialdo-
pyranoside-(1,5) (**132**) in acetonitrile gave the nitroundecose **133**. A single dia-

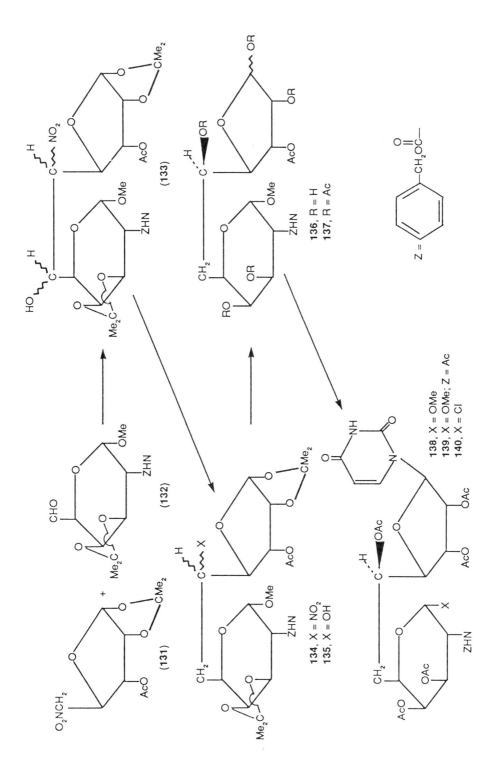

stereomer was isolated in 51% yield. (Since it is the furanose ring that is to be linked to the base, the numbering of the carbon atoms of the undecose will begin from that end in the ensuing discussion.) Dehydration using acetic anhydride followed by treatment with sodium borohydride gave the 6-deoxy compound (**134**). The 5-nitro group was exchanged for a hydroxyl group (**135**) by successive reactions with potassium permanganate and sodium *tert*-butoxide, then sodium borohydride, and finally sodium methoxide. The epimers were separated by silica gel chromatography. One of the epimers was acetylated, the isopropylidene groups were removed with aqueous acetic acid to give **136**, and acetylation afforded **137**. The latter was condensed with bis(trimethylsilyl)uracil under the conditions of Niedballa and Vorbrüggen[68] to give the blocked tunicaminyl nucleoside (**138**). Hydrogenolysis and acetylation provided the hexaacetate **139**, which was shown to be identical to the hexaacetate prepared from a sample of tunicaminyluracil obtained by degradation of tunicamycin. This proved the configuration of the 5-epimer chosen for this synthesis. The other epimer was also subjected to the same reaction pathway to obtain the C-5' analogue of the tunicaminyluracil derivative.[94]

The blocked nucleoside **138** was converted to the 11'-chloro derivative (**140**) by acetolysis followed by treatment with hydrogen chloride in a dioxane–acetyl chloride mixture. Coupling of **140** with 2-acetamido-2-deoxy-4,6-*O*-isopropylidene-3-*O*-propionyl-α-D-glucopyranose (**141**) gave the disaccharide nucleoside derivative **142** in 19% yield.[95] Following O-deacylation and hydrogenolysis, the product was N-acylated with (*E*)-13-methyl-2-tetradecenoic acid with the aid of dicyclohexylcarbodiimide. Careful acidic hydrolysis to remove the isopropylidene group afforded a tunicamycin (**143**) whose physical and spectral properties were identical to those of tunicamycin V. The preparation of several other tunicamycins was announced, but no details were provided.[95]

3.2.4. Miscellaneous

As of this writing, there has been no reported synthesis of an ezomycin or of hikizimycin. Hanessian et al. have outlined some ideas and presented some model experiments that could eventually lead to a practical route to the ezomycins.[96] The synthesis of a 3',7'-anhydroocturonic acid nucleoside has been reported,[97] and a synthesis of ezoaminuroic acid has been accomplished.[98] One of the component sugars of hikizimycin, 3-amino-3-deoxy-D-glucose (kanosamine), was prepared many years ago.[99] More recently, a derivative of hikosamine, methyl peracetyl α-hikosaminide, has been synthesized.[100]

4. Addendum: Synthesis and Properties of Various Disaccharide Nucleosides

The corynetoxins have definitely been shown to be the same group of compounds as the tunicamycins, the only difference being that the corynetoxins have fatty acid components with longer carbon chains, and with some of these having β-hydroxy groups and α,β-unsaturation.[101] The configurations at the anomeric positions were unfortunately changed to α for the galactosamine

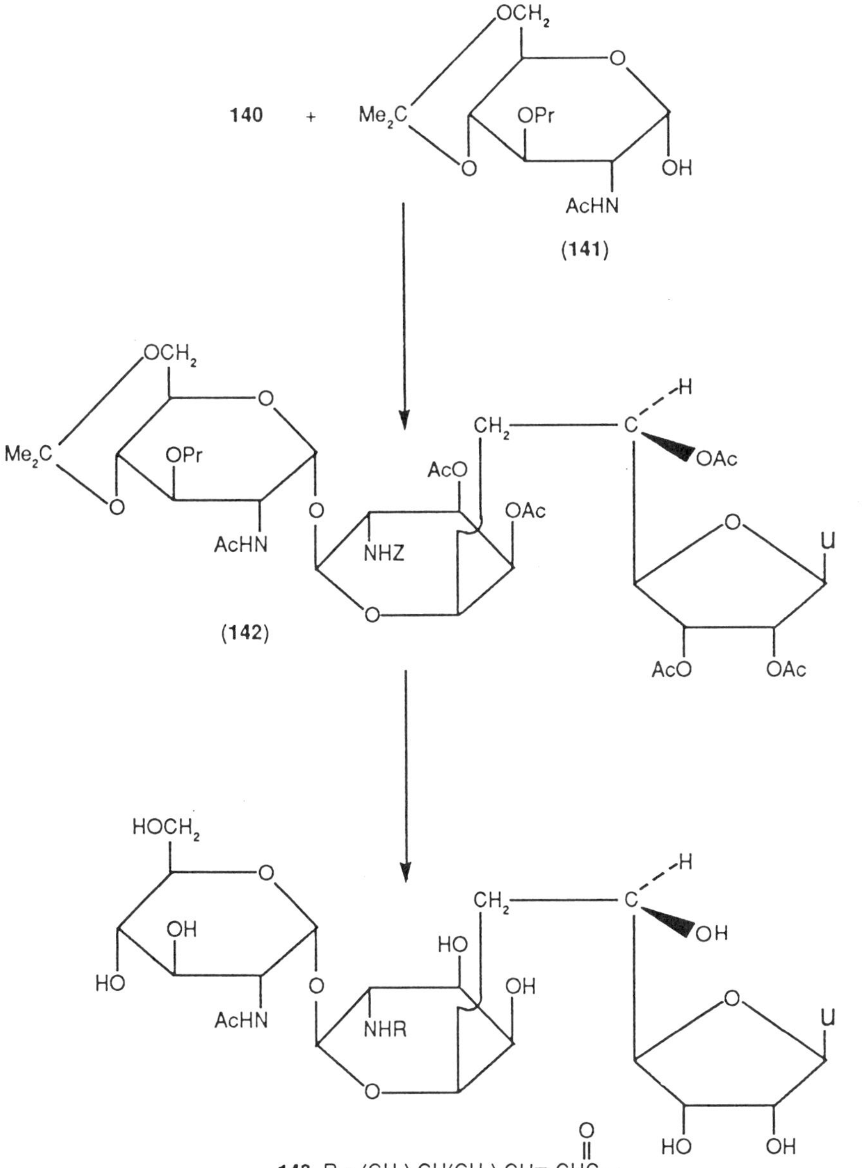

140 + (141)

(142)

143 R = $(CH_3)_2CH(CH_2)_9CH= CHC-$ (with C=O)

$Pr = CH_3CH_2C-$ (with C=O)

$Z =$ [phenyl]$-CH_2OC-$ (with C=O)

$u =$ [uracil]

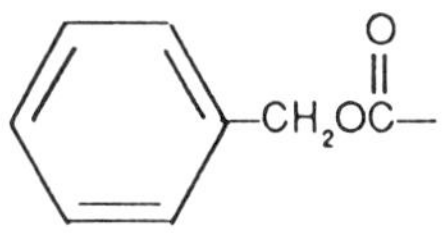

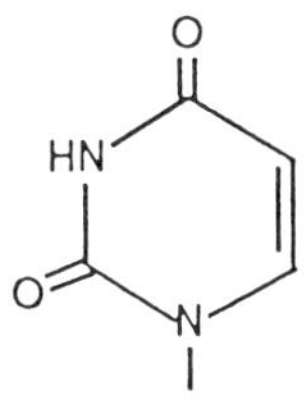

moiety of the 11-carbon sugar and β for the glucosamine residue.[101] Using two-dimensional NMR techniques, Tamura's group has shown that the original assignments of the anomeric linkages were correct; the galactosamine moiety is linked with a β configuration at the C-11' position to N-acetyl-D-glucosamine having an α configuration.[102]

Suami and co-workers have provided experimental details of their synthesis of tunicamycin V (**143**).[103,104] In a different approach from that of this group, tunicamine has been prepared by construction of the galactosamine moiety onto the C-5 position of D-ribose in order to obtain this 11-carbon sugar, which was then used to prepare tunicaminyluracil (**40**).[105] An analogue of tunicamycin has been synthesized by the same reaction sequence outlined in Section 3.2.3 for tunicamycin V.[106] N-Acetyl-D-galactosamine was substituted for N-acetyl-D-glucosamine. What was suspected of being the β,β-linked disaccharide nucleoside was isolated in addition to the desired α,β-linked analogue.

A start has been made toward the synthesis of the ezomycins.[107,108] A major problem has been the failure to obtain the fused 3,7-anhydro ring structure with the correct stereochemistry. Recent approaches to the synthesis of the closely related octosyl acid A should be helpful.[109,110]

The 11-carbon sugar component of hikizimycin, hikosamine (**16**), has been synthesized using asymmetric induction techniques.[111]

A new disaccharide nucleoside antibiotic having antibacterial activity has been isolated from the culture filtrate of *Streptomyces griseus*.[112] This nucleoside has been named capuramycin, based upon the caprolactam part of the molecule. The structure **144** has been proposed on the basis of detailed chemical degradation studies, spectroscopy, and X-ray analysis.[112,113]

A series of disaccharide nucleosides have been prepared by coupling N-acetyl-D-neuraminic acid to various natural or preformed ribosyl nucleosides.[114,115] Methyl N-acetyl-β-D-neuraminate was obtained from N-acetyl-D-neuraminic acid by treatment with methanol and a strong acid ion-exchange

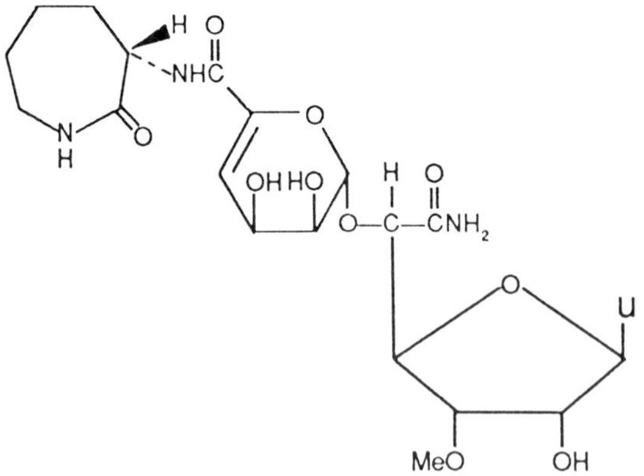

144   U = uracil

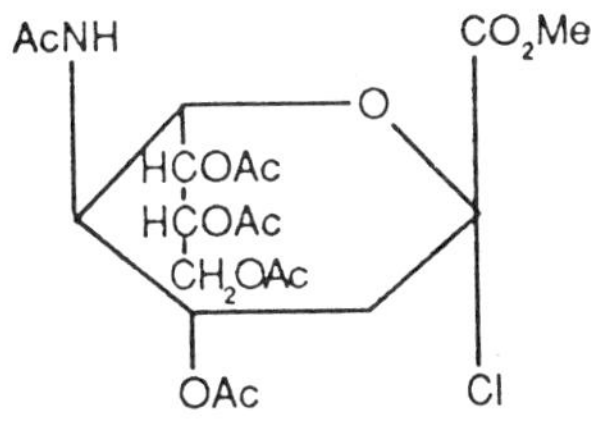

(145)

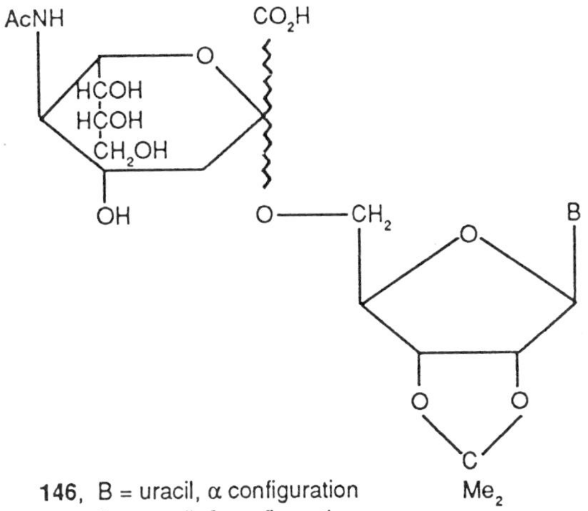

146, B = uracil, α configuration
147, B = uracil, β configuration
148, B = 5-fluorouracil, α configuration
149, B = 5-fluorouracil, β configuration

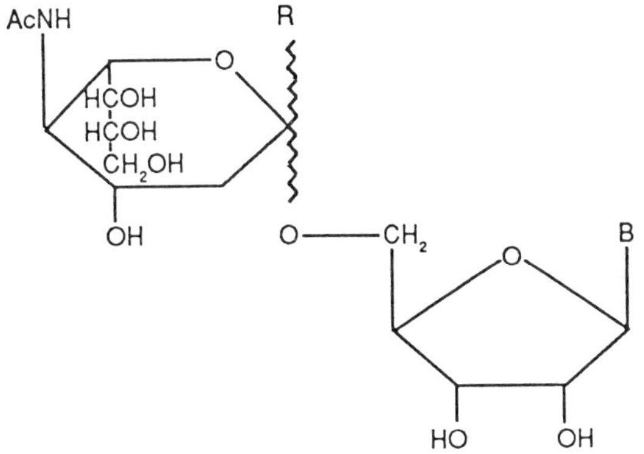

150, B = hypoxanthine, R = CO_2H, α configuration
151, B = hypoxanthine, R = CO_2H, β configuration
152, B = cytosine, R = $CO_2^-NH_4^+$, α configuration
153, B = cytosine, R = $CO_2^-NH_4^+$, β configuration

resin. The sugar was then converted in a single step to methyl 4,7,8,9-tetra-*O*-acetyl-*N*-acetyl-2-chloro-2-deoxy-β-D-neuraminate (**145**) with acetyl chloride, a method which afforded **145** in crystalline form. The chloride (**145**) was coupled with 2′,3′-*O*-isopropylideneuridine and 5-fluoro-2′,3′-*O*-isopropylideneuridine in acetonitrile containing mercuric bromide and mercuric cyanide. Anomeric mixtures were obtained and were separated by column chromatography. The ester groups were hydrolyzed with aqueous sodium hydroxide, and the pH was adjusted to obtain the acidic form of the sugar. However, the isopropylidene groups were left in place. No explanation was given for this. It is possible that their removal under acidic conditions would also break the newly formed glycosidic bonds. The anomeric configurations of the glycosidic linkages of the new nucleosides **146–149** were determined by ¹H- and ¹³C-NMR spectroscopy and by a new method involving differences in the rates of hydrolysis between α- and β-glycosides of this type.[114] Compound **145** was also coupled with 2′,3′-di-*O*-acetylinosine and 2′,3′-di-*O*-acetyl-*N*-benzoylcytidine, this time using silver trifluoromethanesulfonate and *N*,*N*-dimethylformamide.[115] Again, the anomeric mixtures were separated by column chromatography. The ester groups of the inosine derivatives were removed as above to give **150** and **151**. The ester groups of the cytidine derivatives were removed in methanolic ammonia and afforded the α- and β-anomers of ammonium [*N*-acetyl(cytidine-5′-yl)-D-neuraminosid]-onate (**152** and **153**).

5. References

1. M. H. McCormick and M. M. Hoehn, Isolation of a new antibiotic from *Streptomyces fasiculatus* Nov. SP., *Antibiot. Chemother.* **3**, 718–720 (1953).
2. C. DeBoer, E. L. Caron, and J. W. Hinman, Amicetin, a new streptomyces antibiotic, *J. Am. Chem. Soc.* **75**, 499–500 (1953).
3. J. W. Hinman, E. L. Caron, and C. DeBoer, The isolation and purification of amicetin, *J. Am. Chem. Soc.* **75**, 5864–5866 (1953).
4. Y. Hinuma, M. Kuroya, T. Yajima, K. Ishihara, S. Hanada, K. Watanabe, and K. Kikuchi, An amicetin (sacromycin) from *Streptomyces* SP No. 5223, *J. Antibiot. Ser. A* **8**, 148–152 (1955).
5. S. Tatsuoka, K. Nakagawa, M. Inoue, and S. Fujii, Antibiotics V. Extraction and physiochemical properties of allomycin, an antituberculous antibiotic, and its identity with amicetin, *J. Pharm. Soc. Jpn.* **75**, 1206–1208 (1955) [*CA* **50**, 8695b (1956)].
6. E. H. Flynn, J. W. Hinman, E. L. Caron, and D. O. Woolf, Jr., The chemistry of amicetin, a new antibiotic, *J. Am. Chem. Soc.* **75**, 5867–5871 (1953).
7. C. L. Stevens, R. J. Gasser, T. K. Mukherjee, and T. H. Haskell, The structure of amicetin. A new dimethylamino sugar, *J. Am. Chem. Soc.* **78**, 6212 (1956).
8. C. L. Stevens, K. Nagarajan, and T. H. Haskell, The structure of amicetin, *J. Org. Chem.* **27**, 2991–3005 (1962).
9. C. L. Stevens, B. Cross, and T. Toda, The synthesis of DL-*threo*- and *erythro*-amicetose 2,4-dinitrophenylhydrazones, *J. Org. Chem.* **28**, 1283–1286 (1963).
10. F. A. Hochstein and P. P. Regna, Magnamycin IV. Mycaminose, and aminosugar from magnamycin, *J. Am. Chem. Soc.* **77**, 3353–3355 (1955).
11. C. L. Stevens, P. Blumbergs, and F. A. Daniker, Sterochemistry and synthesis of amosamine: 4,6-Dideoxy-4-dimethylamino-D-glucose, *J. Am. Chem. Soc.* **85**, 1552–1553 (1963).
12. S. Hanessian and T. H. Haskell, Configuration of the anomeric linkages in amicetin, *Tetrahedron Lett.* **1964**, 2451–2460.

13. P. Sensi, A. M. Greco, G. G. Gallo, and G. Rolland, Isolation and structure determination of a new amicetin-like antibiotic: Amicetin B, *Antibiot. Chemother.* **7**, 645–652 (1957).

14. T. H. Haskell, A. Ryder, R. P. Frohardt, S. A. Fusari, Z. L. Jakubowski, and Q. R. Bartz, The isolation and characterization of three crystalline antibiotics from *Streptomyces plicatus, J. Am. Chem. Soc.* **80**, 743–747 (1958).

15. T. H. Haskell, Amicetin, bamicetin, and plicacetin. Chemical studies, *J. Am. Chem. Soc.* **80**, 747–751 (1958).

16. R. Evans and G. Weare, Norplicacetin, a new antibiotic from *Streptomyces plicatus, J. Antibiot.* **30**, 604–606 (1977).

17. M. Konishi, M. Kimeda, H. Tsukiura, H. Yamamoto, T. Hoshiya, T. Miyaki, K. Fujisawa, H. Koshiyama, and H. Kawaguchi, Oxamicetin, a new antibiotic of bacterial origin. 1. Production, isolation, and properties, *J. Antibiot.* **26**, 752–756 (1973).

18. M. Konishi, M. Maruishi, T. Tsuno, H. Tsukiura, and H. Kawaguchi, Oxamicetin, a new antibiotic of bacterial origin. II. Structure of oxamicetin, *J. Antibiot.* **26**, 757–764 (1973).

19. K. Uchida, T. Ichikawa, Y. Shimauchi, T. Ishikura, and A. Ozaki, Hikizimycin, a new antibiotic, *J. Antibiot.* **24**, 259–262 (1971).

20. B. C. Das, J. Defaye, and K. Uchida, The structure of hikizimycin. Part 1. Identification of 3-amino-3-deoxy-D-glucose and cytosine as structural components, *Carbohydr. Res.* **22**, 293–299 (1972).

21. K. Uchida and B. C. Das, Hikosamine, a novel C_{11} aminosugar component of the antibiotic hikizimycin, *Biochimie* **55**, 635–636 (1973).

22. K. Uchida, Structure de l'hikizimycine, un antibiotique nucléosidique, *Agric. Biol. Chem.* **40**, 395–404 (1976).

23. K. Uchida, E. Breitmaier, and W. A. Koenig, ^{13}C-NMR investigations of the nucleoside antibiotic hikizimycin and its constituents, *Tetrahedron* **31**, 2315–2317 (1975).

24. M. Vuilhorgne, S. Ennifar, B. C. Das, J. W. Paschal, R. Nagarajan, E. W. Hagaman, and E. Wenkert, Structure analysis of the nucleoside disaccharide antibiotic anthelmycin by carbon-13 nuclear magnetic resonance spectroscopy. A structural revision of hikizimycin and its identity with anthelmycin, *J. Org. Chem.* **42**, 3289–3291 (1977).

25. R. L. Hamill and M. M. Hoehn, Anthelmycin, a new antibiotic with anthelmitic properties, *J. Antibiot. Ser. A.* **17**, 100–103 (1964).

26. M. Vuilhorgne, S. Ennifar, and B. C. Das, Synthèse et étude conformationelle par R.M.N.-^{13}C de β-nucléosides pyrimidiques contenant une ou deux sousunités hexopyranosyles: Application à la conformation de l'anthelmycine, *Carbohydr. Res.* **97**, 19–30 (1981).

27. A. K. Takaoka, T. Kuwayama, and A. Aoki, Jpn. Patent, 615332 (1971).

28. K. Sakata, A. Sakurai, and S. Tamura, Isolation of novel antifungal antibiotics, ezomycins A_1, A_2, B_1, and B_2, *Agric. Biol. Chem.* **39**, 1883–1890 (1974).

29. K. Sakata, A. Sakurai, and S. Tamura, Structures of ezomycins B_1, B_2, C_1, C_2, D_1, and D_2, *Agric. Biol. Chem.* **41**, 2033–2039 (1977).

30. K. Sakata, A. Sakurai, and S. Tamura, L-Cystathionine as a component of ezomycins A_1 and B_1 from a *Streptomyces, Agric. Biol. Chem.* **37**, 697–699 (1973).

31. K. Sakata, A. Sakurai, and S. Tamura, Ezoaminuroic acid, 3-amino-3,4-dideoxy-D-*xylo*-hexopyranuronic acid, as a constituent of ezomycins A_1 and A_2, *Tetrahedron Lett.* **1974**, 1533–1536.

32. N. Harada and K. Nakanishi, A method for determining the chiralities of optically active glycols. *J. Am. Chem. Soc.* **91**, 3989–3991 (1969).

33. K. Sakata, A. Sakurai, and S. Tamura, Structures of ezomycins A_1 and A_2, *Agric. Biol. Chem.* **40**, 1993-1999 (1976).

34. K. Sakata and J. Uzawa, Application of C-13 NMR spectrometry to the structural investigation of the novel bicyclic anhydrooctose uronic acid nucleosides, constituents of ezomycins, *Agric. Biol. Chem.* **41**, 413–415 (1977).

35. K. Isono, P. F. Crain, and J. N McCloskey, Isolation and structure of octosyl acids. Anhydrooctose uronic acid nucleosides, *J. Am. Chem. Soc.* **97**, 943–945 (1975).

36. H. de Barjac and R. Dedonder, Isolement d'un nucléotide identifiable à la "Toxin Thermostable" de *Bacillus thuringiensis* var. Berliner, *Acad. Sci.* **260**, 7050–7053 (1965).

37. H. de Barjac and R. Dedonder, Purification de la toxine thermostable de *B. thuringiensis* var. *thuringiensis* et analyses complémentaires, *Bull. Soc. Chim. Biol* **50**, 941–944 (1968).

38. G. Benz, On the chemical nature of the heat-stable exotoxin of *Bacillus thuringiensis*, *Experientia* **22**, 81–82 (1966).

39. K. Šebasta, K. Horská, and J. Vaňková, Isolation and properties of the insecticidal exotoxin of *Bacillus thuringiensis* var. *gelechiae*, *Collect. Czech. Chem. Commun.* **34**, 891–900 (1969).

40. R. P. M. Bond, The natural occurrence of allomucic (allaric) acid: A contribution to the assignment of structure of an insecticidal exotoxin from *Bacillus thuringiensis* Berliner, *J. Chem. Soc., Chem. Commun.* **1969**, 338–339.

41. R. P. M. Bond, C. B. C. Boyce, and S. J. French, A purification and some properties of an insecticidal exotoxin from *Bacillus thuringiensis* Berliner, *Biochem. J.* **114**, 477–488 (1969).

42. J. Farkas, K. Šebasta, K. Horská, Z. Samek, L. Dolejš, and F. Šorm, The structure of exotoxin of *Bacillus thuringiensis* var. *gelechiae*, *Collect. Czech. Chem. Commun.* **34**, 1118–1120 (1969).

43. L. Kalvoda, M. Prystaš, and F. Šorm, Nucleic acid components and their analogues, CLX. Determination of the structure of the allaric portion of exotoxin from *Bacillus thuringiensis* by means of periodate oxidation, *Collect. Czech. Chem. Commun.* **38**, 2529–2532 (1973).

44. M. Prystaš, L. Kalvoda, and F. Šorm, Structure proof of exotoxin from *Bacillus thuringiensis*, *Collect. Czech. Chem. Commun.* **40**, 1775–1785 (1975).

45. L. Kalvoda, M. Prystaš, and F. Šorm, The structure of the allaric portion of exotoxin from *Bacillus thuringiensis*, *Tetrahedron Lett.* **1973**, 1873–1876.

46. J. Farkas, K. Šebasta, K. Horská, Z. Samek, L. Dolejš, and F. Šorm, Structure of thuringiensin, the thermostable exotoxin from *Bacillus thuringiensis*, *Collect. Czech. Chem. Commun.* **42**, 909–929 (1977).

47. M. Pais and H. deBarjac, Identification par des méthodes chimiques et spectrales de la toxine thermostable isolée d'un *B. thuringiensis* du sérotype 4a-4c, *J. Carbohydr. Nucleosides Nucleotides* **1**, 213–223 (1974).

48. A. Takatsuki, K. Arima, and S. Tamura, Tunicamycin, a new antibiotic. I. Isolation and characterization of tunicamycin, *J. Antibiot.* **24**, 215–223 (1971).

49. A. Takatsuki, K. Kawamura, M. Okina, Y. Kodama, T. Ito, and G. Tamura, The structure of tunicamycin, *Agric. Biol. Chem.* **41**, 2307–2309 (1977).

50. A. Takatsuki, K. Kawamura, Y. Kodama, T. Ito, and G. Tamura, Structure determination of fatty acids from tunicamycin complex, *Agric. Biol. Chem.* **43**, 761–764 (1979).

51. T. Ito, Y. Kodama, K. Kawamura, K. Suzuki, A. Takatsuki, and G. Tamura, Structure determination of tunicaminyl uracil, a degradation product of tunicamycin, *Agric. Biol. Chem.* **43**, 1187–1195 (1979).

52. T. Ito, A. Takatsuki, K. Kawamura, K. Sato, and G. Tamura, Isolation and structures of components of tunicamycin, *Agric. Biol Chem.* **44**, 695–698 (1980).

53. M. Kenig and C. Reading, Holomycin and an antibiotic (MM 19290) related to tunicamycin, metabolites of *Streptomyces clavuligerus*, *J. Antibiot.* **32**, 549–554 (1979).

54. K. Eckardt, W. Ihn, D. Tresselt, and D. Krebs, The chemical structures of streptovirudins, *J. Antibiot.* **34**, 1631–1632 (1981).

55. J. A. Edgar, J. L. Frahn, P. A. Cockrum, N. Anderton, M. V. Jago, C. C. J. Culvenor, A. J. Jones, K. Murray, and K. J. Shaw, Corynetoxins, causative agents of annual ryegrass toxicity; their identification as tunicamycin group antibiotics, *J. Chem. Soc., Chem. Commun.* **1982**, 222–224.

56. A. Takatsuki and G. Tamura, Inhibition of glycoconjugate biosynthesis by tunicamycin, in: *Tunicamycin* (G. Tamura, ed.), pp. 35–70, Japan Scientific Societies Press, Tokyo (1982).

57. M. L. Wolfrom, P. McWain, F. Shafizadeh, and A. Thompson, 9-β-Lactosyladenine and 2,6-diamino-9-β-lactosylpurine, *J. Am. Chem. Soc.* **81**, 6080–6082 (1959).

58. J. Davoll and B. A. Lowy, A new synthesis of purine nucleosides. The synthesis of adenosine, guanosine, and 2,6-diamino-9-β-D-ribofuranosylpurine, *J. Am. Chem. Soc.* **73**, 1650–1655 (1951).

59. M. L. Wolfrom, P. McWain, and A. Thompson, Nucleosides of disaccharides; cellobiose and maltose, *J. Am. Chem. Soc.* **82**, 4353–4354 (1960).

60. B. R. Baker, Stereochemistry of nucleoside synthesis, in: *The Chemistry and Biology of Purines* (G. E. W. Wolstenholme and C. M. O'Connor, eds.) Ciba Foundation Symposium, pp. 120–133, Little, Brown, & Co., Boston (1957).

61. L. M. Lerner, 9-β-Melibiosyladenine. Protection of the 1→6 glycosidic linkage of melibiose with benzoate ester blocking groups, *J. Org. Chem.* **32**, 3663–3665 (1967).

62. L. M. Lerner, 9-(2-Deoxycellobiosyl)adenine, *J. Med. Chem.* **11**, 912 (1968).
63. N. Yamaoka, K. Aso, and K. Matsuda, New synthesis of nucleosides. The synthesis of glyco-pyranosides of purines, pyrimidines, and benzimidazole, *J. Org. Chem.* **30**, 149–152 (1965).
64. D. R. Rao and L. M. Lerner, Disaccharide nucleosides of benzimidazole, *J. Org. Chem.* **37**, 3741–3743 (1972).
65. C. L. Stevens and P. Blumbergs, Pyrimidine disaccharide nucleosides. Synthesis of an amino sugar disaccharide nucleoside, *J. Org. Chem.* **30**, 2723–2728 (1965).
66. G. E. Hilbert and T. B. Johnson, Researches on pyrimidines. CXVII. A method for the synthesis of nucleosides, *J. Am. Chem. Soc.* **52**, 4489–4494 (1930).
67. J. J. Fox, N. Yung, I. Wempen, and I. I. Doerr, Pyrimidine nucleosides. III. On the synthesis of cytidine and related pyrimidine nucleosides, *J. Am. Chem. Soc.* **79**, 5060–5064 (1957).
68. U. Niedballa and H. Vorbrüggen, A general synthesis of *N*-glycosides. III. A simple synthesis of pyrimidine disaccharide nucleosides, *J. Org. Chem.* **39**, 3664–3667 (1974).
69. H. Ogura, F. Furuhatu, K. Iwaki, and H. Takahashi, Synthetic *O*-glycosyl nucleoside analogs, *Nucleic Acids Res. Symp. Ser.* No. 10, 23–26 (1981).
70. T. Halmos, J. Filippi, J. Bach, and K. Antonakis, Unsaturated ketonucleosides: Keto derivatives of benzoylated disaccharide nucleosides, *Carbohydr. Res.* **99**, 180–188 (1982).
71. B. Kraska and F. W. Lichtenthaler, (1→5)-Verknüpfte Glucopyranosyl- und Galactopyranosyl-ribosen, *Chem. Ber.* **114**, 1636–1648 (1981).
72. K. A. Watanabe, A. Matsuda, M. J. Halat, D. H. Hollenberg, J. S. Nisselbaum, and J. J. Fox, Nucleosides. 114. 5′-*O*-Glucuronides of 5-fluorouridine and 5-fluorocytidine. Masked precursors of anticancer nucleosides, *J. Med. Chem.* **24**, 893–897 (1981).
73. H. Ogura, H. Takahashi, and M. Kobayashi, Synthesis of disaccharide isothiocyanates and nucleoside related compounds, *Nippon Kagaku Kaishi* **1982**, 1673–1681, [*CA* **98**, 72644j (1983)].
74. F. W. Lichtenthaler, Y. Sanemitsu, and T. Nohara, Synthesis of 5′-*O*-glycosyl-*ribo*-nucleosides, *Angew. Chem. Int. Ed. Engl.* **17**, 772–774 (1978).
75. W. Eberhard, F. W. Lichtenthaler, and K. A. Kahn, Galactopyranosyl β-(1→3′)-ribonucleo-sides—Structural evidence and synthesis, *Nucleic Acids Res. Symp. Ser.* No. 9, 15–19 (1981).
76. F. W. Lichtenthaler, W. Eberhard, and S. Braun, Assignment of glycosylation sites in *O*-hexo-pyranosyl-ribonucleosides by ¹³C-NMR, *Tetrahedron Lett.* **22**, 4401–4404 (1981).
77. F. W. Lichtenthaler and K. Kitahara, Ribose-glucose exchange in purine nucleosides, *Angew. Chem. Int. Ed. Engl.* **14**, 815–816 (1975).
78. H. Bredereck, A. Wagner, H. Kuhn, and H. Ott, Synthesen von Di- und Trisacchariden des Gentiobiosetyps, *Chem. Ber.* **93**, 1201–1206 (1960).
79. N. D. Chkanikov, V. N. Tolkachev, and M. N. Preobrazhenskaya, Thymidine- and 2′-deoxycyti-dine 5′-*O*-glucopyranosides, *Biorg. Khim.* **4**, 1620–1626 (1978) [*CA* **90**, 204408g (1979)].
80. N. D. Chkanikov and M. N. Preobrazhenskaya, 5′-*O*-(β-D-Glucopyranosyluronate)-6-aza-uridine, *Biorg. Khim.* **6**, 67–69 (1980) [*CA* **92**, 198662h (1980)].
81. N. D. Chkanikov, V. N. Tokachev, M. Z. Kornveits, and M. N. Preobrazhenskaya, Synthesis of 1-(β-D-arabinofuranosyl)cytosine 5′-*O*-glucuronides. Transglycosylation in the experimental synthesis of 1-(β-D-ribofuranosyl)-1,2,4-triazolecarboxamide *O*-glucuronide, *Bioorg. Khim.* **8**, 1143–1148 (1982) [*CA* **97**, 198401j (1982)].
82. D. Arndt, W. Hulsmann, and B. Tschiersch, Darstellung und Eigenschaften von Glycosiden mit zytotoxischen Aglykonen und Disacchariden im Zuckeranteil, *Pharmazie* **33**, 248–250 (1978).
83. C. L. Stevens, G. H. Ransford, J. Nemec, J. M. Cahoon, and P. M. Pillai, Synthesis and reactions of azido halo sugars, *J. Org. Chem.* **39**, 298–302 (1974).
84. C. L. Stevens, J. Nemec, and G. H. Ransford, Total synthesis of the amino sugar nucleoside antibiotic, plicacetin, *J. Am. Chem. Soc.* **94**, 3280–3281 (1972).
85. S. Laland, W. G. Overend, and M. Stacey, Deoxy-sugars. Part X. Some methanesulphonyl and toluene-*p*-sulphonyl derivatives of α-ethyl-2:3-dideoxy-D-glucoside, *J. Chem. Soc.* **1950**, 738–743.
86. C. L. Stevens, P. Blumbergs, and D. H. Otterbach, Synthesis and chemistry of 4-amino-4,6-dideoxy sugars. 1. Galactose, *J. Org. Chem.* **31**, 2817–2821 (1966).
87. M. Prystaš and F. Šorm, Nucleic acid components and their analogues. CXXXIII. Synthesis of

the sugar fragment of exotoxin from *Bacillus thuringiensis*, *Collect. Czech. Chem. Commun.* **36**, 1448–1471 (1971).

88. M. Prystaš and F. Šorm, Nucleic acid components and their analogues. CXXXIV. Synthesis of the nucleosidic moiety of exotoxin from *Bacillus thuringiensis*, *Collect. Czech. Chem. Commun.* **36**, 1472–1481 (1971).

89. L. Kalvoda, M. Prystaš, and F. Šorm, Synthesis of exotoxin produced by *Bacillus thuringiensis*. I. Formation of the ethereal bond between ribose and glucose, *Collect. Czech. Chem. Commun.* **41**, 788–799 (1976).

90. L. Kalvoda, M. Prystaš, and F. Šorm, Synthesis of exotoxin produced by *Bacillus thuringiensis*. II. Formation of the α-glucosidic bond, nucleosidation, phosphorylation, *Collect. Czech. Chem. Commun.* **41**, 800–815 (1976).

91. M. Prystaš, L. Kalvoda, and F. Šorm, Alternative synthesis of exotoxin from *Bacillus thuringiensis*, *Collect. Czech. Chem. Commun.* **41**, 1426–1447 (1976).

92. K. Horská, L. Kalvoda, and K. Šebasta, Inhibition of DNA-dependent RNA polymerase by derivatives and analogues of thuringiensin, *Collect. Czech. Chem. Commun.* **41**, 3837–3841 (1976).

93. L. Kalvoda, K. Horská, and K. Šebasta, Synthesis of the uracil analog of thuringiensin and its inhibitory effect on DNA-dependent RNA polymerase of *Escherichia coli*, *Collect. Czech. Chem. Commun.* **46**, 667–672 (1981).

94. T. Suami, H. Sasai, and K. Matsuno, Synthesis of methyl hexaacetyl-tunicaminyl uracil, *Chem. Lett.* **1983**, 819–822.

95. T. Suami, H. Sasai, K. Matsuno, N. Suzuki, Y. Fukuda, and O. Sakanaka, Synthetic approach toward antibiotic tunicamycins. VI. Total synthesis of tunicamycins, *Tetrahedron Lett.* **25**, 4533–4536 (1984).

96. S. Hanessian, D. M. Dixit, and T. J. Liak, Studies directed toward the total synthesis of the ezomycins, the octosyl acids, and related antibiotics, *Pure Appl. Chem.* **53**, 129–148 (1981).

97. K. S. Kim and W. A. Szarek, Synthesis of 3′,7′-anhydrooctose nucleosides related to the ezomycins and the octosyl acids, *Can J. Chem.* **59**, 878–888 (1981).

98. T. Ogawa, M. Akatsu, and M. Matsui, Synthesis of a sugar occurring in an antibiotic: Ezoaminuroic acid, the first example of a naturally occurring 3-amino-3-deoxyhexuronic acid, *Carbohydr. Res.* **44**, C22–C24 (1975).

99. H. H. Baer, Synthetic kanosamine, *J. Am. Chem. Soc.* **83**, 1882–1885 (1961).

100. J. A. Secrist and K. D. Barnes, Synthesis of methyl peracetyl α-hikosaminide, the undecose portion of the nucleoside antibiotic hikizimycin, *J. Org. Chem.* **45**, 4526–4528 (1980).

101. J. L. Frahn, J. A. Edgar, A. J. Jones, P. A. Cockrum, N. Anderton, and C. C. J. Culvenor, Structure of the corynetoxins, metabolites of *Corynebacterium rathayi* responsible for toxicity of annual ryegrass (*Lolium rigidum*) pastures, *Aust. J. Chem.* **37**, 165–182 (1984).

102. T. Ito, S. Zushi, K. Kawamura, A. Takatsuki, and G. Tamura, Sterochemical configuration of 1–1′ glycosidic linkages in tunicamycin, *Agric. Biol. Chem.* **49**, 2257–2259 (1985).

103. H. Sasai, K. Matsuno, and T. Suami, Synthetic approach toward antibiotic tunicamycins. VII. Synthesis of tunicamine and tunicaminyluracil derivative, *J. Carbohydr. Chem.* **4**, 99–112 (1985).

104. T. Suami, H. Sasai, K. Matsuno, and N. Suzuki, Total synthesis of tunicamycin, *Carbohydr. Res.* **143**, 85–96 (1985).

105. S. Danishefsky and M. Barbachyn, A fully synthetic route to tunicaminyluracil, *J. Am. Chem. Soc.* **107**, 7761–7762 (1985).

106. K. Kominato, S. Ogawa, and T. Suami, Synthesis of a 2-acetamido-2-deoxy-α-D-galacto-pyranosyl analogue of tunicamycin, *Carbohydr. Res.* **174**, 360–368 (1988).

107. O. Sakanaka, T. Ohmori, S. Kozaki, T. Suami, T. Ishii, S. Ohba, and Y. Saito, Synthetic approach toward antibiotic ezomycins. I. Synthesis of 5-amino-5-deoxyoctofuranose-(1,4) derivatives by Henry reaction and their stereochemistry, *Bull. Chem. Soc. Jpn.* **59**, 1753–1759 (1986).

108. O. Sakanaka, T. Ohmori, S. Kazaki, and T. Suami, Synthetic approach toward antibiotic ezomycins. II. Synthesis of 5-amino-3,7-anhydro-5-deoxyoctofuranose-(1,4) derivatives, *Bull. Chem. Soc. Jpn.* **59**, 3523–3528 (1986).

109. S. Hanessian, J. Kloss, and T. Suzawara, Stereocontrolled access to the octosyl acids: Total synthesis of octosyl A, *J. Am. Chem. Soc.* **108**, 2759–2761 (1986).

110. S. J. Danishefsky, R. Hungate, and G. Schutte, Total synthesis of octosyl acid A. Intramolecular Williamson reaction via a cyclic stannylene derivative, *J. Am. Chem. Soc.* **110**, 7434 (1988).

111. S. Danishefsky and C. Maring, A fully synthetic route to hikosamine, *J. Am. Chem. Soc.* **107**, 7762–7764 (1985).

112. H. Yamaguchi, S. Saito, S. Yoshida, H. Seto, and N. Otake, Capuramycin, a new nucleoside antibiotic. Taxonomy, fermentation, isolation, and characterization, *J. Antibiot.* **39**, 1047–1053 (1986).

113. H. Seto, N. Otake, S. Sato, H. Yamaguchi, K. Takada, M. Itoh, H. S. M. Lu, and J. Clardy, The structure of a new nucleoside antibiotic, capuramycin, *Tetrahedron Lett.* **29**, 2343–2346 (1988).

114. H. Ogura, K. Furuhata, M. Itoh, and Y. Shitori, Synthesis of 2-*O*-glycosyl derivatives of *N*-acetyl-D-neuraminic acid, *Carbohydr. Res.* **158**, 37–51 (1986).

115. S. Sato, K. Furuhata, M. Itoh, Y. Shitori, and H. Ogura, Studies on sialic acids. XI. Synthesis of 2-*O*-glycosyl derivatives of *N*-acetylneuraminic acid, *Chem. Pharm. Bull.* **36**, 914–919 (1988).

Chapter 3

The Synthesis, Reactions, and Properties of Nucleoside Mono-, Di-, Tri-, and Tetraphosphates and Nucleotides with Changes in the Phosphoryl Residue

D. W. Hutchinson

1. Introduction

As accounts of early work on nucleotides can be found in a number of monographs,[1,2] early studies on the synthesis and the chemistry of nucleotides will not be discussed in detail in this chapter. Examples of nucleotides are shown in Fig. 1. The first nucleotide to be isolated was inosinic acid, which was obtained from meat hydrolysates by Liebig in 1847. The nucleotide coenzymes were discovered much later; for example, nicotinamide adenine dinucleotide (NAD$^+$) was not detected until 1905,[3] while adenosine triphosphate (ATP), perhaps the most important coenzyme of all, was not discovered until 1929.[4] A large number of nucleotide coenzymes are known.[5,6] These can function, for example, as cofactors for oxidation–reduction and group transfer reactions. In addition, nucleotide coenzymes can play a major role in the biosynthesis of oligo- and polysaccharides. Nucleoside triphosphates, in addition to their function as coenzymes, are also monomeric building blocks in the biosynthesis of nucleic acids. As the biological importance of nucleotides became recognized, the synthesis of phosphate esters of nucleosides began to be studied.[7,8] The field of nucleotide synthesis has expanded rapidly in recent years with the development of commercially important genetic engineering techniques. These techniques use synthetic oligo(deoxyribonucleotides), and the need for the rapid synthesis of oligo(deoxynucleotides) that contain bases joined in a precise sequence has

D. W. Hutchinson • Department of Chemistry, Warwick University, Coventry CV4 7AL, United Kingdom.

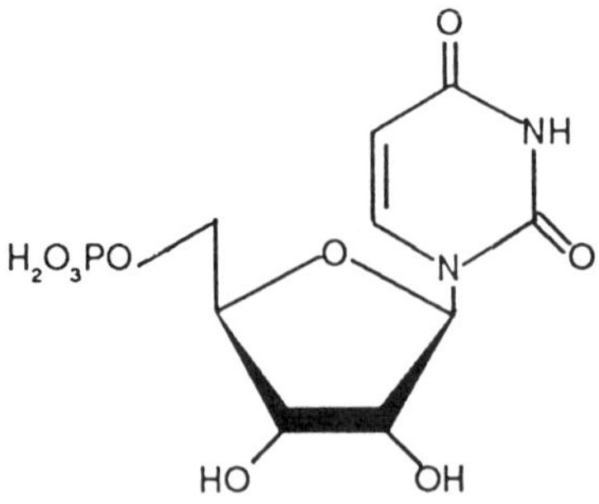

uridine 5'-phosphate

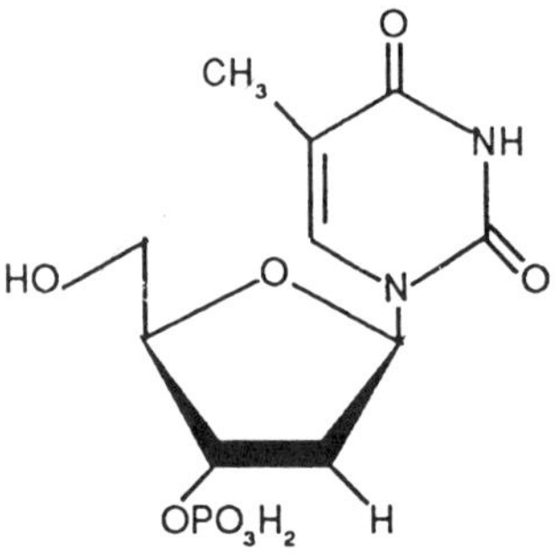

2'-deoxythymidine 3'-phosphate

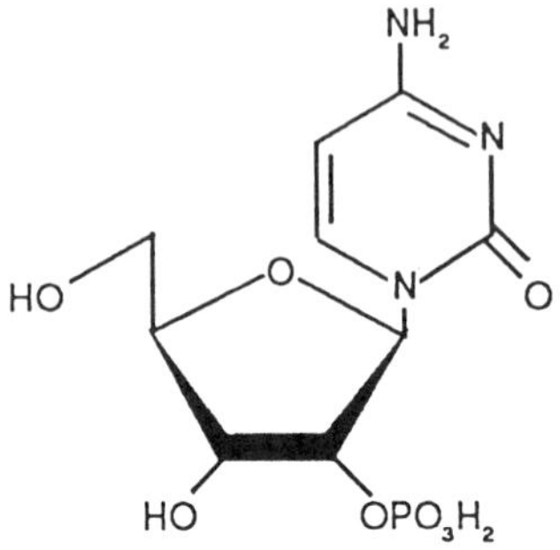

cytidine 2'-phosphate

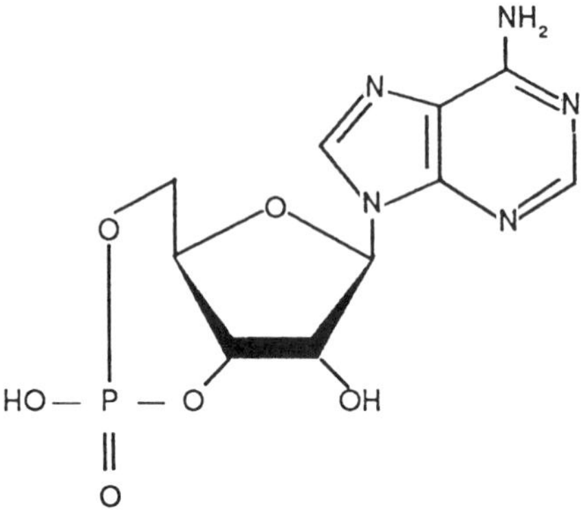

adenosine 3', 5'-cyclic phosphate

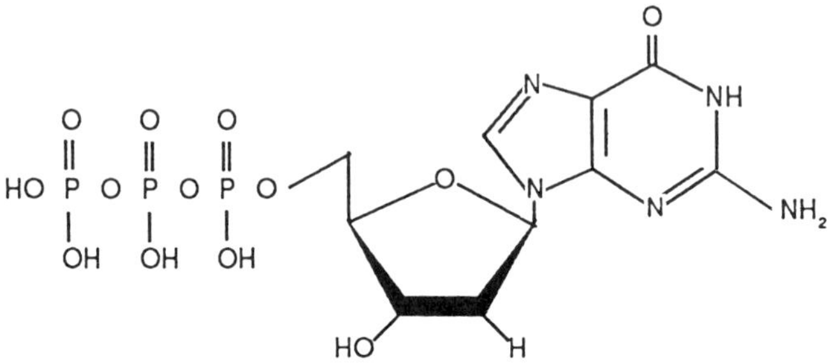

2'-deoxyguanosine 5'-triphosphate

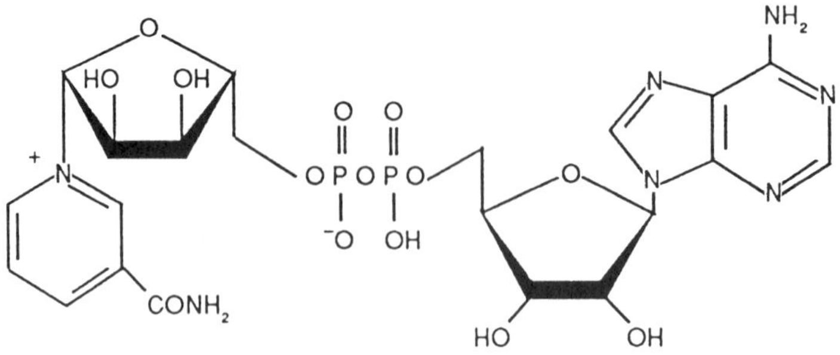

nicotinamide-adenine dinucleotide (NAD+)

Figure 1. Typical nucleoside phosphate esters.

been a great stimulus for the development of new methods for the phosphorylation of nucleosides.

The synthesis of analogues of naturally occurring nucleotides has attracted attention for many years. It has long been realized that chemically modified nucleotides might be used to probe important biochemical processes or have useful chemotherapeutic properties.

Other chapters of this treatise monograph deal with nucleoside analogues and also the synthesis and properties of oligo- and polynucleotides. In this chapter, attention will be focused on general phosphorylation methods applied to the synthesis of nucleoside mono- and polyphosphate esters as well as the synthesis of nucleotide analogues in which the phosphoryl moiety has been altered, such as thiophosphoric and phosphonic acid esters. Examples of nucleoside phosphate esters and their analogues are given in Fig. 2.

2. Synthesis

2.1. General Phosphorylation Methods

All naturally occurring nucleotides are esters of orthophosphoric or condensed polyphosphoric acids. Orthophosphoric acid contains four oxygen atoms, and the synthesis of a nucleotide involves either the nucleophilic attack by an oxygen atom of a phosphate anion on carbon with the concomitant displacement of a suitable leaving group such as halide ion or the transfer of a phosphorus atom and *three* oxygen atoms (or their equivalents) to the hydroxyl group of a nucleoside (Fig. 3).[9] Owing to difficulties in the preparation of suitable substrates, phosphorylation by the nucleophilic attack of a phosphate anion on saturated carbon [pathway (a)] is seldom encountered in nucleotide synthesis. Phosphorylation of a hydroxyl group with an active phosphorus-containing compound is by far the most common chemical route to nucleotides, and these reactions can be regarded as an attack by nucleophilic oxygen on phosphorus. Phosphorylation reactions involving phosphorus can be divided into two main types: (i) the synchronous displacement of a leaving group from a phosphorylating agent by a hydroxyl group of a nucleoside [pathway (b)], and (ii) the initial dissociation of the phosphorylating agent to give a reactive intermediate (metaphosphate) which then phosphorylates the nucleoside [pathway (c)]. This process should be accompanied by racemization at phosphorus. In practice, pathway (c) is seldom encountered except for a few specialized chemical reactions,[9] although metaphosphate has been proposed as an intermediate in several biochemical reactions[10] and appears to be involved in at least one non-enzymatic phosphoryl transfer from ADP.[11]

Phosphoric acid is tribasic, and, if unwanted side products are to be avoided during phosphorylation reactions, it is necessary to protect one or more of the phosphoryl hydroxyl groups. The recent interest mentioned above in the synthesis of oligo(deoxyribonucleotides) of defined sequence has resulted in the development of a large number of protecting groups which can be removed from reaction intermediates with the minimum effect on the rest of the molecule. Examples of these protecting groups will be discussed in the next section. The a phosphodiester   a phosphomonoester   a phosphoramidate

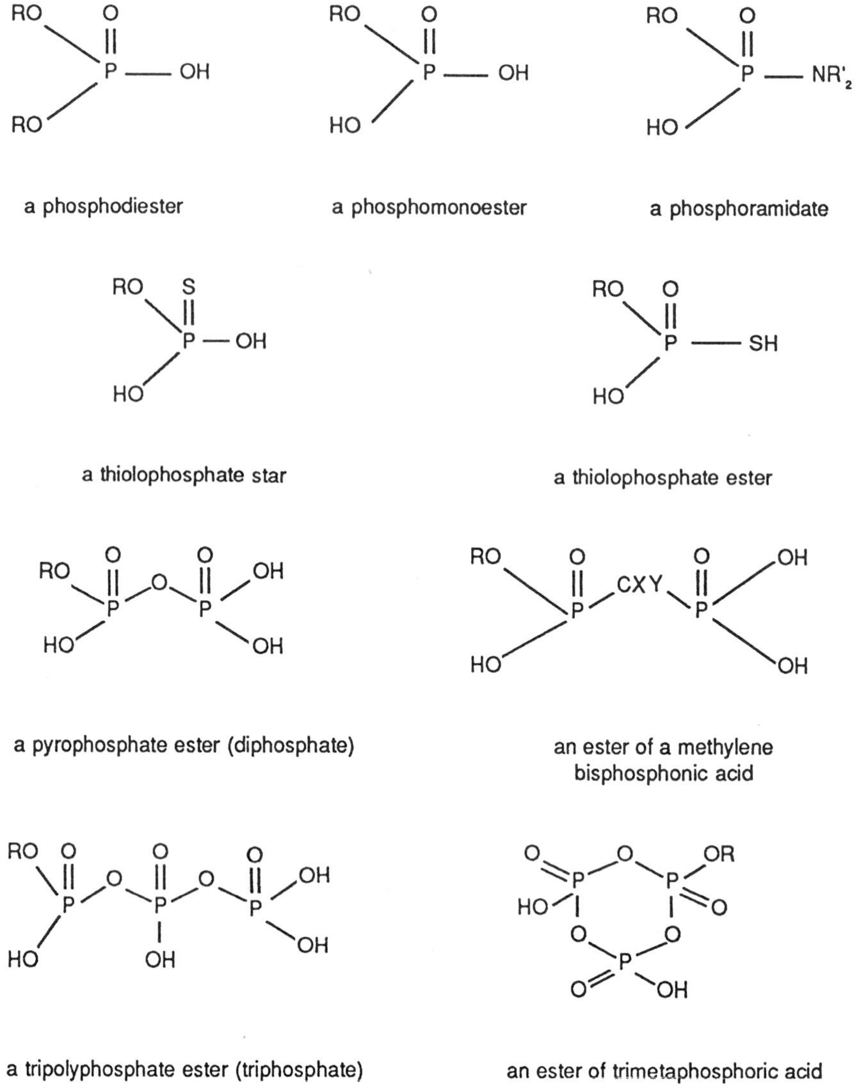

a thiolophosphate star   a thiolophosphate ester

a pyrophosphate ester (diphosphate)   an ester of a methylene bisphosphonic acid

a tripolyphosphate ester (triphosphate)   an ester of trimetaphosphoric acid

Figure 2. Some phosphate analogues.

vast majority of chemical approaches to the synthesis of nucleotides have employed chemical methods, with concomitant problems associated with the protection of reactive groups on the nucleoside and phosphorylating agent. In recent years, enzymatic methods have begun to be used, not only for the synthesis of nucleoside 5′-monophosphates, but also for the synthesis of nucleotide coenzymes. The use of enzymes as synthetic reagents has considerable appeal as problems of regio- and stereospecificity are readily overcome. Some examples of the enzymatic synthesis of nucleotides are given later in this chapter (Section 2.2.8).

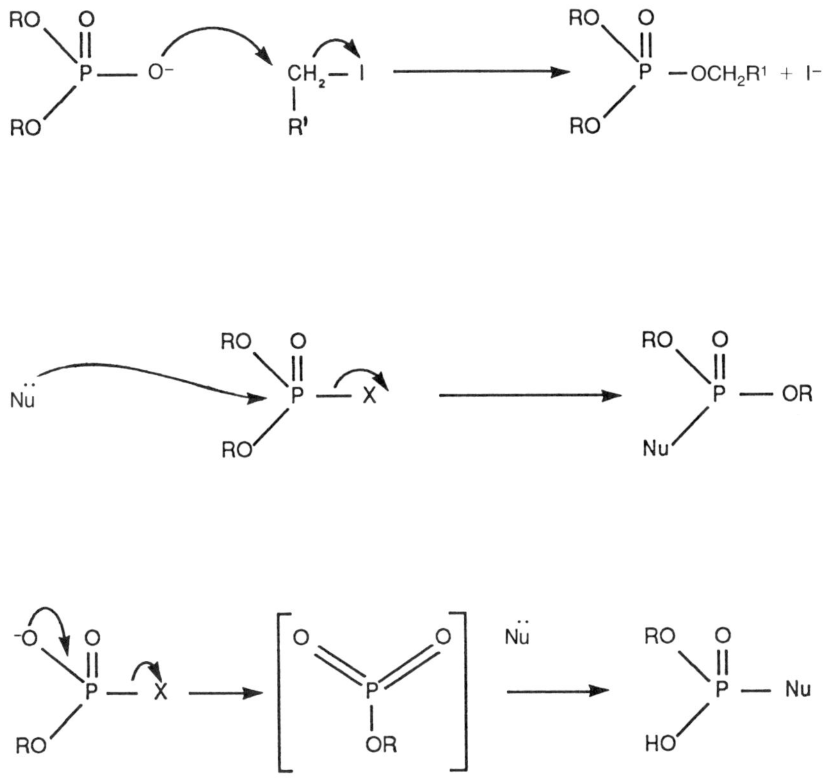

Figure 3. Phosphorylation pathways.

2.2. Synthesis of Nucleoside Monophosphate Esters

2.2.1. Phosphoryl Chloride

Before it can function as a phosphorylating agent by either pathway (b) or (c), phosphoric acid must be activated to form an anhydride. Examples of phosphorylating agents, together with experimental details, are given in recent reviews.[12,13] The most widely used anhydrides are phosphorochloridates, which are anhydrides formed by a reaction between phosphoric and hydrochloric acids (Fig. 4). Phosphoryl chloride (1) has been used extensively to phosphorylate protected nucleosides. It has been observed that selective phosphorylation at the 5'-position in unprotected nucleosides can be achieved if a solution of phosphoryl chloride in a trialkyl phosphate[14] is used. The exact function of the trialkyl phosphate is unclear. It is not only an excellent solvent but it can also react with phosphoryl chloride to form a reactive intermediate (2). The large size of this intermediate may be the cause of the selective attack at the relatively unhindered 5'-hydroxyl of the nucleoside. Examples of the range of nucleosides which have been phosphorylated by this reaction mixture are given in refs. 12 and 13. Purine nucleoside 5'-monophosphates which are labeled with

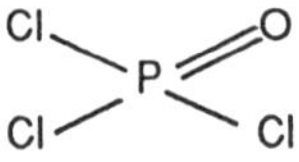

Figure 4. Phosphoryl chloride and related phosphorylating agents.

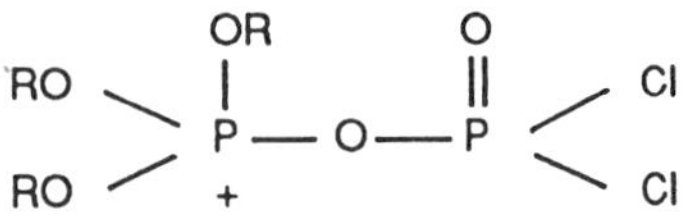

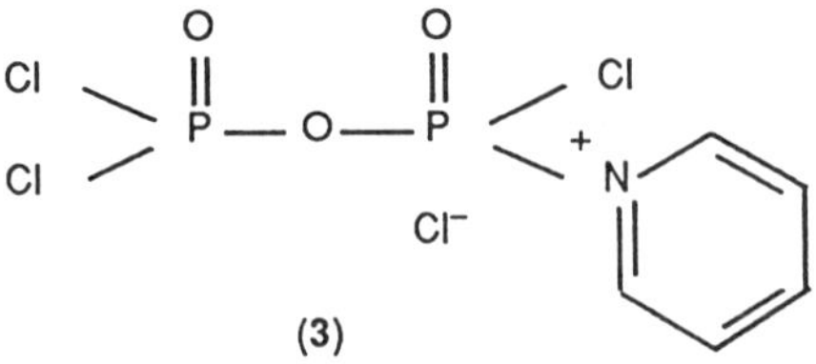

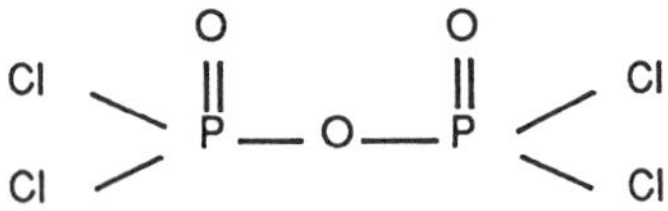

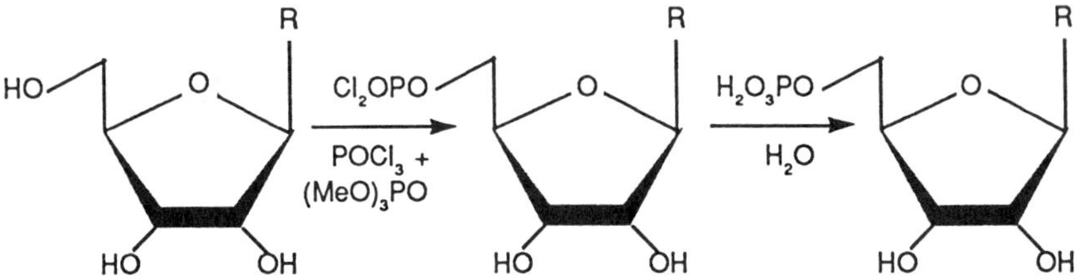

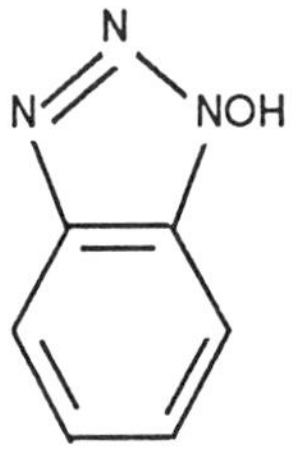

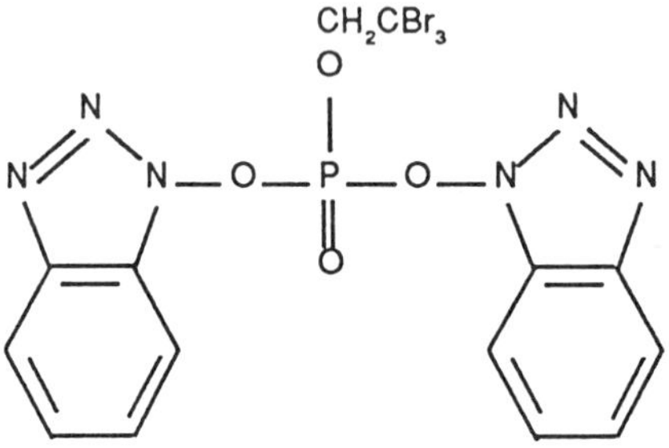

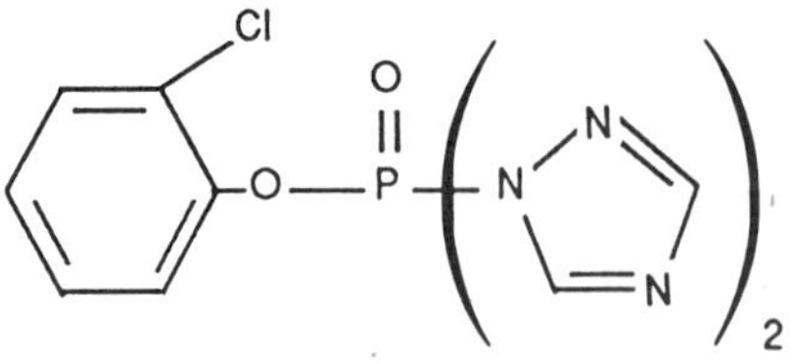

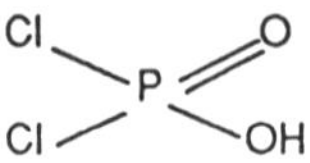

[17]O or [18]O on the phosphate group can be prepared conveniently by treating phosphorus pentachloride in dry triethyl phospate with one equivalent of labeled water to form [17]O- or ^{18}O-labeled phosphoryl chloride *in situ*, which then reacts with the nucleoside in the usual fashion.[15] Selective phosphorylation of the 5'-hydroxyl of nucleosides can also be achieved with phosphoryl chloride in a mixture of acetonitrile, pyridine, and water.[16] In this case, the adduct **3** formed from pyrophosphoryl chloride (**4**) and pyridine has been isolated from the reaction and is probably the phosphorylating species. It has been reported that this reaction mixture is superior to the more conventional phosphoryl chloride–trialkyl phosphate mixtures.[17] Pyrophosphoryl chloride (**4**) itself has been used to phosphorylate unprotected nucleosides.[18] However, this reagent must be used with caution as it has been reported that a large number of other products including polyphosphates can be formed in this reaction. If the hydrolytic workup of phosphorodichloridates is replaced by treatment with ammonia or an amine, then the phosphoromono- and phosphorodiamidates rather than the monophosphates are formed.[19] The phosphoramidates can be used without purification to synthesize polyphosphate esters.

One disadvantage of the phosphoryl chloride–trialkyl phosphate and related phosphorylating agents is that the presence of several reactive P—Cl bonds in the nucleoside phosphorodichloridate (**5**) formed as an intermediate can lead to the formation of phosphodi- and phosphotriesters. Furthermore, large amounts of chloride ion and phosphorylated by-products which must be removed from the reaction introduce the possibility of loss of yield of the target nucleotide. To overcome this problem, mono- and diesters of phosphorochloridic acids are usually employed to phosphorylate nucleosides.

A wide range of protecting groups for the phosphorochloridic acids have been developed, and some of these are listed in Table I. In general, for the synthesis of a nucleotide, a suitably protected nucleoside is treated with the protected phosphorochloridate in the presence of a tertiary base. The nucleoside triester is then isolated and the protecting groups removed. Little stereoselectivity is observed with unprotected nucleosides although it has been reported that selective phosphorylation of the 5'-position can be achieved with the bulky bis(2-*tert*-butylphenyl)[29] and bis(2,2,2-trichloroethyl)[30] phosphorochloridates. In recent years, with the increased interest in the synthesis of oligonucleotides both in solution and using solid-phase techniques, emphasis has been on the development of esterifying groups on phosphorus which can be removed in homogeneous phase, for example, by base-catalyzed hydrolysis. Thus, halogenated phenyl esters of phosphorochloridic acid are frequently used as these protecting groups. These groups are considerably more base labile than the unhalogenated phenyl groups.

As mentioned above, phosphorochloridates are usually employed in the presence of a tertiary base. It was believed initially that the function of the base was merely to remove hydrogen chloride liberated during the phosphorylation reaction. It is now apparent that the tertiary base can also catalyze the phosphorylation reaction; for example, 5-chloro-1-alkylimidazoles are better catalysts than pyridine or 2,6-lutidine for the phosphorylation of nucleosides with phenyl phosphorodichloridate.[36] Perhaps this is due to the formation of phosphoro-

Table I. Some Phosphoryl Protecting Groups Used
with Phosphorochloridates (RO)(R'O)P(O)Cl

Protecting group (R)	Method of removal	Reference
C_6H_5	OH^-/H_2O	20
$2\text{-}ClC_6H_4$	OH^-/H_2O or F^-	21
$4\text{-}ClC_6H_4$	OH^-/H_2O	22
	Aldoximes	23
$3,5\text{-}Cl_2C_6H_3$	*syn*-4-Nitrobenzaldoxime	24
$CNCH_2CH_2$	OH^-/H_2O	22
$4\text{-}NO_2C_6H_4$	$Et_3N/4\text{-}MeC_6H_4SH$	25
$4\text{-}NO_2C_6H_4CH_2CH_2$	DBU	26
$C_6H_5CH_2$	H_2/Pd	27
$4\text{-}NO_2C_6H_4CH_2$	H_2/Pd	28
$2\text{-}Me_3CC_6H_4$	H_2/PtO_2	29
CCl_3CH_2	Zn	30
	Electrochemical	31
CBr_3CH_2	Zn	32
$2\text{-}NO_2C_6H_4CH_2$	$h\gamma$ (>305 nm)	33
4-Chloro-8-quinolyl	$ZnCl_2$	34
ArNH	HNO_2	35
CH_3S	I_2/H_2O	24

imidazolidate intermediates. Replacement of a halogen atom in a phosphoro-chloridate by a heterocyclic residue such as a triazole, tetrazole, or hydroxy-benzotriazole (**6**) leads to improved phosphorylating agents. Thus, **6** can replace chloride from tribromoethyl phosphorodichloridate to give **7**. This reagent (**7**) can phosphorylate relatively hindered hydroxyl groups.[37] The oxybenzotri-azole groups can easily be removed with aqueous triethylamine. Similarly, 2-chlorophenyl phosphorobistriazole (**8**) has been used in place of the corre-sponding phosphorodichloridate in the synthesis of oligo(deoxyribonucleo-tides).[38] Unesterified phosphorochloridic acids are generally too unstable to be useful phosphorylating agents. However, there has been one report of phospho-rodichloridic acid (**9**) being employed to form internucleotide bonds.[39]

2.2.2. Trimetaphosphates

Stable anhydrides of orthophosphoric acid, for example, phosphorus pen-toxide or "polyphosphoric acid," have occasionally been used for nucleotide synthesis although problems can be encountered in the removal of phosphorus-containing by-products from the final target molecule. These anhydrides are useful for the synthesis of radioactively labeled nucleotides, and good experi-mental details for these syntheses are given by Symons.[40] The specific phospho-rylation of the 2'- and 3'-positions in ribonucleosides by sodium trimetaphos-phate (**10**) in aqueous solution has been reported, and similar reactions have been suggested as possible prebiotic routes to nucleotides.[41]

2.2.3. Carbodiimides

Dicyclohexylcarbodiimide [DCC, (**11**)][42] is a highly efficient reagent for promoting phosphorylation reactions and has been used for many years for the preparation of esters of phosphoric acid and polyphosphoric acids. The mechanism of action of DCC is shown in Fig. 5. DCC reacts initially with phosphomonoesters to give imidoyl phosphates (**12**). These compounds can be considered as mixed anhydrides that are formed between the phosphoric acid and the enolic form of dicyclohexylurea.

It was believed that **12** underwent a reaction with excess phosphomonoester to give the fully esterified trimetaphosphate (**13**)[43] which could then react with a variety of nucleophiles. However, recent investigations using ^{31}P-NMR spectroscopy suggest that any metaphosphates which might be present in the reaction are formed at a late stage in the reaction and that the major phosphorylating species is the imidoyl phosphate (**12**).[44] One feature of reactions involving DCC is that the hydrated product [dicyclohexylurea (**14**)] is highly insoluble and can readily be removed from the reactions. If free orthophosphoric acid is condensed with an alcohol in the presence of DCC, significant amounts of polyphosphate esters are formed. Protection of the phosphoric acid as a mono-or diester obviates this problem. 2-Cyanoethyl phosphate is a popular vehicle for the introduction of a phosphoryl group into a molecule with the aid of DCC, and the 2-cyanoethyl group can readily be removed with base at the end of the reaction sequence.[45] DCC has also been used extensively in oligonucleotide synthesis[46] and in the synthesis of nucleotide coenzymes.[6]

2.2.4. Arenesulfonyl Chlorides

One problem which can be encountered when DCC is used to prepare phosphodiesters is that N-acyl ureas are frequently formed as by-products. This problem can be avoided if the phosphomonoester is activated with an arenesulfonyl chloride (**15**), and these reagents have been used interchangeably with DCC to activate mononucleotide building blocks in oligonucleotide synthesis. The course of the reaction between 5'-*O*-trityl-2'-deoxythymidine 3'-(2-chlorophenyl)phosphate and 3'-*O*-acetyl-2'-deoxythymidine promoted by triisopropylbenzenesulfonyl chloride and catalyzed by pyridine or tetrazole has been studied by ^{31}P-NMR spectroscopy. The results are summarized in Fig. 6.[44,47] In this case, the active phosphorylating species is either the pyrophosphate tetraester (**16**) or the phosphorotetrazolidate (**17**) and not a triester of trimetaphosphate. Other heterocycles that are weak acids, for example, 3-nitro-1,2,4,-triazole, can replace tetrazole in this reaction. Arenesulfonyl chlorides have been used extensively to condense phosphomono- and diesters with hydroxyl groups in nucleosides. The groups which have been used to protect the phosphoryl hydroxyl groups in these esters are usually the same as those that have been mentioned above in connection with the protection of phosphorochloridates (Table I). It is worth mentioning in this context that hindered primary amines such as *tert*-butylamine will remove 2-cyanoethyl groups smoothly and rapidly from phosphoryl residues under conditions that leave 4-chlorophenyl groups

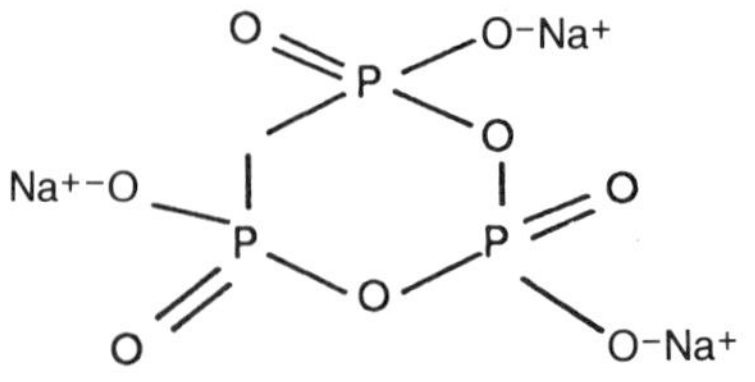

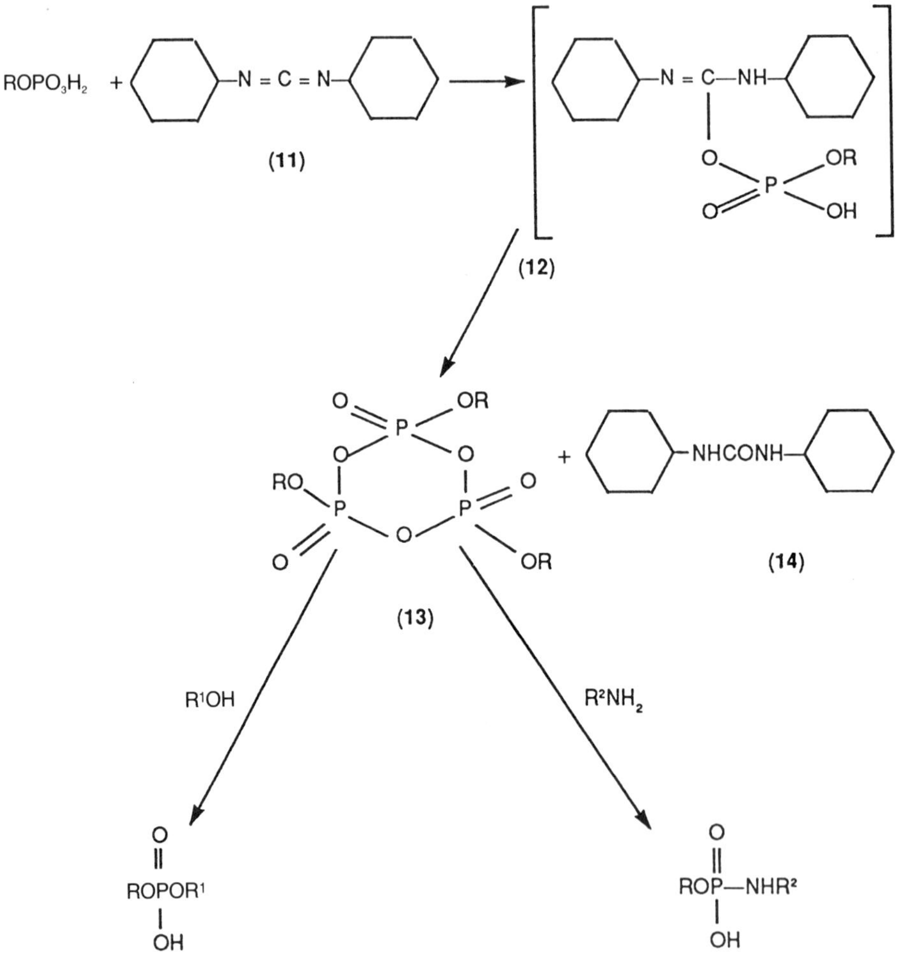

Figure 5. Mechanism of action of dicyclohexylcarbodiimide (DCC).

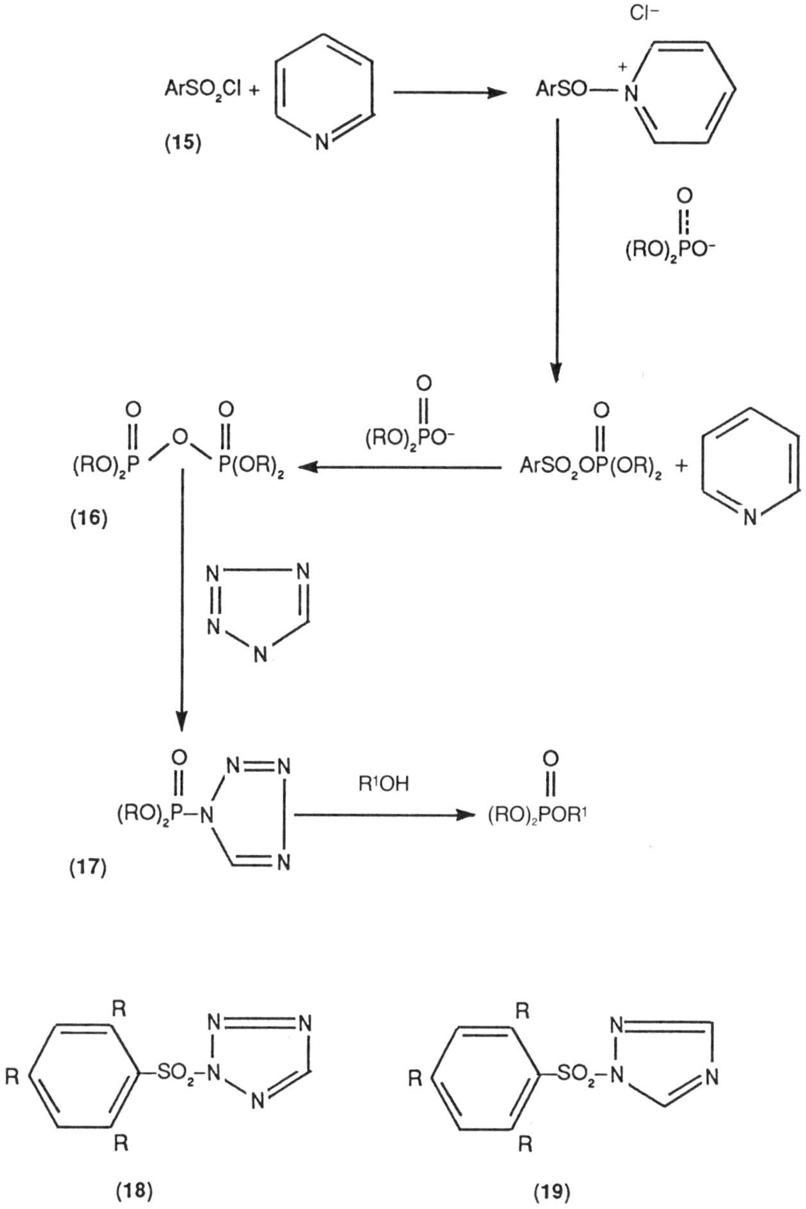

Figure 6. Activation of phosphates by arenesulfonyl chlorides.

untouched.[48] Also, sulfonation of hydroxyl groups by arenesulfonyl chlorides, which is always a problem with nucleosides,[49] can be reduced if bulky substituents, for example, isopropyl, are present in the 2- and 6-positions of the arene ring. The rate of esterification promoted by an arenesulfonyl chloride is increased if the chlorine atom in **15** is replaced by a tetrazole or 1,2,4-triazole residue to give **18** and **19**,[50] and little or no sulfonation of the 5'-hydroxyl of a nucleoside occurs during the internucleotide bond formation if the phosphoryl residue of the esterifying group is protected by an aryl rather than an alkyl group. However, base protection is essential if unwanted side reactions are to be avoided.[51]

2.2.5. *Miscellaneous Phosphorylation Reactions*

Pyrophosphate esters of ene-diols, for example, **20**, will selectively phosphorylate primary alcohols in the presence of secondary alcohols,[52] although the imidazolidate **21** is claimed to be a better phosphorylating agent (Fig. 7).[53] These reagents are very sensitive to traces of water and must be used under strictly anhydrous conditions. They are considerably superior to other mixed anhydrides, such as *O*-benzyl phosphorous *O*-diphenyl phosphoric anhydride (**22**), which have been used to phosphorylate nucleosides in the past.[54]

The use of phosphate anions to carry out displacement of halide at saturated carbon [pathway (a) above] has been used occasionally as a means to prepare nucleotides. Silver dibenzyl phosphate will react with 5'-iodo-5'-deoxy-2',3'-*O*-isopropylideneuridine to give UMP after removal of the protecting groups.[55] However, the formation of cyclonucleotides is a serious competing reaction in this synthesis, severely limiting the yield of nucleotide. The reaction of dibenzyl phosphate with a 5'-diazonucleoside, which gives the dibenzyl ester of the corresponding nucleotide as the sole product, is a variant of this reaction.[56] Dibenzyl phosphate can be activated if it is treated with a mixture of triphenylphosphine and diethyl azocarboxylate (**23**). This combination has been used to phosphorylate selectively the 5'-hydroxyl group of nucleosides.[57] It is proposed that the zwitterion **24** is formed initially and that this reacts with the 5'-hydroxyl group of the nucleoside to give a phosphoroxyphosphoronium ion (**25**). The latter then undergoes an S_N2 attack by dibenxyl phosphate to give the nucleotide dibenzyl ester. These reagents will phosphorylate octan-2-ol with inversion of configuration at C-2, supporting the hypothesis of an S_N2 attack at carbon during the final phosphorylation step.[58] A similar method of activating phosphomonoesters involves the use of triphenylphosphine and 2,2'-dipyridyl disulfide. In this case, *S*-(2-pyridyl)phosphorothioates (**26**) are formed as reaction intermediates.[59]

2.2.6. *Phosphites and Other Tervalent Phosphorus Derivatives*

Letsinger and Lunsford introduced in 1976 a procedure for the synthesis of oligonucleotides by means of phosphite intermediates.[60] The reason for using tervalent rather than pentavalent phosphorus derivatives was the observation that phosphorochloridites (**27**) react very rapidly and cleanly with the hydroxyl

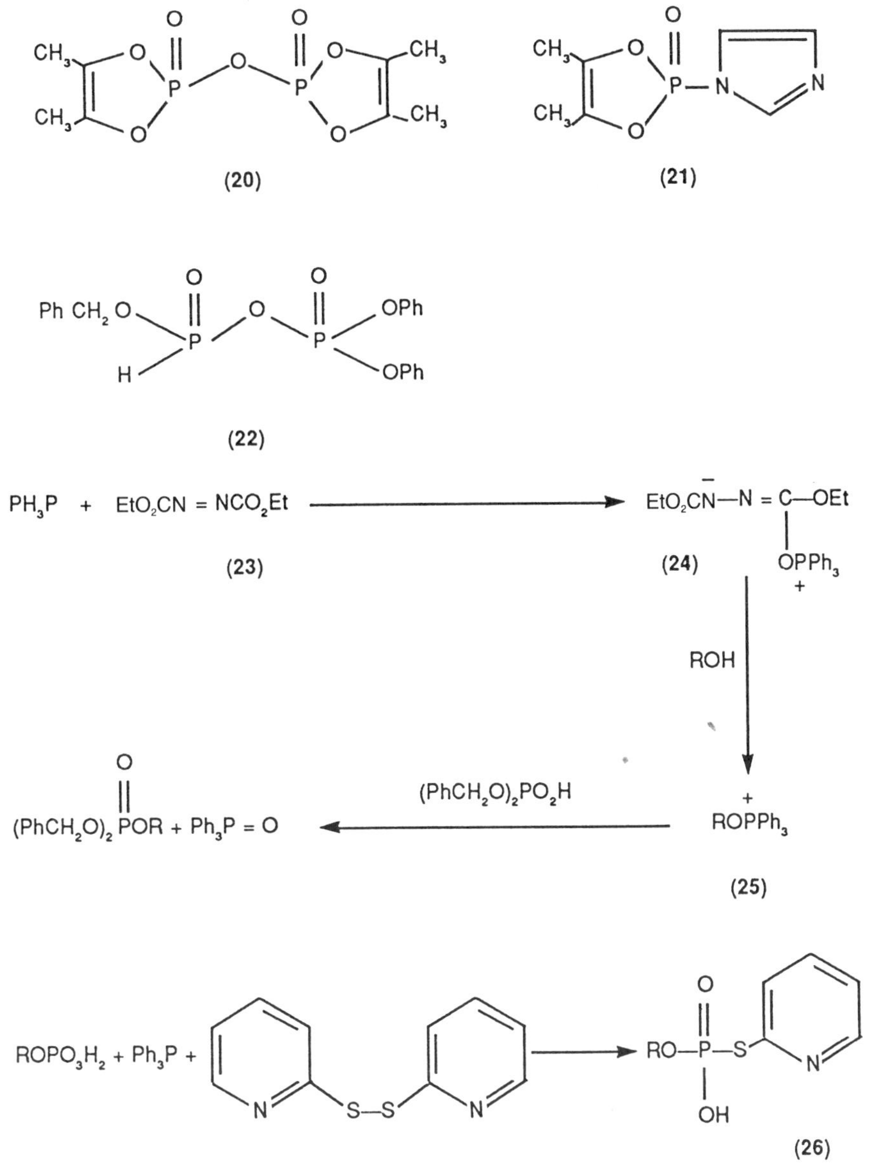

Figure 7. Other phosphorylating agents for nucleosides.

groups of nucleosides in tetrahydrofuran at low temperature. Hence, the synthesis of an oligonucleotide by this method should be less time-consuming than one carried out by standard phosphotriester methods. Dinucleoside monophosphites can be oxidized quantitatively to the corresponding phosphates with iodine in aqueous solution. This approach has been used extensively, and many examples of this approach are given in recent volumes of ref. 12. Drawbacks in the preparation of phosphoromono- and phosphorodichloridites and related analogues such as **28** and **29** have been overcome,[61,62] and this technique has been applied in particular to the solid-phase synthesis of oligonucleotides (Fig. 8).[63]

2.2.7. Solid-Phase Methods

The linear synthesis of oligonucleotides involves the repetitive addition of mononucleotide building blocks to an oligonucleotide. Therefore, solid-phase methods, especially if these can be automated, should be ideally suited to these syntheses. Procedures for the solid-phase synthesis of oligonucleotides have been available for many years, but rapid progress has only recently been made in this field. An illustration of the power of solid-phase synthetic methods is provided by a recent paper describing the preparation of 96 different oligo(deoxyribonucleotide) heptadecamers and their use in the isolation of cDNA clones.[64] The synthesis of only one heptadecamer by solution methods would have attracted considerable attention in the recent past. The major advance in the synthesis of oligonucleotides by this method was made when the phosphodiester approach was abandoned in favor of the phosphotriester approach. Major disadvantages in the former method include incomplete reaction in the phosphorylation and/or condensation stages and the incompatibility of the polymer supports with the solvents necessary for the reactions. The development of the phosphite triester approach with its short reaction times gave an added impetus to the development of solid-phase synthetic methods.

Several descriptions of automated solid-phase oligonucleotide synthesis have appeared,[65-68] and machines to undertake these tasks are commercially available. In general, an activated silica gel which contains succinyl groups is used as the support phase (**30**) (Fig. 8). A base-protected 5′-*O*-dimethoxytrityl-2′-deoxynucleoside is then coupled to **30** using DCC to give the anchored monomer **31**. Detritylation of **31** is followed by coupling with a base-protected 5′-*O*-dimethoxytrityl-2′-deoxynucleoside 3′-(methyl phosphorochloridite) to give **32**. Oxidation of **32** to the phosphotriester gives a product which can then take part in a further cycle of reactions. Any unreacted 5′-hydroxyl groups which were not phosphitylated in this procedure can be blocked by acetylation before starting the next synthetic cycle. The time taken for a single cycle can be as little as 13 minutes using this method.[67] The completed oligonucleotide is demethylated with thiophenol and triethylamine, released from the silica support by the action of ammonia, and purified by HPLC before removal of the final 5′-dimethoxytrityl group. Simple variants of this technique in which the silica is contained in the barrel of a syringe[67] or in which all the steps are performed in conical centrifuge tubes have been reported.[69] Porous glass[70] and synthetic polymers[71] are among other supports that have been used in the solid-phase synthesis of oligonucleotides.

2.2.8. The Enzymatic Synthesis of Nucleotides

The use of enzymes for the selective phosphorylation of nucleosides should have several advantages over corresponding phosphorylations that are carried

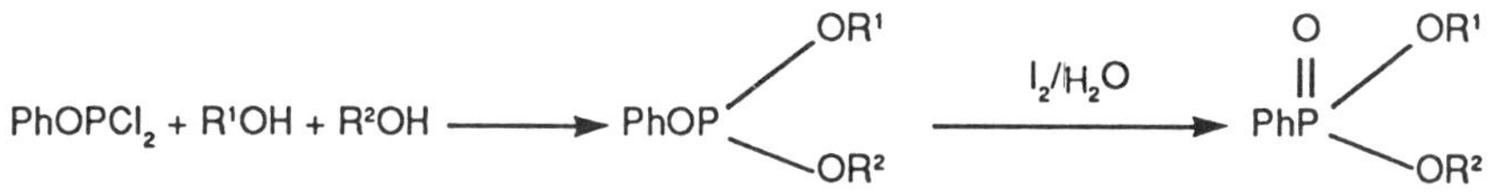

$$PhOPCl_2 + R^1OH + R^2OH \longrightarrow PhOP \Big\langle \begin{matrix} OR^1 \\ OR^2 \end{matrix} \quad \xrightarrow{I_2/H_2O} \quad PhP \Big\langle \begin{matrix} O \\ \| \\ OR^2 \end{matrix} \begin{matrix} OR^1 \end{matrix}$$

(27)

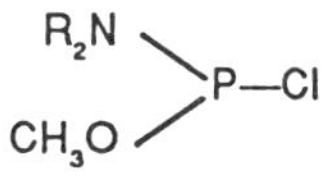

$$\begin{matrix} R_2N \\ CH_3O \end{matrix} \Big\rangle P{-}Cl$$

(28)

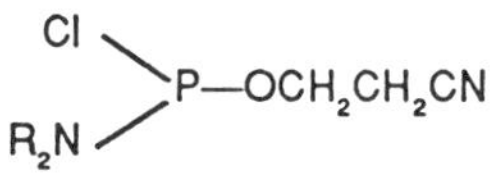

$$\begin{matrix} Cl \\ R_2N \end{matrix} \Big\rangle P{-}OCH_2CH_2CN$$

(29)

$$\text{(Si)} - (CH_2)_3NHCO\,(CH_2)_2COOH$$

(30)

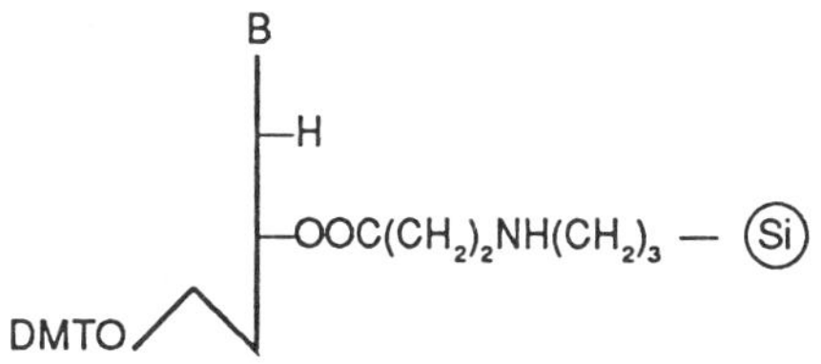

$$-OOC(CH_2)_2NH(CH_2)_3 - \text{(Si)}$$

DMTO

B —H

(31)

where DMT = dimethoxytrityl

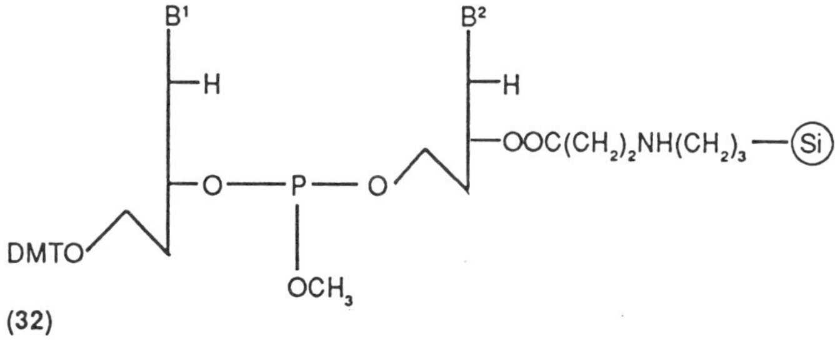

(32)

DMTO ... B¹ —H ... —O—P—O— ... OCH₃ ... B² —H ... —OOC(CH₂)₂NH(CH₂)₃—(Si)

Figure 8. Phosphites and solid-phase oligonucleotide synthesis.

out by chemical means since the need for protecting groups on both nucleoside and phosphorylating agent is avoided. The phosphoryl donor is often a simple phosphomonoester such as a nucleotide or an aryl phosphate, and enzymatic phosphorylation is particularly suited to the syntheses of radiolabeled nucleotides since few steps are involved in the phosphorylation process.[72]

A phosphotransferase which uses 4-nitrophenyl phosphate as phosphoryl donor has been isolated from carrot,[73] and similar enzymes have been obtained from wheat shoots[74] and barley.[75] The carrot enzyme has been used to phosphorylate the 5'-hydroxyl of both natural nucleosides and nucleoside analogues,[76,77] while the wheat shoot enzyme in an immobilized form has been used to phosphorylate the 5'-hydroxyl groups of oligonucleotides.[78] Phosphotransferases from *Escherichia coli*[79] and lupins[80] phosphorylate the 2'- and 3'-hydroxyls of nucleosides and can utilize nucleotides and aryl phosphates as phosphoryl donors.

The enzymatic synthesis of ribonucleoside 5'-triphosphates from RNA[81] and dATP from DNA[82] has been reported. In essence, the nucleic acids are broken down into oligonucleotides either chemically or enzymatically. The oligonucleotides are hydrolyzed to 5'-monophosphates by immobilized nuclease P_1, and the 5'-monophosphates are converted enzymatically to the required 5'-triphosphates.

2.2.9. Synthesis of Cyclic Nucleotides

Ribonucleoside 2',3'-cyclic phosphates (**33**) are intermediates in the chemical and enzymatic degradation of RNA. These compounds can readily be prepared by treating ribonucleoside 2'- (3'-) monophosphates with a number of reagents, including ethyl chloroformate[83] and DCC,[84] which are capable of forming anhydrides with the phosphoryl group. The proximity of the adjacent hydroxyl group in these molecules allows a rapid intramolecular cyclization to take place rather than the much slower intermolecular condensation. The action of aqueous barium hydroxide on nucleoside 5'-triphosphates leads to the 3',5'-cyclic phosphates (**34**) *inter alia*,[85] but yields are low and this is not a suitable method for the preparation of these compounds. However, both ribo- and deoxyribonucleoside 3,5'-cyclic phosphates can be prepared smoothly and efficiently by treating a solution of the nucleoside 5'-monophosphate in pyridine with DCC (Fig. 9).[86]

Although the DCC method is the most widely used route to obtain cyclic phosphates, many other approaches have been reported for the synthesis of 2',3'-[20,87] and 3',5'-[88] cyclic phosphates. Nucleoside 3',5'-cyclic phosphates can be obtained by treating 8-quinolyl esters of nucleosides with cupric chloride in pyridine followed by a removal of the 8-hydroxyquinolyl group.[89] These esters may be prepared from the unprotected nucleosides and 8-quinolyl phosphate (**35**) in the presence of triphenylphosphine and 2,2'-dipyridyl diselenide. Nucleoside 3',5'-cyclic phosphates can also be obtained in high yield by treating unprotected nucleosides with trichloromethanephosphonyl dichloridate (**36**) in triethyl phosphate followed by ring closure of the 5'-trichloromethyl phosphonate with potassium *tert*-butoxide.[90]

Tervalent phosphorus derivatives have also been used to prepare cyclic nucleotides. For example, treatment of 5'-*O*-monomethoxytrityluridine with tris(1-imidazolyl)phosphine gives the nucleoside 2',3'-cyclic phosphorimidazolidite (**37**) in high yield.[91] Oxidation of **37** and removal of the imidazole moiety affords the 2',3'-cyclic phosphate. Unprotected uridine under the same reaction conditions gives polymeric products. Methyl 2'-deoxythymidine 3',5'-cyclic phosphite can be oxidized with $^{18}O_2$, in the presence of azobisisobutyronitrile

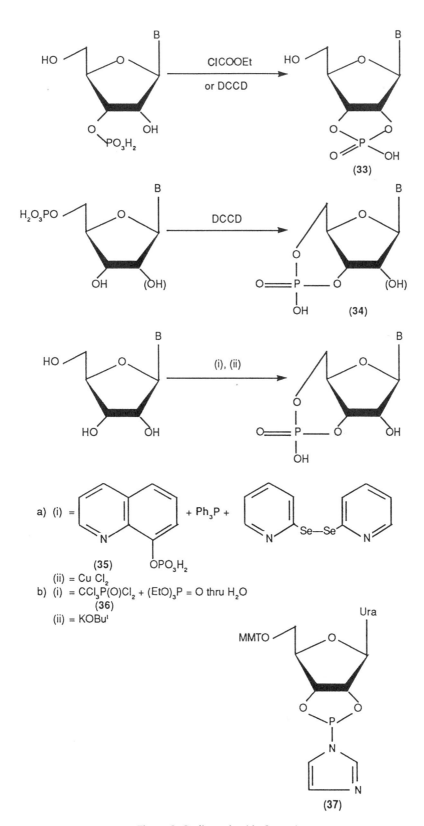

Figure 9. Cyclic nucleotide formation.

(AIBN), to afford the methyl ester of the corresponding 3′,5′-cyclic phosphate which is now labeled with ^{18}O. The diastereomers of this labeled compound can be separated chromatographically and the methyl group removed with *tert*-butylamine to yield the individual diastereomers of 2′-deoxythymidine 3′,5′-cyclic [^{18}O]phosphate.[92]

2.3. Synthesis of Mononucleotide Analogues

2.3.1. Oxygen Isotopes

While nucleotides which contain oxygen isotopes in their phosphoryl residues are not analogues of the naturally occurring nucleotides in the strictest sense, it is convenient to consider their synthesis here. Isotopically labeled nucleotides of this kind have been widely used to study the stereochemistry of enzymatically catalyzed phosphoryl transfer reactions,[93] and excellent stereospecific syntheses have been developed. One of the most successful methods[94] begins with (S)-mandelic acid, which is first converted into (1R,2S)-1,2-[1-^{18}O]-dihydroxydiphenylethane (**38**). Phosphorylation of **38** with [^{17}O]phosphoryl chloride followed by treatment of the reaction product with an alcohol gives the phosphotriester **39**. Hydrogenolysis of the latter gives rise to the (S)-[$^{16}O,^{17}O,^{18}O$]-monophosphate (Fig. 10). In this manner, a number of isotopically labeled nucleoside monophosphates have been prepared, for example, from 2′,3′-*O*-diacetyladenosine.[95] Other chiral $^{16}O,^{17}O,^{18}O$-labeled phosphates have been prepared using the cyclic phosphorochloridate **40**, which can be obtained from (−)-ephedrine and [^{17}O]phosphoryl chloride.[96] Treatment of **40** with an alcohol followed by acidic hydrolysis of the product in [^{18}O]water and catalytic hydrogenolysis leads to the chiral phosphate. Inversion of the order in which the reactions with water and the alcohol are carried out leads to the formation of the chiral phosphate with the opposite configuration.

The 3′,5′-cyclic oxygen-chiral phosphate of adenosine has been prepared by the following route.[97] Adenosine 5′-[$^{16}O,^{17}O,^{18}O$]phosphate was treated with diphenyl phosphorochloridate. The resulting triester of pyrophosphoric acid was cyclized with base to give an equimolar mixture of the three types of oxygen-chiral cAMP. The original assignment of the stereochemistry of [$^{16}O,^{18}O$]cAMP had to be revised due to an incorrect assignment of the stereochemistry of **38**. It is now agreed that the cyclization proceeds with an inversion of configuration at phosphorus. Other syntheses of oxygen-chiral cyclic dAMP or AMP are based on the reaction of the cyclic phosphoranilidate with ^{18}O-labeled carbonyl compounds such as carbon dioxide[98] or benzaldehyde.[99] As will be discussed later in this chapter (Section 3.2), the configuration of $^{16}O,^{18}O$-labeled phosphoryl residues can be determined by high-field ^{31}P-NMR spectroscopy.[100]

2.3.2. Nucleoside Phosphorothioates

Nucleoside phosphorothioates are analogues of nucleotides in which one or more of the phosphoryl oxygen atoms have been replaced by sulfur. While this substitution can affect the acid strength and the metal-chelating ability of the

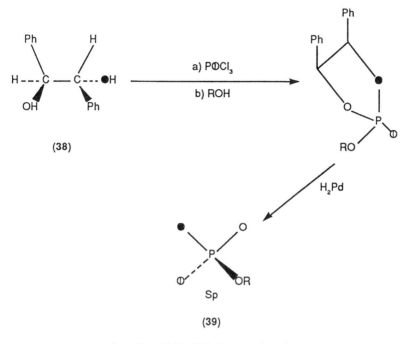

where R = 2′,3′—O,O-diacetyl adenosine-5′

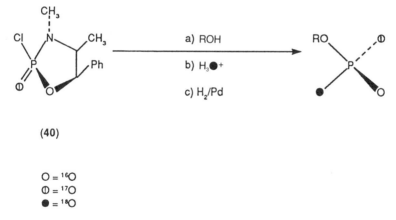

O = ^{16}O
Ⓞ = ^{17}O
● = ^{18}O

Figure 10. Synthesis of nucleotides chirally labeled with oxygen isotopes.

nucleotide analogue, nucleoside phosphorothioates are also used as substrates by a wide range of enzymes involved in the transfer of a nucleotide or phosphoryl group to acceptors other than water. Nucleoside phosphorothioates have been widely used to study enzymatic reactions of this type.[101] One of the most important properties of nucleoside phosphorothioates is that the phosphorus atom is chiral if two nonequivalent groups are attached to it in addition to the phosphoryl oxygen and sulfur atoms. This has allowed studies on the stereochemistry of many enzyme-catalyzed phosphoryl transfer reactions. Furthermore, the presence of a "soft" sulfur atom in the phosphorothioate allows molecules derived from it to be removed selectively from reactions using heavy metal ions attached to a chromatographic support. For example, specific kinds of DNA can readily be removed from biochemical systems by affinity chromatography using mercury(II) ions attached to Sepharose.[102]

Early work on phosphorothioates has been reviewed by Scheit,[103] and only general synthetic methods, together with recent developments, will be discussed here. The earliest method employed for the synthesis of nucleoside phosphorothioates involved the reaction between a protected nucleoside and tris(1-imidazolyl)phosphane sulfur (**41**).[104] This reagent has now been replaced by thiophosphoryl chloride ($PSCl_3$),[105] and ^{35}S-labeled phosphorothioates can be prepared with this reagent. The conversion of a nucleoside phosphoranilidate into a phosphorothioate by treatment with sodium hydride and carbon disulfide has been widely used for the synthesis of phosphorothioates of known stereochemistry at phosphorus.[106] The direct sulfurization of nucleoside 5'-phosphites is an alternative route to diastereomers of nucleoside 5'-phosphorothioates.[107] Reagents related to those employed in oligonucleotide synthesis by the conventional phosphotriester approach, for example, **42** (Fig. 11), have been used to prepare dinucleoside phosphorothioates,[108] and *O,S*-dimethyl thiophosphorochloridate (**43**) has also been used to prepare nucleoside phosphorothioates.[109] Demethylation at oxygen can readily be achieved with pyridine, and the methylthio residue can be removed in its entirety by the action of aqueous iodine. 5'-Deoxy-5'-thio analogues of AMP and IMP have been prepared by reacting trisodium phosphorothioate with the corresponding 5'-iodo-5'-deoxynucleosides. Both analogues are hydrolyzed by weak acid or alkaline phosphatase to afford the 5'-deoxy-5'-thio nucleosides.[110]

Purine nucleoside 2'(3')-phosphorothioates have been obtained from their 2',3'-*O*-di-*n*-butylstannylene derivatives (**44**) by the action of thiophosphoryl chloride followed by alkaline hydrolysis (Fig. 11).[111] Cyclization of the 2'(3')-phosphorothioates with ethyl chloroformate afforded the 2',3'-*O,O*-cyclophosphorothioate. The diastereomers of uridine 2',3'-*O,O*-cyclophosphorothioate can be separated by crystallization of their triethylammonium salts. The structure of the *endo* isomer (**45**) has been determined by X-ray crystallography.[112] The *endo* isomer is a substrate for pancreatic ribonuclease A and has been used to study the steric course of the hydrolysis of ribonucleoside 2',3'-cyclic phosphates by that enzyme. Although the diastereomers of guanosine 2',3'-*O,O*-cyclophosphorothioate cannot readily be separated, the assignment of stereochemistry at phosphorus has been made by ^{31}P-NMR spectroscopy. The mixture has also been used to study the stereochemistry of reactions catalyzed by ribonuclease T_1.[113] Both the (R_P)- and (S_P)-isomers of the dinucleoside phosphorothioate Up(S)A were hydrolyzed by ribonuclease T_2, the former affording the *endo* isomer and the latter the *exo* isomer of uridine 2',3'-*O,O*-cyclophosphorothioate.[114] Thus, RNase T_2 follows an in-line mechanism for the hydrolysis, as does RNase A. The diastereomers of Up(S)A were prepared from the protected uridine using 2-cyanoethyl phosphorothioate and 2,4,6-triisopropylbenzenesulfonyl chloride.

Nucleoside 3',5'-*O,O*-cyclophosphorothioates have been prepared from the

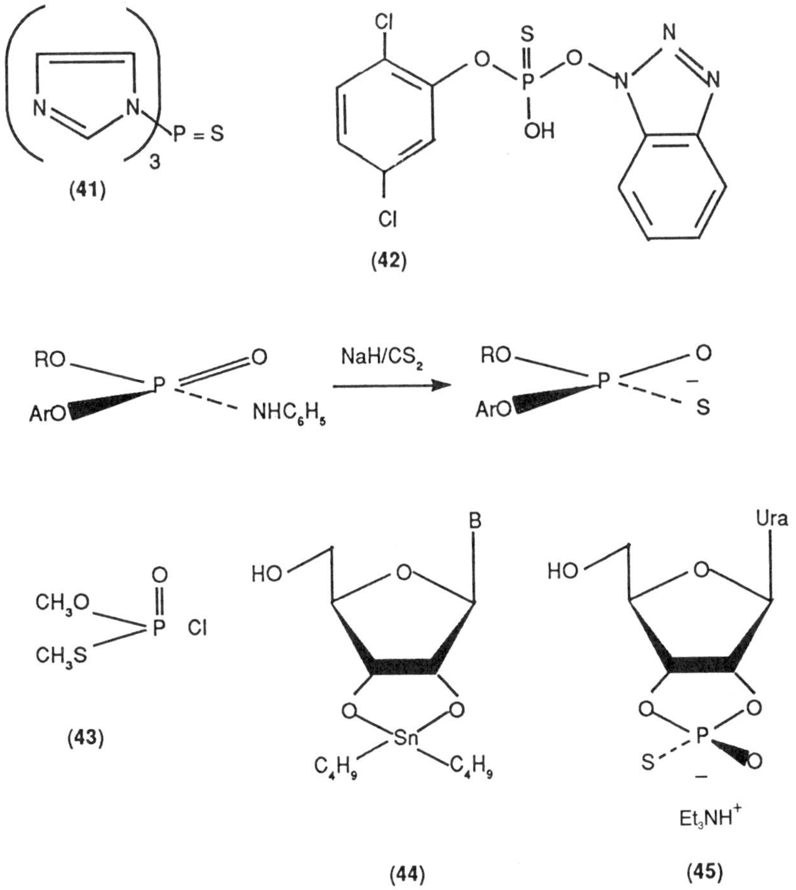

Figure 11. Synthesis of nucleoside phosphorathioates.

corresponding phosphoroanilidates by the action of base and carbon disulfide.[114] An alternative route to adenosine $3',5'$-O,O-cyclophosphorothioate (cAMPS) is provided by the cyclization of adenosine bis(4-nitrophenyl) phosphorothioate in the presence of base.[115] These methods give higher yields than a synthesis which involves the activation of guanosine $5'$-phosphorothioate with diphenyl phosphorochloridate with subsequent cyclization. The assignment of the stereochemistry at phosphorus in cAMPS and related nucleoside phosphorothioates has again been made by ^{31}P-NMR spectroscopy. This technique has also been used to study the bond order and charge localization in nucleoside phosphorothioates.[115]

2.3.3. Nucleoside Phosphoramidates

Nucleoside phosphoramidates, in which one or more phosphoryl oxygen atoms have been replaced by nitrogen, have not been used as widely as nucleoside phosphorothioates to study biological processes since the P—N bond is considerably more labile than the P—S bond even under mildly acidic conditions. The early work on nucleoside phosphoramidates has been reviewed by Scheit.[103] Synthetic routes to these compounds include the phosphorylation of aminonucleosides, for example, **46** with X = NH_2, and the treatment of nucleoside azides, for example, **46** with X = N_3, with triesters of phosphorous acid (Fig. 12).

A number of nucleoside *O,N*-cyclophosphoramidates have been prepared by conventional methods, including compounds **47**[116] and **48**.[117] The cyclophosphoramidates such as **49**[118] and **50**[119] have been prepared as analogues of the anticancer drug cyclophosphamide and show some anticancer activity *in*

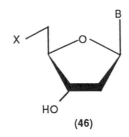

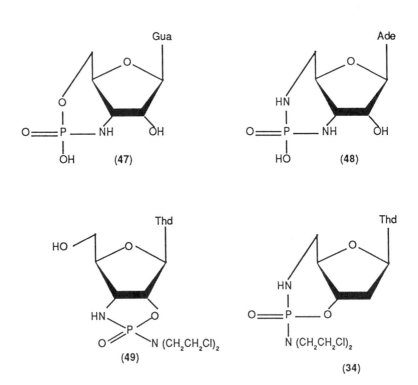

Figure 12. Nucleoside cyclic phosphoramidates.

vitro. The lability of phosphoramidates in aqueous solution has prompted the study of lipophilic nucleoside phosphoramidates as compounds which might be taken up readily into cells and then be hydrolyzed into biologically active compounds.[120]

The use of phosphoranilidates as protecting groups for phosphorochloridates has already been mentioned,[35] as has their conversion to ^{18}O-labeled nucleotides[98] and nucleoside phosphorothioates.[106] Phosphoramidates are excellent intermediates for the synthesis of polyphosphates, and their use in the synthesis of nucleotide coenzymes and related compounds will be mentioned later in this chapter.

2.3.4. Nucleoside Phosphonates

Nucleoside phosphonates have attracted attention for a considerable number of years, ever since it was realized that compounds with a "stable" P—C bond could be of use in the study of metabolic reactions. However, several important physicochemical changes occur when a phosphoryl oxygen is replaced by carbon, and these changes and the effects they produce in biological systems have sometimes been neglected. The first important change is the decrease in acidity of P—OH groups when the phosphorus atom is substituted by an electron-donating alkyl group in place of an oxygen atom. This change in acidity could influence which ionic species is present at physiological pH and hence influence the uptake and reactivity of the analogue. A second major factor is that the analogue might have a different size and shape compared with the natural nucleotide if oxygen is replaced by carbon. For example, the P—C bond is longer than the P—O bond and there are significant differences in the P—X—P bond angle in nucleoside polyphosphate analogues compared to that in the natural nucleoside polyphosphates. These problems are discussed in a major review of phosphonates as analogues of natural phosphates.[121] Phosphonates are easily prepared from nucleoside halides by either the Arbusov or the Michaelis–Becker reactions. Early work on the synthesis of nucleoside phosphonates was carried out with 2′,3′-protected 5′-iodo-5′-deoxynucleosides to give nonisosteric nucleoside 5′-phosphonates (**51**).[103] Isosteric nucleoside 5′-phosphonates in which the 5′-oxygen atom was replaced by a methylene group were prepared later by a different route. A suitably protected nucleoside 5′-aldehyde was coupled with diphenyl triphenylphosphoranylidenemethylphosphonate (**52**) to give an α,β-unsaturated phosphonate diester. Reduction of the latter and deprotection of the phosphoryl residue gave the phosphonate **53** (Fig. 13).[122] The synthesis of isosteric analogues of nucleoside 3′-phosphonates is a much more difficult problem owing to the instability of nucleoside 3′-keto derivatives. The few syntheses that have been reported have started from the phosphonylated ribose-1 chloride, coupling it with the heavy-metal salt of a purine or pyrimidine; for example, an isosteric analogue of 3′-dAMP (**54**) has been prepared in this way.[123]

The difficulties encountered in the synthesis of isosteric nucleoside phosphonates has discouraged an extensive study of their properties, and attention has been turned to the preparation and study of phosphonates in which the

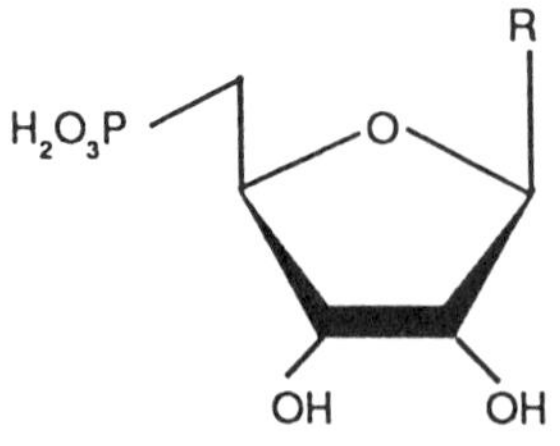

(51)

$Ph_3P = CH–P(O)(OPh)_2$

(52)

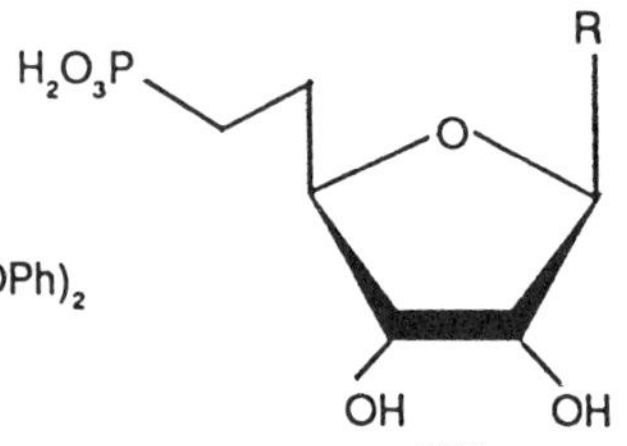

(53)

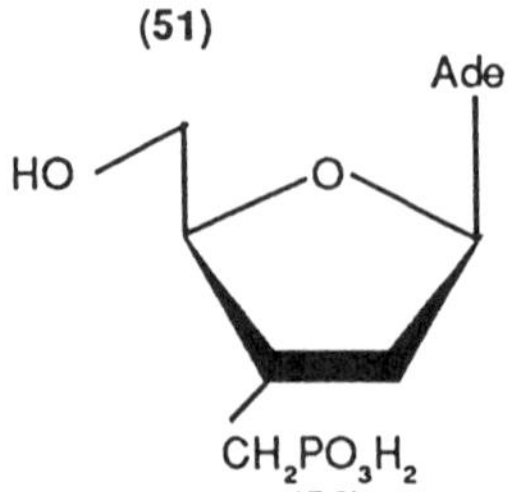

(54)

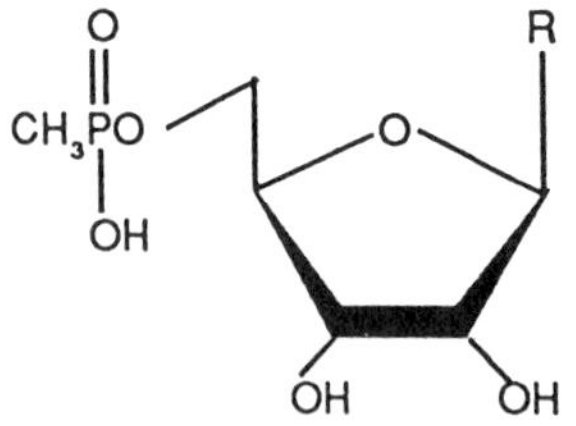

(55)

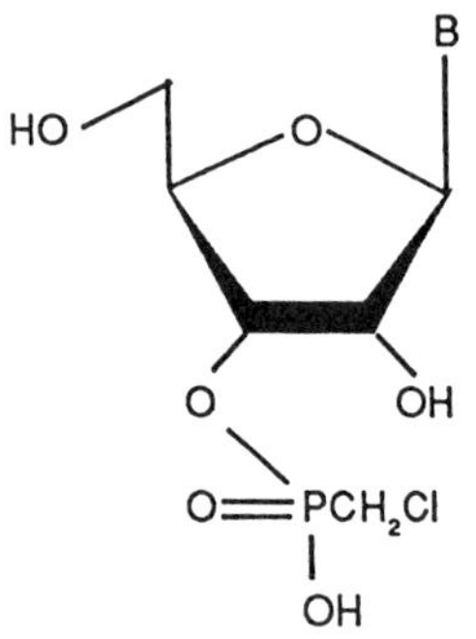

(56)

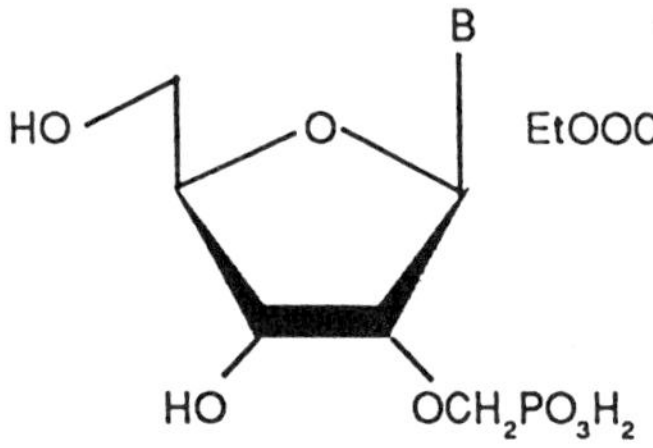

(57)

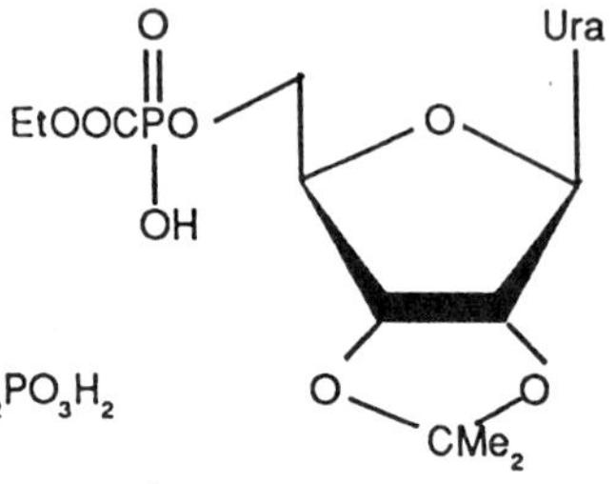

(58)

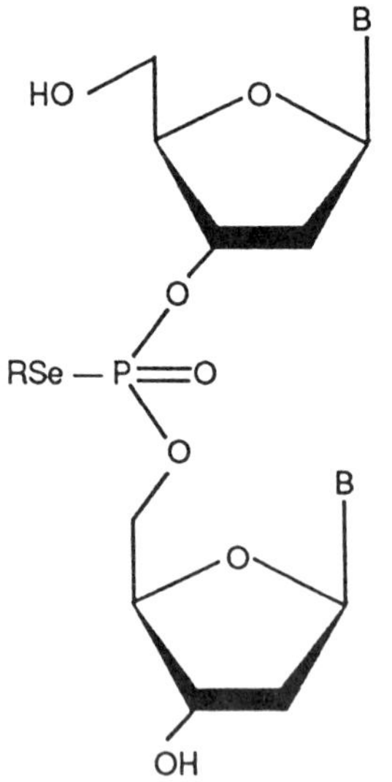

(59)

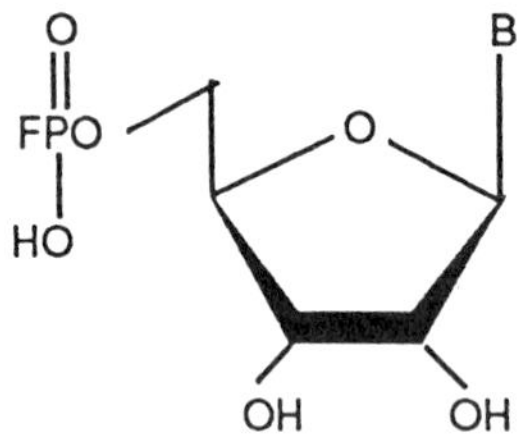

(60)

Figure 13. Nucleoside phosphates.

phosphonate residue is linked through oxygen to the nucleoside moiety as these can be prepared by conventional phosphorylation techniques. A number of such phosphonates have been prepared including phosphonates related to **55**,[124] which are less polar than their phosphate counterparts and hence should be taken up by cells more easily. Oligo(deoxyribonucleotides) derived from **55** have been prepared chemically and are indeed taken up readily by cells.[125] The 2′(3′)- (**56**) and 5′-isomers of ribonucleoside chloromethanephosphonates have been prepared.[126] Whereas compound **56** is isomerized rapidly in alkali to afford **57**, the 5′-isomer is stable under comparable conditions, presumably on steric grounds. The 5′-(ethoxycarbonyl)phosphonate ester of 2′,3′-*O*-isopropyl-ideneuridine (**58**) has been prepared by conventional means[127] and decarboxy-lates smoothly in the presence of sodium hydroxide to afford the nucleoside 5′-phosphite. This can then react *in situ* with a variety of electrophiles.

2.3.5. Nucleoside Phosphites

While nucleoside phosphites have been known for a number of years,[103] their preparation and properties have been studied in detail only since the realization of their importance as intermediates in the synthesis of oligonucleo-tides (see Section 2.2.7), phosphorothioates,[107] and oxygen-chiral phosphates.[92] Mono- and diesters of phosphorous acid exist in aqueous solution mainly in the phosphono form, and monoesters of nucleosides have been studied as nucleotide analogues although their ready oxidation limits their use in biological systems. They can be prepared from a protected nucleoside, phosphorous acid, and a carbodiimide or, better, by transesterification of a nucleoside with triesters of phosphorous acid followed by deprotection of the phosphorous acid residue. The decarboxylation of nucleoside (5′-ethoxycarbonyl)phosphonates, mentioned above, appears to be a promising new route to derivatives to phosphorous acid.[127]

2.3.6. Other Mononucleotide Analogues

A few analogues of mononucleotides in addition to those already discussed have been prepared. For example, treatment of nucleoside phosphites with alkyl or aryl selenides gives rise to the corresponding esters of nucleoside phosphoro-selenic acids (**59**),[128] and dinucleoside monophosphoroselenides have been obtained in a similar manner from the dinucleoside phosphite and elemental selenium in an inert solvent.[129] Treatment of nucleoside 5′-phosphates with 2,4-dinitrofluorobenzene results in the formation of nucleoside 5′-phosphoro-fluoridates (**60**) via the 2,4-dinitrophenylester of the nucleotide.[130] While some enzymatic studies have been carried out with **60**, the properties of these nucleo-tide analogues have been little studied.

2.4. Nucleotide Anhydrides

Nucleotide anhydrides are widely encountered as coenzymes for many important biochemical processes and have been synthetic targets for organic chemists for many years. In this section, synthetic routes to these compounds will be described, and their biological properties will be discussed later in this chapter.

2.4.1. Nucleoside Polyphosphates

Nucleoside polyphosphates can be prepared either by the polyphosphorylation of a hydroxyl function on the sugar residue of the nucleoside or by coupling a nucleotide with another phosphoryl residue. The latter route is the most commonly used for the synthesis of nucleoside polyphosphates. As is the case in the synthesis of mononucleotides and other phosphomonoesters, the phosphoric (or polyphosphoric) acid must be activated before a polyphosphoric acid ester can be formed. The protection of phosphoryl hydroxyl groups was a critical problem, particularly in the synthesis of asymmetrical polyphosphates, until the development of the phosphoramidates as synthetic intermediates. The range of protecting groups used for phosphoryl O—H protection is not as extensive as that developed for the synthesis of mononucleotides, the commonest being benzyl and phenyl. The early work on the synthesis of nucleoside polyphosphates has been reviewed.[5,6] The earliest reported synthesis of ADP involved the coupling of dibenzyl phosphorochloridate with the benzyl ester of adenosine 5'-phosphate,[27] but this route could not readily be extended to more complex polyphosphates owing to the instability of fully esterified polyphosphates. In an early synthesis of ATP, the disilver salt of adenosine 5'-phosphate was treated with an excess of dibenzyl phosphorochloridate followed by a hydrogenolytic removal of the benzyl groups.[131] The resulting ATP is apparently identical with authentic material obtained from natural sources, and no evidence was presented for the presence of the isomeric form (**61**) (Fig. 14). Later work [132] has shown that monoadenosine trimetaphosphate (**62**) is converted rapidly in the presence of water into ATP, and it may be that a trimetaphosphate is formed during the debenzylation process at the end of the synthetic sequence. This route could not, however, be easily adapted for the synthesis of other nucleoside polyphosphates.

The ready cleavage of the P—O—P bond in triesters of pyrophosphoric acids following nucleophilic attack on phosphorus has been exploited extensively for the synthesis of nucleoside polyphosphates.[133] Treatment of a mononucleotide with diphenyl phosphorochloridate leads to the formation of the triester **63**, which can then be made to react *in situ* with nucleophiles such as the anion of a phosphoric acid (Fig. 14). This reaction leads to the expulsion of the anion of diphenyl phosphoric acid. ADP can be prepared in this way using a salt of orthophosphoric acid and AMP. This procedure is a simple method for the preparation of a range of nucleotide anhydrides with the minimum of protection of the starting materials although side reactions can occur.

The pyrophosphorylation of AMP using phosphoryl chloride under phase-transfer conditions has been reported.[134] If an aqueous solution of AMP is stirred with a solution of the dichloride salt of N^1, N^4-bis(*n*-octadecyl)-1,4-diaza-[2,2,2]bicyclooctane (**64**) and the organic phase is separated, treated with phosphoryl chloride in the presence of base, and back-extracted with aqueous so-

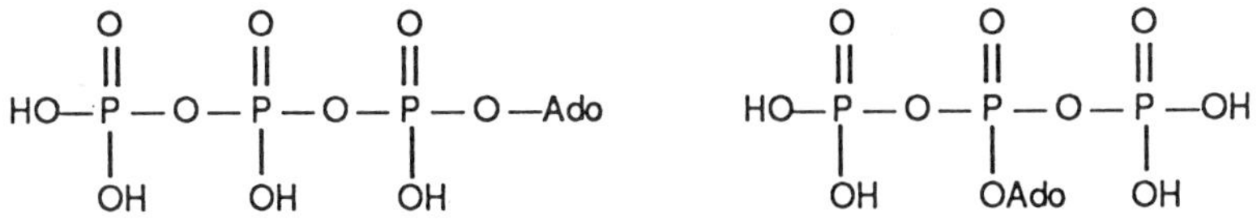

(61)

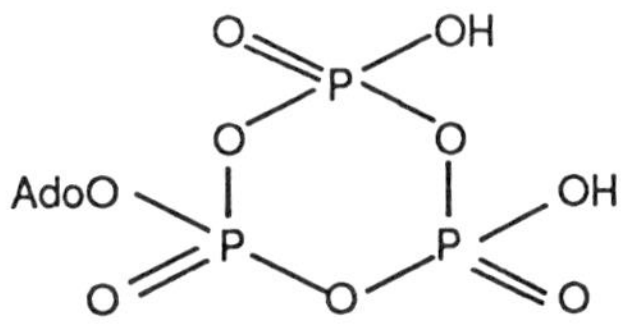

(62)

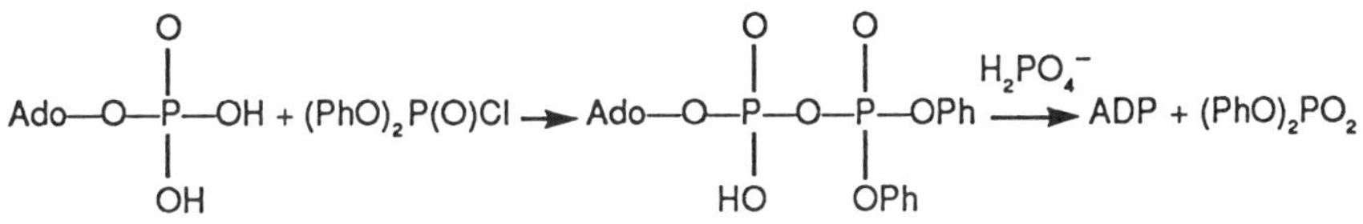

(63)

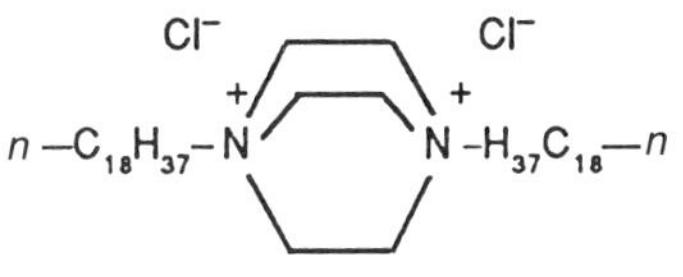

(64)

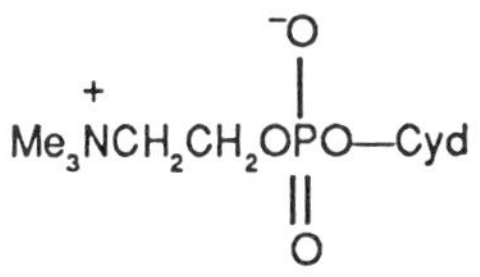

(65)

$$CCl_3C \equiv N$$

(66)

Figure 14. Nucleoside polyphosphates.

dium perchlorate, ADP and ATP are obtained in a moderate yield. The treatment of nucleosides with pyrophosphoryl chloride (**4**) is, as mentioned earlier, an unsatisfactory method for the synthesis of nucleoside polyphosphates. A mixture of polyphosphates is formed, and if the 2′,3′-hydroxyl groups of a ribonucleoside are left unprotected, the cyclic phosphate is also formed. The mixture of polyphosphates can be hydrolyzed to give the mononucleotides in reasonable yield.[18] Also, treatment of the reaction mixture with ammonia leads to phosphoramidates which can be converted into nucleoside di- and triphosphates.[19,20]

A major advance in the synthesis of nucleoside polyphosphates was achieved when carbodiimides were introduced as coupling agents. This allowed the mononucleotides to be used without any protection of the P—OH and C—OH groups. The mechanism of action of carbodiimides is discussed in Section 2.2.3, and the synthesis of nucleoside polyphosphates from a mononucleotide and phosphoric acid follows the mechanism outlined in Fig. 5. As was pointed out in Section 2.2.3, the use of unprotected orthophosphoric acid is unsatisfactory for the synthesis of polyphosphates such as ADP and ATP since there are in orthophosphoric acid three hydroxyl groups available for reaction. This produces a mixture of polyphosphates which must be separated chromatographically. Despite these drawbacks, carbodiimides have been used for the preparation of a number of nucleoside polyphosphates and P^1,P^2-diesters of pyrophosphoric acid. Occasionally, asymmetric esters of pyrophosphoric acid have been obtained in good yield; for example, CDP-choline (**65**) has been prepared from CMP and choline phosphate with the aid of DCC. Choline phosphate is a zwitterion under the conditions of the condensation, and, as it is a stronger acid than CMP, it is not the best nucleophile for carbon. Thus, initial attack on the DCC is by the CMP, but, as phosphoryl groups are good nucleophiles for phosphorus, choline phosphate can attack the CMP–DCC adduct, leading to **65**. CMP can also attack the adduct, and P^1,P^2-dicytidine pyrophosphate is also formed.[135] Trichloroacetonitrile (**66**) is a condensing agent which reacts in a manner similar to DCC but which produces a volatile hydration product, trichloroacetamide.[136] Compound **66** has been particularly useful in the synthesis of ^{32}P-labeled nucleotides and nucleoside polyphosphates.[40]

Organophosphorus compounds containing a P—N bond can function as phosphorylating agents and will react selectively with phosphoryl groups to afford polyphosphates. Phosphoramidic acid and its derivatives are the most synthetically useful reagents of this class, and monoesters of phosphoramidic acid have been used extensively since their introduction over 25 years ago for the synthesis of nucleotide anhydrides. Diesters of phosphoramidic acid are readily available, and they can be prepared in high yield by treating a diester of phosphorochloridic acid with an amine. However, they are poor phosphorylating agents. Selective deesterification of phosphoramidic diesters, for example, debenzylation of the dibenzyl ester with lithium chloride, gives the monoesters **67** (Fig. 15), which are potent phosphorylating agents.[137] A much more convenient route to phosphoramidic monoesters, which is particularly applicable to the preparation of nucleoside phosphoramidates, is the reaction between DCC, a nucleotide, and an amine.[138]

When a monoester of phosphoramidic acid is heated, a pyrophosphate is formed in a reaction which follows second-order kinetics.[139] This observation, taken together with the selectivity of phosphoramidic monoesters for phosphoryl groups rather than alcohol functions, suggests that phosphoryl transfer from phosphoramidates occurs by a direct displacement reaction rather than by a route which involves monomeric metaphosphates. The nature of the departing amine plays an important part in determining the rate of phosphoryl transfer.

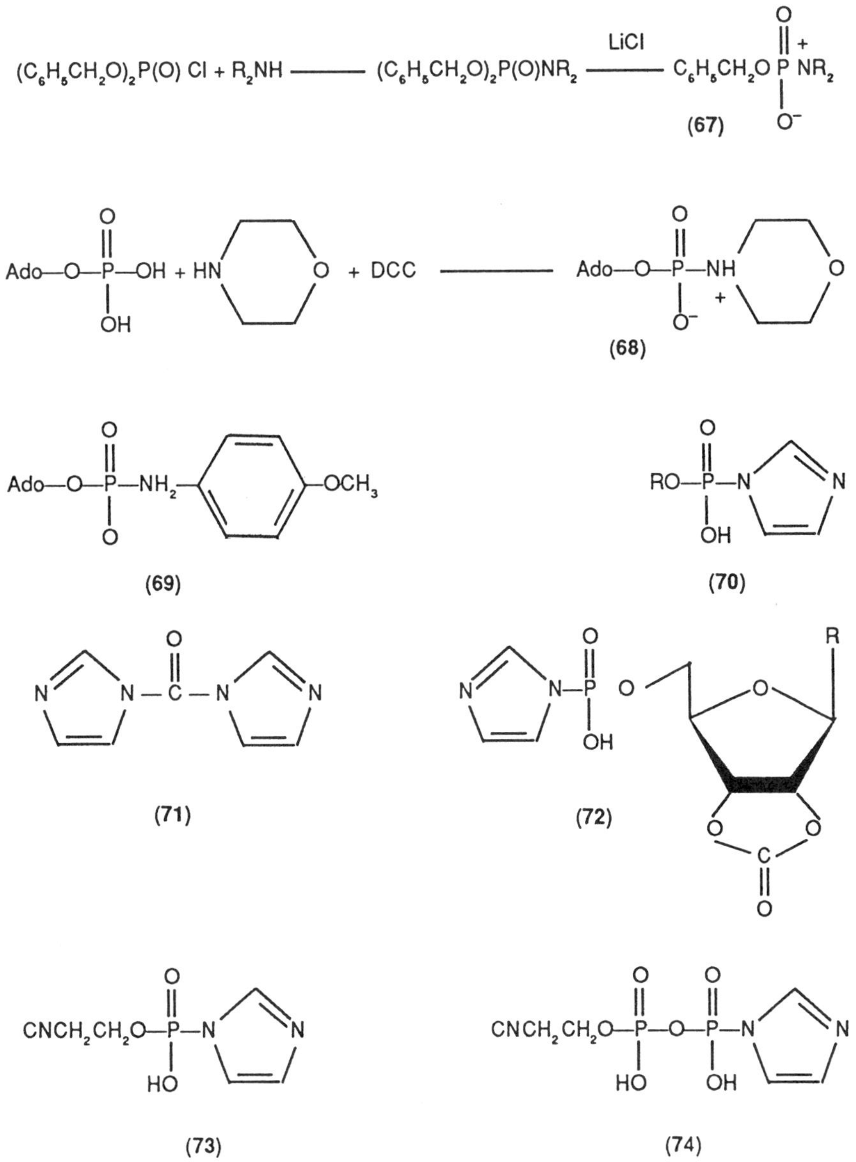

$$(C_6H_5CH_2O)_2P(O)\,Cl + R_2NH \longrightarrow (C_6H_5CH_2O)_2P(O)NR_2 \xrightarrow{\text{LiCl}} C_6H_5CH_2O\overset{\overset{\displaystyle O}{\|}}{\underset{\underset{\displaystyle O^-}{|}}{P}}\overset{+}{N}R_2$$

(67)

Ado—O—P(O)—OH + HN(morpholine) + DCC ⟶ Ado—O—P—NH(morpholine)$^+$

(68)

Ado—O—P—NH₂—⟨C₆H₄⟩—OCH₃

(69)

RO—P—N(imidazole), OH

(70)

(imidazole)N—C(=O)—N(imidazole)

(71)

(imidazole)N—P(=O)(OH)—O—(sugar)

(72)

CNCH₂CH₂O—P(=O)(HO)—N(imidazole)

(73)

CNCH₂CH₂O—P(=O)(HO)—O—P(=O)(OH)—N(imidazole)

(74)

Figure 15. Reactions of nucleoside phosphoramidates and related compounds.

For example, monophenyl phosphate will react almost quantitatively with adenosine 5′-phosphoropiperidate in pyridine after 1 h at room temperature, whereas less than 10% reaction occurs under the same conditions with a phosphoramidate derived from a weaker base, anisidine (**69**).[138] Nucleoside phosphoromorpholidates (**68**) offer the advantages of reactivity and solubility in organic solvents, making them the reagents of choice for the synthesis of symmetrical and unsymmetrical esters of polyphosphoric acid such as ATP,[139] NAD⁺ and analogues,[140] coenzyme A,[141] and nucleoside diphosphate sugars.[142–144] As will be described in Section 2.4.5, many analogues of naturally occurring nucleoside polyphosphates have been prepared with the aid of phosphoromorpholidates and related compounds. While procedures have been developed for the synthesis of nucleoside phosphoramidates in high yield and although some are commercially available, it can be inconvenient to prepare and isolate nucleoside phosphoramidates during a synthetic sequence. Thus, the preparation of nucleoside phosphorimidazolidates (**70**) *in situ* from the nucleotide and carbonyl bis(imidazole) (**71**)[145] is a time-saving development. For example, ATP can be prepared either by activating ADP with **71** and coupling the product with orthophosphate or by activating the orthophosphate with **71** and coupling with ADP.[146] It has been shown that the treatment of ribonucleoside 5′-phosphates with excess **71** can lead to the formation of 2′,3′-*O-O*-cyclic carbonates (**72**).[147] While the cyclic carbonates can readily be destroyed by alkaline hydrolysis, this complication can be avoided by adopting the route which involves the activation of the ortho- or pyrophosphate with **71** prior to a coupling with the nucleotide.

The ATP analogue **61**, in which the adenosine residue is linked to the β-phosphorus atom, has been prepared from adenosine 5′-diphosphoroimidazolidate and orthophosphate.[146] While compound **61** had the predicted chromatographic and electrophoretic properties and had the correct analytical figures, no ³¹P-NMR evidence was presented to confirm the structure. The activation of nucleotides by arenesulfonyl chlorides followed by reaction with morpholine or piperidine affords the corresponding nucleoside phosphoromorpholidates and -piperidates.[148] This synthetic procedure appears to have little advantage over the conventional syntheses of nucleoside phosphoramidates and has been little used. 2-Cyanoethyl phosphoro- (**73**) and pyrophosphoroimidazolidates (**74**) have been prepared from 2-cyanoethyl phosphate or pyrophosphate and **71**, and both **73** and **74** have been used to prepare the α-³²P-labeled nucleoside polyphosphates of high specific radioactivity.[149] A highly active phosphorylating agent prepared by treating 2-cyanoethyl phosphate first with DCC and then with mesitylenesulfonyl chloride will convert 2′-deoxyguanosine 3′,5′-bisphosphate directly into the 3′,5′-bispyrophosphate.[150]

The oxidation of phosphorothioate *S*-esters can generate phosphorylating agents, possibly metaphosphates, and this route has been used for the synthesis of a number of nucleoside polyphosphate esters. For example, di-*n*-butylphosphinothioyl bromide (**75**) reacts smoothly with nucleoside 5′-phosphates in pyridine to form stable adducts (**76**) (Fig. 16). The treatment of **76** with ortho- or pyrophosphate in the presence of silver(I) ions leads to the rapid formation of nucleoside di- and triphosphates.[151] Condensation of a protected guanosine 5′-*S*-phenylthiophosphate (**77**) with stannylated 5′-ribonucleotides (**78**) in the presence of iodine leads to P^1, P^2-(ribonucleoside-5′) diphosphates related to the "cap" structures of eucaryotic mRNA.[152] An alternative synthesis of "cap" structures involves the treatment of methyl phosphorodichloridate with thiophenol and 7-methyl-GMP. Oxidation of the presumed intermediate (**79**) in the presence of DMP or AMP leads to P^1, P^2-(diribonucleoside-5′) triphosphates.[153] Oxidation of *S*-alkyl phosphorothioates with iodine in the presence of pyro-

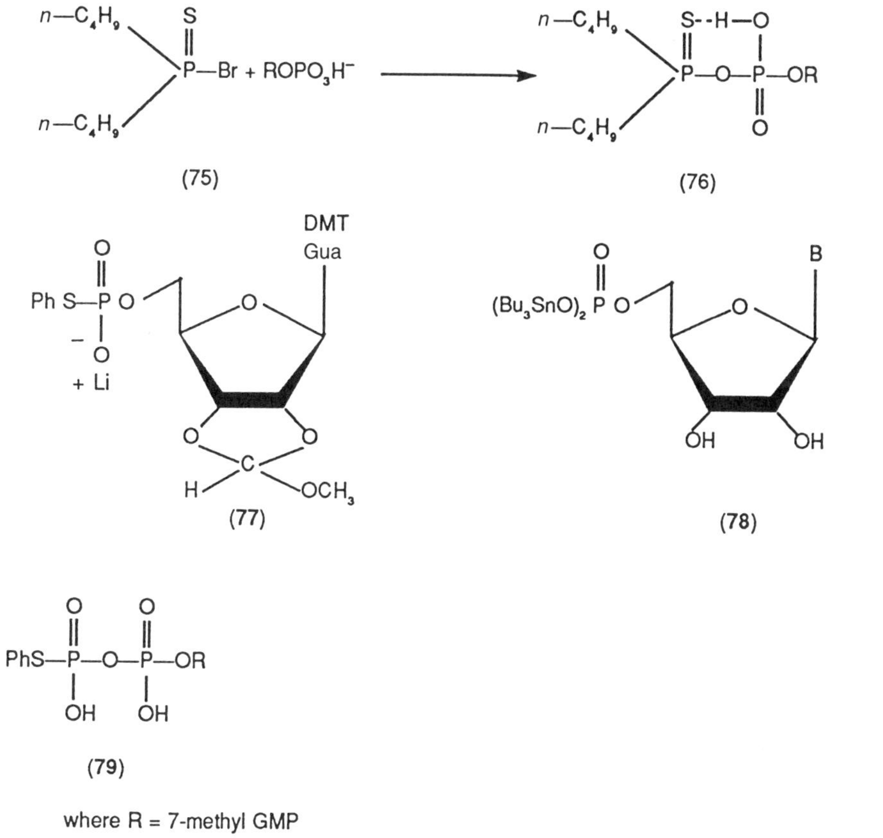

Figure 16. Sulfur-containing reagents used in the chemical synthesis of nucleoside polyphosphates.

phosphate ion is another route to nucleoside 5′-triphosphates.[154,155] P^1P^3-(Diribonucleoside-5′) triphosphates have also been made by condensing a nucleoside 5′-phosphorimidazolidate with the 5′-diphosphate of another nucleoside.[156]

Nucleoside-5′ di- and triphosphates are involved in a very large number of enzymatic reactions, and several procedures have been developed for the synthesis of these compounds that involve either single enzymes or coupled enzyme systems (Table II). Most of these procedures have been specifically developed for the preparation of radioactively labeled nucleoside polyphosphates on a small scale. However, large-scale syntheses of nucleoside triphosphates have been developed and were mentioned in Section 2.2.8. In large-scale synthesis of ribonucleoside triphosphates,[81] RNA is degraded to oligonucleotides with nuclease P_1, and the oligomers are converted into a mixture of nucleoside 5′-diphosphates by immobilized polynucleotide phosphorylase. The diphosphates are then phosphorylated by a mixture of phosphoenolpyruvate and pyruvate kinase. The mixture of 5′-triphosphates obtained in this way proved to be

Table II. Enzymatic Synthesis of Nucleoside Di- and Triphosphates

Nucleotide product	Nucleotide substrate	Enzyme(s) used	Reference
ATP	ADP	3-Phosphoglycerate kinase	157
ATP	ADP	Acetate kinase	158
ATP	ADP	Hexokinase/pyruvate kinase	159
ATP/dATP	AMP/dAMP	Myokinase/pyruvate kinase	160
NTP[a]	NMP	Posphoenolpyruvate synthetase	161
NTP	NMP	Nucleoside monophosphate kinase	162
ATP	Adenosine	Adenosine kinase/myokinase/acetate kinase	163,164
ADP	ATP	Hexokinase	165
Br^{8}ITP	Br^{8}IDP	Nucleoside diphosphate kinase	166
ATP	3'-AMP	Polynucleotide kinase/nuclease P$_1$/pyruvate kinase/ myokinase	167,168
NTP	3'-NMP	Polynucleotide kinase/nuclease P$_1$/pyruvate kinase/ nucleoside monophosphate kinase	169
ATP/GTP	Poly(A)/poly(G)	Polynucleotide phosphorylase/3-phosphoglycerate kinase	170
ATP	RNA	Nuclease P$_1$/acetate kinase/adenylate kinase	171
NTP	RNA	Nuclease P$_1$/polynucleotide phosphorylase/ pyruvate kinase	81
dATP	DNA	DNase I/nuclease P$_1$/adenylate kinase/pyruvate kinase	82

[a]N = nucleoside-5'.

satisfactory for the preparation of UDP-Glc, as the enzyme UDP-Glc pyrophosphorylase will only react with the UTP in the triphosphate mixture. A similar degradation of DNA with DNase I and nuclease P$_1$ followed by selective phosphorylation of the dAMP in the monophosphate mixture using adenylate kinase and pyruvate kinase leads to dATP in high overall yield.[82]

2.4.2. Nucleotide Coenzymes

Most nucleotide coenzymes are P^1,P^2-diesters of pyrophosphoric acid, and hence the chemical methods described in the previous section can be applied to their synthesis. For example, the phosphoramidate route is a good method for the synthesis of these coenzymes, and a large number have been prepared by this route.[138–144] Carbodiimides have not usually been satisfactory for the synthesis of nucleotide coenzymes although reasonable yields of asymmetric pyrophosphate esters can be obtained if one of the components is a zwitterion. The synthesis of CDP-choline from CMP and choline phosphate has already been mentioned (Section 2.4.1),[135] and analogues of NAD$^+$ can be prepared in reasonable yield from the zwitterionic nicotinamide mononucleotide (NMN$^+$, **80**) with this reagent.[172]

Activation of AMP(S), an analogue of AMP in which a phosphoryl oxygen has been replaced by sulfur, with diphenyl phosphorochloridate followed by treatment of the reaction product with a protected nicotinamide mononucleotide is a simple way to prepare NAD(S)$^+$.[173] Di-*n*-butyl phosphinothioyl bromide (**75**) can be used to prepare NAD$^+$ and other coenzymes,[151] while the reaction between phosphorothioates and the disilver salts of phosphoric acids has been used to synthesize FAD, UDP-Glc, and UDP-Gal in high yield.[174]

Large-scale syntheses of NAD$^+$ and NADP$^+$ by a combination of chemical and enzymatic steps have been reported.[175] NMN$^+$ (**80**) is prepared from ribose 5-phosphate by treatment of the latter first with anhydrous ammonia and then with 1-(2,4-dinitrophenyl)-3-carbamyl pyridinium chloride (**81**) (Fig. 17). Pyrophosphorolysis of ATP by **80** leads to NAD$^+$, which can be further phosphorylated with ATP and NAD kinase to afford NADP$^+$. A comparison of the strategies for the *in situ* regeneration of NAD$^+$ from NADH has been published.[176] There have been many reports of the enzymatic synthesis of analogues of nucleotide coenzymes, for example, 3-iodopyridine adenine dinucleotide (**82**),[177] while UDP-[^{14}C]Glc has been prepared enzymatically from ^{14}C-labeled glucose.[178]

Although P^1,P^4-di(nucleoside-5') tetraphosphates (**83**, $n = 2$) and P^1,P^5-di(nucleoside-5') pentaphosphates (**83**, $n = 3$) are not strictly nucleotide coenzymes, it is convenient to consider their synthesis here. The phosphoromorpholidate method is an ideal way to synthesize these α,ω-polyphosphate esters,[179] and the pentaphosphates can be prepared by activating ATP with diphenyl phosphorochloridate and coupling the product with ADP.[180,181]

Treatment of thymidine 5'-phosphorodiamidate (**84**) with pyrophosphate in anhydrous DMF leads to P^1-(thymidine-5')-P^1-amino triphosphate (**85**) in good yield.[182] In DMF at room temperature, the latter exists in equilibrium with the tetraphosphate analogue (**86**). Phosphonate analogues of P^1,P^4-di(adenosine-5') tetraphosphate, for example, **87** have been prepared by activating methylene bisphosphonate with carbonyl bis(imidazole) and coupling the product with AMP.[183]

2.4.3. Nucleoside Polyphosphates That Contain Oxygen Isotopes

Nucleoside polyphosphates that contain oxygen isotopes in the phosphoryl residues have attracted considerable attention since they can be used to study the stereochemistry of phosphoryl transfer by enzymes. Many nucleoside (mainly adenosine) polyphosphates that contain oxygen isotopes have been prepared. The syntheses of mononucleotides that contain oxygen isotopes have been described in Section 2.3.1, and some of these synthetic routes have been adapted for the synthesis of oxygen-labeled polyphosphates. For example, hydrolysis of the ^{17}O-labeled cyclic phosphorochloridate **40** with lithium [^{18}O]hydroxide gives **88**, which, after reaction with ADP and catalytic hydrogenolysis, leads to [$\gamma(S)$-^{16}O,^{17}O,^{18}O]ATP (**89**) (Fig. 18).[184] The synthesis of γ-labeled [^{18}O$_4$]ATP can readily be achieved by coupling adenosine 5'-(β-morpholino)diphosphate[185] with [^{18}O$_4$]orthophosphate. A more complex series of enzymatic reactions can be used to obtain a range of labeled adenosine polyphosphates (Fig. 18).[186] Treatment of [^{18}O$_4$]orthophosphate with potassium cyanate yields the carbamyl [^{18}O$_3$]phosphate. Incubation of this compound with ADP and carbamate kinase produces [γ-^{18}O$_3$]ATP (**90**). When **90** and AMP are used as substrates for

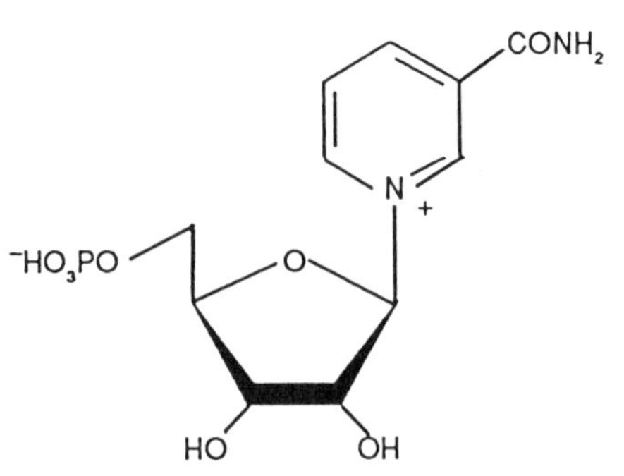

(80)

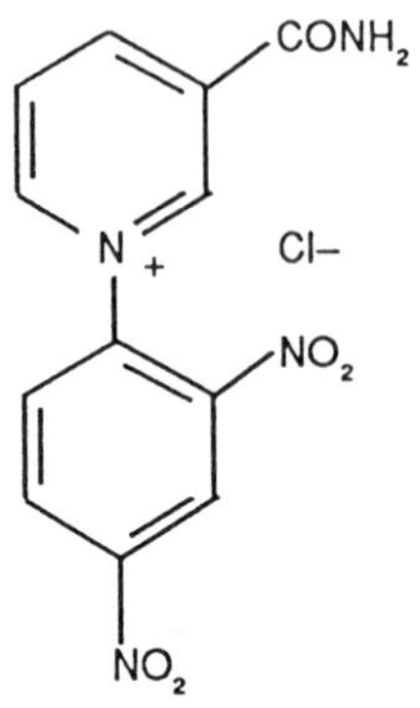

(81)

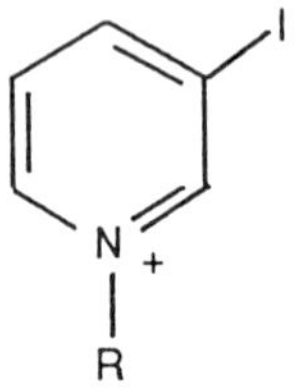

(82)

where R = adenosine 5'-pyrophosphoryl
-5-β-D-ribofuranosyl

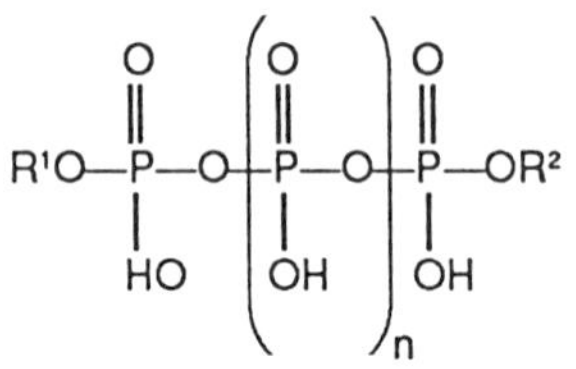

(83)

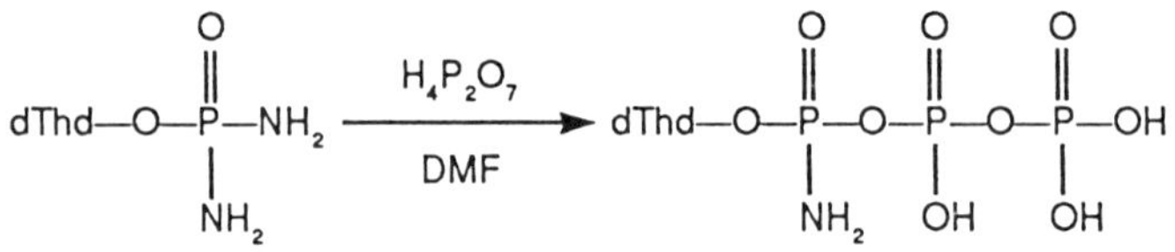

(84)

(85)

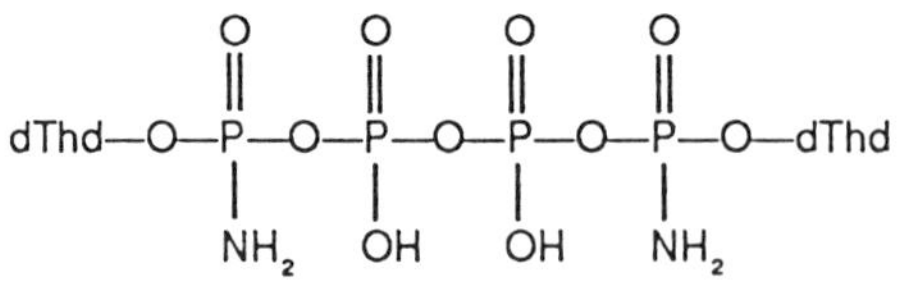

(86)

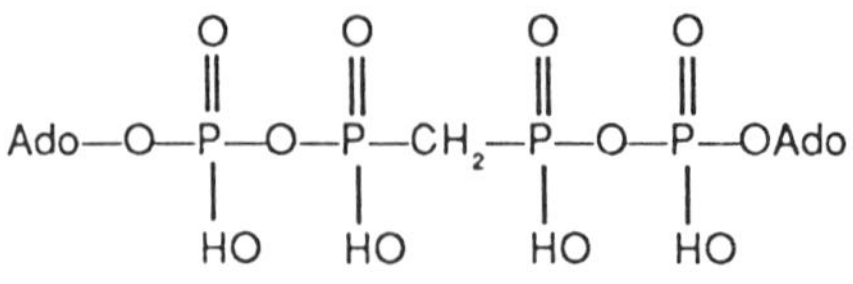

(87)

Figure 17. Nucleoside coenzymes and other analogues of nucleoside polyphosphates.

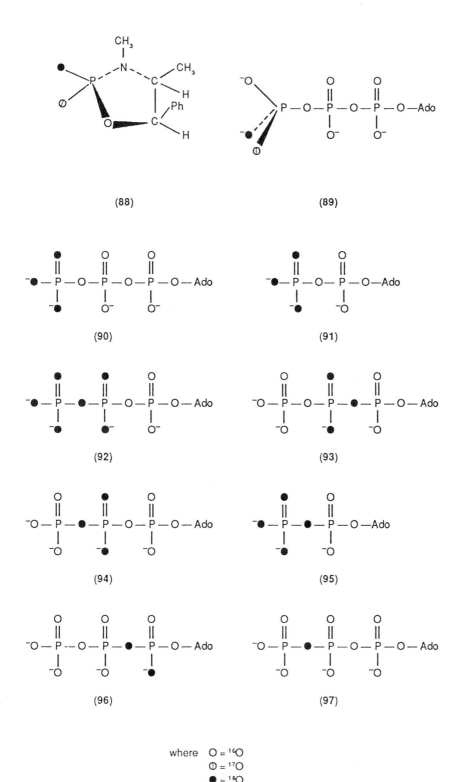

(88)

(89)

(90)

(91)

(92)

(93)

(94)

(95)

(96)

(97)

where $O = {}^{16}O$
$\oplus = {}^{17}O$
$\bullet = {}^{18}O$

Figure 18. Adenosine di- and triphosphates labeled with ^{17}O and ^{18}O.

adenylate kinase, $[\beta\text{-}^{18}O_3]$ADP (**91**) containing three nonbridging oxygen atoms is obtained. Recycling of **91** with carbamyl $[^{18}O_3]$phosphate and carbamate kinase gives $[^{18}O_6]$ATP (**92**). Alternatively, treatment of **90** with valyl-tRNA synthetase results in the randomization of the α,γ-pyrophosphate moiety and the formation of $[\beta\text{-}^{18}O_3]$ATP (**93**).[186] The reaction of **91** with phosphoenol-pyruvate in the presence of pyruvate kinase gives **94**. Alternatively, **93** can be prepared chemically by condensing AMP and $[^{18}O_4]$orthophosphate in the presence of carbonyl bis(imidazole) to give $[\beta\text{-}^{18}O_4]$ADP (**95**).[187] Activation of **95** with carbonyl bis(imidazole) followed by treatment of the product with $[^{16}O_4]$orthophosphate affords **93**. Similarly, activation of a base-protected AMP with diphenyl phosphorochloridate and treatment of the product with $[^{18}O_4]$-orthophosphate, followed by a reaction of the ADP produced with di-n-butyl-phosphinothioyl bromide and $[^{16}O_4]$orthophosphate, gives **93**.[188] (S_P)-Ade-nosine $5'\text{-}[\alpha\text{-}^{17}O,^{18}O_2]$triphosphate (**96**) has been prepared by phosphorylating adenosine with thiophosphoryl chloride, hydrolyzing the resultant $5'$-thiophos-phorodichloridate with ^{18}O-labeled water, and converting enzymatically the AMP(S) produced in this way into $(S_P)\text{-}[^{18}O_2]$ADP(αS). Oxidation of the latter with bromine in ^{17}O-labeled water gave $[\alpha\text{-}^{17}O,^{18}O_2]$ADP, which was finally phosphorylated with pyruvate kinase to **96**.[189]

Two syntheses of adenosine $5'[\beta,\gamma\text{-}^{18}O]$triphosphate (**97**) have been re-ported. One method involves the enzymatic phosphorylation of $(S_P\text{-}[\beta\text{-}^{18}O)]$-ADP$(\alpha S)$.[190] The better method involves the prior synthesis of bridge-labeled pyrophosphate and its reaction with adenosine $5'$-phosphoromorpholi-date.[190,191] The latter route can also be used for the ^{17}O-analogue of **97**.

2.4.4. Nucleoside Polyphosphorothioates

Like the oxygen-labeled nucleoside polyphosphates, nucleoside poly-phosphorothioates have been used to study the stereochemistry of enzyme-catalyzed phosphoryl transfer reactions. The syntheses and properties of these polyphosphorothioates have been the subject of an excellent review by Eck-stein,[101] and only the salient features of their synthesis will be discussed here. The synthesis of nucleoside monophosphorothioates has been discussed in Section 2.3.2, and these compounds have been used as the starting materials for the chemical and enzymatic synthesis of nucleoside polyphosphorothioates. The phosphorothioate residue is chiral, and one important difference between the chemical and the enzymatic synthesis of nucleoside polyphosphorothioates is that a mixture of diastereomers is produced as a result of a chemical reaction whereas, in general, only one diastereomer is produced as a result of an enzy-matic reaction. Furthermore, following the activation of nucleoside monophos-phorothioates by conventional chemical reagents, different products can be produced depending on the reagent. The oxygen atoms of the monophosphoro-thioate are more nucleophilic toward "hard" phosphorus centers than sulfur, and thus activation of the phosphorothioate by diphenyl phosphorochloridate leads to products resulting from the loss of an oxygen atom from the phosphoro-thioate. On the other hand, the sulfur atom is more nucleophilic toward "soft" carbon centers than oxygen. Activation of nucleoside monophosphorothioates with carbodiimides or carbonyl bis(imidazole) leads to a complete loss of sulfur. The synthesis of both diastereomers of ADP(αS) (**98**) and ATP(αS) (**99**) can, therefore, be achieved by activating AMP(S) with diphenyl phosphorochloridate and coupling the product with ortho- or pyrophosphate.[192] The diastereomers can be distinguished by ^{31}P-NMR spectroscopy[193] and can be separated by HPLC.[194] The guanosine analogues can be made in a similar fashion.[195] Where it is not possible to separate the diastereomers chromatographically, selective enzymatic phosphorylation of one isomer can be used to achieve separation. Nucleoside triphosphorothioates in which sulfur is attached to the β- and γ-phosphorus atoms, for example, **100** (Fig. 19), are generally prepared enzymatically. Thus, (S_P)-ATP (**100**) can be prepared using pyruvate kinase and the appropriate isomer of ADP(βS) whereas the (R_P)-isomer (**101**) is obtained with acetate kinase.[196]

Nucleoside polyphosphorothioates containing [17]O and [18]O in which sulfur is attached to the γ-phosphorus atom have also been prepared enzymatically. For example, (R_P)-[β,γ-^{18}O,γ-[18]O$_1$]GTP(γS) (**102**) has been obtained and used to prepare chiral [^{18}O]thiophosphoenolpyruvate (**103**).[197] Other ATP analogues of this kind that have been obtained enzymatically include (S_P)-[β,γ-[17]O,[18]O]-ATP(γS) (**104**)[198] and -GTP(γS).[199] Large-scale enzymatic syntheses of (S_P)-ATP(αS) from AMP(S)[200] and ATP(γS)[201] from ADP and sodium thiophosphate have been reported.

2.4.5. *Other Nucleoside Polyphosphate Analogues*

A number of nucleoside polyphosphate analogues in which the atom between the α,β-phosphorus atoms in nucleoside di- and triphosphates or the β,γ-phosphorus atoms in nucleoside triphosphates has been replaced by atoms other than oxygen have been prepared chemically and used to investigate enzymatic reactions. Some of these syntheses are reviewed by Scheit,[103] and ATP analogues have been reviewed by Yount.[202] These nucleoside polyphosphate analogues are listed in Table III. As the preparation of α,β-substituted polyphosphate analogues is, in general, more difficult than that of β,γ-substituted polyphosphate analogues, fewer of the former have been made. Usually, the α,β-analogues, for example, adenosine 5'-methylene bisphosphonate [**105**, R = Ado-5' (Fig. 20)] are prepared by condensing a 2',3'-*O*-protected nucleoside with the pyrophosphate analogue with the aid of DCC.[203] Nucleophilic displacement reactions have occasionally been used, and the deoxythymidine analogue (**105**, R = dThd-5') has been prepared by the displacement of a toluenesulfonyl (tosyl) residue from the 5'-position of the sugar residue of the tosyl thymidine by methylene bisphosphonate ion.[206] The β,γ-analogues can readily be obtained by conventional methods which have already been discussed in this chapter. Thus, adenylyl 5'-methylene bisphosphonate (**106**)[207] has been obtained by the phosphoramidate route, and activation of either the nucleoside monophosphate or the pyrophosphate analogue with diphenyl phosphorochloridate has been used on several occasions, for example, for the synthesis of adenosine 5'-phosphohypophosphate (**107**).[213] The bismethylene analogue (**108**) of trimetaphosphate is an acid anhydride and will react with alcohols with a

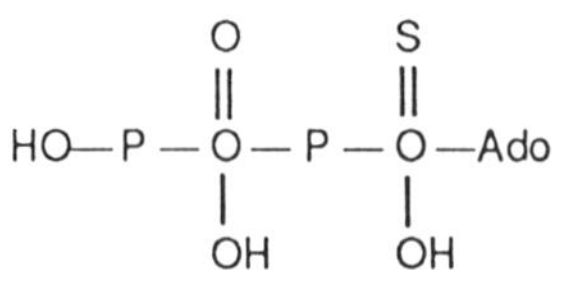

(98)

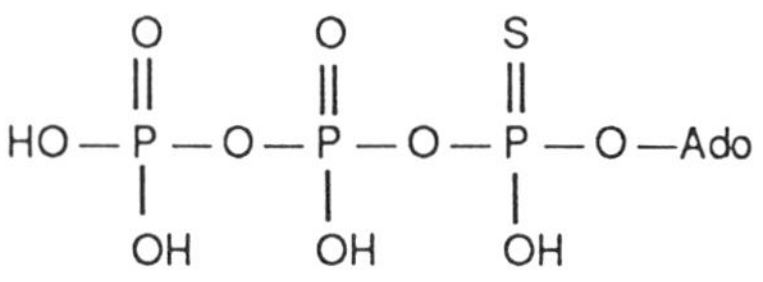

(99)

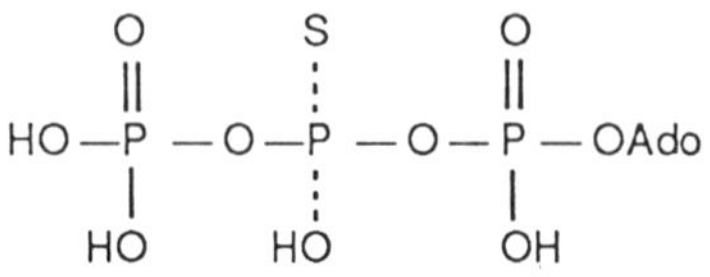

(100)

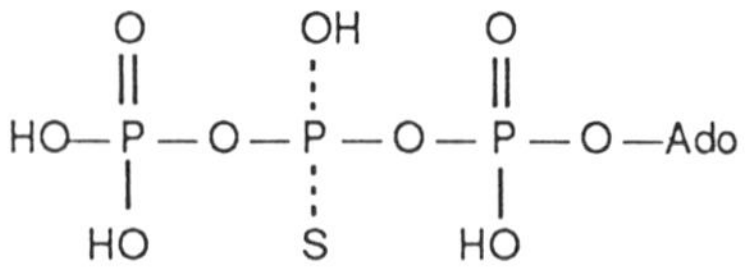

(101)

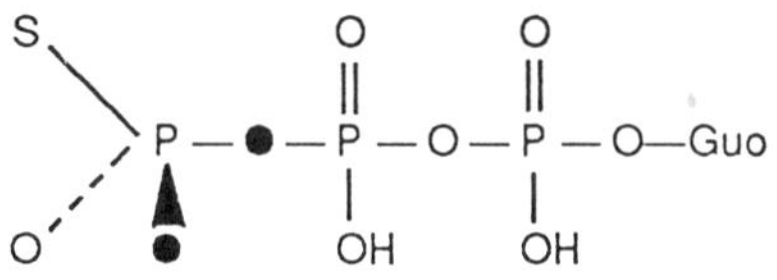

(102)

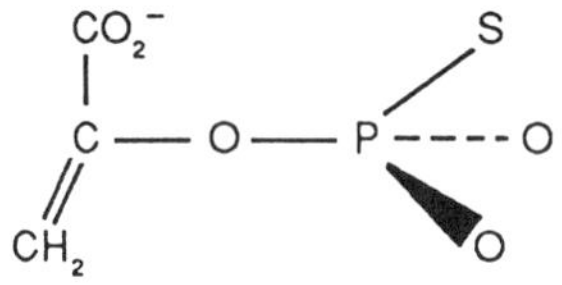

(103)

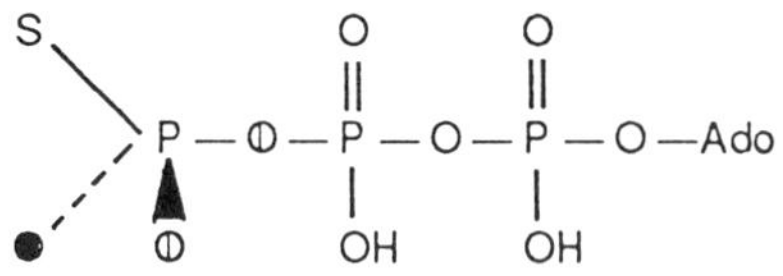

(104)

Figure 19. Nucleoside polyphosphorothioates.

*Table III. Nucleoside Di- and Triphosphate
Analogues with Atoms Other than Oxygen between
the α and β or β and γ Phosphorus Atoms*

Compound[a]	Reference
α, β-*Analogues*	
Ado-5' PCH$_2$P (**105**)	203
Ado-5' PCH$_2$PCH$_2$P (**109**)	204
Guo-5' PCH$_2$POP	205
dThd-5' PCH$_2$P (**105**)	206
Guo-di-3',5' PCH$_2$P	150
β, γ-*Analogues*	
Ado-5' POPCH$_2$P (**106**)	207
Ado-5' POPCCl$_2$P	208
Ado-5' POPCF$_2$P	208
Ado-5' POPCHFP	208
Ado-5' POPC:CP	208
N-5' POPOOP	209
Ado-5' POPNHP	210
Guo-5' POPCH$_2$P	205
Guo-5' POPCCl$_2$P	208
Guo-5' POPCF$_2$P	208
Guo-5' POPCHFP	208
dThd-5' PCH$_2$POP	211
dThd-5' PCH$_2$PCH$_2$P	206
Ino-5' POPNHP	212
Ado-5' POPP (**107**)	213
Guo-5' POPP	214
dThd-3'-POP, 5'-POPCH$_2$P	215

[a]P = a phosphoryl residue; N = a ribo- or deoxyribonucleoside.

subsequent opening of the phosphorus-containing ring. 2',3-*O*-Isopropylidene-adenosine reacts with **108** to give the symmetrical 5'-bis(dihydroxyphosphinyl-methyl)phosphinate ester and **109** (X = O) after removal of the isopropylidene group (Fig. 20).[204] It is of interest that some nucleoside triphosphate analogues of this type, for example, **110** are commercially available although no detailed synthetic procedure has been published for these compounds.

Several analogues of nucleoside polyphosphates have been prepared in which the oxygen atom between the 5'-carbon atom of the nucleoside and the α-phosphorus atom of the polyphosphate has been replaced. These include the analogues with NH,[204,216,217] CH$_2$,[218] and CHCN (**111**)[219] groups. The 5'-amino nucleoside polyphosphates can be made by treating the 5'-amino nucleoside with trimetaphosphate or its analogue (**108**). In the latter instance, **109** (X = NH) is formed.[204] Compound **111** was obtained by activation of the related nucleoside monophosphonate with diphenyl phosphorochloridate followed by a reaction of the product with pyrophosphate.[219] Analogues of adenosine polyphosphates in which there is a carbonyl group at the 5'-position (**112**)[220] have been prepared by activating 9-(β-D-ribofuranosyluronic acid)adenine (**113**) with diphenyl phosphorochloridate and coupling the product with ortho- or tripoly

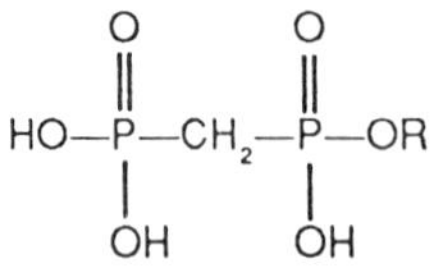

(105)

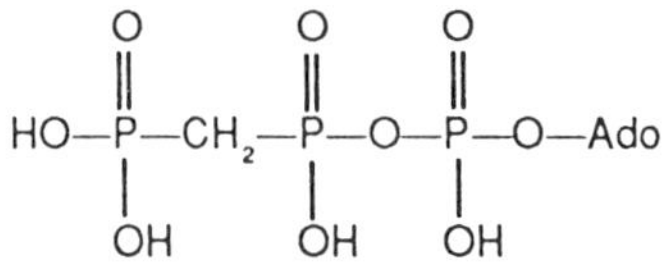

(106)

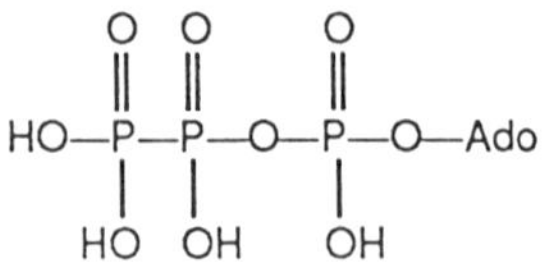

(107)

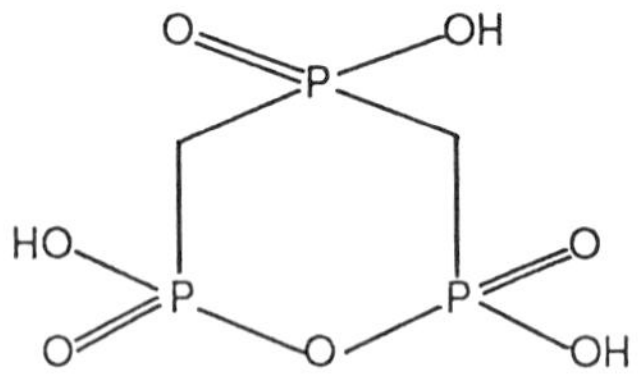

(108)

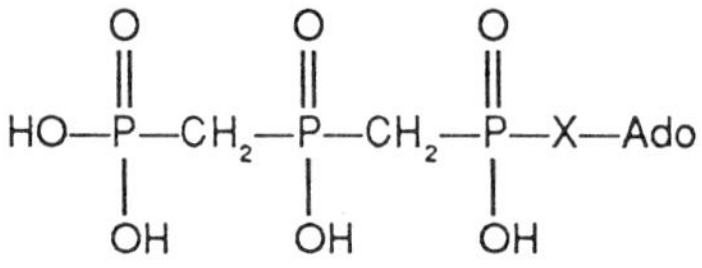

(109)

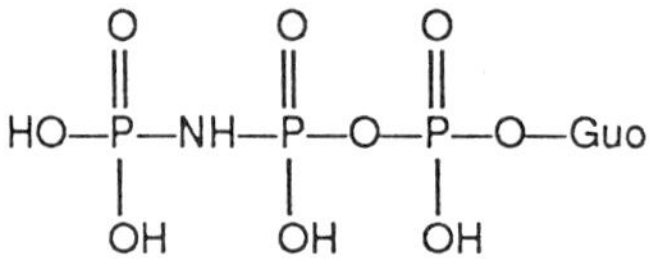

(110)

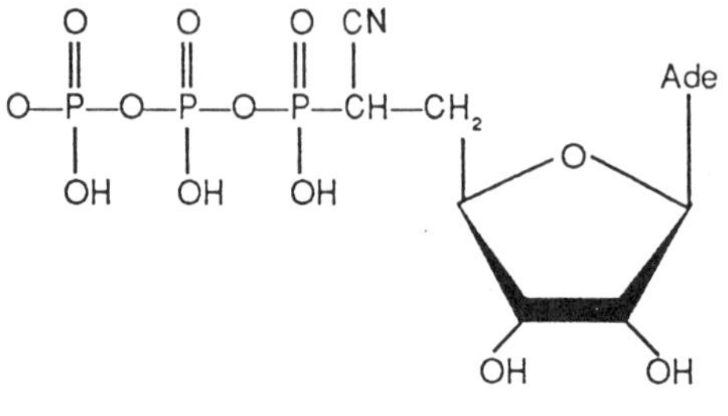

(111)

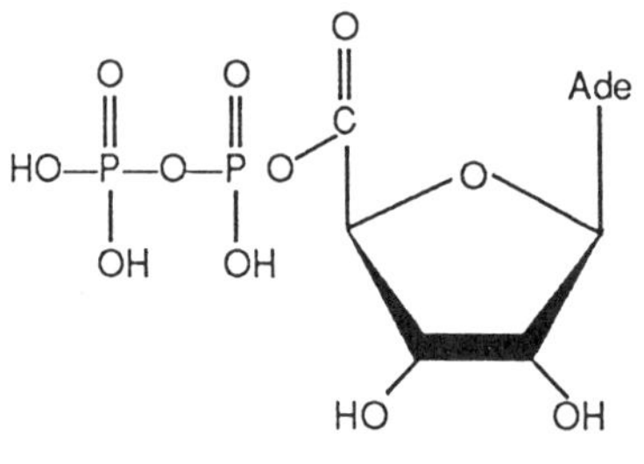

(112)

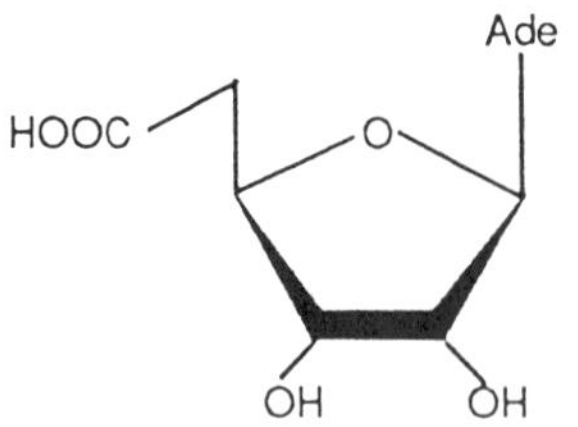

(113)

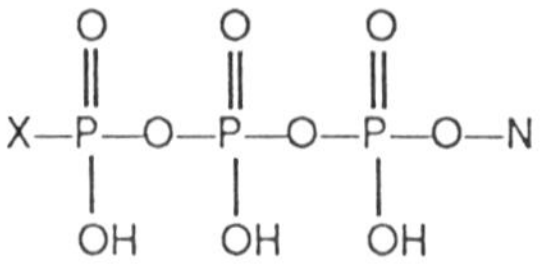

(114)
where N = Ado - 5' or Guo-5'

Figure 20. Nucleoside polyphosphonates and related analogues.

phosphate. These highly reactive carboxylic–phosphoric acid anhydrides have been used to probe the active site of AMP- and ATP-requiring enzymes. γ-Modified ATP and GTP analogues (**114**) have also been used as affinity probes for studying the active sites of enzymes. Two synthetic routes to **114** have been employed. GTP analogues have been prepared by activating GDP or GTP with diphenyl phosphorochloridate and then treating the mixed anhydride with an appropriate nucleophile.[221] Alternatively, cyclization of ATP to the metaphosphate ester (**62**) followed by ring opening with a nucleophile leads to γ-modified ATP.[222] A derivative of ppApU in which the 5′-pyrophosphoryl group is esterified with a 4-azidophenyl residue has been prepared by activating 4-azidophenyl phosphate with carbonyl bis(imidazole) and then coupling the phosphoroimidazolidate with pApU.[223] Carbonyl bis(imidazole) has also been used to prepare ATP analogues which are spin-labeled at the γ-position.[224]

Analogues of ATP in which the terminal or γ-phosphorus atom has been replaced by groups other than phosphorus have also been described. Phosphonoformic and phosphonoacetic acid esters of nucleosides have been prepared and examined as antiviral agents.[225] Condensation of adenosine 5′-phosphoromorpholidate with phosphonoacetic acid leads to the analogue **115** ($n = 1$) (Fig. 21).[226] Interestingly, the analogue **115** with $n = 0$, derived from phosphonoformic acid, cannot be obtained in a similar way and appears to be highly unstable. The -AsO_3H_2 group differs from the -PO_3H_2 group only slightly in shape and size. A number of arsonomethyl and arsonate analogues of AMP, ADP, and ATP have been prepared [**116–118** (Fig. 21)]. The AMP analogue **116** was obtained from the α-bromoacid **119** by treatment with an alkaline solution of arsenite followed by decarboxylation of the product.[227] The ADP analogue **117** was obtained by a conventional route involving the coupling of arsonomethylphosphonic acid with 2′,3′-O-isopropylideneadenosine in the presence of DCC.[228] The ATP analogue **118** can be isolated from the reaction between ADP and arsenate in the presence of a submitochondrial preparation from beef heart.[229] Compound **117** is a poor substrate for ADP-requiring enzymes, but the AMP analogue **116** has some activity as an enzyme substrate.

The enzymatic formation of sulfate esters of phenols and polysaccharides requires a cofactor, adenosine 5′-sulfatophosphate (**120**, X = H) or adenosine 3′-phosphate 5′-sulfatophosphate (**120**, X = PO_3H_2). The latter has been synthesized by treatment of adenosine 2′,3′-cyclic phosphate 5′-phosphate with triethylamine N-sulfonic acid (**121**) and hydrolysis of the intermediate with RNase T_2.[230] Hydrolysis of the intermediate from this reaction with spleen phosphodiesterase leads to the 2′-isomer of **120** (X = PO_3H_2). This route is simpler than an earlier synthesis in which adenosine 2′,3′-cyclic phosphate 5′-phosphoromorpholidate was coupled with pyridine N-sulfonic acid.[231] An enzymatic synthesis for **120** (X = PO_3H_2) using an extract from *Chlorella* has been published.[232]

2.5. Biosynthetic Pathways

The biosynthetic routes to purine and pyrimidine nucleotides have been known for several years and are described in detail in a number of texts.[6] In the biosynthesis of inosine 5′-phosphate, which is shown in Fig. 22, the purine ring is

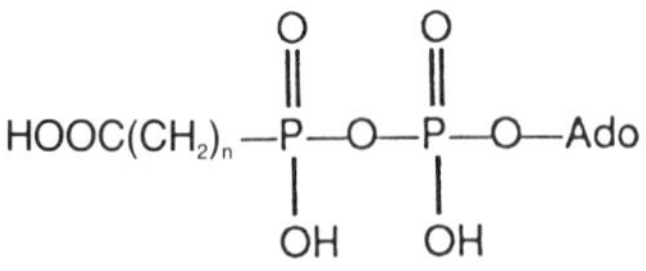

(115)

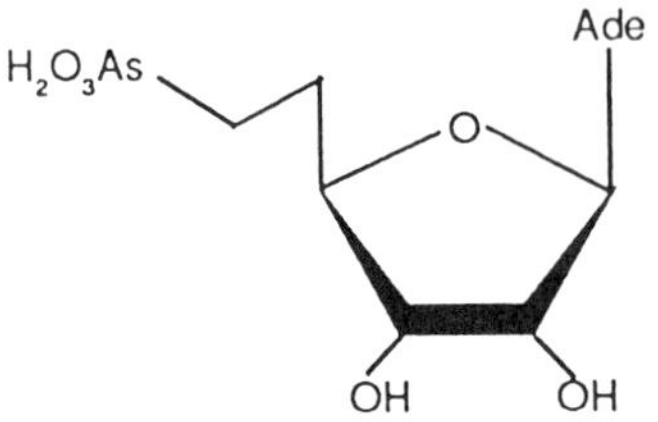

(116)

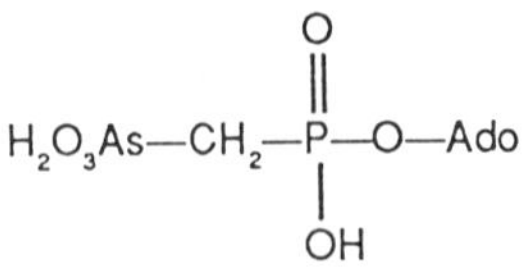

(117)

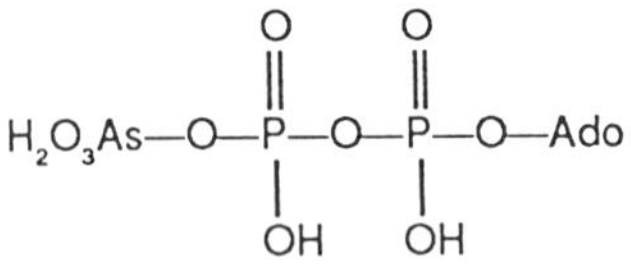

(118)

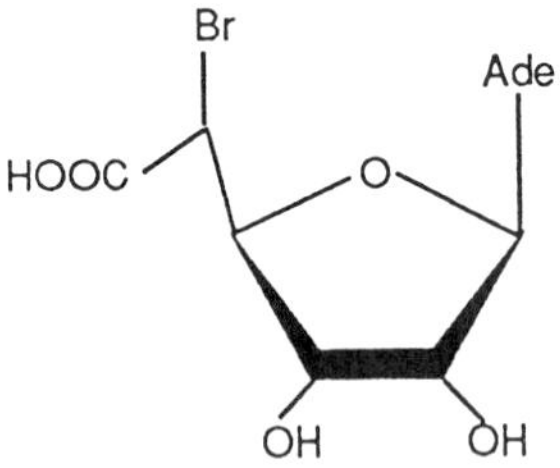

(119)

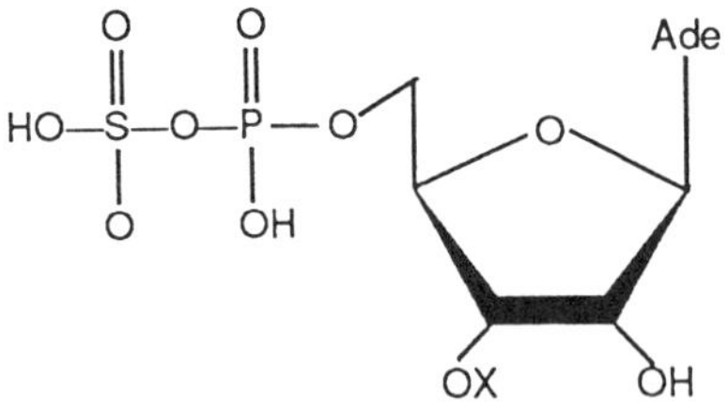

(120)

Et$_3$NSO$_3$H

(121)

Figure 21. Nucleotide analogues containing carbon, arsenic, and sulfur in the phosphoryl moiety.

built onto the ribose 5-phosphate residue. The initial step in this biosynthesis is the direct transfer of a pyrophosphoryl residue from ATP to the 1-position of ribose 5-phosphate. This is unusual as reactions of ATP leading to direct transfer of a pyrophosphoryl residue to a substrate are uncommon. Displacement of the pyrophosphoryl residue from ribose 5-phosphate 1-pyrophosphate (PRPP) (**122**) by an amino group is followed by the construction of the imidazole ring of the purine residue. This is followed by the addition of the pyrimidine ring, leading to IMP. Modification of the hypoxanthine ring in IMP affords AMP and GMP (Fig. 23). A key step in the biosynthesis of pyrimidine nucleosides

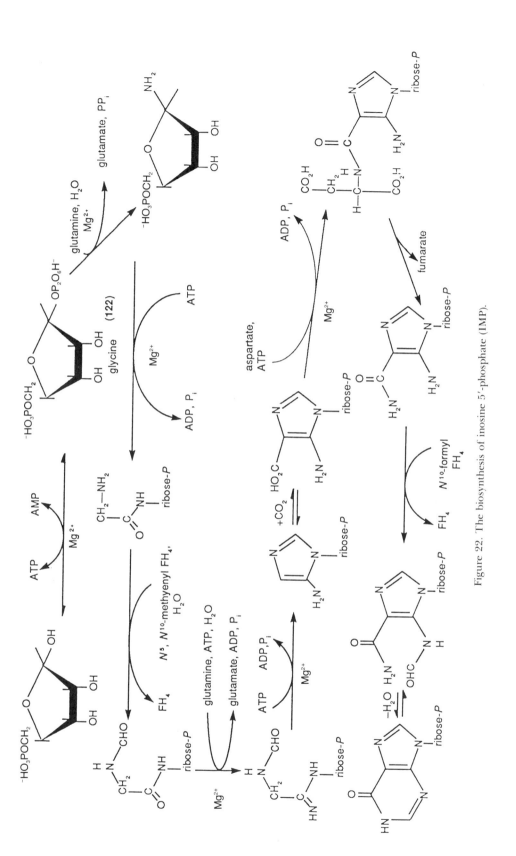

Figure 22. The biosynthesis of inosine 5'-phosphate (IMP).

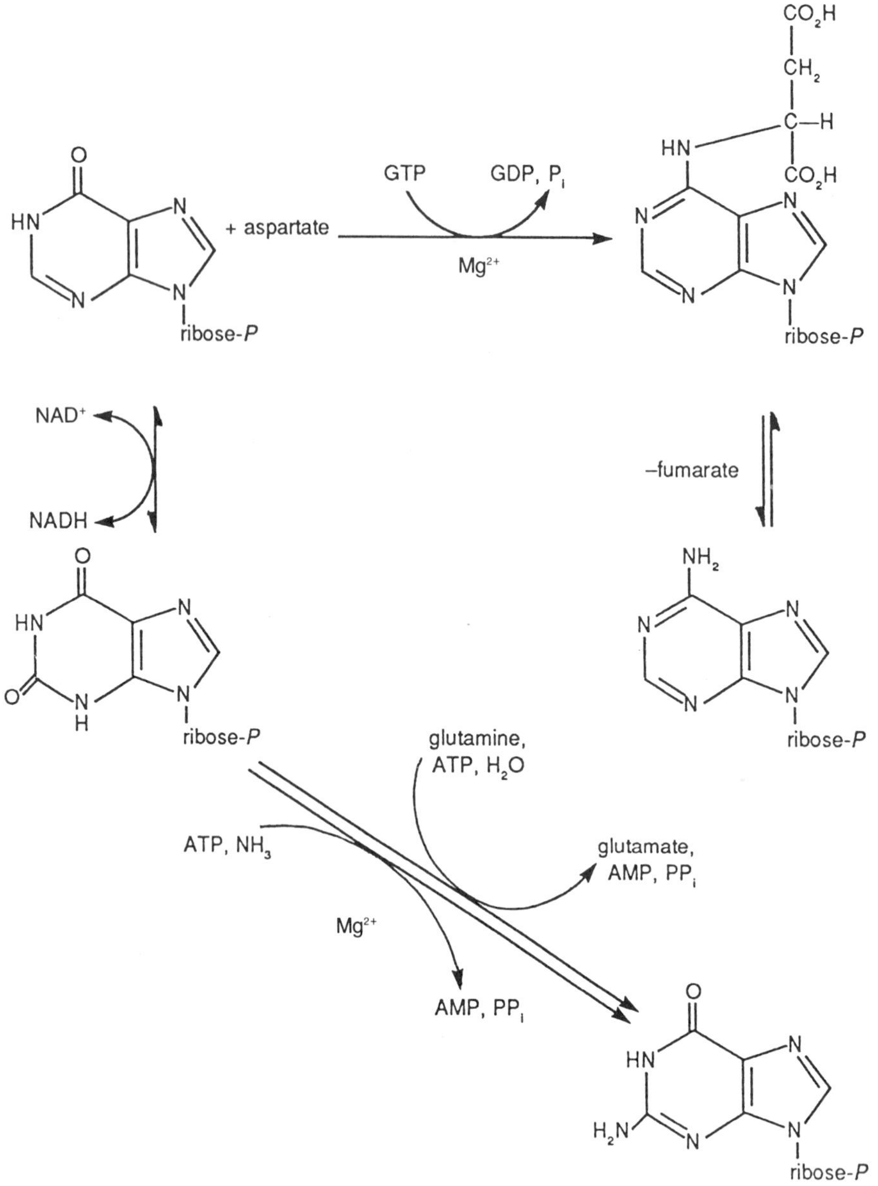

Figure 23. The biosynthetic conversion of IMP into AMP and GMP.

also involves **122**, which loses pyrophosphate following attack by orotic acid to generate **123** (Fig. 24). Decarboxylation of **123** leads to UMP, which is further phosphorylated to UTP and aminated to CTP. Dephosphorylation of the latter to CDP and reduction at the 2'-position generates dCDP, which is hydrolyzed to dCMP and deaminated to dUMP. This deoxyribonucleotide is converted into dTMP by thymidylate synthetase. Another enzyme exists to convert ribonucleo-

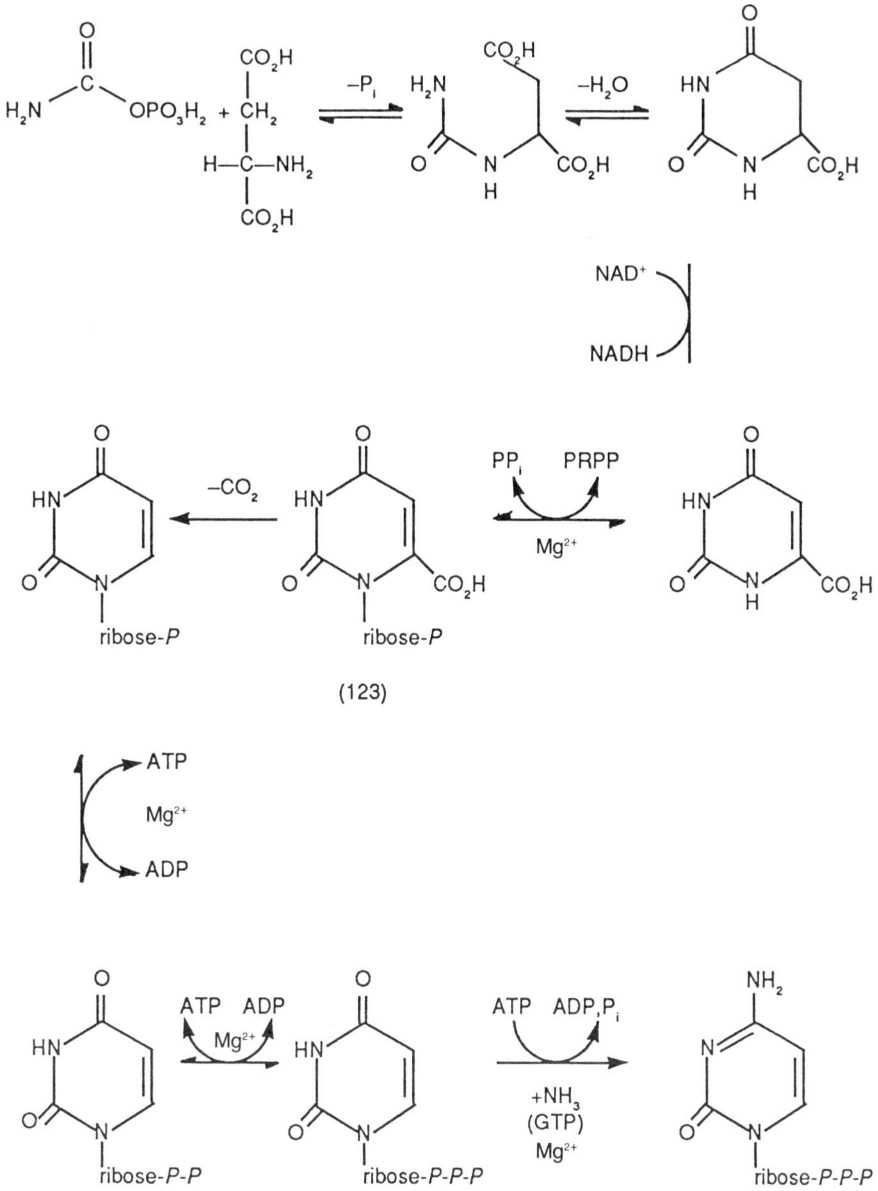

Figure 24. The biosynthesis of pyrimidine nucleotides.

side triphosphates to their 2′-deoxy counterparts. In addition to the above pathways, a "salvage" mechanism exists whereby purine or pyrimidine bases can be converted to the mononucleotide by displacement of pyrophosphate from **122**, for example,

$$\text{purine} + \textbf{122} \leftrightarrows \text{purine ribonucleotide} + PP_i$$

All biosynthetic pathways leading to nucleoside mono- and polyphosphates involve ATP. At neutral pH, ATP exists as the tetraanion, and all enzymatic reactions which involve ATP normally require the presence of divalent metal ions (usually Mg^{2+}). The hydrolysis of ATP is exergonic, and hydrolysis of the β,γ-phosphoryl bridge to ADP and inorganic phosphate has $\Delta G^0 = -7.3$ kcal/mol (-30.6 kJ/mol),[233] while hydrolysis of the α,β-phosphoryl bridge to AMP and pyrophosphate has $\Delta G^0 = -7.7$ kcal/mol (-32 kJ/mol). This exergonic cleavage of the triphosphate residue in ATP has led to the latter being termed "energy-rich," which is a rather misleading term. However, many biochemical reactions take advantage of the energy release that accompanies the cleavage of the triphosphate moiety in ATP. The catalysis of ATP hydrolysis by metal ions will be considered later in this chapter (Section 3.4).

There are three possible pathways for cleavage of the triphosphate residue in ATP (Fig. 25). The vast majority of enzymatic reactions involve attack at the α- or the γ-phosphorus atoms of ATP (Table IV).[5] The number of enzymatic reactions in which a pyrophosphoryl residue is transferred from ATP as a single entity is very limited, despite the favorable energy release which accompanies cleavage of the α,β-phosphoryl bridge in ATP. One reason for the preferential attack at the α- and γ-phosphorus atoms of ATP may be metal complexation. As mentioned above, all reactions which involve ATP also require metal ions which, presumably, can complex with the triphosphate chain of ATP. If the metal ion were to complex with the β- and γ-phosphoryl groups of the triphosphate, a stable metal chelate with a six-membered ring (**124**) would be formed. Attack on the α-phosphorus atom by a nucleophile would cleave the α,β-phosphoryl bridge and release inorganic pyrophosphate in the form of the metal complex, which should make this a good leaving group. Similarly, complex formation between a metal ion and the α- and β-phosphoryl groups would lead to another chelate with a stable six-membered ring (**125**). Attack by a nucleophile on the γ-phosphorus atom would then result in a cleavage of the β,γ-phosphoryl bridge and the expulsion of ADP in the form of a metal complex. Again, the formation of such a metal complex should make ADP a better leaving group than in the uncomplexed form. The formation of a metal ion complex between the α- and the γ-phosphoryl groups, leaving the β-phosphoryl group available for attack by a nucleophile, would result in the formation of a comparatively unstable complex with the metal ion in an eight-membered ring (**126**). Furthermore, attack by a nucleophile on the central phosphorus atom of the triphosphate chain could not be followed by the expulsion of a stable metal complex of ADP or inorganic pyrophosphate. A further factor in assisting reactions which involve the transfer of an adenyl residue from ATP to a substrate with concomitant release of inorganic pyrophosphate is the existence in all living matter of pyrophosphatases which cleave one molecule of pyrophosphate into two molecules of orthophosphate. This hydrolysis of pyrophosphate is exergonic, and this process means that the transfer of an adenyl residue to a substrate is essentially irreversible. The overall transfer reaction can be written:

$$ATP + X \leftrightarrows AMP—X + PP_i \rightarrow AMP—X + 2P_i$$

As may be inferred from Section 2.4.3, ATP can be formed in a large number of reactions, and this fact has been exploited by many workers for the

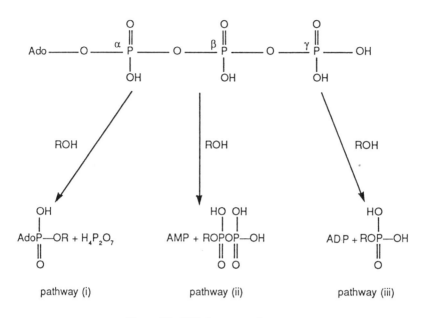

Figure 25. ATP cleavage pathways.

(124)

(125)

(126)

Table IV. Some Enzymatic Phosphoryl and Nucleotidyl Transfer Reactions Involving Nucleoside Polyphosphate Esters[a]

Phosphoryl transfer	Nucleotidyl transfer
ATP:D-fructose 1-phosphotransferase	ATP:NMN$^+$ adenylyltransferase
ATP:riboflavin 5'-phospotrasferase	ATP:FMN adenylyltransferase
ATP:AMP phosphotransferase (adenylate kinase)	ATP:α-D-glucose 1-phosphate uridylyltransferase
ATP:creatine phosphotransferase	GTP:α-D-mannose 1-phosphate guanylyltransferase
ATP:acetate phosphotransferase	
ATP:pyruvate phosphotransferase (pyruvate kinase)	CTP:choline phosphate cytidylyltransferase
ATP:thiamine pyrophosphotransferase	Nucleoside diphosphate:polynucleotide nucleotidyltransferase (polynucleotide phosphorylase)
ATP:D-ribose 5-phosphate pyrophosphotransferase	
ATP:hydroxymethyldihydropterin pyrophosphotransferase	Nucleoside triphosphate:RNA nucleotidyltransferase (RNA polymerase)
ATP:guanosine 5'-pyrophosphatase pyrophosphotransferase	Deoxynucleoside triphosphate:DNA nucleotidyltransferase (DNA polymerase)

[a]Ref. 5.

synthesis of oxygen-labeled and chiral ATP. These have been used to establish the stereochemistry of phosphoryl (or nucleotidyl) transfer from nucleoside polyphosphates. One additional class of phosphorylation reactions that involve ATP for which the stereochemistry of phosphoryl transfer is frequently unknown is the phosphorylation or pyrophosphorylation of hydroxyl groups prior to the formal elimination of water. Hydroxide ion is a poor leaving group in elimination reactions. On the other hand, if a hydroxyl group is mono- or pyrophosphorylated, then the anion of a strong acid (ortho- or pyrophosphoric acid) will be expelled during the elimination reaction, thus rendering this reaction energetically more favorable. One example of many such reactions is the formation of isopentenyl pyrophosphate (**127**) (Fig. 26).[234] Mevalonate (**128**), a molecule containing six carbon atoms, undergoes three successive phosphoryl transfers from ATP to yield the triphosphate **129**. Decarboxylation of **129**, followed by the loss of phosphate from the intermediate, leads to **127**, which contains five carbon atoms. This compound is the building block of terpenes and steroids. During the joining of terpenoid fragments, pyrophosphate is eliminated at every prenyl transfer step when a new carbon–carbon bond is formed. The biosynthesis of CTP from UTP and glutamine is an additional example of the activation by phosphorylation of an oxygen atom prior to its elimination. Phosphorylation of the enolic form of UTP leads to the phosphate **130**, which then undergoes attack by the amido function of the glutamine. Phosphate is eliminated from the intermediate **131**, and CTP is formed.[235]

The biosynthesis of nucleotide coenzymes occurs in a standard manner. Usually, an adenyl (or other nucleotidyl residue) is transferred from the nucleoside triphosphate to a monophosphorylated moiety, for example, FMN, NMN$^+$, or a sugar phosphate. Thus, the biosynthesis of FAD occurs in two steps. First, riboflavin kinase catalyzes the conversion of riboflavin into FMN, and then FMN-adenyl transferase converts this into FAD.[236]

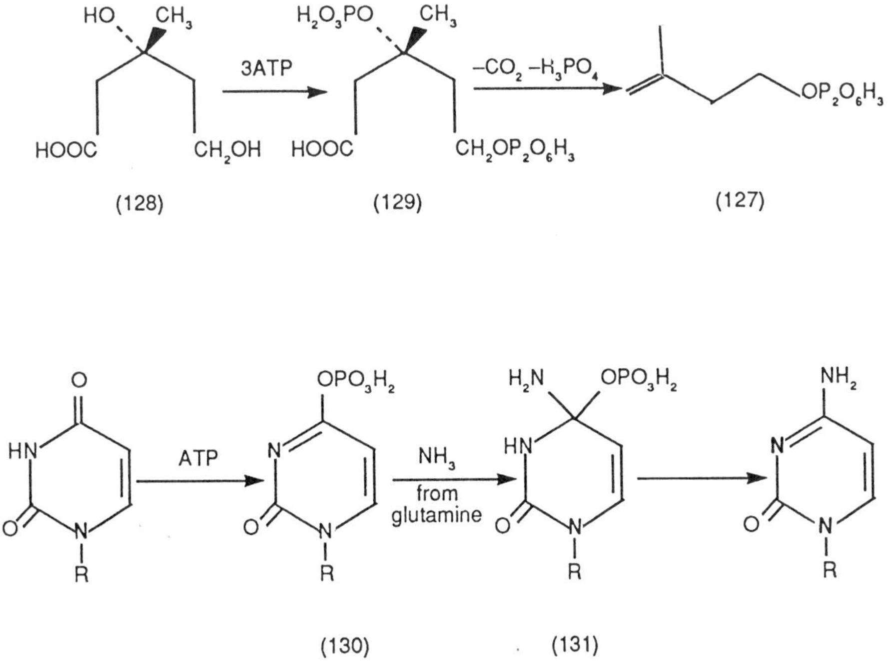

where R = β –D– ribofuranosyl   5'-triphosphate

Figure 26. Biosynthetic pathways to isopentenyl pyrophosphate and CTP.

$$\text{Riboflavin} + \text{ATP} \rightarrow \text{FMN} + \text{ADP}$$

$$\text{FMN} + \text{ATP} \rightarrow \text{FAD} + \text{PP}_i$$

Ribose 5-phosphate 1-pyrophosphate (**122**) is the phosphoribosyl donor in the synthesis of NMN$^+$ (**80**), and **122** reacts with nicotinamide to produce the mononucleotide.[237] an adenyl transfer reaction converts the latter to NAD$^+$, which can then be phosphorylated, if required, by ATP to NADP$^+$.[238]

$$\text{Nicotinamide} + \mathbf{122} \rightarrow \text{NMN}^+ + \text{PP}_i$$

$$\text{NMN}^+ + \text{ATP} \rightarrow \text{NAD}^+ + \text{PP}_i$$

3. Chemical Properties and Reactions of Nucleotides and Their Analogues

3.1. Analytical Methods

Nucleoside mono- and polyphosphates are acidic, water-soluble compounds, and this governs which techniques can be used for their analysis. Paper[239] and thin-layer[240] chromatography are useful techniques for the analysis of simple mixtures but are not always satisfactory for the analysis of complex mixtures owing to poor resolving power. High-performance liquid chromatography (HPLC) is being increasingly used for such analyses in view of the excellent resolution which can be obtained with this technique. Early investigators used ion-exchange HPLC for the separation of nucleotide mixtures,[241] but reversed-phase HPLC is now the common means of separating nucleosides and nucleotides.[242,243] One advantage of this technique is that volatile buffers can be employed, making simple the recovery of samples from column eluates.[244] HPLC can also be used to separate mixtures of nucleotides, even when the latter bear polyphosphate chains.[245] Methods for the separation of nucleotides by the use of ion-exchange resins and polysaccharides are well established and are described in standard texts.[246] Gas–liquid chromatography, however, has been little used for the separation of nucleotides and related compounds owing to volatility problems although these can be overcome to a certain extent by derivatization.[247]

The analysis by mass spectrometry of mixtures which contain nucleotides is also beset by problems of volatility, necessitating extensive derivatization of the compounds being analyzed. The mass spectrometry of phosphorus compounds, including nucleotides, has been well reviewed by Chapman[248] and others.[249] They have covered the literature up to the period 1983–84. The major advance since that time has been the development of sputtering techniques to overcome the problem of low volatility, particularly the application of field desorption (FD) and fast atom bombardment (FAB) to this problem. Improvements in liquid chromatography–mass spectrometry, for example, the use of "thermospray" injection techniques, are another area in which progress has recently been made. Although the thermospray method has mainly been used to identify nucleosides rather than their phosphorylated derivatives.[250] FAB mass spectrometry offers many advantages over FD mass spectrometry, particularly with respect to sample manipulation and the ease of obtaining reproducible spectra. It is to be expected that FAB mass spectrometry will emerge as an important technique for the analysis of nucleotide mixtures. FAB spectra of trimethylsilyl nucleotides,[251] mono- and oligonucleotides,[252] and NAD$^+$[253] have already been published.

3.2. ^{31}P Nuclear Magnetic Resonance Studies

The identification of nucleoside mono- and polyphosphates by ^{31}P-NMR spectroscopy is very well established and is described in texts.[254] This technique has assumed great importance in the determination of the stereochemical arrangement of groups around four-coordinate phosphorus as the chemical shift of the phosphorus resonance is dependent on the mass and magnetic nature of the oxygen atoms to which the phosphorus atom is attached.[100,186] The ^{17}O atom is paramagnetic ($I = \frac{5}{2}$), and when a ^{17}O atom is directly bonded to phosphorus, marked broadening of the phosphorus resonance occurs due to quadrupolar relaxation. Thus, signals due to phosphorus bonded to ^{17}O are, in effect, removed from the ^{31}P-NMR spectrum, leaving signals due to phosphorus bonded only to ^{16}O and ^{18}O. Substitution of ^{16}O bonded to phosphorus by the heavier isotope ^{18}O causes the ^{31}P signal to move upfield, and the magnitude of this shift is related to the fractional difference in mass between the two oxygen isotopes as well as the number of ^{16}O and ^{18}O atoms bonded to phosphorus.[100,255,256] Signals due to all the species in orthophosphate bonded to ^{16}O and ^{18}O can be distinguished, namely, $P^{16}O_4$, $P^{16}O_3{}^{18}O$, $P^{16}O_2{}^{18}O_2$, $P^{16}O^{18}O_3$, and $P^{18}O_4$. The intensities of these signals are related to the amounts of each species present.

Phosphoryl derivatives that bear $^{16}O^{17}O^{18}O$ and sulfur have been extensively used to study the stereochemistry of phosphoryl transfer reactions. In general, oxygen-chiral phosphate and polyphosphate esters are preferable to phosphorothioates for the investigation of enzymatic reactions. The replacement of phosphoryl oxygen by sulfur can lead to a considerable decrease in the rate of the reaction, and in some cases the enzyme does not appear to accept one of the diastereomers of the phosphorothioate. Phosphorothioates do not occur naturally, and it may be that the stereochemistry of an enzymatic phosphoryl transfer reaction that has been deduced using a phosphorothioate is not the same as when a natural substrate is used. However, at least six enzymatic phosphoryl transfer reactions have been shown to have the same stereochemistry whether phosphate or phosphorothioate derivatives are used. Hence, the assumption that phosphorothioates can be used to study enzymatic phosphoryl transfer reactions appears to be valid.[93] However, the development of methods for the stereochemical analysis of oxygen-chiral phosphomonoesters is of extreme importance. The stereochemical analysis of $[^{16}O,^{17}O,^{18}O]$phosphomonoesters has been achieved with the aid of chiral propane-1,2-diol or related compounds.[257] Alkaline phosphatase from *E. coli* will catalyze the transfer of a chiral phosphoryl group from a phosphomonoester in a stereospecific manner to (S)-propane-1,2-diol to give, for example, **132** (Fig. 27). Treatment of **132** with diphenyl phosphorochloridate or -imidazolidate results in ring closure, and three isomers (**133–135**) of the cyclic phosphodiester are formed. Treatment of this mixture of isomers with an excess of diazomethane leads to the formation of six cyclic phosphotriesters. Quadrupolar line broadening of the ^{31}P-NMR signals derived from phosphorus atoms bonded to ^{17}O results in only two (**136** and **137**) of the six isomers being detectable by ^{31}P-NMR spectroscopy. Based on the known isotopic content of the starting material, the intensities of the lines in the ^{31}P-NMR spectra of **136** and **137** can be predicted and compared with those found experimentally. Thus, the stereochemistry of the original phosphomonoester can be deduced. A similar method for the stereochemical analysis of chiral $[^{16}O,^{17}O,^{18}O]$thiophosphates has been devised using (S)-2-iodo-1-phenylethanol (**138**).[258] Alternatively, the six cyclic phosphotriesters obtained from compounds **133–135** can be separated by HPLC and their structures deduced by metastable-ion mass spectrometry.[255] The stereochemical courses of some enzymatic reactions of phosphodiesters are shown in Table V.

3.3. Hydrolysis of Phosphates and Polyphosphates

As mentioned briefly in Section 2.1, there are two main mechanisms for nucleophilic substitution at phosphorus—associative and dissociative. The dissociative mechanism involving metaphosphate is rarely encountered although evidence for this process has been obtained on several occasions.[9,10,11,259] Three

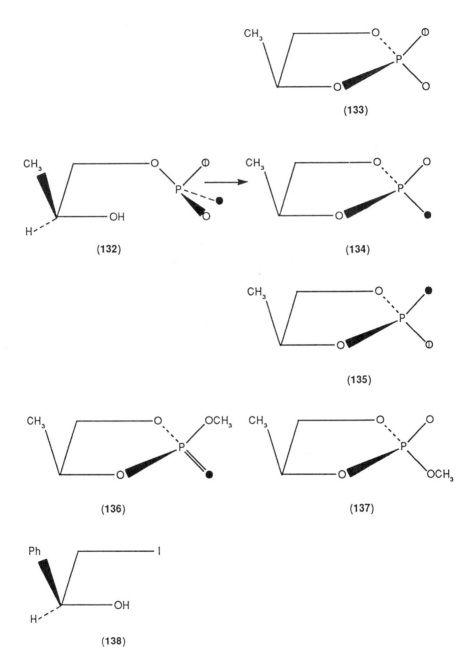

Figure 27. Stereochemical analysis of intramolecular phosphoryl transfer reactions.

Table V. Stereochemical Course of Some Enzymatic
Reactions of Phosphodiesters as Studied
by the Phosphorothioate (A) and Oxygen Isotope (B) Methods[a]

Enzyme	Result	Method
Ribonuclease A		
Transesterification	Inversion	A
Hydrolysis	Inversion	A
Ribonuclease T_1	Inversion	A
Ribonuclease T_2	Inversion	A
DNA-dependent RNA polymerase		
Initiation	Inversion	A
Elongation	Inversion	A
DNA-dependent DNA polymerase	Inversion	A
t-RNA nucleotidyltransferase	Inversion	A
RNA ligase, ligation step	Inversion	A
UDP-glucose pyrophosphorylase	Inversion	A
Galactose-1-phosphate uridylyltransferase	Retention	A
Acetyl-CoA synthetase	Inversion	A
Adenylate cyclase	Inversion	A, B
Bacterial		
Mammalian	Inversion	A
Enterobacter aerogenes phosphohydrolase	Inversion	A
Bovine intestine phosphodiesterase	Retention	A
Aminoacyl-tRNA synthetase	Inversion	A
Snake venom phosphodiesterase	Retention	A, B
Cyclic phosphodiesterase	Inversion	A, B
Phosphofructokinase[b]	Inversion	B
Phosphoglucomutase[c]	Retention	A
Glucose 6-phosphatase[d]	Retention	B

[a]Adapted from ref. 262.
[b]Ref. 264.
[c]Ref. 265.
[d]Ref. 266.

associative mechanisms can be imagined which differ in the occurrence and mode of decomposition of pentacovalent intermediates. Involvement of the latter in the hydrolysis of cyclic phosphodiesters was first demonstrated by Westheimer,[260] who defined the rules followed in the pseudorotation and reaction of these intermediates. The main rules to be observed are (a) the preferred addition and expulsion of groups from apical positions, (b) the preference of electronegative groups to adopt apical positions, and (c) the tendency of cyclic systems to bridge apical and equatorial positions in the trigonal bipyramid.

The first associative mechanism is an S_N2 reaction in which the displacing nucleophile attacks the phosphorus atom on the side opposite the leaving group (Scheme 1 in Fig. 28). Inversion of configuration would be a compulsory result of such a reaction. The other two associative mechanisms involve the formation of pentacovalent intermediates and differ in whether these intermediates undergo pseudorotation or not. In one mechanism (Scheme 2 in Fig. 28), the displacing nucleophile again attacks the phosphorus atom from the side opposite the

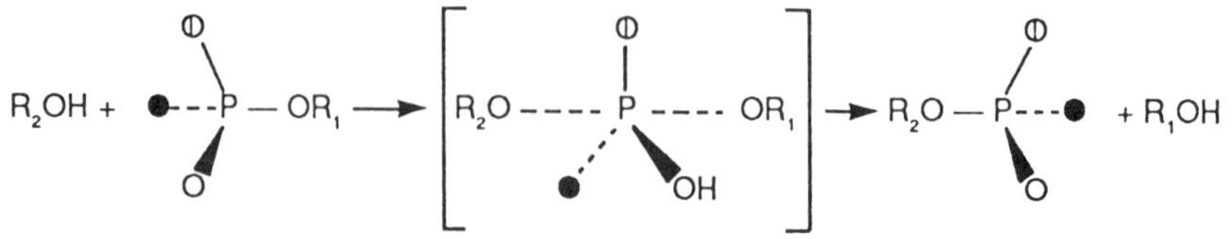

Scheme 1

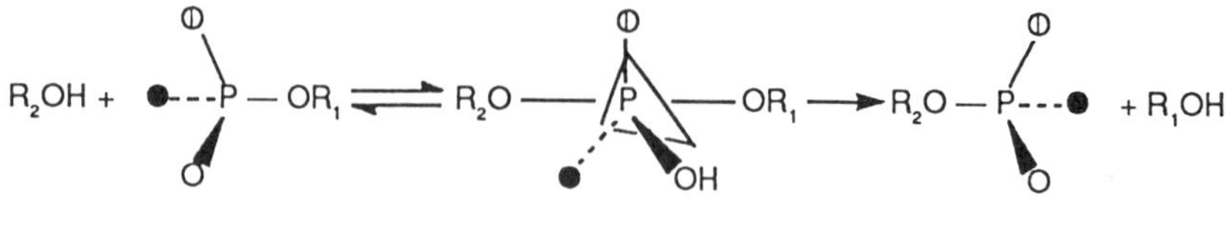

Scheme 2   (139)

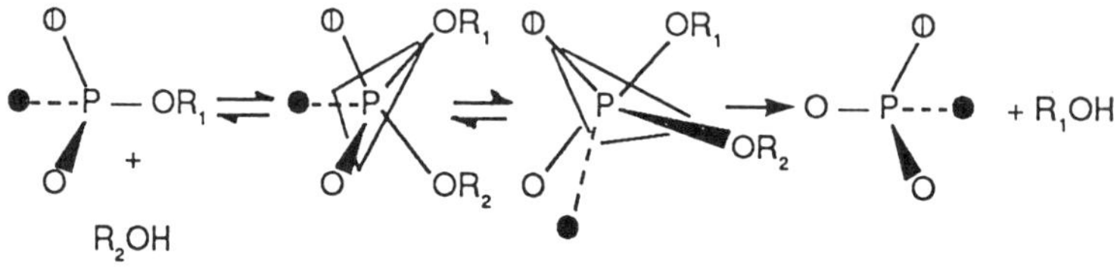

Scheme 3   (140)   (141)

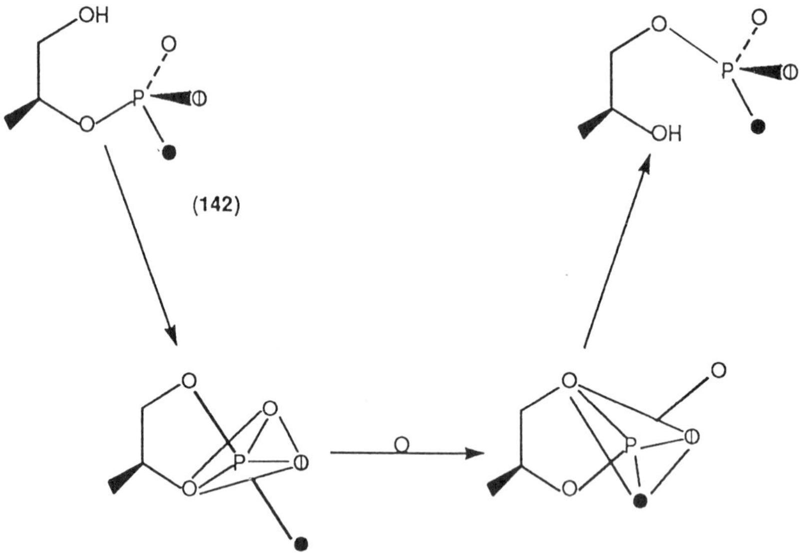

Scheme 4

Figure 28. Stereochemistry of associative mechanisms for nucleophilic displacement at phosphorus.

leaving group. The resulting pentacovalent intermediate (**139**) formed by addition of the nucleophile to the phosphomonoester has the nucleophile and the leaving group in apical positions and the other three substituents on phosphorus in equatorial positions. Decomposition of **139** results in expulsion of the leaving group and inversion of configuration at phosphorus. In the other associative mechanism (Scheme 3 in Fig. 28), the attacking nucleophile is in the apical position in the intermediate (**140**), but the leaving group is in an equatorial position. As groups must leave pentacovalent intermediates from apical positions if possible, pseudorotation of **140** occurs to give **141** before the leaving group is expelled. The net result of this process is retention of configuration at phosphorus.

The phosphate group in ribonucleoside $2'(3')$-phosphates can migrate readily under acidic conditions to the adjacent hydroxyl group. A study, using the methods outlined above to follow the stereochemistry at phosphorus, of the isomerization of the model system 2-[(R)-$^{16}O,^{17}O,^{18}O$]phospho(S)-propane-1,2-diol (**142**) under acidic conditions showed that the reaction proceeds with retention of configuration at phosphorus and that at least two pentacovalent intermediates which are interconvertible by pseudorotation are probably involved.[261] In this case, the entering nucleophile is constrained to attack the phosphorus atom "adjacent" to the leaving group rather than by an "in-line" mechanism (Scheme 4 in Fig. 28). Presumably, the isomerization of ribonucleoside $2'(3')$-phosphates follows a similar route.

The stereochemical course of a large number of enzymatic phosphoryl transfer reactions has been determined (Table V).[262,263] The simplest interpretation of the results obtained so far is that all these phosphoryl transfer reactions occur by "in-line" displacement mechanisms (Scheme 1 or Scheme 2 in Fig. 28). It should be noted here that it is very difficult, if not impossible, to distinguish between these two mechanistic possibilities in enzymatic reactions. When direct displacement of a leaving group by a nucleophile occurs, inversion of configuration results. Where retention of configuration at phosphorus is observed, it may be postulated that a phosphoryl-enzyme intermediate participates in the reaction. In effect, two "in-line" displacement reactions occur, the first from substrate to enzyme, the second from phosphorylated enzyme to attacking nucleophile, leading to an overall retention of configuration. The postulate of a two-stage reaction leading to retention of configuration can often be confirmed by kinetic and other data.[266] The only authenticated example of an "adjacent" mechanism elucidated for a reaction which may have importance in a biological system is that described above when steric constraints force the cyclization of phosphorylated propanediol to follow this route.

Investigations on the nonenzymatic hydrolysis of ATP have been summarized.[10] For the acid $ATPH_4$ and the monoanion $ATPH_3^-$, hydrolysis appears to occur by an addition–elimination route involving pentacovalent intermediates, that is, by an associative process, whereas for the tetraanion ATP^{4-} and the trianion $ATPH^{3-}$ the hydrolysis appears to take place via a metaphosphate, that is, by a dissociative process. The mechanism of hydrolysis of the dianion $ATPH_2^{2-}$ is not so clear-cut, and both routes may be involved. These results were obtained largely from kinetic evidence as studies with oxygen-labeled ATP do not give results that can be readily interpreted when the hydrolytic reactions are carried out in water. Nonenzymatic phosphoryl transfer reactions from ATP in nonaqueous media appear to involve metaphosphate, as has already been mentioned (Section 2.1). Detailed studies on the mechanisms of nonenzymatic phosphoryl transfer reactions from other nucleoside polyphosphates have not been made.

3.4. Metal Complexes of Nucleotides

Divalent metal ions play an important part in many of the enzymatic reactions of nucleoside mono- and polyphosphates that involve the transfer of a phosphoryl or nucleotidyl group to a substrate. Consequently, the properties of metal complexes of nucleotides have attracted attention for some years. The metal ion may play a part in (a) charge neutralization, (b) polarization, (c) strain induction, (d) the alignment of reactants at the active site of an enzyme, or (e) coordination with the leaving group. Several of these factors may come into play during an enzymatic reaction but may not be of importance in model hydrolytic reactions in the absence of a protein. For example, metal ions have a profound influence on the enzymatic reactions of ATP, but the effect of metal ions on the hydrolysis of ATP to ADP in the absence of an enzyme is comparatively small and the hydrolysis of nucleoside 5'-diphosphates is little affected by metal ions.[267] The formation of complexes between metal ions and nucleoside di- and triphosphates has been widely studied both in solution and by crystallographic methods.[268] From a study of the dephosphorylation, in the presence of divalent metal ions, of ATP and other nucleoside triphosphates to be corresponding nucleoside diphosphates, it has been concluded that a labile γ-phosphoryl group can be achieved if a metal ion is coordinated to the α- and β-phosphoryl groups.[269] Migration of the metal ion along the polyphosphate backbone could lead to the formation of a complex between the β- and the γ-phosphoryl groups and would make the molecule susceptible to nucleophilic attack at the α-phosphorus atom. Alternatively, formation of a complex (**143**) between the nucleoside triphosphate and two metal ions may also result in enhanced reactivity toward nucleophilic attack at the γ-phosphorus atom (Fig. 29). With purine nucleoside triphosphates, interaction can also occur between the metal ion and N-7 of the purine base. However, interaction between metal ions and the base in pyrimidine nucleoside triphosphates is very weak. This observation may be of importance in connection with the initiation of RNA synthesis by the DNA-dependent RNA polymerase from *E. coli* in that it is observed that initiation only occurs with ATP or GTP and not with CTP or UTP. The metal ion may play a role in discriminating between the nucleoside triphosphates and orient the first nucleotide in a specific manner that allows nucleic acid synthesis to take place.[270] It is also of interest in this context that metal ions catalyze the dephosphorylation of ATP to a greater extent than that of pyrimidine nucleoside triphosphates under the same conditions. This suggests that base–metal interactions may play a part in this reaction.[271] An example of a metal ion-promoted dephosphorylation of ATP is the zinc ion-catalyzed phosphorylation of 2-hydroxymethyl-1,10-phenanthroline.[272] The phosphoryl transfer is absolutely de-

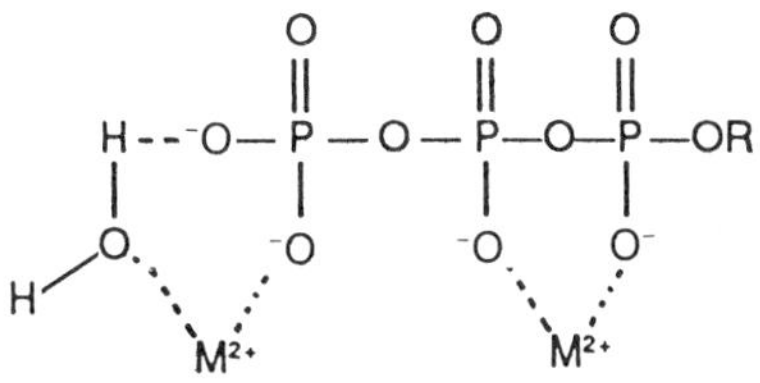

(143)

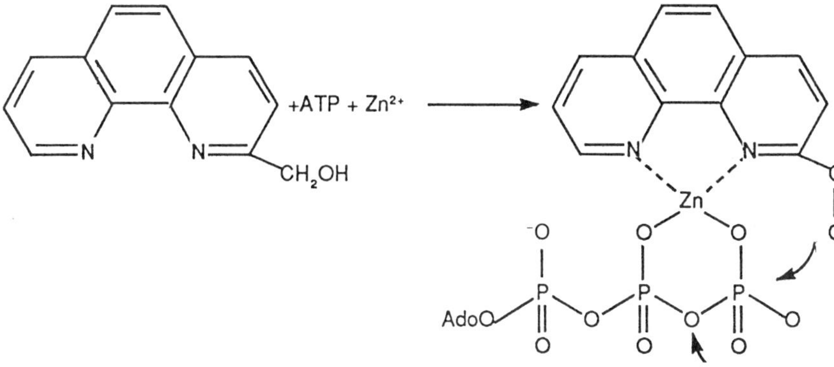

(144)

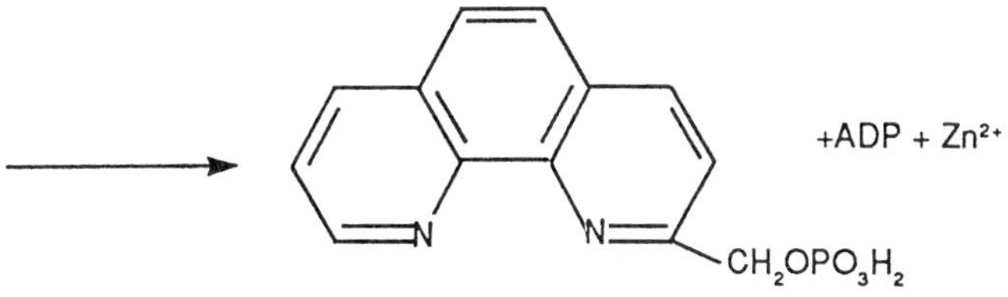

Figure 29. Metal ion-catalyzed reaction of ATP.

pendent on the present of zinc ions and presumably proceeds through a ternary complex (144) involving the phenanthroline, ATP, and a zinc ion (Fig. 29). Once 144 is formed, the stereochemistry of the complex may bring the primary alcohol function of the phenanthroline close to the α-phosphoryl group of the ATP, and hence phosphorylation is assisted. It has been suggested that this reaction could be a model for many biological phosphoryl transfer reactions from ATP.

Although metal ions do not accelerate the hydrolysis of nucleoside diphosphates to an appreciable extent, they can promote their self-association by coordinating with the phosphoryl residues and reducing charge repulsion. In this way, base stacking can be facilitated. Alternatively, some metal ions, for example, zinc, can enhance the stability of dimers by forming an intermolecular link between a phosphoryl residue in one nucleoside diphosphate and the next.[273]

It is difficult to study the coordination isomers of complexes formed between ATP and magnesium(II) ions by NMR spectroscopy as they equilibrate more rapidly than the NMR time scale. However, chromium(III) ions form exchange-inert complexes with ATP, and these complexes can be substrates for enzymatic reactions. ATP and chromium(III) ions can form mono- and tridentate complexes, but when ATP and chromium(III) chloride are allowed to react at pH 5.7 and 4°C, β,γ-bidentate Cr(III)–ATP complexes are obtained.[274] Four stereoisomers can be separated from the reaction mixture, two possessing a left-hand screw configuration (Λ) at the β-phosphate (**145**) and two possessing a right-hand screw configuration (Δ) at the β-phosphate (**146**) (Fig. 30). The two pairs of isomers appear to be due to the α-phosphoryl group adopting either a pseudo-axial or a pseudo-equatorial position in the complex (e.g., **147** and **148** for the Λ-isomer). Hexokinase, glycerokinase, creatine kinase, and arginine kinase will only accept the Λ-conformers of the bidentate Cr(III)–ATP complex, whereas pyruvate kinase, myokinase, and fructose-6-phosphate kinase will only accept the Δ-conformers of the bidentate complex.[275] This suggests that there are steric differences at the active sites of these two groups of enzymes. ATP(βS) diastereomers can also form Λ- and Δ-conformers with metal ions.[276] In this case, the "hardness" or "softness" of the metal ion determines the structure of the complex. For example, a "soft" metal ion such as cadmium(II) will bind preferentially to sulfur at the β-phosphorus atom. The R_p-isomer of ATP(βS) will confer a right-handed screw sense to the direction of the triphosphate chain, giving the Λ-isomer (**149**). A "hard" metal ion such as magnesium(II) will bind preferentially at oxygen at the β-phosphorus atom, giving the Δ-isomer (**150**). By analogy, the S_p isomer of ATP(βS) will give **151** and **152**. Using these guidelines, the substrate specificities of a number of ATP-requiring enzymes have been determined. For example, creatine kinase binds the $\Lambda,\beta\gamma$-bidentate chelate obtained from ATP and a magnesium(II) ion. During the reaction, a $\Lambda,\beta\gamma$-bidentate magnesium(II)–ADP complex is formed, but it is not clear whether the change in coordination occurs before or after the loss of the γ-phosphoryl residue from ATP.[277] The formation of Δ- and Λ-isomers of cobalt(III) complexes has been used to assign the stereochemistry at the α-phosphorus atom of [α-^{18}O]dADP.[278] Incubation of $[Co(NH_3)_4]^+$ ions with dADP, under conditions already established for the cobalt complexes of ADP, leads to the formation of the $\alpha\beta$-bidentate complexes, which could be separated into the Δ- and Λ-isomers by chromatography on cross-linked cycloheptaamylose. If the oxygen atom of a P—O bond is complexed to cobalt, the P—O bond order is reduced, and this can be detected by changes in the ^{31}P-NMR spectra. Thus, the stereochemistry of the various complexes can be determined and correlated with the stereochemistry of oxygen substitution at the α-phosphorus atom.

In a method for determining the interaction of ^{16}O and ^{17}O at phosphorus of [β-^{17}O]ATP with manganese(II) ions at the active site of the enzyme, the effect of the ^{17}O on the ESR spectrum of the manganese was studied.[279] If the [^{17}O]ATP forms a chelate with the manganese(II) ion at the active site of an enzyme, superhyperfine coupling between the nuclear spin of the ^{17}O and the paramagnetic manganese(II) ion produces line broadening in the ESR spectrum of the latter. It has been shown that (R_p)-[β-^{17}O]ATP, and not (S_p)-[β-^{17}O]ATP, gives rise to line broadening in the ESR spectrum of manganese(II)

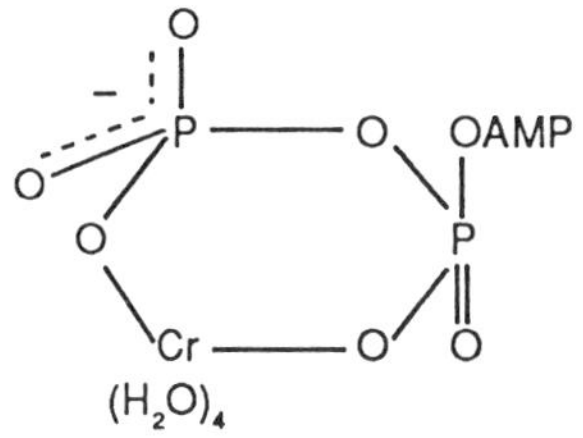

(145)

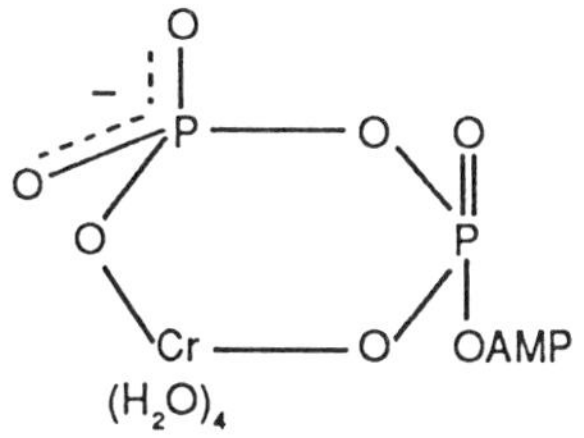

(146)

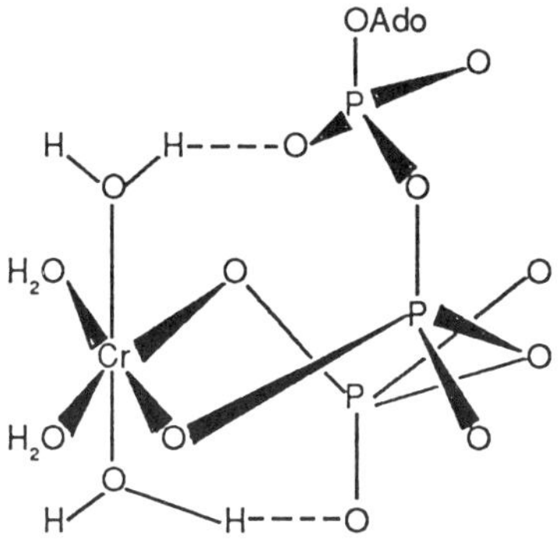

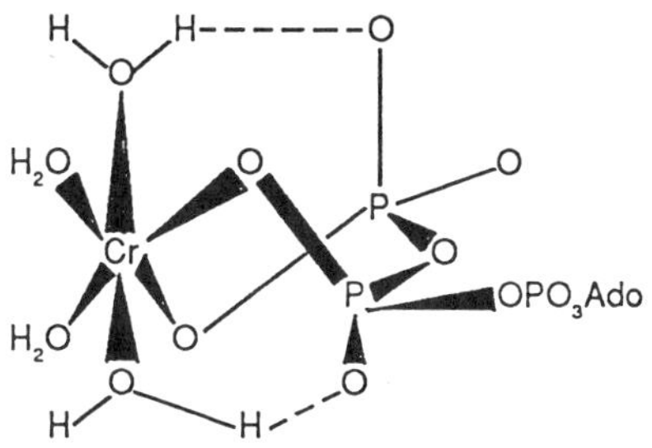

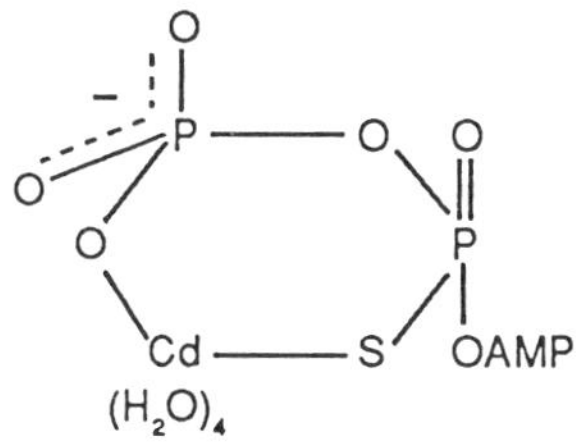

(149)

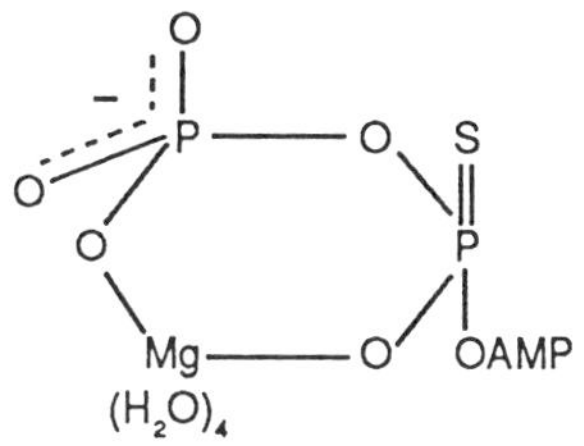

(150)

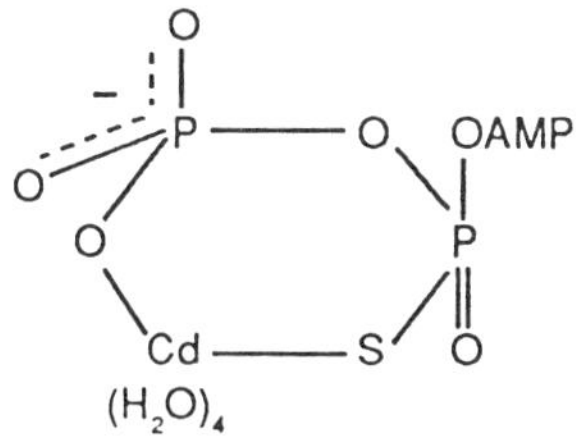

(151)

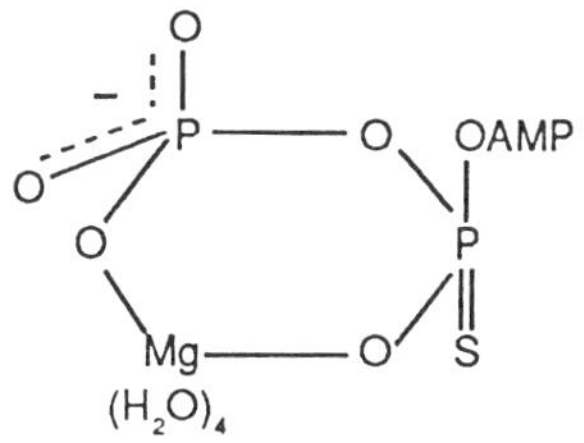

(152)

Figure 30. Bidentate complexes of ATP with metal ions.

ions with porcine liver adenylate kinase. This would indicate that the metal ion is complexed with the *pro-R* oxygen atom of ATP. The perturbation by manganese(II) ions of the longitudinal relaxation rates of the ^{31}P nuclei in the Δ-diastereomer of the $\beta\gamma$-bidentate Co(NH$_3$)$_4$ATP which binds to the (Na$^+$ + K$^+$)-ATPase from kidney has been used to determine the manganese(II)–phosphorus distances for all the phosphorus atoms in the ATP complex.[280]

The structures and properties of nucleotide analogues which have modified phosphoryl residues, other than phosphorothioates, have been little studied. ^{31}P-NMR spectroscopy has, however, been used to study the binding of the [α,β]- and [β,γ]-methylene analogues of ATP to magnesium[281] and to study the binding of a number of isopolar ATP and GTP analogues to divalent metal ions.[208,282]

The crystal structures of a large number of nucleotide–metal ion complexes have been determined,[283] but little effort has been made to correlate the structural information obtained in this manner in the solid state with the biochemical properties of these complexes in aqueous solution.

It has been suggested that metal ions can act as catalysts in a number of prebiotic condensation reactions, and the observation that lead(II) ions will catalyze the condensation of adenosine 5'-phosphoroimidazolidate to oligoadenylic acids supports this view.[284] In this case, however, the thermodynamically more stable 2'–5' link predominates. Any 3'–5' linked oligomers can readily be removed by digestion of the reaction products with nuclease P$_1$, and hence this is a simple route for the synthesis of 2'–5'-linked oligoadenylic acids. The 3'-hydroxyl appears to play a part in this reaction as the 5'-phosphoroimidazolidate of 3'-deoxyadenosine will not oligomerize in the presence of lead(II) ions under comparable conditions. Other metal ions can catalyze the oligomerization of nucleoside 5'-phosphoroimidazolidates, and, for example, zinc ions will catalyze the condensation of guanosine 5'-phosphoroimidazolidate. The efficiency of this reaction is greatly improved by the addition of a poly(C) template to the reaction, and in this case the oligomers are predominantly 3'–5' linked.[285]

3.5. Nucleotide Analogues as Affinity Labels

The technique of affinity labeling involves binding a close analogue of a natural ligand or substrate to an enzyme or other macromolecule. If the analogue possesses a reactive group, a covalent bond may be formed between it and the macromolecule, leading to loss of biochemical activity. Furthermore, if the analogue is radioactive, degradation of the covalently linked ligand–macromolecule complex leads to an oligomer bearing the radioactive label. Further degradation of this oligomer allows, for instance, the amino acid sequence around a binding site in a protein to be determined. The successful use of an affinity label depends on the presence of a suitable amino acid side chain at or very near the binding site and an affinity label (or active-site-directed irreversible inhibitor) which should ideally react only at or near the active site of the macromolecule since it should possess the same features which facilitate the tight binding of the natural ligand. In practice, the introduction of a bulky reactive group into the ligand can have an adverse effect on its binding. Two main classes of affinity label can be envisaged: (a) those which contain electrophilic groups that can react directly with electron-rich amino acid side chains and (b) those which contain functional groups that can be activated once binding has taken place. For example, photoactivation of an azido group in a ligand after binding has taken place can lead to the formation of a nitrene, which can then react with electron-rich centers in the macromolecule.

Much of the early work on the use of nucleotide analogues as affinity labels has been compiled in a volume of *Methods of Enzymology*.[286] Oxidation of a ribonucleoside-5′ di- or triphosphate with periodate is a simple method for the generation of reactive intermediates that can be used as affinity labels (Fig. 31). With ATP, the dialdehyde **153** (R = $P_3O_9H_4$) is formed on oxidation with periodate, and the latter can react with an amino group in the active site of an enzyme, possibly by a two-step reaction involving the formation of **154** (R = $P_3O_9H_4$) and then **155** (R = $P_3O_9H_4$). ADP reacts similarly. However, caution must be exercised with these reagents; for example, **153** (R = $P_3O_9H_4$) loses its tripolyphosphate moiety on standing at neutral pH at room temperature.[287] The ADP analogue **153** (R = $P_2O_7H_3$) was found to lose pyrophosphate after binding to a protein.[288]

General methods for the preparation of γ-modified ATP derivatives,[221–223] suitable for use as affinity labels, have been described in Section 2.4.5. Phosphoramidates can also be employed for these syntheses, and adenosine 5′-phosphoromorpholidate has been coupled with a series of ω-(5-nitrobenzimidazol-2-yl) phosphates to give **156** (R = NO_2). Catalytic reduction of the nitro group in these compounds followed by diazotization gave **156** (R = N_2^+). The latter compound bound covalently to the NAD$^+$ binding site of dehydrogenases.[289] In a similar manner, uridine 5′-phosphoromorpholidate was coupled with 4-nitrophenyl phosphate to produce **157** (R^1 = H, R^2 = NO_2). A catalytic reduction of the nitro group was followed by acylation of the amino group with the bromoacetyl ester of *N*-hydroxysuccinimide to give **157** (R^1 = H, R^2 = $NHCOCH_2Br$) (Fig. 31).[290] This derivative was an affinity label for UDP-galactose-4-epimerase from *E. coli.*

A major disadvantage in the use of electrophilic affinity labels is that they are often too reactive, especially as enzymatic reactions are carried out in 55 *M* water. Thus, inactivation of the reagent by solvent water can occur before it reaches the protein. Alternatively, intramolecular decomposition may occur (see compound **153** above). In complex biological systems, a reaction with other components such as cell walls may also occur. These problems can be largely overcome with photogenerated affinity labels (photoaffinity labels) as they are activated after they have been bound to the macromolecule. The classes of reagents commonly employed as photoaffinity labels are carbenes (generated from diazo compounds) and nitrenes (generated from azido compounds). Of the two, nitrenes have been the most used because of their lower reactivity, which makes them less indiscriminate in their reactions with C—H and O—H. Furthermore, their precursors (azides) are easier to synthesize than diazo compounds. Thus, 8-azidoadenosine nucleotides can be prepared by the action of azide ion on 8-bromoadenosine nucleotide.[291] These derivatives, together with the 2-azidoadenosine nucleotides,[292] have been used extensively as photoaffinity labels. Nucleoside 5′-polyphosphate esters of 4-azidophenol have been popular

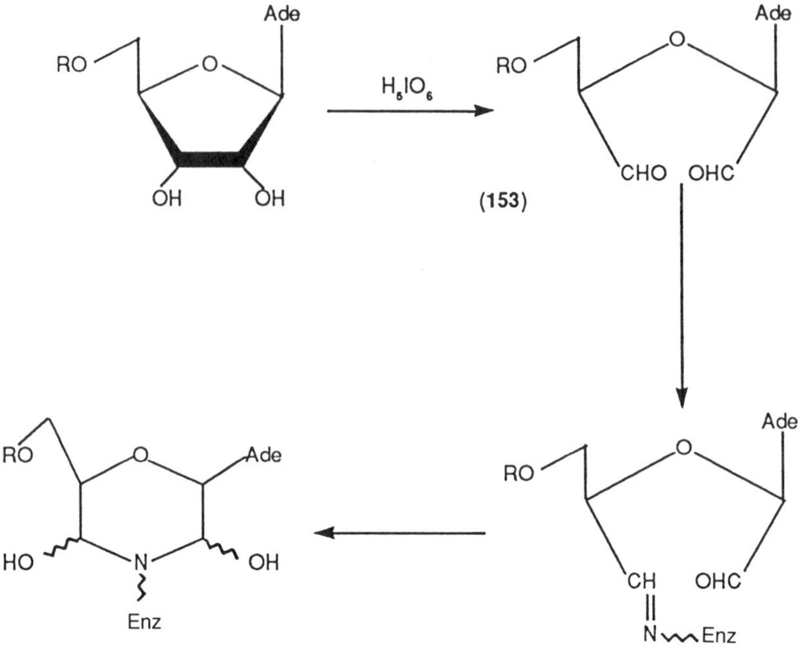

(153)

(155)

(154)

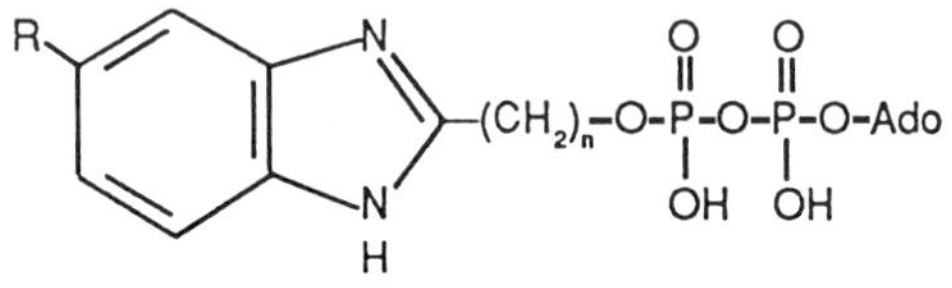

(156)

where n = 2, 4 or 6

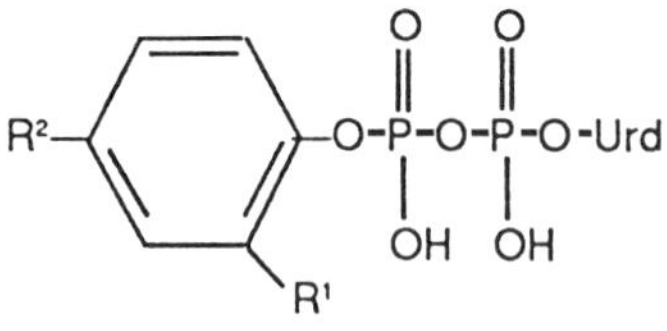

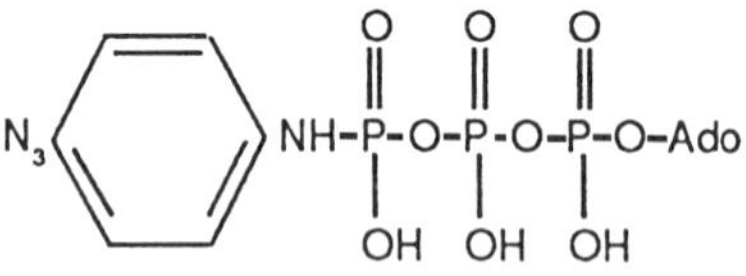

(157)

(158)

Figure 31. Nucleotides as affinity labels.

photoaffinity reagents. These can be synthesized by standard procedures. For example, condensation of uridine 5'-phosphoromorpholidate with 4-azido-2-nitrophenyl phosphate leads to **157** (R^1 = NO_2, R^2 = N_3), a photoaffinity analogue of UDP-galactose which has been used to label the galactosyl transferase subunit of lactose synthase.[293] The γ-4-azidoanilidate of ATP (**158**), obtained by the action of the amine on adenosine trimetaphosphate, has been used to label the leucyl-tRNA synthetases from *Euglena gracilis*.[294] Unmodified nucleotides, for example, CDP, can be used as photoaffinity labels, but in these cases care must be used to avoid radiation damage of the macromolecule as radiation of shorter wavelength than that used with azidonucleotides must be used to activate the cytosine residue.[295]

4. Addendum

Since this manuscript was prepared in 1986, perhaps the major development in the field of nucleotide analogues has been the routine use of commercially available DNA synthesizers (e.g., from Applied Biosystems or Milligen) for the synthesis of oligodeoxynucleotides. The chemistry used for these syntheses is largely that devised by Caruthers and co-workers [296] although H-phosphonates are being increasingly used.[297] The organic chemistry employed in DNA synthesis has been reviewed[298] as has the synthesis of oligonucleotides that contain modified internucleotide links and chemically reactive groups.[299]

Not only naturally occurring oligomers but also those containing phosphorothioate or methylphosphonate linkages are now readily accessible, and the availability of oligonucleotides of defined sequence has transformed many areas of molecular biology. Thus, "antisense" oligo(deoxynucleotides) and oligo(deoxythionucleotides) have been prepared and investigated as inhibitors of the replication of human immunodeficiency virus.[300] With the appropriate use of protecting groups for 2'-hydroxyl groups, the solid-state method has been adapted for the synthesis of oligo(ribonucleotides).[301] With the use of phosphite methodology, 3'-thiothymidylyl(3'-5')thymidine (**159**) has been obtained by coupling 5'-protected 3'-thiothymidine with 3'-*O*-thymidine followed by deblocking.[302] The mass spectrometric analysis of impurities encountered in the synthesis of nucleoside 3'-phosphoramidites has been reported.[303] Oligo(deoxynucleotides) with alternating 3'(N)$\rightarrow$5'P phosphoramidate rather than phosphodiester links have been prepared by the carbodiimide-mediated condensation of $pG_{NH_2}pC_{NH_2}$ (**160**) or $pC_{NH_2}pG_{NH_2}$.[304] Related oligo(deoxynucleotides) with phosphoramidate links, for example, **161** can act as templates for this carbodiimide-mediated condensation.[305]

Isotopically labeled nucleoside polyphosphates have continued to be extensively investigated as probes of the stereochemistry of biochemical and chemical phosphoryl transfer reactions. A rapid method for preparing α-thio analogues of ribo- and deoxyribonucleoside di- and triphosphates, which starts from thiophosphoryl chloride and the unprotected nucleoside, has been developed.[306] When (S_P)-adenosine 5'-[β-$^{16}O,^{17}O,^{18}O$]diphosphate was heated with 2-*O*-benzyl-(*S*)-propane-1,2-diol, the phosphorylated propanediol produced was

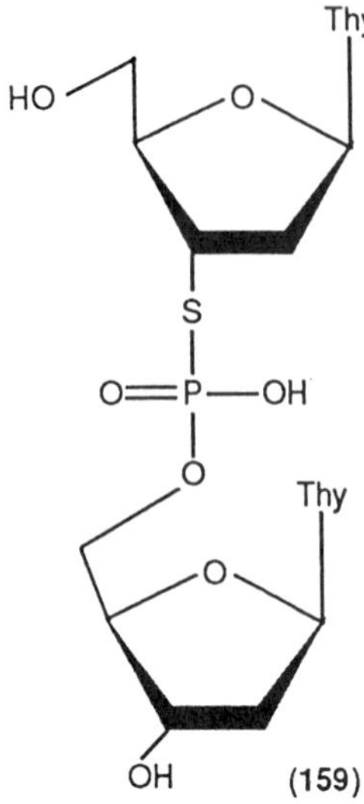

(159)

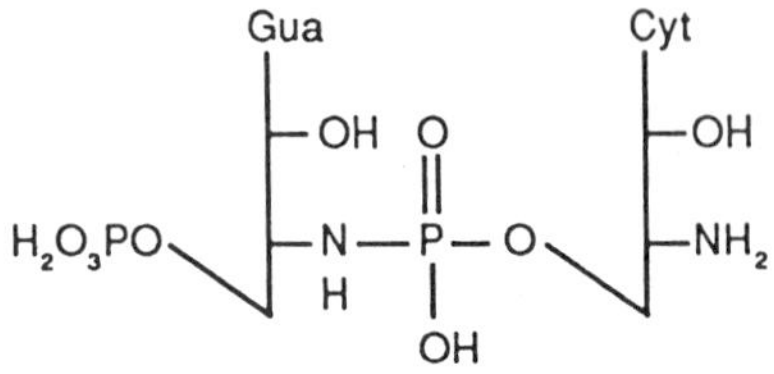

(160)

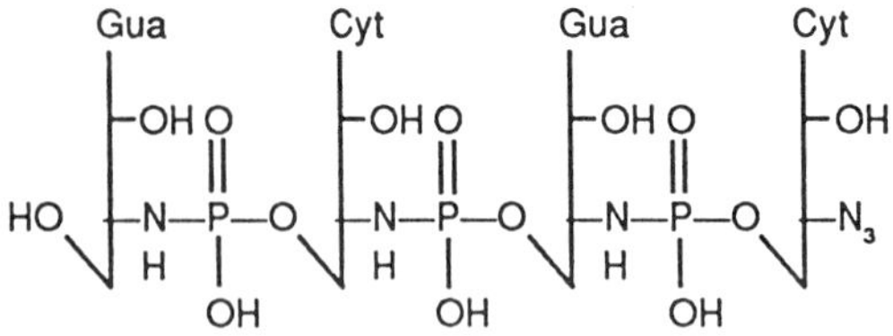

(161)

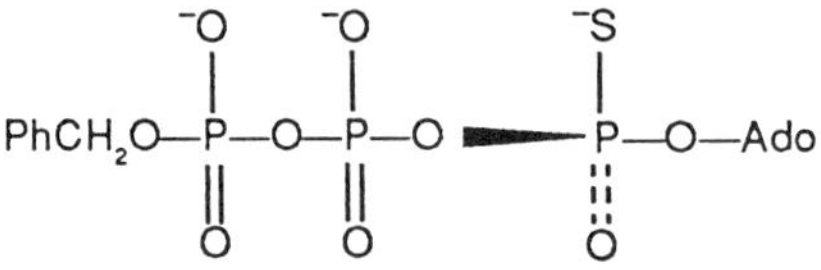

(162)

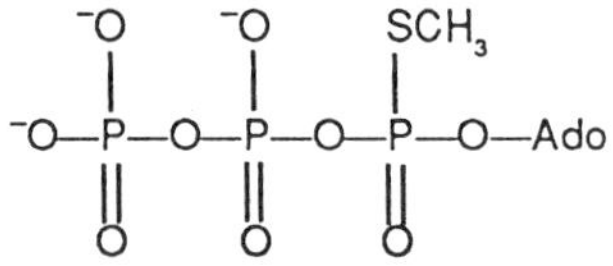

(163)

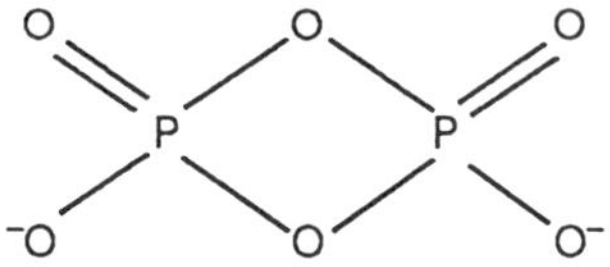

(164)

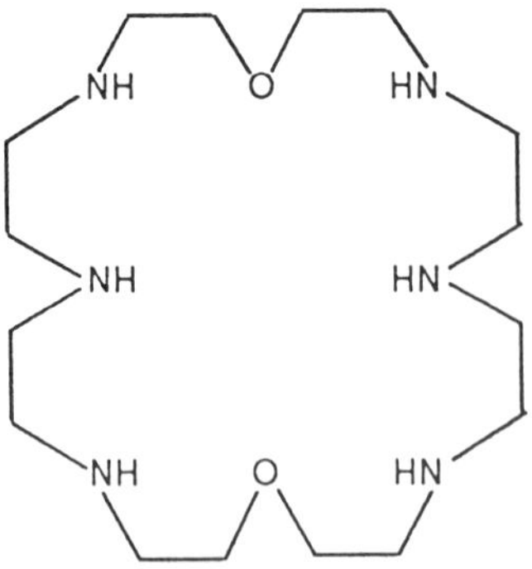

(165)

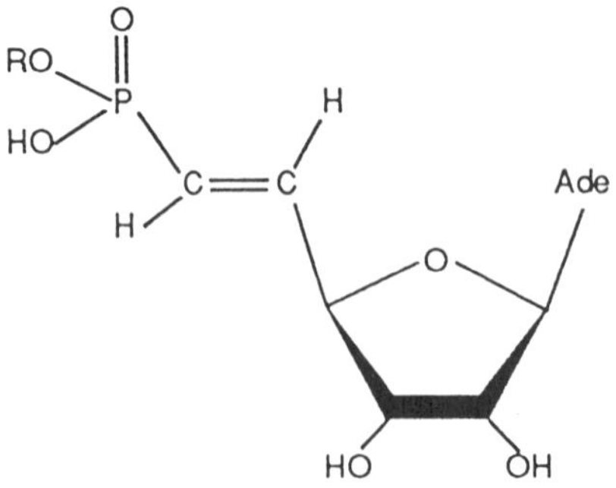

(166)

racemic at phosphorus.[307] Since neither starting material nor the product racemized under the conditions of the experiment, this observation suggests either that free metaphosphate was involved in the phosphoryl transfer or that an alternative acceptor (perhaps acetonitrile) was phosphorylated with rapid racemization before phosphorylation of the propanediol. It has been observed[308] that the possible involvement of species such as $CH_3CN^+ \cdot PO_3^- \cdot AMP$ makes difficult the unambiguous proof of involvement of monomeric metaphosphate in phosphoryl transfer reactions from adenosine polyphosphates in acetonitrile as solvent. On the other hand, phosphoryl transfer reactions of adenosine $5'$-[α,β-^{18}O]diphosphate trianion in acetonitrile, acetonitrile–*tert*-butanol, and neat *tert*-butanol are all accompanied by closely similar bridge-to-nonbridge scrambling in the reisolated starting material. The simplest interpretation of this scrambling involves the participation of monomeric metaphosphate even in hydroxylic solvents.[309] The mechanism of cyanogen bromide-mediated displacement of sulfur from adenosine $5'$-[(S-α-thio-γ-benzyl]-triphosphate (**162**) appears to involve the γ-phosphate and a cyclotriphosphate,[310] whereas the isotopic distribution in the products of hydrolysis of adenosine $5'$-(S-methyl-1-thio)triphosphate (**163**) suggests the participation of a cyclodiphosphate dianion (**164**).[311] The fragmentation of isotopically labeled uridine $5'$-triphosphate determined by fast atom bombardment tandem mass spectrometry can be used to follow positional isotope exchange in this nucleotide.[312] This technique may have uses as an alternative method to ^{31}P-NMR spectroscopy for studying the fate of oxygen isotopes during phosphoryl transfer reactions.

Isotopically and/or sulfur-labeled adenosine triphosphates have continued to be employed to investigate the stereochemistry of enzymatic phosphoryl transfer reactions. Among the reactions that have been shown to proceed with inversion, presumably by an "in-line" mechanism, are those catalyzed by thymidine kinase induced by herpes simplex virus-1[313] and phenylalanyl-tRNA synthetase.[314] Biotin carboxylase from *E. coli*, in the absence of biotin, will hydrolyze ATP, possibly by first phosphorylating bicarbonate to give a mixed anhydride, which is then hydrolyzed. Thus, the hydrolysis of the terminal phosphoryl residue of ATP occurs with a single inversion at phosphorus.[315] Considerable effort has been expended in the study of the enzymatic reactions of bis($5'$-nucleotidyl) polyphosphates. Both their formation[316] and enzymatic hydrolysis[317] appear to involve "in-line" phosphoryl displacements. Phosphonate analogues of Ap_4A have been prepared in which bridging oxygen atoms have been replaced by, for example, CH_2, CF_2, CHF, CCl_2, or $CHCl$ residues, and their hydrolysis by enzymes from *Artemia*, *E. coli*, and lupins has been studied.[318,319] From the products of these hydrolyses, it appears that the position of the Ap_4A analogue in the active sites of the enzymes can depend on the electronegativity of the phosphonate substituent. Other enzymatic reactions that proceed with inversion at phosphorus include deoxyadenosyl transfer to the antibiotic tobramycin, catalyzed by gentamycin nucleotidyl transferase,[320] and the cyclization of ATP to cAMP which is catalyzed by the adenylate cyclase from *Bordetella pertussis*.[321] The stereochemical course of the reaction catalyzed by galactose-1-phosphate uridylyl transferase has also been determined.[321]

Treatment of 5'-*O*-tosyl purine nucleosides with the tetrabutylammonium salts of the corresponding phosphoryl derivatives in acetonitrile affords the nucleoside 5'-diphosphate, 5'-methylene bisphosphates, or ATP in good yields.[322] Protection of the 2',3'-hydroxyl groups of the sugar residues and the amino groups in guanine and cytosine residues improves the yields still further, but only poor yields of pyrimidine 5'-triphosphates were obtained. Macrocyclic polyamines will catalyze the hydrolysis of nucleoside polyphosphates. Thus, 1,4,7,13,16,19-hexaaza-10,22-dioxocyclotetraeicosane (**165**) can catalyze the hydrolysis of ATP with the production of the phosphorylated macrocycle.[323,324] This macrocycle will cleave ATP and analogues with a readily hydrolyzable link between P_β and P_γ, with loss of the terminal phosphoryl group. On the other hand, ATP analogues with a stable link in this position [e.g., adenylyl 5'-(difluoromethylene bisphosphonate)] are hydrolyzed very slowly with the release of AMP.[325] Thus, the macrocycle is beginning to show selectivity resembling that of an enzyme. Heptakis(2,6-dimethyl)-β-cyclodextrin and heptakis(2,3,6-trimethyl)-β-cyclodextrin can catalyze transphosphorylation reactions between AMP, ADP, and ATP in neutral aqueous solution in the presence of magnesium ions. This reaction appears to resemble that catalyzed by adenylate kinase.[326]

A number of studies on phosphonate analogues of nucleotides have been reported. For example, a vinyl analogue of ATP (**166**, R = $P_2O_7H_3$) has been prepared[327] by the phosphoroimidazolidate route from the monophosphate analogue (**166**, R = H).[328] Compound **166** (R = $P_2O_7H_3$) binds to methionine adenosyl transferase in such a manner as to suggest that C-4' and P-1 are in a *trans* relationship. Esters formed between phosphonoformic and phosphonoacetic acids and 2'-deoxyuridine analogues purine nucleosides have anti-herpes virus activity.[329,330]

5. References

1. P. T'so, in: *Basic Principles in Nucleic Acid Chemistry* (P. T'so, ed.), Vol 1, pp. 1–92, Academic Press, New York (1974).
2. E. Chargaff and J. N. Davidson (eds.), *The Nucleic Acids*, Vols. 1–3, Academic Press, New York (1965); P. A. Levene and L. W. Bass, *Nucleic Acids*, Chemical Catalog. Co. (Tudor), New York (1931).
3. A. Harden and W. J. Young, The influence of phosphates on the fermentation of glucose by yeast extract, *Proc. Chem. Soc.* **21**, 189–190 (1905).
4. K. Lohmann, The pyrophosphate fraction in muscle, *Naturwissenschaften* **17**, 624–625 (1929).
5. D. W. Hutchinson, in: *Comprehensive Organic Chemistry* (D. H. Barton and W. D. Ollis, eds.), Vol. 5, pp. 105–145, Pergamon Press, Oxford (1979).
6. A. M. Michelson, *The Chemistry of Nucleosides and Nucleotides*, Academic Press, New York (1963).
7. W. W. Zorbach and R. S. Tipson (eds.), *Synthetic Procedures in Nucleic Acid Chemistry*, Wiley, New York (1973).
8. L. B. Townsend and R. S. Tipson (eds.), *Nucleic Acid Chemistry*, Wiley, New York (1978).
9. F. H. Westheimer, Monomeric metaphosphates, *Chem. Rev.* **81**, 313–326 (1981).
10. F. Ramirez and J. F. Maracek, in: *Organophosphorus Chemistry* (D. W. Hutchinson and J. A. Miller, eds.), Vol. 12, pp. 142–163, Royal Society of Chemistry, London (1981).
11. P. M. Cullis and A. J. Rous, Stereochemical course of the phosphoryl transfer from adenosine 5'-diphosphates to alcohols in acetonitrile and the possible role of monomeric metaphosphate, *J. Am. Chem. Soc.* **108**, 1298–1300 (1986).

12. D. W. Hutchinson, J. A. Miller, S. Trippett, and B. J. Walker (eds.), *Organophosphorus Chemistry*, Vols. 1–17, Royal Society of Chemistry, London (1970–1986).

13. L. A. Slotin, Current methods of phosphorylation of biological molecules, *Synthesis* **1977**, 737–752.

14. M. Yoshikawa, T. Kato, and T. Takenishi, A novel method for phosphorylation of nucleosides to 5'-nucleotides, *Tetrahedron Lett.* **1967**, 5065–5068.

15. R. S. Goody, A simple and rapid method for the synthesis of nucleoside 5'-monophosphates enriched with ^{17}O or ^{18}O on the phosphate group, *Anal. Biochem.* **119**, 322–324 (1982).

16. T. Sowa and S. Ouchi, The facile synthesis of 5'-nucleotides by the selective phosphorylation of a primary hydroxyl group of nucleosides with phosphoryl chloride, *Bull. Chem. Soc. Jpn.* **48**, 2084–2090 (1975).

17. K. Imai, S. Fujii, K. Takanohashi, Y. Furukawa, T. Masuda, and M. Honjo, Selective phosphorylation of the primary hydroxyl group in nucleosides, *J. Org. Chem.* **34**, 1547–1550 (1969).

18. J. Tomasz and A. Simoncits, On the phosphorylation of guanosine by pyrophosphoryl chloride, *J. Carbohydr. Nucleosides Nucleotides* **2**, 315–325 (1975).

19. A. Simoncits and J. Tomasz, Nucleoside 5'-phosphoramidates, synthesis and some properties, *Nucleic Acids Res.* **2**, 1223–1233 (1975).

20. J. H. van Boom, J. F. M. de Rooy, and C. B. Reese, Preparation of 2', 3'-cyclic phosphates of ribonucleosides and diribonucleoside phosphates via phosphotriester intermediates, *J. Chem. Soc., Perkin Trans 1* **1973**, 2513–2517.

21. J. F. M. de Rooij, G. Wille-Hazeleger, P. M. J. Burgers, and J. H. van Boom, Neighbouring group participation in the unblocking of phosphotriesters of nucleic acids, *Nucleic Acids Res.* **6**, 2237–2259 (1979).

22. S. De Bernardini, F. Waldmeir, and C. Tamm, A simple preparation of protected deoxynucleoside-3'-phosphates, *Helv. Chim. Acta* **64**, 2141–2147 (1981).

23. C. B. Reese, R. C. Titmas, and L. Yau, Oximate ion promoted unblocking of oligonucleotide phosphotriester intermediates, *Tetrahedron Lett.* **1978**, 2727–2730.

24. C. B. Reese and L. Yau, *O*-Aryl *S*-methyl phosphorochloridothioates: Terminal phosphorylating agents in the phosphotriester approach to oligonucleotide synthesis, *J. Chem. Soc., Chem. Commun.* **1978**, 1050–1052.

25. C. B. Reese and Y. T. Yan Kui, 4-Nitrophenyl phenyl phosphorochloridate: A new phosphorylating agent for oligonucleotide synthesis, *J. Chem. Soc., Chem. Commun.* **1977**, 802–804.

26. E. Uhlmann and W. Pfleiderer, Substituted β-phenylethyl groups. New protecting groups for oligonucleotide synthesis by the phosphotriester method, *Helv. Chim. Acta* **64**, 1688–1703 (1981).

27. J. Braddiley and A. R. Todd, Muscle adenylic acid and adenosine diphosphate, *J. Chem. Soc.* **1947**, 648–651.

28. B. Jastorff and H. Hettler, Phosphorylation of 5'-amino'5'-deoxynucleosides, *Tetrahedron Lett.* **1969**, 2543–2544.

29. J. Hes and M. P. Mertes, Di(2-*t*-butylphenyl)phosphorochloridate. A new selective phosphorylating agent, *J. Org. Chem.* **39**, 3767–3769 (1974).

30. G. R. Owen, J. P. H. Verheyden, and J. G. Moffatt, Synthesis of some 4'-fluorouridine derivatives, *J. Org. Chem.* **41**, 3010–3017 (1976).

31. J. Engels, Selective electrochemical removal of protecting groups in nucleotide synthesis, *Angew. Chem. Int. Ed. Engl.* **18**, 148–149 (1979).

32. J. Engles and U. Krahmer, Bis[2,2,2-tribromoethyl]phosphorochloridate: A suitable phosphorylating agent for nucleosides, *Synthesis* **1981**, 485–486.

33. M. Rubinstein, B. Amit, and A. Patchornik, The use of a light-sensitive phosphate protecting group for some mononucleotide syntheses, *Tetrahedron Lett.* **1975**, 1445–1448.

34. H. Takaku, K. Kamaike, and K. Kasuga, Synthesis of bis(5-chloro-8-quinolyl) nucleoside 5'-phosphates in oligoribonucleotide synthesis by the phosphotriester approach, *J. Org. Chem.* **47**, 4937–4940 (1982).

35. E. Ohtsuka, A. Yamane, T. Doi, and M. Ikehara, Chemical synthesis of the 5'-half molecule of *E. coli* tRNA$_2^{Gly}$, *Tetrahedron* **40**, 47–57 (1984).

36. J. H. van Boom, P. M. J. Burgers, G. R. Owen, C. B. Reese, and R. Saffhill, Approaches to oligoribonucleotide synthesis via phosphotriester intermediates, *J. Chem. Soc., Chem. Commun.* **1971**, 869–871.

37. J. E. Marugg, M. Tromp, P. Jhurani, C. F. Hoyng, G. A. van der Marel, and J. H. van Boom, Synthesis of DNA fragments by the hydroxybenzotriazole phosphotriester approach, *Tetrahedron* **40**, 73–78 (1984).

38. C. Broka, T. Hozumi, R. Arentzen, and K. Itakura, Simplifications in the synthesis of short oligonucleotide blocks, *Nucleic Acids Res.* **8**, 5461–5471 (1980).

39. M. Rubinstein and A. Patchornik, A novel method for phosphodiester and internucleotide bond synthesis, *Tetrahedron* **31**, 2107–2110 (1975).

40. R. H. Symons, in: *Methods in Enzymology* (J. G. Hardman and B. W. O'Maley, eds.), Vol. 38, pp. 410–420, Academic Press, New York (1974).

41. A. W. Schwartz, Specific phosphorylation of the 2'- and 3'-positions in ribonucleosides, *J. Chem. Soc., Chem. Commun.* **1969**, 1393.

42. A. Williams and I. T. Ibrahim, Carbo-diimide chemistry: Recent advances, *Chem. Rev.* **81**, 589–636 (1981).

43. G. Weimann and H. G. Khorana, On the mechanism of internucleotide bond synthesis by the carbodiimide method, *J. Am. Chem. Soc.* **84**, 4329–4341 (1962).

44. Y. Wang, Z. Yang, Q. Wang, Y. Xu, X. Liu, J. Xu, and C. Chen, The role of metaphosphate in the activation of the nucleotide by TPS and DCC in the oligonucleotide synthesis, *Nucleic Acids Res.* **14**, 2699–2706 (1986).

45. G. M. Tener, 2-Cyanoethyl phosphate and its use in the synthesis of phosphate esters, *J. Am. Chem. Soc.* **83**, 159–168 (1961).

46. M. Ikehara, Synthesis of ribooligonucleotides having sequences of transfer ribonucleic acids, *Acc. Chem. Res.* **7**, 92–96 (1974).

47. V. F. Zarytova and D. G. Knorre, General scheme of the phosphotriester condensation in the oligodeoxyribonucleotide synthesis with arylsulfonyl chlorides and aryl sulfonyl azides, *Nucleic Acids Res.* **12**, 2091–2110 (1984).

48. H. M. Hsiung, Improvements in the phosphotriester synthesis of deoxyribooligonucleotides—the use of hindered primary amines and a new isolation procedure, *Tetrahedron Lett.* **23**, 5119–5122 (1982).

49. S. E. Creasey and R. D. Guthrie, Use of mesitylenesulphonyl ('Mesisyl') chloride as a selective sulphonylating reagent, *J. Chem. Soc., Chem. Commun.* **1971**, 801–802.

50. S. S. Jones, B. Rayner, C. B. Reese, A. Ubasawa, and M. Ubasawa, Synthesis of the 3'-terminal decaribonucleoside nonaphosphate of yeast alanine transfer ribonucleic acid, *Tetrahedron* **36**, 3075–3085 (1980).

51. F. Himmelsbach, B. S. Schultz, T. Trichtinger, R. Charabula, and W. Pfleiderer, The *p*-nitrophenylethyl (NPE) group. A versatile new blocking group for phosphate and aglycone protection in nucleosides and nucleotides, *Tetrahedron* **40**, 59–72 (1984).

52. F. Ramirez and J. F. Marecek, Synthesis of phosphodiesters: The cyclic enediol phosphoryl (CEP) method, *Synthesis* **1985**, 449–488.

53. F. Ramirez, T. E. Gavin, S. B. Mandal, S. V. Kelkar, and J. F. Marecek, Synthesis of deoxyribotetranucleotides by the cyclic enediol phosphorylation method, *Tetrahedron* **39**, 2157–2161 (1983).

54. N. S. Corby, G. W. Kenner, and A. R. Todd, Ribonucleoside-5' phosphites. A new method for the preparation of mixed secondary phosphites, *J. Chem. Soc.* **1952**, 3669–3674.

55. J. Davoll, N. Anand, V. M. Clark, R. Hall, and A. R. Todd, A synthesis of uridine-5' pyrophosphate, a breakdown product of the coenzyme "uridine-diphosphate-glucose," *J. Chem. Soc.* **1952**, 3665–3668.

56. T. M. Chapman, J. M. Simpson, D. C. Kapp, and P. Butch, 5'-Diazonucleosides as alkylating agents in esterification reactions. Model studies and a thymidine-based reagent, *J. Carbohydr. Nucleosides Nucleotides* **7**, 241–256 (1980).

57. O. Mitsunobu, K. Kato, and J. Kimura, Selective phosphorylation of the 5'-hydroxy groups of thymidine and uridine, *J. Am. Chem. Soc.* **91**, 6510–6511 (1969).

58. O. Mitsunobu and M. Eguchi, Preparation of carboxylic esters and phosphoric esters by the activation of alcohols, *Bull. Chem. Soc. Jpn.* **44**, 3427–3430 (1971).

59. T. Mukaiyama and M. Hashimoto, Phosphorylation of alcohols and phosphates by oxidation–reduction condensation, *Bull. Chem. Soc. Jpn.* **44**, 196–199 (1971).

60. R. L. Letsinger and W. B. Lunsford, Synthesis of thymidine oligonucleotides by phosphite triester intermediates, *J. Am. Chem. Soc.* **98**, 3655–3661 (1976).

61. S. L. Beaucage, A simple and efficient preparation of deoxynucleoside phosphoramidites *in situ*, *Tetrahedron Lett.* **25**, 375–378 (1984).

62. M. F. Moore and S. L. Beaucage, Selective activation of phosphines by weak acids and the preparation of deoxynucleoside phosphoramidites, *J. Org. Chem.* **50**, 2019–2025 (1985).

63. T. Atkinson and M. Smith, in: *Oligonucleotide Synthesis: A Practical Approach* (M. J. Gait ed.), IRL Press, Oxford (1984).

64. O. Myklebost, B. Williamson, A. F. Markham, S. R. Myklebost, J. Rogers, D. E. Woods, and S. E. Humphries, The isolation and characterisation of cDNA clones for human apolipoprotein C11, *J. Biol. Chem.* **259**, 4401–4404 (1984).

65. M. D. Matteucci and M. H. Caruthers, Synthesis of deoxyoligonucleotides on a polymer support, *J. Am. Chem. Soc.* **103**, 3185–3191 (1981).

66. A. Elmblad, S. Josephson, and G. Palm, Synthesis of mixed oligodeoxyribonucleotides following the solid phase method, *Nucleic Acids Res.* **10**, 3291–3301 (1982).

67. T. Tanaka and R. L. Letsinger, Syringe method for stepwise chemical synthesis of oligonucleotides, *Nucleic Acids Res.* **10**, 3249–3260 (1982).

68. H. Seliger, C. Scalfi, and F. Eisenbeiss, An improved syringe technique for the preparation of oligonucleotides of defined sequence, *Tetrahedron Lett.* **24**, 4963–4966 (1983).

69. P. L. de Haseth, R. A. Goldman, C. L. Cech, and M. H. Caruthers, Chemical synthesis and biochemical reactivity of bacteriophage lambda P_R promoter, *Nucleic Acids Res.* **11**, 773–787 (1983).

70. G. R. Gough, M. J. Brunden, and P. T. Gilham, 2'(3')-*O*-Benzoyluridine 5'-linked to glass: An all-purpose support for solid phase synthesis of oligodeoxyribonucleotides, *Tetrahedron Lett.* **24**, 5321–5324 (1983).

71. J. L. Barascut, J. Cuartero, and J. L. Imbach, Use of a new type of support for solid-phase oligonucleotide synthesis, *Nucleosides Nucleotides* **2**, 193–202 (1983).

72. B. Preussel, M. Fritz, and P. Langen, An enzymatic procedure for the synthesis of [³H]-thymidine 5'-triphosphate, *J. Labelled Compd. Radiopharm.* **14**, 617–624 (1978).

73. E. Chargaff and E. F. Brunngraber, Purification and properties of a nucleoside phosphotransferase from carrot, *J. Biol. Chem.* **245**, 4825–4831 (1970).

74. J. I. Ademola and D. W. Hutchinson, The rapid purification of a phosphotransferase from wheat shoots, *Biochim. Biophys. Acta* **615**, 283–286 (1980).

75. D. C. Prasher, M. C. Carr, T. C. Tsai, and P. A. Frey, Nucleoside phosphotransferase from barley, *J. Biol. Chem.* **257**, 4931–4939 (1982).

76. C. L. Harvey, E. M. Clericuzio, and A. L. Nussbaum, The small scale preparation of 5'-nucleotides and analogs by carrot phosphotransferase, *Anal. Biochem.* **36**, 413–421 (1970).

77. R. Marutzky, H. Peterssen-Borstel, J. Flossdorf, and M. R. Kula, Large scale enzymatic synthesis of nucleoside 5'-monophospates using a phosphotransferase from carrots, *Biotechnol. Bioeng.* **16**, 1449–1458 (1974).

78. J. I. Ademola and D. W. Hutchinson, Preparation and properties of an insolubilised phosphotransferase, *Biotechnol. Bioeng.* **22**, 2419–2424 (1980).

79. E. F. Brunngraber and E. Chargaff, A nucleotide phosphotransferase from *Escherichia coli*: Purification and properties, *Biochemistry* **12**, 3005–3011 (1973).

80. A. Guranowski, Nucleoside phosphotransferase from yellow lupin seedling cotyledons, *Biochim. Biophys. Acta* **569**, 13–22 (1979).

81. C. H. Wong, S. L. Haynie, and G. M. Whitesides, Preparation of a mixture of nucleoside triphosphates from yeast RNA: Use in enzymatic synthesis requiring nucleoside triphosphate regeneration and conversion to nucleoside diphosphate sugars, *J. Am. Chem. Soc.* **105**, 115–117 (1983).

82. W. E. Ladner and G. R. Whitesides, Enzymatic synthesis of deoxyATP using DNA as starting material, *J. Org. Chem.* **50**, 1076–1079 (1985).

83. A. M. Michelson, Homopolymers of cytidylic and pseudouridylic acid copolymers with repeating subunits, and the stepwise synthesis of polyribonucleotides, *J. Chem. Soc.* **1959**, 3655–3669.

84. M. Smith, J. G. Moffat, and H. G. Khorana, Observations on the reactions of carbodiimides with acids and some new applications in the synthesis of phosphoric acid esters, *J. Am. Chem. Soc.* **80**, 6204–6212 (1958).

85. D. Lipkin, R. Markham, and W. H. Cook, The degradation of adenosine-5'-triphosphoric acid (ATP) by means of aqueous barium hydroxide, *J. Am. Chem. Soc.* **81**, 6075–6080 (1959).

86. M. Smith, G. I. Drummond, and H. G. Khorana, Ribonucleoside-3′,5′ cyclic phosphates. A general method of synthesis and some properties, *J. Am. Chem. Soc.* **83**, 698–706 (1961).

87. D. M. Brown, in: *Advances in Organic Chemistry: Methods and Results*, (R. A. Raphael, E. C. Taylor, and H. Wynberg, eds.), Vol. 3, pp. 75–157, Interscience, New York (1963).

88. J. H. van Boom, P. M. J. Burgers, P. van Deursen, and C. B. Reese, Preparation of nucleoside 3′,5′-cyclic phosphates by phosphotriester intermediates, *J. Chem. Soc., Chem. Commun.* **1974**, 618–619.

89. H. Takaku, M. Kato and T. Hata, Preparation of nucleoside 3′,5′-cyclic phosphates via 8-quinolyl nucleoside 5′-phosphates as useful intermediates, *Chem. Lett.* **1978**, 181–184.

90. R. Marumoto, T. Nishimura, and M. Honjo, A new method for synthesis of nucleoside 3′,5′-cyclic phosphates. Cyclisation of nucleoside 5′-trichloromethylphosphonates, *Chem. Pharm. Bull. Jpn.* **23**, 2295–2300 (1975).

91. T. Shimidzu, K. Yamana, N. Kanda, and S. Kitagawa, Cyclic phosphorylation reactions of diols with tri(1-imidazolyl)phosphine, *Bull. Chem. Soc. Jpn.* **56**, 3483–3485 (1983).

92. T. M. Gajda, A. E. Sopchik, and W. G. Bentrude, O_2/AIBN, free-radical method for the stereospecific oxidation of nucleoside trialkyl phosphites: New preparation of the individual diastereomeric, ^{18}O-labelled thymidine 3′-5′-cyclic monophosphates, *Tetrahedron Lett.* **22**, 4167–4170 (1981).

93. P. A. Frey, Stereochemistry of enzymatic reactions of phosphates, *Tetrahedron* **38**, 1541–1567 (1982).

94. P. M. Cullis, R. L. Jarvest, G. Lowe, and B. V. L. Poter, The stereochemistry of 2-substituted-2-oxo-4,5-diphenyl-1,3,2-dioxaphospholans and related chiral [$^{16}O,^{17}O,^{18}O$]phosphate monoesters, *J. Chem. Soc., Chem. Commun.* **1981**, 245–246.

95. R. Bicknell, P. M. Cullis, R. L. Jarvest, and G. Lowe, The stereochemical course of nucleotidyl transfer catalysed by ATP sulfurlase, *J. Biol. Chem.* **257**, 8922–8927 (1982).

96. S. J. Abbott, S. R. Jones, S. A. Weinman, and J. R. Knowles, Chiral [$^{16}O,^{17}O,^{18}O$]phosphate monoesters. 1. Asymmetric synthesis and stereochemical analysis of [1(*R*)-$^{16}O,^{17}O,^{18}O$]phospho-(*S*)-propane-1,2-diol, *J. Am. Chem. Soc.* **100**, 2558–2560 (1978).

97. R. L. Jarvest and G. Lowe, The stereochemical course of the hydrolysis of cyclic AMP by beef heart cyclic AMP phosphodiesterase, *J. Chem. Soc., Chem. Commun.* **1980**, 1145–1147.

98. J. A. Gerlt and J. A. Coderre, Oxygen chiral phosphodiesters. 1. Synthesis and configurational analysis of cyclic [^{18}O]-2′-deoxyadenosine-3′,5′-monophosphate, *J. Am. Chem. Soc.* **102**, 4531–4533 (1980).

99. J. Baranaik, K. Lesiak, M. Sochacki, and W. J. Stec, Stereospecific synthesis of cyclic adenosine 3′, 5′-(S_p)-[^{18}O]-phosphate, *J. Am. Chem. Soc.* **102**, 4533–4534 (1980).

100. R. D. Sammons, P. A. Frey, K. Bruzik, and M.-D. Tsai, Effects of ^{17}O and ^{18}O on ^{31}P n.m.r.: Further investigation and applications, *J. Am. Chem. Soc.* **105**, 5455–5461 (1983).

101. F. Eckstein, Phosphorothioate analogues of nucleotides—tools for the investigation of biochemical processes, *Angew. Chem. Int. Ed. Engl.* **22**, 423–439 (1983).

102. A. E. Reeve, M. M. Smith, V. Pigiet, and R. C. C. Huang, Incorporation of purine nucleoside 5′-[γ-S]triphosphates as affinity probes for initiation of RNA synthesis in vitro, *Biochemistry* **16**, 4464–4469 (1977).

103. K. H. Scheit, *Nucleotide Analogs, Synthesis and Biological Function*, Wiley, New York (1980).

104. F. Eckstein, Nucleoside phosphorothioates, *J. Am. Chem. Soc.* **92**, 4718–4723 (1970).

105. F. Eckstein and M. Goumet, in: *Nucleic Acid Chemistry* (L. B. Townsend and R. S. Tipson, eds.), Vol. 2, pp. 861–864, Wiley, New York (1978).

106. Z. J. Lesnikowski, W. Niewiarowski, W. S. Zielinski, and W. J. Stec, 2′-Deoxyribonucleoside 3′-aryl phosphoranilidates, *Tetrahedron* **40**, 15–32 (1984).

107. J. T. Chen and S. J. Benkovic, Synthesis and separation of diastereomers of deoxynucleoside 5′-*O*-(1-thio)triphosphates, *Nucleic Acids Res.* **11**, 3737–3751 (1983).

108. Ö. Kemal, C. B. Reese, and H. T. Serafinowska, Use of 2,5-dichlorophenyl phosphorodichloridothioate in the synthesis of diastereoisomeric dinucleoside phosphorothioates, *J. Chem. Soc., Chem. Commun.* **1983**, 591–593.

109. U. Asseline and N. T. Thuong, New simple synthesis of nucleoside phosphorothiolate, *Tetrahedron Lett.* **22**, 847–850 (1981).

110. E. F. Rossomando, G. A. Cordis, and G. D. Markham, 5′-Deoxy-5′-thioanalogs of adenosine

and inosine 5'-monophosphate: Studies with 5'-nucleotidase and alkaline phosphatase, *Arch. Biochem. Biophys.* **220**, 71–78 (1983).

111. A. Holy and P. Kois, Synthesis of 2'(3')-phosphates of 2'(3')-phosphorothioates of 5'-*O*-carboxymethylinosine and related compounds, *Collect. Czech. Chem. Commun.* **45**, 2817–2829 (1980).

112. F. Eckstein, Investigation of enzyme mechanisms with nucleoside phosphorothioates, *Angew. Chem. Int. Ed. Engl.* **14**, 160–166 (1975).

113. F. Eckstein, H. H. Schulz, H. Rüterjans, W. Haar, and W. Maurer, Stereochemistry of the transesterification step of ribonuclease T1, *Biochemistry* **11**, 3507–3512 (1972).

114. P. A. Frey and R. D. Sammons, Bond order and charge localisation in nucleoside phosphorothioates, *Science* **228**, 541–545 (1985).

115. P. M. J. Burgers and F. Eckstein, Diastereomers of 5'-*O*-adenosyl 3'-*O*-uridyl phosphorothioate: Chemical synthesis and enzymatic properties, *Biochemistry* **18**, 592–596 (1979).

116. M. Morr, Synthesis of 3'-amino-3'-deoxyguanosine, 3'-amino-3'-deoxyguanosine 5'-monophosphate and its 3',5'-cyclophosphate, *Liebigs Ann. Chem.* **1982**, 666–674.

117. M. Morr and E. Lugiger, Reactions of aminonucleosides with thiophosphorylating reagents, *Chem. Ber.* **111**, 2152–2172 (1978).

118. A. Okruszek and J. G. Verkade, 2',3'-Bis(2-chloroethyl)-aminophosphoryl-3'-amino-3'-deoxyadenosine: A cyclic nucleotide with antitumour activity, *J. Med. Chem.* **22**, 882–885 (1979).

119. R.-S. Lin, P. H. Fischer, and W. H. Prusoff, Synthesis and antineoplastic activity of a novel series of phosphoramide mustard analogues of pyrimidine deoxyribonucleosides, *J. Med. Chem.* **23**, 1235–1237 (1980).

120. D. Farquhar, N. J. Kuttesch, M. G. Wilkerson, and T. Winkler, Synthesis and biological evaluation of neutral derivatives of 5-fluoro-2'-deoxyuridine 5'-phosphate, *J. Med. Chem.* **26**, 1153–1158 (1983).

121. R. Engel, Phosphonates as analogues of natural phosphates, *Chem. Rev.* **77**, 349–367 (1977).

122. J. A. Montgomery, H. J. Thomas, R. L. Kisliuk, and Y. Gaumont, Phosphonate analogue of 2'-deoxy-5-fluorouridylic acid, *J. Med. Chem.* **22**, 109–111 (1979).

123. M. Morr, T. Ernst, and L. Grotjahn, 2',3'-Dideoxy-3'-*C*-(phosphonomethyl)adenosine, the phosphonate analogue of 2'-deoxyadenosine 3'-phosphate, *Z. Naturforsch., B* **38**, 1665–1668 (1983).

124. P. S. Miller, J. Yano, E. Yano, C. Carroll, K. Jayaraman, and P. O. P. Ts'o, Nonionic nucleic acid analogues. Synthesis and characterisation of dideoxyribonucleoside methylphosphonates, *Biochemistry* **18**, 5134–5143 (1979).

125. P. S. Miller, K. B. McParland, K. Jayaraman, and P. O. P. Ts'o, Biochemical and biological effects of nonionic nucleic acid methylphosphonates, *Biochemistry* **20**, 1874–1880 (1981).

126. I. Rosenberg and A. Holy, Preparation of 2'(3')-*O*-phosphonomethyl derivatives of ribonucleosides, *Collect. Czech. Chem. Commun.* **48**, 778–789 (1983).

127. M. Sekine, H.Mori, and T. Hata, Protecting of phosphonate function by means of ethoxycarbonyl group. A new method for the generation of reactive silyl phosphite intermediates, *Bull. Chem. Soc. Jpn.* **55**, 239–242 (1982).

128. M. Sekine and T. Hata, Selenium-containing nucleotides: Se-ethyl and Se-phenyl thymidine 5'-phosphoroselenoates, *Chem. Lett.* **1979**, 801–802.

129. M. J. Nemer and K. K. Ogilvie, Phosphoramidate analogues of diribonucleoside monophosphates, *Tetrahedron Lett.* **21**, 4153–4154 (1980).

130. P. W. Johnson, R. von Tigerstrom, and M. Smith, Reaction of nucleoside 2,4-dinitrophenyl phosphate with fluoride; a convenient method for the preparation of nucleoside phosphorofluoridate, *Nucleic Acids Res.* **2**, 1745–1749 (1975).

131. A. M. Michelson and A. R. Todd, A novel synthesis of adenosine triphosphate, *J. Chem. Soc.* **1949**, 2487–2490.

132. T. Glonek, R. A. Kleps, and T. C. Myers, Cyclisation of the phosphate side chain of adenosine triphosphate: Formation of monoadenosine 5'-trimetaphosphate, *Science* **1985**, 352–355 (1974).

133. A. M. Michelson, *The Chemistry of Nucleosides and Nucleotides*, Academic Press, New York (1963), pp. 221–222.

134. I. Tabushi and J. Imuta, Biomimetic one-pot synthesis of nucleotide phosphates, *Tetrahedron Lett.* **23**, 5415–5418 (1982).

135. E. P. Kennedy, The synthesis of cytidine diphosphate choline, cytidine diphosphate ethanol-amine and related compounds, *J. Biol. Chem.* **222**, 185–191 (1956).

136. F. Cramer and G. Weimann, Trichloroacetonitrile, a reagent for the selective esterification of phosphoric acids, *Chem. Ber.* **94**, 996–1007 (1961).

137. V. M. Clark, G. W. Kirby, and A. R. Todd, The use of phosphoramidic esters in acylation. A new preparation of adenosine 5'-pyrophosphate and adenosine 5'-triphosphate, *J. Chem. Soc.* **1957**, 1497–1501.

138. J. G. Moffatt and H. G. Khorana, The synthesis and some reactions of nucleoside-5(phosphoro-morpholidates and related compounds. Improved methods for the preparation of nucleoside-5' polyphosphates, *J. Am. Chem. Soc.* **83**, 649–658 (1961).

139. V. M. Clark and S. G. Warren, The formation of pyrophosphates from phosphoramidate monoesters, *J. Chem. Soc.* **1965**, 5509–5513.

140. H. G. Khorana, *Some Recent Developments in the Chemistry of Phosphate Esters of Biological Interest*, Wiley, New York (1961).

141. J. G. Moffatt and H. G. Khorana, The total synthesis of Coenzyme A, *J. Am. Chem. Soc.* **83**, 663–675 (1961).

142. S. Roseman, J. J. Distler, J. G. Moffatt, and H. G. Khorana, An improved general method for the synthesis of nucleotide coenzymes. Synthesis of uridine-5', cytidine-5' and guanosine-5' deriv-atives *J. Am. Chem. Soc.* **83**, 659–662 (1961).

143. J. G. Moffatt, in: *Methods in Enzymology* (E. F. Neufeld and V. Ginsburg, eds.), Vol. 8, pp. 136–142, Academic Press, New York (1966).

144. A. D. Elbein, in: *Methods in Enzymology* (E. P. Neufeld and V. Ginsburg, eds.), Vol. 8, pp. 142–145, Academic Press, New York (1966).

145. D. E. Hoard and D. G. Ott, Conversion of mono- and oligodeoxyribonucleotides into 5'-tri-phosphates, *J. Am. Chem. Soc.* **87**, 1785–1788 (1965).

146. (a) J. W. Kozarich, A. C. Chinault, and S. M. Hecht, Ribonucleoside phosphates via phosphor-imidazolidate intermediates. Synthesis of pseudoadenosine 5'-triphosphate, *Biochemistry* **12**, 4458–4463 (1973); (b) S. M. Hecht and J. W. Kozarich, A chemical synthesis of adenosine 5'-[γ-^{32}P]triphosphate, *Biochim. Biophys. Acta* **331**, 307–309 (1973).

147. M. Maeda, A. D. Patel, and A. Hampton, Formation of ribonucleotide 2',3'-cyclic carbonates during conversion of ribonucleoside 5'-phosphates to diphosphates and triphosphates by the phosphorimidazolidate procedure, *Nucleic Acids Res.* **4**, 2843–2853 (1977).

148. S. N. Zagrebelnyli, V. F. Zarytova, A. S. Levina, L. N. Semenova, F. V. Yarmolinskaya, and S. M. Yasnetskaya, Use of arenesulfonyl chloride for preparing nucleoside 5'-phosphamides, *Bioorg. Khim.* **4**, 729–734 (1978).

149. R. H. Symons, Practical methods for the routine chemical synthesis of ^{32}P-labelled nucleoside di- and triphosphates, *Biochim. Biophys. Acta* **209**, 296–305 (1970).

150. G. N. Bennett, G. R. Gough, and P. T. Gilham, Guanosine tetraphosphate and its analogues, *Biochemistry* **15**, 4623–4628 (1976).

151. K. Furusawa, M. Sekine, and T. Hata, A new method for the synthesis of nucleotide coenzymes by means of di-*n*-butylphosphinothioyl bromide, *J. Chem. Soc., Perkin Trans. 1* **1976**, 1711–1716.

152. M. Sekine, T. Kamimura, and T. Hata, A convenient method for the synthesis of P^1-(7-methyl-guanosine-5')-P^2-(ribonucleoside-5')diphosphates, *J. Chem. Soc., Perkin Trans. 1* **1985**, 997–1000.

153. I. Nakagawa, S. Konya, S. Ohtani, and T. Hata, A "capping" agent: P^1-S-phenyl-P^2-7-methyl-guanosine-5'-pyrophosphorothioate, *Synthesis* **1980**, 556–557.

154. A. F. Cook, M. J. Holman, and A. L. Nussbaum, Nucleoside S-alkyl phosphorothioates II. Preparation and chemical and enzymatic properties, *J. Am. Chem. Soc.* **91**, 1522–1527 (1969).

155. S. S. Jones and C. B. Reese, Chemical synthesis of 5'-O-triphosphoryladenylyl-(2'-5')-ade-nylyl-(2'-5')-adenosine, *J. Am. Chem. Soc.* **101**, 7400–7401 (1979).

156. S. Bornemann and E. Schlimme, Preparation of 5',5''-phosphate linked dinucleosides, *Z. Naturforsch., C* **36**, 135–141 (1981).

157. P. F. Schendel and R. D. Wells, The synthesis and purification of [γ-^{32}P]adenosine triphosphate with high specific activity, *J. Biol. Chem.* **248**, 8319–8321 (1973).

158. P. I. Bauer and G. Varady, A simple method for synthesising [γ-^{32}P]nucleoside triphosphates using [^{32}P]acetyl phosphate and acetate kinase, *Anal. Biochem.* **91**, 613–617 (1978).

159. B. L. Hirschbein, F. P. Mazenod, and G. M. Whitesides, Synthesis of phosphoenolpyruvate and its use in adenosine triphosphate cofactor regeneration, *J. Org. Chem.* **47**, 3765–3766 (1982).

160. R. H. Symons, The rapid, simple and improved preparation of high specific activity α-[^{32}P]dATP and α-[^{32}P]ATP, *Nucleic Acids Res.* **4**, 4347–4355 (1977).

161. D. Johnson, M. MacCoss, and S. Narindrasorasak, The enzymatic synthesis of ATP analogues, *Biochem. Biophys. Res. Commun.* **71**, 144–149 (1976).

162. M. Imazawa and F. Eckstein, Synthesis of sugar-modified nucleoside 5′-triphosphates with partially purified nucleotide kinases from calf thymus, *Biochim. Biophys. Acta* **570**, 284–290 (1979).

163. B. R. Martin and H. P. Voorheis, A simple method for the synthesis of adenosine 5′-[α-^{32}P]triphosphate on a preparative scale, *Biochem. J.* **161**, 555–559 (1977).

164. R. L. Baughn, Ö. Adalsteinsson, and G. M. Whitesides, Large scale enzyme-catalysed synthesis of ATP from adenosine and acetyl phosphate. Regeneration of ATP from AMP, *J. Am. Chem. Soc.* **100**, 304–306 (1978).

165. I. L. Cartwright and D. W. Hutchinson, A simple, rapid preparation of α-[^{32}P]-labelled adenosine diphosphate, *Nucleic Acids Res.* **4**, 2507–2510 (1977).

166. M. Kezdi, L. Kiss, O. Bojan, T. Pavel, and O. Barzu, 8-Bromoinosine 5′-diphosphate, a suitable phosphate acceptor of nucleoside diphosphate kinase, *Anal. Biochem.* **76**, 361–364 (1976).

167. K. Kihara, H. Nomiyama, M. Yukuhiro, and J.-I. Mukai, Enzymatic synthesis of [α-^{32}P]ATP of high specific activity, *Anal. Biochem.* **75**, 672–673 (1976).

168. T. F. Walseth and R. A. Johnson, The enzymatic preparation of [α-^{32}P]nucleoside triphosphates, cyclic [^{32}P]AMP, and cyclic [^{32}P]GMP, *Biochim. Biophys. Acta* **562**, 11–31 (1979).

169. A. E. Reeve and R. C. C. Huang, A method for the enzymatic synthesis and purification of [α-^{32}P]nucleoside triphosphates, *Nucleic Acids Res.* **6**, 81–90 (1979).

170. K. A. Abraham, A novel method for the synthesis of nucleoside triphosphates labelled with inorganic [^{32}P]phosphate specifically in the β-position, *Biochem. J.* **161**, 615–617 (1977).

171. H. J. Leuchs, J. M. Lewis, V. M. Rios-Mercadillo, and G. M. Whitesides, Conversion of the adenosine moieties of RNA into ATP for use in cofactor recycling, *J. Am. Chem. Soc.* **101**, 5829–5830 (1979).

172. H. Vutz, R. Koob, R. Jeck, and C. Woenckhaus, Properties of nicotinamide-adenine dinucleotides with modified nonfunctional adenosine, *Liebigs Ann. Chem.* **1980**, 1259–1270.

173. T. Meyer, K. Wielskens, J. Thiem, and H. Hilz, NAD[S], an NAD analogue with reduced susceptibility to phosphodiesterase, *Eur. J. Biochem.* **140**, 531–537 (1984).

174. I. Nakagawa and T. Hata, A new method for the synthesis of nucleotide coenzymes and their analogues, *Bull. Chem. Soc. Jpn.* **46**, 3275–3277 (1973).

175. D. R. Walt, A. Findeis, V. M. Rios-Mercadillo, J. Auge, and G. M. Whitesides, An efficient chemical and enzymatic synthesis of nicotinamide-adenine dinucleotide, *J. Am. Chem. Soc.* **106**, 234–239 (1984).

176. L. G. Lee and G. M. Whitesides, Enzyme-catalyzed organic synthesis: A comparison of strategies for *in situ* regeneration of NAD from NADH, *J. Am. Chem. Soc.* **107**, 6999–7008 (1985).

177. M. A. Abdallah, J. F. Bielmann, J. P. Samama, and D. Wrixon, Preparation and properties of 3-halopyridine-adenine dinucleotides, NAD$^+$ analogues and of model compounds. *Eur. J. Biochem.* **64**, 351–360 (1976).

178. A. W. H. Tan, A simplified method for the preparation of pure UDP[^{14}C]glucose, *Biochim. Biophys. Acta* **582**, 543–547 (1979).

179. J. R. Reiss and J. G. Moffatt, The synthesis of α,ω-dinucleoside 5′-polyphosphates, *J. Org. Chem.* **30**, 3381–3387 (1965).

180. J. Köhrle, K. S. Boos, and E. Schlimme, Preparation of [^{14}C]-P^1,P^5-di(adenosine-5′)penta-phosphate by direct reaction of ADP with activated ATP, *Annalen* **1977**, 1160–1166.

181. P. Feldhaus, T. Fröhlich, R. S. Goody, M. Isakov, and R. H. Schirmer, Synthetic inhibitors of adenylate kinases in the assays for ATPases and phosphokinases, *Eur. J. Biochem.* **57**, 197–204 (1975).

182. A. Simoncsits and J. Tomasz, A new type of nucleoside 5′-triphosphate analogue: P_1-(nucleoside 5′-)P_1-aminotriphosphates, *Tetrahedron Lett.* **1976**, 3995–3998.

183. N. B. Tarusova, V. V. Shumyantseva, A. C. Krylov, M. Ya Karpeiskii, and R. M. Khomutov,

Synthesis and properties of diadenosine tetraphosphate and its phosphonate analogs, *Bioorg. Khim.* **9**, 838–843 (1983).

184. W. A. Blättler and J. R. Knowles, Stereochemical course of glycerol kinase, pyruvate kinase and hexokinase: Phosphoryl transfer from chiral $[\gamma(S)\text{-}^{16}O,^{17}O,^{18}O]$ATP, *J. Am. Chem. Soc.* **101**, 510–511 (1979).

185. S. Meyerson, E. S. Kuhn, F. Ramirez, and J. F. Maracek, Hydrolysis of adenosine-5′ triphosphate: An isotope-labelling study, *J. Am. Chem. Soc.* **104**, 7231–7239 (1982).

186. M. Cohn and A. Hu, Isotopic ^{18}O shifts with ^{31}P n.m.r. of adenosine nucleotides synthesised with ^{18}O in various positions, *J. Am. Chem. Soc.* **102**, 913–916 (1980).

187. M. R. Webb, A method for determining the positional isotope exchange in a nucleoside triphosphate: Cyclisation of nucleoside triphosphate by dicyclohexyl carbodiimide, *Biochemistry* **19**, 4744–4748 (1980).

188. G. Lowe and B. S. Sproat, Evidence of a dissociative $S_N1(P)$ mechanism of phosphoryl transfer by rabbit muscle pyruvate kinase, *J. Chem. Soc., Perkin Trans. 1* **1978**, 1622–1630.

189. G. Lowe and G. Tansley, An investigation of the mechanism of activation of tryptophan by tryptophanyl-RNA synthetase from beef pancreas, *Eur. J. Biochem.* **138**, 597–602 (1984).

190. P. M. Cullis, Facile syntheses of bridge-oxygen-labelled pyrophosphates: The preparation of adenosine 5′-[β, γ-^{18}O]triphosphate, *J. Am. Chem. Soc.* **105**, 7783–7784 (1983).

191. M. A. Reynolds, N. J. Oppenheimer, and G. L. Kenyon, Enzyme-catalysed positional isotope exchange by phosphorus-31 nuclear magnetic resonance spectroscopy using either ^{18}O- or ^{17}O-β,γ-bridge labelled adenosine 5′-triphosphate, *J. Am. Chem. Soc.* **105**, 6663–6667 (1983).

192. F. Eckstein and R. S. Goody, Synthesis and properties of diastereoisomers of adenosine 5′-(O-1-thiotriphosphate) and adenosine 5′-(O-2-thiotriphosphate), *Biochemistry* **15**, 1685–1691 (1976).

193. E. K. Jaffe and M. Cohn, ^{31}P nuclear magnetic resonance spectra of the thiophosphate analogues of adenine nucleotides; effects of pH and Mg^{2+} binding, *Biochemistry* **17**, 652–657 (1978).

194. F. R. Bryant, S. J. Benkovic, D. Sammons, and P. A. Frey, The stereochemical course of thiophosphoryl group transfer catalyzed by T4 polynucleotide kinase, *J. Biol. Chem.* **256**, 5965–5969 (1981).

195. B. A. Connolly, P. J. Romanius, and F. Eckstein, Synthesis and characterisation of diastereomers of guanosine 5′-O-(1-thiotriphosphate) and guanosine 5′-O-(2-thiotriphosphate), *Biochemistry* **21**, 1983–1989 (1982).

196. D. Yee, V. W. Armstrong, and F. Eckstein, Mechanistic studies on DNA dependent RNA polymerase from *Escherichia coli* using phosphorothioate analogues, *Biochemistry* **18**, 4116–4123.

197. K.-F. Sheu, H.-T. Ho, L. D. Nolan, P. Markovitz, J. P. Richard, M. F. Utter, and P. P. Frey, Stereochemical course of thiophosphoryl group transfer catalysed by mitochondrial phospho-enolpyruvate carboxykinase, *Biochemistry* **23**, 1779–1783 (1984).

198. M. A. Gonzalez, M. R. Webb, K. M. Welsh, and B. S. Cooperman, Evidence that catalysis by yeast inorganic pyrophosphatase proceeds by direct phosphoryl transfer to water and not via a phosphoryl enzyme intermediate, *Biochemistry* **23**, 797–801 (1984).

199. M. R. Webb, G. H. Reed, B. F. Cooper, and F. B. Rudolph, The stereochemical course of phospho transfer catalyzed by adenylosuccinate synthetase, *J. Biol. Chem.* **259**, 3044–3046 (1984).

200. J. R. Moran and G. M. Whitesides, A practical enzymatic synthesis of (S_p)-adenosine 5′-O-(1-thiotriphosphate), *J. Org. Chem.* **49**, 704–706 (1984).

201. O. Abril, D. C. Crans, and G. M. Whitesides, Practical enzymatic synthesis of adenosine 5′-O-(3-thiotriphosphate), *J. Org. Chem.* **49**, 1360–1364 (1984).

202. R. G. Yount, ATP analogs, *Adv. Enzymol.* **43**, 1–56 (1975).

203. T. C. Myers, K. Nakamura, and A. Danielzadeh, Phosphonic acid analogs of nucleoside phosphates. III. The synthesis of adenosine-5′-methylene diphosphonate, a phosphonic acid analog of adenosine-5′-diphosphate, *J. Org. Chem.* **30**, 1517–1520 (1965).

204. D. B. Trowbridge, D. M. Yamamoto, and G. L. Kenyon, Ring openings of trimetaphosphoric acid and its bismethylene analog. Synthesis of 5′-bis(dihydroxyphosphinylmethyl)phosphinate and 5′-amino-5′-deoxyadenosine 5′-triphosphate, *J. Am. Chem. Soc.* **94**, 3816–3824 (1972).

205. A. M. Spiegel, R. W. Downs, Jr., and G. D. Aurbach, Guanosine 5′,α-β-methylene triphosphate,

a novel GTP analog, causes persistent activation of adenylate cyclase: Evidence against a pyrophosphorylation mechanism, *Biochem. Biophys. Res. Commun.* **76**, 758–764 (1977).

206. J. A. Stock, Synthesis of phosphonate analogues of thymidine di- and triphosphate from 5'-O-toluene sulfonylthymidine, *J. Org. Chem.* **44**, 3997–4000 (1979).

207. T. C. Myers, K. Nakamura, and J. W. Flesher, Phosphonic analogs of nucleoside phosphates. I. The synthesis of 5'-adenylyl methylene diphosphonate, a phosphonic analog of ATP, *J. Am. Chem. Soc.* **85**, 3292–3295 (1963).

208. G. M. Blackburn, D. E. Kent, and F. Kolkmann, The synthesis and metal binding characteristics of novel, isopolar phosphonate analogues of nucleotides, *J. Chem. Soc., Perkin Trans. 1* **1984**, 1119–1125.

209. K. J. Gibson and N. J. Leonard, Nucleoside 5'-(β-γ-peroxytriphosphates), *Biochemistry* **23**, 78–85 (1984).

210. R. G. Yount, D. Babcock, D. Ojala, and W. Ballantyne, Adenylyl imidodiphosphate, an adenosine triphosphate analog containing a P–N–P linkage, *Biochemistry* **10**, 2484–2489 (1971).

211. P. T. Englund, J. A. Huberman, T. M. Jovin, and A. Kornberg, Binding of triphosphates to deoxyribonucleic acid polymerase, *J. Biol. Chem.* **244**, 3038–3044 (1969).

212. S. M. Schuster, R. J. Gertschen, and H. A. Lardy, Effect of inosine 5'-(β,γ-imido)triphosphate and other nucleotides on beef heart mitochondrial ATPase, *J. Biol. Chem.* **251**, 6705–6710 (1976).

213. P. Remy, G. Dirheimer, and J. P. Ebel, Synthesis of adenosine 5'-phosphohypophosphate, *Biochim. Biophys. Acta* **136**, 99–107 (1967).

214. P. Remy, M. L. Engel, G. Dirheimer, J. P. Ebel, and M. Revel, Guanosine-5'-phosphohypophosphate: Preparation and effects on protein synthesis *in vitro*, *J. Mol. Biol.* **48**, 173–176 (1970).

215. J. A. J. den Hartog, M. P. Lawson, E. W. P. de Jong, and J. H. van Boom, Chemical synthesis of thymidine 3'-diphosphate 5'-[(phosphonomethyl)phosphonyl-phosphate], an asymmetrically phosphorylated thymidine derivative, *Recl. J. R. Neth. Chem. Soc.* **100**, 317–319 (1981).

216. J. Wilkes, B. Hapke, and R. Letsinger, A 5'-amino analog of adenosine diphosphate, *Biochem. Biophys. Res. Commun.* **53**, 917–922 (1973).

217. R. Letsinger, J. Wilkes, and L. B. Dumas, Enzymatic synthesis of polydeoxyribonucleotides possessing internucleotide phosphoramidate bonds, *J. Am. Chem. Soc.* **94**, 292–293 (1972).

218. T. C. Meyers, Phosphonic acid analogs of nucleoside phosphates, U.S. Patent 3,238,191 (1966) [*CA* **64**, 15972 (1966)].

219. A. Hampton, T. Sasaki, and B. Paul, Synthesis of 6'-cyano-6'-deoxyhomoadenosine-6'-phosphonic acid and its phosphoryl and pyrophosphoryl anhydrides and studies of their interactions with adenine nucleotide utilising enzymes, *J. Am. Chem. Soc.* **95**, 4404–4414 (1973).

220. A. Hampton, P. J. Harper, T. Sasaki, P. Howgate, and R. K. Preston, Carboxylic-phosphonic mixed anhydrides isosteric with AMP and ATP as reagents for AMP and ATP sites, *Biochem. Biophys. Res. Commun.* **65**, 945–950 (1975).

221. F. Eckstein, W. Burns, and A. Parmeggiani, Synthesis of guanosine 5'-di- and triphosphate derivatives with modified terminal phosphates, *Biochemistry* **14**, 5225–5232 (1975).

222. D. G. Knorre, V. A. Kurbatov, and V. V. Samukov, General method for the synthesis of ATP gamma-derivatives, *FEBS Lett.* **70**, 105–108 (1976).

223. L. H. De Riemer and C. F. Meares, Synthesis of mono- and dinucleotide photoaffinity probes for ribonucleic acid polymerase, *Biochemistry* **20**, 1606–1612 (1981).

224. S. A. Kedik and R. I. Zhdanov, Adenosine triphosphate spin labelled by γ-phosphate group, *Izv. Akad. Nauk SSSR, Ser. Khim.* **1978**, 1689.

225. H. Griengl, W. Hayden, and B. Rosenwirth, Phosphonoformic and -acetic acid esters of nucleosides and nucleoside analogs, *Nucleic Acids Res. Symp. Ser.* **14**, 273–274 (1984).

226. D. W. Hutchinson, P. A. Cload, and M. C. Haugh, Organophosphorus compounds as antiviral agents, *Phosphorus and Sulfur* **14**, 285–293 (1983).

227. S. R. Adams, M. J. Sparkes, and H. B. F. Dixon, The arsonomethyl analogue of adenosine 5'-phosphate, *Biochem. J.* **221**, 829–836 (1984).

228. D. Webster, M. J. Sparkes, and H. B. F. Dixon, An arsenical analogue of adenosine diphosphate, *Biochem. J.* **169**, 239–244 (1978).

229. M. J. Gresser, ADP-arsenate, *J. Biol. Chem.* **256**, 5981–5983 (1981).

230. J. P. Horwitz, J. P. Neenan, R. S. Misra, J. Rozhin, A. Huo, and K. D. Philips, Facile synthesis of

3'-phospho- and 2'-phosphoadenosine 5'-phosphosulfate, *Biochim. Biophys. Acta* **480**, 376–381 (1977).

231. R. Cherniak and E. A. Davidson, Synthesis of adenylyl sulfate and adenylyl sulfate 3'-phosphate, *J. Biol. Chem.* **239**, 2986–2990 (1964).

232. M. L. Tsang, J. Lemieux, J. A. Schiff, and T. B. Bojarski, Preparation of adenosine 5'-phosphosulfate (APS) and adenosine 3'-phosphate 5'-phosphosulfate (PAPS) prepared by an improved procedure, *Anal. Biochem.* **74**, 623–626 (1976).

233. C. Walsh, *Enzymatic Reaction Mechanisms*, W. H. Freeman, San Francisco (1979), pp. 132–178.

234. K. Bloch, S. Chaykin, A. H. Phillips, and A. DeWaard, Mevalonic acid pyrophosphate and isopentenyl pyrophosphate, *J. Biol. Chem.* **234**, 2595–2604 (1959).

235. D. E. Koshland, Jr. and A. Levitzki, in: *The Enzymes* (P. D. Boyer, ed.), 3rd ed., Vol. 10, pp. 539–560, Academic Press, New York (1974).

236. M. K. Horwitt and L. A. Witting, in: *The Vitamins* (W. H. Sebrell and R. S. Harris, eds.), Vol. 5, p. 53, Academic Press, New York (1972).

237. J. Preiss and P. Handler, Biosynthesis of diphosphopyridine nucleotide *J. Biol. Chem.* **233**, 493–500 (1958).

238. A. Kornberg, Reversible enzymatic synthesis of diphosphopyridine nucleotide and inorganic pyrophosphate, *J. Biol. Chem.* **182**, 779–793 (1950).

239. G. R. Wyatt, in: *The Nucleic Acids* (E. Chargaff and J. N. Davidson, eds.), Vol. 1, pp. 243–265, Academic Press, New York (1955).

240. K. Randerath, *Thin Layer Chromatography*, 2nd ed., Academic Press, New York (1966).

241. C. Horvath, HPLC ion-exchange chromatography with narrow-bore columns: Rapid analysis of constituents at subnanomole levels, *Methods Biochem. Anal.* **21**, 79–154 (1973).

242. M. W. Taylor, H. V. Hershey, R. A. Levine, K. Coy, and S. Olivelle, Improved method of resolving nucleotides by reversed-phase high performance liquid chromatography, *J. Chromatogr.* **219**, 133–139 (1981).

243. N. E. Hoffman and J. C. Liao, Reversed phase high performance liquid chromatography separations of nucleotides in the presence of solvophobic ions, *Anal. Chem.* **49**, 2231–2234 (1977).

244. C. Y. Ip, D. Ha, P. W. Morris, M. L. Puttemans, and D. L. Venton, Separation of nucleosides and nucleotides by reversed-phase high-performance liquid chromatography with volatile buffers allowing sample recovery, *Anal. Biochem.* **147**, 180–185 (1985).

245. R. E. Brown, P. J. Cayley, and I. M. Kerr, in: *Methods in Enzymology* (S. Pestka, ed.), Vol. 79, pp 208–216, Academic Press, New York (1981).

246. S. Zadrazil, in: *Synthetic Procedures in Nucleic Acid Chemistry* (W. W. Zorbach and R. S. Tipson, eds.), Vol. 2, pp. 533–660, Wiley, New York (1973).

247. A. E. Pierce, in: *Synthetic Procedures in Nucleic Acid Chemistry* (W. W. Zorbach and R. S. Tipson, eds.), Vol. 2, pp. 125–143, Wiley, New York (1973).

248. J. R. Chapman, in: *Oganophosphorus Chemistry* (D. W. Hutchinson and J. A. Miller, eds.), Vol. 14, pp. 278–304, Royal Society of Chemistry, London (1983).

249. A. L. Burlingame and N. Castagnoli (eds.), *Mass Spectrometry in the Health and Life Sciences*, Elsevier, Amsterdam (1985).

250. C. G. Edmonds, M. L. Vestal, and J. A. McCloskey, Thermospray liquid chromatography–mass spectrometry of nucleosides and of enzymatic hydrolystates of nucleic acids, *Nucleic Acids Res.* **13**, 8197–8206 (1985).

251. D. L. Slowikowski and K. H. Schram, Fast atom bombardment mass spectrometry of nucleosides, nucleotides and oligonucleotides, *Nucleosides Nucleotides* **4**, 309–346 (1985).

252. J. L. Aubagnac, F. M. Devienne, R. Combarieu, J. L. Barascut, J. L. Imbach, and H. B. Lazrek, Characterisation of underived nucleotides by mass spectrometry using the MBSAFAB ionisation method, *Org. Mass Spectrom.* **18**, 361–364 (1983).

253. M. Panico, G. Sindona, and N. Uccella, Fast atom bombardment induced zwitterionic oligonucleotide quasimolecular ions sequenced by MS/MS, *J. Am. Chem. Soc.* **105**, 5607–5610 (1983).

254. L. J. Berliner and J. C. Reubens (eds.), *Biological Magnetic Resonance*, Vols. 1–6, Plenum Press, New York (1978–1984).

255. G. Lowe, B. V. L. Potter, B. S. Sproat, and W. E. Hull, The effect of ^{17}O and the magnitude of the ^{18}O-isotope shift in ^{31}P nuclear magnetic resonance spectroscopy, *J. Chem. Soc., Chem. Commun.* **1979**, 733–735.

256. M. D. Tsai, in: *Methods in Enzymology* (D. L. Purich, ed.), Vol. 87, pp. 235–279, Academic Press, New York (1982).

257. S. L. Buchwald, D. E. Hansen, A. Hassett, and J. R. Knowles, in: *Methods in Enzymology* (D. L. Purich, ed.), Vol. 87, pp. 279–301, Academic Press, New York (1982).

258. J. R. P. Arnold and G. Lowe, Synthesis and stereochemical analysis of chiral inorganic [^{16}O,^{17}O,^{18}O]thiophosphate, *J. Chem. Soc., Chem. Commun.* **1986**, 865–867.

259. P. M. Cullis and A. J. Rous, Stereochemical course of phosphoryl transfer reactions of P^1,P^1-disubstituted pyrophosphate in aprotic solvent. A model for the enzyme-catalyzed "dissociative" phosphoryl transfer, *J. Am. Chem. Soc.* **107**, 6721–6723 (1985).

260. F. H. Westheimer, Pseudo-rotation in the hydrolysis of phosphate esters, *Acc. Chem. Res.* **1**, 70–78 (1968).

261. S. L. Buchwald, D. H. Pliura, and J. R. Knowles, Stereochemical evidence for pseudorotation in the reaction of a phosphoric monoester, *J. Am. Chem. Soc.* **106**, 4916–4922 (1984).

262. F. Eckstein, P. J. Romaniuk, and B. A. Connolly, in: *Methods in Enzymology* (D. L. Purich, ed.), Vol. 87, pp. 197–212, Academic Press, New York (1982).

263. H. G. Floss, M.-D. Tsai, and R. W. Woodward, in: *Topics in Stereochemistry* (E. L. Eliel, S. H. Wilen, and N. L. Allinger, eds.), Vol. 15, pp. 253–321, Wiley, New York (1984).

264. R. L. Jarvest, G. Lowe, and B. V. L. Potter, The stereochemical course of phosphoryl transfer catalysed by *Bacillus stearothermophilus* and rabbit skeletal-muscle phosphofructokinase with a chiral [^{16}O,^{17}O,^{18}O]phosphate ester, *Biochem. J.* **199**, 427–432 (1981).

265. G. Lowe and B. V. L. Potter, A stereochemical investigation of phosphoryl transfer catalysed by phosphoglucomutase by the use of α-D-glucose 1-[(S)-^{16}O,^{17}O,^{18}O]phosphate, *Biochem. J.* **199**, 693–698 (1981).

266. G. Lowe and B. V. L. Potter, The stereochemical course of phosphoryl transfer catalysed by glucose 6-phosphatase, *Biochem. J.* **201**, 665–668 (1982).

267. M. Tetas and J. M. Lowenstein, The effect of bivalent metal ions on the hydrolysis of adenosine di- and triphosphate, *Biochemistry* **2**, 350–357 (1963).

268. H. Pezzano and F. Podo, Structure of binary complexes of mono- and polynucleotides with metal ions of the first transition group, *Chem. Rev.* **80**, 365–401 (1980).

269. H. Sigel and P. E. Amsler, On the mechanism of metal ion promoted dephosphorylation of purine nucleoside 5′-triphosphates, *J. Am. Chem. Soc.* **98**, 7390–7400 (1976).

270. D. Chattergi and F. Y. H. Wu, Direct coordination of nucleotide with the intrinsic metal in *E. coli* RNA polymerase, *Biochemistry* **21**, 4657–4664 (1982).

271. H. Sigel and F. Hofstetter, Metal-ion-promoted dephosphorylation of the 5′-triphosphates of uridine and thymidine, and a comparison with the reactivity in the corresponding cytidine and adenosine nucleotide systems, *Eur. J. Biochem.* **132**, 569–577 (1983).

272. D. S. Sigman, G. M. Wahl, and D. J. Creighton, Models for metalloenzymes. Zinc ion catalysed phosphorylation of 1,10-phenanthroline-2-carbinol by adenosine triphosphate, *Biochemistry* **11**, 2236–2242 (1972).

273. K. H. Scheller and H. Sigel, A proton nuclear magnetic resonance study of purine and pyrimidine nucleoside 5′-diphosphates. Extent of macrochelate formation in monomeric metal ion complexes and promotion of self-stacking by metal ions, *J. Am. Chem. Soc.* **105**, 5891–5900 (1983).

274. D. Dunaway-Mariano and W. W. Cleland, Preparation and properties of chromium(III) adenosine 5′-triphosphate, chromium(III) adenosine 5′-diphosphate, and related chromium(III) complexes, *Biochemistry* **19**, 1496–1505 (1980).

275. D. Dunaway-Mariano and W. W. Cleland, Investigation of substrate specificity and reaction mechanism of several kinases using chromium(III) adenosine 5′-triphosphate and chromium(III) adenosine 5′-diphosphate, *Biochemistry* **19**, 1506–1515 (1980).

276. M. Cohn, Some properties of the phosphorothioate analogues of adenosine triphosphate as substrates of enzymic reactions, *Acc. Chem. Res.* **15**, 326–332 (1982).

277. P. M. J. Burgers and F. Eckstein, Structure of the metal–nucleotide complex in the creatine kinase reaction, *J. Biol. Chem.* **255**, 8229–8233 (1980).

278. J. A. Coderre and J. A. Gerlt, Enzymatic synthesis and configurational analysis of [α-^{18}O]-2′-deoxyadenosine 5′-diphosphate, *J. Am. Chem. Soc.* **102**, 6594–6597 (1980).

279. H. R. Kalbitzer, R. Marquetant, B. A. Connolly, and R. S. Goody, Structural investigations of the

Mg.ATP complex at the active site of porcine adenylate kinase using phosphorothioate analogs and electron paramagnetic resonance of Mn(II) with chiral [17]O-labelled ATP analogs, *Eur. J. Biochem.* **133**, 221–227 (1983).

280. C. Klevickis and C. M. Grisham, Phosphorus-31 nuclear magnetic resonance studies of the conformation of an adenosine triphosphate analogue at the active site of (Na$^+$ + K$^+$)-ATPase from kidney medulla, *Biochemistry* **21**, 6979–6984 (1982).

281. L. H. Schliselfeld, C. T. Biert, and R. J. Labotka, ^{31}P nuclear magnetic resonance of phosphonic acid analogues of adenosine nucleotides as function of pH and magnesium ion concentration, *Biochemistry* **21**, 317–320 (1982).

282. H. J. Vogel and W. A. Bridger, Phosphorus-31 nuclear magnetic resonance studies of methylene and fluoro analogues of adenine nucleotides. Effects of pH and magnesium ion binding, *Biochemistry* **21**, 394–401 (1982).

283. O. Kennard, D. G. Watson, F. H. Allen, and S. A. Belard (eds.), *Molecular Structures and Dimensions*, Vols. 1–16, D. Reidel Publishing Co., Dordrecht, The Netherlands (1970–1985).

284. H. Sawai, T. Shibata, and M. Ohno, Preparation of oligoadenylates with 2′–5′ linkage using Pb^{2+} ion catalyst, *Tetrahedron* **37**, 481–485 (1981).

285. P. K. Bridson and L. E. Orgel, Catalysis of accurate poly(C)-directed synthesis of 3′–5′-linked oligoguanylates by Zn^{2+}, *J. Mol. Biol.* **144**, 567–577 (1980).

286. W. B. Jakoby and M. Wilchek (eds.), *Methods in Enzymology*, Vol. 46, Academic Press, New York (1977).

287. P. N. Lowe and R. B. Beechey, Interactions between the mitochondrial adenosine triphosphatase and periodate-oxidised adenosine 5′-triphosphate, an affinity label for adenosine 5′-triphosphate binding sites, *Biochemistry* **21**, 4073–4082 (1982).

288. M. M. King and R. F. Colman, Affinity labelling of nicotinamide adenine dinucleotide dependent isocitrate dehydrogenase by the 2′-3′-dialdehyde derivative of adenosine 5′-diphosphate, *Biochemistry* **22**, 1656–1665 (1983).

289. A. Burkhard, A. Dworsky, R. Jeck, M. Pfeiffer, S. Pundak, and C. Woenckhaus, Diazonium derivatives of ADP for identifying essential amino acid residues in the active centre of dehydrogenases, *Z. Physiol. Chem.* **362**, 1079–1090 (1981).

290. Y.-H. Huang Wong, F. B. Winer, and P. A. Frey, *p*-(Bromoacetamido)phenyl uridyl pyrophosphate: An active-site-directed irreversible inhibitor for uridine diphosphate galactose 4-epimerase, *Biochemistry* **18**, 5332–5341 (1979).

291. I. L. Cartwright, D. W. Hutchinson, and V. W. Armstrong, The reaction between thiols and 8-azidoadenosine derivatives, *Nucleic Acids Res.* **3**, 2331–2339 (1976).

292. J. J. Czarnecki, M. S. Abbott, and B. R. Selman, Localisation of the tight ADP-binding site on the membrane-bound chloroplast coupling factor one, *Eur. J. Biochem.* **136**, 19–24 (1983).

293. T. K. Lee, L.-J. C. Wong, and S. S. Wong, Photoaffinity labelling of lactose synthase with a UDP-galactose analog, *J. Biol. Chem.* **258**, 13166–13171 (1983).

294. R. Krauspe and O. I. Lavrik, Photoaffinity labelling of chloroplastic and cytoplasmic leucyl-tRNA synthetases of *Euglena gracilis* by the α-(*p*-azidoanilidate) of ATP, *Eur. J. Biochem.* **132**, 545–550 (1983).

295. I. W. Caras, T. Jones, S. Eriksson, and D. W. Martin, Jr., Direct photoaffinity labelling of the catalytic site of mouse ribonucleotide reductase by CDP, *J. Biol. Chem.* **258**, 3064–3068 (1983).

296. L. J. McBride, R. Kierzek, S. L. Beaucage, and M. H. Caruthers, Amidine protecting groups for oligo nucleotide synthesis, *J. Am. Chem. Soc.* **108**, 2040–2048 (1986).

297. P. J. Garegg, I. Lindh, T. Regberg, J. Stawinski, R. Stromberg, and C. Henrichson, Nucleoside H-phosphonates III. Chemical synthesis of oligodeoxyribonucleotides by the hydrogen-phosphonate approach, *Tetrahedron Lett.* **27**, 4051–4054 (1986).

298. E. Sonveaux, The organic chemistry underlying DNA synthesis, *Bioorg. Chem.* **14**, 274–325 (1986).

299. N. T. Thoung and U. Asseline, Chemical synthesis of natural and modified oligonucleotides, *Biochimie* **67**, 673–684 (1985).

300. M. Matsukura, K. Shinozuka, G. Zon, H. Mitsuya, M. Reitz, J. S. Cohen, and S. Broder, Phosphorothioate analogs of oligodeoxynucleotides: Inhibitors of replication and cytopathic effects of human immunodeficiency virus, *Proc. Natl. Acad. Sci. U.S.A.* **84**, 7706–7710 (1987).

301. T. S. Rao, C. B. Reese, H. T. Serafinowska, H. Takaku, and G. Zappia. Solid phase synthesis of

the 3′-terminal nonadecaribonucleoside octadecaphosphate sequence of yeast alanine transfer ribonucleic acid, *Tetrahedron Lett.* **28**, 4897–4900 (1987).

302. R. Cosstick and J. S. Vyle, Synthesis and phosphorus–sulphur bond cleavage of 3′-thiothymidylyl (3′-5′) thymidine, *J. Chem. Soc., Chem. Commun.* **1988**, 992–993.

303. P. C. Toren, D. F. Betsch, H. L. Weith, and J. M. Coull, Determination of impurities in nucleoside 3′-phosphoramidites by fast atom bombardment mass spectrometry, *Anal. Biochem.* **152**, 291–294 (1986).

304. W. S. Zielinski and L. E. Orgel, Oligoaminonucleoside phosphoramidates. Oligomerisation of dimers of 3′-amino-3′-deoxynucleotides (GC and CG) in aqueous solution, *Nucleic Acids Res.* **15**, 1699–1715 (1987).

305. W. S. Zielinski and L. E. Orgel, Autocatalytic synthesis of a tetranucleotide analogue, *Nature* **327**, 346–347 (1987).

306. R. S. Goody and M. Isakov, Simple synthesis and separation of the diastereomers of α-thioanalogs of ribo- and deoxyribo- di- and triphosphates, *Tetrahedron Lett.* **27**, 3599–3602 (1986).

307. P. M. Cullis and A. J. Rous, Stereochemical course of phosphoryl transfer from adenosine 5′-diphosphate to alcohols and the possible role of monomeric metaphosphate, *J. Am. Chem. Soc.* **108**, 1298–1300 (1986).

308. G. Lowe and S. P. Tuck, Positional isotope exchange in adenosine 5′-[β-$^{18}O_4$]diphosphate and the possible role of monomeric metaphosphate, *J. Am. Chem. Soc.* **108**, 1300–1301 (1986).

309. P. M. Cullis and D. Nicholls, The existence of monomeric metaphospate in hydroxylic solvent: A positional isotope exchange study, *J. Chem. Soc., Chem. Commun.* **1987**, 783–785.

310. G. Lowe and G. Semple, The mechanism of displacement of sulphur from adenosine 5′-[(S)-α-thio-γ-benzyl]tri-phosphate, *J. Chem. Soc., Chem. Commun.* **1988**, 377–378.

311. P. M. Cullis and M. B. Schilling, The reactivity of adenosine 5′-[S-methyl-1-thio]triphosphate: A facile way of generating a cyclophosphate dianion, *J. Chem. Soc., Chem. Commun.* **1989**, 106–108.

312. L. M. Mallis, F. M. Raushel, and D. H. Russell, Differentiation of isotopically labelled nucleotides using fast atom bombardment tandem mass spectrometry, *Anal. Chem.* **59**, 980–984 (1987).

313. J. R. P. Arnold, M. S. Cheng, P. M. Cullis, and G. Lowe, Stereochemical course of phosphoryl transfer catalysed by herpes simplex virus-1 induced thymidine kinase, *J. Biol. Chem.* **261**, 1985–1987 (1986).

314. S. P. Harnett, G. Lowe, and G. Tansley, Mechanisms of activation of phenylalanine synthetase and synthesis of P^1,P^4-bis(5′-adenosyl)tetraphosphate by yeast phenylalanyl-tRNA synthetase, *Biochemistry* **24**, 2908–2915 (1985).

315. D. E. Hansen and J. R. Knowles, *N*-Carboxybiotin formation by pyruvate carboxylase: The stereochemical consequence at phosphorus, *J. Am. Chem. Soc.* **107**, 8304–8305 (1985).

316. S. P. Harnett, G. Lowe, and G. Tansley, A stereochemical study of the mechanism of activation of donor oligonucleotides by RNA ligase from bacteriophage T4-infected *Escherichia coli*, *Biochemistry* **24**, 7446–7449 (1985).

317. R. M. Dixon and G. Lowe, Synthesis of (Rp, Rp)-P^1,P^4-bis(5′-adenosyl)-1[^{17}O,$^{18}O_2$],4[^{17}O,$^{18}O_2$] tetraphosphate from (Sp, Sp)-P^1,P^4-bis(5′-adenosyl)-1[*thio*-$^{18}O_2$],4[*thio*-$^{18}O_2$]tetraphosphate with retention at phosphorus and the stereochemical course of hydrolysis by the unsymmetrical Ap_4A phosphodiesterase from lupin seeds, *J. Biol. Chem.* **264**, 2069–2074 (1989).

318. A. G. McLennan, G. E. Taylor, M. Prescott, and G. M. Blackburn, Recognition of $\beta\beta'$-substituted and $\alpha\beta,\alpha'\beta'$-disubstituted phosphonate analogues of bis(5′-adenosyl) tetraphosphate by the bis(5′-nucleosidyl)-tetraphosphate pyrophosphohydrolases from *Artemia* embryos and *E. coli*, *Biochemistry* **28**, 3868–3875 (1989).

319. A. Guranowski, E. Starzynska, G. E. Taylor, and G. M. Blackburn, Studies on some specific Ap_4A-degrading enzymes using various methylene analogues of P^1,P^4-bis (5′,5″-adenosyl) tetraphosphate, *Biochem. J.* **262**, 241–244 (1989).

320. J. E. Van Pelt, R. Iyengar, and P. A. Frey, Gentamycin nucleotidyl transferase. Stereochemical inversion at phosphorus in enzymic 2′-deoxyadenylyl transfer to tobramycin, *J. Biol. Chem.* **261**, 15995–15999 (1986).

321. A. Arabshahi, R. S. Brody, A. Smallwood, T.-C. Tsai, and P. A. Frey, Galactose-1-phosphate uridylyl transferase. Purification of the enzyme and stereochemical course of each step of the double displacement mechanism, *Biochemistry* **25**, 5583–5589 (1986).

322. V. J. Davisson, D. R. Davis, V. M. Dixit, and C. D. Poulter, Synthesis of nucleotide 5′-diphosphates from 5′-*O*-tosyl nucleosides, *J. Org. Chem.* **52**, 1794–1801 (1987).

323. M. W. Hosseni, J.-M. Lehn, L. Maggiora, K. B. Mertes, and M. P. Mertes, Supramolecular catalysis in the hydrolysis of ATP facilitated by macrocyclic polyamines: Mechanistic studies, *J. Am. Chem. Soc.* **109**, 537–544 (1987).

324. P. G. Yohannes, M. P. Mertes, and K. B. Mertes, Pyrophosphate formation via a phosphoramidate intermediate in polyammonium macrocycle/metal ion-catalysed hydrolysis of ATP, *J. Am. Chem. Soc.* **107**, 8288–8289 (1985).

325. G. M. Blackburn, G. R. J. Thatcher, M. W. Hosseni, and J.-M. Lehn, Evidence for a protophosphatase catalysed cleavage of ATP by a dissociative-type mechanism within a receptor substrate complex, *Tetrahedron Lett.* **28**, 2779–2782 (1987).

326. K. Hattori and K. Takahashi, Reversible transphosphorylation among ATP, ADP, and AMP in the presence of a methylated cyclodextrin, *Chem. Lett.* **1985**, 985–988.

327. F. Kappeler, T. T. Hai, and A. Hampton, Use of a vinyl phosphonate analog of ATP as a rotationally constrained probe of the C5′–O5′ torsion angle in ATP complexed to methionine adenosyl transferase, *Bioorg. Chem.* **13**, 289–295 (1985).

328. A. Hampton, F. Kappeler, and F. Perini, Evidence for the conformation about the C5′-O5′ bond of ATP complexed to AMP kinase, *Bioorg. Chem.* **5**, 31–35 (1976).

329. M. M. Vaghefi, P. A. McKernan, and R. K. Robins, Synthesis and antiviral activity of certain nucleoside 5′ phosphonoformate derivatives, *J. Med. Chem.* **29**, 1389–1393 (1986).

330. R. W. Lambert, J. A. Martin, G. J. Thomas, I. B. Duncan, M. J. Hall, and E. P. Heimer, Synthesis and antiviral activity of phosphonoacetic and phosphonoformic acid esters of 5-bromo-2′-deoxyuridine and related pyrimidine nucleosides and acyclonucleosides, *J. Med. Chem.* **32**, 367–374 (1989).

Chapter 4

The Synthesis and Chemistry of Heterocyclic Analogues of Purine Nucleosides and Nucleotides

Ganapathi R. Revankar and Roland K. Robins

1. Introduction, Scope, and Nomenclature

The usefulness of heterocyclic analogues of the natural nucleic acid components as biochemical tools and as chemotherapeutic agents has prompted expansion of the term "nucleoside" from its original rigid definition,[1] confined only to the carbohydrate derivatives of purines and pyrimidines isolated from the alkaline hydrolysates of yeast nucleic acids. In this chapter, the term "nucleoside" includes all glycosyl derivatives of nitrogen hyterocycles, of both synthetic and natural origin, in which the heterocyclic base is linked through *nitrogen* to the C-1 position of the carbohydrate, regardless of the nature of the carbohydrate moiety. The extraordinary growth of research related to the nucleoside antibiotics during the past 25 years has convincingly marshaled great interest in exploring other potentially active analogues of the nucleic acid components. Although considerable work has been done in this area, only a few reviews[2–4] have highlighted major developments in this field. An effort has now been made to compile a comprehensive review on this subject.

The present chapter is designed to summarize and complement previous reports on the synthesis and chemistry of heterocyclic analogues of purine nucleosides and nucleotides. This covers the area of both synthetic and natural nucleosides of bicyclic systems in which a five-membered ring is fused to a six-membered ring containing one or more nitrogen atoms in each ring, *except purine*, which has been covered separately in Vol. 1 of this treatise.

Numbering of the carbohydrate moiety originates at the anomeric carbon, which is the carbon involved in the glycosidic linkage. The nomenclature and

Ganapathi R. Revankar • Triplex Pharmaceutical Corporation, The Woodlands, Texas 77380. *Roland K. Robins* • ICN Nucleic Acid Research Institute, Costa Mesa, California 92626.

numbering for the heterocyclic aglycon will be discussed briefly at the beginning of each section.

2. Synthesis, and Chemical and Physical Properties of Heterocyclic Analogues of Purine Nucleosides and Nucleotides

2.1. Five-Membered Ring Fused to a Six-Membered Ring with Two Nitrogen Atoms in the Ring System (One Nitrogen Atom in Each Ring)

The original nomenclature "pyrindole" introduced by Perkin and Robinson[5] for the fused pyrrole–pyridine system persisted until Kruber [6] presented the aza designation for 7-azaindole (4). Although there is an overwhelming general usage of the "azaindole" nomenclature to be found in the literature, the *Ring Index*[7] and the *Chemical Abstracts* designation has been adopted for the present work, and numbering is as shown in (1). Accordingly, 4-azaindole (1) is pyrrolo[3,2-*b*]pyridine, 5-azaindole (2) is pyrrolo[3,2-*c*]pyridine, 6-azaindole (3) is pyrrolo[2,3-*c*]pyridine, and 7-azaindole (4) is pyrrolo[2,3-*b*]pyridine. Even though this convention has been used in the *Chemical Abstracts* since 1925, occasional inconsistencies have appeared in the literature.[8]

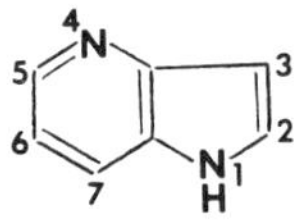

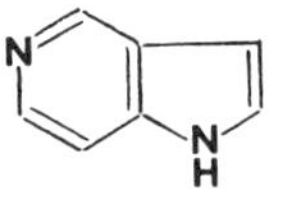

A large number of pyrrolopyridine derivatives, excluding the nucleosides, have found considerable use as chemotherapeutic agents.[9–15] The reason for the absence of reported chemotherapeutic activity of pyrrolopyridine nucleosides is most likely due to the paucity of such derivatives.

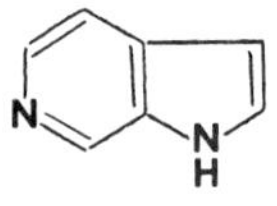

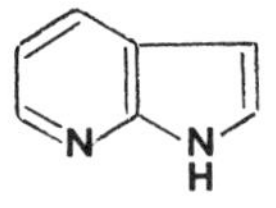

The first chemical synthesis of a pyrrolo[2,3-*b*]pyridine (7-azaindole) nucleoside was accomplished by Preobrazhenskaya and co-workers[16] using the "indoline–indole" method.[17,18] The reaction of either 2,3-dihydro-4-methylpyrrolo[2,3-*b*]pyridine[19] (5, X = H) or 6-chloro-2,3-dihydro-4-methylpyrrolo[2,3-*b*]pyridine[19] (5, X = Cl) with D-glucose in absolute ethanol in the presence of ammonium sulfate led to the corresponding 1-β-D-glucopyranosyl-2,3-dihydropyrrolo[2,3-*b*]pyridines (6). Acetylation of 6 with acetic anhydride in pyridine gave the corresponding tetraacetates (7), which, on subsequent dehydrogenation with 2,3-dichloro-5,6-dicyano-1,4-benzoquinone (DDQ) to afford 9, followed by deacetylation, furnished 1-β-D-glucopyranosyl-4-methylpyrrolo[2,3-*b*]pyridine[16] (8, X = H) and 6-chloro-1-β-D-glucopyranosyl-4-methylpyrrolo[2,3-*b*]pyridine[16] (8, X = Cl). The versatility of this synthetic "indoline–indole" method for the preparation of pyrrolopridine nucleosides was further demonstrated[20,21] by the successful synthesis of several 1-β-D-ribofuranosyl-pyrrolo[2,3-*b*]pyridines. Thus, treatment of 5-*O*-trityl-D-ribofuranose (10) with

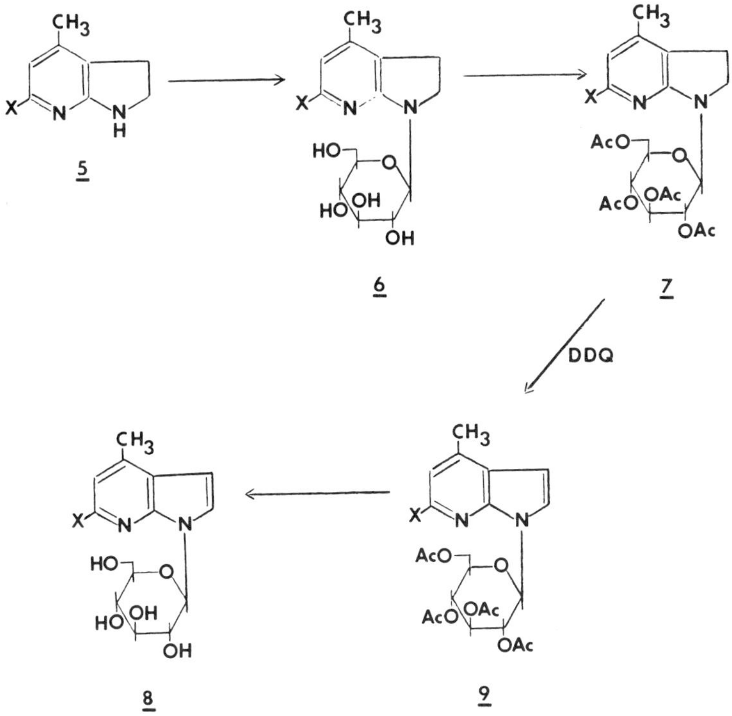

2,3-dihydro-4-methylpyrrolo[2,3-*b*]pyridine (**5**, X = H) or its 6-chloro derivative in boiling ethanol furnished, after proper chemical transformation as described for **8**, 1-β-D-ribofuranosyl-4-methylpyrrolo[2,3-*b*]pyridine (**11**, X = H) and the corresponding 6-chloro derivative (**11**, X = Cl), in 51% and 34% yield, respectively. A similar methodology was employed to obtain 1-α-L-arabinopyranosides of 6-chloro-2,3-dioxo-2,3-dihydro-4-methylpyrrolo[2,3-*b*]pyridine[22] and 6-chloro-4-methyl-5-nitropyrrolo[2,3-*b*]pyridine.[23] Through a combination of ^{1}H-NMR studies, periodate oxidation techniques, and ORD determination, the β-anomeric configuration has been assigned to these nucleosides. The site of glycosylation was determined by UV spectral studies.[16–20]

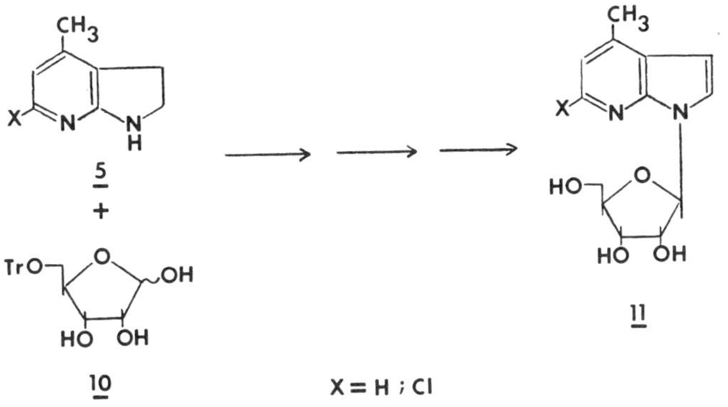

Synthesis of the dideazaadenosine analogue 4-amino-1-β-D-ribofuranosyl-pyrrolo[2,3-*b*]pyridine (**14**) has also been reported.[24] The acid-catalyzed fusion of 4-(acetylamino)-2,3-dihydropyrrolo[2,3-*b*]pyridine (**12**) with 1,2,3,5-tetra-*O*-acetyl-D-ribofuranose gave the *N*-1 glycosyl derivative (**13**). The subsequent oxidation of **13** with DDQ afforded **15**. Deacetylation of **15** with sodium methoxide furnished **14**, which is probably the first example of a dideazaadenosine containing one nitrogen atom in each of the two heterocyclic rings.[24]

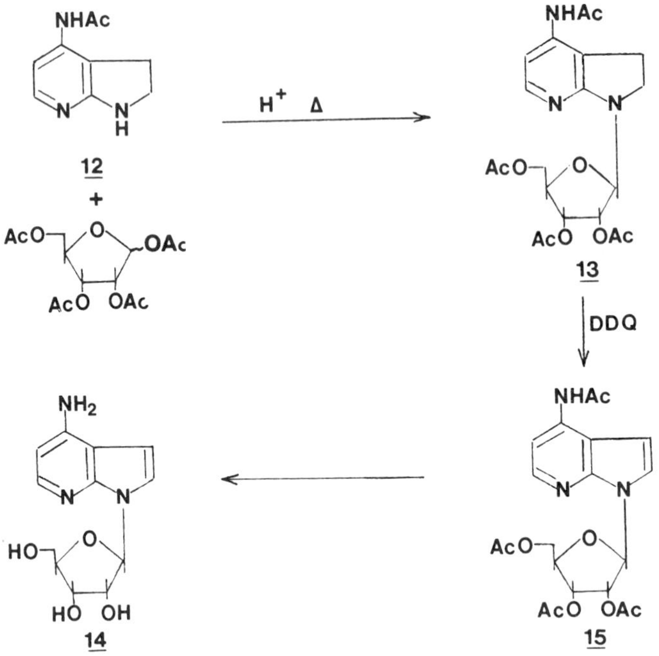

Nucleosides in the pyrrolo[3,2-*c*]pyridine (5-azaindole) series have re-mained relatively unexplored. The chemical synthesis of certain *N*-glycosides of 2,3-dimethylpyrrolo[3,2-*c*]pyridines (**17**) by the condensation of chloromer-curi-2,3-dimethylpyrrolo[3,2-*c*]pyridine (**16**) with protected bromo sugars has been reported.[25] However, no adequate proof was presented for either the anomeric configuration or the site of glycosylation. It is of particular interest to

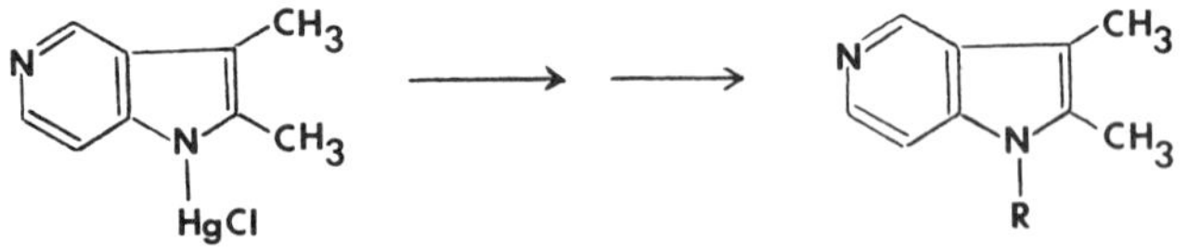

R = β-D-arabinofuranosyl
β-D-galactopyranosyl
β-D-glucopyranosyl

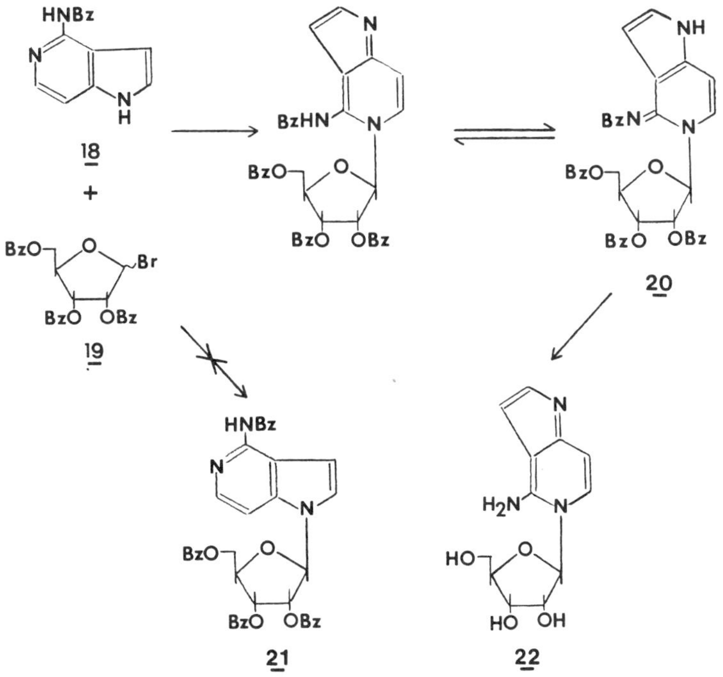

note that the direct alkylation of 4-benzamidopyrrolo[3,2-*c*]pyridine (**18**) with 2,3,5-tri-*O*-benzoyl-D-ribofuranosyl bromide (**19**) in acetonitrile in the presence of mercuric cyanide gave, in 50% yield, 4-benzoylimino-5-(2,3,5-tri-*O*-benzoyl-β-D-ribofuranosyl)pyrrolo[3,2-*c*]pyridine (**20**),[26] which was erroneously thought to be 4-benzamido-1-(2,3,5-tri-*O*-benzoyl-β-D-ribofuranosyl)pyrrolo-[3,2-*c*]pyridine (**21**).[27] Subsequent debenzoylation of **20** under alkaline conditions afforded 4-amino-5-β-D-ribofuranosylpyrrolo[3,2-*c*]pyridine (**22**).[26] The molecular structure of **22** was deduced on the basis of ¹H-NMR and mass spectra,[28] as well as by X-ray crystallographic studies.[29]

Application of the stereospecific sodium salt glycosylation procedure, which was developed recently in our laboratory,[30,31] to 4-chloropyrrolo[3,2-*c*]-pyridine[26] (**23**) furnished a convenient route to 5-azaindole nucleosides.[32,33] Thus, direct glycosylation of the sodium salt of **23** with 1-chloro-2-deoxy-3,5-di-*O*-*p*-toluoyl-α-D-*erythro*-pentofuranose[34] (**24**) in an inert atmosphere gave 4-chloro-1-(2-deoxy-3,5-di-*O*-*p*-toluoyl-β-D-*erythro*-pentofuranosyl)pyr-rolo[3,2-*c*]pyridine (**25**). Treatment of **25** with hydrazine hydrate afforded **26a**, which, on catalytic (Pd/C) hydrogenation, furnished the 2′-deoxydideazaadeno-sine analogue 4-amino-1-(2-deoxy-β-D-*erythro*-pentofuranosyl)pyrrolo[3,2-*c*]py-ridine (**26b**).[32] A similar glycosylation of the sodium salt of **23** with 1-chloro-2,3-,5-tri-*O*-benzyl-α-D-arabinofuranose (**27**)[35] gave the corresponding blocked nucleoside (**28**). Reaction of **28** with hydrazine hydrate, followed by catalytic hydrogenation, gave 4-amino-1-β-D-arabinofuranosylpyrrolo[3,2-*c*]pyridine (**30**).[32]

When the sodium salt of 4,6-dichloropyrrolo[3,2-*c*]pyridine[36] (**31**) was allowed to react with **24** in acetonitrile, only 4,6-dichloro-1-(2-deoxy-3,5-di-*O*-*p*-toluoyl-β-D-*erythro*-pentofuranosyl)pyrrolo[3,2-*c*]pyridine (**32**) was obtained. Treatment of **32** with methanolic ammonia at an elevated temperature gave 4,6-dichloro-1-(2-deoxy-β-D-*erythro*-pentofuranosyl)pyrrolo[3,2-*c*]pyridine (**33**),[31] in which the chloro group is available for nucleophilic displacement reactions[37] to obtain **26b**.[37c] 4-Amino-1-β-D-ribofuranosylpyrrolo[3,2-*c*]pyridine was recently prepared[37c] by this method. Similarly, glycosylation of the sodium salt of ethyl 2-cyanomethylpyrrole-3-carboxylate[38] (**34**) with **24** gave the corresponding blocked nucleoside (**35**). Ring closure of **35** with hydrazine furnished 5-amino-6-hydrazino-1-(2-deoxy-β-D-*erythro*-pentofuranosyl)pyrrolo[3,2-*c*]pyridine-4-one (**37**). Catalytic hydrogenation of **37** in the presence of Raney nickel gave the 2′-deoxyguanosine analogue 6-amino-1-(2-deoxy-β-D-*erythro*-pento-furanosyl)pyrrolo[3,2-*c*]pyridin-4-one (**36**).[33]

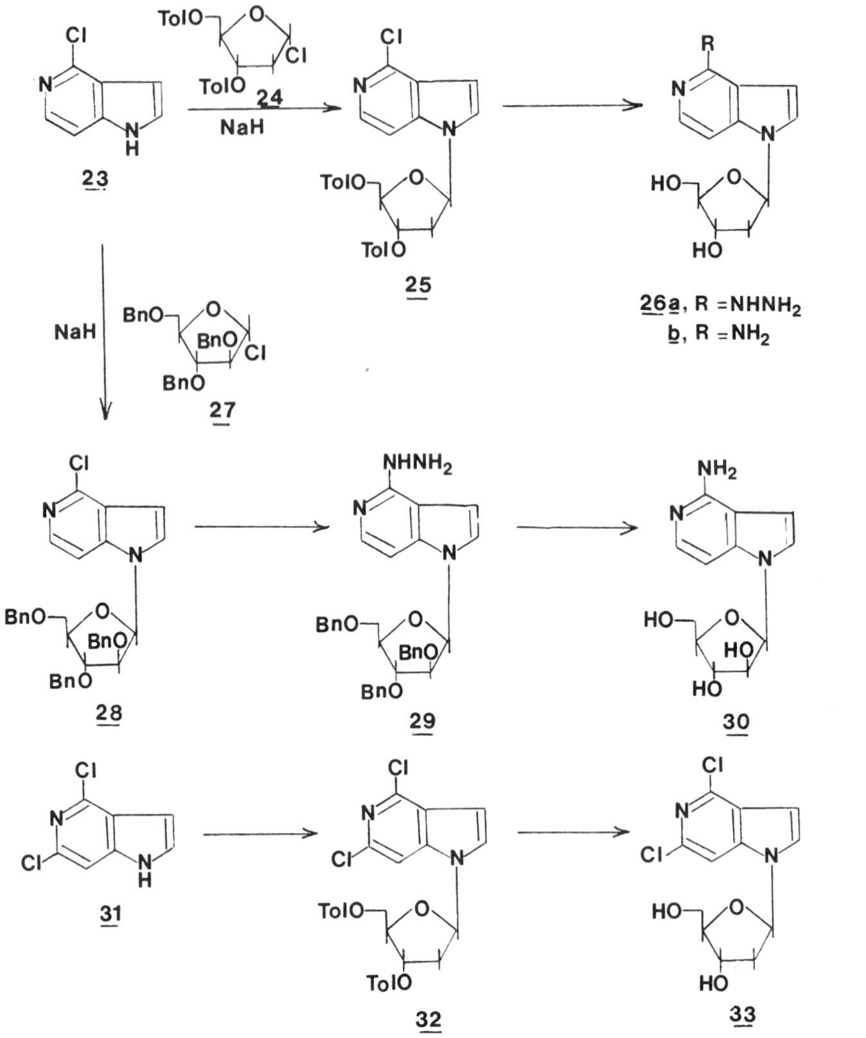

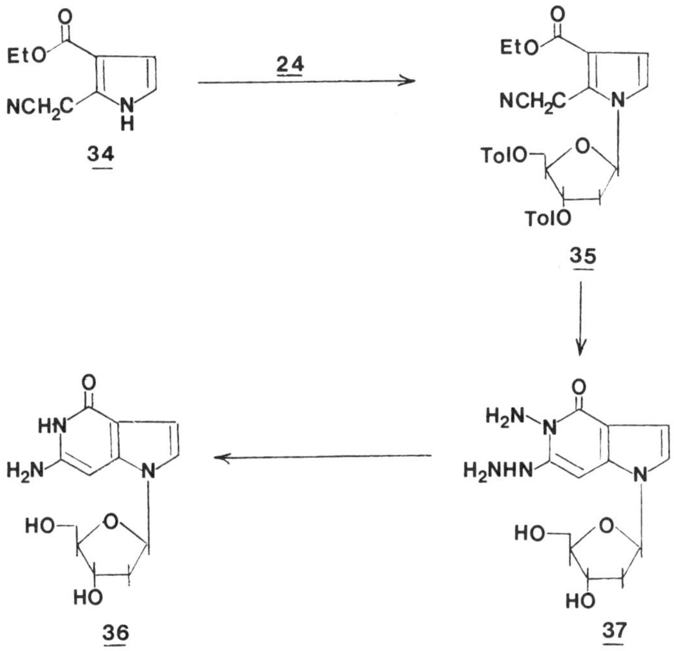

Since the starting halo sugars **24** and **27** have the α-configuration,[35,39] the exclusive formation of the blocked nucleosides **25**, **28**, **32**, and **35** is assumed to be due to a direct Walden inversion (S_N2) at the carbon at the 1-position of the carbohydrate by the anionic heterocyclic nitrogen.[30]

Although several pyrrolo[3,2-b]pyridines[40,41] and pyrrolo[2,3-c]pyridines[40–43] have been reported, the nucleoside derivatives of these ring systems remain unknown.

4-Amino-1-β-D-ribofuranosylpyrrolo[2,3-b]pyridine (**14**) was evaluated against L1210 and S-180 cells in culture and found to be inactive.[24] Similarly, compound **14** did not show any significant inhibitory activity against platelet aggregation induced by ADP and collagen at a concentration of $1 \times 10^{-4} M$.[44]

2.2. Five-Membered Ring Fused to a Six-Membered Ring with Three Nitrogen Atoms in the Ring System (at Least One Nitrogen Atom in Each Ring)

Since the isolation of the antibiotics tubercidin, toyocamycin, and sangivamycin in the early 1960s and subsequent structural elucidation of these antibiotics as 7-deazapurine (pyrrolo[2,3-d]pyrimidine) nucleosides, the developments in the area of deazapurine nucleosides have been so prodigious that it is beyond the scope of this section to attempt a comprehensive review of all the chemical and biochemical reports that are available on this subject. Ensuing interest in the chemical synthesis and the unusual biological properties of such 1-, 3-, 7-, and 9-deazapurine nucleosides represents one of the most well explored areas of research in nucleoside chemistry. The heterocyclic portion of this group of natural nucleoside antibiotics is structurally related to 4-aminopyrrolo[2,3-*d*]pyrimidine and is numbered according to the *Chemical Abstracts* nomenclature along with the other deazapurines and pyrazolopyridines as shown in Fig. 1.

2.2.1. *Imidazo[4,5-b]pyridine (1-Deazapurine) Nucleosides*

Glycosylation of the imidazo[4,5-*b*]pyridine system has been accomplished by the conventional mercuri salt, fusion, direct alkylation, and trimethylsilyl procedures.

   2.2.1.1. Synthesis Employing the Halomercuri Salt of a Preformed Aglycon and an Acylated Sugar Halide.   The first chemical synthesis of an imidazo[4,5-*b*]pyridine nucleoside was accomplished by Chatterjee and co-workers.[45] The next compound in this area was prepared by Mizuno *et al.*[46] by the condensation of the chloromercuri derivative of 1-deazapurine (**38**) with polyacyl glycosyl halide under conditions similar to those employed in the conventional synthesis of purine neucleosides.[47,48] Although Chatterjee and co-workers assigned the site of glycosylation as N-1 in their original paper[45] by a comparison[49,50] of the UV absorption spectra of these nucleosides with that of the model 1-methylimidazo-[4,5-*b*]pyridine, the glycosyl attachment was later[51] shown to be on N-3. By analogy with similar results observed in the purine series, the β-configuration was assumed on the basis of the *trans* rule.[52,53] Since exceptions to the *trans* rule have been reported,[54] Mizuno *et al.*[46,55] obtained conclusive evidence for both the site of ribosylation and the anomeric configuration by exhaustive spectroscopic comparison of the 1- and 3-methyl isomers of imidazo[4,5-*b*]pyridines with **39** (R = β-D-ribofuranosyl) and by the synthesis of the cyclonucleoside **42** from **39**. The formation of **42** by an intramolecular quaternization is feasible only with the 2′,3′-*O*-isopropylidene-5′-tosylate of 3-β-D-ribofuranosylimidazo[4,5-*b*]pyridine (**41**).

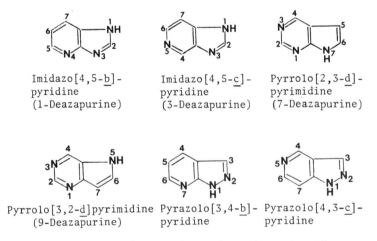

Figure 1. Numbering scheme for deazapurines and pyrazolopyridines.

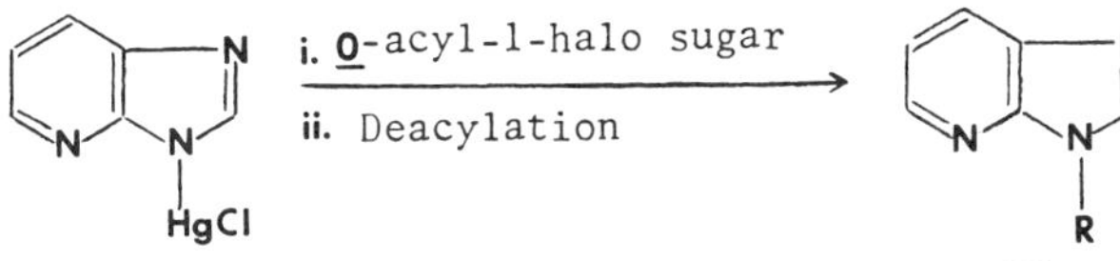

$$38 \xrightarrow{\substack{\text{i. } \underline{O}\text{-acyl-1-halo sugar} \\ \text{ii. Deacylation}}} 39$$

$$R = \beta\text{-}\underline{D}\text{-galactopyranosyl}$$
$$\beta\text{-}\underline{D}\text{-glucopyranosyl}$$
$$\beta\text{-}\underline{D}\text{-ribofuranosyl}$$
$$\beta\text{-}\underline{D}\text{-xylopyranosyl}$$

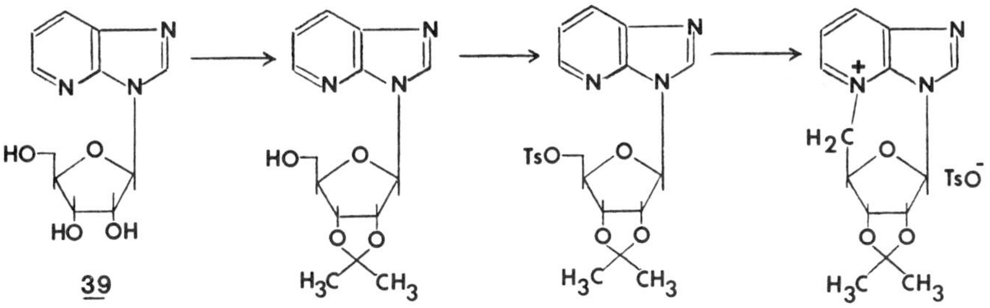

39 → **40** → **41** → **42**

The synthesis of 6-nitro-3-β-D-ribofuranosylimidazo[4,5,-*b*]pyridine (**44**) was also accomplished[56] from the chloromercuri derivative of 6-nitroimidazo-[4,5-*b*]pyridine (**43**) and 2,3,5-tri-*O*-benzoyl-D-ribofuranosyl chloride. By using a similar methodology, a number of 6-bromo-3-β-D-glycosylimidazo[4,5-*b*]pyridines have been prepared.[57]

Product distribution in the ribosylation of deazapurines by the halomercuri procedure was studied by Mizuno *et al.*[58] using various chromatographic techniques. This resulted in the isolation of 1-β-D-ribofuranosylimidazo[4,5-*b*]pyridines (**45**). The results are summarized in Table I. It was observed that ribosylation of the bases in class A gave either one of two possible isomers in predominant amounts or failed to afford a detectable amount of one of the two possible isomers. Ribosylation of bases in class B gave each of two possible isomers in about equal amounts.

The halomercuri procedure was also employed[59] for the synthesis of

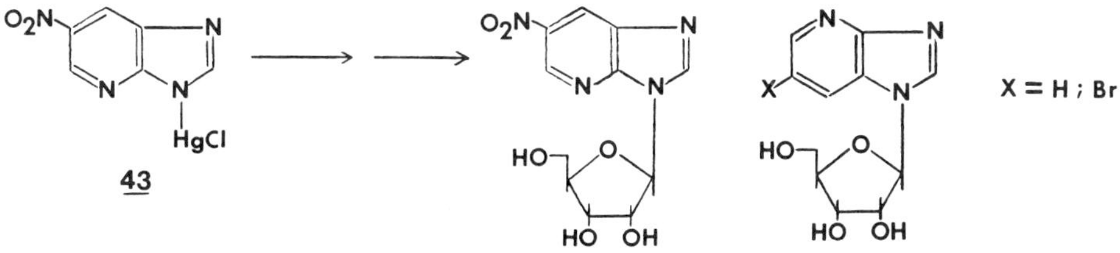

43 → → **44**   **45**   X = H ; Br

Table I. Product Distribution in Ribosylation
of Deazapurines by Chloromercuri Method

Aglycon	Yield of isomers (%)		Isomeric ratio	Method of separation[a]
	3H	1H		
Class A				
(imidazo[4,5-b]pyridine structure)	68	—	—	A
	68	3.3	20	B
(5-bromo structure)	42	13	3.2	C
(4-chloro imidazo[4,5-c]pyridine structure)	84	16	5.3	D
Class B				
(imidazo[4,5-c]pyridine structure)	41	36	1.1	A
	45	55	0.82	D
(N-oxide structure)	23.7	20.2	1.17	A

[a] A, Alumina column chromatography; B, a combination of alumina column chromatography and paper chromatography; C, a combination of fractional crystallization and subsequent paper chromatography; D, gas chromatography.

7-amino-3-β-D-ribofuranosylimidazo[4,5-*b*]pyridine (1-deazaadenosine, **48**). Condensation of the chloromercuri-7-acetamidoimidazo[4,5-*b*]pyridine (**46**) with 2,3,5-tri-*O*-benzoyl-D-ribofuranosyl chloride (**47**) in boiling xylene, followed by deacylation, gave **48**. It is of particular interest, however, that chloromercuri-7-benzamidoimidazo[4,5-*b*]pyridine failed to react with **47** in boiling xylene under these same conditions. The benzamido heterocycle was recovered quantitatively,[60] whereas the chloromercuri derivative of 5-formamido-7-chloroimidazo[4,5-*b*]pyridine[61] did react with **47** to furnish a respectable yield of 5-formamido-3-β-D-ribofuranosyl-7-chloroimidazo[4,5-*b*]pyridine (**49**)[62] after debenzoylation. The conformation of the structure of **48** was elucidated through a combination of ^{1}H-NMR and UV spectral studies,[60] ORD determination (by observing a negative Cotton effect), and the formation of a $N^{4},5'$-cyclo-nucleoside.

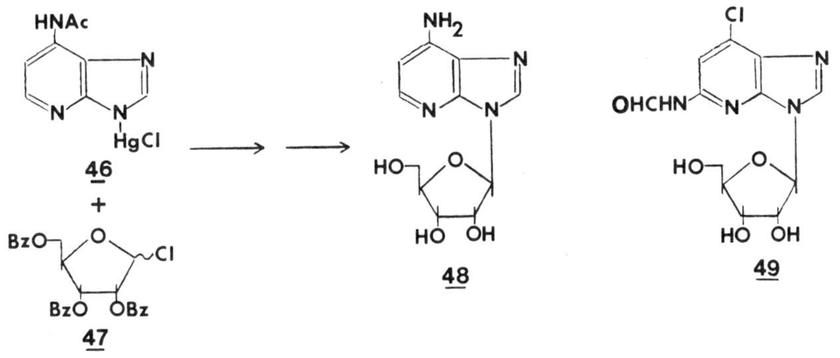

2.2.1.2. Synthesis Employing the Acid-Catalyzed Fusion of a Preformed Aglycon and an Acyl Sugar. The synthesis of 6-methylthio-3-β-D-ribofuranosylimidazo-[4,5-*b*]pyridine[53], the 1-deaza analogue of the cytotoxic 6-methylthio-9-β-D-ribofuranosylpurine, was first reported by Montgomery and Hewson,[63] employing the acid-catalyzed fusion procedure. Fusion of 7-chloroimidazo[4,5-*b*]-pyridine[64] (**50**) with 1,2,3,5-tetra-*O*-acetyl-β-D-ribofuranose (**51**) in the presence of either *p*-toluenesulfonic acid[63,65] or monochloroacetic acid,[66] followed by deacetylation with methanolic ammonia, gave 7-chloro-3-β-D-ribofuranosyl-imidazo[4,5-*b*]pyridine (**52**), along with a small amount of the α-anomer.[66]

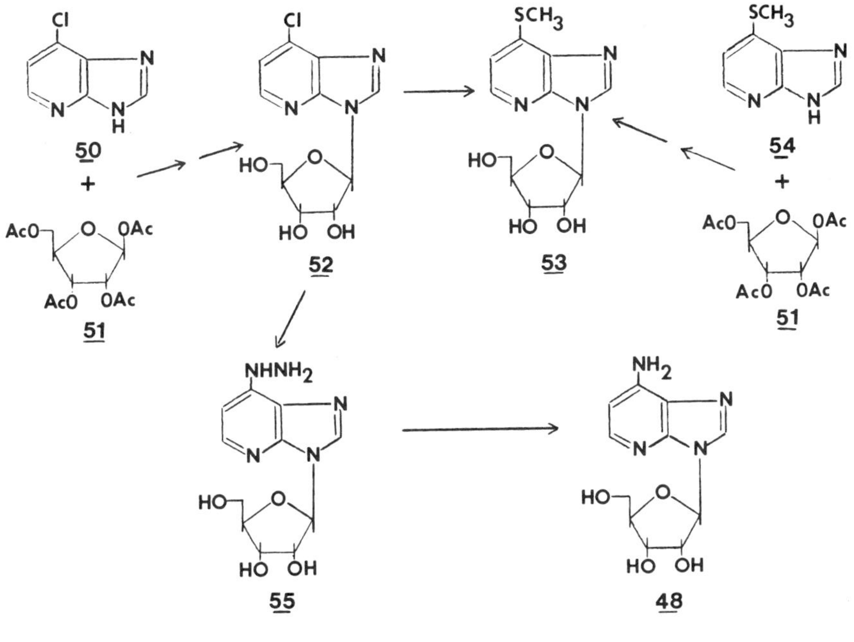

Catalytic hydrogenation of **52** afforded the nebularine analogue **39**[45,55] of established structure. Displacement of the chloro group of **52** with methanethiol gave **53** in a rather low yield (30%).[63] Alternatively, the acid-catalyzed fusion of 7-methylthioimidazo[4,5-*b*]pyridine (**54**) with **51**, followed by deacetylation of the intermediate blocked nucleoside, gave **53** in 67% yield along with a small amount of the α-anomer. Nucleophilic displacement of the chlorine atom of **52** with hydrazine hydrate and subsequent reduction of the resultant 7-hydrazino compound (**55**) with Raney nickel furnished 1-deazaadenosine (**48**) in 64% yield. This represents[66] the most practical method for the synthesis of **48**. A displacement of the halogen from **52** by dimethylamine proceeded in good yield (69%), but the yields from the reaction of **52** with methylamine or furfurylamine were rather low (~30%).

A similar fusion[60,62] of imidazo[4,5-*b*]pyridine with 1-*O*-acetyl-2,3,5-tri-*O*-benzoyl-β-D-ribofuranose (**57**), in the presence of *p*-toluenesulfonic acid, was followed by debenzoylation to give 3-β-D-ribofuranosylimidazo[4,5-*b*]pyridine (**39**) in 10% yield. However, when the fusion occurred in the presence of zinc chloride, the yield of **39** increased to 30%.[62] Interestingly enough, the fusion of 5-amino-7-chloroimidazo[4,5-*b*]pyridine (**56**) with **57** in the presence of zinc chloride and subsequent debenzoylation has been reported[62] to give 3-β-D-ribofuranosyl-5-ribosylamino-7-chloroimidazo[4,5-*b*]pyridine (**58**). The presence of two sugar residues was shown by ribose estimation. One of the ribosyl moieties was assigned to the imidazole nitrogen at the 3-position while the other ribosyl group was presumed to be on the exocyclic amino group. However, the actual structural assignment of **58** awaits further work. Similar fusion of 5,7-dichloroimidazo[4,5-*b*]pyridine (**59**)[67,68] with tetra-*O*-acetyl-β-D-ribofuranose (**51**) in the presence of a mixture of monochloroacetic acid and *p*-toluenesulfonic acid at 175°C afforded, after column chromatography, 5,7-dichloro-3-(2,3,5-tri-*O*-acetyl-β-D-ribofuranosyl)imidazo[4,5-*b*]pyridine (**60a**). Subsequent deacetylation of **60a** with methanolic ammonia gave 5,7-dichloro-3-β-D-ribofuranosyl-imidazo[4,5-*b*]pyridine[69] (**60b**) in a 25% overall yield. Catalytic hydrogenation of **60b** gave a dehalogenated product, identical with **39**, which established the site of ribosylation in **60b** as N-3 and the anomeric configuration as β. The reaction of **60b** with various nucleophiles afforded the 7-substituted 5-chloro derivatives, which, on hydrogenation in the presence of Pd/C, furnished a series of N[6]-substituted 1-deazaadenosines with significant adenosine receptor agonist properties.[70]

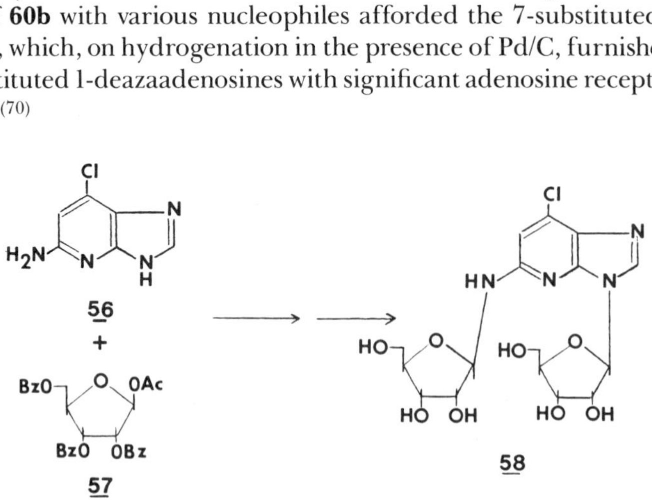

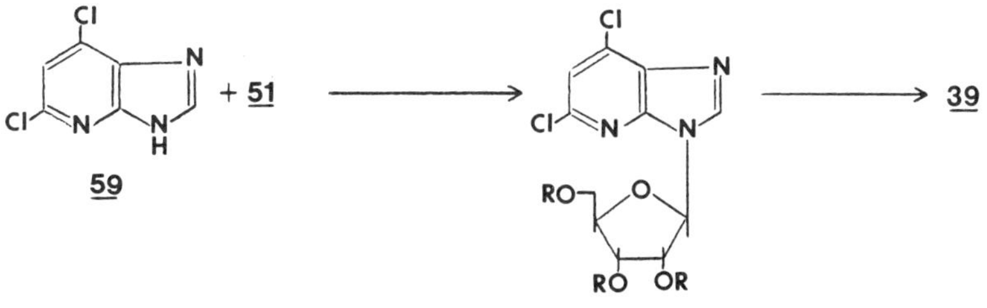

59 + **51** ⟶ **60a**, R = Ac

b, R = H

⟶ **39**

2.2.1.3. Synthesis Employing the Direct Condensation of a Preformed Aglycon and an Acyl Sugar Halide in the Presence of Mercuric Cyanide. The "mercuric cyanide procedure" developed by Yamaoka and co-workers[71] has been used as an alternative route to 1-deazapurine nucleosides.[72] In a typical reaction, Mizuno and co-workers[60] treated 7-aminoimidazo[4,5-*b*]pyridine hydrochloride (**61**)[73] with **47** in the presence of mercuric cyanide in nitromethane or nitrobenzene.[74] This was followed by debenzoylation of the crude product, which furnished a mixture of **48** (3%), 7-amino-4-β-D-ribofuranosylimidazo[4,5-*b*]pyridine (**62**, 30%), and 4-β-D-ribofuranosyl-7-ribosylaminoimidazo[4,5-*b*]pyridine (**63**, 1%). The individual nucleosides were separated on an alumina column; however, the unequivocal structural determination of **63** needs further work. Although in this case the "mercuric cyanide procedure" failed to afford a reasonably good yield of 1-deazaadenosine, it is an example in which alkylation of 1-deazaadenine in aprotic solvents takes place predominantly at N-4 of the imidazopyridine ring system.[75]

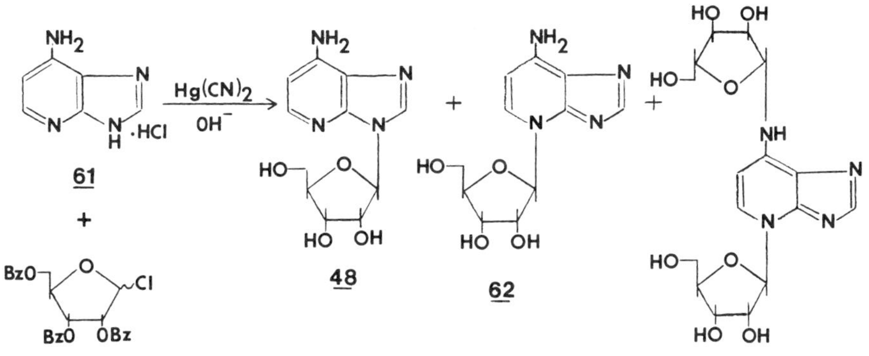

5-Aminoimidazo[4,5-*b*]pyridin-7-one (1-deazaguanine) was synthesized by Markees and Kidder[61] as early as 1956 and subsequently by Gorton and Shive.[76] Its biological activity as a purine antagonist proved to be rather disappointing. Better results were to be expected from the corresponding ribonucleoside[77]; therefore, Schelling and Salemink[69] prepared 1-deazaguanosine (**68**) by a direct condensation of 5-acetamido-7-benzyloxyimidazo[4,5-*b*]pyridine (**64**) with **47** in nitromethane in the presence of potassium cyanide as a hydrogen

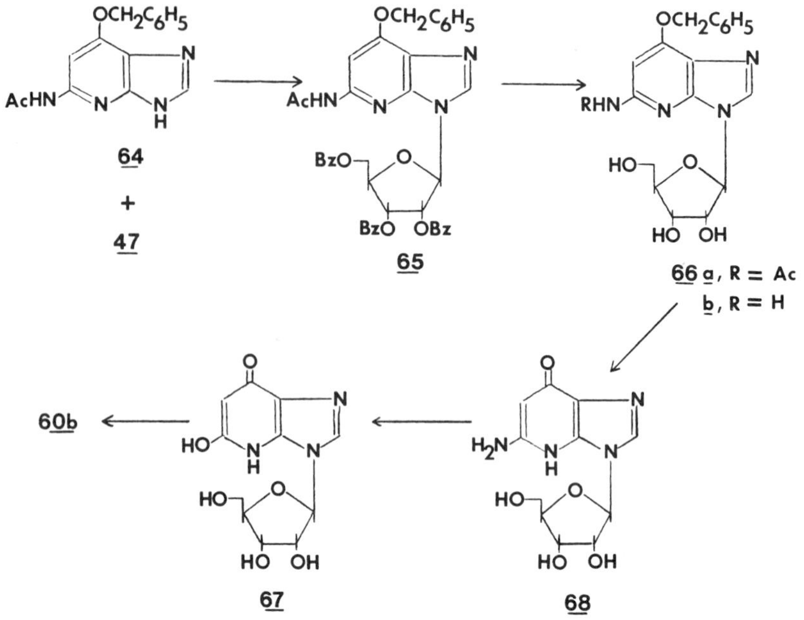

OCH$_2$C$_6$H$_5$ OCH$_2$C$_6$H$_5$ OCH$_2$C$_6$H$_5$

AcHN   **64**   AcHN   BzO   **65**   RHN   HO   HO   OH   **66 a**, R = Ac   **b**, R = H

+

47

60b ←   HO   HO   HO   OH   **67**   ← H$_2$N   HO   HO   OH   **68**

halide acceptor. Under these conditions, 5-acetamido-7-benzyloxy-3-(2,3,5-tri-*O*-benzoyl-β-D-ribofuranosyl)imidazo[4,5-*b*]pyridine (**65**) was isolated. Subsequent debenzoylation of **65** with sodium methoxide in boiling methanol gave a 1:1 mixture of 5-acetamido-7-benzyloxy-3-β-D-ribofuranosylimidazo[4,5-*b*]-pyridine (**66a**) and its deacetylated product (**66b**). Further treatment of the mixture with an aqueous sodium hydroxide solution gave **66b** as the sole product. Catalytic hydrogenation of **66b** gave 5-amino-3-β-D-ribofuranosylimidazo-[4,5-*b*]pyridin-7-one (**68**) in almost quantitative yield. Diazotization of **68** in dilute acetic acid furnished 1-deazaxanthosine[69] (**67**). Compound **67** was chlorinated with phosphorus oxychloride in the presence of *N*,*N*-dimethylaniline after acetylation of the ribosyl moiety,[78] to give the 5,7-dichloronucleoside (**60a**). Subsequent deacetylation of **60a** led to **60b**, identical with the product prepared by the fusion procedure by the same workers.[69] The structures of the compounds involved in the sequence leading to **68** were confirmed by IR, ^{1}H-NMR, and UV spectroscopic studies and were substantiated by the above transformation leading to **60b**.

The synthesis of 5-amino-3-β-D-ribofuranosylimidazo[4,5-*b*]pyridine-7-thione (1-deaza-6-thioguanosine, **72a**) was also accomplished[79] by the direct alkylation procedure. Condensation of 5-acetamido-7-chloroimidazo[4,5-*b*]pyridine[67] (**69**) with 2,3,5-tri-*O*-acetyl-D-ribofuranosyl chloride (**70**) in 1,2-dichloroethane containing Linde 4A molecular sieve gave the tetraacetyl nucleoside **71a**, which was deacetylated with sodium methoxide in methanol to give 5-amino-7-chloro-3-β-D-ribofuranosylimidazo[4,5-*b*]pyridine (**71b**). The reaction of **71b** with hydrogen sulfide in ethanol containing sodium methoxide in a sealed tube gave **72a**.[79] A similar reaction of **71b** with methanethiol gave 1-deaza-6-methylthioguanosine (**72b**).[79]

5-Aminoimidazo[4,5-*b*]pyridine ribonucleosides (**76a** and **77a**) have also been prepared by the direct alkylation procedure.[80] Ribosylation of 5-acet-amidoimidazo[4,5-*b*]pyridine (**73**) with **47** in nitromethane containing mercuric cyanide gave 5-acetamido-3-(2,3,5-tri-*O*-benzoyl-β-D-ribofuranosyl)imidazo-[4,5-*b*]pyridine (**74a**) together with a considerable amount (18.3%) of the N-1 glycosyl isomer (**75a**).[80] Treatment of each of the nucleosides (**74a** and **75a**) with methanolic sodium methoxide[81] effected a removal of the *O*-benzoyl protecting groups to give 5-acetamidoimidazo[4,5-*b*]pyridine nucleosides (**74b** and **75b**). Further treatment of **74b** and **75b** with hydrazine–acetic acid in pyridine at 70–75°C gave the corresponding 5-aminoimidazo[4,5-*b*]pyridine ribonucleosides (**76a** and **77a**).[80] Nucleosides **76a** and **77a** were treated with sodium nitrite in hypophosphorous acid to yield the imidazopyridine derivatives **76b** and **77b**,[80] respectively.

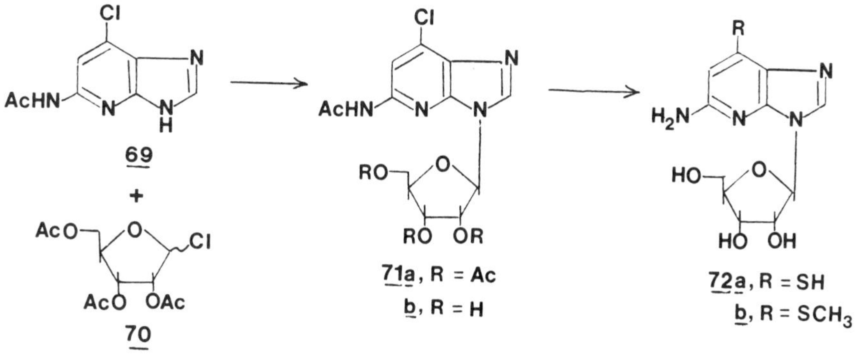

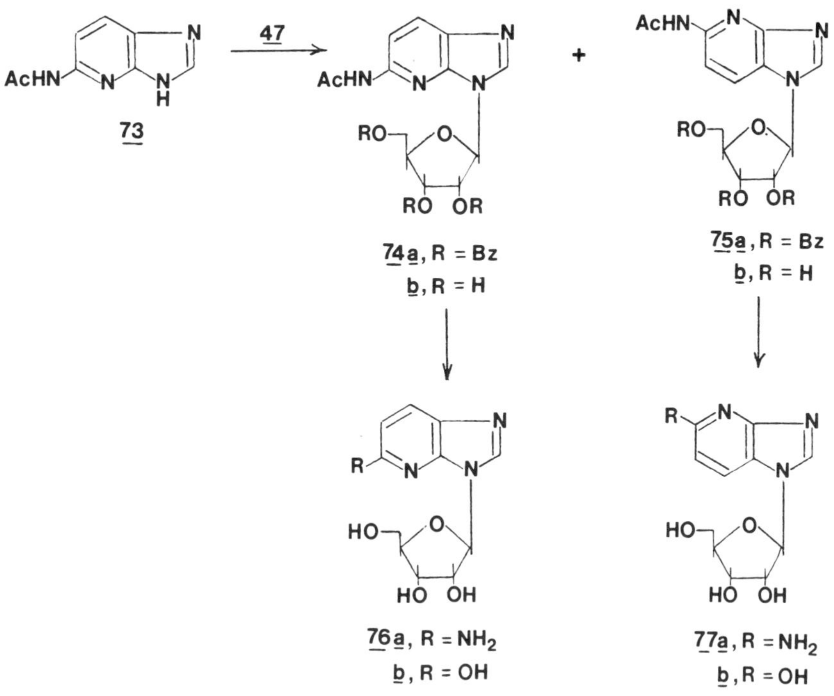

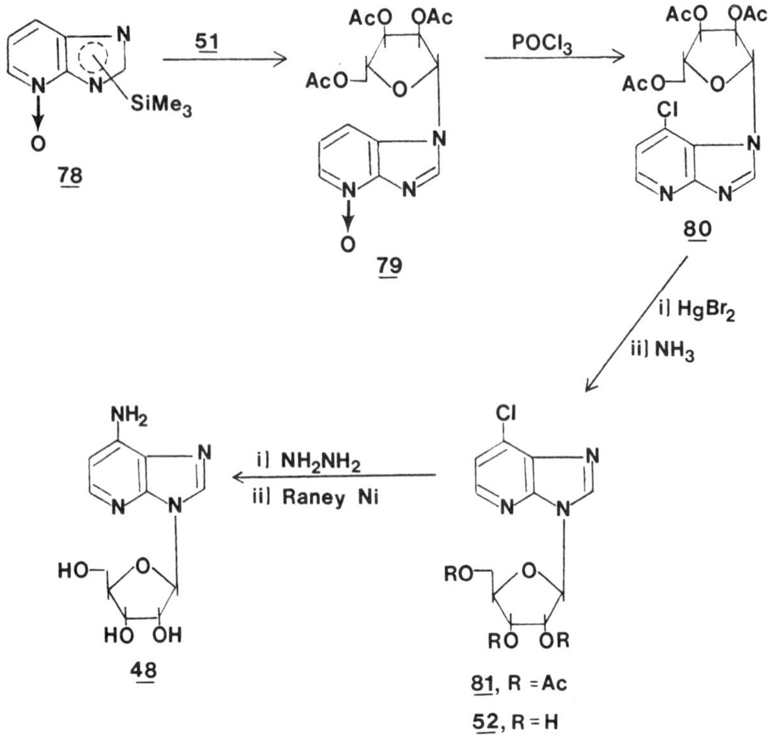

2.2.1.4. Synthesis Employing the Trimethylsilyl Derivative of a Preformed Aglycon and an Acylated Sugar Halide.   A new and efficient synthesis of 1-deazaadenosine (**48**) has recently been described using the trimethylsilyl procedure,[82,83] utilizing the recent advances in both aromatic amine *N*-oxides[84] and nucleoside chemistry.[85] Condensation of the trimethylsilyl derivative (**78**) of imidazo-[4,5-*b*]pyridine 4-oxide[84] with tetra-*O*-acetyl-β-D-ribofuranose (**51**) in the presence of stannic chloride gave an 82% yield of 1-(2,3,5-tri-*O*-acetyl-β-D-ribofuranosyl)imidazo[4,5-*b*]pyridine 4-oxide (**79**). Treatment of **79** with either phosphorus oxychloride or the Vilsmeier reagent (POCl₃/DMF) gave the chlorinated nucleoside **80**. When compound **80** was heated in toluene containing mercuric bromide, glycosyl migration occurred to afford the thermodynamically more stable product **81**.[82,83] 1-Deazaadenosine (**48**) was obtained from the deblocked nucleoside (**52**) by treatment with hydrazine hydrate, followed by Raney nickel reduction, in an overall yield of 25%.

Ribosylation of imidazo[4,5-*b*]pyridine or 7-nitroimidazo[4,5-*b*]pyridine[86] with **51**, catalyzed by an equimolar amount of stannic chloride in acetonitrile, is reported to yield exclusively 3-(2,3,5-tri-*O*-acetyl-β-D-ribofuranosyl)imidazo-[4,5-*b*]pyridine (57% yield)[85] and the corresponding 7-nitro derivative[86] (82% yield), respectively. In sharp contrast, when the trimethylsilyl derivative of imidazo[4,5-*b*]pyridine was used and the reaction time was limited, the kinetically controlled 4-(2,3,5-tri-*O*-acetyl-β-D-ribofuranosyl)imidazo[4,5-*b*]pyridine was obtained as a major product (60–70%), which readily rearranged under more drastic conditions of prolonged reaction time and a large excess of stannic chloride to furnish the N-1 and N-3 glycosyl isomers.[85] Similarly, glycosylation of the trimethylsilyl derivative of 6-nitroimidazo[4,5-*b*]pyridine with **57** in the presence of stannic chloride in acetonitrile gave the corresponding N-1 (10.8%) and N-3 (35%) glycosyl derivatives. Removal of the benzoyl blocking groups with sodium methoxide, followed by a catalytic reduction of the nitro group, gave 6-amino-1-β-D-ribofuranosylimidazo[4,5-*b*]pyridine and the corresponding N-3 ribofuranosyl isomer.[80]

1-Deazaguanosine (**68**) was also prepared by the trimethylsilyl procedure[87,88] in good yield starting with 5-acetamido-7-chloro-3-β-D-ribofurano-sylimidazo[4,5-*b*]pyridine (**83b**). Condensation of the trimethylsilyl derivative of 5-acetamido-7-chloroimidazo[4,5-*b*]pyridine (**82**) with **19** in the presence of mercuric cyanide gave an 80% yield of 5-acetamido-7-chloro-3-(2,3,5-tri-*O*-benzoyl-β-D-ribofuranosyl)imidazo[4,5-*b*]pyridine (**83a**). Treatment of **83a** with sodium methoxide in methanol gave **83b**, which, when heated with sodium benzylate in benzyl alcohol, afforded 5-benzylamino-7-benzyloxy-3-β-D-ribo-furanosylimidazo[4,5-*b*]pyridine (**84**). Hydrogenolysis of **84** using 10% palladium on carbon furnished 1-deazaguanosine (**68**).[87,88] A similar glycosylation of the trimethylsilyl derivative of 5-amino-7-chloroimidazo[4,5-*b*]pyridine with **19** and subsequent debenzylation of the condensation product with methanolic ammonia gave the versatile intermediate 5-amino-7-chloro-3-β-D-ribofura-nosylimidazo[4,5-*b*]pyridine, which has been used for further transformation reactions.[88] The versatility of the trimethylsilyl procedure has also been demonstrated by the preparation of the 3-β-D-xylofuranosyl and 3-α-L-rhamnopyra-nosyl derivatives of imidazo[4,5-*b*]pyridine.[89]

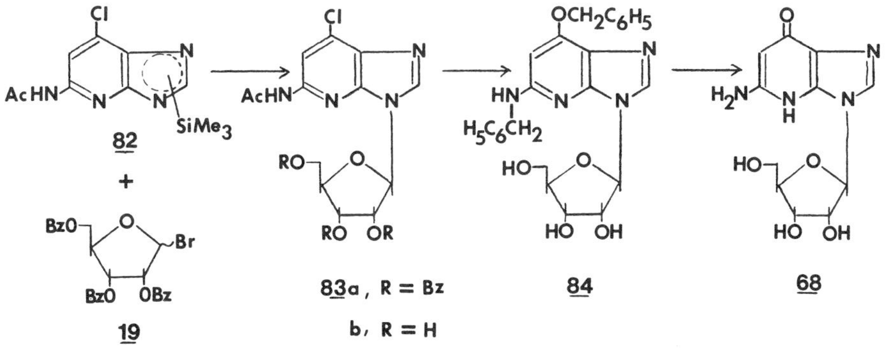

2.2.1.5. Chemical Transformations. A Dimroth-type ring transformation of a 1-methylpurine nucleoside to a 1-deazapurine nucleoside has been reported.[90] Sulfur methylation of 1-methyl-6-thioinosine (**85**), followed by treatment with diethyl sodiomalonate, gave 1-methyl-6-bis(ethoxycarbonyl)methyl-ene-1,6-dihydro-9-β-D-ribofuranosylpurine (**87**). Compound **87** underwent the Dimroth-type rearrangement in methanolic potassium hydroxide. This gave a 1-deazapurine derivative (**88**), which, on heating in aqueous alkali, afforded 7-methylamino-3-β-D-ribofuranosylimidazo[4,5-*b*]pyridin-5(4*H*)-one (**89**).[90]

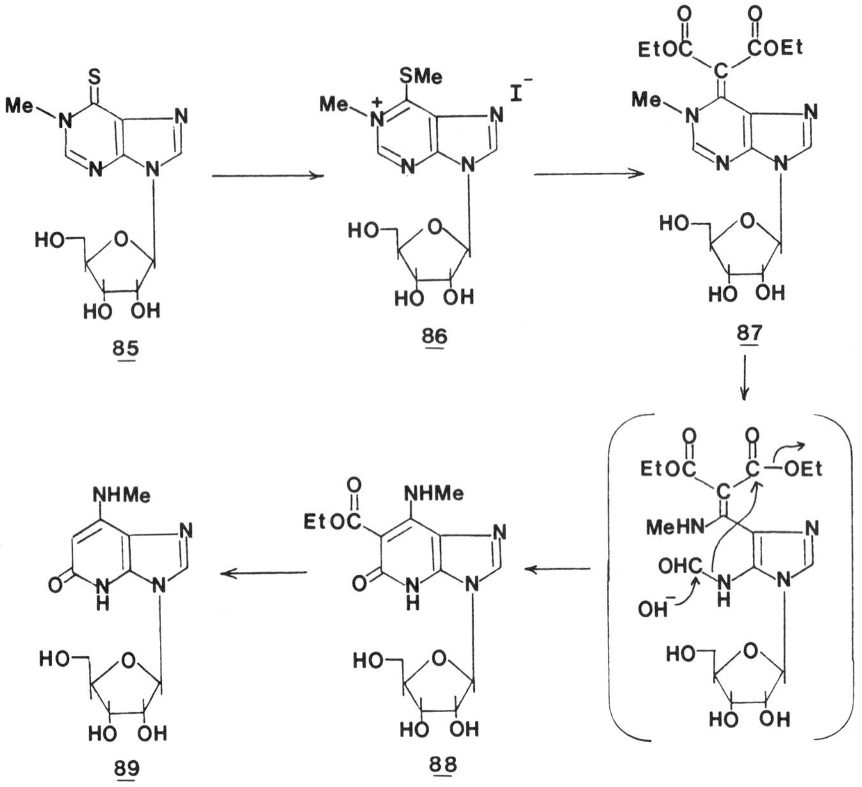

The use of recent advances in the transformation of ribofuranosides to arabinofuranosides[91] and 2'-deoxyribofuranosides[92,93] was found to be very fruitful for the preparation of 1-deaza-araA (**93**) and 2'-deoxy-1-deazaadenosine (**98**). The 3',5'-O-silylated 7-chloro-1-deazapurine ribonucleoside **90**, prepared from **52** and 1,3-dichloro-1,1,3,3-tetraisopropyldisiloxane, was converted into the corresponding trifluoromethanesulfonyl derivative (**91**).[94] Nucleophilic displacement of the trifyloxy group with sodium acetate in hexamethylphosphoramide gave the 2'-O-acetyl-arabinofuranoside **92**. Treatment of **92** with methanolic ammonia was followed by tetrabutylammonium fluoride to afford **94**. Compound **93** was obtained by the treatment of **94** with hydrazine hydrate and subsequent Raney nickel reduction.[94] Similarly, 3',5'-O-silylated 1-deazaadenosine (**96a**) was converted into the 2'-O-phenoxythiocarbonyl derivative **96b**, which was then reduced with tributyltin hydride to afford **97**. Treatment of **97** with tetrabutylammonium fluoride gave **98**.[94]

7-Furfurylamino-3-β-D-ribofuranosylimidazo[4,5-b]pyridine (1-deazakinetin ribofuranoside, **100**),[66] a potent cytokinin agent, was synthesized by condensing 1-deazaadenosine (**48**) with furfural, followed by reduction of the intermediate **99** with sodium borohydride. Alternatively, compound **100** was also prepared[65,87] from 7-chloro-3-β-D-ribofuranosylimidazo[4,5-b]pyridine (**52**) and furfurylamine in 52% yield.

1-Deazaadenosine was successfully converted to the 5'-alkylthio-5'-deoxy

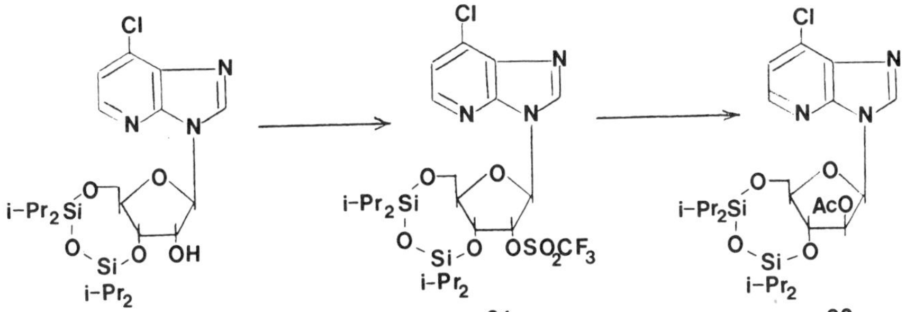

90 91 92

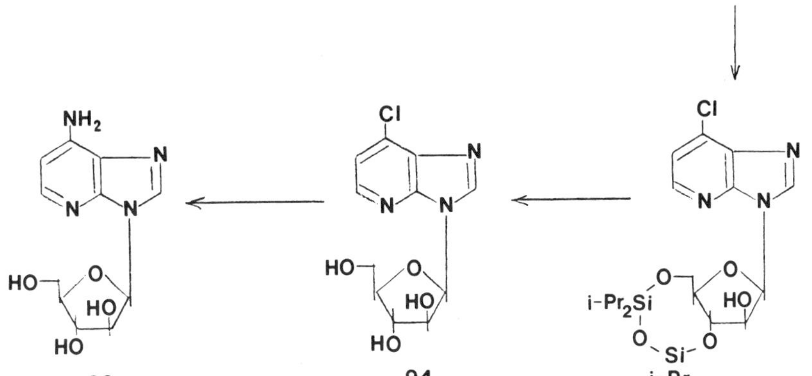

93 94 95

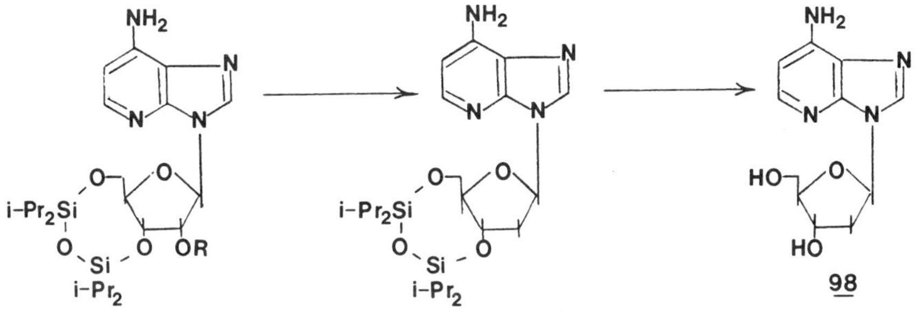

96a, R = H

b, R = C(S)OC$_6$H$_5$

97 98

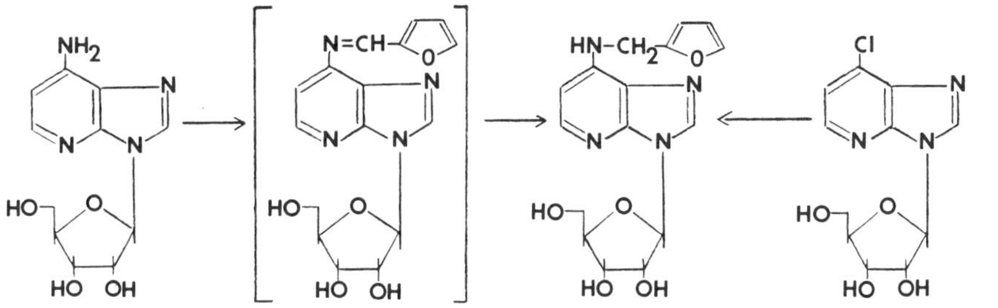

48 99 100 52

derivatives **102** via 5′-chloro-5′-deoxy-1-deazaadenosine (**101**). Treatment of
1-deazaadenosine (**48**) with thionyl chloride in hexamethylphosphoramide[95]
gave **101**. The reaction of **101** with isobutanethiol in the presence of sodium
hydride in dimethylformamide afforded 5′-deoxy-5′-isobutylthio-1-deazaadeno-
sine (1-deazaSIBA, **102a**).[83] Similarly, *S*-(1-deazaadenosyl)homocysteine (1-
deazaSAH, **102b**) was obtained from **101** and L-homocysteine in liquid ammonia.[83]

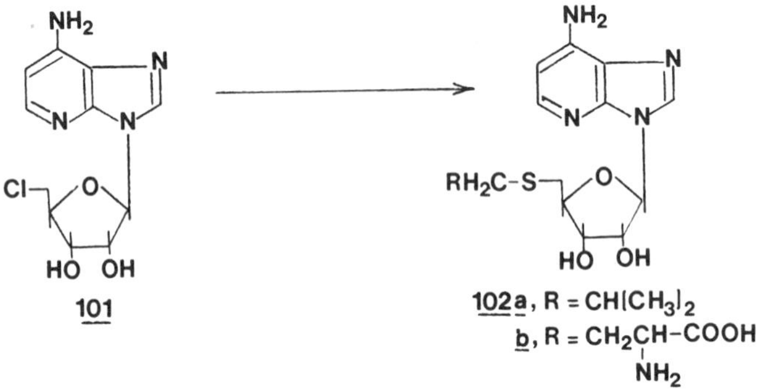

101

102a, R = CH(CH₃)₂
b, R = CH₂CH–COOH
                 |
                NH₂

Kitano *et al.*[96] studied the conformation of 1- and 3-deazaadenosines in
solution by ¹H-NMR spectroscopy. Features of coupling constants indicate that
the furanose rings of adenosine and 1- and 3-deazaadenosine have similar
conformational preferences and that conformations around the C(4′)—C(5′)
bond are preferentially *gauche–gauche*. Nuclear Overhauser effect and spin–
lattice relaxation time measurements demonstrate that 1-deazaadenosine predominantly adopts the *syn* conformation similar to that of adenosine, whereas
3-deazaadenosine has a greater *anti* component. These results would suggest
that the *syn* conformation in 1-deazaadenosine, as well as in adenosine, is stabilized presumably through a hydrogen bond between N-3 (purine nomenclature)
and the 5′-hydroxyl group.

2.2.1.6. Imidazo[4,5-b]pyridine Nucleotides. The first chemical synthesis of a
3-β-D-ribofuranosylimidazo[4,5-*b*]pyridine 5′-monophosphate by the conventional method was reported by Woenckhaus and Pfleiderer.[97] Subsequently,
mono- and diphosphates of 1-deazaadenosine have been synthesized by both
chemical[98,99] and enzymatic[100] methods. The synthesis of 1-deazaadenosine
5′-monophosphate (**105**) was also reported by Mizuno *et al.*[101] by the treatment
of 2′,3′-*O*-ethoxymethylene-1-deazaadenosine (**103**) with pyrophosphoryl tetrachloride. This blocking group was used since 1-deazaadenosine is more acid
labile than 3-deazaadenosine with respect to the N-glycosyl bond cleavage.
When 2′,3′-*O*-isopropylidene-1-deazaadenosine was used for the synthesis of
105, a removal of the protecting group under acidic conditions was always
accompanied by considerable cleavage of the N-glycosyl bond. By employing the
general procedure of Khorana and co-workers[102] for the synthesis of nucleoside
3′,5′-cyclic phosphates from 5′-phosphates, 1-deazaadenosine 3′,5′-cyclic phosphate (**107**) was prepared[101] from **105** in a 58% yield. Hydrolysis of **107** with

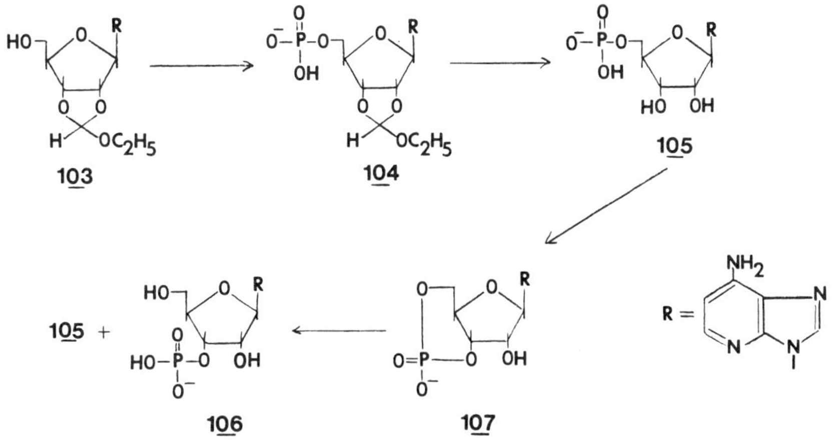

103 → **104** → **105**

R = (7-amino-imidazo[4,5-b]pyridine structure with NH$_2$)

105 + **106** ← **107**

barium hydroxide gave **105** and 1-deazaadenosine 3′-phosphate (**106**) in a ratio of 27:73.

For the synthesis of the nucleoside 2′,3′-cyclic phosphate, a number of methods are available, among which Ueda and Kawai's procedure[103] was found to be successful. Using this procedure, 1-deazaadenosine 2′,3′-cyclic phosphate (**108**) was prepared[101] in 23% yield, whereas only a 3% yield of 3-deazaadenosine 2′,3′-cyclic phosphate was obtained using the same approach. The nucleotide **108** could be hydrolyzed to afford the corresponding nucleoside 3′-phosphate (**106**) and 2′-phosphate (**109**) by the enzyme ribonuclease M.

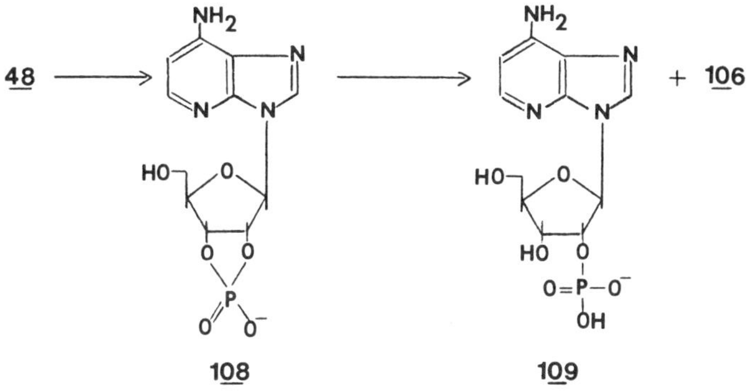

48 → **108** → **109** + **106**

Extension of this work led Mizuno *et al.*[104] to prepare the dinucleoside phosphates[105] containing 1-deaza- and 3-deazaadenosine. It is important to protect both the 5′-hydroxy and 7-amino group of **48** with the alkali-labile dimethylaminomethylene group.[106,107] 1-Deazaadenosine (**48**) reacted smoothly with dimethylformamide dimethylacetal to give 7-(2-dimethylaminomethylene-amino)-3-β-D-ribofuranosylimidazo[4,5-*b*]pyridine,[108] which, in turn, was treated with dimethoxytrityl chloride (DMTrCl) to afford the 7-(2-dimethyl-aminomethyleneamino)-5′-*O*-dimethoxytrityl compound (**110**). Treatment of **110** with 2′,3′,N^6-triacetyladenosine 5′-monophosphate in the presence of 2,4,6-triisopropylbenzenesulfonyl chloride furnished the corresponding blocked dinucleoside monophosphate. This compound was then deblocked by the conventional procedure to give **111**. Similarly, adenylyldeazaadenosine (**112**) has been prepared.[104] Hypochromicity and CD spectra of some of these dinucleoside monophosphates were determined.[99,104]

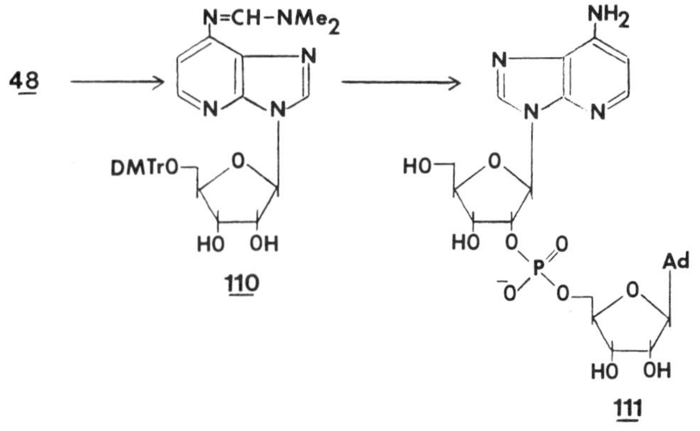

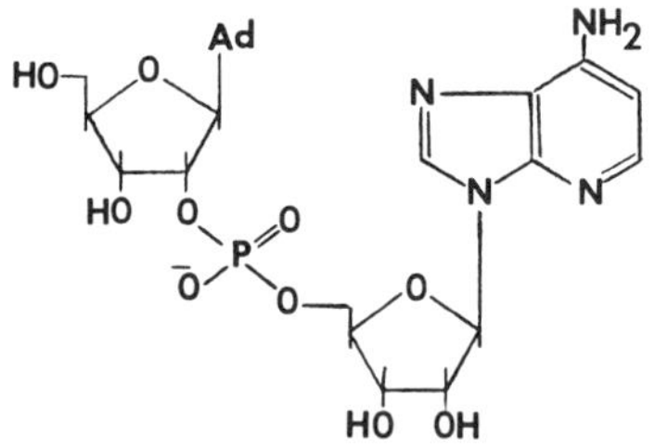

2.2.1.7. Biochemical Properties of 1-Deazapurine Nucleosides and Nucleotides. 3-β-D-Ribufuranosylimidazo[4,5-b]pyridine completely inhibited the Ranikhet disease virus multiplication at concentrations of 2.5, 1.25, and 0.625 mg/ml,[109–111] but the corresponding 6-halo (chloro or bromo) derivatives are completely inactive.[112] 3-β-D-Glucopyranosyl- and 3-β-D-galactopyranosyl-6-bromoimidazo[4,5-b]pyridine showed strong antibacterial activity against both gram-positive and gram-negative bacteria.[113] 1-Deazaadenosine showed good activity *in vitro* as an inhibitor of HeLa, KB, P388, and L1210 leukemia cell line growth, with ID_{50} values ranging from 0.34 μM (KB) to 1.8 μM (P388).[86]

1-Deazakinetin ribofuranoside (**100**) and 7-(3-methyl-2-butenylamino)-3-β-D-ribofuranosylimidazo[4,5-b]pyridine have been tested for their growth-promoting activity in the tobacco bioassay.[65,114] These nucleosides, in the concentration range from 320 μM to 125 μM, were found to be less active than kinetin itself.

1-Deazaadenosine was not deaminated by calf intestine adenosine deaminase[115]; however, it showed an inhibitory effect toward the deamination of adenosine.[116] 1-Deazaadenosine is a competitive inhibitor of adenosine deaminase, and its inhibitory effect is of the same order of magnitude as that reported for 2-aminopurine riboside.[117] 1-Deazaadenosine phosphate derivatives show the ability to act as substrates or activators for a number of enzymes, including cAMP-dependent protein kinase and 5′-nucleotidase.[101] In sharp contrast, 3-deazaadenosine phosphate derivatives were found to be poor (nearly inactive) substrates or activators in these enzyme systems. 1-Deazaadenosine 3′,5′-cyclic phosphate was shown to be a better activator of protein kinase than cAMP.[118] 1-Deazaadenosine diphosphate is a good substrate for *E. coli* polynucleotide phosphorylase.[119] Both 1-deaza- and 3-deazaadenosine 5′-phosphates were completely hydrolyzed by snake venom 5′-nucleotidase, but at different rates.[101]

2.2.2. Imidazo[4,5-c]pyridine (3-Deazapurine) Nucleosides

2.2.2.1. Synthesis Employing the Halomercuri Salt of a Preformed Aglycon and an Acylated Sugar Halide. The first synthesis of an imidazo[4,5-c]pyridine (3-deazapurine) nucleoside was reported simultaneously by two groups.[46,51] Condensation of the chloromercuri-imidazo[4,5-c]pyridine derivative **113a** with the chlorobenzoyl sugar derivative **47** in boiling xylene, followed by debenzoylation of the blocked nucleoside (**114a**), gave 3-β-D-ribofuranosylimidazo[4,5-c] pyridine (**117a**). The other positional isomer (**115a**) escaped the attention of Jain *et al.*[51] A more careful study by Mizuno and co-workers[46,120] provided **115a** in a 27% yield; on debenzoylation, **115a** furnished 1-β-D-ribofuranosylimidazo-[4,5-c]pyridine (3-deazanebularine, **118a**). 4-Chloroimidazo[4,5-c]pyridine[121-123] under similar conditions gave two isomeric blocked nucleosides (**114b** and **115b**), which, on debenzoylation with methanolic ammonia, afforded **117b** and **118b**, respectively.

The structural assignment of **114b** was made by converting **114b** to **117a**. 3-(2,3,5-Tri-O-benzoyl-β-D-ribofuranosyl)imidazo[4,5-c]pyridine (**114a**) was

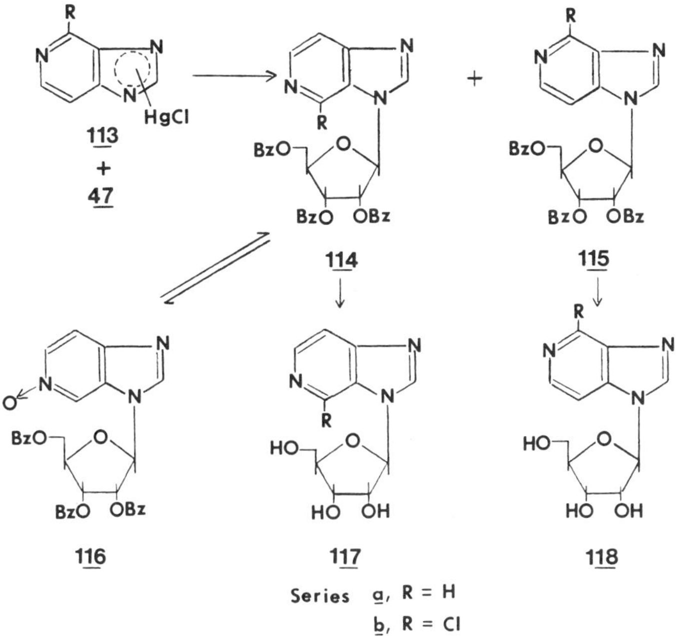

Series **a**, R = H
b, R = Cl

converted into the corresponding N_5-oxide **116** with monoperphthalic acid. Compound **116**, on treatment with phosphoryl chloride, gave 3-(2,3,5-tri-*O*-benzoyl-β-D-ribufuranosyl)-4-chloroimidazo[4,5-*c*]pyridine in 59% yield, identical with **114b**. Catalytic dehalogenation of **118b** gave **118a**. Although no unequivocal proof was obtained for the β-configuration, the site of ribosylation was assigned by the comparison of UV absorption spectra of these nucleosides with those of the corresponding model methyl compounds.

The halomercuri salt procedure has also been used[124,125] for the preparation of various 2-substituted imidazo[4,5-*c*]pyridine nucleosides.

2.2.2.2. Synthesis Employing the Acid-Catalyzed Fusion of a Preformed Aglycon and an Acyl Sugar.   The synthesis of 4-methylthio-1-β-D-ribofuranosylimidazo-[4,5-*c*]pyridine (**124a**) and 4-amino-1-β-D-ribofuranosylimidazo[4,5-*c*]pyridine (3-deazaadenosine, **122**) was reported almost simultaneously by two groups[123,126] using the acid-catalyzed fusion procedure.[127] 4-Chloroimidazo[4,5-*c*]pyridine (**119**)[121–123] was fused with tetra-*O*-acetylribofuranose (**51**) in the presence of either *p*-toluenesulfonic acid[123] or monochloroacetic acid[126] at 160°C to furnish crystalline 4-chloro-1-(2,3,5-tri-*O*-acetyl-β-D-ribofuranosyl)imidazo[4,5-*c*]-pyridine (**120a**)[126] as the major product. A small amount of the α-anomer was also isolated, which, on treatment with alcoholic ammonia, gave 4-chloro-1-α-D-ribofuranosylimidazo[4,5-*c*]pyridine. Similar deacetylation of **120a** with ethanolic ammonia afforded 4-chloro-1-β-D-ribofuranosylimidazo[4,5-*c*]pyridine (**120b**).[120] The chloro group of **120b** was found to be less reactive than that of 6-chloropurine ribonucleoside. However, the chloro group of **120b** could be replaced by a mercapto group by the treatment of **120b** with sodium hydrosulfide in methanol at 90°C for 18 h in a pressure vessel to obtain 3-deaza-6-

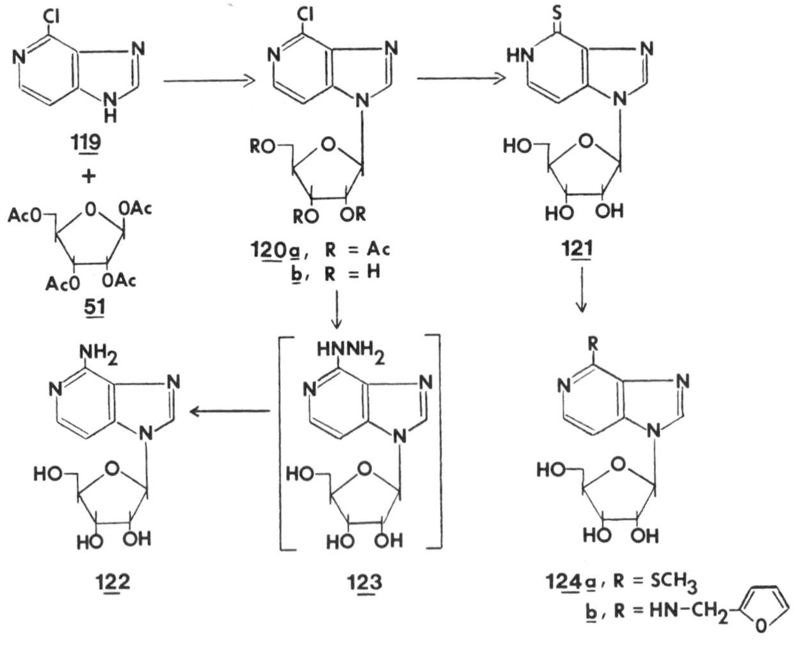

thiopurine ribonucleoside (**121**). Methylation of **121** with methyl iodide gave **124a**, which was found to be far less cytotoxic in HEp-2 cell lines than 6-methyl-thiopurine ribonucleoside.[123]

Treatment of **120b** with hydrazine in a nitrogen atmosphere yielded the corresponding 4-hydrazine derivative **123**, which was then converted with Raney nickel to the desired 4-amino-1-β-D-ribofuranosylimidazo[4,5-*c*]pyridine (**122**) in good yield. This represented the first reported[126] synthesis of 3-deazaadeno-sine of melting point 225–226°C. Although the yield of 3-deazaadenosine was later claimed[128] to be improved to 85%, the reported low melting point (145–147°C) raises doubts about the authenticity of the product. It is interesting to note that the halomercuri salt coupling of 4-chloroimidazo[4,5-*c*]pyridine gave[120] approximately equal amounts of nucleoside products with the D-ribo-furanose moiety at position 1 and at position 3. However, the acid-catalyzed fusion procedure gave no evidence whatsoever of sugar attachment at any position other than position 1. The anomeric configuration of **120b** and its α-anomer was unequivocally established by subjecting this anomeric pair to periodate oxidation followed by reduction with sodium borohydride.[126]

Nucleophilic substitution of the 4-chloro group of **120b** with furfurylamine gave 4-furfurylamino-1-β-D-ribofuranosylimidazo[4,5-*c*]pyridine (**124b**, 3-de-azakinetin ribofuranoside).[65]

In contrast to the trimethylsilyl procedure, the acid-catalyzed fusion of 4,6-dichloroimidazo[4,5-*c*]pyridine (**125**) with **51** gave 4,6-dichloro-1-(2,3,5-tri-*O*-acetyl-β-D-ribofuranosyl)imidazo[4,5-*c*]pyridine (**126**) in 88% yield, along with a trace amount of another nucleoside product.[129,130] Treatment of **126** with ethanolic ammonia at 140°C gave 4-amino-6-chloro-1-β-D-ribofuranosylimid-azo[4,5-*c*]pyridine (**127**),[129] which, on reductive dechlorination, afforded 3-deazaadenosine (**122**).

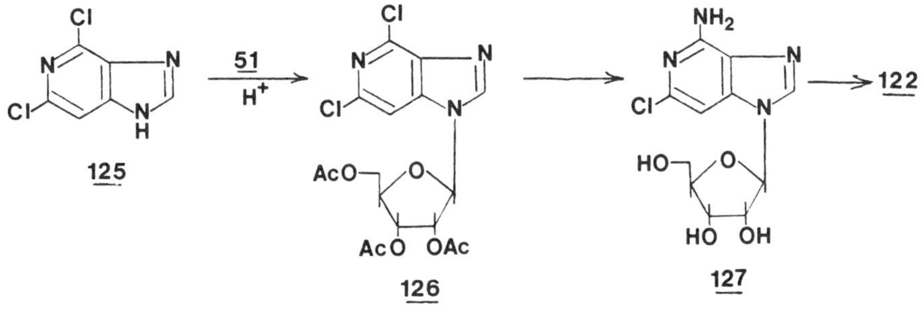

2.2.2.3. Synthesis Employing the Direct Condensation of a Preformed Aglycon and an Acyl Sugar Halide. The "mercuric cyanide procedure"[71] was used by Mi-zuno *et al.*[131] for the preparation of 4-chloro-1-(2,3,5-tri-*O*-benzoyl-β-D-ribo-furanosyl)imidazo[4,5-*c*]pyridine (**128**). Thus, direct condensation of 4-chloro-imidazo[4,5-*c*]pyridine (**119**) with 2,3,5-tri-*O*-benzoyl-D-ribofuranosyl chloride (**47**) in the presence of mercuric cyanide gave crystalline **128** in a 47% yield.[131] However, the yield of **128** was no better than that attained through the acid-catalyzed fusion procedure (72%). Debenzoylation of **128** furnished **120b**, which, upon treatment with methylamine at 120°C, gave the corresponding 4-methylamino nucleoside (**129**, R = NHCH$_3$).[128,131] Reaction of **120b** with aqueous hydrazine gave the dehalogenated nucleoside, 1-β-D-ribofuranosyl-imidazo[4,5-c]pyridine (**118a**), rather than the hydrazino derivative. However, treatment of **120b** with anhydrous ethanolic hydrazine at 120°C in a sealed tube gave a 68% yield of the *crystalline* 4-hydrazino derivative, which, on subsequent hydrogenation, gave 3-deazaadenosine (**122**).[131] Compound **120b** in boiling aqueous acetic acid or acetic acid in the presence of sodium acetate gave 3-deaza-inosine (**129**, R = OH) along with several by-products.

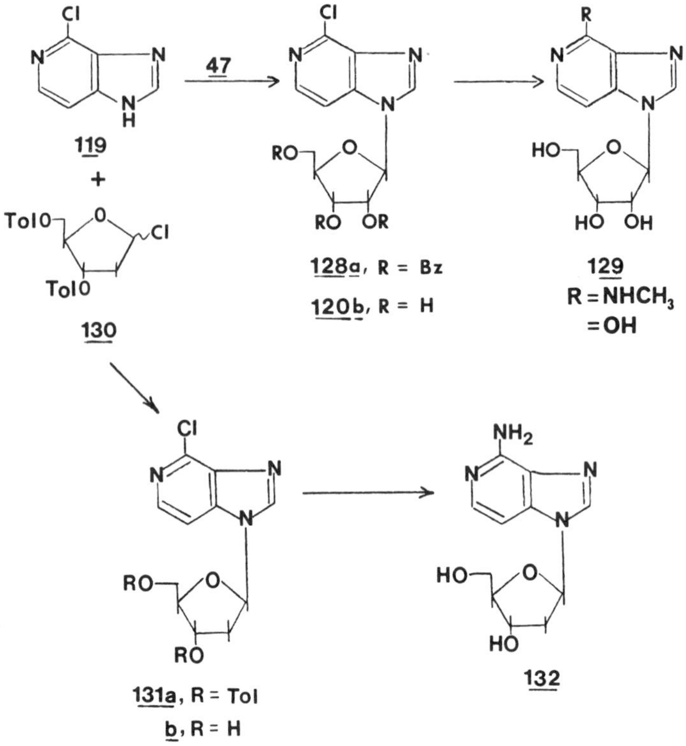

This direct condensation procedure has been found to be applicable to the preparation of 2'-deoxy-3-deazaadenosine (**132**).[132] Treatment of **119** with 2-deoxy-3,5-di-O-*p*-toluoyl-α-D-*erythro*-pentofuranosyl chloride (**130**) in nitro-methane containing mercuric cyanide gave a mixture of the blocked α- and β-anomers of **131a**. Deacylation of **131a** with methanolic ammonia gave a 33.2% yield of 4-chloro-1-(2-deoxy-β-D-*erythro*-pentofuranosyl)imidazo[4,5-c]pyridine (**131b**) and its α-anomer.[132] Further treatment of **131b** with anhydrous hydra-zine followed by Raney nickel reduction gave **132**.

The synthesis of 1-β-D-arabinofuranosylimidazo[4,5-c]pyridine (ara-3-deazaadenine, **133**) was also accomplished by the direct condensation proce-dure.[133] When 4,6-dichloroimidazo[4,5-c]pyridine (**125**) was allowed to react with 2,3,5-tri-O-benzyl-α-D-arabinofuranosyl chloride in refluxing 1,2-di-chloroethane in the presence of molecular sieves, a mixture of four blocked

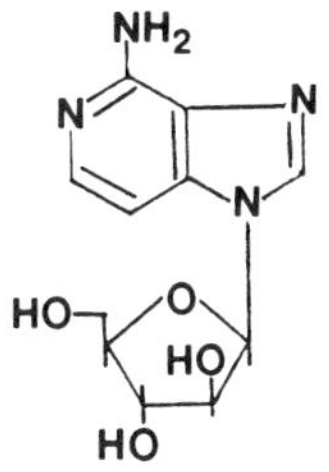

133

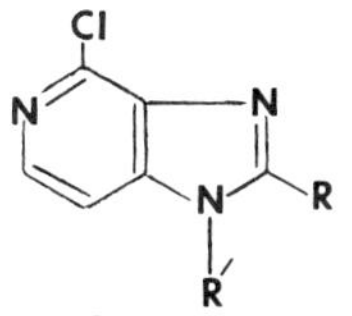

134

R = H or CH₃ : R′ = β-D-galactopyranosyl

β-D-glucopyranosyl

β-D-ribofuranosyl

nucleosides was formed. After resolving the mixture by silica gel column chromatography, 4,6-dichloro-1-(2,3,5-tri-*O*-benzyl-β-D-arabinofuranosyl)imidazo-[4,5-*c*]pyridine was obtained in 34–37% yield. Treatment of this blocked nucleoside with ethanolic ammonia, as patterned after the procedure used for preparation of **122** from **125**, gave, after dechlorination and debenzylation, the desired ara-3-deazaadenine.[133]

By using this direct condensation method, Stetsenko and Miroshnichenko[134] have prepared various 2-methyl-4-chloro-1-β-D-glycosylimidazo[4,5-*c*]pyridines (**134**) in good yields.

2.2.2.4. Synthesis Employing the Trimethylsilyl Derivative of a Preformed Aglycon and an Acylated Sugar Halide. Both the acid-catalyzed fusion procedure[126] and the mercuric cyanide method[131] furnished 3-deazaadenosine in rather low yield. The major disadvantage of these syntheses was the low susceptibility to nucleophilic displacement of the chloro group of 4-chloro-1-β-D-ribofuranosyl-imidazo[4,5-*c*]pyridine (**120b**) in comparison with that of 6-chloropurine riboside, a result of the higher electron density at C-4 in the former nucleoside. This prompted May and Townsend[135,136] to synthesize 4,6-dichloro-1-β-D-ribo-furanosylimidazo[4,5-*c*]pyridine (β-**136b**), in which the presence of an electron-withdrawing group at C-6 decreases the electron density at C-4, thereby increasing the reactivity of the 4-chloro group toward nucleophilic displacement. Thus, the condensation of the trimethylsilyl derivative (**135**) of 4,6-dichloroimid-azo[4,5-*c*]pyridine with 2,3,5-tri-*O*-benzoyl-D-ribofuranosyl bromide (**19**) gave a mixture of four nucleosidic products, which were separated by silica gel column chromatography. Their identity was established by UV and ¹H-NMR[137] spectroscopy as 4,6-dichloro-1-(2,3,5-tri-*O*-benzoyl-α- and -β-D-ribofuranosyl)imid-azo[4,5-*c*]pyridine (α-**136a**, 30%, and β-**136a**, 43%) and 4,6-dichloro-3-(2,3,5-tri-*O*-benzoyl-α- and -β-D-ribofuranosyl)imidazo[4,5-*c*]pyridine (α-**137a**, 11%, and β-**137a**, 16%). In contrast, the acid-catalyzed fusion of 4,6-dichloro-imidazo[4,5-*c*]pyridine and tetra-*O*-acetyl-β-D-ribofuranose gave essentially a single nucleoside, 4,6-dichloro-1-(2,3,5-tri-*O*-acetyl-β-D-ribofuranosyl)imid-azo[4,5-*c*]pyridine (**126**), in 88% yield with only traces of other nucleosidic material.[129] A removal of the benzoyl groups from **136a** and **137a** with sodium methoxide gave the free nucleosides, **136b** and **137b**.[136,138] The reaction of β-**136b** with various nucleophiles afforded the 4-substituted 6-chloro derivatives, which, on further catalytic hydrogenation, afforded a series of 4-substituted

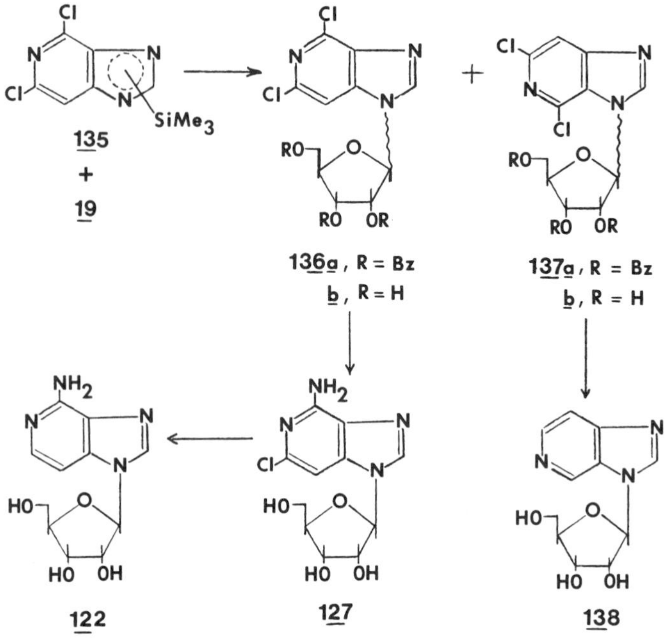

1-β-D-ribofuranosylimidazo[4,5-c]pyridines. Treatment of β-**136b** with ethanolic ammonia at 140°C gave **127**, which, on subsequent reductive dehalogenation, gave 3-deazaadenosine (**122**) in good yield.

The above route was found to be unrewarding when applied to the synthesis
of 6-amino-1-β-D-ribofuranosylimidazo[4,5-c]pyridine-4-thione (3-deaza-6-
thioguanosine, **143a**), a structural analogue of 6-thioguanosine with strong
growth-inhibitory activity against transplanted tumors in animals.[139,140] However, the successful synthesis of **143a** was accomplished[104,105] as follows. In order
to increase the susceptibility of the bromo group of 6-amino-4-bromoimidazo-
[4,5-c]pyridine[141] toward nucleophilic displacement, as well as to prevent
ribosylation of the exocyclic amino group,[60,62] 6-acetamido-4-bromoimidazo-
[4,5-c]pyridine (**139**) was selected[142] for glycosylation studies. The silylation of
139 with N,O-bis(trimethylsilyl)acetamide gave the crystalline bis-trimethylsilyl
derivative (**140**). Compound **140** was glycosylated with 2,3,5-tri-O-benzoyl-D-
ribofuranosyl bromide (**19**) in the presence of sodium iodide under silyl fusion
conditions[143] (method A) to provide, after removal of protecting groups, a
mixture of three nucleosides. These compounds were identified as 6-acetamido-
4-bromo-1-α- and -β-D-ribofuranosylimidazo[4,5-c]pyridine (α-**142** and β-**142**)
and 6-acetamido-4-bromo-3-β-D-ribofuranosylimidazo[4,5-c]pyridine (**141**).
However, coupling of **140** with **19** under Wittenburg conditions[144] [in boiling
toluene in the presence of mercuric bromide and mercuric oxide, (method B)]
afforded a mixture of two nucleosides, **141** and β-**142**, with no evidence for the
formation of α-**142**. A similar condensation of **140** with **19** in the presence of
mercuric cyanide in boiling nitromethane again provided a mixture of **141** and
β-**142**, with no observed formation of the α-anomer.[142] The yield of the individual nucleosides varied with the glycosylation conditions employed. Thiation of

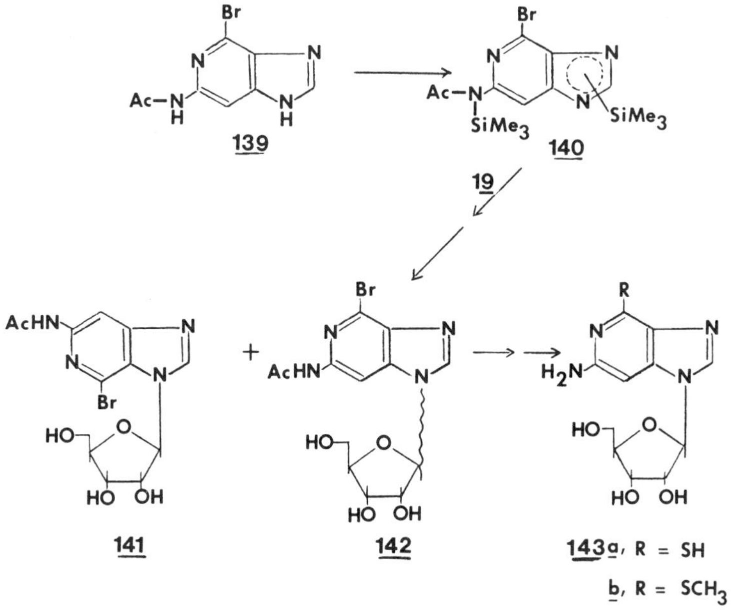

β-**142** with a concomitant deacetylation was accomplished with methanolic sodium hydrogen sulfide to provide 6-amino-1-β-D-ribofuranosylimidazo[4,5-*c*]-pyridine-4-thione (**143a**, 3-deaza-6-thioguanosine). Treatment of an aqueous solution of 3-deaza-6-thioguanosine with methyl iodide in the presence of an equivalent amount of sodium hydroxide provided the corresponding 4-methyl-thio derivative (**143b**). The structures of all the nucleosides were established by the conventional methods, including NOE measurements.[145]

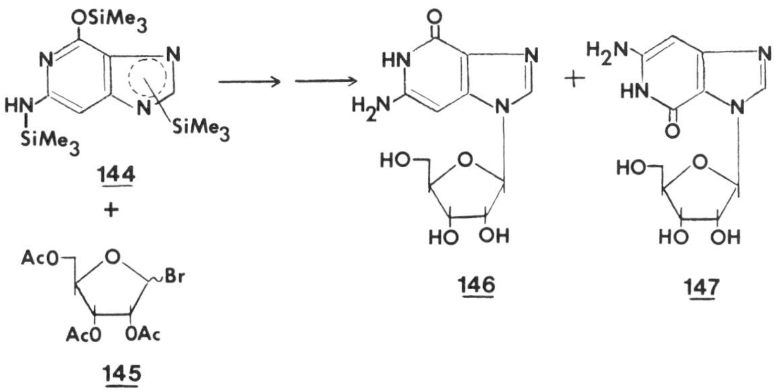

The synthesis of 3-deazaguanosine (**146**) was also accomplished by employing this procedure.[146,147] The condensation of the trimethylsilyl derivative of 6-aminoimidazo[4,5-*c*]pyridin-4-one (**144**) with 2,3,5-tri-*O*-acetyl-D-ribofuranosyl bromide (**145**) in acetonitrile for three days afforded a mixture of several nucleoside products from which, after deacetylation, 6-amino-1-β-D-ribofuranosylimidazo[4,5-*c*]pyridin-4-one (3-deazaguanosine, **146**) and the corresponding N-3 glycosyl isomer (**147**) were isolated in crystalline form in rather low yields. These compounds were found to be identical with the ones synthesized unambiguously by the ring-closure procedure (Section 2.2.2.5).

A similar glycosylation of the trimethylsilyl derivative of 1,3-dihydroimidazo[4,5-*c*]pyridin-2-one with 1-*O*-acetyl-2,3,5-tri-*O*-benzoyl-β-D-ribofuranose in the presence of SnCl$_4$ in a mixture of 1,2-dichloroethane and acetonitrile gave N-1 and N-3 monoribonucleosides in a 0.55:0.45 ratio in 42% overall yield. Deprotection of these nucleosides with methanolic ammonia furnished 1,3-dihydro-1-β-D-ribofuranosylimidazo[4,5-*c*]pyridin-2-one and the N-3 glycosyl isomer in 85% and 90% yield, respectively.[148]

2.2.2.5. Synthesis Employing the Ring Closure of an Imidazole Nucleoside Precursor.

The prerequisite imidazole nucleoside, methyl 5-cyanomethyl-1-(2,3,5-tri-*O*-acyl-β-D-ribofuranosyl)imidazole-4-carboxylate (**150**), for the synthesis of 3-deazaguanosine (**146**), was obtained by Robins and co-workers[147] by three independent procedures: (A) stannic chloride-catalyzed condensation of methyl 5(4)-cyanomethyl-*N*-trimethylsilylimidazole-4(5)-carboxylate (**149**) with a fully acylated ribofuranose (**51** or **57**)[149]; (B) condensation of **149** with **19** in acetonitrile[150]; and (C) acid-catalyzed fusion of **149** with **51**.

It was found that the yield and ratio of positional isomers of imidazole nucleosides in method A markedly depends on the ratio of stannic chloride to **149** and the protected sugars, **51** or **57**. Thus, the treatment of 1 equivalent of **149** in 1,2-dichloroethane or acetonitrile with 1 equivalent of **51** or **57** and 1.44 molar equivalent of stannic chloride gave a quantitative yield of the blocked imidazole nucleoside, **150** or **151**. However, condensation of 1 equivalent of **149** with 1 equivalent of **57** and 0.72 molar equivalent of stannic chloride afforded a 29.5% yield of **150** and a 34.5% yield of methyl 4-cyanomethyl-1-(2,3,5-tri-O-benzoyl-β-D-ribofuranosyl)imidazole-5-carboxylate (**153**). Further reduction of the amount of stannic chloride to 0.36 molar equivalent effected a slow and incomplete reaction to form **150** and **153** in 10% and 20% yields, respectively. The ribosylation of **149** with the bromo sugar **19** in acetonitrile (method B) provided a mixture of the positional isomers **150** and **153** in 6% and 48% yields, respectively. Interestingly enough, fusion of the imidazole (**148**) with the fully protected sugar (**51**) in the presence of bis(*p*-nitrophenyl)phosphate[147] (method C) or chloroacetic acid[151] provided **151** (19%), methyl 5-cyanomethyl-1-(2,3,5-tri-*O*-acetyl-α-D-ribofuranosyl)imidazole-4-carboxylate (α-**151**, 4.7%), and methyl 4-cyanomethyl-1-(2,3,5-tri-*O*-acetyl-β-D-ribofuranosyl)imidazole-5-carboxylate (**154**, 47%) and the corresponding α-anomer (α-**154**, 11.8%).

Although 3-deazaguanosine (**146**) could be obtained[152,153] directly from the blocked imidazole nucleosides **150** or **151** and liquid ammonia in 20–30% yield, a more convenient route to **146** was developed[147] which provided the pure nucleoside in 69% yield. Treatment of **150** with liquid ammonia gave the intermediate 5-cyanomethyl-1-β-D-ribofuranosylimidazole-4-carboxamide (**152**) in 81% yield; when **152** was heated in aqueous sodium carbonate in ethanol, **146** was formed in 85% yield. On the other hand, the corresponding positional isomer 4-cyanomethyl-1-β-D-ribofuranosylimidazole-5-carboxamide (**155**) could not be isolated and was directly converted to 6-amino-3-β-D-ribofuranosylimidazo[4,5-*c*]pyridin-4-one (7-β-D-ribofuranosyl-3-deazaguanine, **147**) in 80%

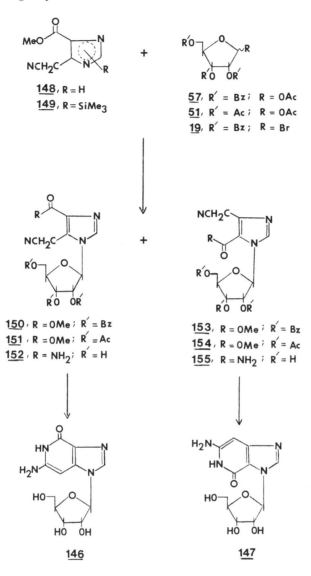

yield.[147] The structure of these imidazole nucleosides (**150–154**), as well as **146** and **147**, was established by using ¹H- and ¹³C-NMR spectroscopy.

The versatility of this unique ring closure of appropriate imidazole nucleosides under basic conditions to provide the requisite 3-deazaguanine nucleosides was further demonstrated by the preparation of 6-amino-1-(5-deoxy-β-D-ribofuranosyl)imidazo[4,5-*c*]pyridin-4(5*H*)-one (5′-deoxy-3-deazaguanosine, **156**),[154] 6-amino-1-(2-deoxy-β-D-*erythro*-pentofuranosyl)imidazo[4,5-*c*]pyridin-4(5*H*)-one (2′-deoxy-3-deazaguanosine, **157**),[155,156] and 6-amino-1-β-D-arabinofuranosylimidazo[4,5-*c*]pyridin-4(5*H*)-one (**158**).[157] The corresponding α-anomers as well as N-3 glycosyl derivatives of 6-aminoimidazo[4,5-*c*]pyridin-4(5*H*)-one have also been reported.[154–157]

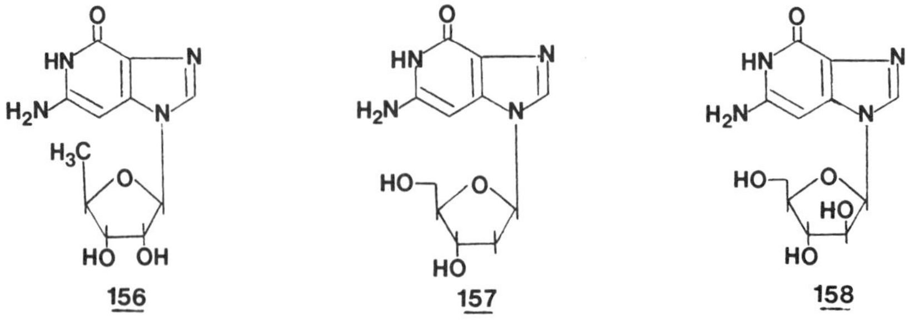

Ring closure of the imidazole nucleoside **150** with hydrazine hydrate was found to give 1-amino-3-deazaguanosine (**160**),[156] which, on hydrogenation in the presence of Raney nickel, provided a new synthesis of **146**. A plausible mechanism for this ring closure has been suggested,[156] involving the direct attack of hydrazine on the ester carbonyl carbon to give the intermediate **159**, which then cyclizes immediately to form **160**.

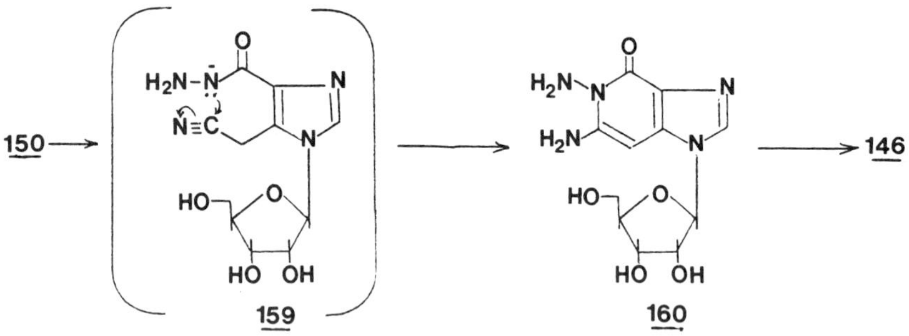

The methodology of this ring closure of appropriate imidazole nucleosides to 3-deazapurine nucleosides was further extended by Cook and Robins[158] and also by Gupta and Bhakuni[159] and Tanaka and co-workers.[160–162] Treatment of 4(5)-cyano-5(4)-cyanomethyl-1-trimethylsilylimidazole (**161**) with **57** in 1,2-dichloroethane containing 1.44 molar equivalent of stannic chloride gave two blocked imidazole nucleosides, 5-cyano-4-cyanomethyl-1-(2,3,5-tri-*O*-benzoyl-β-D-ribofuranosyl)imidazole (**164**, 72%) and 4-cyano-5-cyanomethyl-1-(2,3,5-tri-*O*-benzoyl-β-D-ribofuranosyl)imidazole (**165**, 18%). A reduction in the amount of stannic chloride used in this glycosylation procedure, to 0.72 molar equivalent, significantly affected the yield and the isomeric ratio (50% of **164** and 25% of **165**). When the blocked imidazole nucleoside **164** was treated with ethanolic hydrogen sulfide, in the presence of triethylamine, a 94% yield of 6-amino-3-(2,3,5-tri-*O*-benzoyl-β-D-ribofuranosyl)imidazole[4,5-*c*]pyridine-4(5*H*)-thione (**168a**) was obtained. Debenzoylation of **168a** with sodium methoxide provided 6-amino-3-β-D-ribofuranosylimidazo[4,5-*c*]pyridine-4(5*H*)-thione (3-deaza-7-β-D-ribofuranosyl-6-thioguanine, **168b**). Hydrogenolysis of **168b** with Raney nickel provided 6-amino-3-β-D-ribofuranosylimidazo[4,5-*c*]pyridine (**167**). In a similar manner, treatment of compound **165** with hydrogen sulfide provided 6-amino-1-(2,3,5-tri-*O*-benzoyl-β-D-ribofuranosyl)imidazo[4,5-*c*]pyridine-4-

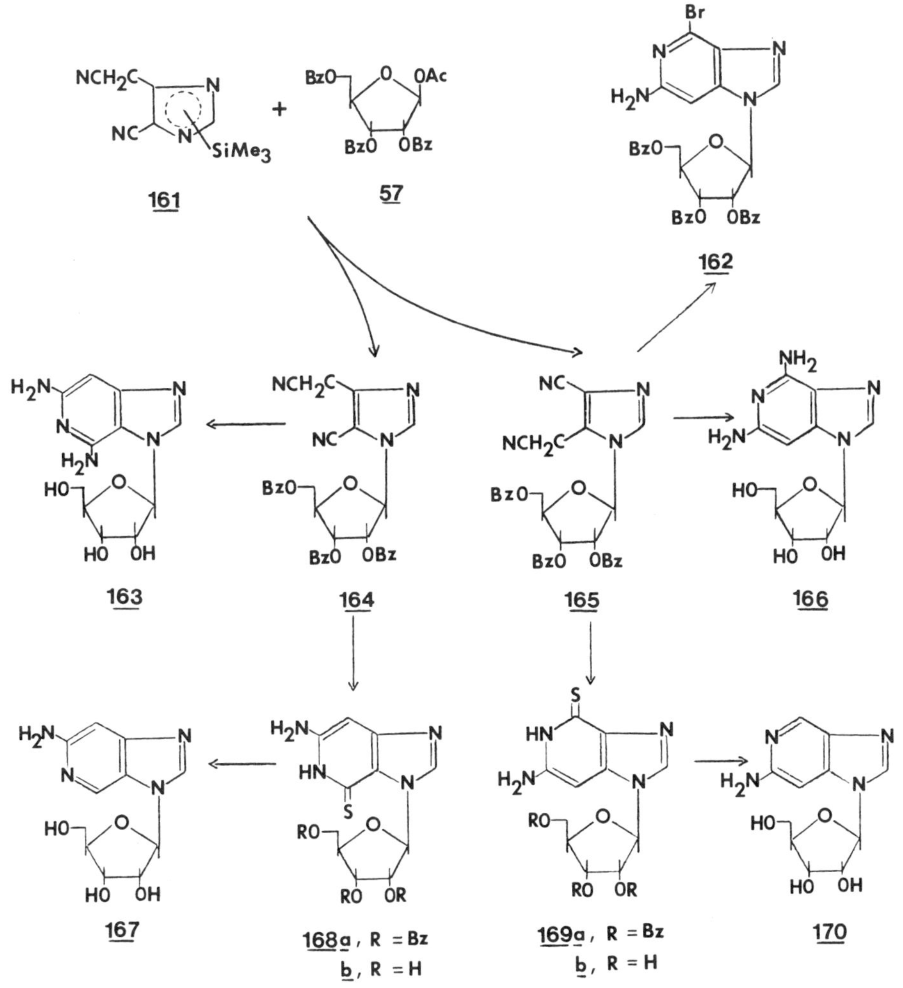

(5*H*)-thione (**169a**) in almost quantitative yield. Debenzoylation of **169a** with sodium methoxide in methanol gave 6-amino-1-β-D-ribofuranosylimidazo[4,5-*c*]pyridine-4(5*H*)-thione (**169b**), which was shown to be identical with 3-deaza-6-thioguanosine.[145] Dethiation of **169b** with Raney nickel provided 6-amino-1-β-D-ribofuranosylimidazo[4,5-*c*]pyridine (**170**). 4,6-Diamino-3-β-D-ribofuranosylimidazo[4,5-*c*]pyridine (**163**) was obtained by reacting **164** with methanolic ammonia at 130°C. The corresponding positional isomer 4,6-diamino-1-β-D-ribofuranosylimidazo[4,5-*c*]pyridine (**166**) was obtained in excellent yield when **165** was treated with methanolic ammonia at room temperature. Although the attempts to cyclize **164** with anhydrous hydrogen bromide failed, compound **165** under identical conditions provided a moderate yield of 6-amino-4-bromo-1-(2,3,5-tri-*O*-benzoyl-β-D-ribofuranosyl)imidazo[4,5-*c*]pyridine (**162**). The structures of all these nucleosides were established by ¹H-NMR spectroscopy or transformation to compounds of established structure, or both. The mechanistic implications of these novel cyclizations were also discussed.[158] The carbocyclic analogue of 3-deazaadenosine {(±)-4-amino-1-[(1α,2β,3β,4α)-2,3-dihydroxy-4-(hydroxymethyl)cyclopentyl]imidazo[4,5-c]pyridine (C-c³Ado)} has been synthesized by the reductive ring has been synthesized by the reductive ring annulation of (±)-1(1,4/2,3)-4-(3-nitro-2-chloro-4-pyridylamino)-2,3-dihydroxy-1-cyclopentanemethanol.[163]

2.2.2.6. Synthesis Employing the Sodium Salt of a Preformed Aglycon and an Acylated Sugar Halide. By using a general and stereospecific glycosylation procedure, developed recently in our laboratory,[31] 2'-deoxy-3-deazapurine nucleosides have been prepared. Treatment of the sodium salt of 4,6-dichloroimidazo[4,5-c]pyridine[164] (125), generated *in situ* by sodium hydride in acetonitrile, with 1-chloro-2-deoxy-3,5-di-O-p-toluoyl-α-D-*erythro*-pentofuranose[34] (24) in an inert atmosphere gave crystalline 4,6-dichloro-1-(2-deoxy-3,5-di-O-p-toluoyl-β-D-*erythro*-pentofuranosyl)imidazo[4,5-c]pyridine (171). Ammonolysis of 171 with methanolic ammonia at elevated temperature and pressure gave 4-amino-6-chloro-1-(2-deoxy-β-D-*erythro*-pentofuranosyl)imidazo[4,5-c]pyridine (172).[31]

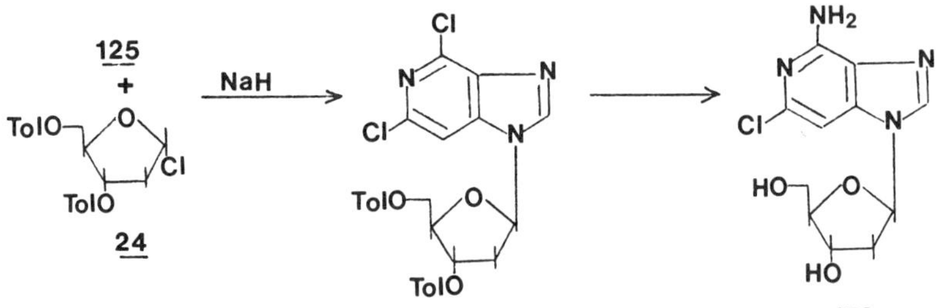

The sodium salt glycosylation procedure has also been used[165] for the preparation of a number of 3-deazapurine arabinosides of general formula 173. For the selective preparation of 7-β-D-ribofuranosyl and 7-(2-deoxy-β-D-ribofuranosyl) derivatives of 3-deazaguanine, this procedure was found to be very fruitful.[166] Recently, a new inhibitor of S-adenosylhomocysteine hydrolase, 3-deazaneplanocin, has been prepared by employing the sodium salt of 4-chloroimidazo[4,5-c]pyridine.[167]

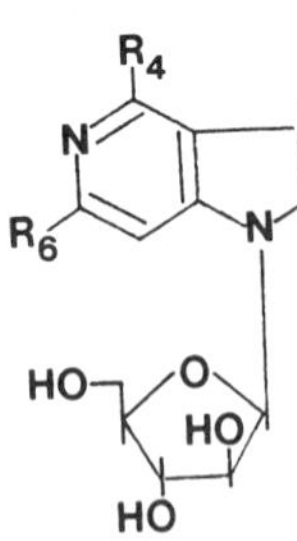

R_4 = NH_2, OH, SH

R_6 = H, NH_2

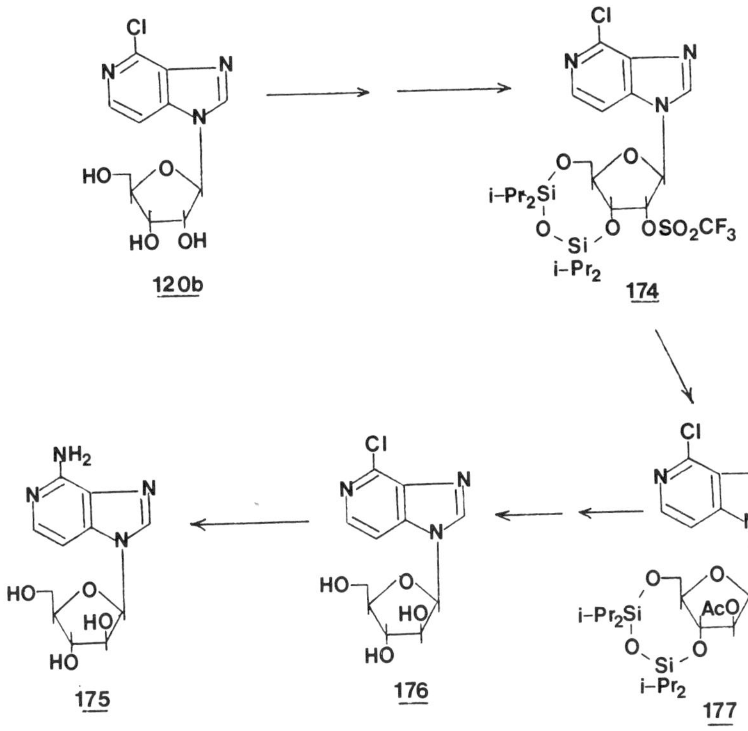

2.2.2.7. *Chemical Transformations.* Application of the procedure developed by Fukukawa *et al.*[91] to 4-chloro-1-β-D-ribofuranosylimidazo[4,5-*c*]pyridine (**120b**) provided an alternate route to the arabinosyl nucleosides.[94] The 3′,5′-*O*-silylated ribonucleoside, prepared from **120b** and 1,3-dichloro-1,1,3,3-tetra-isopropyldisiloxane, was converted into the corresponding trifluoromethane-sulfonyl derivative (**174**). Nucleophilic displacement of the trifloxy group with sodium acetate gave **177** in an 87% yield. Treatment of **177** with methanolic ammonia, followed by tetrabutylammonium fluoride, furnished 4-chloro-1-β-D-arabinofuranosylimidazo[4,5-*c*]pyridine (**176**). Compound **176** was converted to 3-deaza-araA (**175**) by the treatment of **176** with hydrazine hydrate, followed by Raney nickel reduction.[94] Similarly, 4-amino-1-β-D-xylofuranosylimidazo-[4,5-*c*]pyridine, 4-amino-1-(3-deoxy-β-D-*glycero*-pent-3-enofuranosyl)imidazo-[4,5-*c*]pyridine, and 4-amino-1-(5-deoxy-β-D-*erythro*-pent-4-enofuranosyl)imid-azo[4,5-*c*]pyridine have been prepared[168] from 4-chloro-1-β-D-ribofuranosyl-imidazo[4,5-*c*]pyridine.

3′-Deoxy-3-deazaadenosine was prepared by the chemical transformation of the corresponding ribonucleoside with 2-acetoxyisobutyryl bromide and was condensed with *N,N′*-bis-trifluoroacetyl-L-homocysteine dimethyl ester. Subsequent deprotection of the resulting *N*-trifluoroacetyl-*S*-3′-deoxyadenosyl-L-homocysteine gave *S*-3′-deoxy-3-deazaadenosyl-L-homocysteine.[169]

2.2.2.8. *Enzymatic Synthesis.* A number of 1-β-D-ribofuranosylimidazo-[4,5-*c*]pyridines and their 4-substituted derivatives have been prepared by an enzymatic process. These 3-deazapurine ribonucleosides are produced from the corresponding 3-deazapurines with a ribose donor and the enzyme purine nucleoside phosphorylase.[151,170–172] Similarly, by using 4-aminoimidazo[4,5-*c*]-pyridine, thymidine, purine nucleoside phosphorylase, and thymidine phosphorylase, 4-substituted 1-(2-deoxy-β-D-*erythro*-pentofuranosyl)imidazo[4,5-*c*]pyridines (**132**) have been obtained.[173,174]

Several laboratories have reported the synthesis of 5'-modified 3-deazaadenosines, including 5'-deoxy-5'-isobutylthio-3-deazaadenosine,[175] 4-amino-1-β-D-ribofuranosylimidazo[4,5-*c*]pyridinyl-5'-*S*-L-homocysteine [3-deaza-*S*-adenosylhomocysteine (3-deaza-SAH)],[129] N^6-methyl-3-deaza-SAH,[176,177] N^6-dimethyl-3-deaza-SAH,[176] and their corresponding *S*-adenosylmethionine (SAM) analogues.[178]

Singh *et al.*[179] reported an X-ray crystallographic investigation, a molecular orbital calculation of atomic charge densities, and a detailed conformational analysis of the rotational barrier around the glycosyl bond, using both the molecular orbital method and the van der Waals contact method with 3-deazaadenosine. It was observed that 3-deazaadenosine crystallizes in space group $P2_12_12_1$ of the orthorhombic system with four formula units in a cell of dimensions $a = 4.637(2)$, $b = 12.549(6)$, and $c = 19.878(10)$ Å. The observed and calculated densities are 1.54(2) and 1.535 g/cm^3, respectively. The solid-state conformation around the glycosyl linkage is *anti*, with a X value of +4.5°. Recently, Revankar *et al.*[156] have reported single-crystal X-ray diffraction analysis data for 2'-deoxy-3-deazaguanosine (**157**), its α-anomer, and 7-(2-deoxy-β-D-*erythro*-pentofuranosyl)-3-deazaguanine.

The conformation of 1-β-D-ribofuranosylimidazo[4,5-*c*]pyridine has been studied by Eyring and co-workers.[180] An analysis of the circular dichroism data shows that 3-deazanebularine has strong, if not exclusive, preference for the *syn* conformation in aqueous solution. The solution conformation of 3-deazaadenosine has been determined by nuclear magnetic resonance methods in aqueous and ammonia solutions.[181] 3-Deazaadenosine exists preferentially in the S-*syn-g+/t* (70%) and the N-*anti-g+/t* (30%) conformation.[181]

2.2.2.9. Imidazo[4,5-c]pyridine Nucleotides. 4-Amino-1-β-D-ribofuranosylimidazo[4,5-*c*]pyridine 5'-monophosphate (**178**, R = NH$_2$) was synthesized[99] chemically from 3-deazaadenosine by phosphorylation with phosphorus oxychloride,[182] and the corresponding diphosphate was obtained by the morpholidate method.[183] Subsequently, the 5'-monophosphates of 4-chloro-[131,184,185] and 4-hydroxy-1-β-D-ribofuranosylimidazo[4,5-*c*]pyridine[131] (**178**, R = Cl or OH) have been prepared using Tener's general procedure.[186]

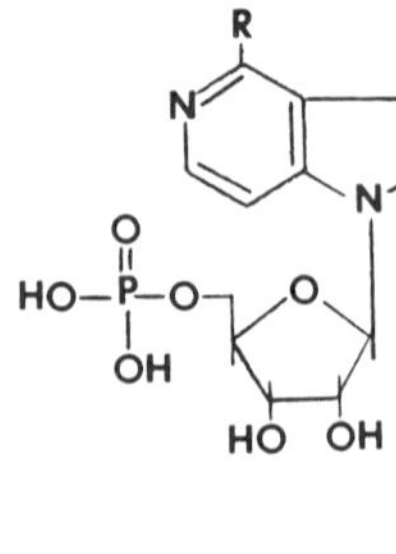

Mizuno and co-workers[101,131] described the synthesis of a variety of phosphates of 3-deazaadenosine. 3-Deazaadenosine 5′-phosphate (**178**, R = NH_2) was prepared by the treatment of 2′,3′-*O*-isopropylidene-3-deazaadenosine (**179**) with prophosphoryl tetrachloride in acetonitrile, followed by hydrolysis. The general procedure of Khorana and co-workers[102] for the synthesis of nucleoside 3′,5′-cyclic phosphates from 5′-monophosphates, however, failed to give satisfactory results when applied to **178**. With this approach, the yield of the requisite 3′,5′-cyclic phosphate (**180**) was discouragingly low, and the product was found to be always contaminated with 3-deazaadenosine 2′,3′-cyclic phosphate (**181**). However, by adopting the procedure of Borden and Smith,[187] **178** was converted to the corresponding *p*-nitrophenyl ester, which, in the presence of potassium *tert*-butyl alkoxide in DMSO, gave **180**. Recently, a low-yield synthesis of **180** has been reported[188] from the trichloromethyl phosphonate intermediate. By using tri-*n*-butylammonium pyrophosphate in DMF,[103] **181** was obtained from **122**, but only in 3% yield.[101] Compound **181** was enzymatically hydrolyzed[189] to the corresponding nucleoside 2′-phosphate (**182**) and 3′-phosphate (**183**) by ribonuclease M.

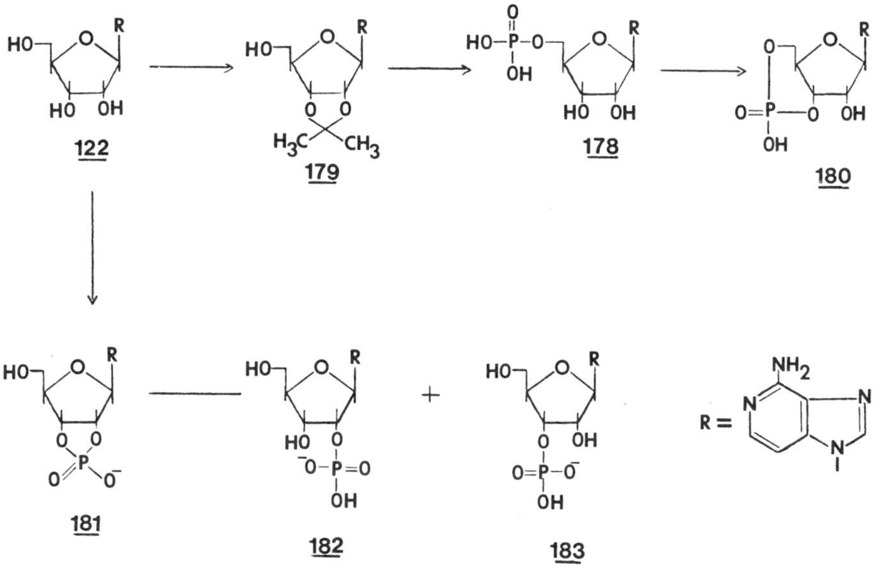

The method described by Ueda and Kawai[103] was found to be convenient for the preparation of 4-chloro-1-β-D-ribofuranosylimidazo[4,5-*c*]pyridine 2′,3′-cyclic phosphate (51.2% yield) and compares favorably with the direct cyclic-phosphorylation procedure.[190] 3-Deazaguanosine 5′-monophosphate (**185**) was obtained directly from the unprotected 3-deazaguanosine (**146**) and also from the cyclization of the corresponding imidazole nucleotide precursor (**184**).[147] Ring closure of **184** in the presence of aqueous sodium carbonate (pH 10.0, 100°C, 40 min) gave **185** in over 75% yield. Phosphorylation of **146** with phosphorus oxychloride in trimethyl phosphate[191] provided a mixture of products, from which **185** was isolated[147] in 30% yield. Treatment of the *N*,

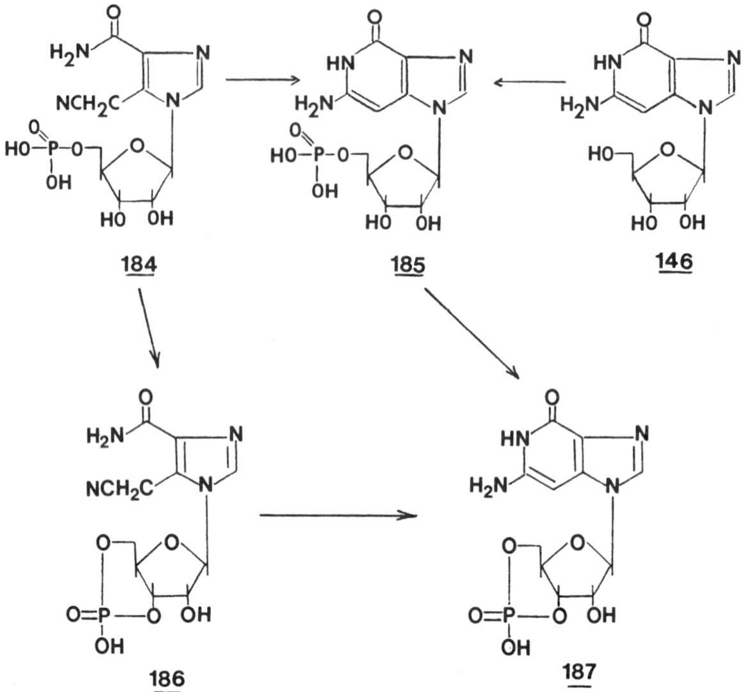

N'-dicyclohexylmorpholinecarboxamidine salt of **184** in anhydrous pyridine with dicyclohexylcarbodiimide (DCC) gave **186**. Ring closure of **186** in the presence of aqueous sodium carbonate provided 3-deazaguanosine $3',5'$-cyclic phosphate (**187**).[156] However, direct DCC-mediated cyclization of **185** to obtain **187** gave a complex reaction mixture from which pure **185** was isolated only in poor yield.[156]

2.2.2.10. Biochemical Properties of 3-Deazapurine Nucleosides and Nucleotides. 1-β-D-Ribofuranosylimidazo[4,5-*c*]pyridine (**118a**) was found to be active against vaccinia virus in chick embryo cells.[112] 6-Chloro-4-dimethylamino-1-β-D-ribofuranosylimidazo[4,5-*c*]pyridine showed good antitumor activity against L1210 lymphoid leukemia *in vivo*.[136] 1-(2,3,4,6-Tetra-*O*-acetyl-β-D-galactopyranosyl)- and 1-β-D-glucopyranosyl-4-chloro-2-methylimidazo[4,5-*c*]pyridines inhibited the growth of sarcoma-37 in mice.[113] Both 3-deazaguanine and 3-deazaguanosine exhibited broad-spectrum antiviral activity against a variety of DNA and RNA viruses,[47,192–195] as well as potent antitumor activity against L1210 leukemia and mammary adenocarcinomas in mice.[196–198] 3-Deazaguanine showed potent cytotoxic activity *in vitro* against a variety of tumor cells including L1210,[196,198,199] HeLa,[200] human KB cells,[200] human myeloid leukemia cells (HL-60),[201] Ehrlich ascites tumor cells,[200] Chinese hamster ovary cells,[202] primary Chinese hamster embryo cells,[197] and mammary carcinoma EMT-6 cells.[203]

    3-Deazaguanine has shown inhibitory action against a broad spectrum of animal breast tumor models, including rat mammary adenocarcinomas, mouse mammary adenocarcinomas,[196,197] slow- and fast-growing mammary tumors in mice, and the human breast xenograft subrenal capsule implant system. These observations are very interesting since mammary adenocarcinoma (e.g., R3230AC) is a non-hormone-dependent, estrogen-sensitive, slow-growing tumor model for postoperative breast carcinomas and is not particularly susceptible to conventional cancer chemotherapeutic agents. Mammary adenocarcinoma 13762 is a fast-growing, non-hormone-dependent tumor which is not responsive to estrogen treatment. These results suggest a possible use of 3-deazaguanine in chemotherapy of human breast cancer. 3-Deazaguanine also inhibited the growth of *Escherichia coli* B3 *in vitro* but did not exhibit any antibacterial activity *in vivo*.[204,206,207] The 7-glycosyl derivatives of 3-deazaguanine (e.g., **147**) showed antibacterial activity against several gram-negative strains *in vivo* without any appreciable toxicity to the host.[204,205] The antibacterial action of **147** has been ascribed to its cleavage to 3-deazaguanine in *E. coli* B-infected cells.[208]

6-Thio-3-deazaguanine (TDG) exhibited significant antitumor activity against a variety of experimental animal tumor models including C_3H mammary adenocarcinoma, Lewis lung carcinoma, adenocarcinoma 755, and leukemias L1210 and P388.[209] However, TDG was ineffective against 3-deazaguanine-resistant L1210, both *in vitro* and *in vivo*, and CEM cells *in vitro*. The spectrum of antitumor activity and mechanism of action of TDG appears to be different from that of 3-deazaguanine.[209]

Recently, 3-deazaguanosine (**146**) was shown to be a potent antileishmanial agent,[210] at least 20 times more active than 3-deazaguanine or allopurinol ribonucleoside against *Leishmania tropica in vitro*. Like **146**, 2′-deoxy-3-deazaguanosine (**157**) also demonstrated significant broad-spectrum antiviral activity against certain DNA and RNA viruses *in vitro*, as well as moderate activity against L1210 and P388 leukemia in cell culture.[156] 3-Deazaguanosine is metabolized to the 5′-triphosphate derivative in Chinese hamster ovary cells deficient in hypoxanthine-guanine phosphoribosyltransferase,[211] without prior conversion to the base, 3-deazaguanine. The identity of the kinase responsible for the phosphorylation of 3-deazaguanosine is unknown.[212]

The biological effects of 3-deazaadenosine (**122**) are very varied. 3-Deazaadenosine has been shown to inhibit replication of RNA type C virus, HL-23, and Rous sarcoma virus.[175,213–215] Avian sarcoma virus B77, grown in chicken embryo fibroblasts, was inhibited by **122** at 100 μM concentration. 3-Deazaadenosine is also active *in vitro* against the vesicular stomatitis virus (VSV), parainfluenza 3 virus, and reovirus at less than 10 $\mu g/ml$, which is 5–20-fold lower than the concentration cytotoxic to the host cell. The mode of antiviral action of 3-deazaadenosine has been attributed to the inhibition of *S*-adenosylhomocysteine hydrolase (SAHase),[128,176,178,216–222] which results in an inhibition of methylation required for viral RNA transcription, more specifically, by inhibiting the methylation of the 5′-cap of the viral mRNA. Since the accumulation of *S*-adenosylhomocysteine (AdoHcy) inhibits the methyltransferase reaction from *S*-adenosylmethionine (AdoMet) by a feedback mechanism, the RNA viral mRNA is not transcribed due to inhibition of the methylation of the guanylate molecule at N-7 in the 5′-terminal position. 3-Deazaadenosine is not enzymatically deaminated or phosphorylated and is capable of acting both as an inhibitor of SAHase and as a substrate of this enzyme to yield 3-deazaadenosyl-homocysteine,[223] which has been shown to inhibit methylations *in vivo*, itself acting as a feedback inhibitor of the viral methyltransferases. It has been postu-lated[214] that the accumulation of *S*-adenosylhomocysteine and 3-deazaadeno-sylhomocysteine leads to an inhibition of the methylation reactions essential to viral mRNA transcription. 3-Deazaadenosine inhibits induction of murine erythroleukemia cell differentiation.[224]

The carbocyclic analogue of 3-deazaadenosine (C-c^{3}Ado) has been re-ported to inhibit the *in vitro* replication of several DNA and RNA viruses such as vaccinia, vesicular stomatitis, measles, parainfluenza, and reo type 1, at 0.75–3.8 μM, giving a selectivity index of 400–2000.[163,225–229] However, it is less potent than araA against herpes simplex virus type 1 (HSV-1). C-c^{3}Ado, administered intraperitoneally (i.p.), protected newborn mice against a lethal infection of VSV,[226] and its antiviral activity spectrum is similar to that of 2,3-dihydroxy-propyladenine, [(*S*)-DHPA]. In a recent study,[230] it has been shown that the inhibitory effects of C-c^{3}Ado and 3-deazaneplanocin on murine L929 cell AdoHcy hydrolase are closely correlated with their antiviral activity against vaccinia and vesicular stomatitis viruses. The *in vitro* antiviral activity of C-c^{3}Ado, in general, is superior to that of 3-deazaadenosine. Like 3-deazaadenosine, C-c^{3}Ado was also found to be a potent reversible inhibitor of SAHase[163,231] and is neither deaminated by ADA nor phosphorylated by AK in L1210 cells.[163] Although 3-deazaadenosine and C-c^{3}Ado are known to be potent inhibitors of SAHase, the effects on nucleotide pools apparently are not mediated via this inhibition.[232]

The coenzyme nicotinamide-3-deazapurine dinucleotide[185] showed simi-lar activities to NAD with various dehydrogenases. 4-Chloro-1-β-D-ribofurano-sylimidazo[4,5-*c*]pyridine 5′-phosphate (**178**, R = Cl) was examined[184] for activation of the threonine deaminase from *E. coli*, in terms of the K_a (activation constant), and exhibited almost the same K_a value (0.14) as AMP (0.10). In a similar study[233] with a regulatory enzyme of mammalian origin, glycogen phosphorylase b, compound **178** (R = Cl) showed a K_a value 120 times higher than that of AMP, although the effect on the activation constant might be due not only to the chlorine atom, but also to the carbon atom by which N-3 of the purine ring was replaced, indicating that the amino group at position 6 makes a significant contribution to the binding and also that the 5′-phosphate is required for the activation.[233] Dihydronicotinamide-3-deazapurine dinucleotide forms more or less stable complexes with pig heart lactate dehydrogenase,[234] and these complexes do not markedly differ from the complex formed with the natural cofactor. Fluorometric studies indicate[234] a change in conformation of the coenzyme by forming the coenzyme–enzyme complex, as proposed by Velick.[235]

2.2.3. Pyrrolo[2,3-d]pyrimidine (7-Deazapurine) Nucleosides

The pyrrolo[2,3-*d*]pyrimidine ring system, consisting of pyrrole fused with a pyrimidine ring, is found in nature. This ring system is sometimes referred to as 7-deazapurine and differs from purine only in that the N-7 atom of purine has been replaced by a —CH=. Although the parent heterocyclic ring system was first prepared in 1911,[236] prior to 1955 very few references were cited in the literature on this class of compounds. However, their structural resemblance to purines, the natural occurrence of their derivatives, and their unusual biological properties have prompted a great deal of activity directed toward the synthesis and biological evaluation of this ring system in recent years. In fact, the volume of literature pertaining to this class of compounds has been phenomenal. The chemistry and biochemical properties of the pyrrolo[2,3-*d*]pyrimidine nucleosides have been treated in considerable detail in recent reviews.[2,3,237]

The first nucleoside antibiotic that was shown to contain the pyrrolo-[2,3-*d*]pyrimidine ring system was isolated in 1955 from *Streptomyces toyocaensis* by Nishimura *et al.*[238] and subsequently by Ohkuma[239,240] from *Streptomyces* strain and was named toyocamycin. The antibiotic isolated from a culture of *Streptomyces* sp. No. E-212[241,242] and *Streptomyces* strain No. 1037[243] has been shown[244] to be identical with toyocamycin. Toyocamycin was also isolated together with mannosidohydroxystreptomycin and hydroxystreptomycin from the cultures of *Streptomyces* 86.[245] Ohkuma[240] first proposed the chemical structure of toyocamycin as 4-amino-7-β-D-ribofuranosylpyrrolo[2,3-*d*]pyrimidine-5-carbonitrile (**188b**) without adequate proof for the site of glycosylation and anomeric configuration, on the basis of preliminary degradation studies which provided the known 4-aminopyrrolo[2,3-*d*]pyridine.[246]

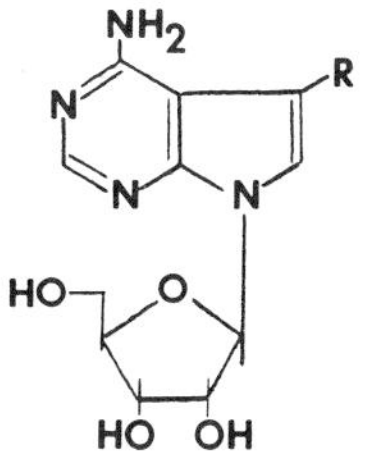

188a, R = H
  b, R = CN
  c, R = CONH₂

The degradation studies[247] of toyocamycin and tubercidin with sodium periodate finally resulted in the isolation and characterization of the aglycon of these nucleoside antibiotics as 4-aminopyrrolo[2,3-*d*]pyrimidine-5-carbonitrile, **190** (R = CN); 4-aminopyrrolo[2,3-*d*]pyrimidine, **190** (R = H), and carbons 1' and 2' of the ribose moiety (as glyoxal bisphenylosazone, **189**) provided some proof for the structures of these nucleosides. A total synthesis of the pyrrolo-[2,3-*d*]pyrimidine ribonucleoside toyocamycin, as well as sangivamycin and tubercidin, was subsequently accomplished.[248,249] This study finally furnished proof that antibiotic 1037, antibiotic E-212, unamycin B,[250,251] vengicide,[252,253] and toyocamycin all possessed the same structure.

The second pyrrolo[2,3-*d*]pyrimidine nucleoside antibiotic to be isolated

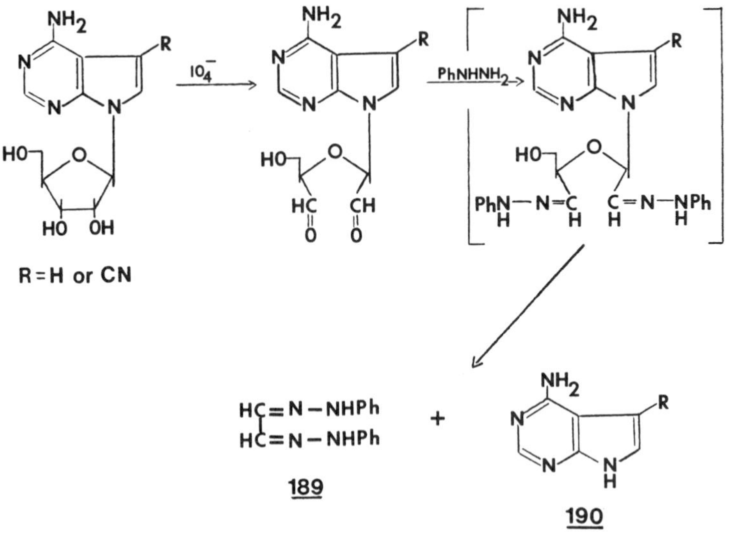

was tubercidin (**188a**). It was isolated in 1957 by Anzai and co-workers[254,255] from the fermentation broth of *Streptomyces tubercidicus*[256,257] and subsequently by Pike *et al.*[258] and Wechter and Hanze[259,260] from *Streptomyces sparsogenes.* The antibiotic isolated from *Streptomyces cuspidosporus*[261,262] was found to be identical with tubercidin. Shirato and co-workers[263,264] reported an improved fermentation procedure (using glycerol medium) and a new isolation technique[265] (using Amberlite IRC-50 or IR-200 H⁺ ion-exchange resins at pH 5.0) for tubercidin from *Streptomyces* mutant T-2-17. In a manner similar to that by which the structure of toyocamycin was elucidated,[240] Suzuki and Marumo[266,267] proposed the chemical structure for tubercidin as 4-amino-7-β-D-ribofuranosyl-pyrrolo[2,3-*d*]pyrimidine (**188a**). The ribosyl linkage was determined as on N-7 by comparing the UV absorption spectrum of tubercidin with that of 4-amino-7-methylpyrrolo[2,3-*d*]pyrimidine. The information of the anomeric configuration of tubercidin as β was established by Mizuno and co-workers[268,269] by the intramolecular quaternization of the 5'-*O*-*p*-toluylsulfonyl derivative of tubercidin (**192**) to the cyclonucleoside tosylate (**193**). This type of cyclization is feasible only if the 5'-carbon of the carbohydrate moiety is in the

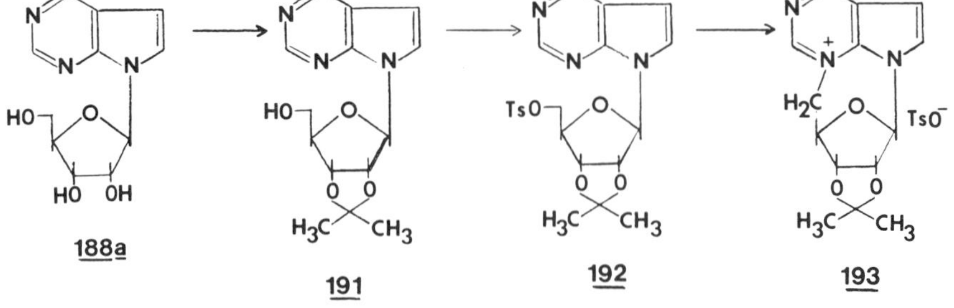

proximity of the N-1 atom of a pyrrolo[2,3-*d*]pyrimidine, which could occur only when the nucleoside antibiotic possesses the β configuration. A more rigorous proof of the structure of tubercidin was obtained[270] by the total chemical synthesis of this nucleoside antibiotic and the synthesis of the inosine analogue, 7-β-D-ribofuranosylpyrrolo[2,3-*d*]pyrimidine-4-one.[268,271]

The other naturally occurring pyrrolo[2,3-*d*]pyrimidine nucleoside antibiotic, sangivamycin (BA-90912), was subsequently isolated in 1963 by Rao and Renn[272] from a species of *Streptomyces*, which was later identified as *Streptomyces rimosus*.[273] The antibiotic isolated from *Streptomyces purpureofuscus*[274] was also found to be identical with sangivamycin. The structure of sangivamycin (**188c**) was deduced by hydrolytic studies.[275] Alkaline hydrolysis of toyocamycin (**188b**) gave an acid (**194**), identical with the product isolated following alkaline hydrolysis of sangivamycin (**188c**). The tetraacetyl derivative of sangivamycin (**196**), on dehydration with phosphorus oxychloride, gave the tetraacetyl derivative of toyocamycin (**195**). Selective acid hydrolysis of toyocamycin furnished sangivamycin and proved that the nucleoside antibiotic is 4-amino-7-β-D-ribofuranosylpyrrolo[2,3-*d*]pyrimidine-5-carboxamide.[275]

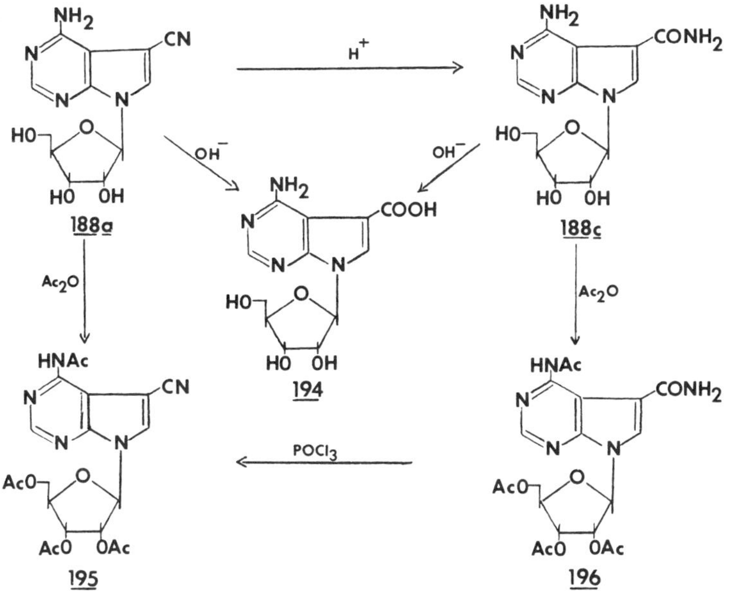

Subsequently, a number of naturally occurring pyrrolo[2,3-*d*]pyrimidine nucleoside antibiotics[276] have been isolated, including nucleoside Q (**197**),[277–280] cadeguomycin (**198**),[281–284] Antibiotic AB-116 (Kanagawamicin, **199**),[285,286] dapiramicin (**200**),[287–289] 5-iodo-5'-deoxytubercidin (**201**),[290,291] and mycalisines A and B.[292,293]

2.2.3.1. Synthesis Employing the Ring Closure of a Pyrimidine Nucleoside Precursor. Since the isolation and structural elucidation of tubercidin, toyocamycin, sangivamycin, and cadeguomycin, considerable interest has been generated in

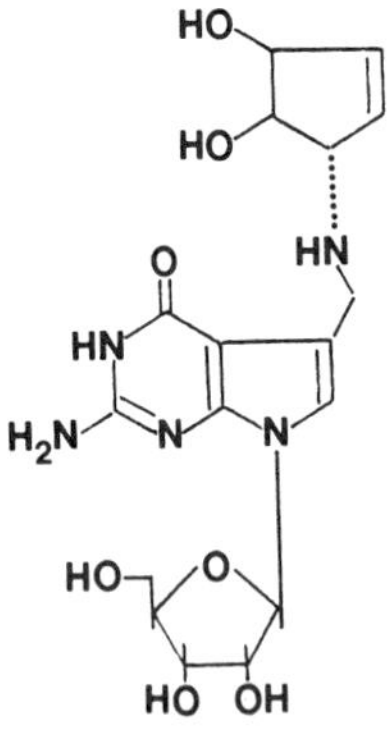

197

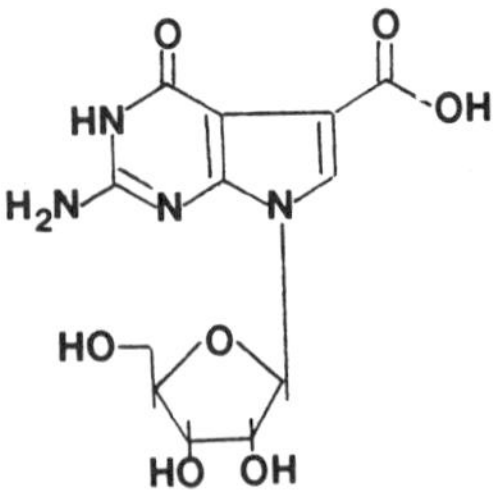

198

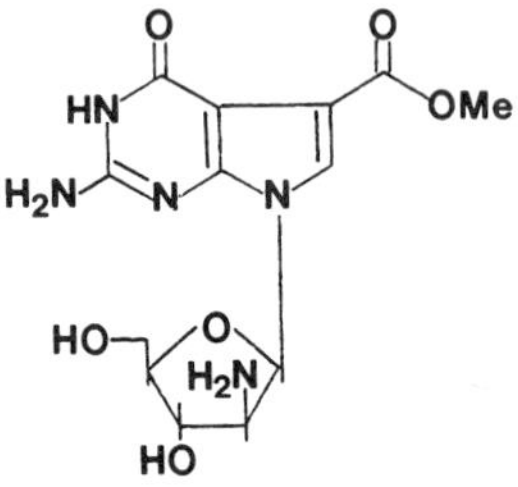

199

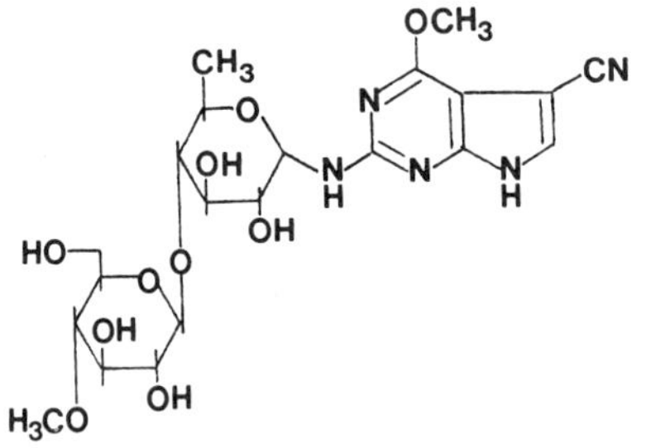

200

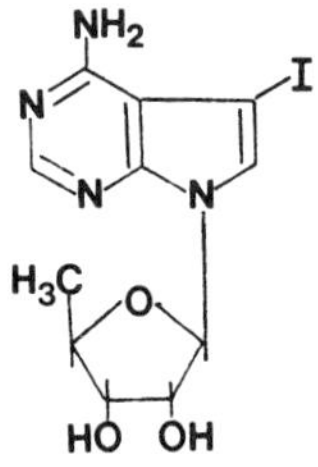

201

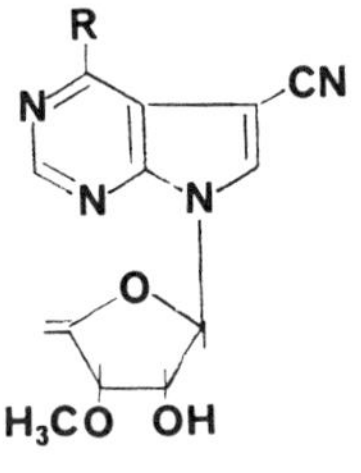

R = NH$_2$,   **Mycalisine A**
R = OH,   **Mycalisine B**

the synthesis of pyrrolo[2,3-*d*]pyrimidine nucleosides. The first chemical synthesis of such pyrrolopyrimidine nucleosides was reported in 1963 by Mizuno and co-workers.[268,271] The condensation of 4-amino-5-(2,2-diethoxyethyl)-6-pyrimidone (**202**) with 5-*O*-trityl-2,3,4-tri-*O*-acetyl-*aldehydo*-D-ribose (**203**) in the presence of ammonium chloride afforded the Schiff base intermediate **204**. Deacetylation of **204** with methanolic ammonia effected a cyclization of the carbohydrate moiety to the furanose form. Reacetylation of the nucleoside, followed by ring annulation with dilute acid, gave 7-(5-*O*-trityl-2,3-di-*O*-acetyl-β-D-ribofuranosyl)pyrrolo[2,3-*d*]pyrimidin-4(3*H*)-one (**205**) in 21% yield from **203**. Subsequent deacetylation with methanolic ammonia and detritylation with 80% acetic acid afforded 7-β-D-ribofuranosylpyrrolo[2,3-*d*]pyrimidin-4-(3*H*)-one (**206**). Tubercidin (**188a**) was also deaminated with nitrous acid to give the deamino compound,[258,268,271,294] which was shown to be identical with 7-deazainosine (**206**). The site of ribosylation in tubercidin was, by the above synthetic scheme, unambiguously shown to be N-7 and the anomeric configuration was established to be β on the basis of cyclonucleoside formation.[269]

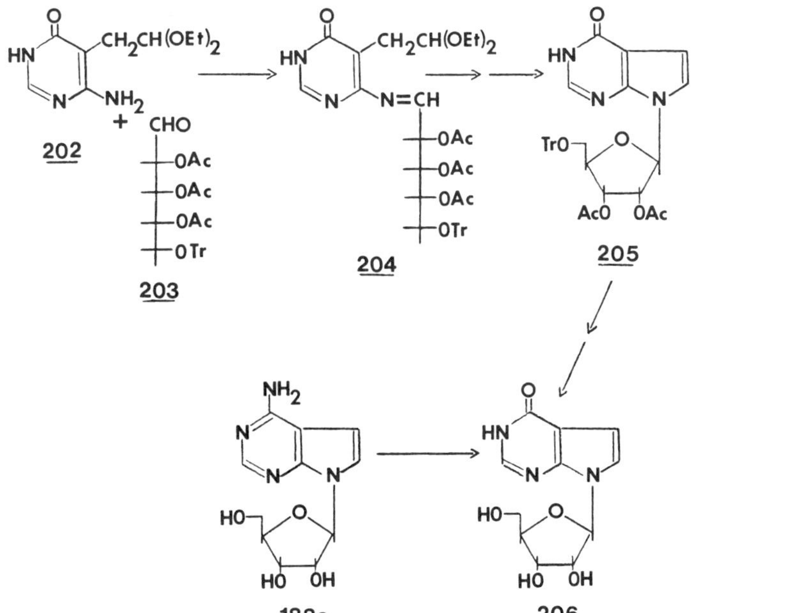

By exploiting the nucleophilic nature of the pyrimidine 5-position in 6-amino-3-β-D-ribofuranosylpyrimidine-2,4(1*H*)-dione (**207**)[295] with chloro-acetaldehyde[296], Winkley[297] has prepared 3-β-D-ribofuranosylpyrrolo[2,3-*d*]-pyrimidine-2,4(1*H*,7*H*)-dione (**208**) in 9% yield.

By using this ring-closure procedure, a number of 7-alkyl analogues of tubercidin (**209**),[298] as well as the carbocyclic analogues of 6-methyltubercidin

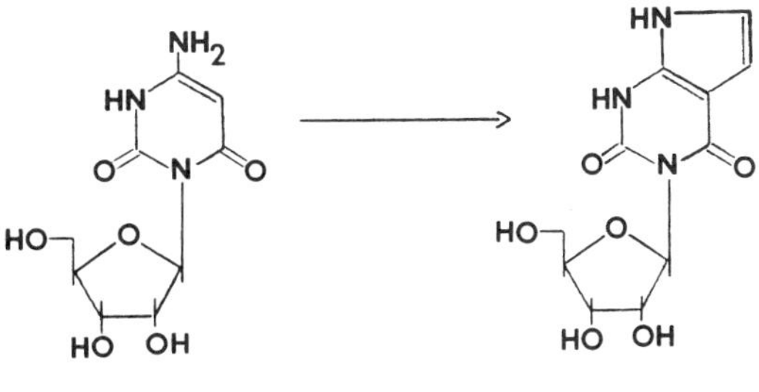

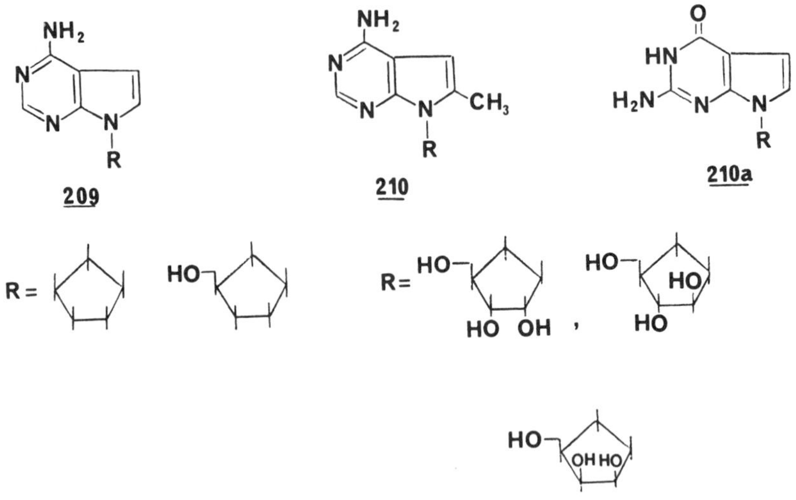

209   210   210a

R =

(210)[299–301] and 7-deazaguanosine (210a),[302] have been prepared. The latter nucleosides exhibited selective inhibitory activities against the multiplication of HSV-1 and HSV-2 in cell culture. Repeated administration of these compounds at 10 mg/kg i.p. to mice infected with HSV-2 increased the number of survivors and lengthened significantly the mean survival time.[302]

2.2.3.2. Synthesis Employing the Halomercuri Salt of a Preformed Aglycon and an Acylated Sugar Halide. An unsuccessful attempt to synthesize toyocamycin was made in 1965 by Taylor and Hendess.[303] Ribosylation of 4-amino-5-cyanopyrrolo[2,3-*d*]pyrimidine[304] by the chloromercuri salt procedure[47] and subsequent deacetylation of the reaction product with alcoholic ammonia furnished less than a 1% yield of a nucleoside product, which was not completely characterized. The nonidentity of this product and toyocamycin was established by paper chromatography. On the basis of available spectral data, the compound was assigned the structure 4-amino-1-β-D-ribofuranosylpyrrolo[2,3-*d*]pyrimidine-5-carbonitrile (211). The halomercuri salt procedure was also used in an attempt to synthesize tubercidin, but the desired nucleoside was not obtained.[296]

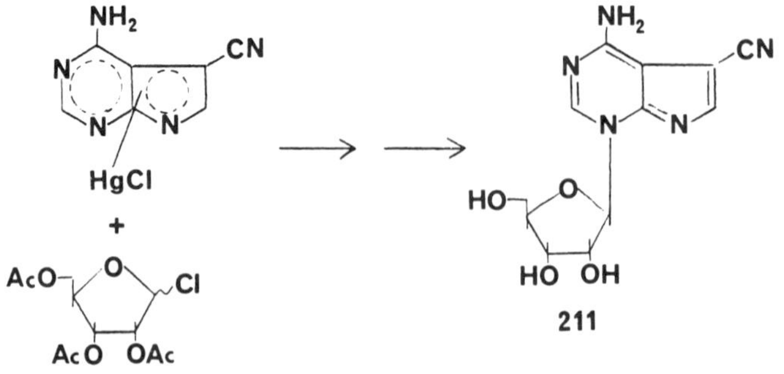

211

However, the chloromercuri procedure has been successfully employed by Bobek *et al.*[305] for the preparation of 4'-thio analogues of toyocamycin. Treatment of 2,3,5-tri-*O*-acetyl-4-thio-D-ribofuranosyl chloride (**213**) with the chloromercuri derivative of 4-acetamido-6-bromopyrrolo[2,3-*d*]pyrimidine-5-carbonitrile (**212**, R = NHAc) furnished a 20% yield of the acetylated nucleoside (**214**, R = NHAc). The structure was established on the basis of its spectroscopic characteristics and the *trans* rule. Deacetylation of **214** (R = NHAc) with methanolic ammonia gave 4-amino-6-bromo-7-(4-thio-β-D-ribofuranosyl)pyrrolo-[2,3-*d*]pyrimidine-5-carbonitrile (**215a**, R' = Br), which, on subsequent dehalogenation, afforded 4'-thiotoyocamycin (**215b**, R' = H). As a result of substituting S for O in the ribofuranosyl moiety, the UV absorption maxima for 4'-thiotoyocamycin exhibit small bathochromic shifts relative to those observed for toyocamycin.[305]

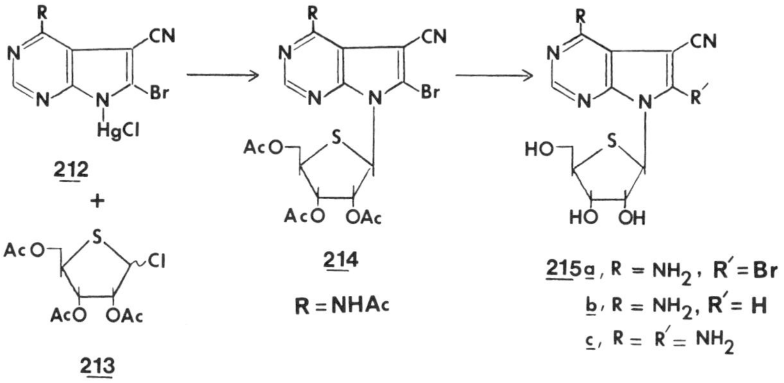

212
+
213

214
R = NHAc

215a, R = NH₂, R' = Br
b, R = NH₂, R' = H
c, R = R' = NH₂

A similar condensation[305] of the chloromercuri derivative of 4-chloro-6-bromopyrrolo[2,3-*d*]pyrimidine-5-carbonitrile (**212**, R = Cl) with **213** gave crystalline 4-chloro-6-bromo-7-(2,3,5-tri-*O*-acetyl-4-thio-β-D-ribofuranosyl)pyrrolo[2,3-*d*]pyrimidine-5-carbonitrile (**214**, R = Cl) in a 25.4% yield. Treatment of **214** with methanolic ammonia at 5°C effected a removal of the sugar protecting groups with a concomitant displacement of the 6-bromo group by NH₂ to furnish 4-chloro-6-amino-7-(4-thio-β-D-ribofuranosyl)pyrrolo[2,3-*d*]pyrimidine-5-carbonitrile (**215b**, R = Cl, R' = NH₂). The susceptibility of the 6-bromo group of **214** (R = Cl) to nucleophilic substitution is presumably due to the ability of the S atom to accommodate both positive and negative charges. However, treatment of **214** (R = Cl) with methanolic ammonia at 115°C produced 4,6-diamino-7-(4-thio-β-D-ribofuranosyl)pyrrolo[2,3-*d*]pyrimidine-5-carbonitrile (**215c**).

2.2.3.3. Synthesis Employing the Fusion of a Preformed Aglycon and an Acyl Sugar.
(i) Noncatalyzed. Since previous attempts[303] to synthesize toyocamycin by the halomercuri salt procedure had proven unsuccessful, Robins and co-workers [270] explored the possibility of utilizing the fusion procedure.[127,306] This procedure had been used very successfully in the purine and pyrimidine series. The

first glycosylation of a pyrrolo[2,3-*d*]pyrimidine by the fusion procedure was reported in 1967.[270] A successful fusion has been shown to be influenced by several factors, but presumably the most critical factor is the proper selection of the aglycon utilized. It was observed that the presence of electron-withdrawing substituents on the aglycon facilitates the reaction. Fusion of 4-chloro-6-methyl-thiopyrrolo[2,3-*d*]pyrimidine-5-carbonitrile (**216**) and 1,2,3,5-tetra-*O*-acetyl-ribose (**51**) at 165°C in the absence of a catalyst gave 4-chloro-6-methylthio-7-(2,3,5-tri-*O*-acetyl-β-D-ribofuranosyl)pyrrolo[2,3-*d*]pyrimidine-5-carbonitrile (**217**). Nucleophilic displacement of the 4-chloro group from **217** was effected by methanolic ammonia to furnish the 6-methylthio derivative of toyocamycin (**218**).[270,307] However, efforts to obtain toyocamycin from **218** by dethiation, using a variety of reaction conditions, were unsuccessful.[270]

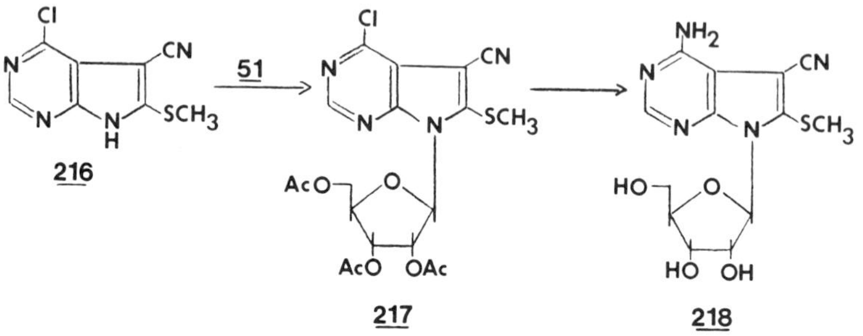

(ii) Acid-catalyzed. A route to the synthesis of toyocamycin via 4-acet-amido-6-bromopyrrolo[2,3-*d*]pyrimidine-5-carbonitrile (**219**) was then initiated[248,249] on the basis of the observation that electron-withdrawing substituents on a purine moiety facilitates the fusion condensation[308] with an acetylated sugar. Fusion of **219** with 1,2,3,5-tetra-*O*-acetylribose (**51**) in the presence of bis(*p*-nitrophenyl)phosphate[309] gave the blocked nucleoside (**222**) in a 9% yield. Complete deacetylation of **222** with methanolic ammonia furnished 4-amino-6-bromo-7-β-D-ribofuranosylpyrrolo[2,3-*d*]pyrimidine-5-carbonitrile (**225**). A removal of the 6-bromo group of **225** with pd/C in a hydrogen atmosphere gave an 88% yield of **188b**. This provided the first total synthesis of toyocamycin.[248] 6-Bromosangivamycin (**226**) was prepared, in 65% yield, by the treatment of **225** with 30% hydrogen peroxide in ammonium hydroxide. A removal of the 6-bromo group from **226** gave **188c**, which was found to be identical with an authentic sample of sangivamycin. The site of ribosylation in synthetic toyocamycin and related compounds was assigned as N-7 on the basis of UV absorption studies, and the β configuration was assigned by the ¹H-NMR studies of 2′,3′-*O*-isopropylidene-5′,1-toyocamycin cyclonucleoside. Final proof for the structural assignment of toyocamycin and sangivamycin was obtained by the conversion of sangivamycic acid to tubercidin (**188a**), the structure of which has been unequivocally established by Mizuno and co-workers.[269]

Since the above fusion reaction with **219** proceeded in a rather poor yield, efforts were then directed[249] toward an alternative synthesis of toyocamycin. The fusion of 6-bromo-4-chloropyrrolo[2,3-*d*]pyrimidine-5-carbonitrile (**220**)

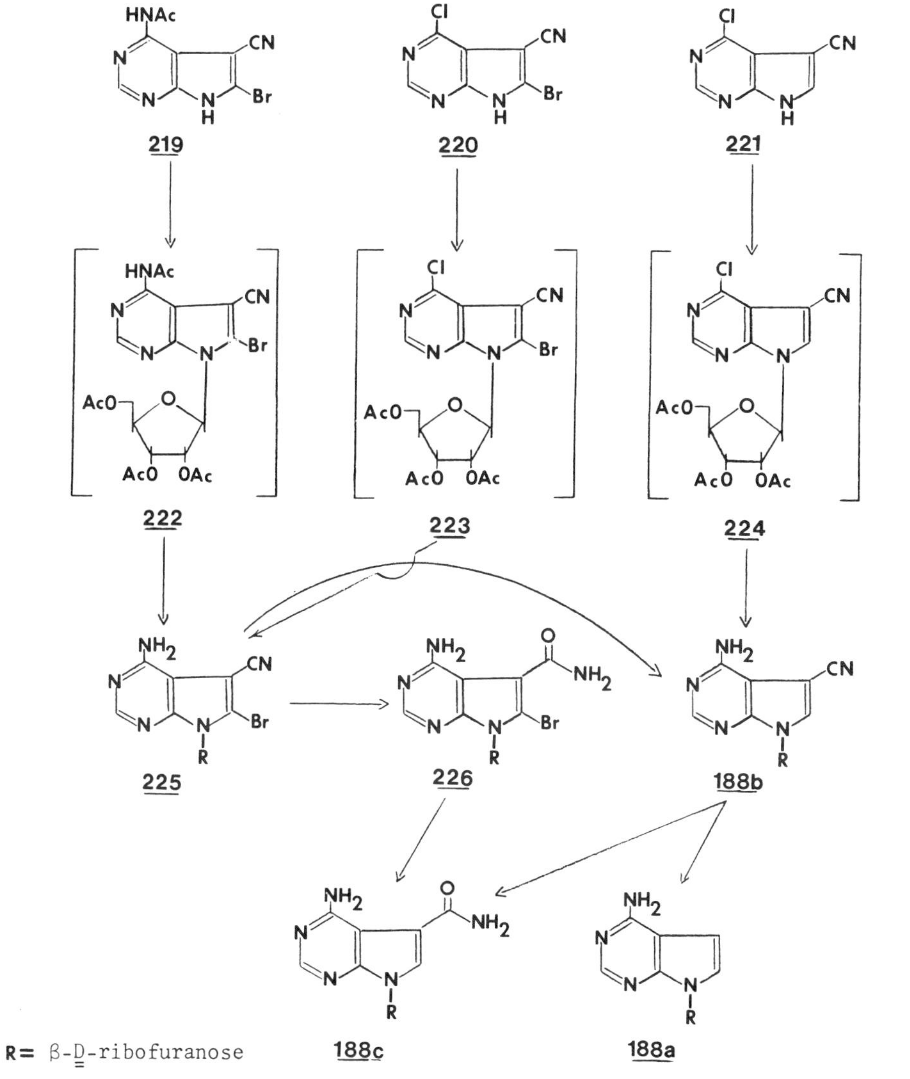

R= β-D-ribofuranose

with **51** in the presence of dichloroacetic acid[310] furnished a 31% yield of a blocked nucleoside (**223**). A similar fusion of 4-chloropyrrolo[2,3-*d*]pyrimidine-5-carbonitrile (**221**) furnished a 28% yield of the acetylated nucleoside **224**. Variation of the acid catalyst produced little or no change in yield, and fusion without an acid catalyst proceeded in nearly the same yield. Removal of the blocking groups with concomitant amination of **223** and **224** proceeded smoothly to give toyocamycin. Likewise, fusion of 4-acetamido-6-methylthio-7-acetylpyrrolo[2,3-*d*]pyrimidine-5-carbonitrile (**227**) with **51** in the presence of bis(*p*-nitrophenyl)phosphate gave a 46% yield of the blocked nucleoside **228**. Subsequent deacetylation of **228** with methanolic ammonia afforded **218**.[307] The 7-β-D-ribofuranosyl analogue of **218** has also been prepared[307] in a similar manner.

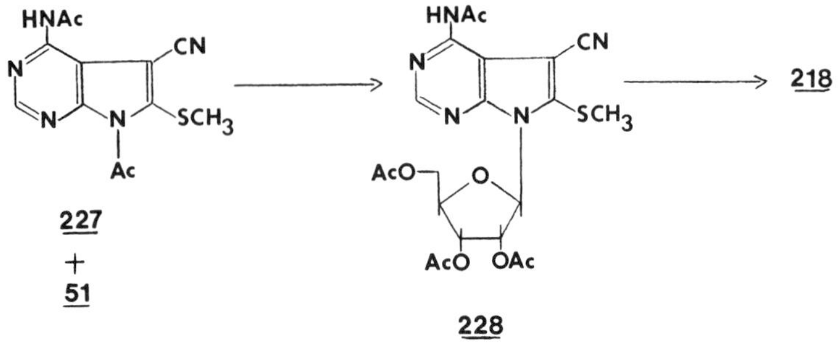

Application of the "indoline–indole" method for the synthesis of tuberci-din has also been found to be successful.[311] Fusion of 4-chloro-5,6-dihydropyr-rolo[2,3-*d*]pyrimidine (**229**) with **51** in the presence of iodine gave a 60% yield of 4-chloro-7-(2,3,5-tri-*O*-acetyl-β-D-ribofuranosyl)-5,6-dihydropyrrolo[2,3-*d*]-pyrimidine (**230**), along with a 5% yield of the corresponding α-anomer. Subse-quent dehydrogenation of **230** with MnO$_2$ and then ammonolysis furnished tubercidin (**188a**).[311] A similar fusion of 4-methylthio-5,6-dihydropyrrolo-[2,3-*d*]pyrimidine gave the corresponding 7-ribosyl derivative in rather low yield.[311]

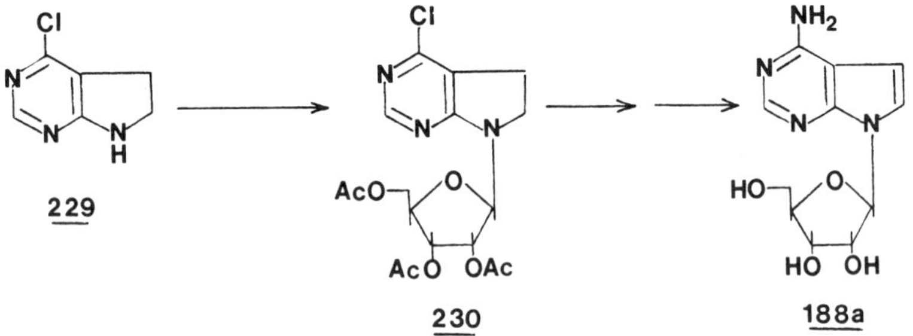

2.2.3.4. Synthesis Employing the Trimethylsilyl Derivative of a Preformed Aglycon and an Acylated Sugar Halide. Even though the fusion procedure has been used successfully to obtain tubercidin, toyocamycin, and sangivamycin,[249] the application of this procedure to other mono-, di-, and trisubstituted pyrrolo-pyrimidines was rather disappointing. Extensive decomposition of the carbohy-drate and aglycon derivatives made product isolation very laborious. Application of the trimethylsilyl procedure[150,312,313] was found to be more rewarding. Using a trimethylsilyl alkylation procedure (silylated heterocycle and an appropriate halosugar at room temperature in an inert solvent, such as acetonitrile)[150,313] or a silyl fusion procedure (fusion of silylated aglycon and a halosugar under reduced pressure)[312,314] and utilizing *N*-trimethylsilyl-4-chloro-6-methylthio-pyrrolo[2,3-*d*]pyrimidine-5-carbonitrile (**231**), poor yields of nucleoside mate-rial were obtained.[315] However, under the Wittenburg conditions[316] using **231** and 2,3,5-tri-*O*-acyl-D-ribofuranosyl bromide (**19**) in a nonpolar solvent, in the presence of an acid acceptor (such as mercuric oxide), a mixture of three isomeric nucleosides was obtained[315]; after separation, these were identified as 4-chloro-6-methylthio-7-(2,3,5-tri-*O*-benzoyl-β-D-ribofuranosyl)pyrrolo[2,-3-*d*]pyrimidine-5-carbonitrile (**232**, 17%) and its N-1 (**233**, 46%) and N-3 (**234**, 31%) ribosyl isomers.[315] Structural assignments were made on the basis of UV absorption studies and further confirmed by a chemical conversion to derivatives of established structure.[315]

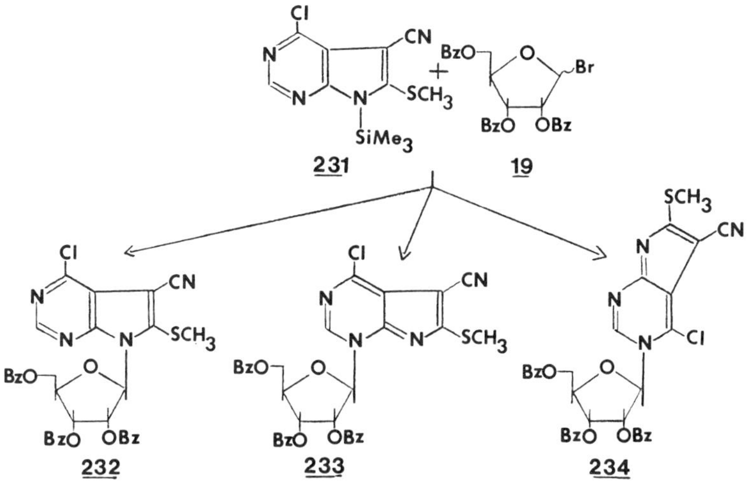

Ribosylation of 4-chloro-5-nitropyrrolo[2,3-*d*]pyrimidine (**235**, X = NO₂) by the Wittenburg procedure using mercuric oxide furnished a 23% yield of 4-chloro-5-nitro-7-(2,3,5-tri-*O*-benzoyl-β-D-ribofuranosyl)pyrrolo[2,3-*d*]pyrimidine (**237**, X = NO₂). A similar ribosylation of **235** (X = H) furnished two nucleosides, identified as 4-chloro-1-(2,3,5-tri-*O*-benzoyl-β-D-ribofuranosyl)-pyrrolo[2,3-*d*]pyrimidine (**236**, 11%) and its N-7 ribosyl isomer (**237**, X = H), 18%). The structure of **237** (X = H) was established by treatment with sodium methoxide in methanol, which furnished the well-characterized 4-chloro-7-β-D-ribofuranosylpyrrolo[2,3-*d*]pyrimidine.[317] However, the anomeric configuration of the isomeric 1-ribofuranoside (**236**) was only assumed to be β on the basis of the *trans* rule.[53]

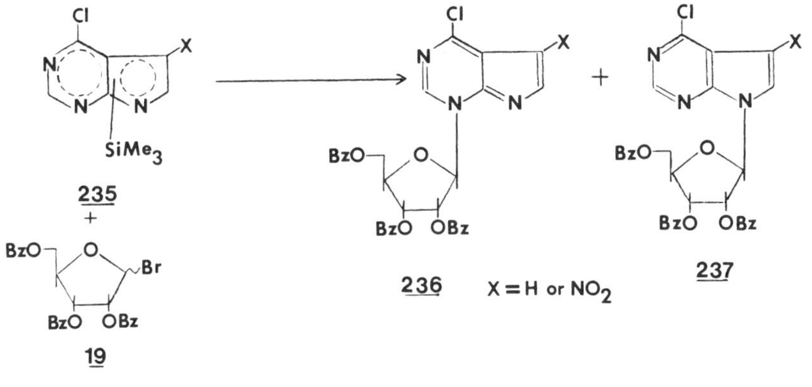

Glycosylation[315] of the bis-trimethylsilyl derivative (**238**) of 2-methylthio-pyrrolo[2,3-*d*]pyrimidin-4-one,[296] by the above procedure, gave three nucleoside products, which were isolated in yields of 8%, 64%, and 7% and were assigned the structures 239, 240, and 241, respectively. However, similar glycosylation of the trimethylsilyl pyrrolo[2,3-*d*]pyrimidin-4(3*H*)-one has been reported[318] to give exclusively the N-3 nucleoside, whereas glycosylation of the trimethylsilyl 5-methyl-2-methylthiopyrrolo[2,3-*d*]pyrimidin-4(3*H*)-one has been reported[318] to give exclusively the N-3 nucleoside, whereas glycosylation of the trimethylsilyl 5-methyl-2-methylthiopyrrolo[2,3-*d*]pyrimidin-4(3*H*)-one under Wittenburg conditions gave the *O*-glycoside.[319] Attempts to rearrange the *O*-riboside (241) to an *N*-ribonucleoside with mercuric bromide in acetonitrile at reflux temperature were unsuccessful.[320] Deacetylation and dethiation of 239 gave 3-β-D-ribofuranosylpyrrolo[2,3-*d*]pyrimidin-4-one (1-ribosyl-7-deazahypoxanthine, 242). However, attempts to dethiate 240 were unsuccessful.[315] Deacetylation of 240 furnished 2-methylthio-7-β-D-ribofuranosyl-pyrrolo[2,3-*d*]pyrimidin-4-one (243), which was also obtained from 4-chloro-2-methylthio-7-(2,3,5-tri-*O*-benzoyl-β-D-ribofuranosyl)pyrrolo[2,3-*d*]pyrimidine (244) by basic hydrolysis.

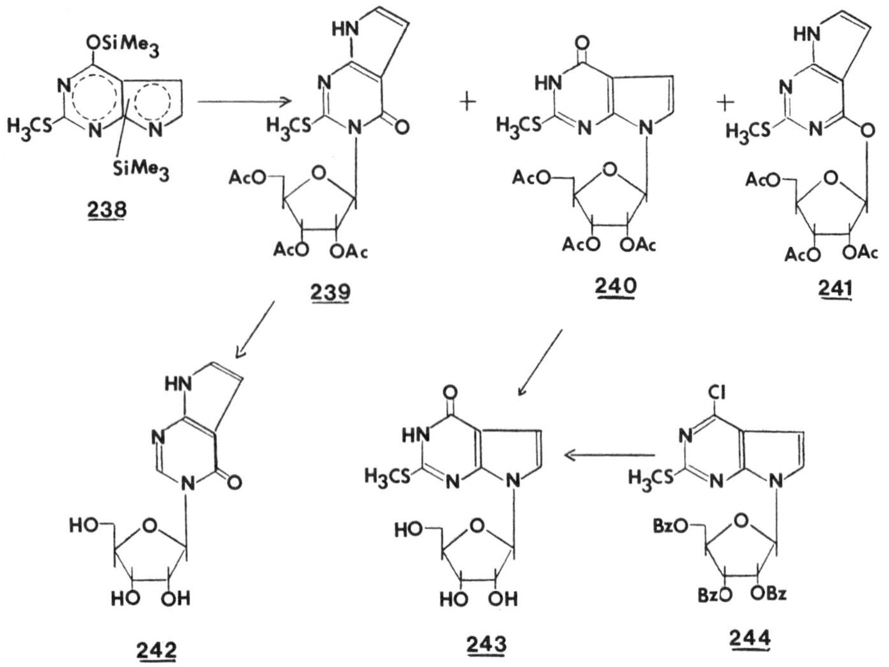

The Wittenburg procedure for ribosylation, which requires mercuric oxide in a nonpolar solvent, is a modification of the Koenigs–Knorr reaction for the preparation of aliphatic and aromatic glycosides. Silver salts have been utilized with excellent results in the Koenigs–Knorr synthesis,[321] and this has prompted[315] use of similar conditions in the ribosylation procedure.

The ribosylation of 2-methylthiopyrrolo[2,3-*d*]pyrimidin-4-one in the presence of one equivalent of silver oxide again furnished three nucleosides, 239 (11%), 240 (7%), and 241 (78%). A similar ribosylation of silylated 5-bromo-4-chloro-2-methylthiopyrrolo[2,3-*d*]pyrimidine in the presence of silver oxide gave a

48% yield of the corresponding N-7 ribosyl derivative. Likewise, 4-chloro-2-methylthiopyrrolo[2,3-*d*]pyrimidine furnished an 82% yield of 4-chloro-2-methylthio-7-(2,3,5-tri-*O*-benzoyl-β-D-ribofuranosyl)pyrrolo[2,3-*d*]pyrimidine (**244**).[315]

2-Methylthio-7-(2,3,5-tri-*O*-acetyl-β-D-ribofuranosyl)pyrrolo[2,3-*d*]pyrimidin-4-one (**240**) was found to be a viable intermediate for the synthesis of 7-deazaguanosine (**246**).[322] Treatment of **240** with 30% hydrogen peroxide in methanol gave the 2-methylsulfonyl derivative **245**, which was not isolated but, on subsequent amination with liquid ammonia at 85°C, afforded 2-amino-7-β-D-ribofuranosylpyrrolo[2,3-*d*]pyrimidin-4-one (**246**). The rather poor yield of **246** prompted[322] an investigation for an alternate synthesis of **246**. Coupling of the bis-trimethylsilyl derivative of 2-acetamidopyrrolo[2,3-*d*]pyrimidin-4-one (**247**) with **145** in the presence of mercuric oxide and mercuric bromide furnished two major nucleosides, **248** and **249**, in 31% and 28% yields, respectively. Deacetylation of **248** with methanolic sodium methoxide gave a 47% yield of 7-deazaguanosine (**246**).

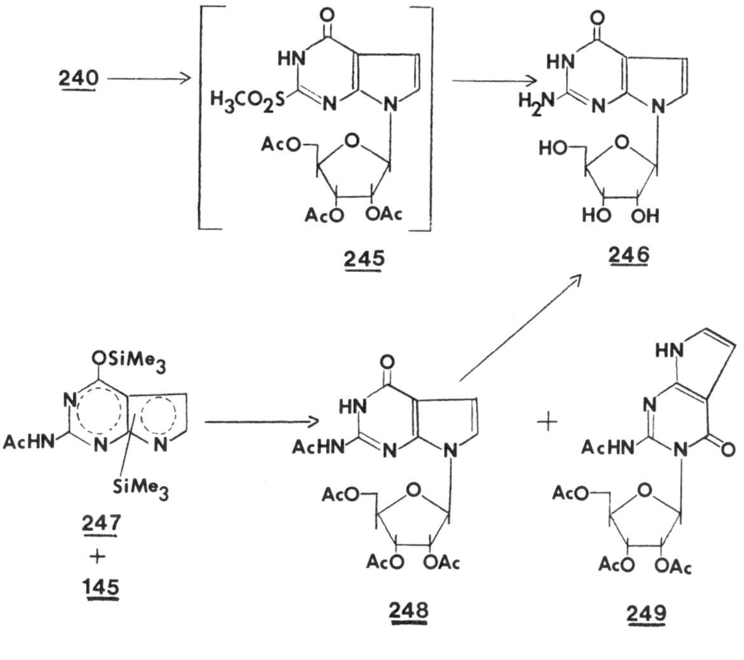

Glycosylation of the trimethylsilyl derivative of 2-mercapto-5,6-dihydropyrrolo[2,3-*d*]pyrimidin-4-one (**250**) with either tetra-*O*-acetylglucopyranosyl bromide or tri-*O*-benzoyl-D-ribofuranosyl bromide gave[323] the corresponding 2-β-D-glycosylthio derivatives (**251**). Deacylation of **251** with sodium methoxide furnished the free glycosides. The formation of *S*-glycosides is not uncommon since similar observations were made in the pyrimidine[295] and purine[324] series, in which the *S*-glycosidic bond is extremely susceptible to hydrolysis. However, ribosylation of the trimethylsilyl 4-hydroxy-5,6-dihydropyrrolo[2,-3-*d*]pyrimidine gave the corresponding N-2 and N-7 ribofuranosides, as well as the N-3 ribofuranoside of 7-acetyl-4-oxo-3,4,5,6-tetrahydropyrrolo[2,3-*d*]pyrimidine.[325]

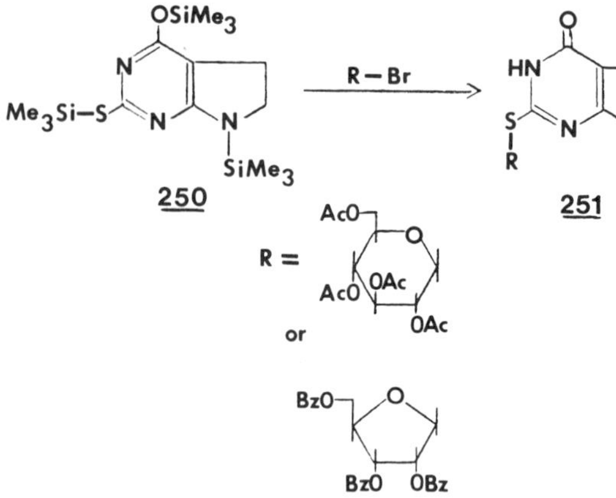

2.2.3.5. Synthesis Employing the Sodium Salt of a Preformed Aglycon and an Acylated Sugar Halide. Seela *et al.*[326] have reported the use of sodium hydride in the glycosylation of 4-chloro-5-methyl-2-methylthiopyrrolo[2,3-*d*]pyrimidine, but no yield of the isolated blocked nucleoside was reported. Goto and co-workers[327] had previously reported the use of sodium hydride in the glycosylation of 4-methoxy-5-methyl-2-methylthiopyrrolo[2,3-*d*]pyrimidine with 2,3,5-tri-*O*-benzyl-D-ribofuranosyl bromide. However, this procedure gave a complex anomeric mixture and an abnormal by-product (**252**). The desired 4-methoxy-5-methyl-2-methylthio-7-(2,3,5-tri-*O*-benzyl-β-D-ribofuranosyl)pyrrolo[2,3-*d*]pyrimidine was isolated in less than 25% yield.

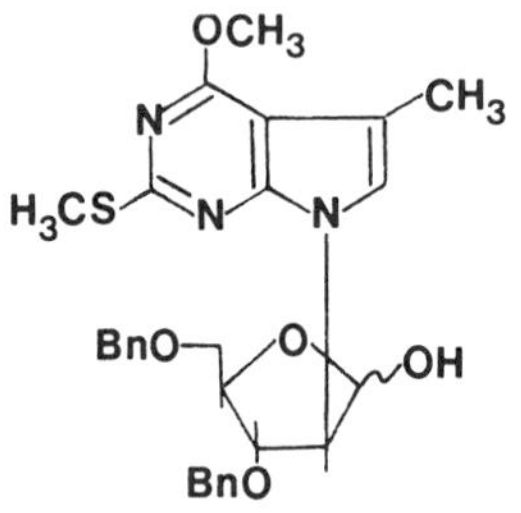

A general and stereospecific sodium salt glycosylation procedure was developed recently in our laboratory,[30,31,328] and found to be remarkably successful for the preparation of halogenated pyrrolo[2,3-*d*]pyrimidine nucleosides. Glycosylation of the sodium salt of 2-methylthio- (or 2-methyl-) 4,6-dichloropyrrolo[2,3-*d*]pyrimidine (**253**) with **19** gave the corresponding blocked N-7 glycosyl derivative (**254**) as the major product.[30] Debenzoylation and concomitant amination of **254** with methanolic ammonia at elevated temperature gave 4-amino-6-chloro-2-methylthio-7-β-D-ribofuranosylpyrrolo[2,3-*d*]pyrimidine

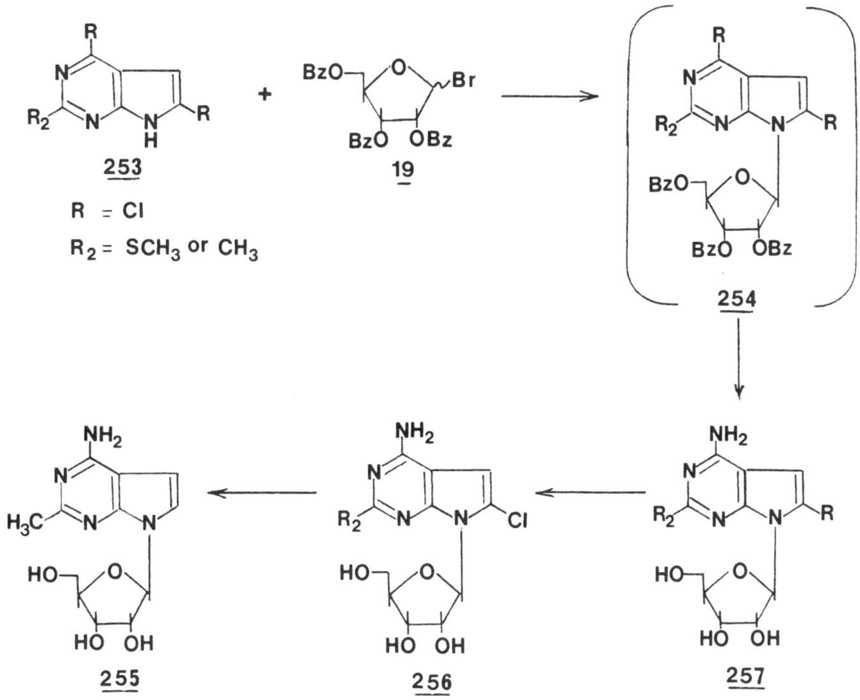

(**257**, R = Cl, R$_2$ = SCH$_3$) and the corresponding 2-methyl derivative (**257**, R = Cl, R$_2$ = CH$_3$). Treatment of **257** (R = Cl, R$_2$ = SCH$_3$) with Raney nickel gave 6-chlorotubercidin (**256**, R$_2$ = H), which, on catalytic dehalogenation, provided a new synthetic route to tubercidin.[30] A similar dehalogenation of **256** (R$_2$ = CH$_3$) with Pd/C in a hydrogen atmosphere provided 2-methyltubercidin (**255**).[30]

This stereospecific glycosylation procedure was also found to be applicable to the direct preparation of 2'-deoxytubercidins.[329,330] Treatment of the sodium salt of 2,4-dichloropyrrolo[2,3-*d*]pyrimidine (**258a**) with **24** in acetonitrile gave a 60% yield of 2,4-dichloro-7-(2-deoxy-3,5-di-*O*-*p*-toluoyl-β-D-*erythro*-pento-furanosyl)pyrrolo[2,3-*d*]pyrimidine (**259a**), which, when allowed to react with methanolic ammonia at 100°C, furnished 2-chloro-2'-deoxytubercidin (**260a**). Dehalogenation of **260a** with Pd/C provided 2'-deoxytubercidin (**260b**).[31]

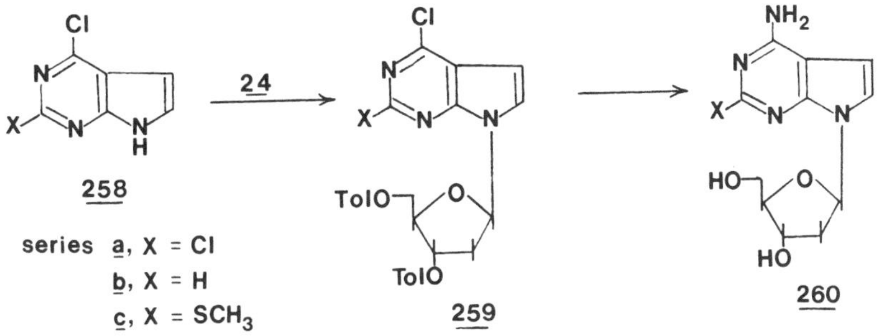

2'-Deoxytubercidin was also prepared in good yield by the direct glycosylation of the sodium salt of 4-chloropyrrolo[2,3-*d*]pyrimidine (**258b**)[246] with **24** to obtain 4-chloro-7-(2-deoxy-3,5-di-*O*-*p*-toluoyl-β-D-*erythro*-pentofuranosyl)pyrrolo[2,3-*d*]pyrimidine (**259b**), followed by ammonolysis.[31] Similar 2'-deoxyribosylation of the sodium salt of 2-methylthio-4-chloropyrrolo[2,3-*d*]pyrimidine (**258c**)[296] with **24** gave **259c**, which, on ammonolysis, gave 2-methylthio-4-amino-7-(2-deoxy-β-D-*erythro*-pentofuranosyl)pyrrolo[2,3-*d*]pyrimidine (**260c**). Dethiation of **260c** with Raney nickel furnished yet another route to 2'-deoxytubercidin.[31] Essentially the same methodology was used to prepare[329] 6-chloro-2'-deoxytubercidin (**261a**), 6-methyl-2'-deoxytubercidin (**261b**), and 2-methyl-6-chloro-2'-deoxytubercidin (**262**). Since the starting halosugar **24** has the α configuration in the solid state,[39] the exclusive formation of the blocked 2'-deoxy-β-nucleosides in this sodium salt glycosylation procedure is believed to be due to a Walden inversion (S_N2 at the C-1 carbon by the anionic heterocyclic nitrogen.

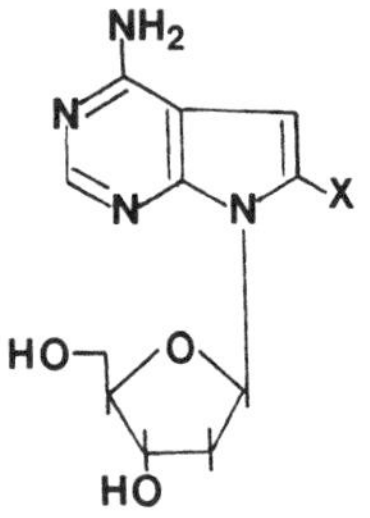

261a, X = Cl
b, X = CH₃

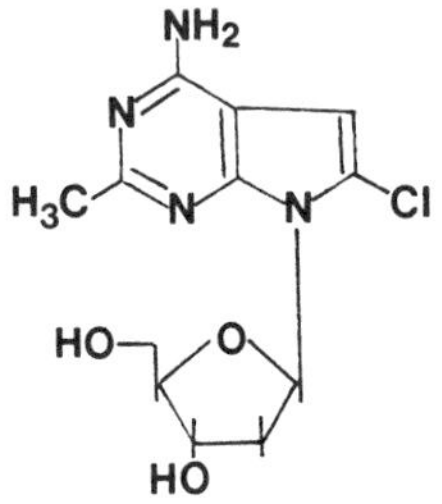

262

   This versatile procedure provided a stereospecific method for the synthesis of 2'-deoxytoyocamycin, *ara*-sangivamycin, and *ara*-cadeguomycin, as well as the ribonucleosides tubercidin, 7-deazaguanosine, and cadeguomycin and a host of other pyrrolo[2,3-*d*]pyrimidine nucleosides of biological significance. For the synthesis of 2'-deoxytoyocamycin (**266**) and *ara*-sangivamycin (**269**), 2-amino-5-bromopyrrole-3,4-dicarbonitrile[331] (**263**) served as a useful starting material. The protection of the amino group was effected by the treatment of **263** with diethoxymethyl acetate in acetonitrile to give the 2-ethoxymethyleneamino compound, which, on treatment with **24** in the presence of NaH, gave a 75% yield of the corresponding blocked nucleoside **264**. Compound **264** cleanly cyclized to 4-amino-6-bromo-7-(2-deoxy-β-D-*erythro*-pentofuranosyl)pyrrolo-[2,3-*d*]pyrimidine-5-carbonitrile (**265**) on treatment with methanolic ammonia. Selective acetylation of **265** with acetic anhydride in pyridine, followed by reductive debromination[332] with *N,O*-bis(trimethylsilyl)acetamide (BSA) in the presence of "naked" fluoride ion and deacetylation, gave 2'-deoxytoyocamycin[333] (**266**), in which the nitrile function was available for further transformation reactions to give **270**.

   The 2-ethoxymethyleneaminopyrrole was also used in the glycosylation studies to obtain *ara*-sangivamycin (**269**) and certain related compounds. Treat-

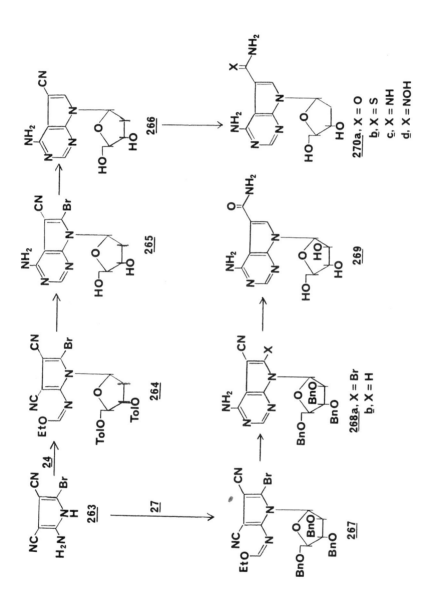

ment of the sodium salt of the pyrrole with 1-chloro-2,3,5-tri-*O*-benzyl-α-D-ara-binofuranose[35] (**27**) gave the corresponding benzyl blocked nucleoside **267**, which was readily ring closed in the presence of methanolic ammonia to yield 4-amino-6-bromo-7-(2,3,5-tri-*O*-benzyl-β-D-arabinofuranosyl)pyrrolo[2,3-*d*]-pyrimidine-5-carbonitrile (**268a**). Reductive debromination of **268a** with BSA gave **268b**, which, on oxidative hydration by the treatment with ammonium hydroxide and hydrogen peroxide, followed by debenzylation with palladium hydroxide in the presence of cyclohexene, gave *ara*-sangivamycin[333] (**269**). Compound **268b** was used for further transformation reactions[334] to obtain potent antiviral[335] nucleosides which are totally free of contamination with the ribonucleoside antibiotics toyocamycin or sangivamycin.

The naturally occurring cytotoxic[336] nucleoside antibiotic tubercidin was prepared by the direct glycosylation of the sodium salt of 4-chloropyrrolo-[2,3-*d*]pyrimidine[246] (**258b**). Reaction of **258b** with 1-chloro-2,3-*O*-isopropyli-dene-5-*O*-(*tert*-butyl)dimethylsilyl-α-D-ribofuranose[337] (**272**) in CH$_3$CN in the presence of NaH gave a 67% yield of the blocked nucleoside **273a**, which, on deprotection with aqueous trifluoroacetic acid, furnished 4-chloro-7-β-D-ribo-furanosylpyrrolo[2,3-*d*]pyrimidine (**274**) in almost quantitative yield. Treatment of **274** with MeOH/NH$_3$ gave tubercidin in 81% yield.[338,339] A similar glycosy-lation of the sodium salt of 2-amino-4-chloropyrrolo[2,3-*d*]pyrimidine[340] (**271**) with **272** gave the blocked nucleoside **273b**, which, on sequential conversion of the C-4 chloro group into a methoxy, deisopropylidenation of the glycon moiety, and cleavage of the ether linkage without affecting the glycosidic cleavage, gave

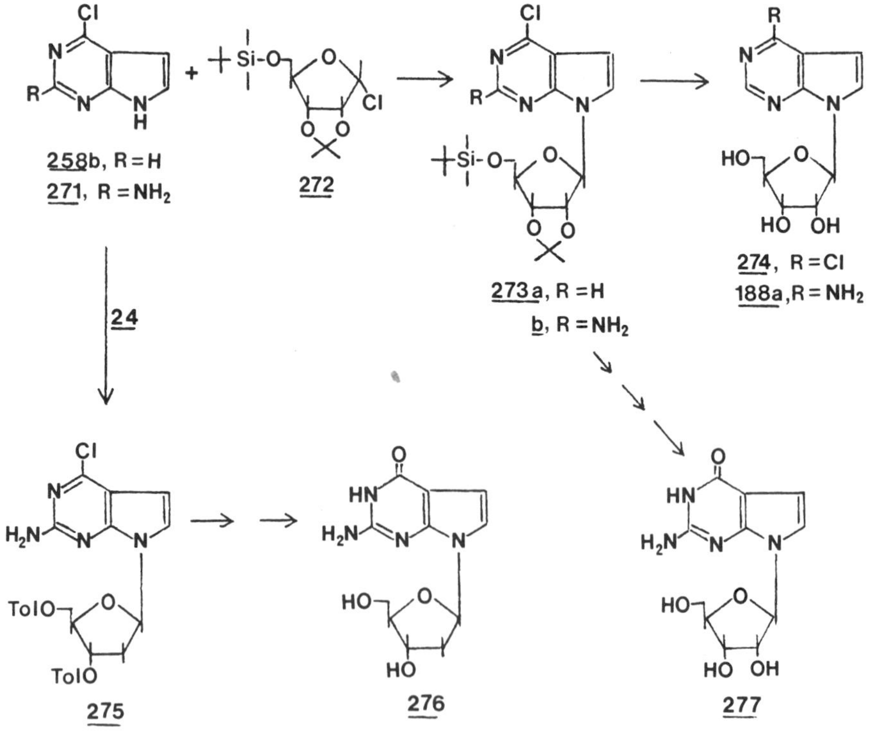

7-deazaguanosine (**277**)[338,339] in good yield. In a similar manner, condensation of the sodium salt of **271** with **24** and subsequent treatment of the condensation product **275** with NaOCH$_3$, followed by Me$_3$SiI treatment, provided a straightforward route to the preparation of 2'-deoxy-7-deazaguanosine (**276**).[338,339]

Cadeguomycin (**198**) is a novel pyrrolo[2,3-*d*]pyrimidine nucleoside antibiotic, isolated[281] in 1982 from the culture broth of *Streptomyces hygroscopicus* IM7912T as a minor component together with tubercidin and characterized as 2-amino-4(3*H*)-oxo-7-β-D-ribofuranosylpyrrolo[2,3-*d*]pyrimidine-5-carboxylic acid.[282] This interesting antibiotic inhibited the growth of solid IMC carcinoma and pulmonary metastasis of Lewis lung carcinoma in mice with low toxicity.[341] It also enhanced cell-mediated immunity and macrophage activity.[341] Cadeguomycin displayed a unique property of enhancing uptake of pyrimidine nucleosides into K562 human myelogenous leukemic cells and YAC-1 murine lymphoma cells, and it potentiated cytotoxicity of *ara-C*[341–343] as well as 5-fluoro-2'-deoxycytidine[344] both *in vitro* and *in vivo*. This interesting and potent activity resulted in the syntheses of cadeguomycin,[283,284,345,346] *ara*-cadeguomycin,[346,347] and 2'-deoxycadeguomycin.[346,348] Utilization of the sodium salt glycosylation method was found to be very successful for the synthesis of these nucleosides.

The sodium salt of 2-amino-4-chloropyrrolo[2,3-*d*]pyrimidine-5-carbonitrile[346] (**279**), generated *in situ* by treatment with NaH, was reacted with **24** in anhydrous CH$_3$CN to give the protected nucleoside, which, on deprotection with 2 *N* NaOH in dioxane, followed by neutralization, afforded 2'-deoxycadeguomycin[346,348] (**284**). The formation of **284** is of particular interest since the base treatment of the protected nucleoside not only converted the nitrile function to the carboxylic acid but also concomitantly hydrolyzed the chloro group to a keto function. A similar glycosylation of the sodium salt of methyl 2-amino-4-chloropyrrolo[2,3-*d*]pyrimidine-5-carboxylate[346,348] (**280**) with **24** and further functional group manipulation of the intermediate **281** furnished a second route to **284**.

A total synthesis of cadeguomycin (**198**) and *ara*-cadeguomycin (**283**) has been accomplished by reacting the appropriate α-halogenoses (**272** and **27**, respectively) with the sodium salt of **279** and subsequent deprotection of the glycon moiety, followed by saponification.[346]

The sodium salt glycosylation procedure has recently been used for the preparation of a number of 4-substituted 7-β-D-arabinofuranosylpyrrolo[2,3-*d*]-pyrimidines[349] (**285**), as well as 7-iodo-2',3'-dideoxy-7-deazaguanosine[350] (**286**), which is a key intermediate in the preparation of reagents for the automated sequencing of DNA. This procedure was also found to be useful for the preparation of certain substituted 7-[(2-hydroxyethoxy)methyl[pyrrolo[2,3-*d*]-pyrimidine[351,352] structurally related to the antiviral agent acyclovir.

2.2.3.6. Synthesis of Pyrrolo[2,3-d]pyrimidine Nucleosides by the Phase-Transfer Glycosylation Procedure. A regiospecific glycosylation procedure for the preparation of pyrrolo[2,3-*d*]pyrimidine nucleosides, known as the "phase-transfer glycosylation method," has been reported by Seela and co-workers.[330] Normally, the glycosylation reaction is carried out in a biphasic mixture of dichloro-

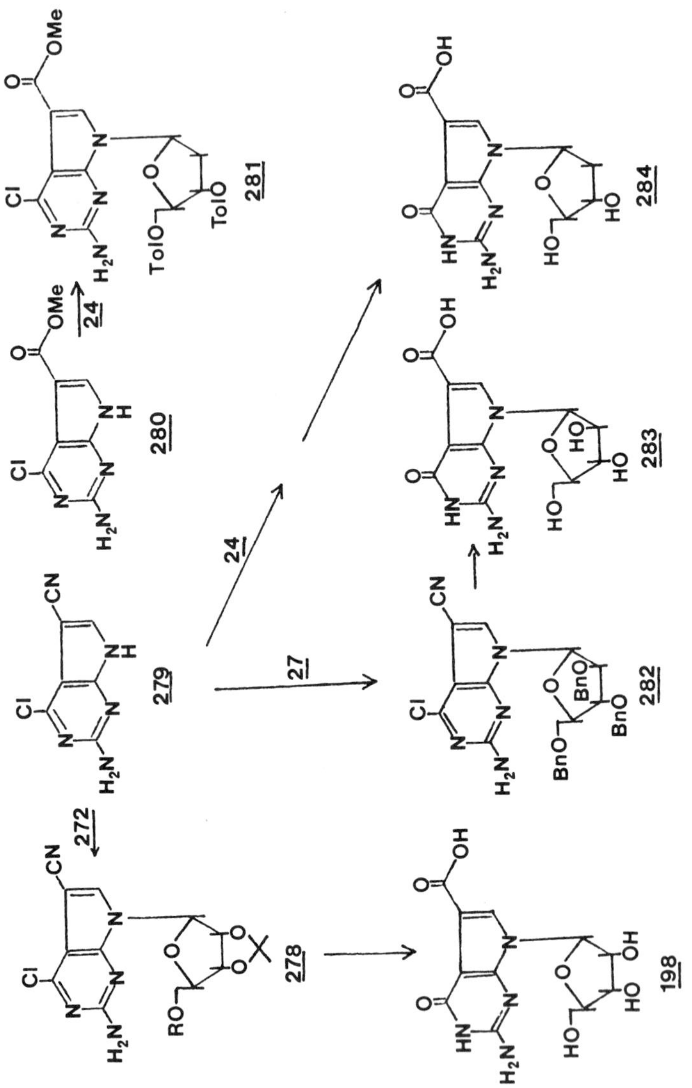
278
279
280
281
282
283
284
198
272
24
27
24

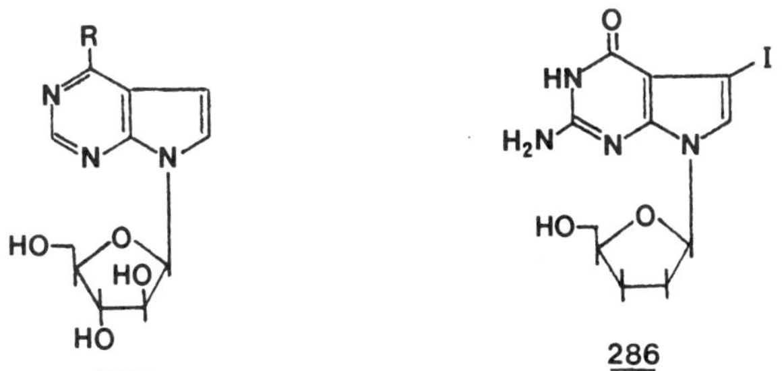

285

R = NH$_2$, OH, SH

286

methane and 50% aqueous sodium hydroxide in the presence of a phase-transfer catalyst,[353–355] such as tetrabutylammonium bisulfate or the cryptand tris[2-(2-methoxyethoxy)ethyl]amine (TDA-1).[356] Under these conditions, 4-methoxy-2-methylthiopyrrolo[2,3-*d*]pyrimidine (287) was reacted with 2,3,5-tri-*O*-benzyl-D-ribofuranosyl bromide (288)[357] to give selectively N-7 glycosyl derivatives (289 and 290).[358] Hydrolysis (methyl ether), desulfurization (methylthio group), and catalytic hydrogenation (benzyl ether) of 289 gave 7-deaza-inosine (206).[358]

The ratio of anomers 289 and 290 was influenced by a change in the concentration of sodium hydroxide. A low concentration of sodium hydroxide

287

288

289

290

291

277

292

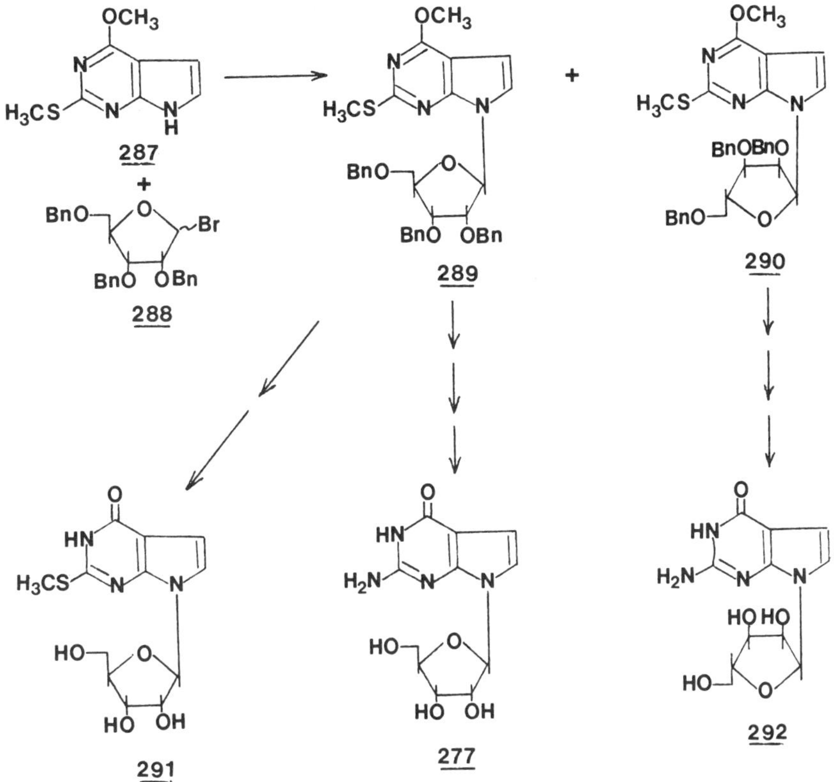

(~10%) resulted in a low yield of glycosylation products, while 50% aqueous sodium hydroxide was found to be optimal. It has been postulated that a relatively solvation-free anion, accompanied by a highly lipophilic solvated tetra-butylammonium cation generated at a 50% sodium hydroxide concentration, accounts for the high reactivity. A reactive aglycon is a basic requirement for an effective glycosylation since **288** gradually decomposes under the strong alkaline reaction conditions of phase-transfer glycosylation.[330] Like the sodium salt glycosylation procedure, the S_N2 displacement of the anomeric bromine of **288** by the anion of the aglycon in the phase-transfer glycosylation procedure has been proposed.[330]

Debenzylation of 2-methylthio-7-(2,3,5-tri-O-benzyl-β-D-ribofuranosyl)-pyrrolo[2,3-d]pyrimidin-4(3H)-one[358] with boron trichloride in dichloromethane at −78°C gave 2-methylthio-7-deazainosine (**291**).[359] Alkylation of 2-methyl-thio-7-(2,3,5-tri-O-benzyl-β-D-ribofuranosyl)pyrrolo[2,3-d]pyrimidin-4(3H)-one[358] at N-3 was accomplished with chloromethyl isopropyl ether. This nucleoside was treated with a mixture of sodium hydride and acetamide, followed by removal of the isopropoxymethyl and benzyl groups with boron trichloride and subsequent ammonolysis, to give 7-deazaguanosine (**277**).[360] The α-anomer of 7-deazaguanosine (**292**) was also similarly prepared[361] by a multistep conversion of **290**.

Phase-transfer catalysis of 4-chloro-2-methylthiopyrrolo[2,3-d]pyrimidine (**258c**) with **288** gave the anomers **293a** and **294a**. Debenzylation of these blocked nucleosides with boron trichloride gave the free nucleosides **293b** and **294b**,[362] which were converted to the corresponding 2-methylthiotuberci-dins.[363] Contrary to these observations, glycosylation of 2-amino-4-methoxy-pyrrolo[2,3-d]pyrimidine[364] with **24** under the conditions of phase-transfer catalysis gave exclusively, after ammonolysis, 2-amino-7-(2-deoxy-β-D-*erythro*-pentofuranosyl)-4-methoxypyrrolo[2,3-d]pyrimidine.[365] No α-anomer was observed. Cleavage of the 4-methoxy group of this blocked nucleoside with sodium *p*-thiocresolate in hexamethylphosphoramide/toluene gave 2′-deoxy-7-deaza-guanosine,[365] which, on further Barton deoxygenation, provided 2′,3′-di-deoxy-7-deazaguanosine.[366] Hydrogen bromide/acetic acid in tetrahydrofuran was found to be a better reagent for the cleavage of the 4-methoxy group.[367]

The yield of phase-transfer glycosylation of 2-amino-4-methoxypyrrolo-[2,3-d]pyrimidine with **24** is limited under liquid–liquid conditions (50%

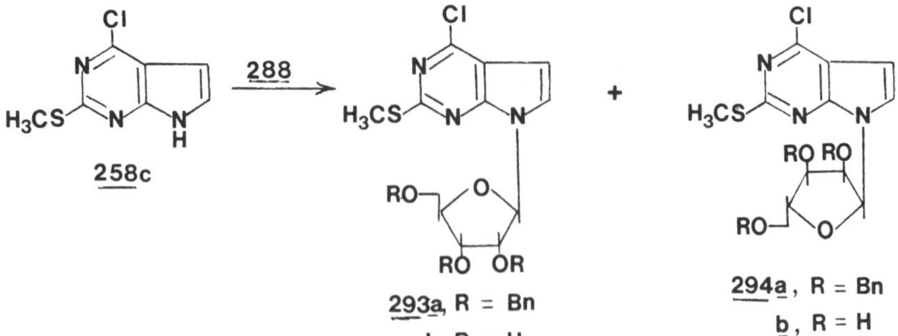

258c

293a, R = Bn
b, R = H

294a, R = Bn
b, R = H

aqueous NaOH, Bu_4NHSO_4 in CH_2Cl_2) due to deprotection of alkali-labile protecting groups in the halogenose. To overcome these difficulties, a solid–liquid phase-transfer glycosylation method employing an aprotic solvent, solid KOH, and the cryptand TDA-1 has been developed.[368] This procedure leads stereospecifically in high yield to 2-amino-4-alkoxy- as well as 2-amino-4-chloro-pyrrolo[2,3-*d*]pyrimidine 2-deoxy-β-D-*erythro*-pentofuranosides.[368] Nucleophilic displacement of the 4-chloro group of 2-amino-4-chloro-7-(2-deoxy-β-D-*erythro*-pentofuranosyl)pyrrolo[2,3-*d*]pyrimidine thus generated a route to 2-amino-2′-deoxytubercidin and related nucleosides.[369]

Similar glycosylation of 4-amino-2-methylthiopyrrolo[2,3-*d*]pyrimidine with **24** gave the corresponding blocked N-7 glycoside with the β configuration.[370] Alkaline hydrolysis of the protecting groups afforded 2′-deoxyy-2-methylthiotubercidin, and desulfurization gave 2′-deoxytubercidin (**260b**).[370]

Several 4-substituted pyrrolo[2,3-*d*]pyrimidine 2′,3′-dideoxyribofuranosides, including 2′,3′-dideoxytubercidin, were prepared from the corresponding 4-chloro derivative.[371] The 4-chloro derivative was in turn prepared by Barton deoxygenation of 4-chloro-7-(2-deoxy-β-D-*erythro*-pentofuranosyl)pyrrolo-[2,3-*d*]pyrimidine.

The phase-transfer glycosylation technique has proved to be very successful for the preparation of arabinofuranosyl nucleosides. By employing this procedure, *ara*-tubercidin (**285**, R = NH_2),[374–378] *ara*-7-deazainosine (**285**, R = OH),[379] *ara*-7-deazaguanosine (**295a**),[380–382] 2-amino-4-methoxy-7-β-D-arabinofuranosyl-pyrrolo[2,3-*d*]pyrimidine (**295b**),[382] 7-β-D-arabinofuranosyl-2,4-dichloropyr-rolo[2,3-*d*]pyrimidine (**296a**),[383] and *ara*-7-deazaxanthosine (**296b**)[384] have been obtained. This procedure was also found to be useful for the preparation of acyclo-7-deazaguanosine[364] and stereo-selective synthesis of α-D-ribonucleo-sides of substituted pyrrolo[2,3-*d*]pyrimidines.[372,373]

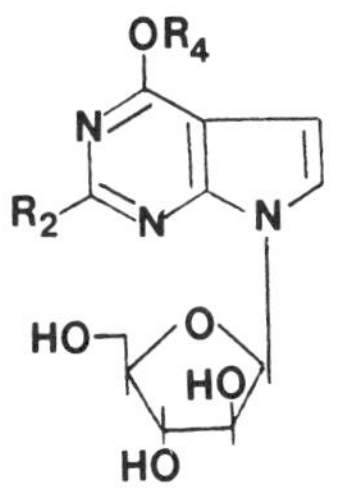

295a, R_2= NH_2, R_4 =H
b, R_2= NH_2, R_4 = CH_3

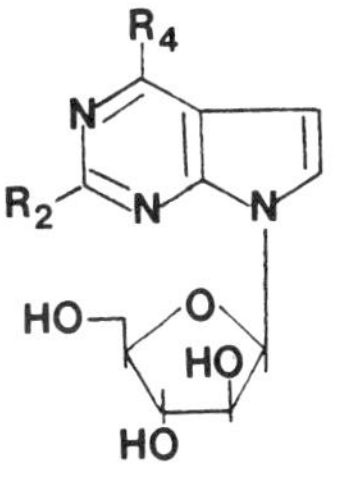

296a, R_2= R_4=Cl
b, R_2= R_4=OH

2.2.3.7. Chemical Transformations of the Aglycon of Pyrrolo[2,3-d]pyrimidine Nucleosides. The unusual biological properties[238,239,385] and antitumor activity[386–391] of naturally occurring pyrrolo[2,3-*d*]pyrimidine nucleosides have stimulated a great deal of interest in the synthesis of various derivatives of these nucleoside antibiotics. The synthesis of several 4-substituted pyrrolo[2,3-*d*]pyrimidine nucleoside derivatives was described by Gerster *et al.*[317] The synthesis of 4-chloro-7-β-D-ribofuranosylpyrrolo[2,3-*d*]pyrimidine (**298**) was accomplished

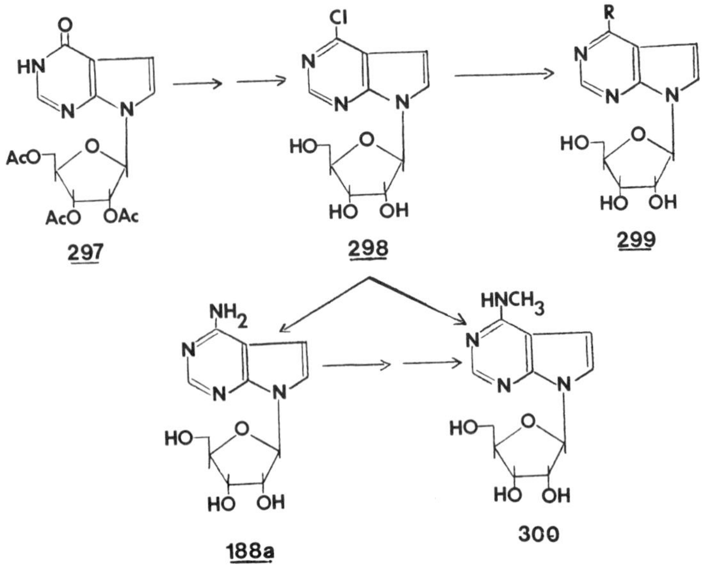

by the treatment of 7-(2,3,5-tri-*O*-acetyl-β-D-ribofuranosyl)pyrrolo[2,3-*d*]pyri-
midin-4-(3*H*)-one (**297**) with phosphorus oxychloride, followed by deacetylation. Nucleophilic displacement of the 4-chloro group of **298** resulted in the
synthesis of various analogues of tubercidin, including 7-deazanebularine (**299**,
R = H). Treatment of **298** with methanolic ammonia at 150°C provided the first
total chemical synthesis of tubercidin (**188a**). Direct methylation of **188a** with
methyl iodide gave 4-amino-3-methyl-7-β-D-ribofuranosylpyrrolo[2,3-*d*]pyrimi-
dine, which rearranged to 4-methylamino-7-β-D-ribofuranosylpyrrolo[2,3-*d*]-
pyrimidine (**300**) when treated with aqueous sodium hydroxide.[392,393] The
synthesis of **300** was also accomplished directly from **298** with methylamine. A
similar treatment of **298** with hydroxylamine gave a 41% yield of 4-hydroxyl-
amino-7-β-D-ribofuranosylpyrrolo[2,3-*d*]pyrimidine.[394] The α-anomer of tubercidin was prepared by ammonolysis of 4-chloro-7-(2,3,5-tri-*O*-acetyl-α-D-
ribofuranosyl)pyrrolo[2,3-*d*]pyrimidine.[395]

   When tubercidin was treated with 2,4-dinitrophenoxyamine[396] in methanolic DMF, crystalline 3-aminotubercidin (**301**)[397] was obtained in a 93% yield.
Similarly, when tubercidin was allowed to react with ethylene oxide in aqueous
acidic media, followed by treatment with alkali, 4-(2-hydroxyethyl)tubercidin
(**302**) was obtained.[398] The synthesis of 5-hydroxymethyltubercidin (**303**)[399]
was achieved by starting with methyl tubercidin-5-carboxylate[275] as well as

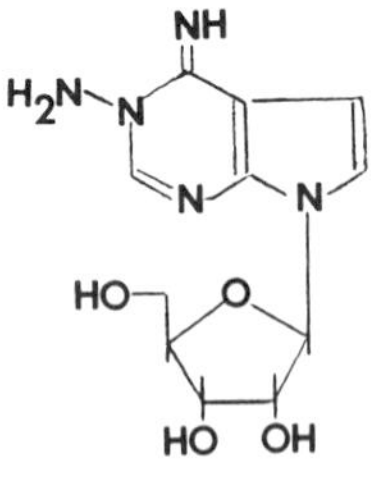

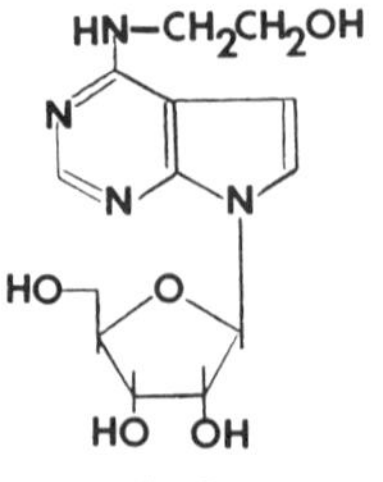

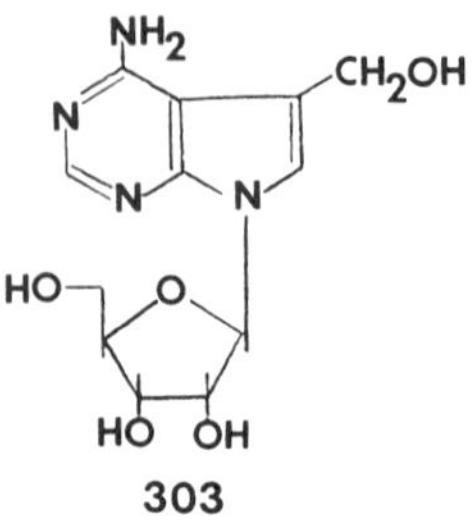

from 5-mercuritubercidin.[400] Isopropylidenation of methyl tubercidin-5-carb-oxylate, followed by reduction with lithium aluminum hydride and subsequent saponification, gave **303** in good yield.[399]

7-β-D-Ribofuranosylpyrrolo[2,3-*d*]pyrimidine-4-carboxamide (**306**), a structural analogue of the broad-spectrum antiviral agent 9-β-D-ribofuranosylpurine-6-carboxamide,[401] has recently been reported by Revankar and co-workers.[402] The direct treatment of 4-chloro-7-(2,3,5-tri-*O*-acetyl-β-D-ribofuranosyl)pyrrolo[2,3-*d*]pyrimidine (**308**)[317] with methanethiol in the presence of a stoichiometric amount of potassium *tert*-butoxide provided 4-methylthio-7-(2,3,5-tri-*O*-acetyl-β-D-ribofuranosyl)pyrrolo[2,3-*d*]pyrimidine (**304**). The KMnO$_4$ oxidation of **304** in acetic acid gave the corresponding 4-methylsulfonyl derivative (**305**). Treatment of **305** with sodium cyanide in dimethylformamide provided the corresponding 4-carbonitrile derivative (**307**), which, on treatment with ammonium hydroxide in the presence of H$_2$O$_2$, gave **306**.[402]

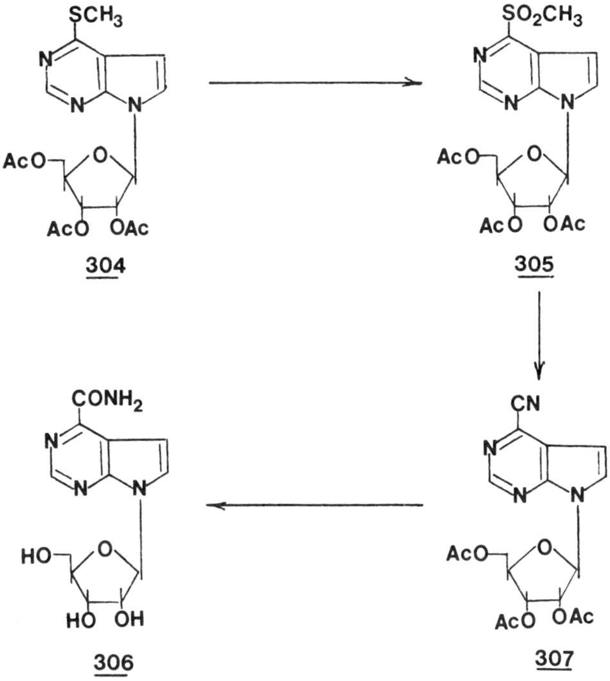

Several interesting 4,5-disubstituted 7-β-D-ribofuranosylpyrrolo[2,3-*d*]pyrimidines have been described.[403] The synthesis of these compounds involved the direct introduction of a functional group via electrophilic substitution in the pyrrole moiety at C-5. Thus, the treatment of 4-chloro-7-(2,3,5-tri-*O*-acetyl-β-D-ribofuranosyl)pyrrolo[2,3-*d*]pyrimidine (**308**) with *N*-bromoacetamide furnished the 5-bromo derivative of **308**, which, on deacetylation, provided 5-bromo-4-chloro-7-β-D-ribofuranosylpyrrolo[2,3-*d*]pyrimidine (**309**). Nucleophilic displacement of the 4-chloro group of **309** gave various 4,5-disubstituted pyrrolo[2,3-*d*]pyrimidine nucleosides (**310**). The substitution of a bromo, chloro,

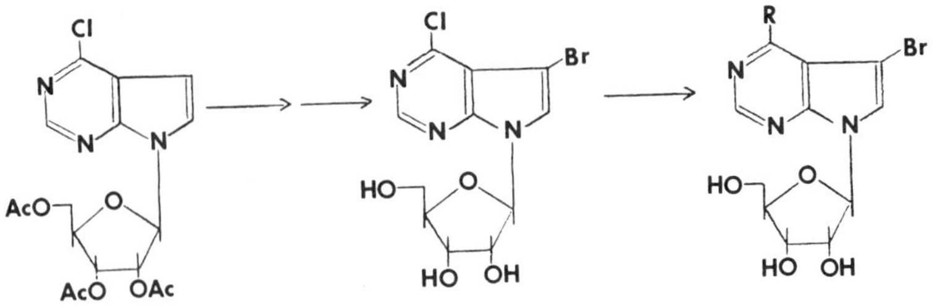

308   309   310a R = NH$_2$
                              b R = OH
                              c R = SH

or iodo group for the 5-cyano group of toyocamycin gave rise to compounds with remarkable biological activity against Rous sarcoma virus.[404] This methodology has recently been used to prepare the naturally occurring 5'-deoxy-5-iodotubercidin (**201**).[405]

Direct bromination of tubercidin with *N*-bromosuccinimide in DMF gave 5-bromotubercidin (**310a**).[406] When the reaction medium was buffered with potassium acetate, the major product was 6-bromotubercidin. 5,6-Dibromotubercidin was also formed in minor amounts under both conditions. Treatment of tubercidin with *N*-chlorosuccinimide gave 5-chlorotubercidin.[406] The unique activity[407–409] of 5-bromotubercidin as a reversible inhibitor of tRNA and heterogeneous nuclear RNA, as well as an inhibitor of the formation of mRNA, prompted[410] the synthesis of additional 5-halosubstituted pyrrolo[2,3-*d*]pyrimidine nucleosides. Catalytic dehalogenation of **308** gave 7-(2,3,5-tri-*O*-acetyl-β-D-ribofuranosyl)pyrrolo[2,3-*d*]pyrimidine (**311**). Alternatively, acetylation of 7-deazanebularine[403] using acetic anhydride in pyridine produced **311**. Halogenation of **311** with *N*-halo (chloro- or bromo-) succinimide or iodine monochloride, followed by deacetylation with methanolic ammonia, furnished the corresponding 5-halogenopyrrolo[2,3-*d*]pyrimidine nucleosides (**313**).[410]

Further electrophilic substitution reactions with 4,5-disubstituted pyrrolo[2,3-*d*]pyrimidine nucleosides furnished interesting nucleoside derivatives.[411] The reactivity of the pyrrolopyrimidine ring at position 6 with another exocyclic group (viz., CN or CONH$_2$) already present at position 5 has been established, and it was observed that hydrogen and bromine are susceptible to

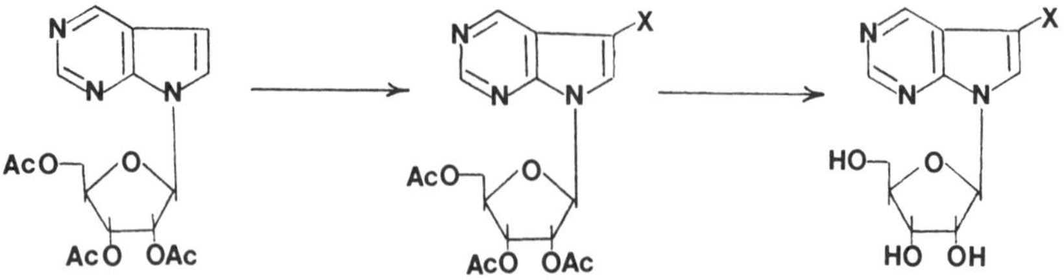

311   312   313

X = Cl, Br, I

electrophilic and acid-catalyzed nucleophilic substitutions,[412] respectively. Treatment of toyocamycin (**188b**) with bromine water gave 6-bromotoyocamycin (**314**). Attempts to nitrate toyocamycin with cold nitric acid, acetyl nitrate, nitronium tetrafluoroborate, or tetranitromethane in pyridine were unsuccessful.[411] Sulfonation attempts with Sulfan B, chlorosulfonic acid, or trimethylamine–sulfur trioxide complex were likewise unsuccessful, as was attempted formylation (DMF/POCl$_3$), cyanation with phenyl cyanate,[413] and diazo coupling.

Treatment of **314** with 30% hydrogen peroxide under basic conditions gave 6-bromosangivamycin (**318**), and subsequent attempts to displace the bromo group from **318** with thiourea in ethanol gave only the glycosidic cleavage product[414] 4-amino-6-thiopyrrolo[2,3-*d*]pyrimidine-5-carboxamide (**319**). However, nucleophilic displacement of the bromo group by sulfide was accomplished without appreciable glycosidic cleavage by the treatment of **318** at 100°C with aqueous methanol saturated with hydrogen sulfide gas to give **320**. This was the first example of *acid-catalyzed* nucleophilic substitution in a pyrrole or condensed pyrrole system.[411] A similar treatment of **318** with methanolic methanethiol in the presence of formic acid afforded 6-methylthiosangivamycin (**321**). Treatment of 6-bromotoyocamycin (**314**) with hydroxylamine resulted in a reaction with the cyano group to give 4-amino-6-bromo-7-β-D-ribofuranosylpyrrolo-[2,3-*d*]pyrimidine-5-carboxamidoxime (**315**). However, treatment of **314** with the strong nucleophile hydrazine resulted in a nucleophilic substitution at position 6

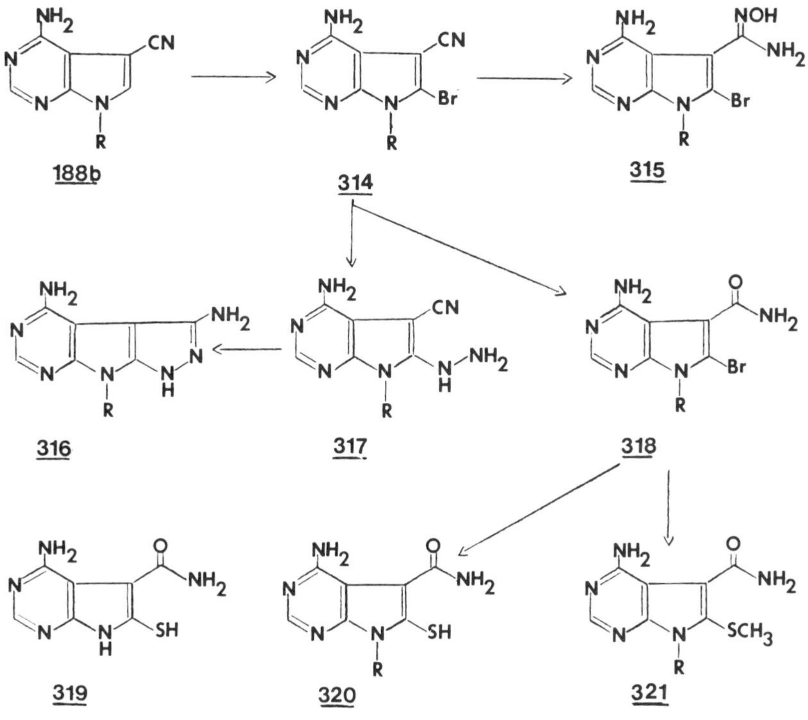

R = β-D-ribofuranosyl

(**317**), which was then followed by a reaction with the cyano group to afford the tricyclic nucleoside 4,5-diamino-8-β-D-ribofuranosylpyrazolo[3′,4′:5,4]pyrrolo[2,3-*d*]pyrimidine (**316**).[411,415]

When the silyl derivative of 6-bromotoyocamycin (**314**) was exposed to "naked" fluoride ion (potassium fluoride and dicyclohexyl-18-crown-6)[416] in acetonitrile, a facile reductive debromination occurred to give toyocamycin.[332] Similarly, 6-bromosangivamycin (**318**) afforded sangivamycin. It has been suggested that the "naked" fluoride ion may interact with the silylating agent to produce the trimethylsilyl anion, which may be the species responsible for the debromination.[332] Thiocyanation[417,418] and cyanation[419] of tubercidin have recently been reported. Treatment of 2′,3′,5′-tri-O-acetyltubercidin (**322**) with thiocyanogen chloride gave the monothiocyanate derivative **323**. When **323** was allowed to react with ethanethiol and methyl iodide in the presence of sodium bicarbonate, the corresponding methylthio derivative (**324a**) was formed. Deacetylation of **324a** gave 5-methylthiotubercidin (**324b**).[417,418] Tubercidin was also converted to toyocamycin by a multistep synthesis.[419]

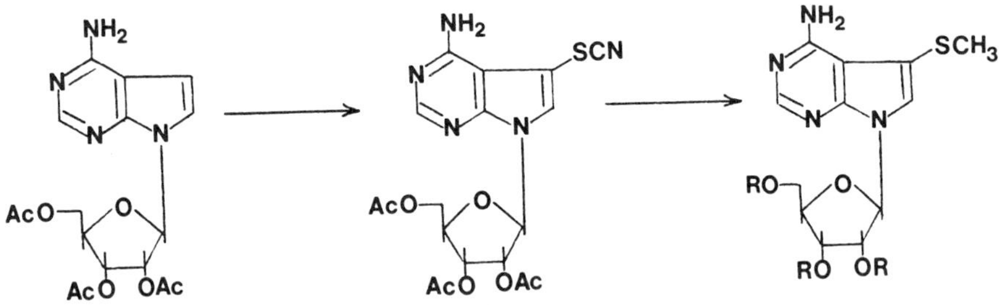

Townsend and co-workers[420,421] have studied the relative chemical reactivity of the cyano group in the pyrrole moiety of toyocamycin by changing the functional group at position 4 in the pyrimidine ring. The differences in reactivity appear to be a function of the pH at which the reaction is conducted, as well as the reagent and conditions employed. The reactivity of the cyano group at position 5 could be greatly decreased by the introduction of a group capable of supporting an anion in position 4. The amino group in position 4 allowed the 5-cyano group to undergo nucleophilic attack; however, a strong nucleophile was required for this reaction to proceed. Although both exocyclic groups (4-chloro, 4-methylseleno, and 5-cyano) reacted with various nucleophiles, the initial reaction was always a nucleophilic displacement of the 4-chloro or 4-methylseleno group, and the 5-cyano group reacted only after this had occurred.[421–424] This interesting study has furnished a number of 4,5-disubstituted 7-β-D-ribofuranosylpyrrolo[2,3-*d*]pyrimidines with carboxamido-type groups (amidoxime, amidrazone, thiocarboxamide) at position 5, which are amenable toward the synthesis of sangivamycin derivatives with a receptor-complement feature[425] (a triangulation of three electronegative atoms, one nitrogen and two oxygen atoms, with very distinct parameters and at least one lone pair of electrons on each atom). The triangulation proposed for sangivamycin (Fig. 2)[425] prompted

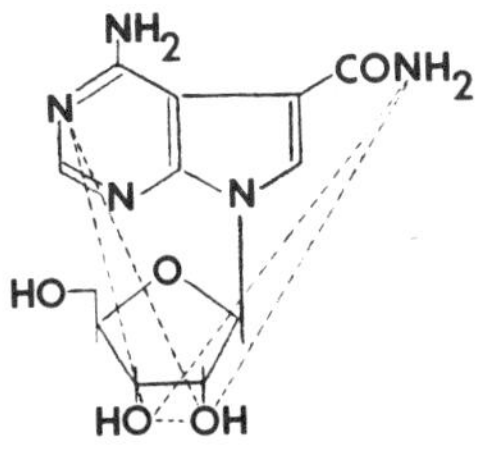

Figure 2. Triangulation for sangivamycin.

Schram and Townsend[426] to synthesize a series of sangivamycin derivatives with various five- and six-membered, aromatic and nonaromatic rings at position 5. Thus, the ring annulation of the carboxamidrazone[427] moiety of 4-amino-7-β-D-ribofuranosylpyrrolo[2,3-*d*]pyrimidine-5-carboxamidrazone (**325**)[420] with ninhydrin, diketones, formic acid, and other reagents afforded good yields of **326**. Similar ring annulation of toyocamycin, thiosangivamycin, and 4-amino-7-β-D-ribofuranosylpyrrolo[2,3-*d*]pyrimidine-5-carboxamidoxime afforded some interesting compounds.[426] Nucleophilic addition to the 5-cyano group of both 6-amino- and 4,6-diamino-7-β-D-ribofuranosylpyrrolo[2,3-*d*]pyrimidine-5-carbonitrile was found to be much more difficult under both acidic and basic conditions[428] than addition to the nitrile group of toyocamycin. This would indicate that an amino group on the pyrrole unit of a pyrrolo[2,3-*d*]pyrimidine substantially deactivates the adjacent 5-cyano group.

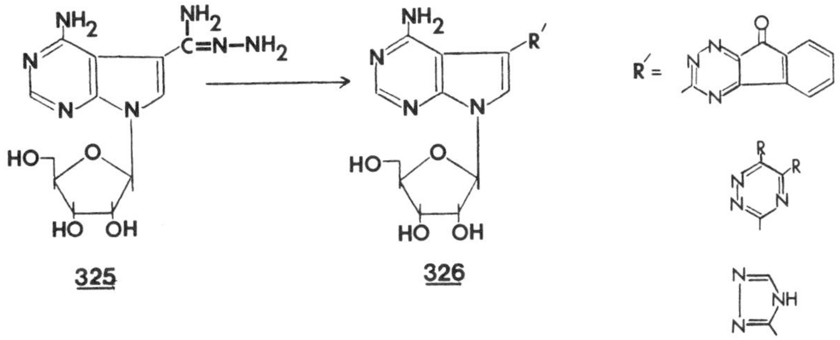

325   **326**

A chemical transformation of the aglycon of preformed pyrrolo[2,3-*d*]-pyrimidine nucleosides provided a convenient route to the synthesis of the nucleoside antibiotic cadeguomycin (**198**). Alkaline hydrolysis of 2-amino-5-cyano-7-β-D-ribofuranosylpyrrolo[2,3-*d*]pyrimidin-4(3*H*)-one (**327**)[429] or the acid hydrolysis (3 *N* HCl at 80°C for 44 h) of 2-amino-7-β-D-ribofuranosylpyrrolo[2,3-*d*]pyrimidin-4(3*H*)-one-5-carboxamide (**328**)[284] gave cadeguomycin (**198**). Cadeguomycin was also synthesized from 2-diacetylamino-3-methoxymethyl-5-methyl-6-bromo-7-(2,3-*O*-isopropylidene-5-*O*-acetyl-β-D-ribofuranosyl)pyrrolo[2,3-*d*]pyrimidin-4-one via a five-step sequence.[283] The chemical transformation of the appropriately substituted heterocyclic ring in the pyrrolo[2,3-*d*]pyrimidine nucleosides also furnished[430–439] a number of tricyclic nucleosides.

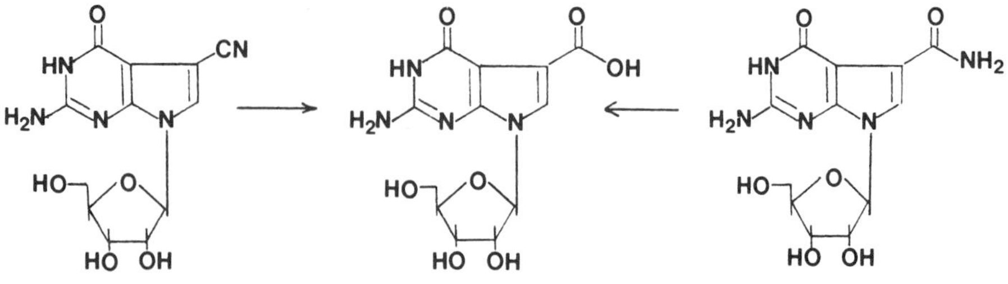

The potent cytokinin agent N^6-(Δ^2-isopentenyl)-7-deazaadenosine (**329**, R = H) was synthesized[440] using the general procedure of Jones and Robins.[441] Tubercidin was treated with an alkyl halide to yield the N^1-quaternized nucleoside, which was subsequently converted quantitatively to the N^6-substituted derivative by heating in aqueous ammonia. Similarly, the N^6-(Δ^2-isopentenyl) derivatives of toyocamycin and sangivamycin (**329**, R = CN, CONH$_2$) have been prepared by Lewis and Townsend.[442] The fragmentation pattern in the mass spectrum of **329** (R = H) was analogous to that of N^6-(Δ^2-isopentenyl)adenosine.[443] It is of particular interest to note that, although tubercidin has an N-ribosyl linkage which is chemically and enzymatically more stable than that of adenosine,[385] the mass spectra of **329** (R = H) and N^6-(Δ^2-isopentenyl)adenosine, determined at 70 eV, do not reflect this difference.

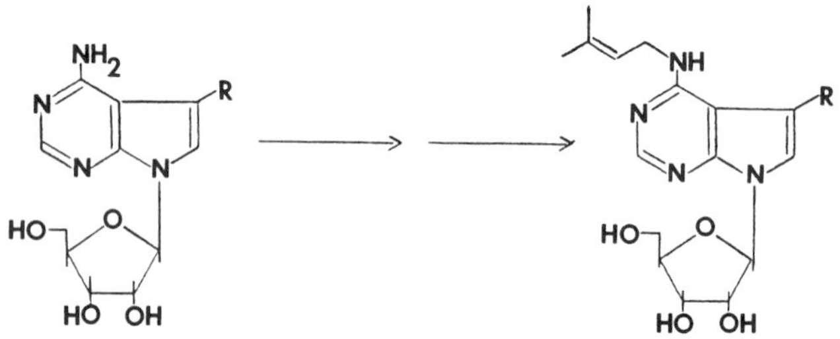

The 7-deaza analogue of kinetin riboside, 4-furfurylamino-7-β-D-ribofuranosylpyrrolo[2,3-d]pyrimidine (**330**), was prepared by Iwamura and co-workers[444–446] by reacting **298** with furfurylamine. This compound was found to be a potent anticytokinin in the tobacco callus bioassay.

2.2.3.8. Chemical Transformations of the Carbohydrate Moiety of Pyrrolo[2,3-d]-pyrimidine Nucleosides. Although the first chemical synthesis of tubercidin 5'-monophosphate (**335**) was reported in 1964 by Pike *et al.*,[294] the yield was low and the isolation was rather difficult. By protecting the amino function, Hanze[448] was able to double the overall yield of crystalline **335**. Benzoylation of 2',3'-O-isopropylidenetubercidin (**331**) gave $N^4,N^4,5'$-tribenzoylisopropylidenetubercidin (**332**), which, on selective debenzoylation with cold sodium methoxide, afforded N^4-benzoylisopropylidenetubercidin (**333**). Phosphorylation of

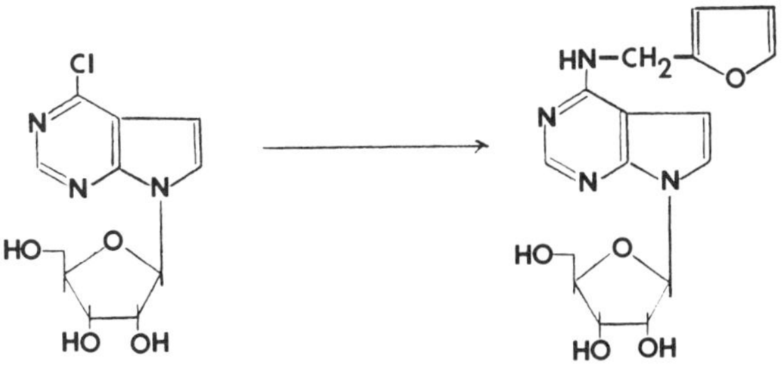

298 → **330**

333 by the Tener procedure[186] gave the crystalline 5'-cyanoethylphosphate, which, when treated with aqueous methanolic ammonium hydroxide, was converted into 2',3'-*O*-isopropylidenetubercidin 5'-phosphate (**336**). A removal of the isopropylidene group of **336** gave crystalline **335**. However, direct phosphorylation of unprotected tubercidin with phosphorus oxychloride in trimethyl phosphate led to **335** in >85% yield.[449,450] The synthesis of the tubercidin 5'-di- and triphosphates was accomplished by conventional procedures.[450–452] The 5'-phosphates of toyocamycin and sangivamycin were also obtained by phosphorylation of the unprotected nucleosides with phosphorus oxychloride in trimethyl phosphate.[453,454] Synthesis of the 5'-di- and triphosphates of toyocamycin and sangivamycin was achieved by conventional procedures.[454,455] 2'-Deoxy-7-deazaguanosine 5'-triphosphate has recently been prepared[456] and shown to enhance resolution in M13 dideoxy sequencing of G—C-rich regions of DNA.[457]

Treatment of **335** with DCC in pyridine in the presence of 4-morpholine-

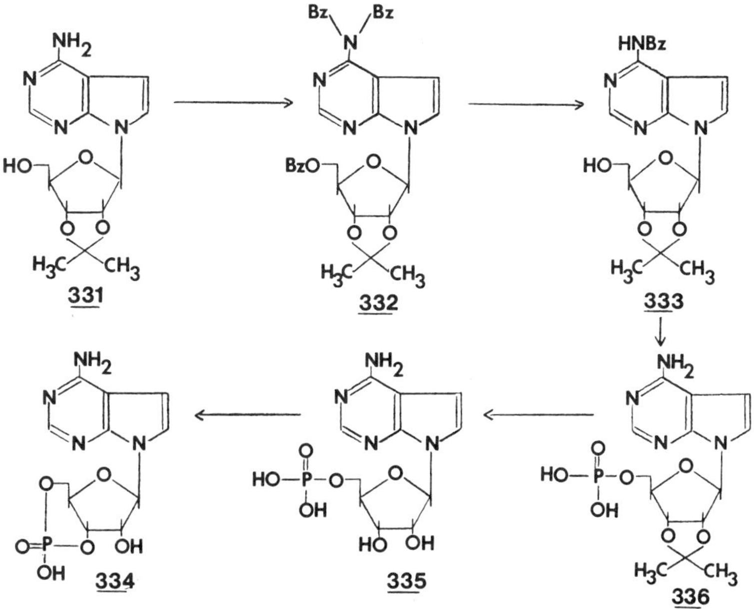

331 → **332** → **333** → **336** → **335** → **334**

N, N'-dicyclohexylcarboxamidine gave tubercidin $3', 5'$-cyclic phosphate (**334**).[448] Compound **334** was also obtained via the tubercidin $5'$-(trichloromethyl phosphonate) intermediate, in good yield.[188] Intramolecular cyclization of toyocamycin $5'$-phosphate with DCC in pyridine/N, N-dimethylformamide (1:1) gave toyocamycin $3', 5'$-cyclic phosphate.[453] Alkaline hydrolysis of toyocamycin $3', 5'$-cyclic phosphate with ammoniacal hydrogen peroxide gave sangivamycin $3', 5'$-cyclic phosphate.[453] DCC-mediated cyclization of 4-substituted 7-β-D-ribofuranosylpyrrolo[2,3-d]pyrimidine $5'$-phosphates in a large excess of pyridine gave the corresponding $3', 5'$-cyclic phosphates.[260] Tubercidin $2', 3'$-cyclic phosphate was also prepared directly from tubercidin in 50% yield, employing pyrophosphoric acid and tri-n-butylamine in N, N-dimethylformamide.[458] Treatment of $5'$-monomethoxytrityltubercidin with p^1-diphenyl-p^2-morpholino-pyrophosphorochloridate also gave tubercidin $2', 3'$-cyclic phosphate in rather low yield (1.8%).[459] The preparation of labeled tubercidin $5'$-phosphate by thermal phosphorylation of tubercidin with an alkali metal phosphate at 155–165°C was reported in a Czech patent.[460] However, the $2'(3')$-phosphate was also formed and was separated by chromatography. The phosphorylation of $2', 3'$-O-isopropylidene-7-deazainosine by the modified procedure of Kochetkov *et al.*[461] gave 7-deazainosine $5'$-monophosphate[462] in an 82% yield. The monophosphate was converted to the 7-deazainosine $5'$-diphosphate employing the morpholidate procedure.[463] The synthesis of tubercidin $3'$-phosphate was accomplished[464] by reacting tubercidin with monomethoxytrityl chloride[465] to give $5'$-monomethoxytrityltubercidin. This compound was subsequently phosphorylated[466] to give a mixture of the $2'$- and $3'$-phosphates. Deprotection by acid treatment, followed by ion-exchange chromatography, furnished tubercidin $2'$- and $3'$-monophosphates.[464]

The synthetic sequence for the preparation of mono- and diphosphates of 7-deazaguanosine also started with regioselective phosphorylation. Phosphorylation of 7-deazaguanosine (**277**) with $POCl_3$ in trimethyl phosphate gave 7-deazaguanosine $5'$-phosphate (**337a**).[467] Compound **337a** was converted to the diphosphate (**337b**) by activation of the tri-n-butylammonium salt of **337a** with 1,1'-carbonyldiimidazole to obtain the imidazolidate intermediate, which,

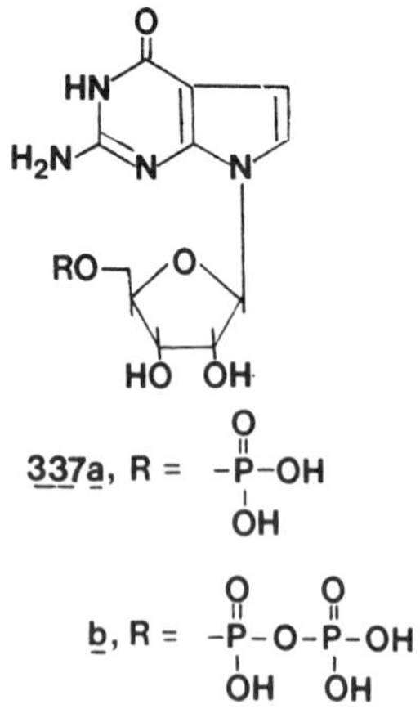

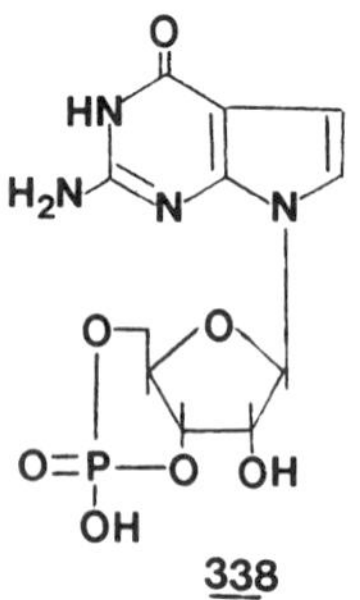

$\underline{337a}$, R = $-\overset{\overset{\text{O}}{\|}}{\underset{\underset{\text{OH}}{|}}{\text{P}}}$-OH

$\underline{b}$, R = $-\overset{\overset{\text{O}}{\|}}{\underset{\underset{\text{OH}}{|}}{\text{P}}}$-O-$\overset{\overset{\text{O}}{\|}}{\underset{\underset{\text{OH}}{|}}{\text{P}}}$-OH

338

upon condensation with tri-*n*-butylammonium phosphate, afforded 7-deaza-guanosine 5'-diphosphate (**337b**).[468] Cyclization of **337a** with DCC in pyridine gave 7-deazaguanosine 3',5'-cyclic phosphate (**338**).[467]

A series of 5'-*O*-monoesters of tubercidin (**340**) have been prepared[469] by the treatment of **331** with chlorosulfonic acid or an acyl chloride or anhydride, followed by mild acid hydrolysis to remove the isopropylidene group selectively. *S*-Tubercidinylhomocysteine (**341**), a potent inhibitor of several *S*-adenosyl-methionine-dependent methylases, was synthesized[470] by reacting 2',3'-*O*-iso-propylidene-5'-*O*-tosyltubercidin (**192**) with homocysteine anion in liquid am-monia, followed by deacetonation. Alternatively, **341** was obtained by coupling

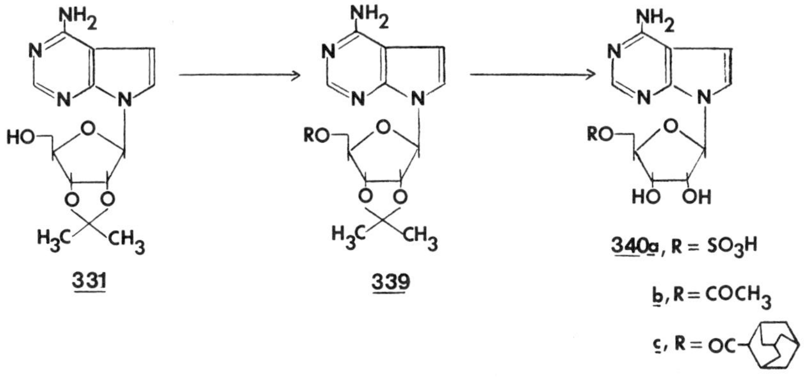

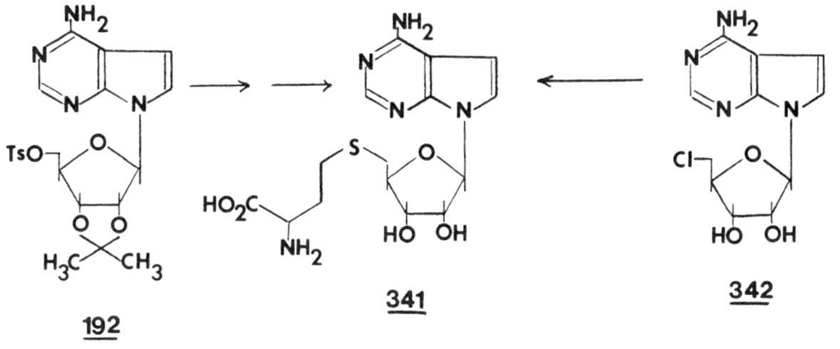

5'-chloro-5'-deoxytubercidin (**342**) with homocysteine.[177,178,471,472] The latter procedures were found to afford cleaner reaction products in higher yields. 3'-Deoxytubercidin, prepared from tubercidin by the treatment with 2-acetoxy-isobutyryl bromide, was converted into 5'-chloro-3',5'-dideoxytubercidin, which in turn was condensed with L-homocysteine sodium salt to afford *S*-3'-deoxytubercidinyl-L-homocysteine.[169] A convenient and high-yield procedure for the sugar monomethylation of tubercidin was described by Robins and co-workers.[473,474] A methanolic solution of tubercidin, containing $10^{-3}\,M$ stannous chloride dihydrate, was treated with a solution of diazomethane in glyme. This gave the 2'-*O*-methyl (**343**, 53%) and 3'-*O*-methyl (**344**, 39%) derivatives, which were separated on a Dekkar column.[475] In addition to the normal ^{1}H-NMR and mass spectral evidence, conclusive proof for the isomeric structures was obtained[473] by ^{1}H-NMR double-resonance analysis of the product acylated with trifluoroacetic anhydride.

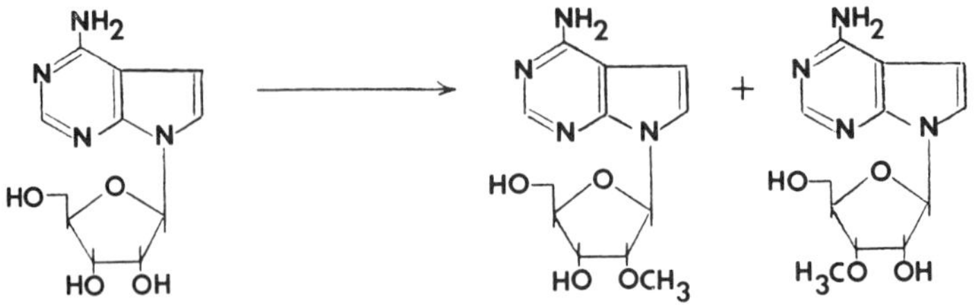

The modification of the ribose moiety of tubercidin has been the subject of research by Anzai and Matsui[476] in recent years. After a series of unfruitful attempts, the successful synthesis of 5-deoxytubercidin (**350**) from tubercidin through a route involving catalytic hydrogenation of N^4,N^4-dibenzoyl-5′-deoxy-5′-iodo-2′,3′-O-isopropylidenetubercidin (**349**) was achieved.[476] Mesylation of isopropylidenetubercidin (**331**), followed by benzoylation with an excess of benzoyl chloride, gave **346**,[477] along with a minor amount of the tribenzoyl compound **351**. Treatment of **346** with sodium iodide in methyl ethyl ketone gave three products, the major product being **349**. Catalytic hydrogenation of **349** afforded N^4-benzoyl-5′-deoxy-2′,3′-O-isopropylidenetubercidin (**347**) and the N^4,N^4-dibenzoyl derivative (**348**), with the former being predominant. Treatment of **347** with dilute sodium hydroxide, followed by aqueous acetic acid, furnished 5′-deoxytubercidin (**350**).[476]

Recently, Robins and co-workers[478] have reported a facile, two-step synthesis of certain 5′-deoxyribonucleosides from the parent ribonucleosides, via the 5′-deoxy-5′-chloro intermediates. This involved a chlorination of the unprotected nucleoside antibiotics (**352**) with the hexamethylphosphoramide–thionyl chloride reagent[95,479] to yield 5′-chloro-5′-deoxytubercidin (**353**, R = H)[472] and 5′-chloro-5′-deoxytoyocamycin (**353**, R = CN).[478] Dehalogenation of **353** with tributyltin hydride in tetrahydrofuran using α,α'-azobisisobutyronitrile as initiator[480] gave the desired **350** and 5′-deoxytoyocamycin (**354**, R = CN) in excellent yield. Oxidation of **354** (R = CN) with hydrogen peroxide in ammonium hydroxide gave 5′-deoxysangivamycin (**354**, R = CONH$_2$) in a 53% yield. Treatment of **353** (R = H) with NaSCH$_3$ under basic conditions furnished 5′-deoxy-5′-methylthiotubercidin.[472]

Anzai and Matsui[481,482] treated **346** with sodium azide in hexamethylphosphoramide to obtain N^4-benzoyl-5′-azido-5′-deoxy-2′,3′-O-isopropylidenetubercidin (**355**) in 68% yield, along with a 7% yield of the ring-opened cyclonucleoside (**356**). The formation of **356** was of particular interest as the reaction could probably be initiated by a nucleophilic attack of the azide ion on the benzoyl group. The benzazide thus formed may then be converted to various products[483] through the Curtius rearrangement, as shown below. Alkaline hydrolysis of **355** afforded the 5′-azido-5′-deoxy derivative (**357**), which, on catalytic hydrogenation, gave a quantitative yield of 5′-amino-5′-deoxy-2′,3′-O-isopropylidenetubercidin (**358**). Deacetonation of **358** with boiling aqueous alcoholic hydrochloric acid furnished 5′-amino-5′-deoxytubercidin (**359**).[483]

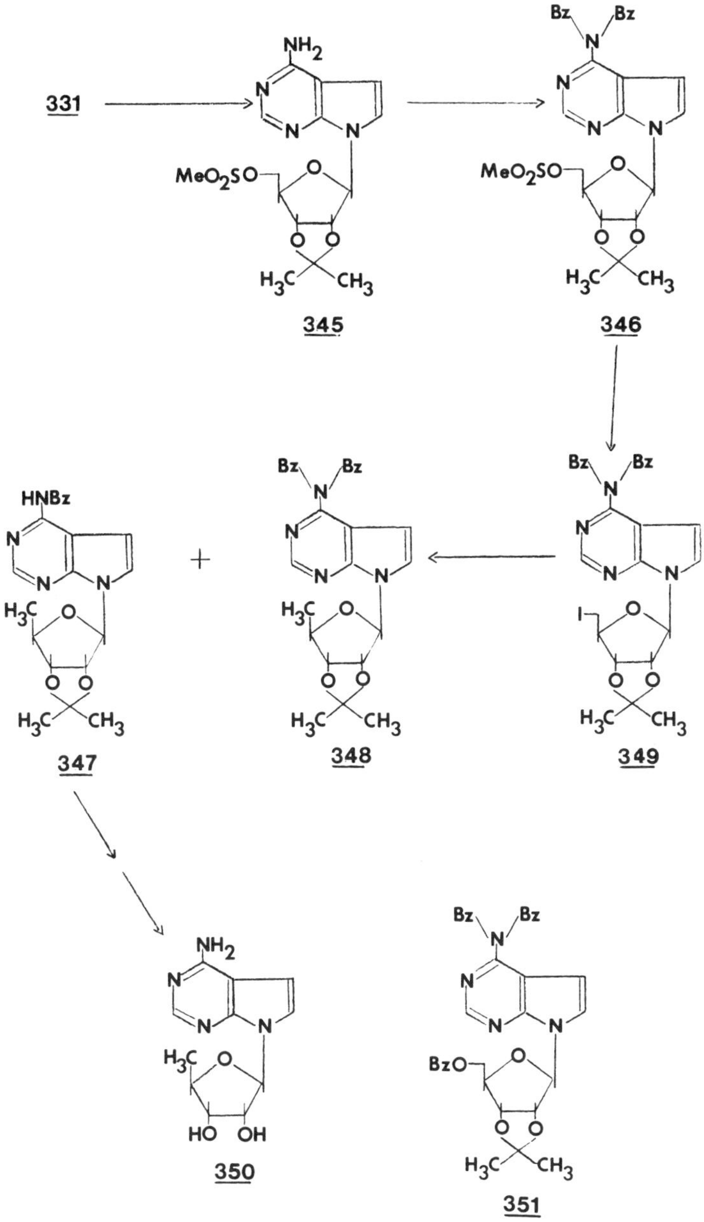

Even though the enzymatic reduction of tubercidin to a biologically inter-esting 2'-deoxy derivative[484] was reported [386] as early as 1964, the chemical synthesis of 2'- and 3'-deoxytubercidin was reported only recently. The first chemical synthesis of 3'-deoxytubercidin (**362**) was reported simultaneously by Robins *et al.*[485] and Moffatt and co-workers[486] using the "abnormal" reaction of α-acetoxyisobutyryl halides with diols. The reaction of tubercidin (**188a**) with α-acetoxyisobutyryl halide[487] in acetonitrile gave the acetylated intermediate

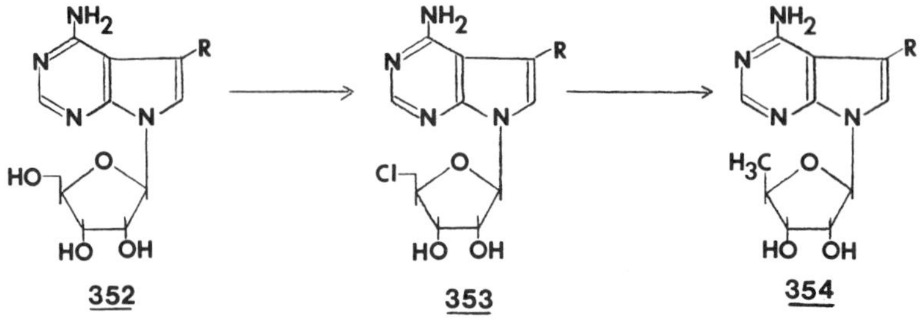

R = H, CN, CONH₂

$$R = H,\ CN,\ CONH_2$$

352 → **353** → **354**

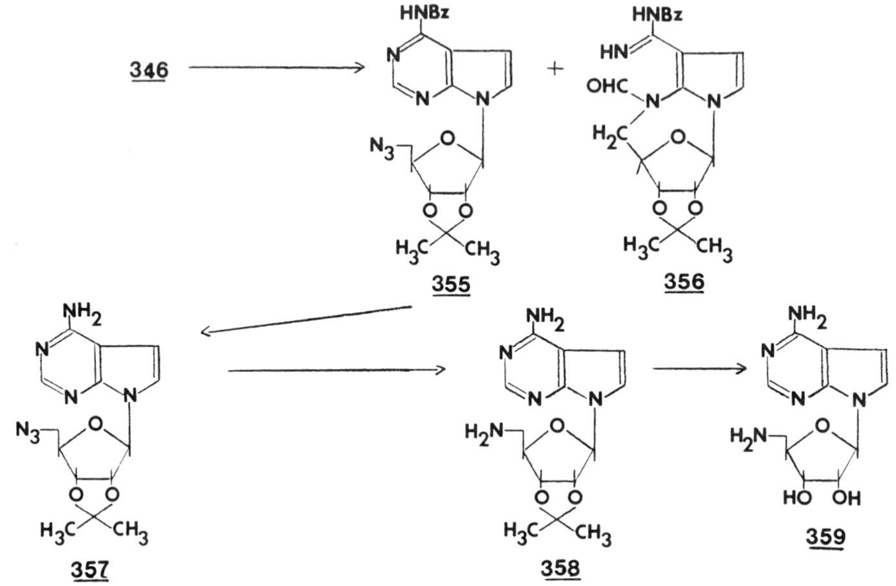

346 → **355** + **356**

357 → **358** → **359**

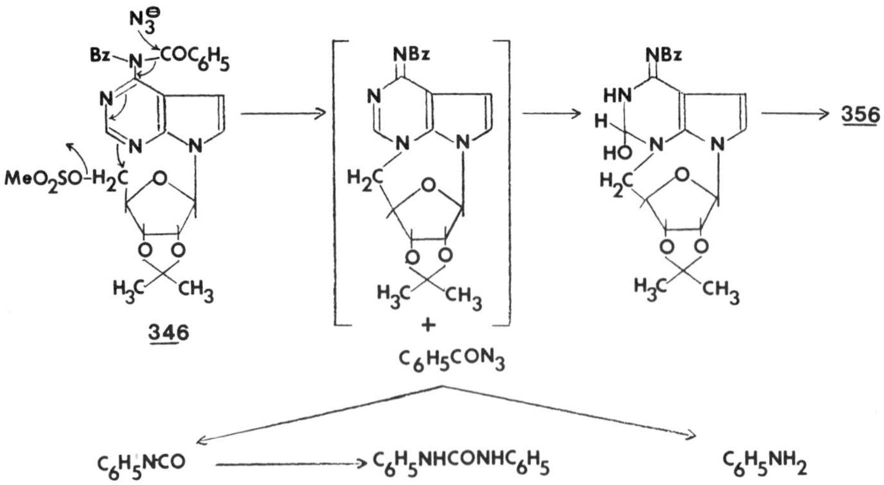

346 → [...] → → **356**

C₆H₅CON₃

$$C_6H_5CON_3$$

C₆H₅NCO ──→ C₆H₅NHCONHC₆H₅   C₆H₅NH₂

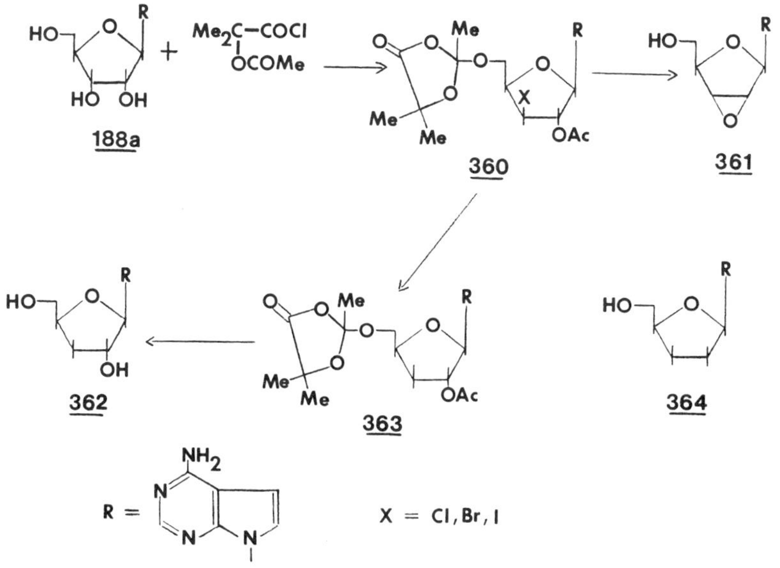

360, which could arise by the mechanism depicted[485] in Fig. 3. Attack of the 5′-oxygen on the less sterically hindered acetyl carbonyl carbon of α-acetoxyisobutyryl halide with concurrent attack of the acetyl carbonyl oxygen of the carbonyl halide and subsequent deprotonation of the 5′-oxygen would lead to **i**. A corresponding attack of either the 2′- or 3′-hydroxyl oxygen of **i** would lead to the intermediate **ii**. Protonation of the ether-type oxygen of **ii** and attack of the 3′-oxygen on the orthoacetate carbon would lead to the mixed orthoester-anhydride, **iii**. This intermediate (**iii**) could be protonated at the carbonyl function of the mixed anhydride and concurrently suffer $S_N 2$ attack by halide at the 3′-carbon to give **360** directly or lose α-hydroxyisobtuyric acid to give the 2′,3′-acetoxonium ion intermediate **iv**,[487,488] which could then undergo an $S_N 2$ attack by halide at the 3′-carbon to give **360**. This reaction seems to be completely regiospecific.

Catalytic hydrogenation of **360** gave the blocked 3′-deoxy intermediate **363**, which, on saponification with methanolic ammonia, furnished 3′-deoxytubercidin (**362**) in good yield. However, catalytic hydrogenolysis of **360** (X = Br) followed by hydrolysis gave an equal amount of 2′,3′-dideoxytubercidin (**364**) along with **362**. The formation of **364** was explained through a palladium-catalyzed *trans* elimination of the acetate group, giving a 2′,3′-olefin, which was concomitantly reduced.[488] A facile conversion of **360** into crystalline 2′,3′-anhydrotubercidin (**361**) was also noted[485,489] upon treatment with sodium methoxide. Compound **361** served as an excellent starting material for the synthesis of 2′-deoxytubercidin (**260b**).[490] The synthesis of **260b** was accomplished by the intramolecular migration of *S*-benzyl from C-3′ to C-2′ via an episulfonium ion rearrangement. The benzoylation of **361** gave the $N^4,N^4,5'$-O-tribenzoyl derivative (**365**) in almost quantitative yield. Treatment of **365** with sodium benzylthiolate in hot THF gave **366**, which on mesylation furnished the 2′-mesylate derivative (**367**). Reaction of **367** with sodium benzoate in hot DMF

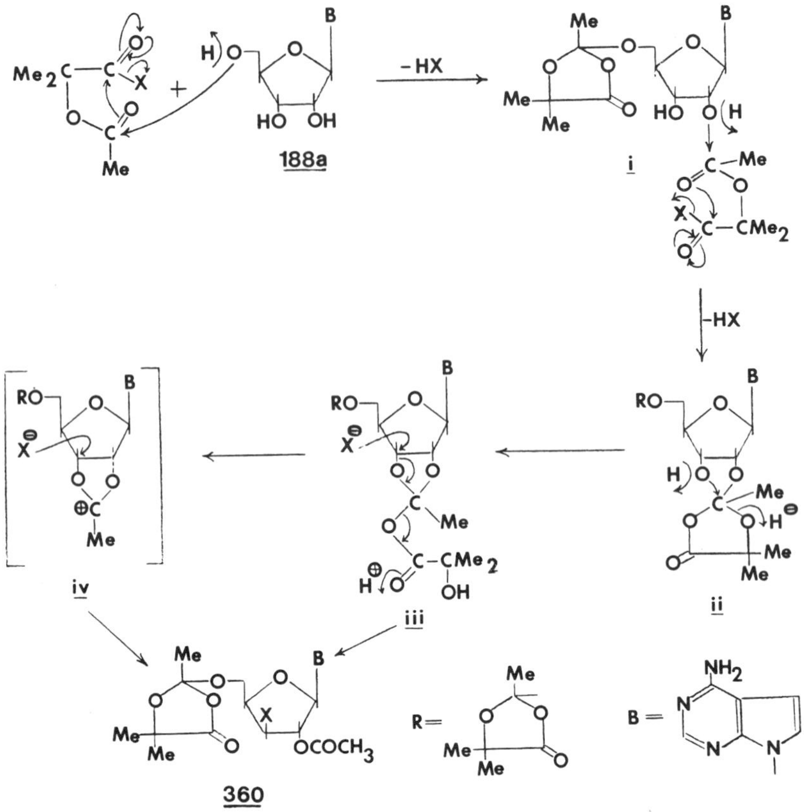

Figure 3. Reaction mechanism of tubercidin with α-acetoxyisobutyryl halide.

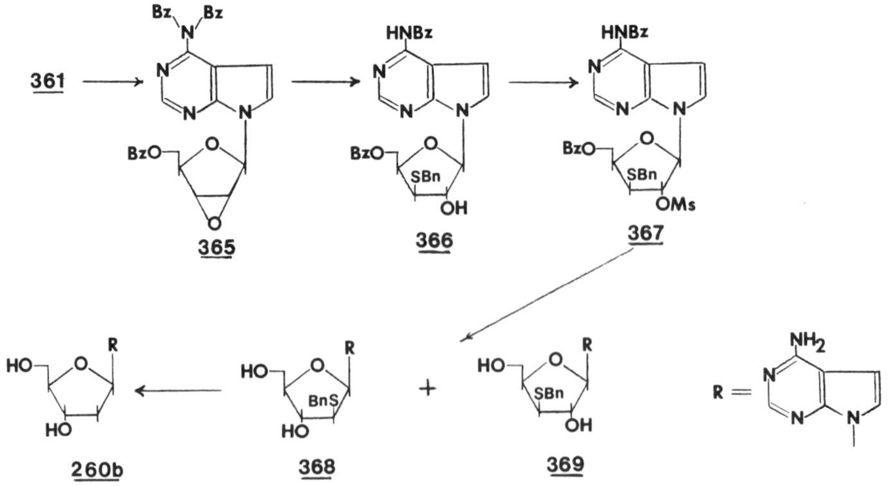

and subsequent deblocking with sodium methoxide afforded **369** and its 2'-*S*-benzylthioarabino isomer (**368**) in a ratio of 2:3. Desulfurization of **368** with Raney nickel furnished 2'-deoxytubercidin (**260b**) in an overall yield of 27%. However, the synthesis of 2'-deoxytubercidin by the sodium salt glycosylation procedure[31] is more practical.

A four-step deoxygenation procedure to provide the requisite 2'-deoxynucleoside, using phenoxythiocarbonylation[92,491,492] or imidazolylthiocarbonylation[93,493,494] of the 2'-hydroxyl group of the corresponding 3',5'-protected β-D-ribonucleoside, has recently been developed. Using this methodology, 2'-deoxytubercidin, 2'-deoxytoyocamycin, and 2'-deoxysangivamycin have been prepared[92] from the corresponding ribonucleosides. A multistep synthesis of 2'-deoxysangivamycin[495a] and 2',3'-dideoxysangivamycin[495b] from toyocamycin has recently been described. Tubercidin 5'-triphosphate has been reduced to 2'-deoxytubercidin 5'-triphosphate enzymatically in the presence of the ribonucleotide reductase from *Lactobacillus leichmannii* in 76% yield.[496] Similar enzymatic reduction of toyocamycin 5'-triphosphate in the presence of an allosteric activator dGTP gave 2'-deoxytoyocamycin 5'-triphosphate.[496]

A general procedure of access into a "nucleophilically functionalized" sugar derivative of acyloxonium ion intermediacy via 2',3'-*O*-ortho esters[497] to tubercidin was found to be very successful.[498] Reaction of tubercidin (**188a**) and trimethyl orthoacetate in the presence of trichloroacetic acid gave the 2',3'-*O*-methoxyethylidene derivative (**370**) in high yield.[498] Treatment of **370** with pivaloyl chloride in pyridine provided a mixture primarily containing 4-(N-pivalamido)-7-(3-chloro-3-deoxy-2-*O*-acetyl-5-*O*-pivalyl)pyrrolo[2,3-*d*]pyrimidine (**371a**) and the corresponding 2'-*O*-(4,4-dimethyl-3-pivaloxypent-2-enoyl) derivative (**371b**). Interestingly, no 2'-chloroarabino isomers were detected. Reaction of the purified mixture of **371** with *n*-tributyltin hydride under free-radical-generating conditions,[499] followed by deblocking, produced the dehalogenated product, 3'-deoxytubercidin (**362**). Treatment of **371** with sodium methoxide gave the *ribo*-epoxide (**361**). The epoxide **361** was found to be very sensitive to intramolecular cyclization/degradation as well as to higher temperatures. However, benzoylation of **361** gave the stable *N,N,*5'-*O*-tribenzoyl derivative (**376**), which, when treated with benzoate, underwent a nucleophilic opening at C-3' of the epoxide to give **378**. Debenzoylation of **378** furnished 4-amino-7-β-D-xylofuranosylpyrrolo[2,3-*d*]pyrimidine (**379**).[498,500] Treatment of **378** with methanesulfonyl chloride in pyridine followed by debenzoylation gave the cyclized product, *lyxo*-epoxide **377**. Nucleophilic attack by benzoate at C-3' of **377**, followed by deblocking, gave 4-amino-7-β-D-arabinofuranosylpyrrolo[2,3-*d*]-pyrimidine (**285**).[498,500]

When **370** was allowed to react with excess sodium iodide in boiling pivaloyl chloride/pyridine, a good yield of 4-(*N*-pivalamido)-7-[3-iodo-3-deoxy-2-*O*-(4,4-dimethyl-3-pivaloxypent-2-enoyl)-5-*O*-pivalyl-β-D-xylofuranosyl]pyrrolo-[2,3-*d*]pyrimidine (**372**) was obtained.[498] Catalytic hydrogenolysis of **372**, followed by deblocking, also gave **362**. Oxidative removal of the 2'-*O*-enol ester function from **372**, followed by treatment of the resulting product with methanesulfonyl chloride in pyridine and then iodide, promoted elimination of iodo and mesylate from the 2'- and 3'- positions. The resulting 2',3'-unsaturated product was deblocked to give 4-amino-7-(2,3-dideoxy-β-D-*glycero*-pent-2-enofuranosyl)pyrrolo[2,3-*d*]pyrimidine (**373**).[501] Silver acetate effected an elimination of the elements of HI from **372**, and subsequent deblocking gave 4-amino-7-(3-deoxy-β-D-*glycero*-pent-3-enofuranosyl)pyrrolo[2,3-*d*]pyrimidine (**374**).[490] Catalytic reduction of **374** gave **362** and its 4′-epimer (**375**).

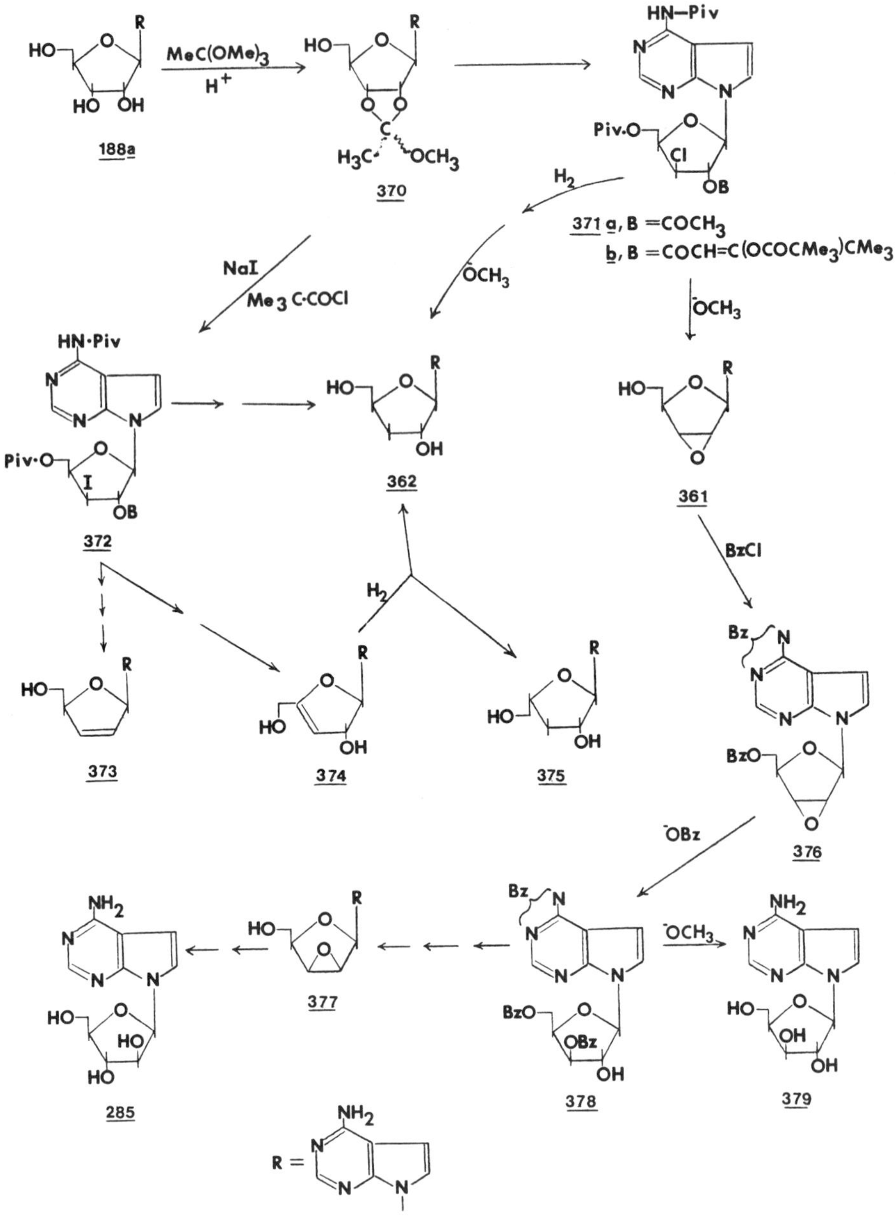

When $N^4,5′$-*O*-dibenzoyltubercidin (**380**)[501] was treated with Corey–Winter reagent,[502] the 2′,3′-thionocarbonate (**381**) was obtained, which, in the presence of trimethyl phosphite, produced $N^4,5'$-O-dibenzoyl-2',3'-didehydro-2',3'-dideoxytubercidin (**382**).[501]

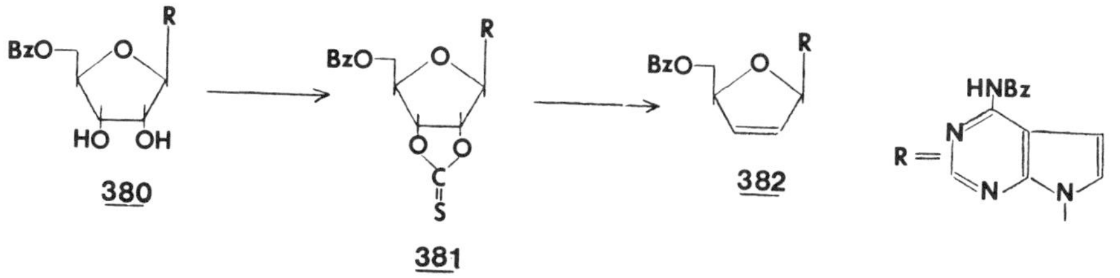

Reaction of 2',3'-O-benzylidenetubercidin (**383**, R = H, R' = C$_6$H$_5$) with N-bromosuccinimide (NBS) in a mixture of carbon tetrachloride and tetrachloroethane at reflux temperature gave a mixture of two diastereomeric products. The minor product was identified as **384**, and the major product **385**, on catalytic reduction, afforded C$_6$,5'-cyclonucleoside (**386**).[503] Analogously, when 2',3'-isopropylidenetubercidin (**383**, R = R' = CH$_3$) was treated with NBS, the cyclonucleoside **385** (R = R' = CH$_3$) was formed, which, on reduction, gave **386** (R = R' = CH$_3$).[503]

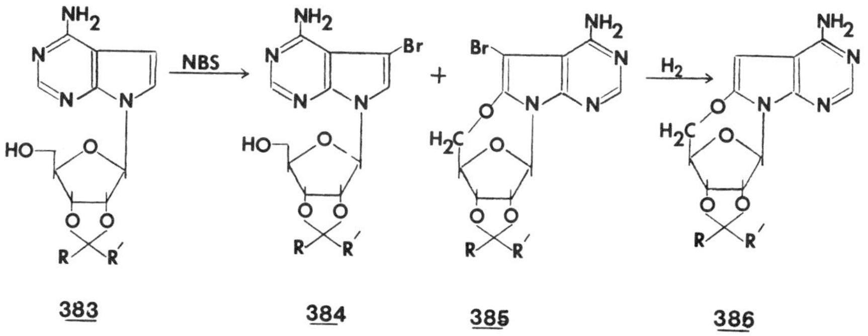

Extension of the chemical transformation of the sugar moiety of pyrrolo-[2,3-d]pyrimidine nucleosides led to the synthesis of several interesting (2'→5')- (**387**) and (3'→5')- (**388**) linked dinucleoside monophosphates.[504–508] The synthesis of these dinucleoside monophosphates generally involved the condensation of a suitably protected 7-deazapurine nucleoside with an appropriate heterocyclic nucleoside 5'-phosphate, according to the procedure of Jacob and Khorana.[509] In addition to these chemical methods, an enzymatic process for the synthesis of 2',5'-oligonucleotides of tubercidin, toyocamycin, sangivamycin,[454] poly(7-deazaguanylic acid),[468,510] and poly(adenylic, 7-deazaadenylic acids)[450] has been reported. Synthesis of oligodeoxynucleotides containing tubercidin has recently been reported.[511] Several enzymatic phosphorylation techniques for the preparation of 7-deazapurine nucleotides are also reported in the literature. Thus the enzymatic phosphorylation of tubercidin using p-nitrophenyl phosphate as the phosphate donor in the presence of *Serratia marcescens* IAM 1223[512] or a phosphotransferase from carrots[513] has been reported to yield tubercidin 5'-monophosphate. The enzymatic phosphorylation of san-

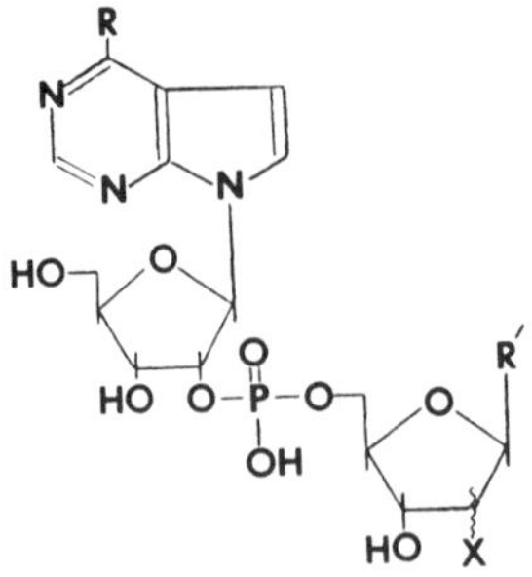

387

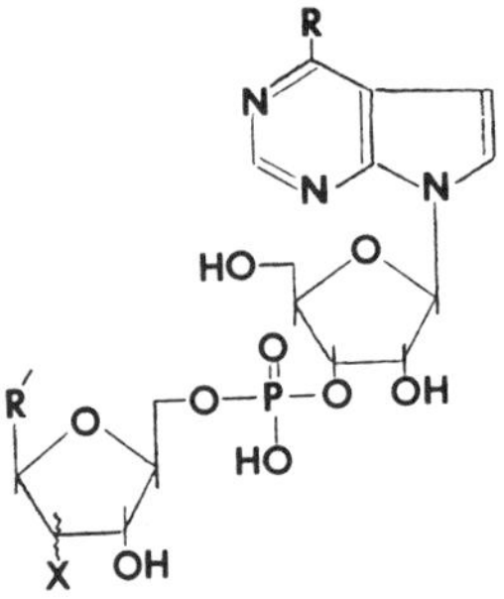

388

givamycin (by enzyme extracts from mouse liver) has also been described by Hardesty *et al.*[514] When sangivamycin was added along with adenosine, the adenosine monophosphate formation was completely inhibited.

The crystal structure,[515–518] conformational analysis,[519,520] and mass spectra[521,522] of tubercidin, 2'-deoxytubercidin, and toyocamycin have been reported. The crystal structure of toyocamycin[523] and the mass spectrum of 4-methylseleno-7-β-D-ribofuranosylpyrrolo[2,3-*d*]pyrimidine[524] have been described. The basic parameters for the investigation of tautomerism in pyrrolo[2,3-*d*]pyrimidine nucleosides has been established by Townsend and co-workers,[525] by employing carbon-13 magnetic resonance spectroscopic methods. The effects of substitution of an N-7 nitrogen atom of purine for a methine group at C-5 in pyrrolo[2,3-*d*]pyrimidines appear mainly additive and are not very dependent on the nature of the substituent at the C-4 (C-6 of purine) position. It is also of considerable interest to note[525] that nitrogen substitution only slightly influences the resonance positions of the β-D-ribofuranosyl carbons, with the largest perturbations being observed at C-1' and C-4'. Analysis of the physical properties of poly-7-deazaadenylic acid suggested[526] that no Hoogsteen-type[98] base pairing was involved in the codon–anticodon recognition process[527] and in the protein-synthesizing systems.[528]

2.2.3.9. Nucleoside Q (Queuosine) and Nucleoside Q*.   The chemical structure of nucleoside Q[277–279] was elucidated [280] to be 2-amino-5-(4,5-*cis*-dihydroxy-1-cyclopenten-3-yl-*trans*-aminomethyl)-7-β-D-ribofuranosylpyrrolo[2,3-*d*]pyrimidin-4(3*H*)-one (**197**). The structure of nucleoside Q is unique since it is a derivative of 7-deazaguanosine having a cyclopentenediol in the side chain at the C-7 position. The relative stereochemistry of the cyclopentene substituents has been determined as 3,5-*trans*;4,5-*cis* on the basis of ¹H-NMR comparisons with synthetic models.[529] This hypermodified nucleoside[530] occupies the first portion of the anticodon[531] of *E. coli* tRNA^Try, tRNA^His, tRNA^Asn, and tRNA^Asp and has been shown to be widely distributed in tRNA from animal as well as plant sources.[532,533] The aglycon portion of nucleoside Q (Q-base or quenine) as also been isolated from bovine amniotic fluid[534] and chemically synthesized.[535] The presence of nucleoside Q in the tRNAs of such a wide range of organisms may reflect its importance in the process of cell growth or growth regulation, although, at present, its actual function is unknown. Biosynthetic studies[536,537] have revealed that nucleoside Q arises from a guanine residue. During the course of modification (G→Q), the C-8 carbon and N-7 nitrogen of the guanine precursor are expelled in a fashion very similar to that observed in the biosynthesis of toyocamycin.[538]

It was also demonstrated[532] that the tRNAs of various animal tissues (including several mammalian tissues) contain a new derivative of nucleoside Q, designated as Q*, and that the Q* content is generally higher than the Q content (e.g., hepatoma cells have larger quantities of Q* than normal cells). It has been shown[539] that nucleoside Q*, isolated from rabbit liver tRNA, consists of a mixture of two O-glycosides, **389** (major, R = β-D-mannosyl, ~75%) and **390** (minor, R = β-D-galactosyl, ~25%). It is probably the first nucleoside analogue isolated from tRNA to contain a carbohydrate moiety in the side chain.[540]

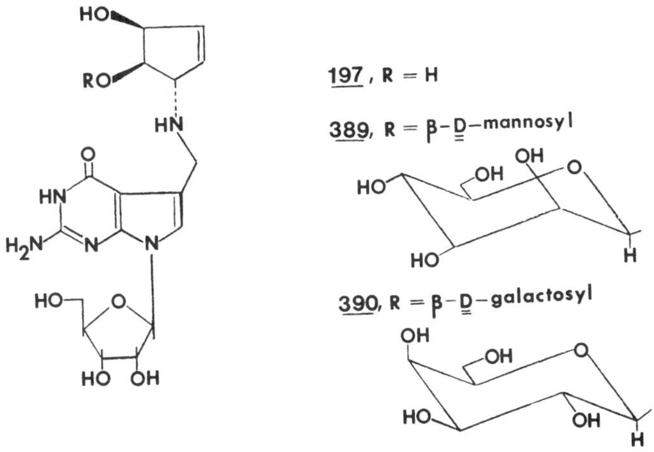

The successful synthesis of 2-amino-5-cyano-7-β-D-ribofuranosylpyrrolo-[2,3-d]pyrimidin-4(3H)-one (**393**), a precursor of nucleosides Q and Q*, was described by Townsend and co-workers.[429] Oxidation of toyocamycin (**188b**) with m-chloroperoxybenzoic acid gave toyocamycin 3-N-oxide (**391**), which, on deamination with nitrous acid, afforded 5-cyano-3-hydroxy-7-β-D-ribofura-nosylpyrrolo[2,3-d]pyrimidin-4-one (**392a**). Acetylation of **392a** provided a near-quantitative yield of **392b**, which, on chlorination with phosphorus oxychloride, gave the 2,4-dichloro compound **395**. Controlled hydrolysis of **395** with aqueous barium hydroxide occurred with a concomitant deacetylation to afford 2-chloro-5-cyano-7-β-D-ribofuranosylpyrrolo[2,3-d]pyrimidin-4(3H)-one (**394**). Treatment of **394** with liquid ammonia at 100°C gave the guanosine analogue **393**. Compound **393** was also obtained at a later date from 5-methyl-2-methylthio-7-(2,3,5-tri-O-benzyl-β-D-ribofuranosyl)pyrrolo[2,3-d]pyrimidin-4(3H)-one[541] by a multistep synthesis.[542] The synthesis of 7-aminomethyl-7-deazaguanosine has also been described.[543,544] The total synthesis of two optically pure diastereo-mers of nucleoside Q, namely, 2-amino-5-(3S,4R,5S-4,5-dihydroxycyclopent-1-en-3-ylaminomethyl)-7-β-D-ribofuranosylpyrrolo[2,3-d]pyrimidin-4(3H)-one and its $3R,4S,5R$ isomer, using optically pure (+)- and (−)-3,4-*trans*-4,5-*cis*-4,5-dihydroxycyclopent-1-en-3-ylamine 4,5-O-acetonide and an appropriately protected pyrrolo[2,3-*d*]pyrimidine nucleoside has recently been accomplished.[541,545] The enzymatic synthesis of nucleoside Q* containing mannose, by a cell-free system from rat liver using purified *E. coli* tRNA^Asp as an acceptor and GDP-mannose as a donor molecule, has been reported.[546]

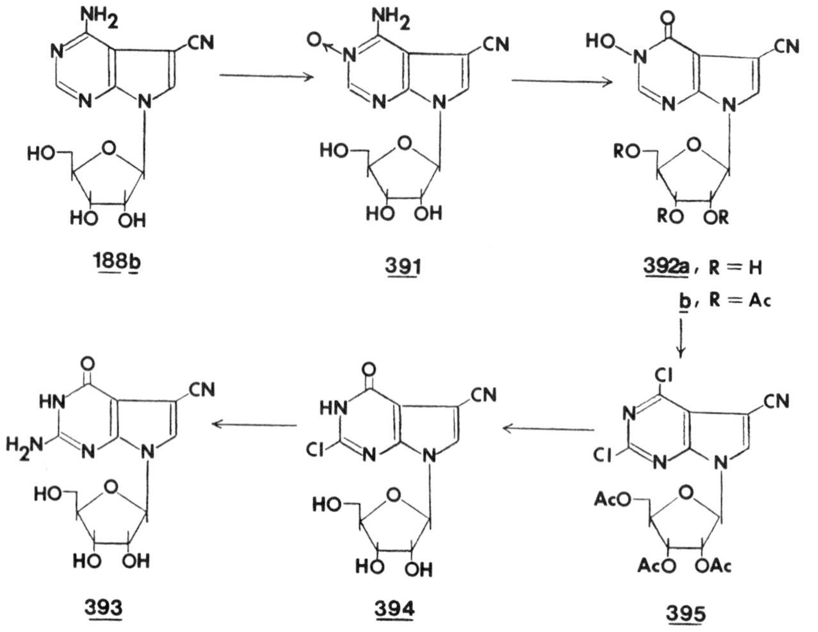

The conformational properties in aqueous solution[547] and determination of the molecular structure of queuosine 5′-monophosphate by X-ray crystallography[548,549] have been reported. Queuosine 5′-monophosphate was isolated from a nuclease P_1 digest of *E. coli* tRNA by Dowex 1 column chromatography.[532]

2.2.3.10. Biochemical Properties of 7-Deazapurine Nucleosides and Nucleotides. Several reviews[2,3,237,550–552] which deal with the biochemical properties of the pyrrolopyrimidine nucleosides in considerable detail are available. The biochemical properties of tubercidin and certain related derivatives in animals have been studied by Smith and co-workers.[553–555] Tubercidin is not deaminated but is rapidly phosphorylated to the triphosphate derivative in cells.[2] Both tubercidin and tubercidin 5′-monophosphate are readily absorbed by mammalian blood cells.[553,554] Tubercidin, toyocamycin, and sangivamycin 5′-triphosphates are effective substrates for rabbit liver tRNA with adenosine nucleoside terminus pyrophosphorylase, and the corresponding nucleosides are incorporated at the 3′-termini of tRNA.[529,556–559] None of the sugar-modified derivatives (e.g., arabinosyl, xylosyl, 2′- and 3′-deoxyribosyl) of tubercidin are as active as tubercidin against cultured mouse L1210 leukemia cells.[560] However, sangivamycin, one of the few pyrrolo[2,3-*d*]pyrimidine nucleosides that have been selected for clinical studies, has cytotoxic activity in a variety of experimental systems *in vitro*[561–563] and *in vivo*.[275,564–569] Phase II studies sponsored by the National Cancer Institute are now under way against colon cancer, gall bladder cancer, and acute myelogenous leukemia (AML) in humans,[570] in which sangivamycin is given by slow infusion to prevent hypotension. Sangivamycin is also active against L1210 leukemia and P388 leukemia (T/C of 167 and 190, respectively

$$\left[\%\,T/C = \frac{\text{Mean life span of treated mice}}{\text{Mean life span of control mice}} \times 100 \right]$$),[391] in contrast to tubercidin,

which is negative against L1210 in animal studies. Significant activity is also shown by sangivamycin against the Lewis lung carcinoma.[391] Thiosangiva-mycin exhibits a T/C of 175 against L1210 leukemia, which is greater than that for sangivamycin.[391] At 3.12 mg/kg per day on a 1 × 9 daily dosage treatment schedule, 4-amino-7-β-D-ribofuranosylpyrrolo[2,3-*d*]pyrimidine-5-carboxami-doxime exhibited a T/C of 204 against L1210 leukemia, which is greater than that for both sangivamycin and thiosangivamycin.[391] Against P388 leukemia at 5 mg/kg per day on a similar treatment schedule, a T/C of 172 was obtained. Sangivamycin does not exhibit appreciable antibacterial activity.[275,567]

5-Iodotubercidin is a useful inhibitor of adenosine kinase.[571–573] Studies have indicated that tubercidin 3′,5′-cyclic phosphate is a significantly better activator of bovine brain, bovine heart, and rat liver protein kinases than cAMP itself,[117,453] whereas toyocamycin 3′,5′-cyclic phosphate is only about half as active as cAMP, and sangivamycin 3′,5′-cyclic phosphate is only about one-fifth as active.[453]

Tubercidin inhibits the growth of several organisms,[255,574] including *Streptococcus faecalis*,[575,576] *Schistosoma mansoni*,[577–581] *Myobacterium tuberculosis*,[582] and *Plasmodium knowles*.[583] Tubercidin also inhibits the egg-laying ability in the housefly.[584,585] The antifungal activity of several substituted pyrrolo[2,3-*d*]-pyrimidine nucleosides has been described.[586] Toyocamycin is a potent inhibitor of *Candida albicans*, *Cryptococcus neoformans*, and *M. tuberculosis* but has little or no effect on other bacteria, fungi, or yeast.[250,251,587] Toyocamycin and several 4-substituted pyrrolo[2,3-*d*]pyrimidine nucleosides showed significant growth-retarding activity in plants.[588–590]

The nucleoside antibiotic cadeguomycin is reported[281] to have inhibitory effects on transplantable animal tumors but no significant activity against bacteria and fungi. Deamination of tubercidin or toyocamycin markedly reduced their antitumor activity and also decreased their toxicity.[591] 7-Deazainosine inhibited the growth of sarcoma-180 cells *in vitro*, as well as Ehrlich ascites and leukemia P388 cells *in vivo*.[592] 7-Deazainosine is active against the parasites *Leishmania tropica*,[210] *L. donovani*, and *Trypanosoma cruzi*[593] *in vitro*. A combination of tubercidin and nitrobenzylthioinosine 5′-monophosphate (NBMPR-P) has been used successfully for the treatment of *Schistosomiasis* in mice.[594]

Tubercidin and some 5-substituted tubercidins have inhibited both DNA and RNA viruses[386,595–602] as well as RNA, DNA, and protein synthesis in mouse fibroblasts,[386,603] human KB cells,[388,554] and human tumors.[604] Of considerable interest is the xylofuranosyl analogue of tubercidin,[500] which, unlike tubercidin, exhibited decreased cytotoxicity and potent activity against HSV-1 and HSV-2 *in vitro*.[605] When xylotubercidin was administered i.p. at 50 mg/kg per day, a significant reduction of the mortality rate of mice infected with HSV-2 was achieved.[606] Acyclovir was not effective in prolonging survival rate in this model, even up to 250 mg/kg per day. Against intracutaneous HSV-2 in hairless mice, xylotubercidin applied topically in 1% DMSO gave 100% survivors without observable toxicity.[606] This nucleoside would appear to have considerable potential for topical treatment of HSV-2 infections. Toyocamycin inhibited rRNA synthesis.[607–611] Tubercidin is as active as adenine as a feedback inhibitor of *de novo* purine biosynthesis,[612,613] and this inhibition probably takes place at the nucleotide level (in the presence of a purine nucleoside kinase).[614,615] The pyrrolo[2,3-*d*]pyrimidine nucleotides cannot function as an energy source for activation of amino acids or esterification of tRNA by mammalian or bacterial enzymes.[529]

The carbocyclic analogue of tubercidin was resistant to ADA and required phosphorylation to exert its cytotoxic effect. Its antiviral activity was marginal. The arabinofuranosyl and lyxofuranosyl analogues of carbocyclic tubercidin are inactive as antitumor or antiviral agents.[299,300] However, the carbocyclic analogue of 7-deazaguanosine exhibited antiherpes properties.[616] At 25 μg/ml [minimum inhibitory concentration (MIC)], the antiviral activity appeared to be selective[616] in culture, whereas antiviral activity was displayed in mice at 10 mg/kg, administered twice a day for 2 days beginning 2 h post infection. The *ara*-carbocyclic analogue of 7-deazaguanosine at 100 mg/kg, given i.p. over a 48-h period, resulted in 4/15 survivors,[302] as compared to 0/30 for the controls.[616]

2.2.4. Pyrrolo[3,2-d]pyrimidine (9-Deazapurine) Nucleosides

Although the syntheses of the pyrrolo[3,2-*d*]pyrimidine (9-deazapurine) *C*-nucleosides, such as 9-deazainosine,[617] 9-deazaadenosine,[618,619] and 9-deazaguanosine,[619,620] have been reported, the preparation of the corresponding *N*-nucleosides is not well documented in the literature. The only synthesis of *N*-nucleosides of this ring system has recently been reported from our laboratory[621,622] by the sodium salt glycosylation procedure.

Glycosylation of the sodium salt of 2,4-dichloropyrrolo[3,2-*d*]pyrimidine (**396**)[623] with **24** in acetonitrile gave the key intermediate 2,4-dichloro-5-(2-deoxy-3,5-di-*O*-*p*-toluoyl-β-D-*erythro*-pentofuranosyl)pyrrolo[3,2-*d*]pyrimidine (**397**). When **397** was treated with methanolic ammonia at an elevated temperature, deprotection of the sugar with a concomitant nucleophilic displacement of the 4-chloro function occurred to give 4-amino-2-chloro-5-(2-deoxy-β-D-*erythro*-pentofuranosyl)pyrrolo[3,2-*d*]pyrimidine (**398a**). Dehalogenation of **398a** with Pd/C in a hydrogen atmosphere provided 7-(2-deoxy-β-D-*erythro*-pentofuranosyl)-9-deazaadenine (**398b**).[621] Similarly, treatment of **397** with aqueous sodium hydroxide gave **401a**, which, on catalytic dehalogenation, furnished 7-(2-deoxy-β-D-*erythro*-pentofuranosyl)-9-deazahypoxanthine (**401b**). 7-(2-Deoxy-β-D-*erythro*-pentofuranosyl)-9-deazaguanine (**399**) was also obtained in good yield from **397** via the intermediates **400a** and **400b**.

The preparation of the corresponding β-D-arabinofuranosyl[621] and β-D-ribofuranosyl[622] nucleosides was accomplished in the same general manner as that of the 2'-deoxyribofuranosylnucleosides, employing the sodium salt of

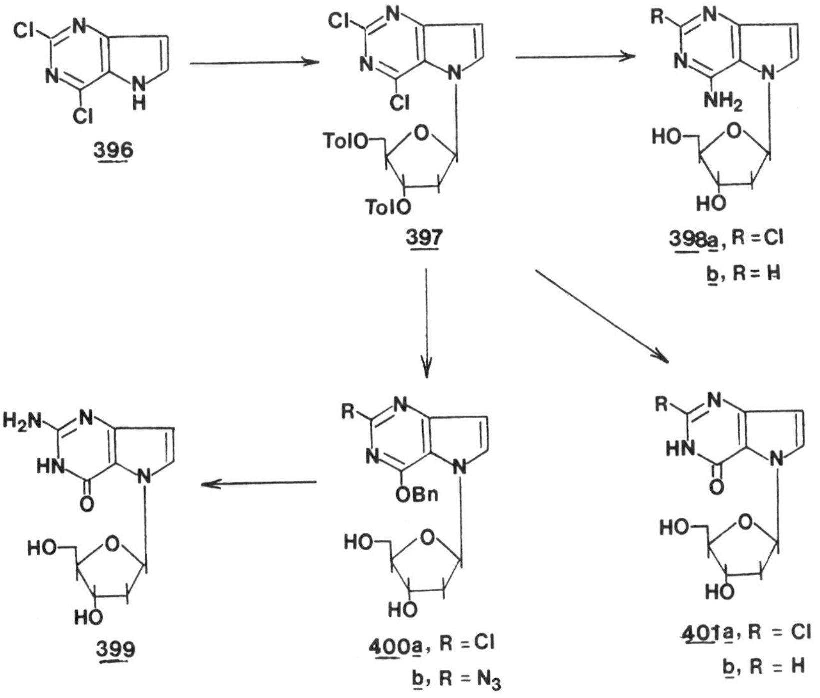

396

397

 TolO

TolO

398a, R = Cl

b, R = H

399

400a, R = Cl

b, R = N$_3$

401a, R = Cl

b, R = H

OBn

396 and the α-halogenose **27** or **272**. The structural assignment of these nucleosides was made on the basis of the single-crystal X-ray diffraction analysis of 2-chloro-5-β-D-arabinofuranosylpyrrolo[3,2-d]pyrimidin-4(3H)-one.[621]

It is of particular interest that 9-deazaadenosine has shown exceptional cytotoxicity against several cell lines of mouse and human leukemia.[618,624] 9-Deazaadenosine was found to be a potent growth inhibitor of nine human solid tumors *in vitro* and exhibits antitumor activity against a human pancreatic carcinoma xenograft.[625,626] 9-Deazaadenosine has also been found to inhibit lymphocyte-mediated cytolysis in a time-dependent manner.[627] However, it did not inhibit the enzyme adenosine kinase.[628] 9-Deazainosine was active against *Leishmania donovani*, *Trypanosoma cruzi*, and *Trypanosoma gambiense* and has very little toxicity for mouse L cells.[593]

2.2.5. Pyrazolo[3,4-b]pyridine Nucleosides

The first chemical synthesis of a pyrazolo[3,4-b]pyridine nucleoside was reported by Preobrazhenskaya and co-workers,[629,630] who investigated this class of compounds as potential antimetabolites of purine nucleosides. 6-Chloropyrazolo[3,4-b]pyridine (**402**, X = Cl)[631] was fused with 1,2,3,5-tetra-O-acyl-D-ribofuranose (**51** or **47**) in the presence of iodine to obtain 6-chloro-1-(2,3,5-tri-O-acyl-β-D-ribofuranosyl)pyrazolo[3,4-b]pyridine (**403**, X = Cl).[629] However, fusion without the catalyst gave N-1 and N-2 isomeric nucleosides in a 1:1 ratio. Deacylation of **403** with methanolic ammonia afforded 6-chloro-1-β-D-ribofuranosylpyrazolo[3,4-b]pyridine (**404a**).[632,633] Nucleophilic displacement of the 6-chloro group from **404a** with ammonia, hydrazine, and dimethylamine furnished the corresponding 6-substituted 1-β-D-ribofuranosylpyrazolo[3,4-*b*]-pyridines (**404b–d**). Dehalogenation of **404a** in the presence of palladium chloride afforded the unsubstituted pyrazolo[3,4-*b*]pyridine riboside (**404**, X = H).

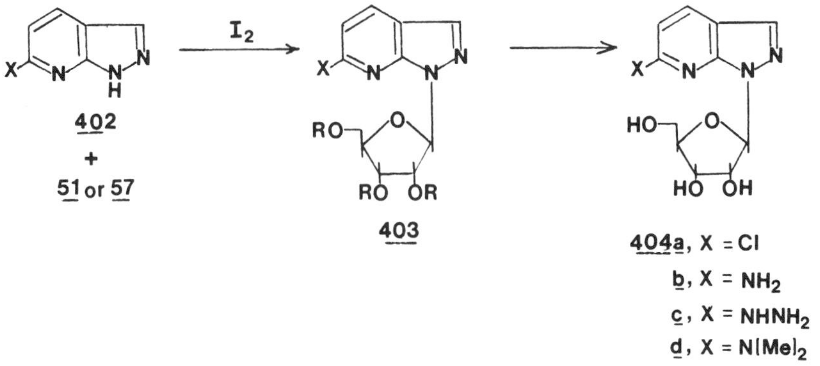

A similar iodine-catalyzed fusion of 6-methylthiopyrazolo[3,4-*b*]pyridine (**402**, X = SCH₃) with **51** or **57** and subsequent deacylation gave 6-methylthio-1-β-D-ribofuranosylpyrazolo[3,4-*b*]pyridine (**404**, X = SCH₃).[563] Dethiation of **404** (X = SCH₃) with Raney nickel afforded the nebularine analogue, which in turn was obtained through the direct fusion of pyrazolo[3,4-*b*]pyridine with **57**, followed by debenzoylation. However, this latter fusion procedure also furnished a small amount of pyrazolo[3,4-*b*]pyridine-2-ribonucleoside. The structures of these nucleosides were established through a combination of UV and ¹H-NMR spectroscopic studies,[634] periodate oxidation studies, and circular dichroism measurements.[630]

A successful synthesis of various C-4-substituted pyrazolo[3,4-*b*]pyridine nucleosides, including the adenosine analogue 4-amino-1-β-D-ribofuranosyl-pyrazolo[3,4-*b*]pyridine, has recently been reported from our laboratory.[635] This involved the nucleophilic displacement by a suitable nucleophile at the C-4 position of 4-chloro-1*H*-pyrazolo[3,4-*b*]pyridine,[636] followed by glycosylation of the sodium salt of the C-4-substituted pyrazolo[3,4-*b*]pyridines with a protected α-halopentofuranose. Use of this methodology furnished a simple and direct route to the β-D-ribofuranosyl, β-D-arabinofuranosyl, and 2-deoxy-β-D-*erythro*-pentofuranosyl nucleosides of C-4-substituted pyrazolo[3,4-*b*]pyridines, wherein the C-4 substituent was azido, amino, methoxy, chloro, or oxo. The structural assignment of these nucleosides was made by single-crystal X-ray diffraction studies.[637]

4-Chloro-1-β-D-ribofuranosylpyrazolo[3,4-*b*]pyridine and 1-β-D-ribofuranosyl-4,7-dihydro-4-oxopyrazolo[3,4-*b*]pyridine were moderately cytotoxic to L1210 and WI-L2 cell lines in culture.[635] The 6-substituted pyrazolo[3,4-*b*]pyridine ribonucleosides are strong inhibitors of xanthine oxidase[18] and IMP dehydrogenase.[638] Compound **404c** showed significant antitumor activity against L1210 in cell culture[18]; however, in animals with transplanted tumors, no dramatically active compound in this series was found.[18]

2.2.6. Pyrazolo[4,3-c]pyridine Nucleosides

The only synthesis of pyrazolo[4,3-c]pyridine nucleosides reported so far is from our laboratory[639] and was accomplished by the cyclization of the corresponding glycosyl derivatives of methyl 3(5)-cyanomethylpyrazole-4-carboxylate.[640] Contrary to the observations with the imidazole counterpart (**149**), glycosylation of the trimethylsilyl derivative of methyl 3(5)-cyanomethylpyrazole-4-carboxylate (**405**) with **57** in the presence of 1.44 molar equivalent of SnCl₄ gave predominantly the N-2 glycosyl isomer methyl 3-cyanomethyl-1-(2,3,5-tri-*O*-benzoyl-β-D-ribofuranosyl)pyrazole-4-carboxylate (**407**) and only a trace amount of the corresponding N-2 isomer (**406**).[639] Subsequent treatment of **407** with ammonia at 100°C provided **410**. 6-Amino-2-β-D-ribofuranosylpyrazolo[4,3-c]pyridine-4(5*H*)-one (**409**) was formed when **410** was heated under reflux in aqueous Na₂CO₃–ethanol mixture. Direct phosphorylation of the unprotected **409** with phosphorus oxychloride in trimethyl phosphate provided the 5′-phosphate derivative **408**.[639]

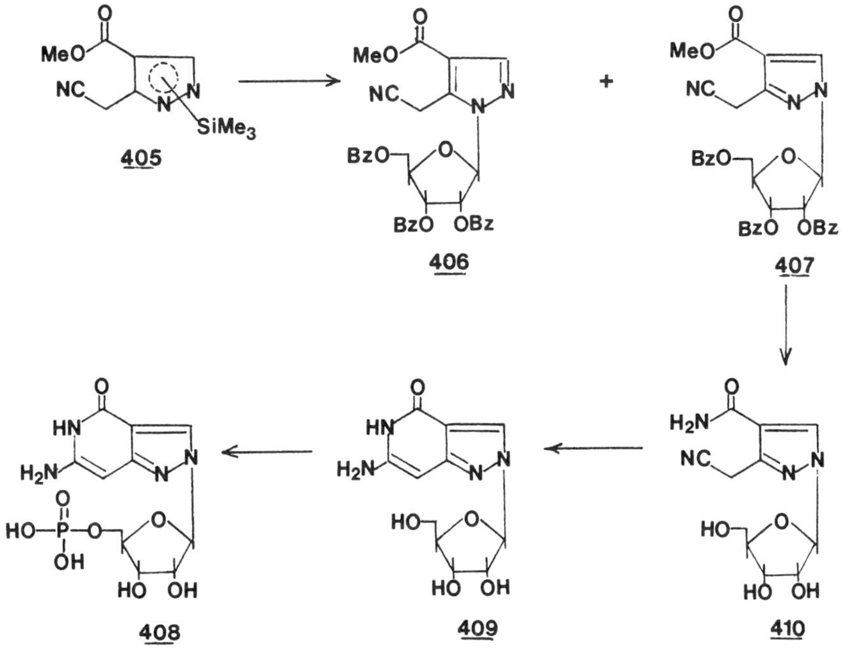

Treatment of the sodium salt of methyl 3(5)-cyanomethylpyrazole-4-carboxylate (**411**) with **24** gave the N-1 and N-2 isomeric blocked nucleosides (**412** and **413**) in a 1:1 ratio. Further treatment of **412** with methanolic ammonia gave **415**, which, on base-catalyzed cyclization, produced the 2′-deoxyguanosine analogue 6-amino-1-(2-deoxy-β-D-*erythro*-pentofuranosyl)pyrazolo[4,3-c]pyridin-4(5*H*)-one (**414**).[639] In a similar manner, **413** was transformed into **416**. Compound **409** exhibited significant activity against Rift Valley Fever virus in mice.[639]

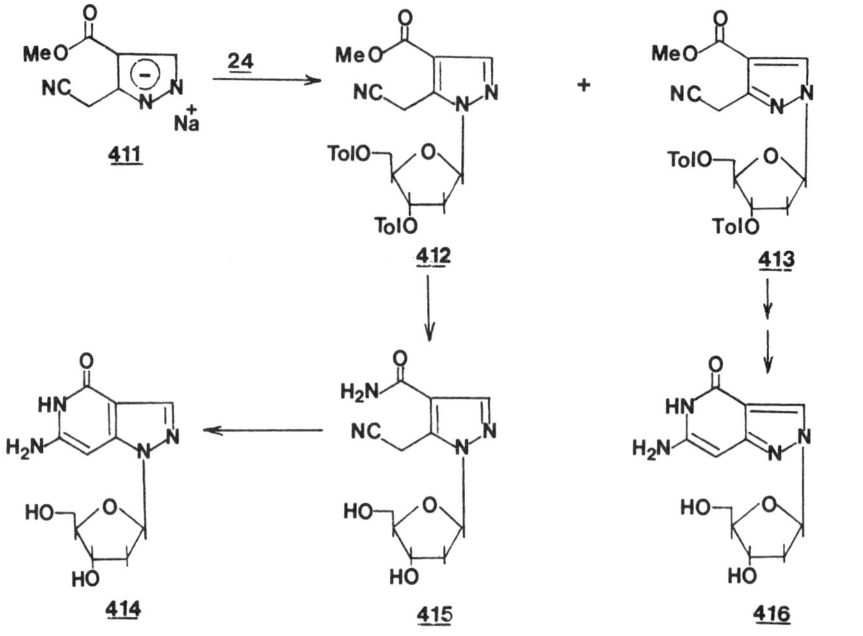

2.3. Five-Membered Ring Fused to a Six-Membered Ring with Four Nitrogen Atoms in the Ring System (at Least One Nitrogen Atom in Each Ring)

Nucleosides that are structurally related to the naturally occurring purine ribonucleosides have generated considerable interest as potential chemotherapeutic agents. These nucleosides usually differ from the natural purine ribosides in the position of attachment of the glycosyl bond (e.g., 3- or 7-β-D-ribofuranosylpurines[641–643]), in the nature of the sugar moiety (e.g., 9-β-D-arabinofuranosyladenine[35,644]), or in the arrangement of the four nitrogen atoms of the bicyclic aglycon. Only a limited amount of research has been done on members of this last category (probably with the exception of pyrazolo[3,4-*d*]pyrimidine). The ring systems included in this section are shown in Fig. 4; nucleosides of only these systems are reported in the literature at the present time.

2.3.1. Imidazo[4,5-b]pyrazine Nucleosides

2.3.1.1. Synthesis Employing the Ring Closure of an Imidazole Nucleoside Precursor. The syntheses of several imidazo[4,5-*b*]pyridines have appeared in the literature[645–647] and usually involve the ring closure of a 2,3-diaminopyrazine with either ethyl orthoformate, acetic anhydride, acyl halides, or urea. Because of the relative inaccessibility of the appropriate diaminopyrazine intermediates, Panzica and Townsend[648] explored an alternate route via ring annulation of the appropriate 4,5-diaminoimidazole derivative and reported the first chemical synthesis of an imidazo[4,5-*b*]pyrazine nucleoside.

Treatment of 5-bromo-4-nitro-1-(2,3,5-tri-*O*-acetyl-β-D-ribofuranosyl)imidazole (**417**)[641] with liquid ammonia at room temperature resulted in a nucleo-

Imidazo[4,5-<u>b</u>]-
pyrazine

Imidazo[4,5-<u>c</u>]-
pyridazine

Imidazo[4,5-<u>d</u>]-
pyridazine

Pyrazolo[3,4-<u>b</u>]-
pyrazine

Pyrazolo[3,4-<u>d</u>]-
pyrimidine

Pyrazolo[4,3-<u>d</u>]-
pyrimidine

<u>v</u>-Triazolo[4,5-<u>b</u>]-
pyridine

<u>v</u>-Triazolo[4,5-<u>c</u>]-
pyridine

Figure 4. Numbering scheme for imidazopyrazine and -pyridazine, pyrazolopyrazine and -pyrimidine, and triazolopyridine ring systems.

philic displacement of the bromo group with a concomitant deacetylation to provide 5-amino-4-nitro-1-β-D-ribofuranosylimidazole (**418**). Catalytic reduction of **418** with Raney nickel gave the diamino compound **420** *in situ*, which was immediately reacted with diacetyl under an atmosphere of nitrogen to obtain 5,6-dimethyl-1-β-D-ribofuranosylimidazo[4,5-*b*]pyrazine (**419**) in an overall yield of 68.2%.

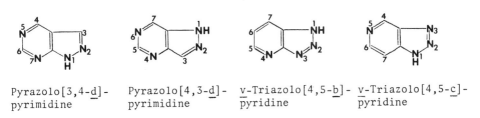

417   418

419   420   421a, R = H
                   <u>b</u>, R = CH$_3$

Similarly, condensation of **420** with glyoxal trimer in 0.1 *N* hydrochloric acid gave the nebularine analogue 1-β-D-ribofuranosylimidazo[4,5-*b*]pyrazine (**421a**). Ring annulation of **420** with aqueous pyruvaldehyde in the presence of an acid gave 5(6)-methyl-1-β-D-ribofuranosylimidazo[4,5-*b*]pyrazine (**421b**).

2.3.1.2. Synthesis Employing the Acid-Catalyzed Fusion of a Preformed Aglycon and an Acyl Sugar. The acid-catalyzed fusion procedure was also examined by Panzica and Townsend[649] for the synthesis of imidazo[4,5-*b*]pyrazine nucleosides. The fusion of 5,6-dimethylimidazo[4,5-*b*]pyrazine (**422**) and **57** at 190–195°C in the presence of bis(*p*-nitrophenyl)phosphate gave the blocked nucleoside (**423**), which, on subsequent debenzoylation, afforded **419** in an overall yield of 64.2%. The site of ribosylation and the anomeric configuration were established by a comparison with an authentic sample (**419**) prepared by the above ring-closure procedure.

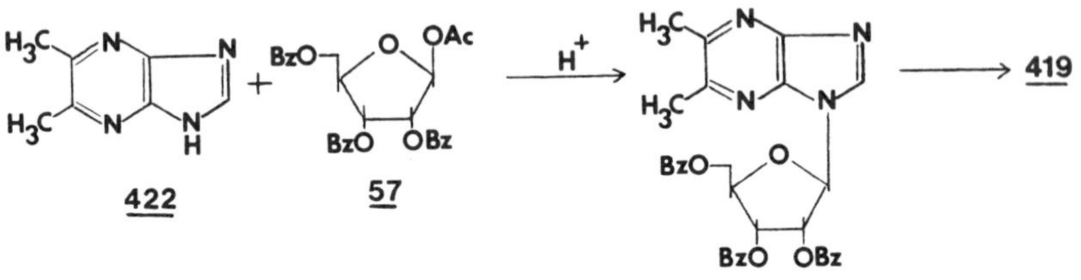

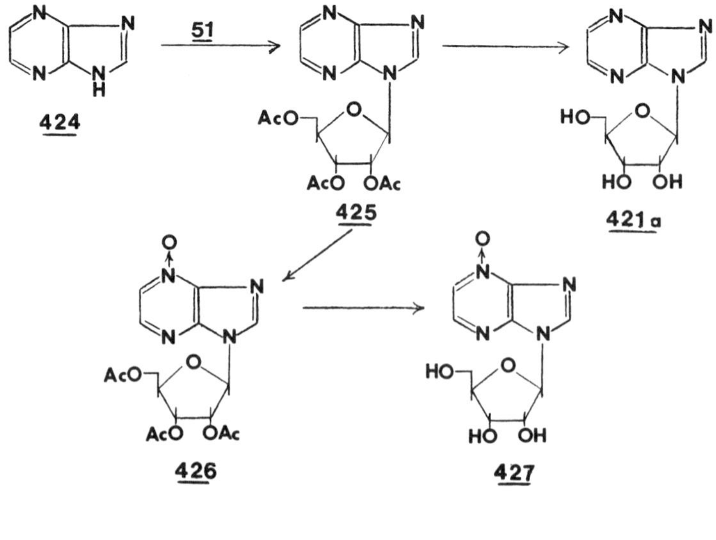

Imidazo[4,5-*b*]pyrazine (**424**)[646] was similarly fused with 1,2,3,5-tetra-*O*-acetyl-D-ribofuranose (**51**) by Bloch and co-workers[650] to obtain the blocked nucleoside **425**. Subsequent deacetylation of **425** gave **421a**. Treatment of **425** with *m*-chloroperoxybenzoic acid afforded the N^4-oxide (**426**), and subsequent deacetylation of **426** furnished 1-β-D-ribofuranosylimidazo[4,5-*b*]pyrazine-4-oxide (**427**).[650] The site of ribosylation was established as N-1 by a comparison of the UV absorption spectrum of **427** with that of 1-methylimidazo[4,5-*b*]-pyrazine-4-oxide and was substantiated by X-ray crystallography. In the crystalline state, **427** exists in the *anti* configuration, with a torsion angle of 7.96° defined by the ring oxygen, C-1 of the carbohydrate, and N-1 and C-2 of the aglycon. The N→O bond distance is 1.302 Å, and in the crystal the *N*-oxide forms a hydrogen bond with the proton of the 2'-hydroxyl group of a neighboring molecule, with a bond distance of 2.85 Å.

2.3.1.3. Synthesis Employing the Trimethylsilyl Derivative of a Preformed Aglycon and an Acylated Sugar Halide. Another approach for the synthesis of imidazo[4,5-*b*]pyrazine nucleosides studied by Panzica and Townsend[649] involved the silylation procedure. Treatment of trimethylsilyl-5,6-dimethylimidazo[4,-5-*b*]pyrazine (**428**) with 2,3,5-tri-*O*-benzoyl-D-ribofuranosyl bromide (**19**) in acetonitrile provided a single protected nucleoside (**423**), which, on subsequent debenzoylation, gave 5,6-dimethyl-1-β-D-ribofuranosylimidazo[4,5-*b*]pyrazine in a 44.6% yield. This compound was found to be identical with **419** prepared by a ring annulation procedure. The site of ribosylation was also corroborated by a comparison of the UV absorption spectrum of **419** with that of 1,5,6-trimethyl-imidazo[4,5-*b*]pyrazine.[649]

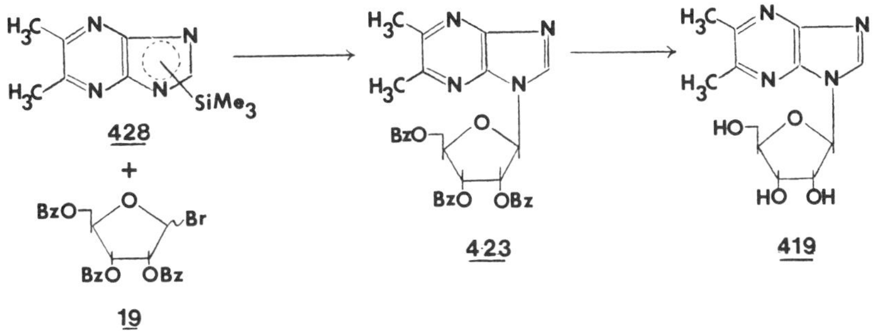

1-β-D-Ribofuranosylimidazo[4,5-*b*]pyrazine-4-oxide (**427**) inhibited the *in vitro* growth of *E. coli* K$_{12}$ by 50% at $6 \times 10^{-6} M$. The nebularine analogue (**421a**) was inactive.[650] The inhibition of growth by **427** was reversed competitively by the natural purines.[651] None of these nucleoside analogues interfered with the *in vitro* growth of L1210 leukemia, mammary carcinoma TA-3, and Burkitt's lymphoma cells. In a cell-free extract from *E. coli*, **427** underwent a glycosidic cleavage.

2.3.2. Imidazo[4,5-c]pyridazine Nucleosides

The imidazo[4,5-*c*]pyridazine ring system was reported for the first time in 1964 by Kuraishi and Castle.[652] The first synthesis of an imidazo[4,5-*c*]pyridazine nucleoside was reported by Halverson and Castle[653] in 1974. The trimethylsilyl procedure was employed. The trimethylsilyl derivative (**429**) of 7-chloroimidazo[4,5-*c*]pyridazine[652] was condensed with 2,3,5-tri-*O*-benzoyl-D-ribofuranosyl bromide (**19**) in the presence of mercuric cyanide to obtain

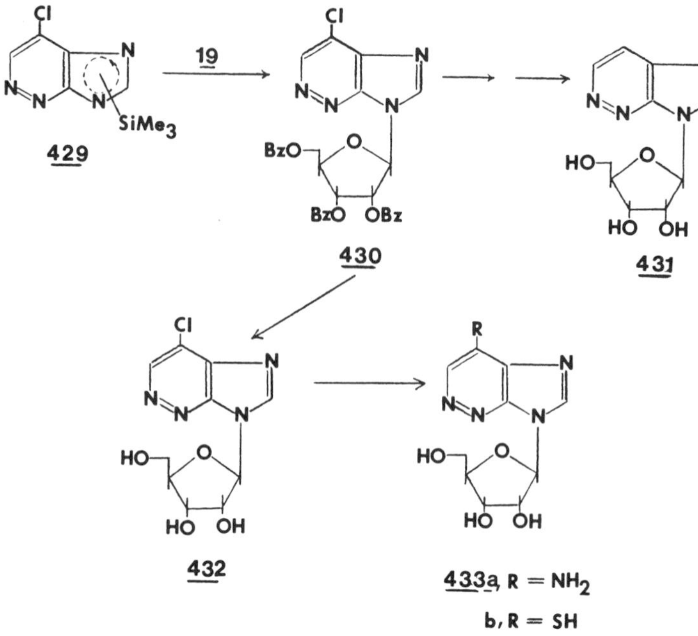

7-chloro-3-(2,3,5-tri-*O*-benzoyl-β-D-ribofuranosyl)imidazo[4,5-*c*]pyridazine
(**430**). Catalytic dehalogenation of **430**, followed by debenzoylation, furnished
the nebularine analogue 3-β-D-ribofuranosylimidazo[4,5-*c*]pyridazine (**431**).
Debenzoylation of **430** with methanolic ammonia gave 7-chloro-3-β-D-ribofura-
nosylimidazo[4,5-*c*]pyridazine (**432**). Treatment of **432** with ethanolic ammonia
at elevated temperature gave the adenosine analogue 7-amino-3-β-D-ribofura-
nosylimidazo[4,5-*c*]pyridazine (**433a**). Likewise, treatment of **432** with methanolic sodium hydrosulfide gave 3-β-D-ribofuranosylimidazo[4,5-*c*]pyrid-
azine-7-thione (**433b**).

The site of ribosylation was established by UV spectral comparisons with
model methyl compounds,[654] and the β-anomeric configuration was assigned
by ^{1}H-NMR spectral studies.[137]

2.3.3. *Imidazo[4,5-d]pyridazine Nucleosides*

*2.3.3.1. Synthesis Employing the Halomercuri Salt of a Preformed Aglycon and an
Acylated Sugar Halide.* The chemistry of the imidazo[4,5-*d*]pyridazine ring
system has been thoroughly investigated,[655] because of the marked similarity
of this ring system to purine; however, only a few imidazo[4,5-*d*]pyridazine
nucleosides are to be found in the chemical literature. The first synthesis of an
imidazo[4,5-*d*]pyridazine nucleoside was reported by Carbon[656] in 1960. A
ribosylation of the chloromercuri derivative of 4(7)-benzamidoimidazo[4,5-*d*]-
pyridazine (**434**) with 2,3,5-tri-*O*-benzoyl-D-ribofuranosyl chloride (**47**) gave
a mixture of blocked nucleosides (**435a,b**). Debenzoylation of **435a,b** with
sodium methoxide afforded the free nucleosides, which were separated by
fractional crystallization to obtain the adenosine analogue 4-amino-1-β-D-

434 + **47**

435a, R_1 = NHBz; R_2 = H
b, R_1 = H; R_2 = NHBz

436

437

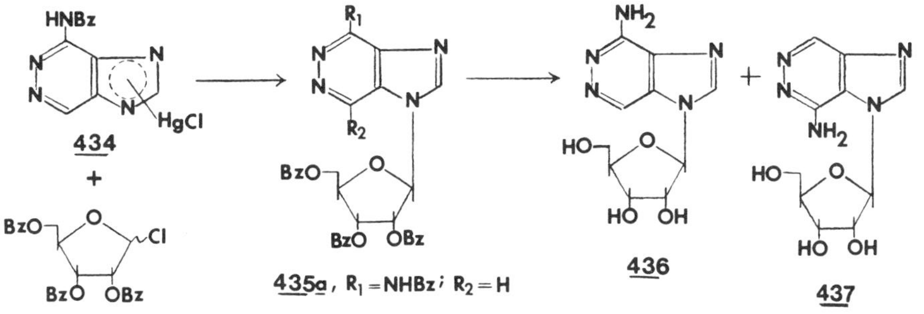

ribofuranosylimidazo[4,5-*d*]pyridazine (**436**) and its 7-amino isomer (**437**). Although the site of ribosylation was assigned on the basis of a comparison of the UV absorption spectra of these nucleosides with those of model 4- and 7-amino-1-methylimidazo[4,5-*d*]pyridazines,[657] the isomer **437** has recently been shown to be the N-6-glycosylated product.[658] The β-anomeric configuration was assigned on the basis of Baker's *trans* rule.[53]

2.3.3.2. Synthesis Employing the Ring Closure of an Imidazole Nucleoside Precursor. The versatility of imidazole nucleosides as precursors for the synthesis of imidazo[4,5-*d*]pyridazine nucleosides was demonstrated by Townsend and co-workers[659] and Robins and co-workers.[660] Treatment of dimethyl 1-(2,3,5-tri-*O*-acyl-β-D-ribofuranosyl)imidazole-4,5-dicarboxylate (**438a** or **438b**) or dimethyl 1-β-D-ribofuranosylimidazole-4,5-dicarboxylate (**438c**)[660] with hydrazine hydrate in ethanol gave crystalline 1-β-D-ribofuranosylimidazole-4,5-dicarboxhydrazide (**439**) in >91% yield. Ring annulation of **439** by heating in 97% hydrazine[661,662] furnished 1-β-D-ribofuranosylimidazo[4,5-*d*]pyridazine-4,7-dione (**440**) in almost quantitative yield.

438a, R = Ac
b, R = Bz
c, R = H

439

440

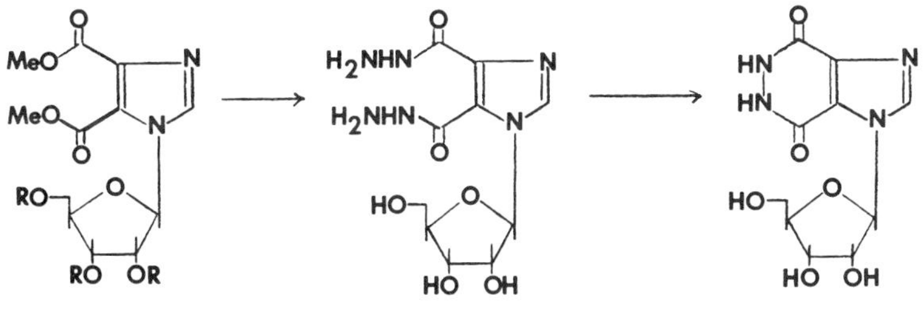

2.3.3.3. Synthesis Employing the Trimethylsilyl Derivative of a Preformed Aglycon and an Acylated Sugar Using a Lewis Acid Catalyst. The Lewis acid catalyzed condensation of the silylated imidazo[4,5-*d*]pyridazin-4(5*H*)-one (**441**) with 1-*O*-acetyl-2,3,5-tri-*O*-benzoyl-β-D-ribofuranose (**57**) was reported[660] to yield the blocked monoribofuranosyl (**442a**, 74%) and the bis-ribofuranosyl (**443a**, 20%) derivatives. Debenzoylation of **442a** and **443a** with sodium methoxide in methanol gave 6-β-D-ribofuranosylimidazo[4,5-*d*]pyridazin-4(5*H*)-one (**442b**) and 1,5-bis(β-D-ribofuranosyl)imidazo[4,5-*d*]pyridazin-7-one (**443b**), respectively. A similar ribosylation of tris(trimethylsilyl)imidazo[4,5-*d*]pyridazine-4,7-dione (**444**) with **57** gave the blocked monoribofuranosyl (**445a**, 38%) and bis-ribofuranosyl (**446a**, 58%) derivatives. Subsequent deblocking of these derivatives provided 1-β-D-ribofuranosylimidazo[4,5-*d*]pyridazine-4,7-dione (**445b**) and the 1,5(6)-bis(β-D-ribofuranosyl) isomer (**446b**), respectively. Nucleoside **445b** was found to be identical with **440** prepared by a ring annulation of the imidazole nucleoside precursor.[660]

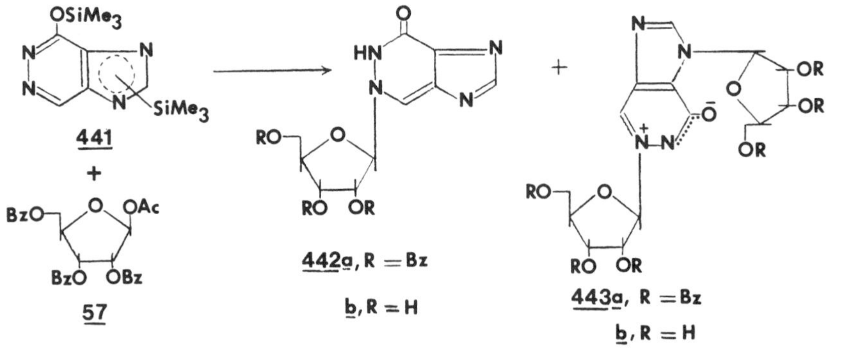

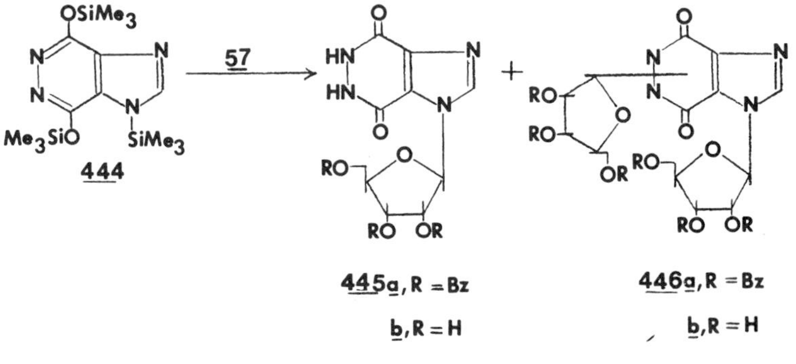

In an attempt[660] to prepare the imidazo[4,5-*d*]pyridazine analogue of inosine, 4,7-dichloro-1-trimethylsilylimidazo[4,5-*d*]pyridazine (**447**) was ribosylated with tetra-*O*-acetyl-β-D-ribofuranose (**51**) in the presence of stannic chloride. This afforded the versatile intermediate 4,7-dichloro-1-(2,3,5-tri-*O*-acetyl-β-D-ribofuranosyl)imidazo[4,5-*d*]pyridazine (**448a**). 4,7-Dichloro-1-β-D-ribofuranosylimidazo[4,5-*d*]pyridazine (**448b**) was obtained by the sodium methoxide deblocking of **448a**. However, when **448a** was treated with sodium benzyloxide, 7-benzyloxy-4-chloro-1-β-D-ribofuranosylimidazo[4,5-*d*]pyridazine **449** was obtained. The formation of **449** was contrary to the reported[657,663] observation that the least sterically hindered chlorine atom (chlorine atom in the 4-position)

of 1-substituted 4,7-dichloroimidazo[4,5-*d*]pyridazines was displaced in a nucleophilic displacement reaction. Hydrogenolysis of **449** furnished 1-β-D-ribofuranosylimidazo[4,5-*d*]pyridazin-7(6*H*)-one (**451**), the structure of which was established by ¹H- and ¹³C-NMR spectroscopic studies. The nebularine analogue, 1-β-D-ribofuranosylimidazo[4,5-*d*]pyridazine (**450**) was obtained by dehalogenation of **448a** with palladium on charcoal in the presence of sodium acetate, followed by subsequent deacetylation. The desired inosine analogue 1-β-D-ribofuranosylimidazo[4,5-*d*]pyridazin-4(5*H*)-one was reported only recently.[58,664] By adopting the strategy of using a removable N-substituent to influence the site of glycosylation,[665] Otter and co-workers[658,664] have ribosylated the trimethylsilyl derivative of 5-benzyloxymethylimidazo[4,5-*d*]pyridazin-4-one (**452**), obtaining a separable mixture of blocked nucleosides **453** and **454**. Removal of the N-5 blocking group from **453**, by treatment with boron trichloride, afforded the intermediate **456a**. Compound **456a** was debenzoylated to give the inosine analogue **455a**. Thiation of **456a**, followed by debenzoylation, afforded 1-β-D-ribofuranosylimidazo[4,5-*d*]pyridazine-4(5*H*)-thione (**455b**).[658]

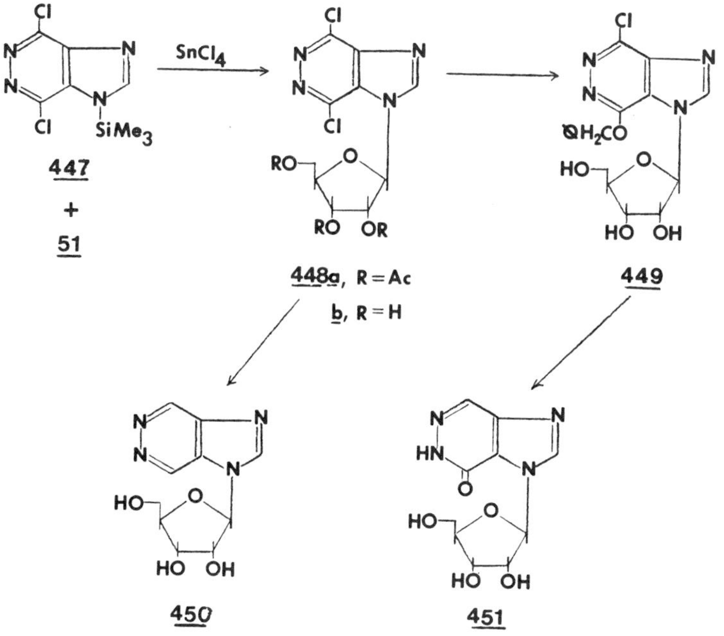

The adenosine analogue (**436**) has been shown to be an excellent substrate but a moderate inhibitor of *S*-adenosylhomocysteine hydrolase from beef liver.[666] However, **436** was not significantly cytotoxic to HEp-2 cells,[614] implying that it was not a substrate for adenosine kinase. The N-6 glycosyl isomer, previously thought to be **437**, was moderately cytotoxic to HEp-2 cells, and it was suggested[614] that this activity may be due to a cleavage to the free base. The inactivity of **436** suggests that it is neither phosphorylated by adenosine kinase nor cleaved to the free base by nucleoside phosphorylase. Compound **436** exhibits a feedback inhibition of *de novo* purine biosynthesis.[667]

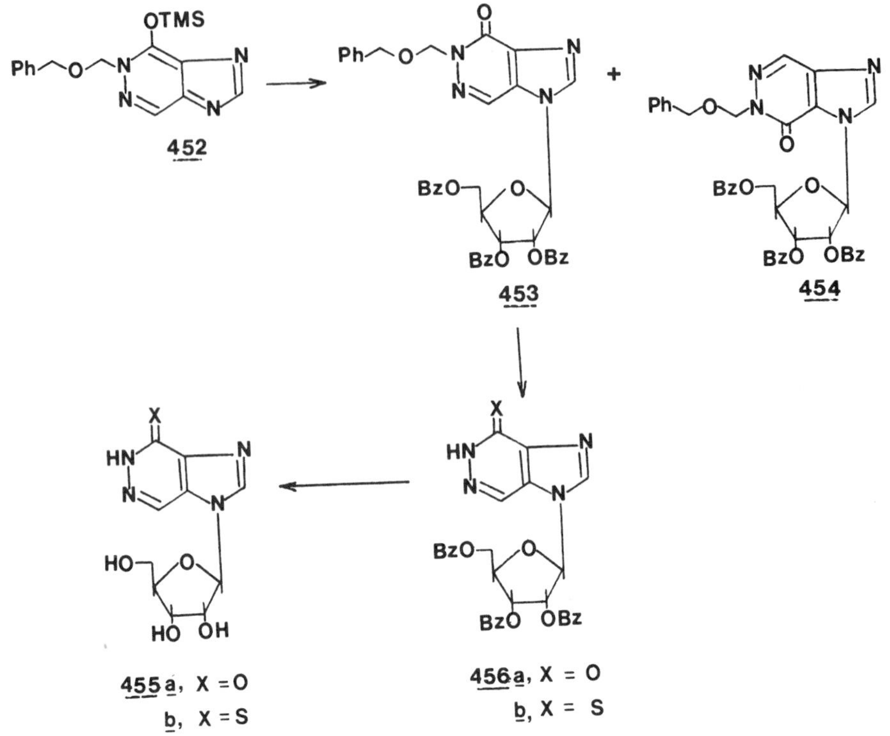

455a, X = O
b, X = S

456a, X = O
b, X = S

2.3.4. Pyrazolo[3,4-b]pyrazine Nucleosides

2.3.4.1. Synthesis Employing the Fusion of a Preformed Aglycon and an Acyl Sugar.
The fusion of pyrazolo[3,4-*b*]pyrazine (**457**)[668] with 1,2,3,5-tetra-*O*-acetyl-β-D-ribofuranose (**51**) at 160°C was reported[669] to yield the corresponding N-1 substituted blocked nucleoside (**458a**), along with a small amount of the N-2 isomer (**460a**). The yield of the N-2 isomer was increased when the fusion was carried out under reduced pressure and temperature for a shorter reaction time.

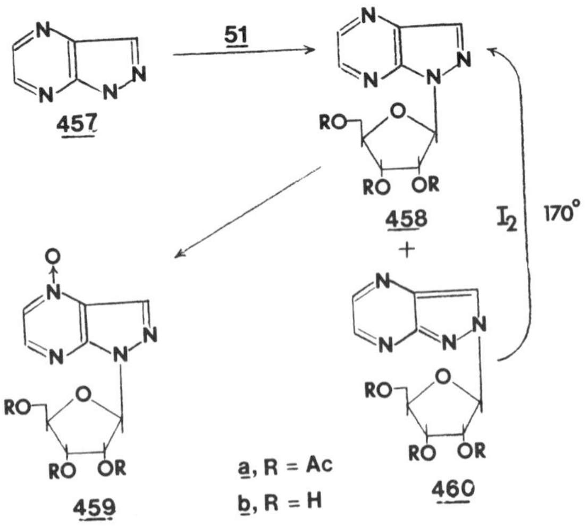

a, R = Ac
b, R = H

The addition of iodine also affected the ratio of N-1 and N-2 isomers, in favor of **458a**. Thus, the fusion of **457** with **51** at 150°C, without iodine, gave a 60% yield of a mixture of approximately equal amounts of **458a** and **460a**. When **460a** was heated at 170°C in the presence of iodine, it completely isomerized[669] to the thermodynamically more stable **458a**. Under similar fusion conditions, 3-phenyl-pyrazolo[3,4-*b*]pyrazine behaved analogously.[670] Deacetylation of the blocked nucleosides **458a** and **460a** with methanolic ammonia furnished the nucleosides **458b** and **460b**, respectively. Oxidation of **458a** with hydrogen peroxide in acetic acid afforded a 35% yield of a compound identified as 1-(2,3,5-tri-*O*-acetyl-β-D-ribofuranosyl)pyrazolo[3,4-*b*]pyrazine-N^4-oxide (**459a**).[671] Subsequent deacetylation of **459a** furnished 1-β-D-ribofuranosylpyrazolo[3,4-*b*]pyrazine-N^4-oxide (**459b**). On the basis of UV, ^{1}H-NMR, and CD spectra, the oxide function was established to be on the N-4 atom of the pyrazolopyrazine ring.[671]

The site of ribosylation in **458** and **460** was established by a comparison of the UV absorption spectra of these nucleosides with those of the 1- and 2-methyl-pyrazolo[3,4-*b*]pyrazines. The β-anomeric configuration was confirmed by ^{1}H-NMR as well as periodate oxidation and sodium borohydride reduction studies.[669]

2.3.4.2. Synthesis Employing the Direct Alkylation of a Preformed Aglycon with an Acyl Sugar Halide. In addition to the fusion procedure for the glycosylation of pyrazolo[3,4-*b*]pyrazines, the direct alkylation method with acetobromoglucose in nitromethane in the presence of mercuric cyanide has also bee employed.[669] Under these conditions, a 70% yield of a mixture of blocked N-1- and N-2-(β-D-glucopyranosyl)pyrazolo[3,4-*b*]pyrazines in a ratio of 1:3 was obtained. No biological data on these pyrazolo[3,4-*b*]pyrazine nucleosides have been reported.

2.3.5. Pyrazolo[3,4-d]pyrimidine Nucleosides

Since the pyrazolo[3,4-*d*]pyrimidine ring is isomeric with purine, the derivatives of pyrazolopyrimidine exhibit many properties similar to those of the corresponding purines.[672] Because of the known increase in effectiveness of some nucleosides against certain tumor lines in comparison with the corresponding aglycons,[673,674] there has been considerable activity in the synthesis of pyrazolo[3,4-*d*]pyrimidine nucleosides.

2.3.5.1. Synthesis Employing the Halomercuri Salt of a Preformed Aglycon and an Acylated Sugar Halide. The first chemical synthesis of a pyrazolo[3,4-*d*]pyrimidine nucleoside was accomplished by Davoll and Kerridge[675] by the condensation of a chloromercuri derivative of an appropriate pyrazolo[3,4-*d*]pyrimidine with 2,3,5-tri-*O*-benzoyl-D-ribofuranosyl chloride (**47**) in boiling toluene. A condensation of the chloromercuri derivative of 4-benzamido- (**461a**) or 4,6-dibenzamido- (**461b**) pyrazolo[3,4-*d*]pyrimidine with halosugar **47**, followed by debenzoylation with methanolic ammonia, gave a mixture of the positional isomers **462** and **463**. These isomers were separated by fractional crystallization. However, only one isomer of type **462** was obtained from a reaction of the chloromercuri salt of the parent pyrazolo[3,4-*d*]pyrimidine with **47**. The guanosine analogue 6-amino-1-β-D-ribofuranosylpyrazolo[3,4-*d*]pyrimidin-4(5*H*)-one (**466**)

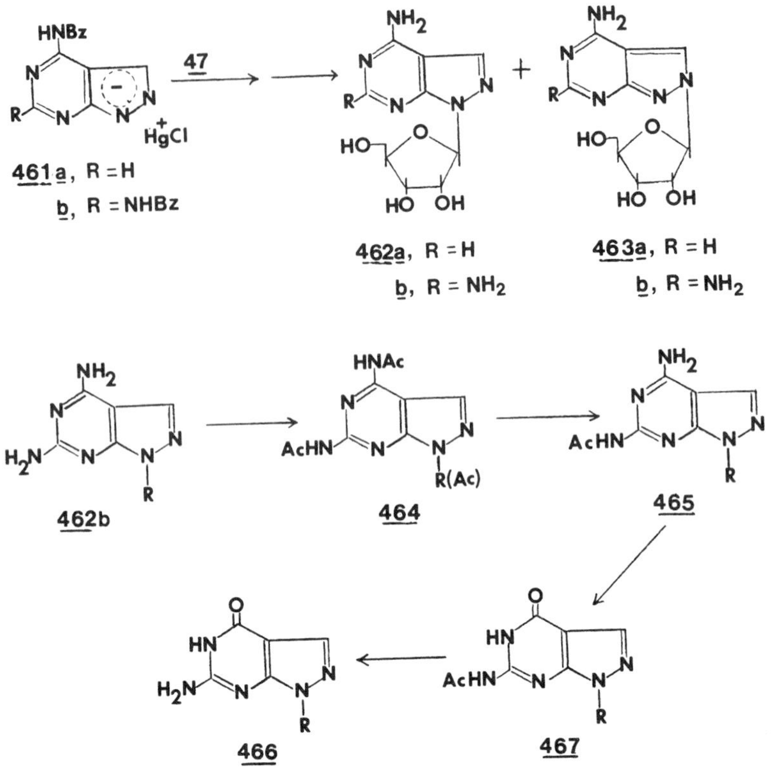

R = β-D-ribofuranosyl

was obtained from **462b** by some conventional transformation reactions.[675] The site of glycosylation of these nucleosides was assigned by a comparison of the UV absorption spectra of the products with those of the corresponding model methyl compounds.[676–678]

2.3.5.2. Synthesis Employing the Acid-Catalyzed Fusion of a Preformed Aglycon and an Acyl Sugar. The fusion of an appropriately substituted pyrazolopyrimidine with a fully protected ribofuranose in the presence of catalysts, such as *p*-toluenesulfonic acid, iodine, and bis(*p*-nitrophenyl)phosphate, has been documented. The ratio of the positional isomers and the anomeric preference depends upon the catalyst selected (**147**). The adenosine analogue 4-amino-1-β-D-ribofuranosylpyrazolo[3,4-*d*]pyrimidine (**469**) was initially prepared[679] by employing the acid-catalyzed fusion procedure. 4-Benzamidopyrazolo-[3,4-*d*]pyrimidine (**468**)[675] was fused with tetra-*O*-acetylribofuranose (**51**) in the presence of *p*-toluenesulfonic acid. This was followed by a deacetylation with sodium methoxide to give a mixture of **469** (12.5% yield) and its N-2 isomer (**470**). The mixture was separated by fractional crystallization.[679] The less soluble (in water) N-1 isomer crystallized first. Although the structures were assigned on the basis of chromatographic behavior, optical rotations, and spec-

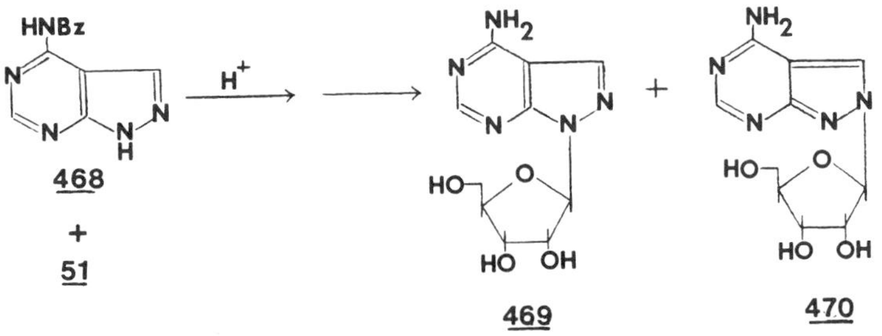

trometric studies, the authors[679] provided no unequivocal proof for the anomeric configuration.

The first chemical synthesis of 4-amino-1-β-D-ribofuranosylpyrazolo[3,-
4-*d*]pyrimidine-3-carbonitrile (6-azatoyocamycin, **476**) and 4-amino-1-β-D-
ribofuranosylpyrazolo[3,4-*d*]pyrimidine-3-carboxamide (6-azasangivamycin,
473) was accomplished by Earl and Townsend[680] using the direct condensation
procedure. Subsequently, the synthesis of these aza analogues of toyocamycin
and sangivamycin has been accomplished by Hecht *et al.*[681] by using the acid-
catalyzed fusion procedure. The fusion of 4-acetamidopyrazolo[3,4-*d*]pyrimi-
dine-3-carbonitrile (**471**, R = CH$_3$) with **51** in the presence of *p*-toluenesulfonic
acid gave predominantly 4-acetamido-1-(2,3,5-tri-*O*-acetyl-β-D-ribofuranosyl)-
pyrazolo[3,4-*d*]pyrimidine-3-carbonitrile (**472**). Heating **472** with concentrated
ammonium hydroxide effected a deacylation with a concomitant hydration of
the nitrile function to afford 6-azasangivamycin (**473**). A similar fusion of
4-trifluoroacetamidopyrazolo[3,4-*d*]pyrimidine-3-carbonitrile (**471**, R = CF$_3$)
with **51** provided 4-amino-1-(2,3,5-tri-*O*-acetyl-β-D-ribofuranosyl)pyrazolo[3,-

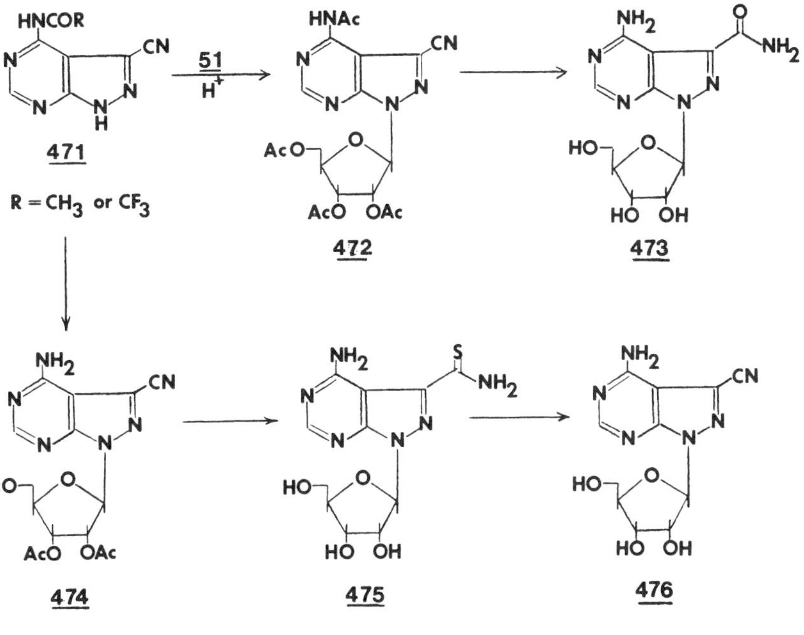

4-*d*]pyrimidine-3-carbonitrile (**474**)[681] in a 54% yield. Since the cyano group in **474** was too reactive to permit a successful removal of the protecting *O*-acyl groups, **474** was converted to 6-azatoyocamycin (**476**) via the intermediate 4-amino-1-β-D-ribofuranosylpyrazolo[3,4-*d*]pyrimidine-3-thiocarboxamide (**475**).[680]

The acid-catalyzed fusion procedure was also successful[682] with 4-methyl-thiopyrazolo[3,4-*d*]pyrimidine (**477**). Compound **477**[683] was fused with **51** in the presence of iodine or bis(*p*-nitrophenyl)phosphate to give 1-(2,3,5-tri-*O*-acetyl-β-D-ribofuranosyl)-4-methylthiopyrazolo[3,4-*d*]pyrimidine (**478**) as the major product.[684] Treatment of **478** with methanolic ammonia, under mild conditions (20°C, 18 h), gave 4-methylthio-1-β-D-ribofuranosylpyrazolo[3,4-*d*]-pyrimidine (**479a**). However, under drastic conditions (150°C, 7 h), the adenosine analogue **469** was obtained. A similar acid-catalyzed [bis(*p*-nitrophenyl)phos-phate] fusion of **477** with 1-*O*-acetyl-2,3,5-tri-*O*-benzoyl-α-D-xylofuranose or with 1-*O*-acetyl-2,3,5-tri-*O*-benzoyl-D-arabinofuranose gave predominantly the N-1-glycosylated products.[685] Further treatment of these protected nucleosides with methanolic ammonia under controlled temperature furnished a route to the preparation of 1-β-D-xylofuranosyl- and 1-α-D-arabinofuranosyl-4-methyl-thio- and 4-aminopyrazolo[3,4-*d*]pyrimidines.[685] Compound **479a**, as well as 1-β-D-ribofuranosylpyrazolo[3,4-*d*]pyrimidine-4-thione (**479b**), prepared[686] from allopurinol ribofuranoside, proved to be useful intermediates for the synthesis of other 4-substituted 1-β-D-ribofuranosylpyrazolo[3,4-*d*]pyrimidines by nucleophilic displacement reactions.[172,687,688]

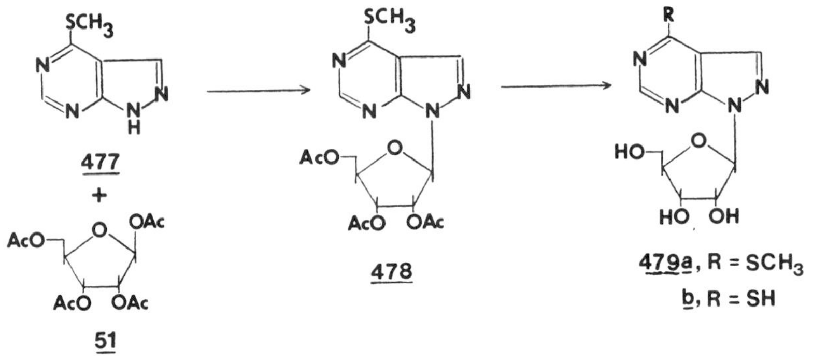

This fusion procedure was later extended to the preparation of various 4- and 4,6-disubstituted pyrazolo[3,4-*d*]pyrimidine-3-carbonitrile nucleosides (**481**).[689–693] The fusion with tetra-*O*-acetyl-β-D-ribofuranose (**51**) was carried out at 170–180°C in the presence of 15–20% of bis(*p*-nitrophenyl)phosphate or iodine and proceeded with high regioselectivity to give **481** in 70–75% yield. The N-2 (**482**) and N-1-α (α-**481**) isomers were also isolated as minor prod-ucts.[689–693] It was also observed that the N-2 isomers (**482**) were converted to the N-1 isomers (**481**) by heating in the presence of iodine. The nitrile function in **481** was converted to $CSNH_2$, $C(NOH)NH_2$, $C(NH)OCH_3$, and other groups.[18,692] A similar fusion of 6-methyl-5-phenyl- (or *o*-methoxyphenyl-)pyrazolo[3,4-*d*]-pyrimidin-4-one with **51** also gave the corresponding N-1 and N-2 glycosides.[694]

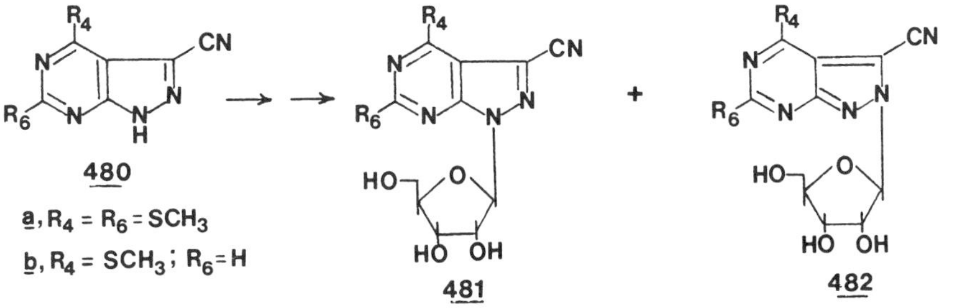

480

a, $R_4 = R_6 = SCH_3$

b, $R_4 = SCH_3$; $R_6 = H$

481

482

2.3.5.3. Synthesis Employing the Direct Condensation of a Preformed Aglycon and an Acyl Sugar Halide. The synthesis of 3,4-disubstituted pyrazolo[3,4-*d*]pyrimidine nucleosides related to the nucleoside antibiotics toyocamycin and sangivamycin was first reported by Earl and Townsend.[680] 4-Acetamidopyrazolo[3,4-*d*]pyrimidine-3-carbonitrile (**471**, R = CH$_3$) was condensed with crystalline 2,3,5-tri-*O*-acetyl-β-D-ribofuranosyl chloride (**483**)[695] in boiling nitromethane in the presence of KCN (as the acid acceptor[71]) to provide a 56% yield of **472**. Treatment of **472** with sodium methoxide in methanol resulted in the formation of crystalline methyl 4-amino-1-β-D-ribofuranosylpyrazolo[3,4-*d*]pyrimidine-3-formimidate (**484a**) in good yield. The cyano group of **472**, as well as the methyl formimidate function of **484a**, was shown[680] to be highly reactive toward nucleophilic substitution displacements to afford a variety of 3-substituted 4-aminopyrazolo[3,4-*d*]pyrimidine nucleosides. Thus, the reaction of **472** with liquid ammonia at room temperature produced a good yield of 4-amino-1-β-D-ribofuranosylpyrazolo[3,4-*d*]pyrimidine-3-carboxamidine (**484b**), which was also obtained from **484a** and liquid ammonia. Both hydroxylamine and hydrazine hydrate reacted very readily with **484a** to give the corresponding 3-carboxamidoxime and 3-carboxamidrazone derivatives (**484c** and **484d**, respectively). Sodium hydrosulfide reacted with **484a** to give **475**, which, on subsequent treatment with mercuric chloride and triethylamine in DMF, furnished the 6-azatoyocamycin analogue (**476**).[680] Treatment of **484a** with aqueous sodium hydroxide gave the 6-azasangivamycin analogue (**473**). Conver-

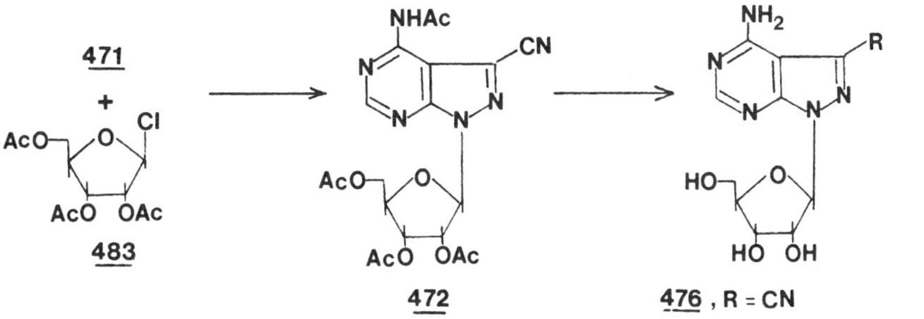

471

+

483

472

476, R = CN

484a, R = C(NH)OCH$_3$

b, R = C(NH)NH$_2$

c, R = C(NH)NOH

d, R = C(NH)NHNH$_2$

e, R = COOH

sion of **473** to 4-amino-1-β-D-ribofuranosylpyrazolo[3,4-*d*]pyrimidine (**469**), via the carboxylic acid derivative (**484e**), established the structural assignment of these nucleosides.

A high-temperature glycosylation procedure has been developed[696–698] recently for the synthesis of 3,4-di- and 3,4,6-trisubstituted pyrazolo[3,4-*d*]-pyrimidine ribonucleosides. This procedure involves the direct condensation of a preformed aglycon with the fully protected sugar, in the presence of the catalyst boron trifluoride diethyl etherate in a boiling polar aprotic solvent such as nitromethane or benzonitrile, for a short period of time. Under these conditions, the formation of an N-5 glycosyl isomer was not observed, and only a minor amount of the N-2 isomer was detected. Table II summarizes the reaction conditions and product distribution of substituted pyrazolopyrimidine nucleosides.

This high-temperature glycosylation procedure has made available certain biologically significant 3-substituted allopurinol ribonucleosides (**487** and **489**), 3-substituted 4-aminopyrazolo[3,4-*d*]pyrimidine nucleosides (**488**),[696] and 3-alkoxy-6-amino-1-β-D-ribofuranosylpyrazolo[3,4-*d*]pyrimidin-4(5*H*)-ones.[698] This procedure was also found to be useful for the preparation of 6-azacadeguoymcin.[697] Oxidation of **486g** with *m*-chloroperoxybenzoic acid gave the corresponding methylsulfonyl derivative (**490**). Treatment of **490** with ammonia at an elevated temperature afforded 6-amino-1-β-D-ribofuranosyl-4(5*H*)-oxopyrazolo[3,4-*d*]pyrimidine-3-carboxamide (**492**). When **492** was heated in 6 *N* NaOH under gentle reflux, 6-azacadeguomycin [6-amino-1-β-D-ribo-furanosyl-4(5*H*)-oxopyrazolo[3,4-*d*]pyrimidine-3-carboxylic acid, (**491**)] was produced.[697]

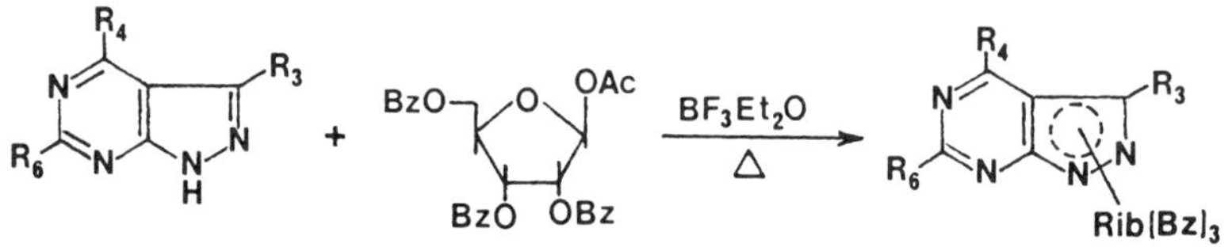

Table II. Reaction Conditions and Product Distribution
in High-Temperature Glycosylation of Substituted Pyrazolopyrimidines[a]

Compound 485	Substituents			Solvent	Time of reaction	% Yield	Isomeric ratio of 486	
	R_3	R_4	R_6				N-1	N-2
a	Br	OH	H	CH_3NO_2	1.5 h	45	100	0
b	Br	NH_2	H	CH_3NO_2	1.5 h	87	100	0
c	Br	OH	Br	CH_3NO_2	20 min	54	100	0
d	H	OH	SCH_3	CH_3NO_2	1.0 h	65	90	10
e	Br	OH	NH_2	CH_3NO_2	1.0 h	63	65	35
f	Br	OH	$SOCH_3$	CH_3NO_2	30 min	40	55	45
g	$CONH_2$	OH	SCH_3	CH_3NO_2	30 min	66	100	0
h	CN	OH	H	C_6H_5CN	15 min	52	80	20

[a]Reference 696.

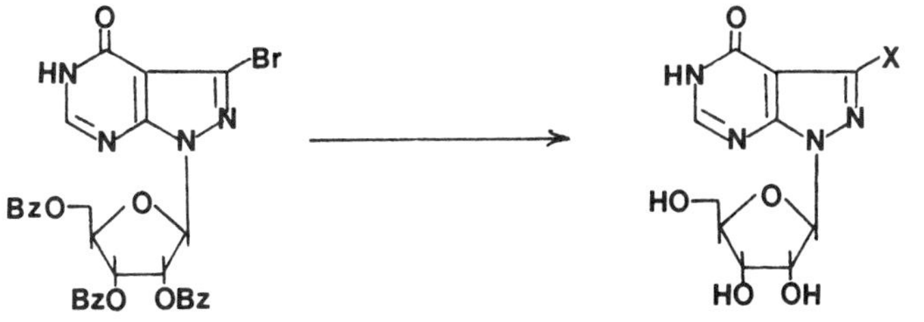

486a

487**a**, X = NH₂

b, X = NHMe

c, X = N(Me₂)₂

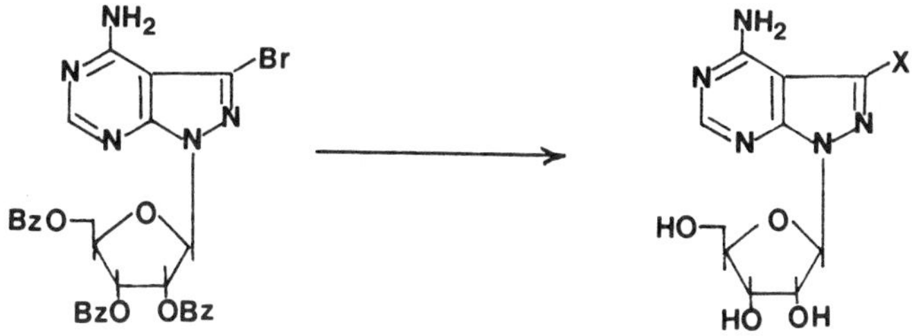

486b

488**a**, X = Br

b, X = NH₂

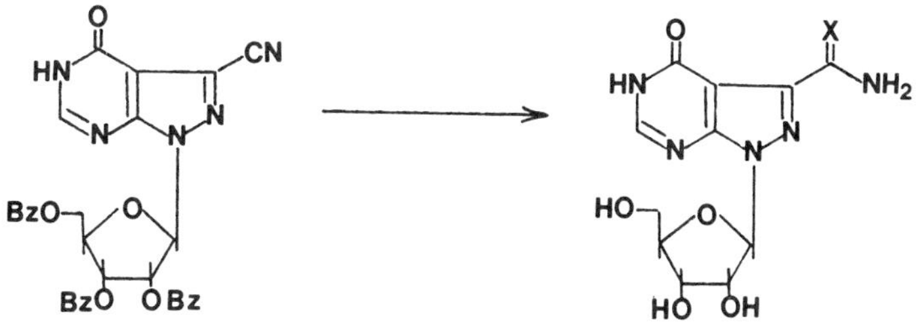

486h

489**a**, X = O; **b**, X = S

c, X = NH; **d**, X = NOH

It has been observed that when the glycosylation is carried out by this procedure at low temperature, the N-2 glycosyl isomer predominates but can be eventually converted to the thermodynamically more stable N-1 isomer at high temperature.[697] This thermal isomerization is analogous to that catalyzed by iodine.[692]

2.3.5.4. Synthesis Employing the Trimethylsilyl Derivative of a Preformed Aglycon and an Acylated Sugar Halide. Revankar and Townsend[699] were the first to use the trimethylsilyl procedure successfully for the synthesis of pyrazolo[3,4-*d*]pyrimidine nucleosides. A sodium iodide catalyzed fusion of trimethylsilyl-4-chloropyrazolo[3,4-*d*]pyrimidine (**493**) with tri-*O*-acetyl-D-ribofuranosyl bromide (**145**), followed by deacetylation under mild conditions, gave 4-chloro-1-β-D-ribofuranosylpyrazolo[3,4-*d*]pyrimidine (**494**). Treatment of **494** with methanolic ammo-

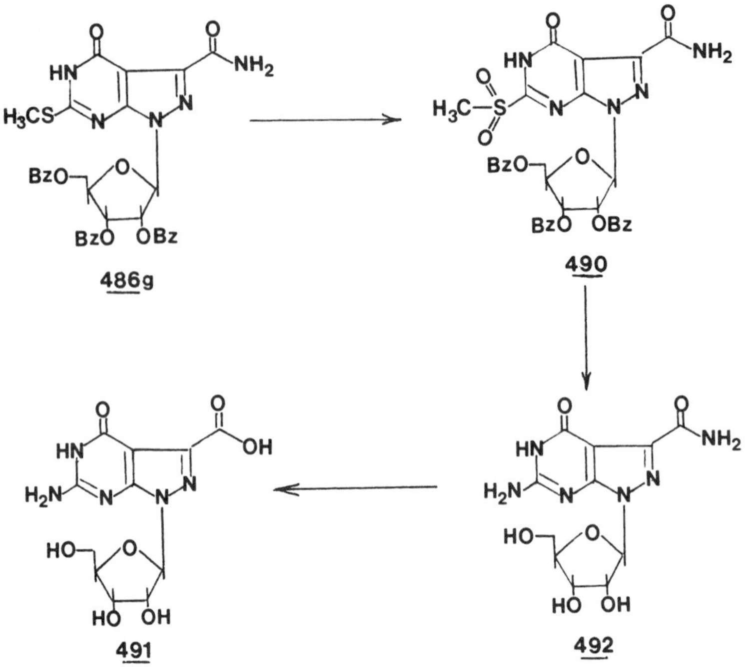

486g → 490

491 ← 492

nia at elevated temperature and pressure furnished the corresponding 4-amino derivative (**469**), identical in all respects with **469** prepared by the acid-catalyzed fusion procedure.[679] The evidence for the β-anomeric configuration of **469** was obtained for the first time[699] by the intramolecular quaternization of the 5'-*O*-tosylate of **469** to afford the corresponding N-7→C-5' cyclonucleoside.

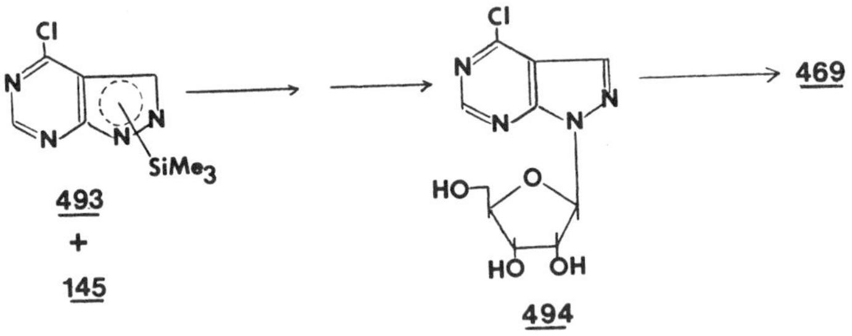

493 + 145 → 494 → 469

Extension of this trimethylsilyl procedure using a Lewis acid catalyst was found to be much more rewarding. The feasibility of such a reaction was demonstrated by Steinmaus[700–702] by the conversion of pyrazolo[3,4-*d*]pyrimidin-4-one (allopurinol) and pyrazolo[3,4-*d*]pyrimidine-4,6-dione (oxoallopurinol) to their corresponding ribonucleosides. A condensation of the tris-trimethylsilyl derivative of pyrazolo[3,4-*d*]pyrimidine-4,6-dione (**495**) with 1,2,3,5-tetra-*O*-acetyl-β-D-ribofuranose (**51**) in the presence of stannic chloride, followed by deacetylation, furnished 7-β-D-ribofuranosylpyrazolo[3,4-*d*]pyrimi-

495 + 51 → 496

dine-4,6-dione (**496**).[700,701] This nucleoside was identical with the one prepared by an enzymatic synthesis.[703] A similar reaction of bis(trimethylsilyl)pyrazolo-[3,4-*d*]pyrimidin-4-one (**497**) with **51** in the presence of boron trifluoride diethyl etherate afforded allopurinol riboside along with several other products.[702,704]

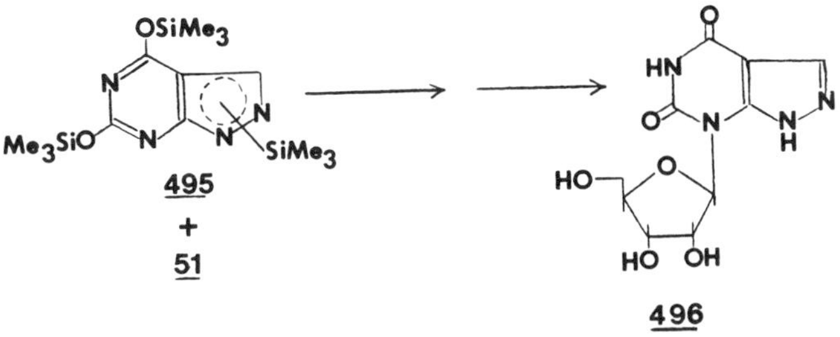

However, a less uniform course was observed[705,706] in the ribosylation of **497** in the presence of stannic chloride. The reaction of crystalline **497** with **57** in the presence of stannic chloride (1 molar equivalent) in 1,2-dichloroethane gave a mixture of N-1 (**498**), N-2 (**500**), and N-5 (**502**) isomers, as well as the 1,5- (**499**) and 2,5- (**501**) bis-ribosylated derivatives. The yield of the desired blocked N-1 isomer under the best of conditions was 17%. A similar reaction catalyzed by trimethylsilyl trifluoromethanesulfonate (TMS-triflate) gave the O^4-glycosylated product (22%), along with **498**, **502**, and trace amount of bis-ribosides.[706] The use of **51** gave an identical product distribution.[707] Deprotection of these blocked nucleosides with sodium methoxide afforded the corresponding free allopurinol ribonucleosides.

a series, R = Bz

b series, R = Ac   Rib = β-D-ribofuranosyl

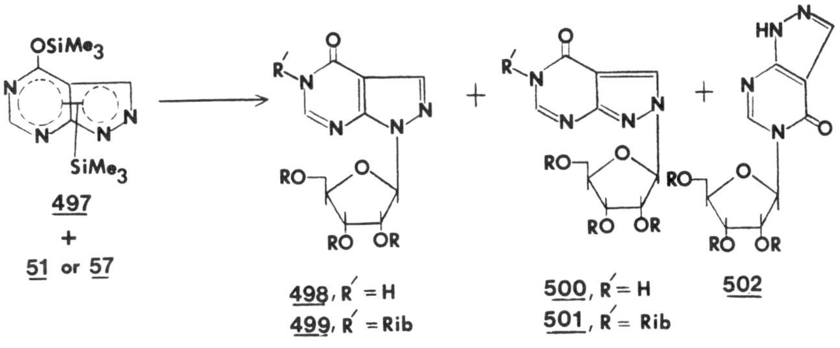

By using the catalyst TMS-triflate,[708] a number of 3,4- and 4,6-disubstituted pyrazolo[3,4-*d*]pyrimidine nucleosides have been recently prepared.[696,709,710] Glycosylation of the trimethylsilyl derivative of 3-bromoallopurinol (**485a**) with **57** in the presence of TMS-triflate produced a major product within the first few hours of the reaction, presumed to be the N-5 glycosyl isomer, which, upon heating for a week, rearranged to the thermodynamically more stable **486a**.[696] The isolated yield of **486a** was 29%. A similar glycosylation of the trimethylsilyl

derivative of 3-cyanoallopurinol (**485h**) or 3-methylthioallopurinol gave the
corresponding blocked N-1 ribonucleosides.[696]

The glycosylation of 4,6-dichloropyrazolo[3,4-*d*]pyrimidine and 4-chloro-
6-methylthiopyrazolo[3,4-*d*]pyrimidine via the corresponding trimethylsilyl intermediate (**503a,b**) with **51**, in the presence of TMS-triflate, gave exclusively the
N-1 glycosyl isomer (**504**).[709] The intermediates **504a** and **504b** provided convenient synthetic routes to the adenosine (**469**), guanosine (**466**), and isoguanosine
(**506**) analogues.[709] The treatment of **504a** with methanolic ammonia gave
4-amino-6-chloro-1-β-D-ribofuranosylpyrazolo[3,4-*d*]pyrimidine (**505a**), which,
on catalytic dehalogenation, afforded **469**. The reaction of **504a** with hot
aqueous sodium hydroxide in methanol furnished the intermediate 6-chloro-
allopurinol riboside, which, on subsequent ammonolysis, gave the guanosine
analogue **466**. Similarly, treatment of **504b** with methanolic ammonia gave
4-amino-6-methylthio-1-β-D-ribofuranosylpyrazolo[3,4-*d*]pyrimidine (**505b**).
Oxidation of **505b** with *m*-chloroperoxybenzoic acid and subsequent treatment
with aqueous sodium hydroxide gave the isoguanosine analogue (**506**).[709] It is
of interest to note that the glycosylation of the trimethylsilyl derivative of
6-chloroallopurinol with 2,3,5-tri-*O*-acetyl-D-ribofuranosyl bromide gave the
N-2 glycosyl isomer as the major product.[709]

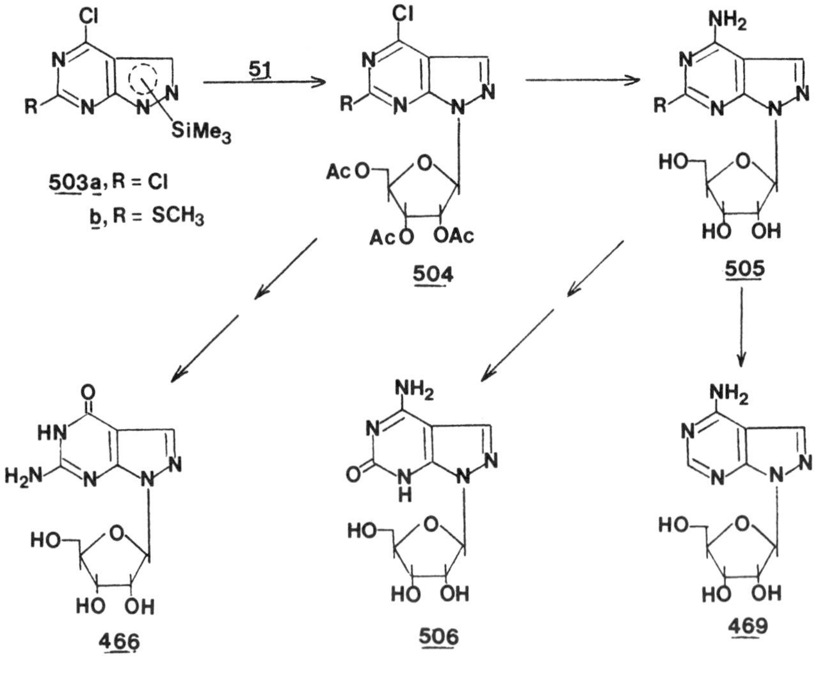

Application of the trimethylsilyl glycosylation procedure to 4-chloro-6-
methylpyrazolo[3,4-*d*]pyrimidine (**507**) gave 4-chloro-6-methyl-1-(2,3,5-tri-*O*-
acetyl-β-D-ribofuranosyl)pyrazolo[3,4-*d*]pyrimidine (**508**) as the only nucleoside product.[710] Ammonolysis of **508** with methanolic ammonia gave 4-amino-
6-methyl-1-β-D-ribofuranosylpyrazolo[3,4-*d*]pyrimidine (**509**). When **508** was
treated with 1 *N* sodium hydroxide in aqueous dioxane, 6-methylallopurinol

507 508 509

510a, X = O
b, X = S
c, X = Se

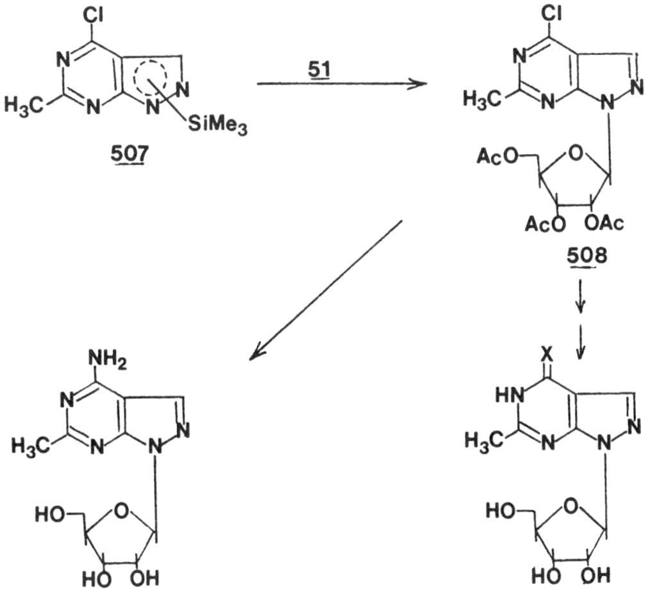

ribonucleoside (**510a**) was obtained. The reaction of **508** with thiourea and selenourea, followed by deacetylation, afforded 6-methylpyrazolo[3,4-*d*]pyrimidine-4(5*H*)-thione ribonucleoside (**510b**) and 6-methyl-1-β-D-ribofuranosylpyrazolo[3,4-*d*]pyrimidine-4(5*H*)-selone (**510c**), respectively.[710]

The 4-chloro group in pyrazolo[3,4-*d*]pyrimidine nucleosides was found to be amenable to nucleophilic displacements, and this provided a convenient route to some biologically significant 4-alkylamino-1-β-D-ribofuranosylpyrazolo[3,4-*d*]pyrimidines.[704,711–714] The 4-methylthio substituent of **478** was also susceptible to nucleophilic displacements, providing a pathway to 1-β-D-ribofuranosylpyrazolo[3,4-*d*]pyrimidine-4-carboxamide (**513**).[402] Oxidation of **478** with *m*-chloroperoxybenzoic acid provided the methyl sulfone (**511**), which, on treatment with sodium cyanide in DMF, gave 1-(2,3,5-tri-*O*-acetyl-β-D-ribofuranosyl)pyrazolo[3,4-*d*]pyrimidine-4-carbonitrile (**512**). Treatment of **512** with ammonium hydroxide containing hydrogen peroxide provided the desired **513**.[402]

511 512 513

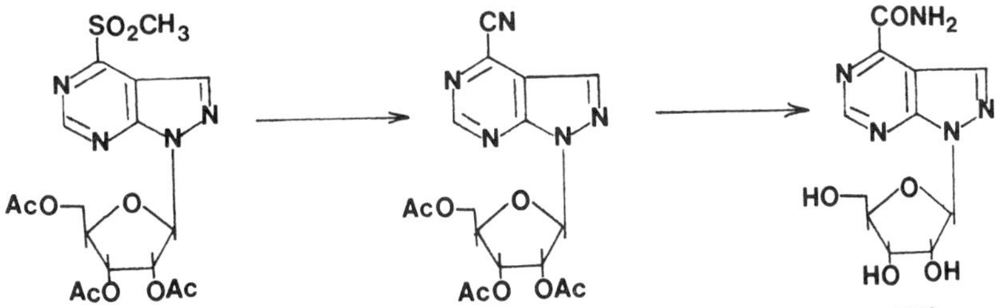

2.3.5.5. Synthesis Employing the Ring Closure of a Pyrazole Nucleoside Precursor. A route for the preparation of pyrazolo[3,4-*d*]pyrimidine nucleosides via the ring annulation of certain pyrazole nucleoside precursors has been described.[715] The crystalline bis-trimethylsilyl derivative (**514**) of 3-(3,3-di-methyl-1-triazeno)pyrazole-4-carboxamide[716,717] was reacted with the bromo-acetyl sugar **145** in anhydrous acetonitrile to obtain an isomeric mixture of two blocked nucleosides. Deacetylation of these compounds afforded 1-β-D-ribo-furanosyl-5-(3,3-dimethyl-1-triazeno)pyrazole-4-carboxamide (**515**) and its 3-(3,3-dimethyl-1-triazeno) isomer (**517**). Catalytic hydrogenation of **515** with Raney nickel, followed by ring annulation of the intermediate (**516**)[718] with diethoxymethyl acetate, gave 1-β-D-ribofuranosylpyrazolo[3,4-*d*]pyrimidin-4 (5*H*)-one (**498**) of established structure.[706] Similarly, **517** was converted to 2-β-D-ribofuranosylpyrazolo[3,4-*d*]pyrimidin-4(5*H*)-one (**500**).

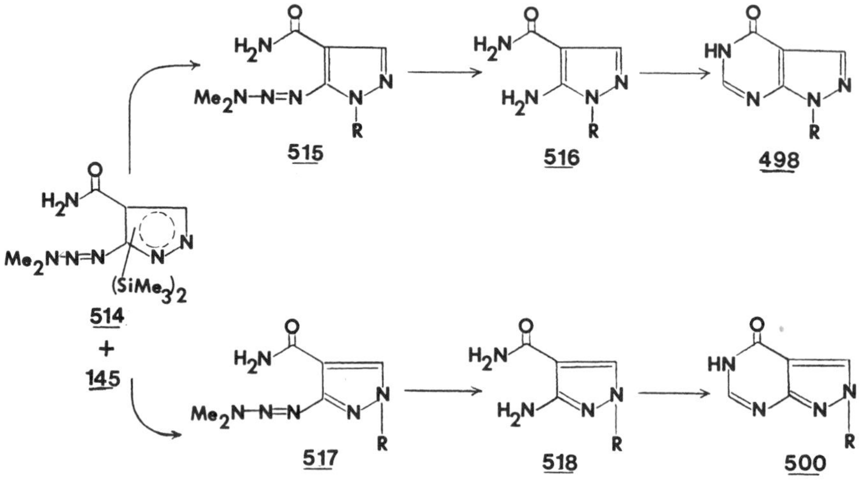

R = β-D-ribofuranosyl

A rather unique method for the specific formation of pyrazolo[3,4-*d*]pyri-midine nucleosides was developed by Schmidt and co-workers.[719–722] This technique involved a synthesis of the viable intermediates **521** and **522** from hydrazinoribose. 2,3-*O*-Isopropylidene-D-ribose was transformed quantitatively into hydrazine derivative **519** by treatment with hydrazine. It was shown[719] that the hydrazine derivative equilibrates to a mixture of **519** and **520**, in which **520** predominates (~90%). A thermodynamically controlled condensation of **519** with an ethoxymethyleneacrylamide derivative[723] gave the pyrazole nucleoside **521** in a 58% yield. Reaction of **521** (R = H) with ethyl formate gave the 2′,3′-*O*-isopropylideneallopurinol riboside (85% yield), which, in the presence of an acid, afforded the allopurinol riboside **498**. Likewise, 2,3,5-tri-*O*-benzyl-D-ribose was reacted with hydrazine to give 1-hydrazino-2,3,5-tri-*O*-benzyl-D-ribose.[722] The latter compound or **519** with ethoxymethylene malonodinitrile and subsequent debenzylation gave 4-amino-1-β-D-ribofuranosylpyrazolo[3,4-*d*]pyrimidine (**469**).[722]

Similar methodology was used to synthesize stereoselectively the 1-(*N*-

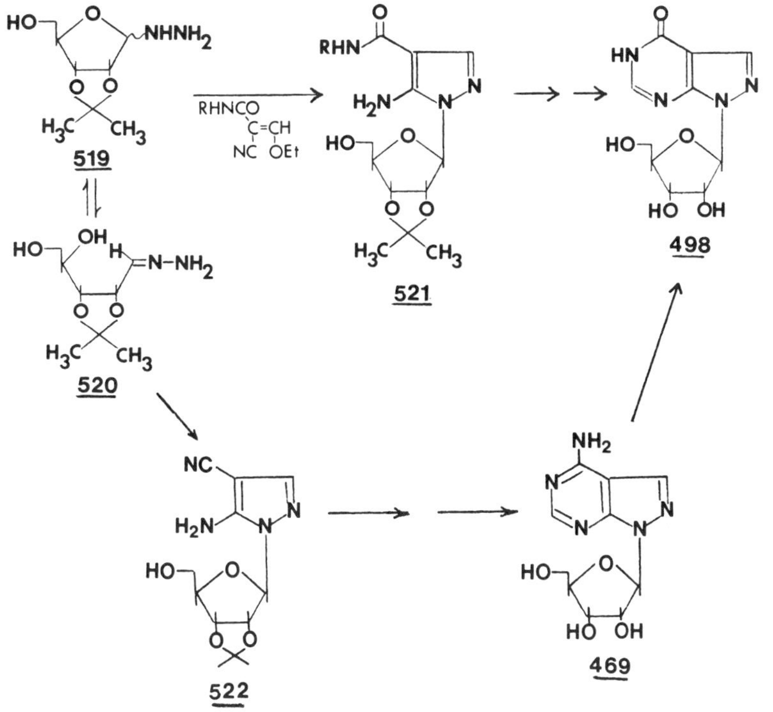

alkyl-β-D-ribofuranuronamidosyl) derivatives of allopurinol and 4-aminopyr-azolo[3,4-*d*]pyrimidine (**524**) from the riburonamide hydrazones (**523**).[724] Likewise, 1-β-D-glucopyranosyl derivatives of allopurinol and 4-aminopyr-azolo[3,4-*d*]pyrimidine have been prepared.[725] However, application of this technique to 2,3,5-tri-*O*-benzyl-D-arabinose hydrazone gave only the 1-α-D-arabinofuranosyl derivatives of allopurinol and 4-aminopyrazolo[3,4-*d*]pyrimi-dine.[726] Although some confidence can be placed in the positional and ano-meric structures of the nucleosides formed, this procedure is lengthy and can be accomplished on a rather limited scale.

2.3.5.6. Synthesis Employing the Alkali Metal Salt of a Preformed Aglycon and a Protected Sugar. The synthesis of pyrazolo[3,4-*d*]pyrimidine nucleosides em-ploying the alkali metal salt of a preformed aglycon has been disclosed in a

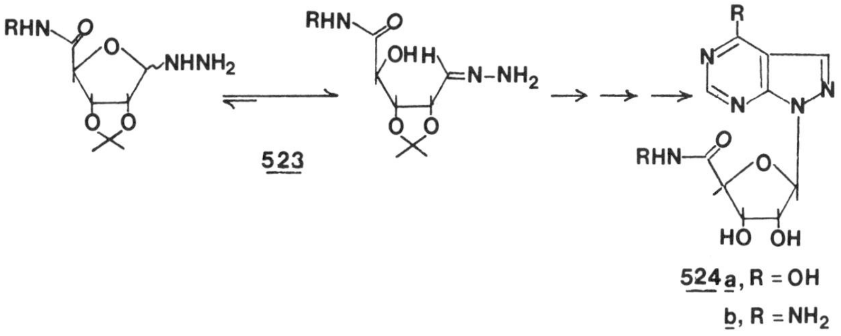

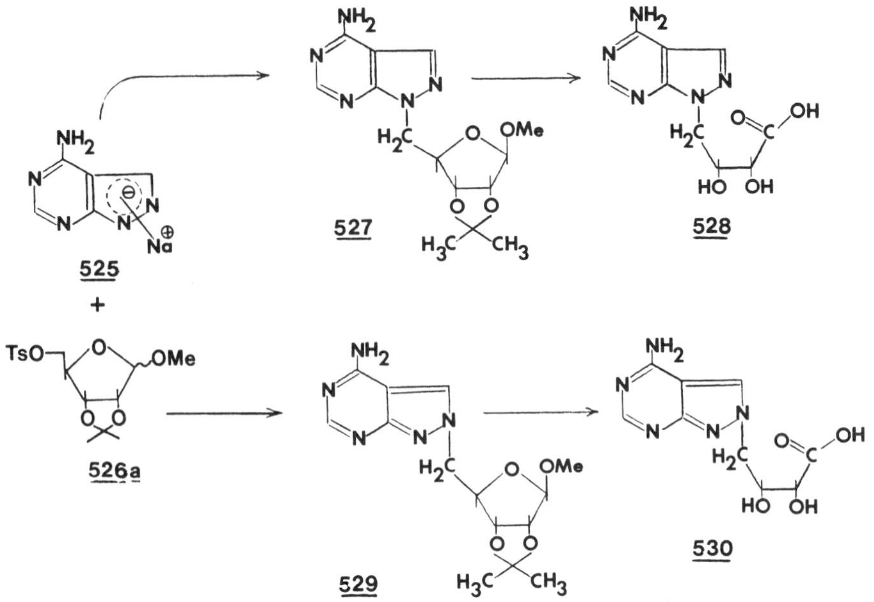

Japanese patent.[727] The sodium salt of 4-aminopyrazolo[3,4-d]pyrimidine (**525**) was condensed with methyl 2,3-O-isopropylidene-5-O-tosyl-D-ribofurano- side (**526a**) in DMF at 90°C to give an isomeric mixture of two protected nucleosides, **527** (minor) and **529** (major). Acid hydrolysis of **527** and **529**, followed by oxidation, furnished 4-(4-amino-1*H*- or -2*H*-pyrazolo[3,4-d]pyri- midin-1-yl or -2-yl)-D-*erythro*-2,3-dihydroxybutyric acid (**528** and **530**).[727]

Similarly, when 3-aminopyrazole-4-carbonitrile (**531**) was treated with methyl 2,3-O-isopropylidene-5-O-tosyl- (or 5-deoxy-5-chloro-) D-ribofurano- side (**526**) in DMF containing an alkali metal hydride, a mixture of N-1 and N-2 glycosylpyrazoles (**532** and **534**) was obtained.[728] These positional isomers were separated by taking advantage of their different solubilities. Treatment of **532** and **534** with an appropriate one-carbon source such as formamide, for- mamidine, or triethyl orthoformate furnished the cyclized products **533** and **535**, respectively.[728] However, glycosylation of the sodium salt of preformed 4-chloropyrazolo[3,4-d]pyrimidine (**536**) with **24** gave 4-chloro-1-(2-deoxy-3,5- di-O-p-toluoyl-β-D-*erythro*-pentofuranosyl)pyrazolo[3,4-d]pyrimidine (**537**) as the major product, along with a minor amount of the N-2 positional isomer. Treatment of **537** with methanolic ammonia afforded 4-amino-1-(2-deoxy- β-D-*erythro*-pentofuranosyl)pyrazolo[3,4-d]pyrimidine (**538**).[31] This simple stereospecific attachment of the 2-deoxy-β-D-ribofuranosyl moiety makes 2'-de- oxypyrazolo[3,4-d]pyrimidine nucleosides readily available.

2.3.5.7. Synthesis Employing the Ring Annulation of Glycosyl Isothiocyanates. A convenient route for the synthesis of 5-glycosylpyrazolo[3,4-d]pyrimidines by the ring annulation of glycosyl isothiocyanates has been reported.[729] Treatment of **539a–c** with 3-aminopyrazole-4-carboxylic acid in the presence of zinc chloride gave the corresponding 5-glycosylpyrazolo[3,4-d]pyrimidines (**540**)[729] in excellent yields. Similarly, a reaction of the glycosyl isothiocyanates (**539**) with 6-hydrazino-1,3-dimethyluracil in acetonitrile afforded *N*-glycosyl-2-(1,3-dimethyl- 2,4-dioxopyrimidin-6-yl)hydrazine thiocarboxamides (**542**).[730,731] The reaction was regioselective, and no other isomeric product was formed. Ring closure of **542** with *N*-bromosuccinimide in methanol or chloroform proceeded well to give 2-glycosylamino-6,8-dioxopyrimido[4,5-*e*]-1,3,4-thiadiazines (**543**) in good yields. Ring contraction of **543** by boiling in toluene readily occurred to give 3-glycosylaminopyrazolo[3,4-*d*]pyrimidine-4,6-diones (**541**).[730,731]

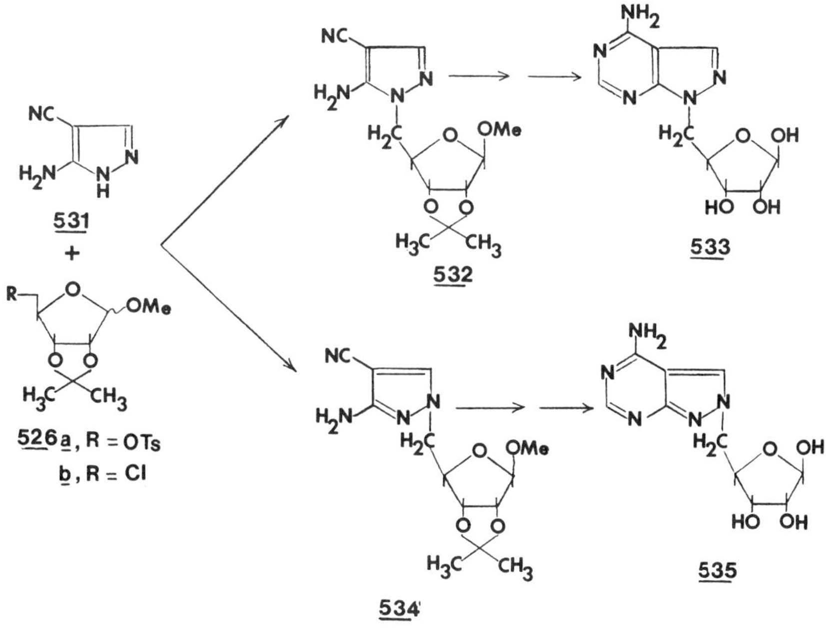

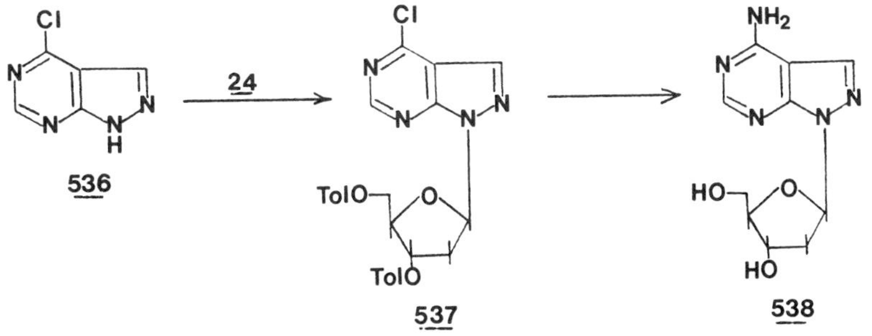

2.3.5.8. Synthesis of Pyrazolo[3,4-d]pyrimidine Nucleosides by the Phase-Transfer Glycosylation Procedure. Utilization of the phase-transfer glycosylation procedure for the preparation of 2'-deoxyribonucleosides of pyrazolo[3,4-*d*]pyrimidines was found to be very successful. The phase-transfer glycosylation was carried out in a biphasic mixture of dichloromethane and 50% aqueous sodium hydroxide solution containing the catalyst Bu_4NHSO_4. Thus, a reaction of

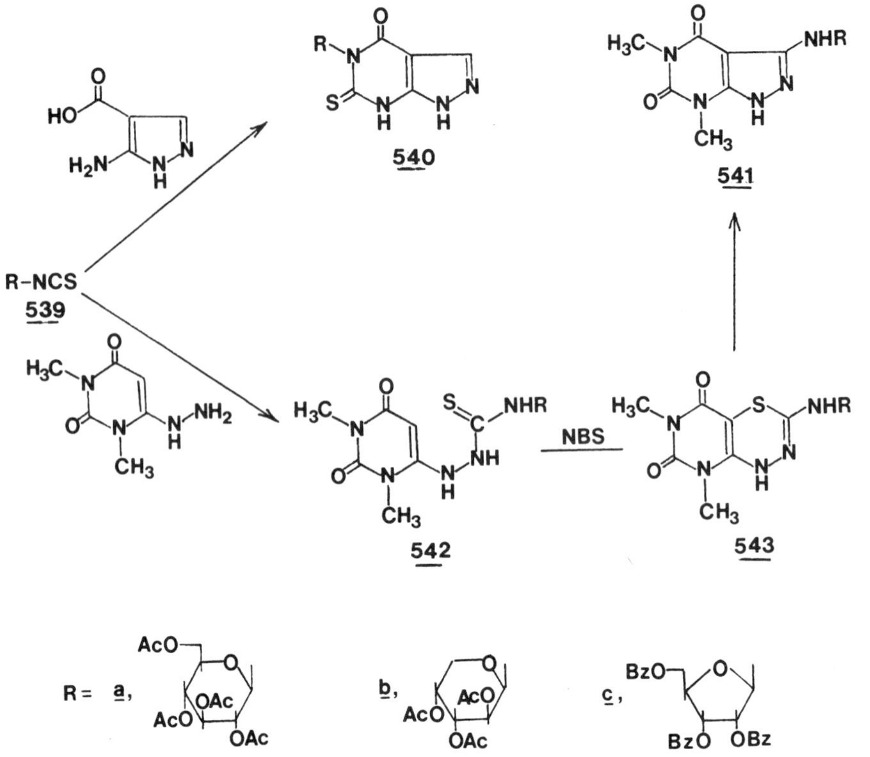

4-methoxypyrazolo[3,4-*d*]pyrimidine (**544a**) with the α-halogenose **24** under the phase-transfer conditions gave a mixture of three nucleoside products.[732] After separation, these nucleosides were identified as 4-methoxy-1-(2-deoxy-3,5-di-*O*-*p*-toluoyl-β-D-*erythro*-pentofuranosyl)pyrazolo[3,4-*d*]pyrimidine (**545a**, 39% yield), the corresponding N-2 glycosyl isomer (**547a**, 18% yield), and the α-anomer of **545a** (7% yield). Treatment of **545a** and **547a** with sodium methoxide in methanol gave the deprotected nucleosides **546a** and **548a**, respectively. Further treatment of **546a** and **548a** with 2 *N* NaOH for 2 h furnished allopurinol 2'-deoxyribonucleoside (**550a**) and the corresponding N-2 glycosyl isomer **551a**.[732] By replacing the aqueous NaOH by NH_4OH, compound **546a** was converted into **538**.[732]

A similar phase-transfer glycosylation of 6-amino-4-methoxypyrazolo-[3,4-*d*]pyrimidine (**544b**) with the halogenose **24** in a biphasic mixture (CH_2Cl_2/50% aqueous NaOH, benzyltriethylammonium chloride) gave **545b** in a 47% yield.[733,734] The N-2 isomer (**547b**) and the α-anomer were formed in low yield. Deprotection of **545b** and **547b** was accomplished with $NaOCH_3$ to give **546b** and **548b**, respectively. Nucleophilic displacement (2 *N* KOH) of the methoxy group of these nucleosides afforded 6-amino-1-(2-deoxy-β-D-*erythro*-pentofuranosyl)pyrazolo[3,4-*d*]pyrimidin-4(5*H*)-one (**550b**) and the N-2 glycosyl isomer **551b**, respectively. Ammonolysis of **546b** readily gave 4,6-diamino-1-(2-deoxy-β-D-*erythro*-pentofuranosyl)pyrazolo[3,4-*d*]pyrimidine (**549**).[735]

However, application of the regular conditions of phase-transfer glycosylation, by using dichloromethane and 50% NaOH in the presence of tetrabutyl-

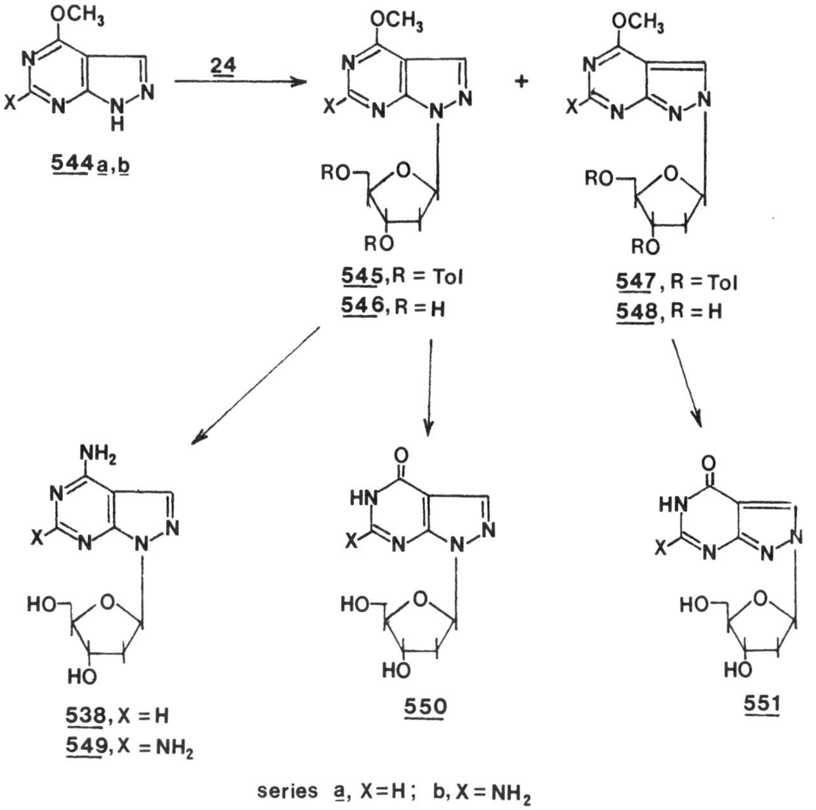

544a,b → 545, R = Tol / 546, R = H + 547, R = Tol / 548, R = H

538, X = H / 549, X = NH₂   550   551

series a, X = H ; b, X = NH₂

ammonium hydrogen sulfate, to 4-chloropyrazolo[3,4-*d*]pyrimidine (**536**) failed.[736] This failure may be due to the instability of the 4-chloro group. When solid potassium carbonate was used instead of aqueous NaOH in the absence of a phase-transfer catalyst, the N-1 (**537**) and the N-2 glycosylation products were formed. In contrast to the phase-transfer glycosylation of 4-methoxypyrazolo[3,4-*d*]pyrimidine (**544a**), the reaction of **536** with the halogenose **24** did not lead to an α-anomer.[736] The protected nucleoside **537** was converted in two steps to 1-(2-deoxy-β-D-*erythro*-pentofuranosyl)pyrazolo[3,4-*d*]pyrimidine-4(5*H*)-thione (**554**)[736] by the reaction with thiourea in methanol. Similarly, 6-amino-4-chloropyrazolo[3,4-*d*]pyrimidine (**552**) was reacted with **24** to yield

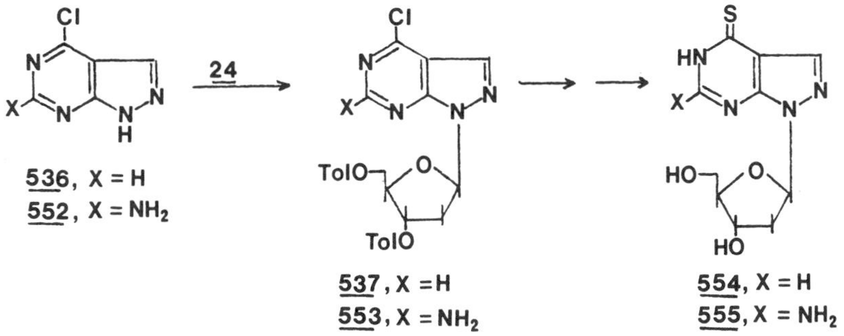

536, X = H / 552, X = NH₂   537, X = H / 553, X = NH₂   554, X = H / 555, X = NH₂

553, which was subsequently converted into 6-amino-1-(2-deoxy-β-D-*erythro*-pentofuranosyl)pyrazolo[3,4-*d*]pyrimidine-4(5*H*)-thione (**555**).[734]

8-Aza-7-deaza-2',3'-dideoxyadenosine (**556**) was prepared from **538**.[737] Benzoylation of the 4-amino group of **538**, followed by 4,4'-dimethoxytritylation of the 5'-hydroxyl function and subsequent Barton deoxygenation of the phenoxythiocarbonyl intermediate, (obtained by the reaction with *O*-phenyl carbonochloridothioate) gave **556**, after removal of the protecting groups. Nucleoside **556** was converted into allopurinol 2',3'-dideoxyribofuranoside (**557**) by adenosine deaminase.[737] Compound **556** exhibited no activity against human immunodeficiency virus *in vitro*.[737]

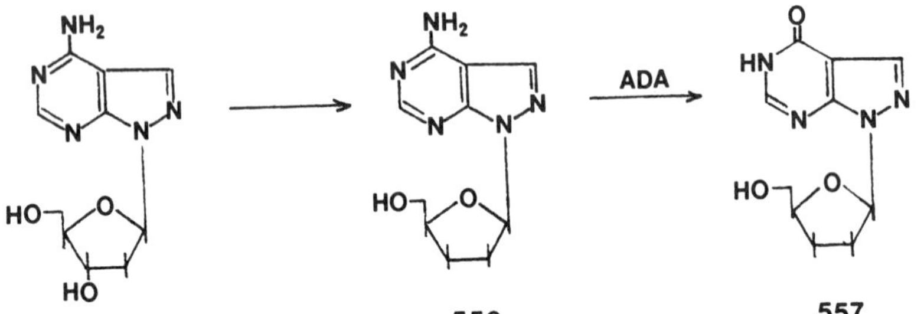

The synthesis of 6-amino-1-(2',3'-dideoxy-β-D-*glycero*-pentofuranosyl)pyrazolo[3,4-*d*]pyrimidin-4(5*H*)-one (8-aza-7-deaza-2',3'-dideoxyguanosine, **558**) from **550b** by a five-step deoxygenation route as described above was also reported recently.[735] Compound **558** was less acid-sensitive at the glycosidic bond than 2',3'-dideoxyguanosine.[735] No antiviral property of **558** is reported.

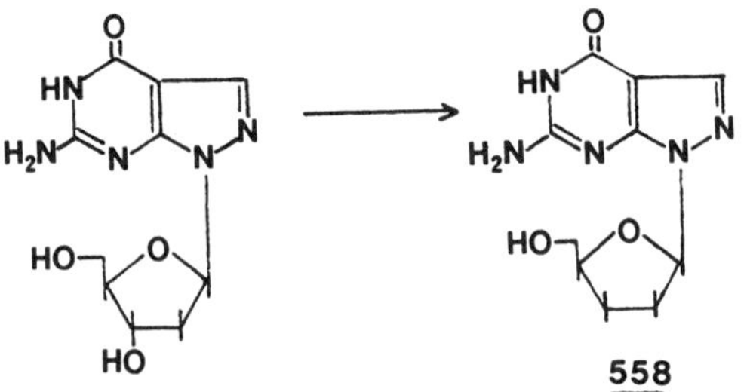

2.3.5.9. Preparation of Pyrazolo[3,4-d]pyrimidine Nucleosides by an Enzymatic Process. 5-Phosphoribosyl-1-pyrophosphate has been shown to react enzymatically with the natural purines,[738,739] pyrimidines,[740] and certain pyridines,[741] as well as with precursors involved in the *de novo* synthesis of purines.[742,743] Way and Parks[744] and subsequently Auscher *et al.*[745] demonstrated that this sugar phosphate also reacts with pyrazolo[3,4-*d*]pyrimidines to give the corresponding nucleoside 5'-phosphates. Exhaustively dialyzed hog-liver acetone powder extracts formed nucleotides from 4-hydroxy-[746] and 6-amino-4-hydroxypyrazolo-

[3,4-*d*]pyrimidines but did not react with 4-aminopyrazolo[3,4-*d*]pyrimidine. However, the hog liver, beef liver, and yeast enzymes purified by only brief dialysis reacted with all of the above pyrazolo[3,4-*d*]pyrimidines,[744] indicating the existence of at least two different purine nucleotide pyrophosphorylases.[747] The difficulties encountered in synthetic glycosylation procedures led Krenitsky and co-workers[172,703] to employ the enzymatic process to attach the sugar at N-1 of pyrazolo[3,4-*d*]pyrimidines.

Pyrazolo[3,4-*d*]pyrimidin-4(5*H*)-one, pyrazolo[3,4-*d*]pyrimidine-4(5*H*)-thione, and pyrazolo[3,4-*d*]pyrimidine-4,6-dione were converted by purine nucleoside phosphorylase (PNPase)[748] to the corresponding N-1 ribosyl derivatives.[172,703,714,749] Ribose 1-phosphate, inosine, xanthosine, and uridine served as ribosyl donors in this enzymatic reaction. 1-β-D-Ribofuranosyl pyrazolo-[3,4-*d*]pyrimidin-4(5*H*)-one prepared enzymatically[714] was found to be identical with the synthetic product, thus proving the N-1 ribosyl attachment and β-anomeric configuration. However, pyrazolo[3,4-*d*]pyrimidine-4,6-dione, in the presence of uridine phosphorylase, was transformed into 7-β-D-ribofuranosyl-pyrazolo[3,4-*d*]pyrimidine-4,6(1*H*,5*H*)-dione and was found to be identical with 7-ribosyloxoallopurinol isolated from the urine of patients treated with allopurinol.[703] Both 1- and 7-β-D-ribofuranosylpyrazolo[3,4-*d*]pyrimidine-4,6-dione 5'-monophosphates were prepared enzymatically.[750–752]

Japanese workers[753–757] have described a new procedure for the production of the ribonucleotides of 4-amino-, 4-hydroxy-, and 6-amino-4-hydroxypyrazolo[3,4-*d*]pyrimidines, using a wide variety of microorganisms, particularly *Brevibacterium ammoniagenes* (ATCC 6872), through the salvage pathway, without the addition of 5-phosphoribosyl-1-pyrophosphate (PRPP). In some cases, crystalline nucleotide was isolated directly from the culture broth. Several fungi have also been shown[758,759] to convert the 4-amino- and 4-hydroxypyrazolo[3,4-*d*]-pyrimidines into their ribonucleosides. 1-β-D-Ribofuranosylpyrazolo[3,4-*d*]pyri-midin-4(5*H*)-one 5'-monophosphate could be dephosphorylated[756,760,761] to the corresponding ribonucleoside by hydrolysis in aqueous solution (pH 4.0 to 9.0) at 140°C for 6 h. Such a dephosphorylation could also be accomplished with various microorganisms.[756] Microbial (*Erwinia carotovora*) ribosylation of allo-purinol through an N-ribosyl transfer reaction from uridine to allopurinol was also described.[762] The structures of these crystalline nucleosides were confirmed through spectroscopic studies.[763]

The enzyme nucleoside 2'-deoxyribosyltransferase (EC 2.4.2.6), prepared from the bacterial species *Lactobacillus leichmannii* (ATCC 7830), converted 4-aminopyrazolo[3,4-*d*]pyrimidine into crystalline 4-amino-1-(β-D-*erythro*-pen-tofuranosyl)pyrazolo[3,4-*d*]pyrimidine[764] in an 83% yield. The deoxyribo-nucleosides of 4-aminopyrazolo[3,4-*d*]pyrimidine and 4-amino-6-hydroxypyra-zolo[3,4-*d*]pyrimidine have been prepared[765,766] from the corresponding bases using thymidine and nucleoside deoxyribosyltransferase from *Lactobacillus helveticus*. However, no experimental details were given.

Recently, the chemical phosphorylation of 4-methylthio-1-β-D-ribofura-nosylpyrazolo[3,4-*d*]pyrimidine (**479a**) using phosphorus oxychloride in tri-methyl phosphate to give the corresponding 5'-monophosphate has been reported.[767] Phosphorylation of the 2',3'-*O*-ethoxymethylidene derivative of

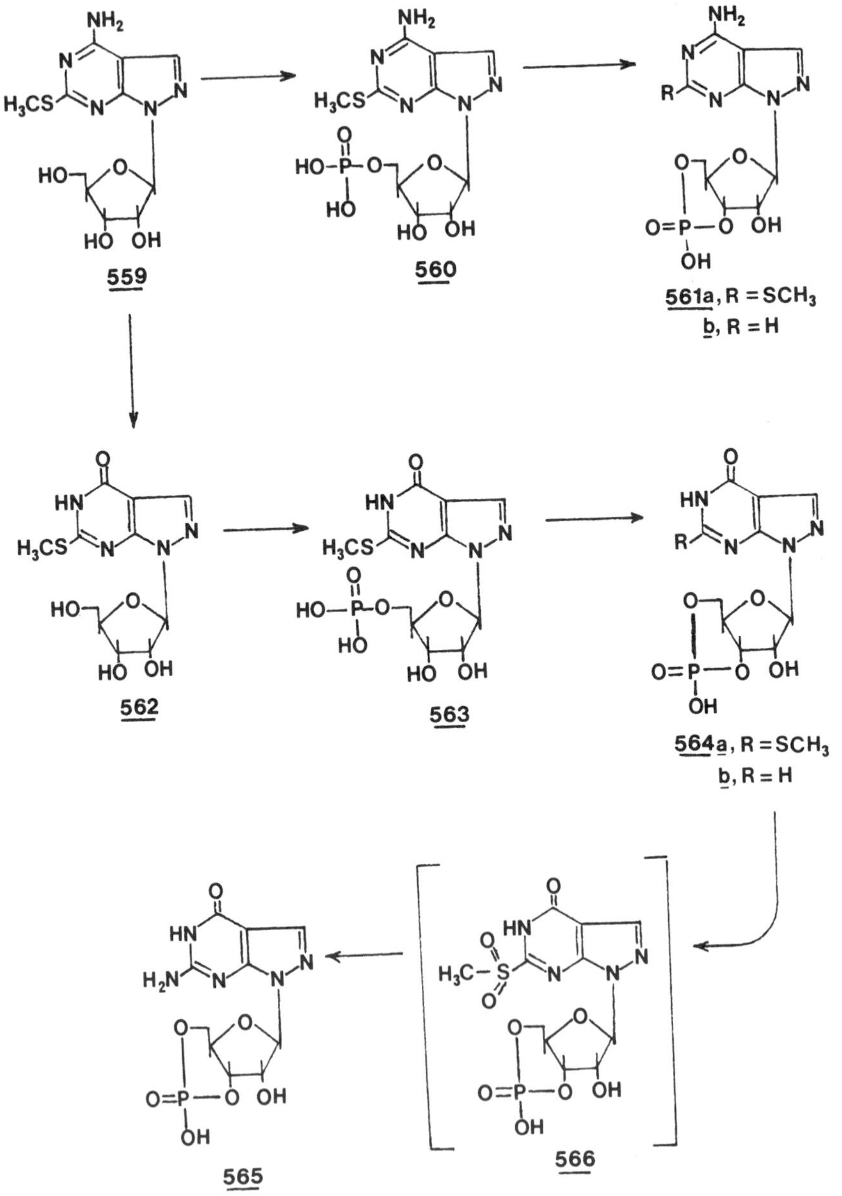

479a with 2-cyanoethyl phosphate in the presence of DCC and a subsequent removal of the blocking groups gave the required 5'-monophosphate of **479a**. Condensation of the 5'-monophosphate imidazolide with pyrophosphoric acid afforded the corresponding 5'-triphosphate.[767] Phosphorylation of 4-amino-6-methylthio-1-β-D-ribofuranosylpyrazolo[3,4-d]pyrimidine[709] (**559**) with $POCl_3$ in trimethyl phosphate gave the corresponding 5'-monophosphate (**560**).[768] N,N'-Dicyclohexylcarbodiimide-mediated intramolecular cyclization of **560** gave the corresponding 3',5'-cyclic phosphate (**561a**), which, on subsequent dethiation, provided the cAMP analogue 4-amino-1-β-D-ribofuranosylpyrazolo[3,4-*d*]pyrimidine 3′,5′-cyclic phosphate (**561b**). A similar phosphorylation of 6-methylthio-1-β-D-ribofuranosylpyrazolo[3,4-*d*]pyrimidin-4(5*H*)-one[768]   (**562**), followed by cyclization with DCC, gave the 3′,5′-cyclic phosphate of **562** (**564a**). Dethiation of **564a** with Raney nickel gave the cIMP analogue 1-β-D-ribofuranosylpyrazolo[3,4-*d*]pyrimidin-4(5*H*)-one 3′,5′-cyclic phosphate (**564b**).[768] Oxidation of **564a** with *m*-chloroperoxybenzoic acid, followed by ammonolysis, provided the cGMP analogue 6-amino-1-β-D-ribofuranosylpyrazolo[3,4-*d*]pyrimidin-4(5*H*)-one 3′,5′-cyclic phosphate (**565**).[768] Recently, several base-modified oligonucleotides derived from 2′-deoxyallopurinol ribonucleoside,[769] 8-aza-7-deaza-2′-deoxyadenosine,[770] and 8-aza-7-deaza-2′-deoxyguanosine[771] have been prepared by solid-phase synthesis employing P(III) chemistry.

The X-ray crystallographic analyses of 4-amino-1-β-D-ribofuranosylpyrazolo[3,4-*d*]pyrimidine (**469**)[772] and 4-amino-6-methylthio-1-β-D-ribofuranosylpyrazolo[3,4-*d*]pyrimidine 5′-monophosphate (**560**)[773] have been reported. In **469**, the glycosyl conformation was shown to be in the "high-*anti*" range.[772]

2.3.5.10. Biochemical Properties of Pyrazolo[3,4-d]pyrimidine Nucleosides and Nucleotides.

4-Amino-1-β-D-ribofuranosylpyrazolo[3,4-*d*]pyrimidine (**469**) is a potent cytotoxic agent to several cell lines,[681,774] but it is not as cytotoxic as tubercidin.[775] Several 3,4-disubstituted pyrazolo[3,4-*d*]pyrimidine ribonucleosides[696,711,713] (e.g., **473**, **476**, **484c**, **489b–d**) were found to be potent inhibitors of L1210 leukemia *in vitro* and *in vivo*.[776] Allopurinol ribonucleoside 5′-phosphate also prevented sarcoma-180 tumor growth in mice and prolonged the survival rate.[777] 4-Amino-1-β-D-ribofuranosylpyrazolo[3,4-*d*]pyrimidine-3-carboxamide (**473**) exhibited the ability to inhibit ADP-induced aggregation of human blood platelets, the inhibitory activity approaching that of adenosine.[778] Compound **469** inhibited the utilization of adenosine by adenosine deaminase and was utilized by the enzyme (from calf intestinal mucosa) as a substrate,[779,780] affording the deaminated product allopurinol riboside (**498**). The adenosine analogue **469** was also shown to be a substrate for adenosine kinase.[781,782]

6-Methyl-1-β-D-ribofuranosylpyrazolo[3,4-*d*]pyrimidine-4(5*H*)-thione (**510b**),[710] 1-β-D-ribofuranosyl-4(5*H*)-thioxopyrazolo[3,4-*d*]pyrimidine-3-carbonitrile,[696] and 4(5*H*)-oxo-1-β-D-ribofuranosylpyrazolo[3,4-*d*]pyrimidine-3-thiocarboxamide (**489b**)[783] exhibited significant broad-spectrum antiviral activity *in vitro*. Compound **489b** had no therapeutic effect on rodents.[783] 4-Amino-6-methylthio-1-β-D-ribofuranosylpyrazolo[3,4-*d*]pyrimidine-3-thiocarboxamide has been shown to be a potent antiherpes agent in cell culture.[784,785]

Pyrazolo[3,4-*d*]pyrimidines have received renewed attention in recent years owing to the discovery of certain derivatives possessing antiparasitic activity.[786,787] Since there is an absence of *de novo* purine biosynthesis in most parasites,[788–792] these organisms are wholly dependent on the salvage pathway for purine nucleoside metabolism and will accept certain pyrazolo[3,4-*d*]pyrimidines in place of purines.[593] Allopurinol {pyrazolo[3,4-*d*]pyrimidin-4(5*H*)-one} was the first such analogue found to be active against several *Leishmania*[793–796] and *Trypanosoma*[796–798] species *in vitro*. However, allopurinol ribonucleoside (**498**) has been shown to be 10-fold more active against *Leishmania braziliensis* and 300-fold more active against *Leishmania donovani* than allopurinol in inhibiting the growth of *Leishmania* promastigotes *in vitro*.[795,796,799] Both allopurinol and **498** are equally effective in preventing the transformation of the intracellular form (amastigote) of *L. donovani* to the extracellular promastigote form.[795] In both *Leishmania*[795] and *Trypanosoma*[800] species, allopurinol is converted to allopurinol ribonucleoside 5′-phosphate (HPPR-MP) by the parasitic nucleoside phosphoribosyltransferase. Sequential conversion of HPRR-MP by the parasite enzymes adenylosuccinate synthetase and succino-AMP lyase gives 4-aminopyrazolo[3,4-*d*]pyrimidine (4-APP) ribonucleoside 5′-phosphate, which is eventually incorporated into the cellular RNA of the parasite as the 5′-triphosphate, resulting in lethality to the parasite.[801–803] This conversion is analogous to the conversion of IMP to AMP in mammalian cells.[792,804] The incorporation of 4-APP-TP into the RNA of the parasite is unique, since mammalian cells do not show either this conversion or the incorporation of allopurinol or its metabolic products into RNA.[795,796] These unusual metabolic transformations of HPPR-MP reveal significant biochemical differences between the host and the parasite.[805]

Recently, 4-APP has been shown to be effective in the treatment of experimental Chagas' disease in mice.[806,807] The beneficial results obtained with allopurinol on human cutaneous leishmaniasis suggest that this compound may be a candidate for a successful treatment of this disease.[807] Moreover, the ribonucleosides of allopurinol (**498**) and 4-APP were shown to be severalfold more active than allopurinol against promastigotes of the isolates of American *L. brazilienses*[808] and *L. mexicana*[808] and *L. tropica*[210] *in vitro*. Pyrazolo[3,4-*d*]pyrimidine-4(5*H*)-thione[683] and the corresponding ribonucleoside (**479b**) are effective *in vitro* against the intracellular forms of *L. donovani* and against the extracellular forms of *L. brazilienses* and *L. mexicana*.[809] Spector and Jones[810] have found that both **498** and **479b** are strong inhibitors of leishmanial guanosine 5′-monophosphate (GMP) reductase and they are weak inhibitors of human GMP reductase.

Ribonucleosides of saturated 4-alkythio (e.g., ethylthio) or the unsaturated 4-allylthio, 4-crotylthio, and (*E*)-4-cinnamylthio derivatives of pyrazolo[3,-4-*d*]pyrimidines are very effective anticoccidials *in vivo* against *Eimeria tenella*.[172,628,811] 4-Cyclopentylamino-1-β-D-ribofuranosylpyrazolo[3,4-*d*]pyrimidine was also shown to be an active anticoccidial agent *in vivo*.[714]

The metabolism of pyrazolo[3,4-*d*]pyrimidines and their ribonucleosides has been studied in great detail.[779,812–818] 4-Aminopyrazolo[3,4-*d*]pyrimidine was converted to nucleotides (mono-, di-, and triphosphates) in all normal and neoplastic tissues, and these derivatives are incorporated into the nucleic acids.[812] Unlike 4-aminopyrazolo[3,4-*d*]pyrimidine, allopurinol or its metabolites are not incorporated into nucleic acids.[819] Although allopurinol 1-ribonucleotide is formed under appropriate conditions *in vitro*,[752,820] where it is a potent inhibitor of the *de novo* purine biosynthesis,[821] its formation *in vivo* has not been demonstrated.[822] The inhibitory effects of allopurinol and oxipurinol on purine and pyrimidine biosynthesis in human fibroblasts were explained[823] by the formation of their respective ribonucleotides either by HGPRTase (allopurinol) or OPRTase (oxipurinol). The enzymatic reactions responsible for these interconversions are summarized in Fig. 5.[824,825] Although allopurinol riboside was found to be a urinary metabolite of allopurinol,[703] it is not excreted by patients with an HGPRTase deficiency.[826] For a detailed discussion of the metabolic transformations of allopurinol, the reader is directed to ref. 824.

2.3.6. *Pyrazolo[4,3-d]pyrimidine Nucleosides*

The pyrazolo[4,3-*d*]pyrimidine nucleosides formycin A (**567**),[827] formycin B[828] (laurusin[829] or oyamycin,[830] **568**), and oxoformycin B (**569**)[831] are a group of naturally occurring *C*-nucleoside antibiotics which are stable to purine nucleoside phosphorylase.[832–834] This resistance to glycosidic cleavage provides a distinct advantage for these nucleosides. A number of derivatives of **567** and **568** modified at the 5- and 7- positions of the heterocycle,[835–844] as well as many derivatives with a modified glycon moiety,[845–849] have been reported. These carbon-linked nucleosides exhibit diverse biological activities.[2,3,210,850] However, only a few reports[851–854] have dealt with the preparation of *N*-nucleoside analogues of formycin A and formycin B.

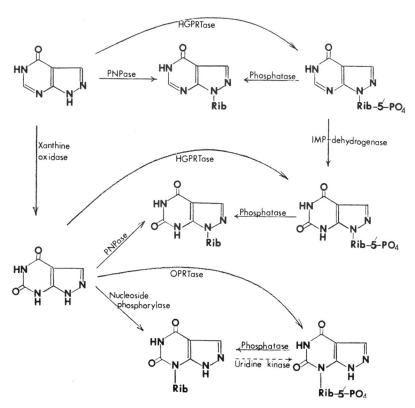

Figure 5. Metabolic transformations of allopurinol.

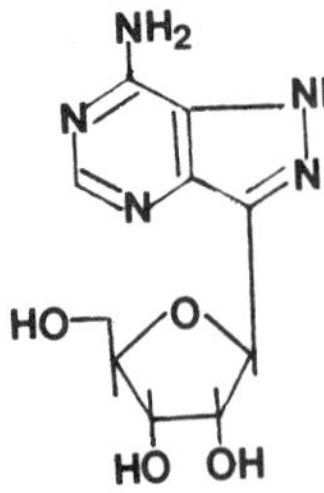

567 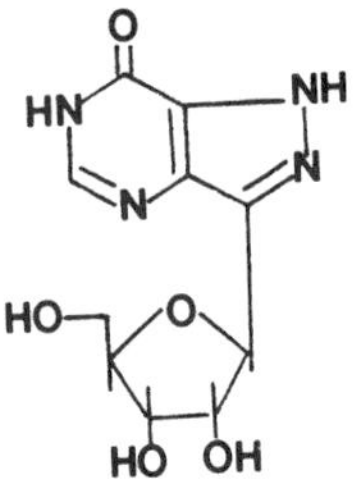   568 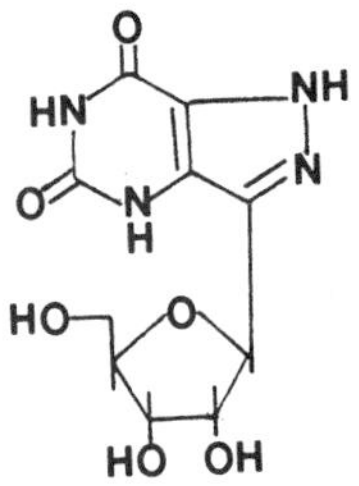   569

2.3.6.1. Synthesis Employing the Acid-Catalyzed Fusion of a Preformed Aglycon and an Acyl Sugar. One of the methods employed for the preparation of *N*-nucleosides of the pyrazolo[4,3-*d*]pyrimidine ring system was by an acid-catalyzed fusion procedure.[851,852] When 7-methylthiopyrazolo[4,3-*d*]pyrimidine (**570**)[855] was fused with a fully acylated ribofuranose (**51**) at 165°C in the presence of bis(*p*-nitrophenyl)phosphate, glycosylation occurred at both ring nitrogens of the pyrazole. This produced a mixture of positional isomers with α-

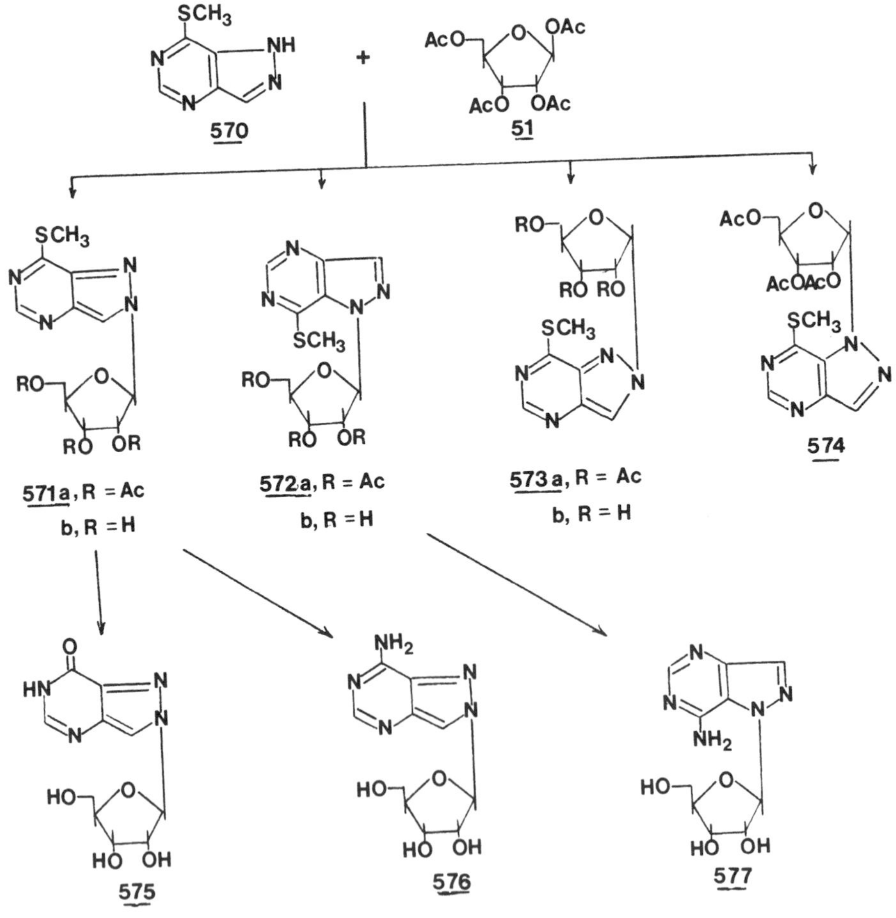

571a, R = Ac   572a, R = Ac   573a, R = Ac
   b, R = H   b, R = H   b, R = H

575   576   577

and β- anomeric configurations (**571a**, **572a**, **573a**, and **574**).[851,852] The N-1 β-anomer (**572a**) was obtained in low yield (5%). The methylthio derivatives were transformed into the corresponding amino and oxopyrazolo[4,3-*d*]pyrimidine nucleosides. Deacylation of **571a** and **572a** with methanolic ammonia at room temperature gave the free nucleosides **571b** and **572b**, respectively. Treatment of either **571a** or **571b** with hydrogen peroxide in acetic acid gave **575**, whereas treatment with methanolic ammonia at 150°C for 4.5 h afforded **576**.[852] Similarly, the reaction of **572b** with MeOH/NH$_3$ furnished **577**.[852] The site of ribosylation and the anomeric configuration of these nucleosides were established by a combination of ^{1}H-NMR and UV spectroscopic studies.[852]

In an effort to increase the yield of the desired **572a**, compound **570** was reacted with 2,3,5-tri-*O*-acetyl-D-ribofuranosyl bromide (**145**) in boiling nitromethane in the presence of molecular sieves, which again led to a preferential formation of the thermodynamically more stable N-2 isomer **571a**.[852] Attempts to isomerize the N-2 glycosyl isomer to the N-1 glycosyl isomer in this series failed.[852]

2.3.6.2. Synthesis Employing the Trimethylsilyl Derivative of a Preformed Aglycon and an Acyl Sugar in the Presence of SnCl$_4$. A number of *N*-ribonucleoside derivatives of 5-methylthio- and substituted 3-methyl pyrazolo[4,3-*d*]pyrimidin-7-one have been prepared by the stannic chloride-catalyzed trimethylsilyl procedure.[853,854] Glycosylation of the bis-trimethylsilyl derivative of 5-methylthiopyrazolo[4,3-*d*]pyrimidin-7(6*H*)-one (**578**) with 1,2,3,5-tetra-*O*-acetyl-β-D-ribofuranose (**51**) in the presence of SnCl$_4$ gave the N-1 and N-2 glycosylated products (**579a** and **580a**) in almost equal yields (30%).[853] Subsequent ammonolysis of these protected nucleosides afforded the free nucleosides **579b** and **580b**, respectively. The structures of these nucleosides were established by ^{13}C-NMR studies.[853] This procedure has been used to prepare 6-substituted 3-methyl-1-β-D-ribofuranosylpyrazolo[4,3-*d*]pyrimidin-7-ones.[854]

Compounds **571b**, **572b**, **575**, and **576** did not exhibit any cytostatic activity against several cell lines *in vitro*,[852] whereas 6-*p*-chlorophenyl- (or *p*-methoxybenzyl-) 3-methyl-1-β-D-ribofuranosylpyrazolo[4,3-*d*]pyrimidin-7-ones showed moderate antiviral activity against HSV-1 in cell culture.[854]

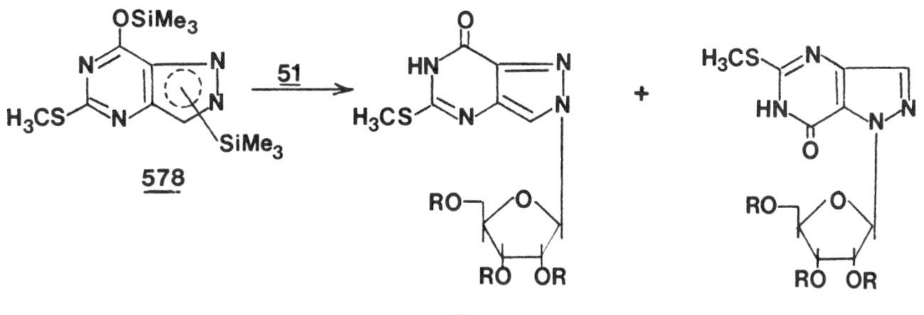

579**a**, R = Ac   580**a**, R = Ac
b, R = H   **b**, R = H

2.3.7. *v-Triazolo[4,5-b]pyridine Nucleosides*

2.3.7.1. Synthesis Employing the Halomercuri Salt of a Preformed Aglycon and an Acylated Sugar Halide. The exclusive formation of a *v*-triazolo[4,5-*b*]pyridine nucleoside (**582**) by a condensation of the chloromercuri derivative (**581**) of *v*-triazolo[4,5-*b*]pyridine[857] with 2,3,5-tri-*O*-benzoyl-D-ribofuranosyl chloride (**47**) in boiling xylene was reported by Jain and co-workers.[56,856] However, a reinvestigation of the reaction by De Roos and Salemink[858] under identical conditions afforded a mixture of three isomeric blocked nucleosides (**582, 583,** and **584**; R = Bz) in a ratio of 12:5:1. Debenzoylation of these blocked nucleosides with methanolic sodium methoxide gave 3-β-D-ribofuranosyl-*v*-triazolo-[4,5-*b*]pyridine (**582,** R = H) and the N-2 and N-1 isomers **583** and **584** (R = H), respectively. Both the anomeric configuration and the site of ribosylation were assigned on the basis of ¹H-NMR spectral studies and UV absorption spectral comparisons using model methyl compounds.[858]

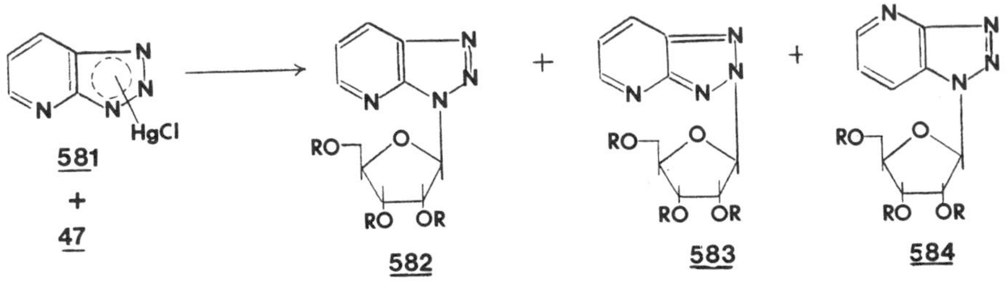

2.3.7.2. Synthesis Employing the Direct Alkylation of a Preformed Aglycon with an Acyl Sugar Halide. In contrast to the halomercuri salt procedure, the direct alkylation of 7-chloro-*v*-triazolo[4,5-*b*]pyridine (**585**),[859] with 2,3,5-tri-*O*-benzoyl-D-ribofuranosyl chloride (**47**) in nitromethane containing potassium cyanide as the hydrogen halide acceptor, provided a 33% yield of 7-chloro-3-(2,3,5-tri-*O*-benzoyl-β-D-ribofuranosyl)-*v*-triazolo[4,5-*b*]pyridine (**586a**) and a 21% yield of the corresponding 2-ribosyl isomer (**587a**).[858] Debenzoylation of **586a** with sodium methoxide in methanol gave 7-chloro-3-β-D-ribofuranosyl-*v*-triazolo[4,5-*b*]pyridine (**586b**), whereas **587a** under identical conditions furnished 7-methoxy-2-β-D-ribofuranosyl-*v*-triazolo[4,5-*b*]pyridine (**590**). However, debenzoylation of **587a** under carefully controlled conditions (pH 8) afforded 7-chloro-2-β-D-ribofuranosyl-*v*-triazolo[4,5-*b*]pyridine (**587b**) in over 65% yield.

Dehalogenation of **586b** and **587b** (via a B—Cl→B—NHNH₂→B—H reaction sequence, because of the sensitivity of the triazole ring to catalytic hydrogenation[860]) to the corresponding dehalogenated nucleosides **582** (R = H) and **583** (R = H) confirmed the structural assignment of **586** and **587**. The 7-chloro group of **586b** was found to be considerably more reactive toward nucleophilic displacement than that of the corresponding 7-chloro-3-β-D-ribofuranosylimidazo[4,5-*b*]pyridine (**52**).[66] Thus, treatment of **586b** with hydrazine hydrate, followed by a reaction of the resultant hydrazino compound with nitrous acid, gave the 7-azido nucleoside (**589**). Subsequent reduction of **589** with sodium dithionite in a base furnished 7-amino-3-β-D-ribofuranosyl-*v*-triazolo[4,5-*b*]pyridine (1-deaza-8-azaadenosine, **588**) in an overall yield of 58%.

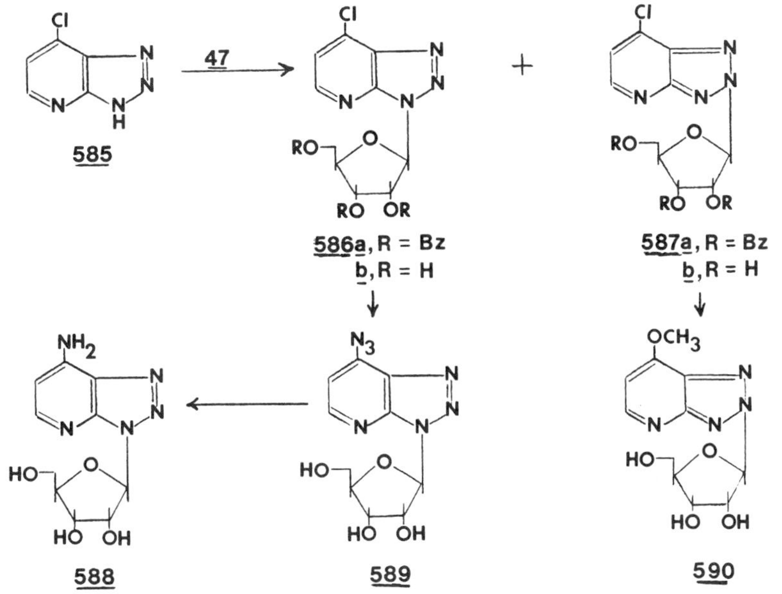

Due to the rapid decomposition of **585** at high temperatures, the fusion of **585** with tetra-*O*-acetylribofuranose gave poor yields of nucleoside material. Coupling under milder conditions (using nitromethane or chlorobenzene) did not improve the yields.[858]

2.3.7.3. Synthesis Employing the Trimethylsilyl Derivative of a Preformed Aglycon and an Acylated Sugar Halide. Although the trimethylsilyl procedure had been claimed to be unsuited[858] for the glycosylation of 7-chloro-*v*-triazolo[4,5-*b*]pyridine, Townsend and co-workers[861,862] successfully employed this procedure for the synthesis of 1-deaza-8-azaguanosine (**596**). The trimethylsilyl derivative of ethyl 7-chloro-*v*-triazolo[4,5-*b*]pyridine-5-carbamate (**591**) was reacted with 2,3,5-tri-*O*-benzoyl-D-ribofuranosyl bromide (**19**) in boiling benzene in the presence of mercuric cyanide to afford an 81% yield of ethyl 7-chloro-3-(2,3,5-tri-*O*-benzoyl-β-D-ribofuranosyl)-*v*-triazolo[4,5-*b*]pyridine-5-carbamate (**592a**) along with a 1% yield of the N-2 positional isomer **593**.[862] Subsequent debenzoylation of **592a** gave ethyl 7-chloro-3-β-D-ribofuranosyl-*v*-triazolo[4,5-*b*]pyridine-5-carbamate (**592b**). The structure of **592b** was assigned on the basis of ¹³C- and ¹H-NMR studies.[863] Treatment of **592b** with sodium benzylate in benzyl alcohol provided 5-amino-7-benzyloxy-3-β-D-ribofuranosyl-*v*-triazolo-[4,5-*b*]pyridine (**595**). Hydrogenolysis of **595** gave 5-amino-3-β-D-ribofuranosyl-*v*-triazolo[4,5-*b*]pyridin-7-(4*H*)-one (1-deaza-8-azaguanosine, **596**[861,862]) in 83% yield. Thiation of **592b** with sodium hydrogen sulfide in DMF at room temperature, followed by hydrolysis of the 5-ethylcarbamate intermediate with 1 *N* KOH, gave 1-deaza-8-aza-6-thioguanosine (**594**) in good yield.[862]

2.3.7.4. Synthesis via Ring Closure of a Pyridine Nucleoside Precursor. A rather unique stereospecific and regioselective method for the synthesis of *v*-triazolo[4,5-*b*]pyridine nucleosides by the ring annulation of pyridine nucleosides

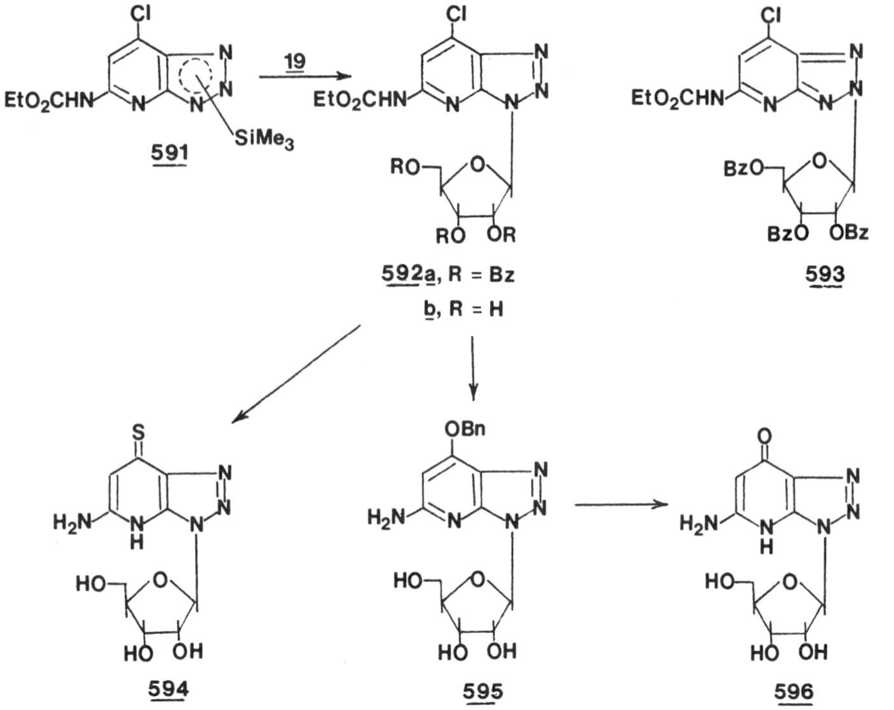

591   592a, R = Bz   b, R = H   593

594   595   596

(analogous to the route developed by Fox and co-workers[864,865] for the synthesis of 8-azapurines) has been reported by Lunch and Sharma.[866] 5-Nitro-1-β-D-ribofuranosylpyridin-2-one (**597**) was converted into the corresponding 4-substituted *v*-triazolo[4,5-*b*]pyridin-5-one (**601**) using the successive substitution and cyclization effected by azide ion in aprotic dipolar media. The dihydro intermediate **599** was isolated in a crystalline form, which, on treatment with an

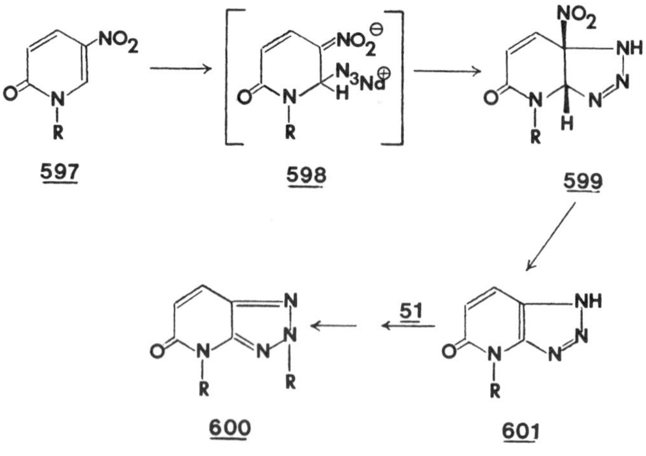

597   598   599

600   601

R = β-D-ribofuranose

acid, gave **601**. The yield of **601** was shown to be significantly better if a protected nitropyridine nucleoside was used.[866]

Further treatment of the trimethylsilyl derivative of **601** (preferably the protected form) with 1,2,3,5-tetra-*O*-acetylribofuranose (**51**), in the presence of a Lewis acid catalyst, effectively introduced the carbohydrate moiety into the 2-position of the *v*-triazolo[4,5-*b*]pyridone nucleus. Subsequent deacylation of this compound provided the bis-riboside (**600**).[866] The structural assignments of these nucleosides were made by a comparison of spectroscopic properties (¹H-NMR, UV) with those of the appropriate model compounds. It was also proposed[866] that the conformation of 4-β-D-ribofuranosyl-*v*-triazolo[4,5-*b*]pyridin-5-one (**601**) assumed "high-*anti*" sugar–base torsional angles.

2.3.8. *v*-Triazolo[4,5-c]pyridine Nucleosides

2.3.8.1. Synthesis Employing the Halomercuri Salt of a Preformed Aglycon and an Acylated Sugar Halide. The synthesis of a *v*-triazolo[4,5-*c*]pyridine nucleoside was first reported by Jain *et al.*[856] in 1965 using the halomercuri salt procedure. The chloromercuri derivative (**602**) of *v*-triazolo[4,5-*c*]pyridine[867] was reacted with 2,3,5-tri-*O*-benzoyl-D-ribofuranosyl chloride (**47**) to obtain 2-β-D-ribofuranosyl-*v*-triazolo[4,5-*c*]pyridine (**603**) in a 52% yield, after debenzoylation of the blocked nucleoside intermediate. A second nucleoside product was also obtained, but it was not adequately characterized. Although the site of ribosylation in **603** was established by UV absorption comparisons with appropriate model methyl compounds, no data were provided for the anomeric configuration.[856]

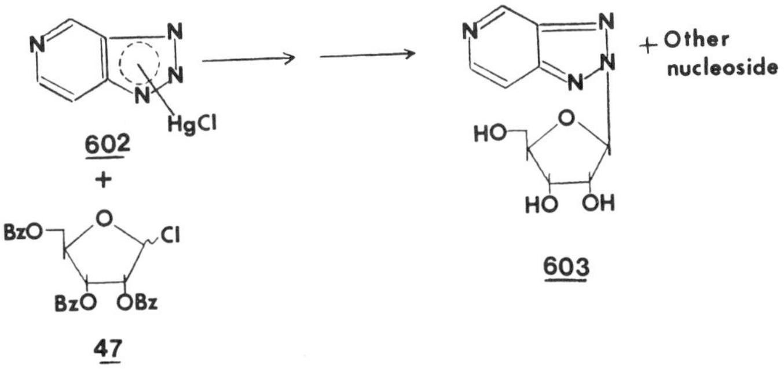

2.3.8.2. Synthesis Employing the Direct Alkylation of a Preformed Aglycon with an Acylated Sugar Halide. May and Townsend[868] treated 4-chloro-*v*-triazolo-[4,5-*c*]pyridine (**604**)[869] with 2,3,5-tri-*O*-benzoyl-D-ribofuranosyl chloride (**47**) under the conditions reported by Yamaoka *et al.*[71] and isolated the N-1 [4-chloro-1-(2,3,5-tri-*O*-benzoyl-β-D-ribofuranosyl)-*v*-triazolo[4,5-*c*]pyridine (**605**, 11%)], N-2 (**606**, 15%), and N-3 (**607**, 8%) isomers. Debenzoylation of **605** with sodium methoside in methanol gave 4-chloro-1-β-D-ribofuranosyl-*v*-triazolo-[4,5-*c*]pyridine (**608**, R = Cl), which, on ammonolysis, provided 4-amino-1-β-D-ribofuranosyl-*v*-triazolo[4,5-*c*]pyridine (3-deaza-8-azaadenosine; **608**, R = NH₂).[868] Similarly, **608** (R = Cl) with methanethiol furnished the 4-methylthio derivative, and subsequent dethiation of the 4 methylthio derivative with Raney

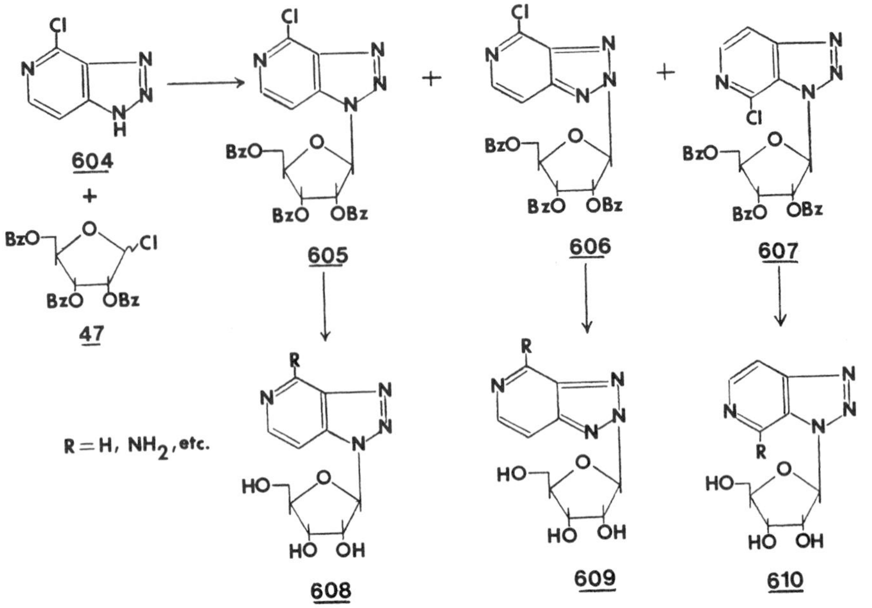

nickel gave the nebularine analogue 1-β-D-ribofuranosyl-v-triazolo[4,5-c]pyri-
dine (**608**, R = H). The debenzoylation of **606** gave 4-chloro-2-β-D-ribofura-
nosyl-v-triazolo[4,5-c]pyridine (**609**, R = Cl). Thiation and then deprotection of
606 gave **609** (R = SH). Subsequent dethiation of **609** afforded **603**, identical
with the compound obtained through the halomercuri procedure.[856] By a
similar sequence of reactions, 4-chloro-3-β-D-ribofuranosyl-v-triazolo[4,5-c]py-
ridine (**610**, R = Cl) and the 4-thione derivative (**610**, R = SH) as well as the
isonebularine analogue (**610**, R = H) were obtained. The structures of these
nucleosides were assigned on the basis of ^{13}C-NMR, ^{1}H-NMR, and UV spectral
studies.[868]

2.3.8.3. Synthesis Employing the Trimethylsilyl Procedure. The trimethylsilyl
procedure was found[868] to give a better yield of 4-chloro-1-(2,3,5-tri-O-ben-
zoyl-β-D-ribofuranosyl)-v-triazolo[4,5-c]pyridine (**605**). The silylation of **604**
with N,O-bis(trimethylsilyl)acetamide at 110°C gave the crystalline silyl derivative. Glycosylation of this silyl derivative with tri-O-benzoyl-D-ribofuranosyl
bromide (**19**) in acetonitrile, in the presence of mercuric cyanide, at reflux
temperature furnished **605**, **606**, and **607** in 25%, 12%, and 4% yield, respectively. The formation of **605** as the major product indicates that the thermodynamically more stable monosilyl derivative of **604** was formed.[868]

*2.3.8.4. Synthesis Employing the Ring Closure of a v-Triazole Nucleoside Precur-
sor.* The synthesis of 8-aza-3-deazaguanosine (**614**) by a methodology involving
the ring closure of a v-triazole nucleoside precursor has been reported by two
groups of investigators.[870,871] The prerequisite v-triazole nucleoside, methyl
5-cyanomethyl-1-(2,3,5-tri-O-benzoyl-β-D-ribofuranosyl)-1,2,3-triazole-4-car-

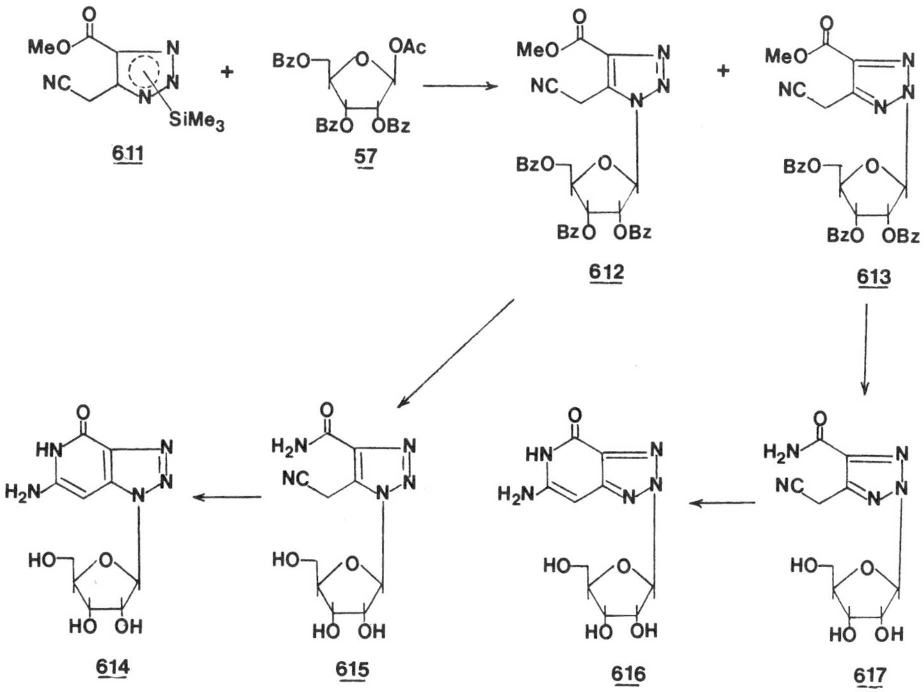

boxylate (**612**), along with the corresponding N-2 glycosyl isomer **613**, was obtained from the SnCl$_4$-catalyzed condensation of the trimethylsilyl derivative of methyl 4(5)-cyanomethyl-1,2,3-triazole-5(4)-carboxylate (**611**) and **57**.[870] A small amount of N-3 glycosyl isomer was also obtained. Compound **612** was also obtained by a 1,3-dipolar cycloaddition reaction of 2,3,5-tri-*O*-benzoyl-β-D-ribofuranosyl azide, followed by a series of functional group transformation reactions.[871] Treatment of **612** with liquid ammonia effected not only a smooth removal of the blocking groups, but also an ammonolysis of the ester function to give **615**. A base-catalyzed ring annulation of **615** provided **614**. A similar sequence of reactions with **613** gave 6-amino-2-β-D-ribofuranosyl-*v*-triazolo-[4,5-*c*]pyridin-4(5*H*)-one (**616**).[870] The structures of these nucleosides were established on the basis of ^{13}C-NMR, ^{1}H-NMR, and nuclear Overhauser enhancement studies.[870,871]

The conformation of 3-deaza-8-azaadenosine (**608**, R = NH$_2$) in solution was studied,[181] and it was determined that the base exists in the *anti* range with some destabilization of the *g*+ rotamer.

2.4. Five-Membered Ring Fused to a Six-Membered Ring with Five Nitrogen Atoms in the Ring System (at Least Two Nitrogen Atoms in Each Ring)

The class of compounds considered in this section includes the aza analogue of purine bases. The 8-azapurines were the first to be studied. Before 1945, the chemistry of azapurines had not been fully explored, and the entire chemistry of 8-azapurines was developed in connection with a study of antimetabolites of nucleic acid components. However, the 2-aza analogues have received relatively little attention by comparison.

The name 2-azapurine was derived by a formal substitution of the methine group in the 2-position of purine by a nitrogen atom. Since this position is substituted in some purine bases, only the aza analogues of adenine and hypoxanthine are amenable to such formal derivation. The *Chemical Abstracts* indices use the more systematic nomenclature imidazo[4,5-*d*]-*v*-triazine, with numbering as shown in **618**. The 8-aza analogues are also formally derived by substitution of the methine group in position 8 of purine by a nitrogen atom. The name 8-azapurine is frequently used in chemical and biochemical literature. However, in compliance with the general principles of the *Ring Index* and IUPAC nomenclature, the systematic nomenclature *v*-triazolo[4,5-*d*]pyridine will be used here, with numbering as shown in **620**. The other ring system included in this section is pyrazolo[4,3-*d*]-*v*-triazine[619].

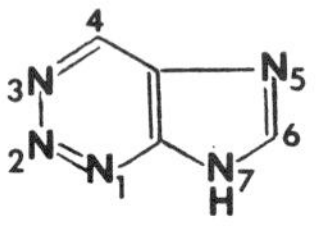

618	**619**	**620**
Imidazo[4,5-d]-v-triazine (2-Azapurine)	Pyrazolo[3,4-d]-v-triazine	v-Triazolo[4,5-d]-pyrimidine (8-Azapurine)

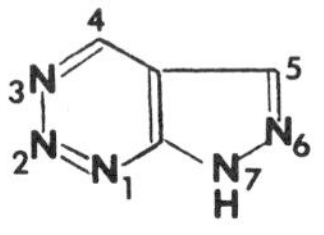

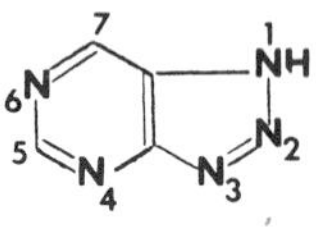

2.4.1. *Imidazo[4,5-d]-v-triazine (2-Azapurine) Nucleosides*

2.4.1.1. Synthesis Employing the Ring Closure of an Imidazole Nucleoside Precursor. Although 2-azaadenine and 2-azahypoxanthine have long been known to inhibit growth of both microbial and mammalian cells,[873–878] little about the biological activity of 2-azapurine nucleosides is reported, probably because of the paucity of such nucleoside derivatives. The first chemical synthesis of a 2-azapurine nucleoside was accomplished by Stevens *et al.*[879] by ring annulation of an imidazole nucleoside. 1-β-D-Ribofuranosyl-5-aminoimidazole-4-carboxamidoxime (**622**) was obtained by the alkaline hydrolysis[880] of adenosine N^1-oxide (**621**).[881] Treatment of **622** with nitrous acid gave 2-azaadenosine N^1-oxide (**623**) in an overall yield of 13%. However, attempts to reduce **623** to 2-azaadenosine were unsuccessful. A similar intramolecular cyclization of diazotized *N*-(5-amino-1-β-D-ribofuranosylimidazole-4-carbonyl)-β-alanine 5′-

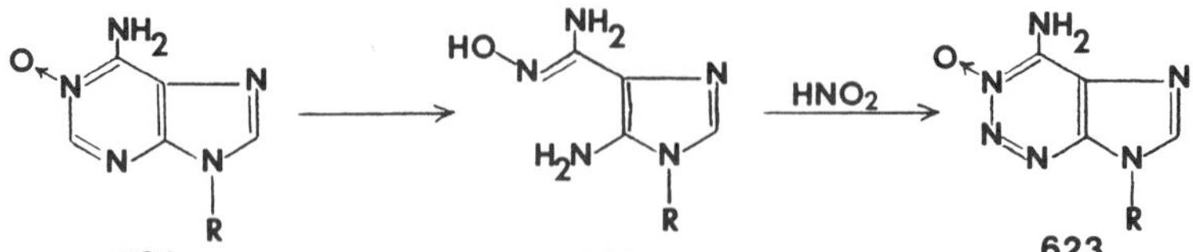

R = β-D-ribofuranosyl

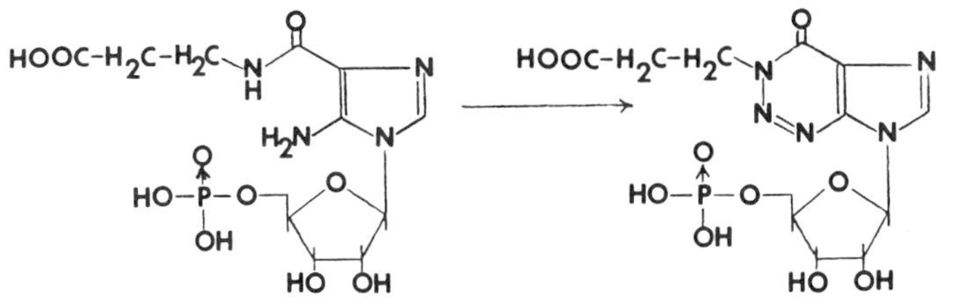

phosphate (CEAICAR, **624**) was reported[882] to yield 1-(2-carboxyethyl)-2-aza-inosinic acid (**625**) in essentially a quantitative yield.

A more convenient synthesis of 2-azapurine nucleosides was later reported by Montgomery and Thomas.[883–885] Benzylation of adenosine N^1-oxide (**626a**) with benzyl bromide in dimethylacetamide gave 1-benzyloxyadenosine hydro-bromide (**627a**). The reaction of **627a** with methanolic ammonia at 80°C opened the pyrimidine ring. A concomitant deformylation of the intermediate N-benzyl-oxy-5-formamido-1-β-D-ribofuranosylimidazole-4-carboxamidine (**628a**) gave, despite the precedent of the Dimroth rearrangement, the 5-aminoimidazole derivative **631a**. Raney nickel catalyzed hydrogenolysis of the benzyloxy group of **631a** afforded 5-amino-1-β-D-ribofuranosylimidazole-4-carboxamidine **630a**, and treatment of **630a** with sodium nitrite in aqueous acetic acid gave 2-aza-adenosine (**629a**). This relatively simply four-step procedure was used to obtain the various 5-amino-1-glycosylimidazole-4-carboxamidines (**630b–e**) required

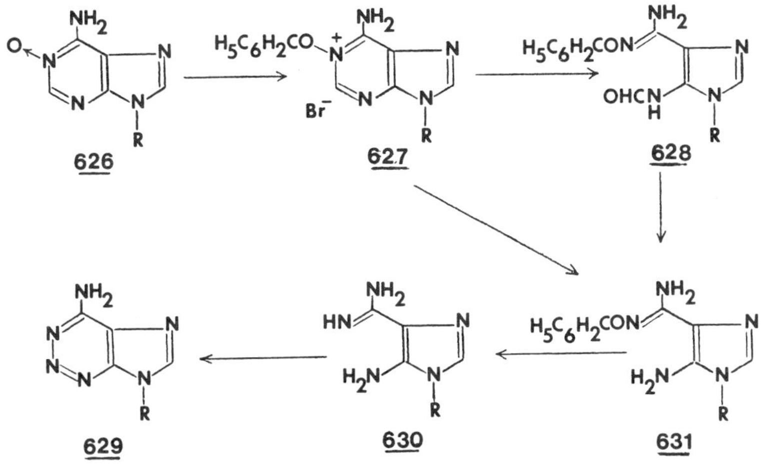

series **a**, R = β-D-ribofuranosyl

b, R = 2-deoxy-β-D-ribofuranosyl

c, R = 3-deoxy-β-D-ribofuranosyl

d, R = α-/β-D-arabinofuranosyl

e, R = β-D-xylofuranosyl

for the preparation of the 2'-deoxy (**629b**) and 3'-deoxy (**629c**)[886] as well as the 9-β-D-arabinofuranosyl (**629d**) and 9-β-D-xylofuranosyl (**629e**)[884] derivatives of 2-azaadenine.

Although earlier attempts[879] to prepare 7-β-D-ribofuranosylimidazo-[4,5-*d*]-*v*-triazin-4-one (2-azainosine, **633**) by the ring annulation of 5-amino-1-β-D-ribofuranosylimidazole-4-carboxamide (**632**, R = H; AICA-riboside) in the presence of nitrous acid were unsuccessful, by employing stronger acidic conditions (6 *N* HCl, to suppress intermolecular diazo coupling[887]) and low temperature (−25°C, to arrest the cleavage of the glycosyl linkage[888]), 2-azainosine was synthesized in excellent yield.[889] Thus, the treatment of **632** (R = H) or the tri-*O*-acetyl derivative (**632**, R = Ac)[890] in 6 *N* HCl or hypophosphorous acid[891] at −25°C with sodium nitrite gave an 86% and a 63% yield of **633** (R = H) and its tri-*O*-acetyl derivative, respectively. Similarly, α-2-azainosine,[892,893] 8-methyl-2-azainosine,[894] 2-azainosine 5'-phosphate (**634**),[889,895] 2-azainosine cyclic 3',5'-phosphate (**635**, X = O),[896] and 2-azaadenosine cyclic 3',5'-phosphate (**635**, X = NH)[896,897] were prepared by the ring closure of an appropriate imidazole precursor. A convenient method for the preparation of **635** (X = NH) involves[898,899] the ring opening of 1,*N*⁶-etheno-cyclic AMP (**636**)[900] with 1 *N* sodium hydroxide solution to afford crystalline 5-amino-1-β-D-ribofuranosyl-

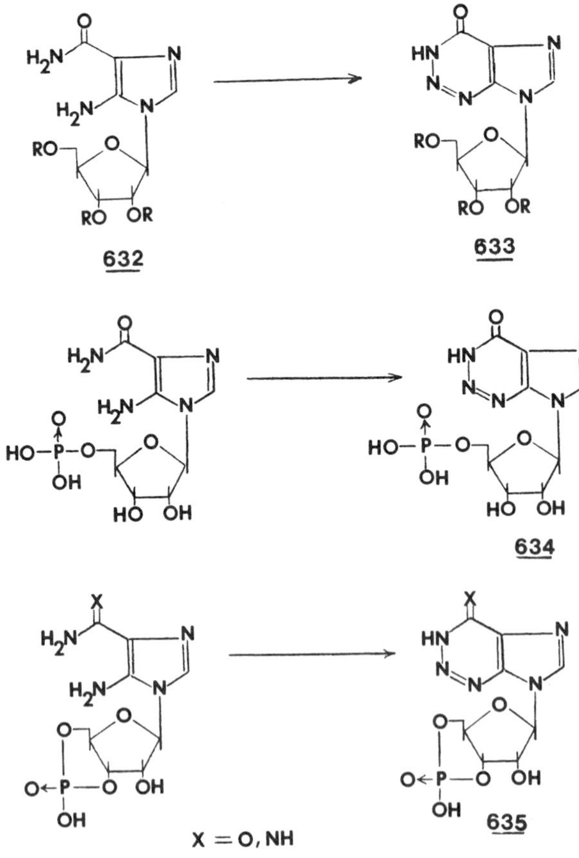

632   633

634

X = O, NH   635

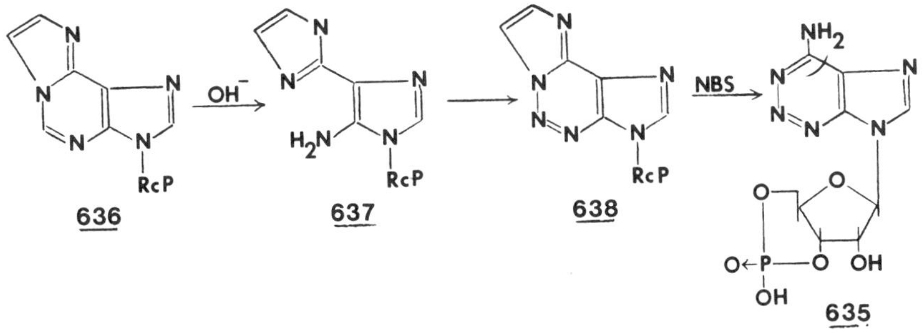

4-(imidazol-2-yl)imidazole cyclic 3′,5′-phosphate (**637**).[901] Nitrosation of **637** at room temperature gave 2-aza-1,*N*⁶-etheno-cyclic AMP (**638**) in 94% yield, which, when stirred overnight in aqueous DMF containing *N*-bromosuccinimide, furnished 2-azaadenosine cyclic 3′,5′-phosphate (**635**) in over 80% yield.

The versatility of this ring annulation procedure as a means to obtain interesting 2-azapurine nucleosides was further demonstrated by Robins and co-workers.[902] Treatment of 5-amino-2-bromo-2-(2,3,5-tri-*O*-acetyl-β-D-ribofuranosyl)imidazole-4-carboxamide (**639a**) or the corresponding free nucleoside (**639b**) with nitrous acid under the conditions reported by Kawana *et al.*[889] furnished 6-bromo-7-(2,3,5-tri-*O*-acetyl-β-D-ribofuranosyl)imidazo[4,5-*d*]*v*-triazin-4-one (**640a**) or the free nucleoside **640b** in good yield. Nucleophilic displacement of the 6-bromo group from either **640a** or **640b** with various nucleophiles gave the corresponding 6-substituted imidazo[4,5-*d*]-*v*-triazine nucleosides (**641**, R = NH₂, OH, SH, etc.).[902] Similarly, the diazotization of 5-amino-1-(2,3-di-*O*-acetyl-5-deoxy-5-iodo-β-D-ribofuranosyl)imidazole-4-carboxamide (**642**) gave the cyclized product **643**, which, on catalytic hydrogenation followed by deacetylation, afforded 5′-deoxy-2-azainosine (**644**).[903] The synthesis of 5-β-D-ribofuranosylimidazo[4,5-*d*]-*v*-triazin-4(3*H*)-one (**646**) was accomplished by Panzica and Townsend[890] by the nitrosative ring annulation of 4-amino-1-β-D-ribofuranosylimidazole-5-carboxamide (**645**)[641] in 4 *N* hydrochloric acid.

2.4.1.2. Synthesis Employing the Direct Alkylation of a Preformed Aglycon with an Acylated Sugar Halide. The synthesis of a 2-azapurine nucleoside by the direct alkylation of a preformed aglycon was reported by Ofitserov and co-workers.[904,905]

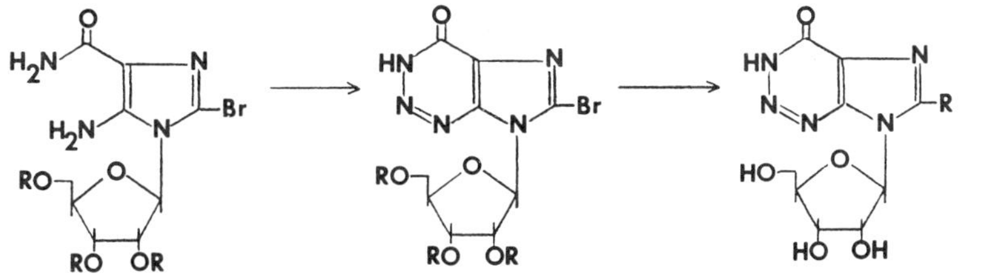

639a, R = Ac
b, R = H

640a, R = Ac
b, R = H

641, R = NH₂, SH, OH, etc.

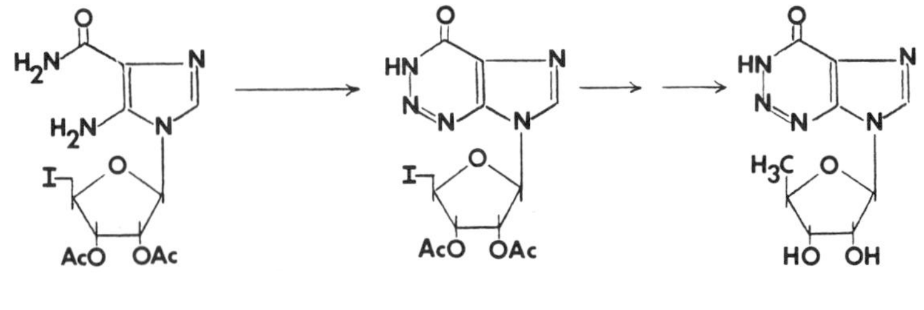

642   643   644

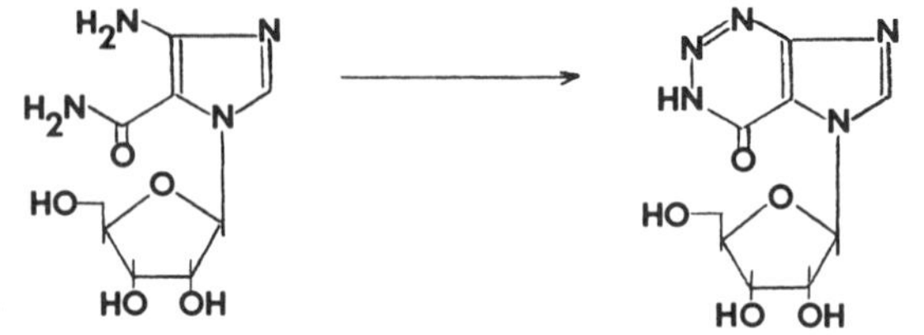

645   646

Treatment of 4-methylthioimidazo[4,5-*d*]-*v*-triazine (**647**)[906] with 2,3,5-tri-*O*-acetyl-D-ribofuranosyl bromide (**145**) in boiling nitromethane containing Hg(CN)₂ gave a 40.5% yield of 7-(2,3,5-tri-*O*-acetyl-β-D-ribofuranosyl)-4-methylthio-imidazo[4,5-*d*]-*v*-triazine, which, on subsequent deacetylation with methanolic ammonia, afforded 4-methylthio-7-β-D-ribofuranosylimidazo[4,5-*d*]-*v*-triazine (**648**). The structure of **648** was established by a combination of ¹H-NMR and UV spectroscopic studies. However, the direct glycosylation of 6-methylimid-azo[4,5-*d*]-*v*-triazin-4(3*H*)-one (**649**) with 2,3,5-tri-*O*-acetyl-D-ribofuranosyl chloride has been reported[894] to give 6-methyl-5-(2,3,5-tri-*O*-acetyl-β-D-ribofuranosyl)imidazo[4,5-*d*]-*v*-triazin-4(3*H*)-one (**650a**) in 10% yield. Deacety-lation of **650a** with MeOH/NH₃ gave 6-methyl-5-β-D-ribofuranosylimidazo-[4,5-*d*]-*v*-triazin-4(3*H*)-one (**650b**).[894]

In solution, 2-azaadenosine has a conformational behavior similar to that of the common purine β-D-ribonucleosides, which is in agreement with the confor-mations observed in the solid state.[907]

2.4.1.3. Biochemical Properties of Imidazo[4,5-d]-v-triazine Nucleosides and Nu-cleotides.   Even though the chemical deamination of 2-azaadenosine failed to yield 2-azainosine, the enzymatic deamination of 2-azaadenosine with adeno-

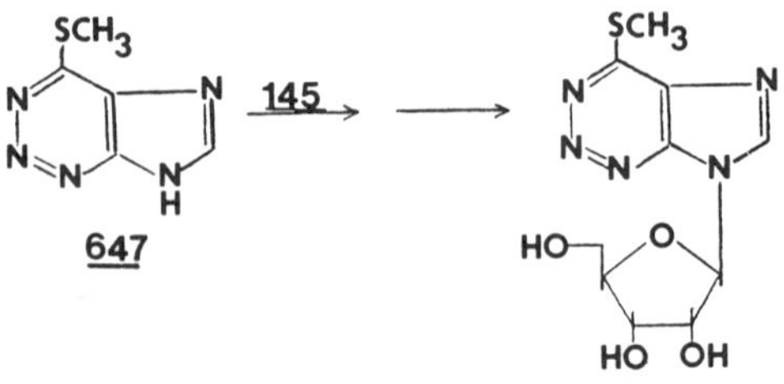

647   648

649

650**a**, R = Ac
b, R = H

sine deaminase was rapid and complete.[886] The enzymatic synthesis of 2-aza-adenosine 5'-triphosphate has been reported,[908] and this triphosphate was shown to be as effective as ATP in the active transport of potassium across the erythrocyte membrane.[909,910] Bennett and co-workers[614,911] studied the cyto-toxicities of certain 2-azapurine nucleosides as substrates for enzymes metaboliz-ing purine nucleosides. 2-Azaadenosine was cytotoxic to human epidermoid carcinoma cells in culture[614] and shows consistent activity against L1210 leuke-mia in both chronic and single-dose schedules.[912] 2-Azaadenosine was found to be much more cytotoxic than 2-azaadenine and was a good substrate for both adenosine kinase and adenosine deaminase. The cytotoxicity of 2-azaadenosine may be due both to its direct phosphorylation and to its conversion to 2-aza-hypoxanthine, as indicated in Fig. 6. 2-Azainosine, 2'-deoxy-2-azainosine, and 2-azahypoxanthine had similar cytotoxicities, and, for both nucleosides, the cytotoxicity resulted from their apparent conversion to 2-azahypoxanthine. 3'-Deoxy-2-azaadenosine and the α-arabinosyl, β-arabinosyl, and β-xylosyl de-rivatives of 2-azaadenine were not cytotoxic. 3'-Deoxy-2-azaadenosine and β-xylofuranosyl-2-azaadenine were moderately good substrates for adenosine deaminase, but α-arabinofuranosyl-2-azaadenine was not deaminated. Al-though β-arabinofuranosyl-2-azaadenine was deaminated, it was at a rate less than 5% of that of araA.[911]

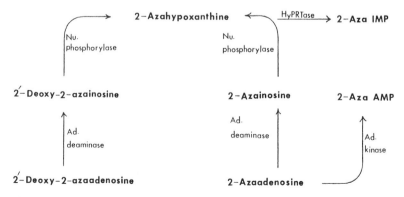

Figure 6. Metabolic conversion of 2-azaadenine nucleosides to 2-azahypoxanthine.

2-Azainosine and its 2′,3′,5′-tri-O-acetyl derivative exhibited significant antiviral activity[883,889] against herpes simplex type 1 and parainfluenza virus in cell culture.

2.4.2. Pyrazolo[3,4-d]-v-triazine Nucleosides

The preparation of the adenosine (657) and the inosine (654) analogues of pyrazolo[3,4-d]-v-triazine nucleosides was reported by Montgomery and co-workers.[884,886] This was accomplished by the ring opening and reclosure sequence as described for 2-azapurine nucleosides (Section 2.4.1.1). 5-Benzyloxy-4-aminopyrazolo[3,4-d]pyrimidine ribonucleoside hydrobromide (653), prepared by the benzylation of the N^5-oxide (652) of 4-amino-1-β-D-ribofuranosylpyrazolo[3,4-d]pyrimidine (469), was ring opened by treatment with methanolic ammonia to obtain 5-amino-N-benzyloxy-1-β-D-ribofuranosylpyrazole-4-carboxamidine (655) in 69% yield. Raney nickel catalyzed hydrogenolysis of the benzyloxy group of 655 gave 5-amino-1-β-D-ribofuranosylpyrazole-4-carboxamidine (656). Ring annulation of 656 with aqueous nitrous acid gave 4-amino-7-β-D-ribofuranosylpyrazolo[3,4-d]-v-triazine (657)[884] in 32% yield. Similarly, 1-β-D-ribofuranosylpyrazolo[3,4-d]pyrimidin-4(5H)-one (498) was converted into 5-amino-1-β-D-ribofuranosylpyrazole-4-carboxamide (651) by the above sequence of reactions. However, 651 could be prepared much more conveniently from 498 by base treatment.[718] Nitrosative ring closure of 651 gave 7-β-D-

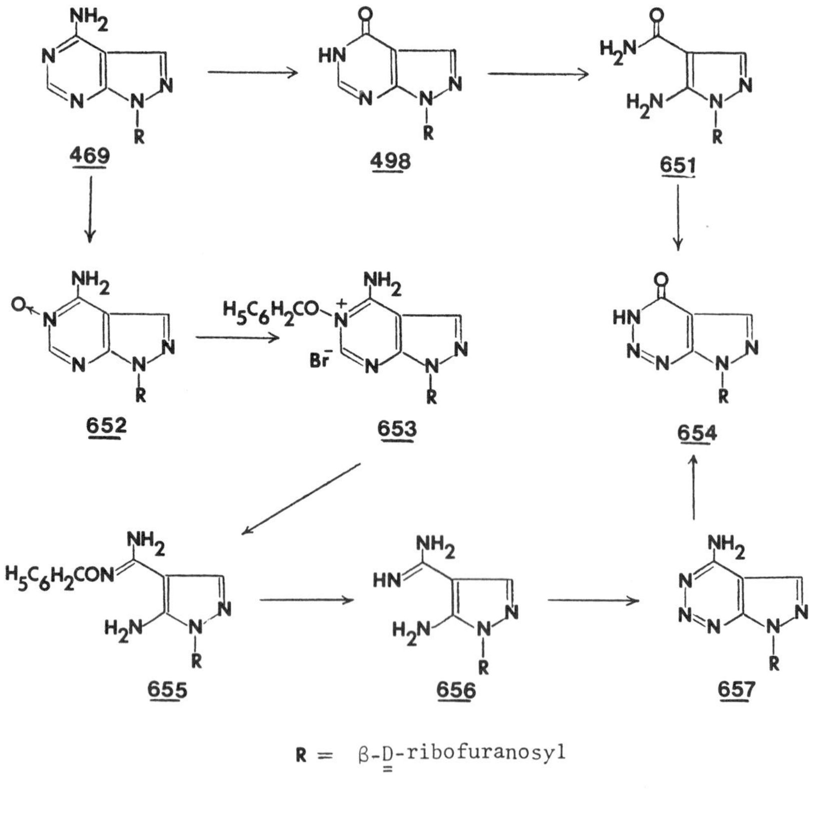

R =   β-D-ribofuranosyl

ribofuranosylpyrazolo[3,4-*d*]-*v*-triazin-4(3*H*)-one (**654**) in an 81% yield. Compound **654** was also obtained by the deamination of **657** with adenosine deaminase.

4-Amino-7-β-D-ribofuranosylpyrazolo[3,4-*d*]-*v*-triazine (**657**) was only 1/200th as toxic as 8-azaadenosine to HEp-2 cells in culture,[614] whereas its synthetic precursor (**656**) was quite cytotoxic, having an ED_{50} value of 1.9 μmol/liter,[884] but the inosine analogue **654** showed no inhibition even at 75 μmol/liter.[886]

2.4.3. *v-Triazolo[4,5-d]pyrimidine (8-Azapurine) Nucleosides*

The synthesis of 5-amino-*v*-triazolo[4,5-*d*]pyrimidin-7(6*H*)-one (8-azaguanine) in 1945 by Roblin *et al.*[872] and the discovery of its antibacterial activity[872] generated considerable interest in the synthesis and biological evaluation of 8-azapurines. 8-Azaguanine later came into prominence through the pioneering works of Kidder[913,914] and Lüdemann.[915] Pathocidin, an antifungal antibiotic isolated from the culture filtrate of an unidentified species of streptomyces, was found to be identical with 8-azaguanine.[916,917] This compound has proved to be a valuable tool for the molecular biologist since it is readily incorporated into the ribonucleic acids.[918–922] Several reviews dealing with the chemistry,[672,923] biochemistry, and pharmacology of 8-azapurines have appeared.[924–927]

v-Triazolo[4,5-*d*]pyrimidin-7(6*H*)-one (8-azahypoxanthine) was found to exhibit[928–930] activity against adenocarcinoma 755 in the mouse. The ribonucleoside 8-azainosine, first prepared by Davoll[931] in 1958, was shown to be active *in vivo* against both Ca 755 and L1210 leukemia. It was also active against a strain of L1210 resistant to 6-mercaptopurine.[929,930] These early observations generated tremendous interest in the synthesis of other 8-azapurine nucleosides.

2.4.3.1. *Synthesis Employing the Halomercuri Salt of a Preformed Aglycon and an Acylated Sugar Halide.*

The glycosylation procedure utilizing a halomercuri salt of an 8-azapurine derivative was first reported by Davoll.[931] Treatment of the chloromercuri derivative of 7-acetamido-*v*-triazolo[4,5-*d*]pyrimidine (**658**) with tetra-*O*-acetyl-α-D-glucopyranosyl bromide (**659**) and subsequent deacetylation of the reaction product furnished 7-amino-3-β-D-glucopyranosyl-*v*-triazolo-[4,5-*d*]pyrimidine (**660**) together with the N-1 glycosyl isomer (**661**). Compound **660** was found to be identical to the 3-glucosyl derivative of 8-azaadenine prepared by the cyclization of an appropriate pyrimidine with nitrous acid. This established the site of glycosylation. However, an analogous reaction with tri-*O*-benzoyl-D-ribofuranosyl chloride (**47**) gave only one ribosyl derivative, to which the structure 7-amino-3-β-D-ribofuranosyl-*v*-triazolo[4,5-*d*]pyrimidine (8-azaadenosine, **662**) was assigned on the basis of UV spectral similarity with **660**. Deamination of **662** with nitrous acid gave 8-azainosine (**663**).

By using this procedure, several 9-β-D-ribofuranosyl and 9-α-L-rhamnopyranosyl derivatives of 2-alkythio-8-azaadenines have been prepared.[932] However, the preparation of 8-azaguanosine (**668**) proved to be rather cumbersome.[931] Application of the chloromercuri salt procedure to 5,7-diacetamido-*v*-triazolo[4,5-*d*]pyrimidine resulted in the formation of 5-amino-1-β-D-ribofuranosyl-*v*-triazolo[4,5-*d*]pyrimidin-7(6*H*)-one.[931] In contrast, when the chloro-mercuri derivative of 7-acetamido-5-methylthio-*v*-triazolo[4,5-*d*]pyrimidine (**664**) was coupled with the chlorobenzoyl sugar (**47**), a mixture of the two isomeric ribosyl derivatives **665** and **666** was obtained. The structure of compound **665** was confirmed by desulfurization with Raney nickel to afford the known 8-azaadenosine (**662**), and compound **666** was assumed to be the 1-β-D-ribofuranosyl derivative.[931] Deamination of **665** with aqueous nitrous acid furnished **669**, which, when heated with aqueous-ethanolic ammonia, afforded

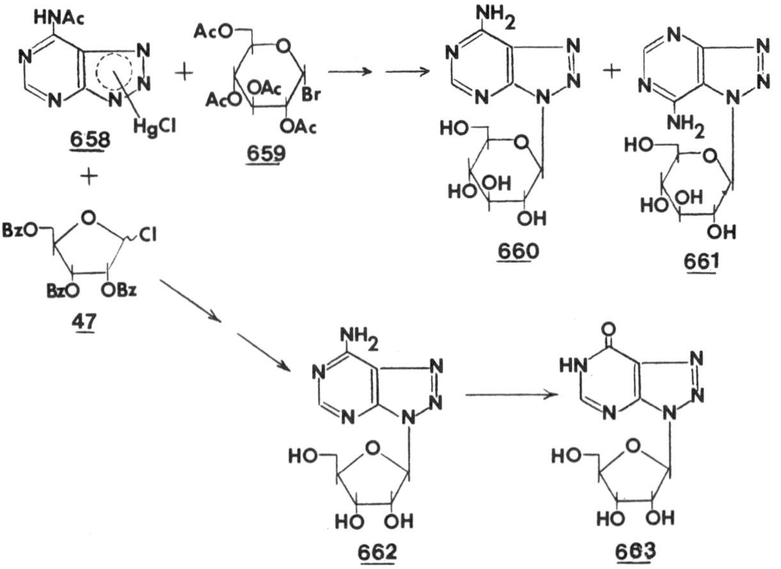

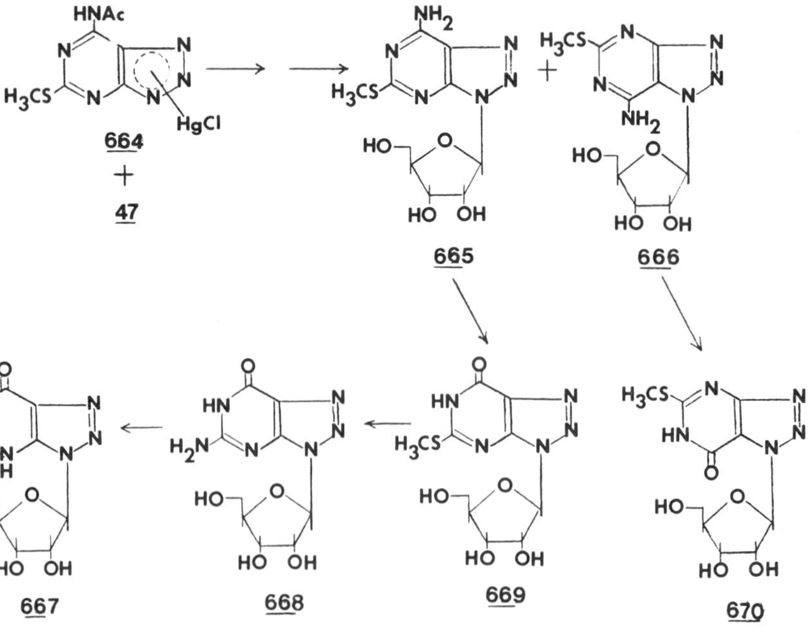

5-amino-3-β-D-ribofuranosyl-v-triazolo[4,5-d]pyrimidin-7(6H)-one (**668**),[931] identical with 8-azaguanosine which had been obtained enzymatically.[933] Deamination of **668** gave 8-azaxanthosine (**667**) in a 47% yield. Similar deamination of **666** with aqueous nitrous acid gave **670**.

The reaction of 2,3,5-tri-O-acetyl-D-ribofuranosyl chloride with the chloromercuri derivative of 7-dimethylamino-5-methylthio-v-triazolo[4,5-d]pyrimidine (**671**) gave the ribofuranosyl derivative. The site of ribosylation was originally assigned as N-1[934] but subsequently established[935] as N-2 (**672**) by a comparison of the UV spectra with that of the 2-alkyl derivative of **671**, prepared by a straightforward synthesis. Desulfurization of **672** afforded 7-dimethylamino-2-β-D-ribofuranosyl-v-triazolo[4,5-d]pyrimidine (**673**).[935]

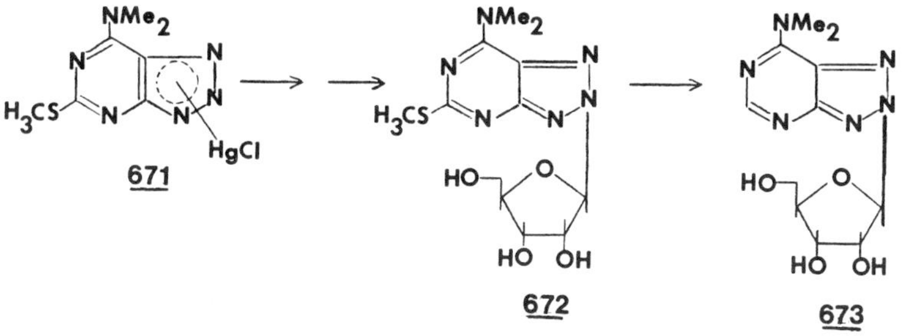

The direct synthesis of 2'-deoxy-8-azaadenosine (**676**) by the condensation of 2-deoxy-3,5-di-O-p-toluoyl-D-$erythro$-pentofuranosyl chloride (**130**) with chloromercuri N-nonanoyl-8-azaadenine (**674**) was described by Tong *et al.*[936] This reaction gave a complex mixture of blocked nucleosides consisting of **675** and several other products. After purification of the complex mixture and deblocking, 2'-deoxy-8-azaadenosine (**676**) and its α-anomer were isolated in an overall yield of 2.7%. Compound **676** was identical with an enzymatically prepared sample.[937] Deamination of **676** with nitrous acid gave 2'-deoxy-8-azainosine (**677**).

2.4.3.2. Synthesis Employing the Acid-Catalyzed Fusion of a Preformed Aglycon and an Acyl Sugar. The acid-catalyzed fusion reaction[938] of N-nonanoyl-8-azaadenine (**678**) with 1,2,3,5-tetra-O-acetyl-D-ribofuranose gave a 14% yield of 8-azaadenosine, after deblocking of the blocked intermediate. Although tetra-O-acetyl-D-ribofuranose with **678** gave only the β-anomer (isolated), the fusion

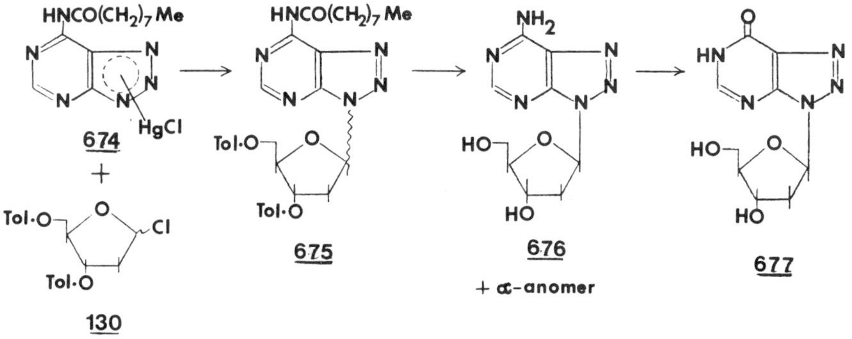

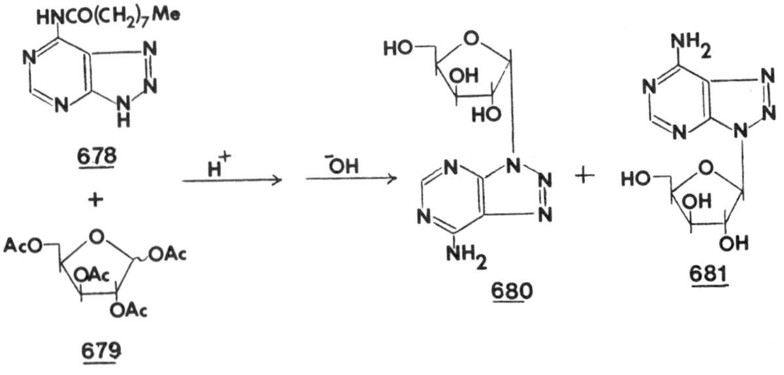

of **678** with 1,2,3,5-tetra-*O*-acetyl-D-xylofuranose (**679**) in the presence of
p-toluenesulfonic acid afforded both the α- and β-nucleosides (**680** and **681**),
along with other unidentified components. The lack of stereospecificity resulting from the use of **679** implies a sugar intermediate that permits, or perhaps
encourages, an approach of the aglycon to C-1 from either side of the ring. In the
case of the reaction with tetra-*O*-acetyl-D-ribofuranose, from which only the
β-anomer has been isolated, the approach of the aglycon is sterically favored
from the top of the ring (i.e., *trans* to the C-2 and C-3 acetyl groups) whereas for
679, with the C-2 acetyl group below the ring and the C-3 acetyl group above the
ring, there is less restriction of approach and thus both anomers are formed.[938]

The fusion of 7-methylthio-*v*-triazolo[4,5-*d*]pyrimidine (**682**) with 1-*O*-
acetyl-2,3,5-tri-*O*-benzoyl-β-D-ribofuranose (**57**), in the presence of bis(*p*-nitro-
phenyl)phosphate, gave a mixture of two nucleoside products. This mixture
was separated into 7-methylthio-3-(2,3,5-tri-*O*-benzoyl-β-D-ribofuranosyl)-*v*-

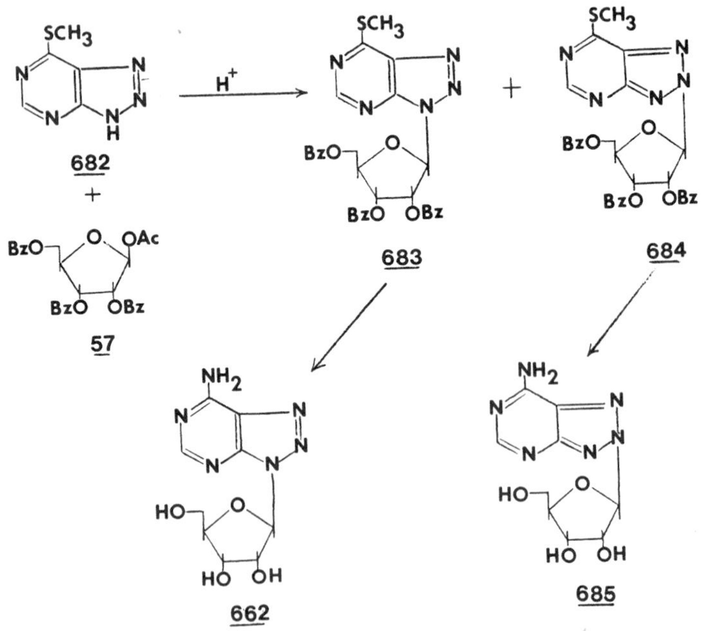

triazolo[4,5-*d*]pyrimidine (**683**, 45.5% yield) and the corresponding N-2 gly-cosyl isomer (**684**, 23% yield.).[939] The structures of these protected nucleosides were confirmed by a conversion to the adenosine analogues (**662** and **685**) with ammonia. However, the fusion of **682** with 1-*O*-acetyl-2,3,5-tri-*O*-benzoyl-D-arabinofuranose (**686a** + **686b**, α/β = 4:1), with or without an acid catalyst, at 200°C gave a four-component mixture. This mixture was resolved by silica gel column chromatography,[940,941] and three of the nucleosides were characterized as 3- (37%), 2- (24%), and 1- (5%) α-D-arabinofuranosyl-7-methylthio-*v*-triazolo-[4,5-*d*]pyrimidines (**687**).[942] The fourth component was found to be unchanged starting sugar, from which **686b** (11%) was isolated. These results indicate that the β-anomeric sugar (**686b**) was unreactive in the fusion reaction, presumably owing to the inability of the 2-*O*-benzoyl group to provide anchimeric assistance in the breaking of the C-1 *O*-acetyl bond.[943]

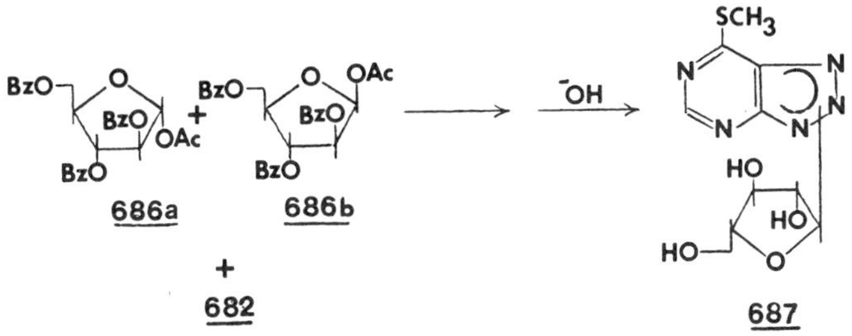

2.4.3.3. Synthesis Employing the Direct Alkylation of a Preformed Aglycon with an Acylated Sugar Halide.

2.4.3.3. Synthesis Employing the Direct Alkylation of a Preformed Aglycon with an Acylated Sugar Halide. The low yield of 8-azaadenosine obtained by the conventional halomercuri procedure[931] (<16%) and acid-catalyzed fusion reaction[938] (14%) prompted Montgomery and co-workers[944] to explore a better and more convenient method for the preparation of 8-azapurine nucleosides. The reaction of *N*-nonanoyl-8-azaadenine (**678**) with 2,3,5-tri-*O*-acyl-D-ribofuranosyl chloride in boiling benzene containing Linde molecular sieve AW500[945] for 4 h was followed by deacylation of the resulting blocked nucleoside to furnish a 48% yield of 8-azaadenosine (**662**). None of the undesired isomeric or anomeric product was detected. This general procedure was then applied to the preparation of other pentofuranosyl-8-azaadenines.[946] The isolated yield of anomerically pure 7-amino-3-β-D-xylofuranosyl-*v*-triazolo[4,5-*d*]pyrimidine (9-β-D-xylofu-ranosyl-8-azaadenine; **688**, R = NH$_2$) from 2,3,5-tri-*O*-acetyl-D-xylofuranosyl chloride was 63%. In the reaction of 3,5-di-*O*-(*p*-chlorobenzoyl)-2-deoxy-D-ribofuranosyl chloride with **678**, triethylamine as the acid acceptor proved superior to the use of molecular sieve. This gave a 44% yield of 7-amino-3-(2-deoxy-D-ribofuranosyl)-*v*-triazolo[4,5-*d*]pyrimidine (**690**, R = NH$_2$), which was separated into the pure anomers prior to deacylation by silica gel column chromatography. The coupling of 2,3,5-tri-*O*-benzyl-α-D-arabinofuranosyl chloride with **678** was followed by treatment with sodium methoxide to give an anomeric mixture of the 2,3,5-tri-*O*-benzyl-D-arabinofuranosyl derivatives of 8-azaadenine in a 40% yield. This is in contrast to the reaction of this sugar with *N*-benzoyladenine, from which only the β-anomer was obtained.[35,947] Debenzylation of the above blocked nucleoside with palladium chloride gave 9-β-D-arabinofuranosyl-8-azaadenine (β-**689**) and the α-anomer α-**689**. The α-anomer (α-**689**) was also obtained[946] in a 78% yield by the reaction of 2,3,5-tri-*O*-acetyl-D-arabinofuranosyl chloride with **678**, followed by sodium methoxide deacylation. Deamination of **688** (R = NH$_2$) and α- and β-**690** (R = NH$_2$) with aqueous nitrous acid gave the corresponding inosine analogues. 9-α-D-Arabino-furanosyl-8-azaadenine-2-^{14}C was also prepared[948] by this procedure.

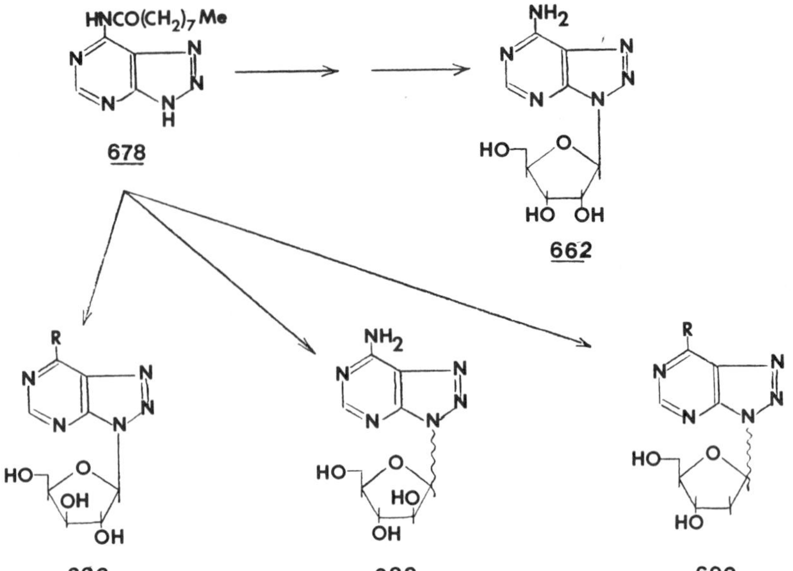

When this glycosylation procedure was applied to 7-methylthio-*v*-triazolo-[4,5-*d*]pyrimidine (**691**),[949] an 89% yield of a 2:1 mixture of 7-methylthio-3- and -2-(2,3,5-tri-*O*-acetyl-β-D-ribofuranosyl-*v*-triazolo[4,5-*d*]pyrimidine (**692** and **693**) was obtained.[950–952] Nucleosides **692** and **693** were proved to be very valuable intermediates for the preparation of a variety of 7-substituted *v*-tri-azolo[4,5-*d*]pyrimidine ribonucleosides[952–954] by the treatment with appropriate nucleophiles. Compound **693** could be rearranged to the thermodynamically more stable **692** with a simple increase in the reaction time or elevated temperature in the presence of molecular sieves.[950] Both 8-aza-6-thioinosine (**694a**)

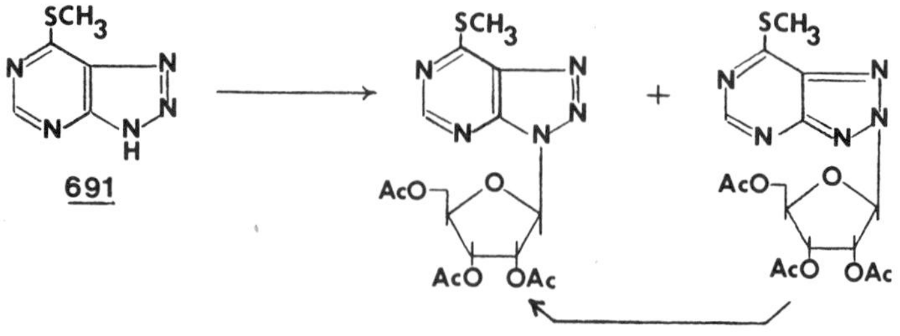

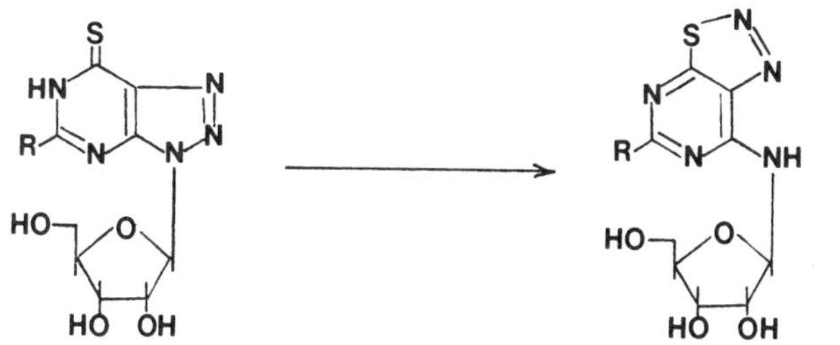

694a, R = H

b, R = NH₂

695a, R = H

b, R = NH₂

and 8-aza-6-thioguanosine (**694b**) were found to be unstable and to rearrange slowly at room temperature to afford N^7-β-D-ribofuranosyl[1,2,3]thiadiazolo-[5,4-*d*]pyrimidine-7-amine (**695a**) and N^7-β-D-ribofuranosyl[1,2,3]thiadiazolo-[5,4-*d*]pyrimidine-5,7-diamine (**695b**).[952,954]

The molecular sieve-catalyzed condensation of 5-acetamido-7-methyl-thio-*v*-triazolo[4,5-*d*]pyrimidine (**696**) with 2,3,5-tri-*O*-acetyl-D-ribofuranosyl chloride gave a rather low yield of a mixture of three nucleosides (N-1, N-2, and N-3 isomers, **697**).[955] Treatment of **697** with hydrogen peroxide in acetic acid, followed by deacetylation, gave the corresponding 8-azaguanine ribonucleosides (**698**, R = H). A similar treatment of **697** with sodium methoxide gave the corresponding 7-methoxy analogues (**698**, R = CH₃). A more convenient synthesis of 7-amino-3-α-D-arabinofuranosyl-*v*-triazolo[4,5-*d*]pyrimidine (**702**, X = H) involved[956] the reaction between 7-methylthio-*v*-triazolo[4,5-*d*]pyrimidine (**699**, R = H) and 2,3,5-tri-*O*-benzoyl-D-arabinofuranosyl bromide (**700**) in the presence of molecular sieves. Treatment of the blocked nucleoside precursor

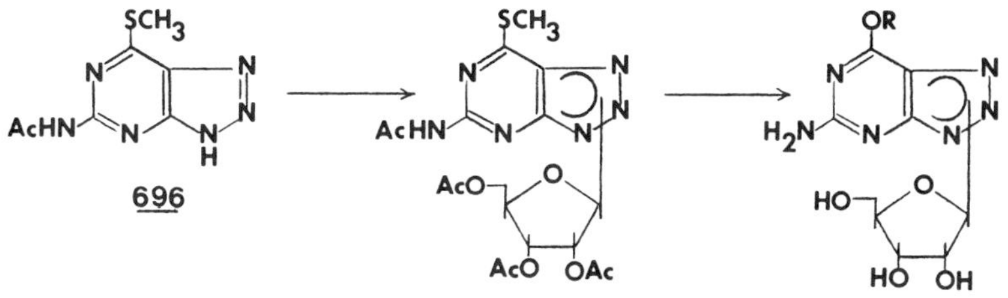

696

697

698

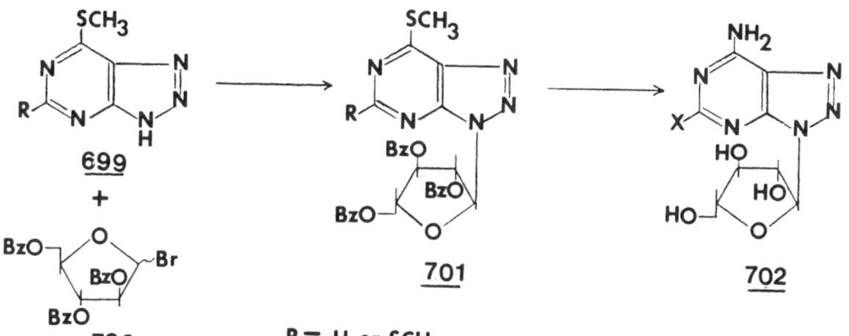

699

+

700

701

702

R = H or SCH₃

701 (R = H) with methanolic ammonia gave **702** (R = H) in an overall yield of 52%. This procedure appears to be superior to the fusion method,[942,957,958] which with **699** (R = H) gives lower yields of **701** (R = H) along with other positional isomers and unreacted sugar. The success of using the bromo sugar **700** for the synthesis of 2-substituted 9-α-D-arabinofuranosyl-8-azaadenine (**702**) was also further demonstrated[956] by reacting 5,7-bis(methylthio)-*v*-triazolo[4,5-*d*]pyrimidine (**699**, R = SCH$_3$) with **700** in the presence of molecular sieves. The blocked nucleoside **701** (R = SCH$_3$) thus obtained, when treated with methanolic ammonia at room temperature, gave **702** (X = SCH$_3$) along with traces of the corresponding N-2 isomer. Displacement of the methylthio group with various nucleophiles gave the corresponding 5-substituted 7-amino-3-α-D-arabinofuranosyl-*v*-triazolo[4,5-*d*]pyrimidines.[956]

In the preparation of 2-fluoro-8-azaadenosine (**706**), some problems were encountered in that the viable starting material 2,6-dichloro-8-azapurine is too unstable[959] to be used in glycosylation studies. However, 2,6-bis(methylthio)-8-azapurine (**699**, R = SCH$_3$) was found to be a key starting material; on treatment with 2,3,5-tri-*O*-acetyl-D-ribofuranosyl chloride (**70**) in toluene in the presence of molecular sieves, **699** (R = SCH$_3$) gave the desired N-3 glycosyl isomer (**703**) in 70–90% yield, along with a minor amount of the N-2 isomer (**708**).[960] Although the separation of these positional isomers was not convenient, 2-methylthio-8-azaadenosine (**704**) obtained by ammonolysis of the above mixture of compounds could be separated chromatographically. Oxidation of **704** with *m*-chloroperoxybenzoic acid gave the methylsulfone (**705**), which, on treatment with ethanolic ammonia, gave 2-amino-8-azaadenosine (**707**). Treatment of **707** in 48% aqueous fluoboric acid with potassium nitrite gave the fluorocompound

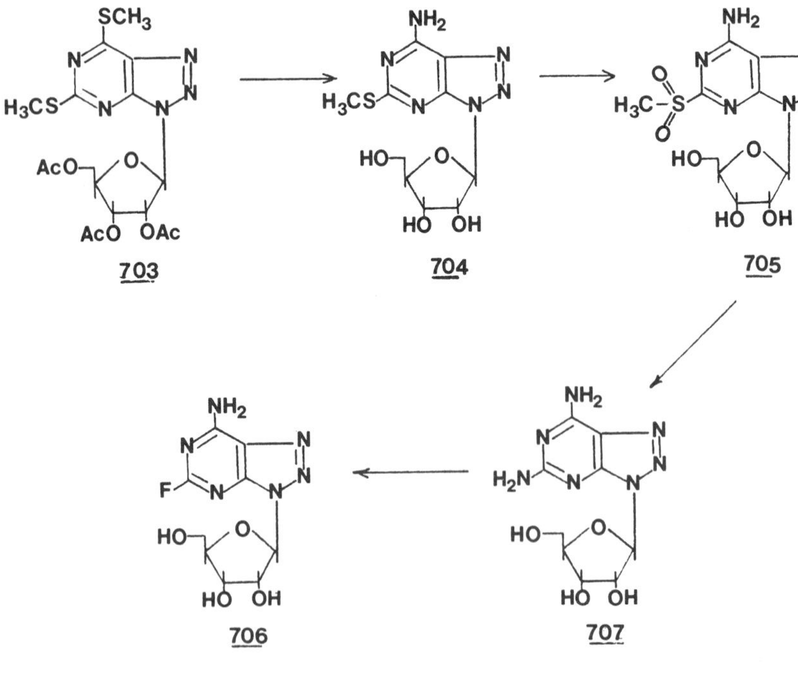

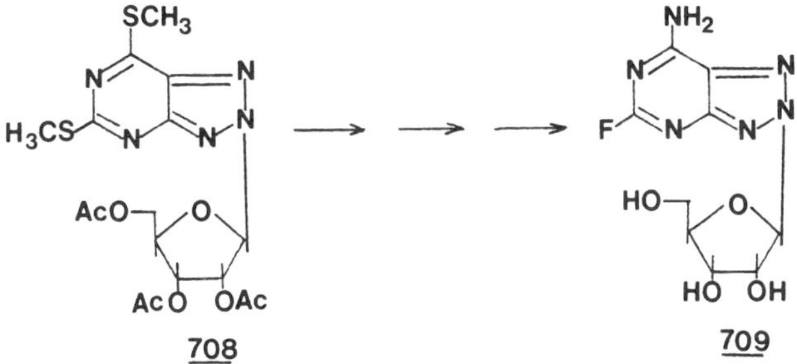

708 **709**

706.[(960,961)] By similar functional group manipulation reactions, the minor component **708** was converted to **709** with the glycon moiety attached at N-2 of the *v*-triazolo[4,5-*d*]pyrimidine ring system.

2.4.3.4. Synthesis Employing the Trimethylsilyl Derivative of a Preformed Aglycon and an Acylated Sugar Halide.   The synthesis of 5-amino-3-(2-deoxy-β-D-*erythro*-pentofuranosyl)-*v*-triazolo[4,5-*d*]pyrimidin-7(6*H*)-one (**711**) and an improved synthesis of 8-azaguanosine (**668**) by the use of the trimethylsilyl procedure were reported by Robins and co-workers.[(939)] A condensation of the tris-trimethylsilyl derivative of 8-azaguanine (**710**) with 2,3,5-tri-*O*-benzoyl-D-ribofuranosyl bromide (**19**) in acetonitrile gave a crystalline nucleoside, which, on deprotection, afforded 8-azaguanosine (**668**) in an overall yield of 35%.

When 2-deoxy-3,5-di-*O*-*p*-toluoyl-D-*erythro*-pentofuranosyl chloride (**130**) was fused with **710** at 100°C, an 80% yield of a mixture of anomeric nucleosides was obtained, which, on deblocking with methanolic ammonia, furnished **711** and the corresponding α-anomer (**712**). A similar fusion[(957)] of the trimethylsilyl derivative **713** of 7-methylthio-*v*-triazolo[4,5-*d*]pyrimidine with 2,3,5-tri-*O*-ben-

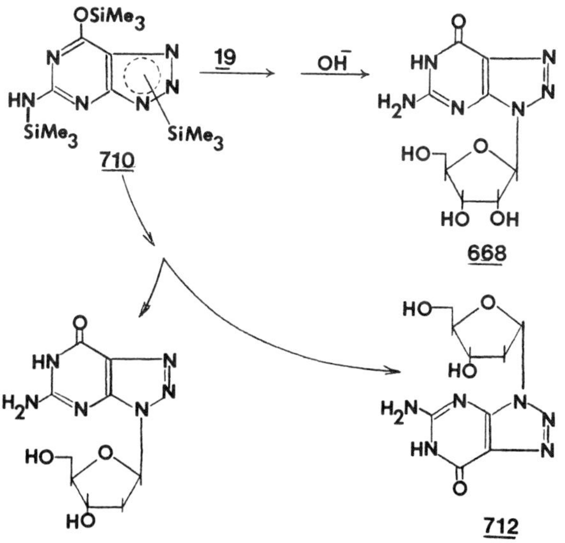

710   **668**

711   **712**

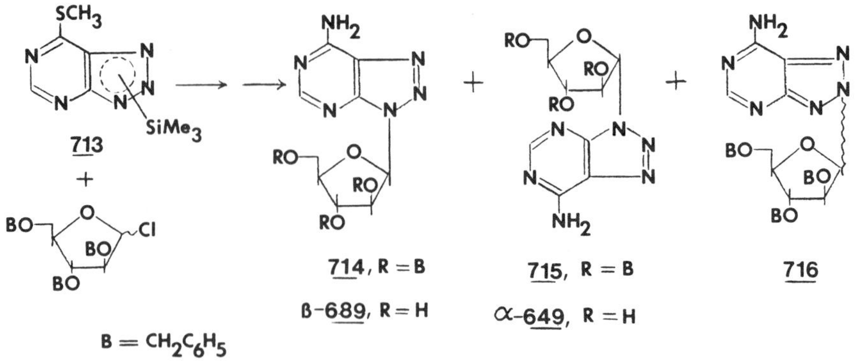

713 + 714, R = B / β-689, R = H + 715, R = B / α-649, R = H + 716

B = CH₂C₆H₅

zyl-D-arabinofuranosyl chloride, followed by ammonolysis, gave a mixture of several nucleosides from which 7-amino-3-(2,3,5-tri-*O*-benzyl-β-D-arabinofura-nosyl)-*v*-triazolo[4,5-*d*]pyrimidine (**714**), the corresponding α-anomer (**715**), and 7-amino-2-(2,3,5-tri-*O*-benzyl-D-arabinofuranosyl)-*v*-triazolo[4,5-*d*]pyrimidine (**716**) were isolated, albeit in low yields. Debenzylation of **714** and **715** with Pd/C in 2-methoxyethanol gave the nucleosides β-**689** and α-**689**, respectively.

2.4.3.5. Synthesis Employing a Ring Closure of a v-Triazole Nucleoside Precursor. The synthesis of *v*-triazolo[4,5-*d*]pyrimidine nucleosides from *v*-triazole derivatives is similar to the synthesis of purines from imidazole precursors. Baddiley *et al.*[962,963] were the first to use this concept for the synthesis of 8-azapurine nucleosides. Starting with a reaction of methyl acetylenedicarboxyl-ate (**717**) with a protected glycosyl azide, the diester (**718**) and further the diamide (**719**) of substituted triazoledicarboxylic acid were obtained. Employing the procedure developed for analogous purine derivatives,[964,965] a Hofmann

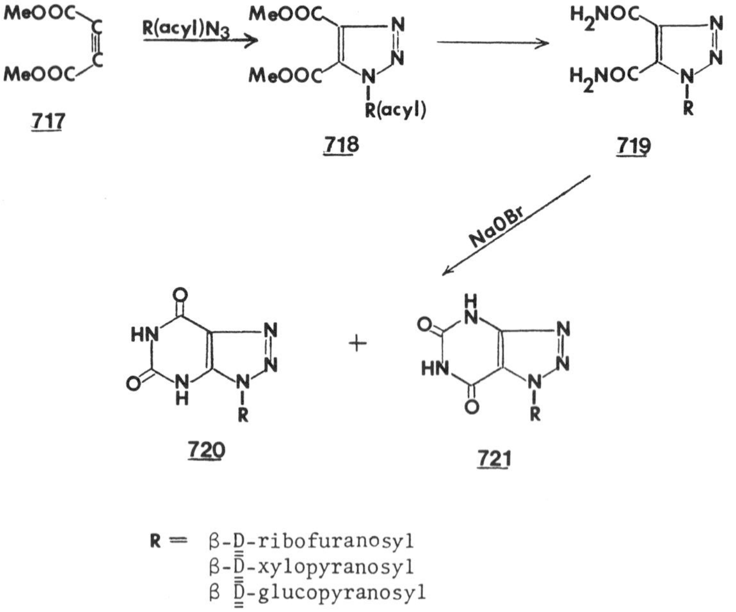

717 → 718 → 719

NaOBr

720 + 721

R = β-D-ribofuranosyl
     β-D-xylopyranosyl
     β D-glucopyranosyl

reaction (alkaline hypobromite) gave a mixture of the corresponding 1- and 3-glycosyl-5,7-dioxo-*v*-triazolo[4,5-*d*]pyrimidines (**720** and **721**).

Ring closure of 2,3,5-tri-*O*-benzoyl-β-D-ribofuranosyl azide (**722**)[966] with cyanoacetamide in aqueous DMF containing KOH gave a 46% yield of 5-amino-1-β-D-ribofuranosyl-*v*-triazole-4-carboxamide (**723**).[939] Ring annulation of **723** with diethoxymethyl acetate furnished 8-azainosine (**663**) in 60% yield, which constitutes an improvement over other procedures.[967] Over the years, this procedure has been improved considerably, and the overall yield of 8-azainosine from the sugar azide has been more than doubled (13% to 30%).[968]

722 723 663

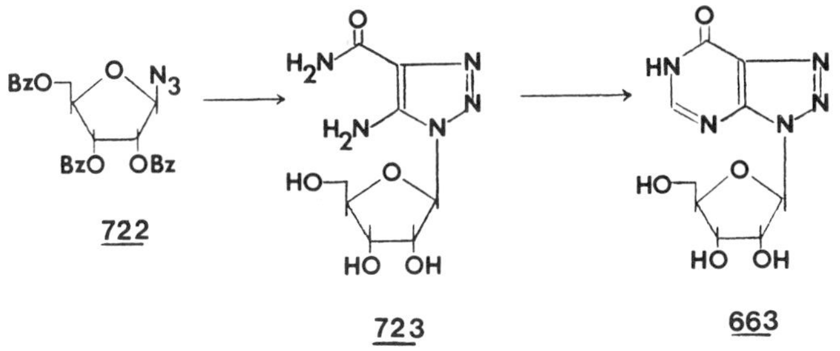

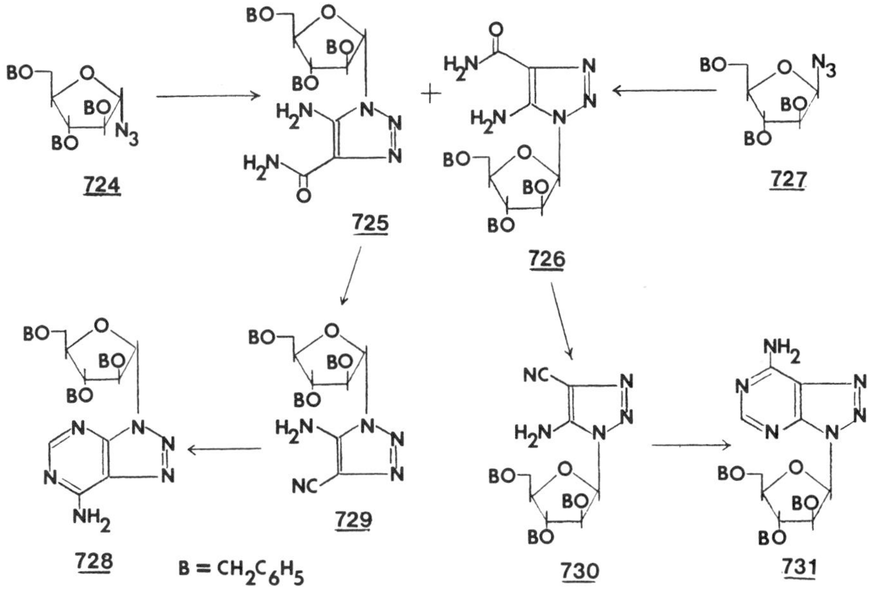

Similarly, when 2,3,5-tri-*O*-benzyl-α-D-arabinofuranosyl azide (**724**)[957] was treated with KOH and cyanoacetamide in aqueous DMF, the sole product was 5-amino-1-(2,3,5-tri-*O*-benzyl-α-D-arabinofuranosyl)-*v*-triazole-4-carboxamide (**725**).[957] A similar cycloaddition with the β-azide **727** gave two products, (**725**) and its β-anomer **726**, in a ratio of 14:1. A mechanism for this unusual rearrangement has been proposed.[969] Dehydration[957] of **725** and **726** with *p*-toluenesulfonyl chloride in pyridine gave the corresponding nitriles, **729** and **730**. Ring annulation of **725** and **726** with diethoxymethyl acetate gave the

724 725 726 727 728 729 730 731

benzyl-blocked α-arabinofuranosyl-8-azahypoxanthine and its β-anomer, respectively. Similar cyclization of either 5-amino-4-cyano-1-(2,3,5-tri-*O*-benzyl-α-D-arabinofuranosyl)-*v*-triazole (**729**) or the corresponding β-anomer (**730**) with diethoxymethyl acetate and methanolic ammonia gave the respective 7-amino-3-(2,3,5-tri-*O*-benzyl-D-arabinofuranosyl)-*v*-triazolo[4,5-*d*]pyrimidines (**728** and **731**).[957,958] Debenzylation of **728** and **731** furnished 7-amino-3-α-D-arabino-furanosyl-*v*-triazolo[4,5-*d*]pyrimidines (α-**689**) and its β-anomer (β-**689**), respectively. The anomeric configuration of these unprotected nucleosides was unequivocally established by the periodate oxidation and sodium borohydride reduction techniques.[969] Additional supporting evidence was also derived from the fact that 8-aza-α-araA (α-**689**) was *not* a substrate for calf intestine deaminase, whereas the β-anomer (β-**689**) was a substrate.[117]

2.4.3.6. Synthesis Employing the Ring Closure of a Pyrimidine Nucleoside Precursor. The general applicability of the ring-closure method for the synthesis of 8-azapurines by the ring annulation of substituted 4,5-diaminopyrimidine by nitrous acid[970,971] was shown by the preparation of glycosyl derivatives of 8-azaguanine (**732**, R^1 = OH),[972–975] where the glycosyl residue was unambiguously attached to position 3 of the *v*-triazolo[4,5-*d*]pyrimidine ring. An analogous cyclization of 5-aminocytidine (**733**) gave 6-β-D-ribofuranosyl-*v*-triazolo[4,5-*d*]pyrimidin-5-one (**734**).[976] This procedure was used successfully for the preparation of the 8-azaadenine-D-glucuronic acid derivatives (**735**).[977]

A novel, one-step synthesis of 2-oxo-8-azapurines by treatment of certain 5-

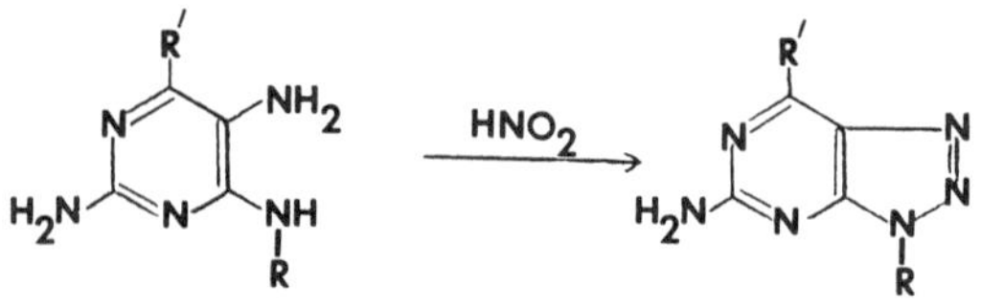

R =   D-ribityl
         D-sorbityl
         tetraacetyl-β-D-glucopyranosyl

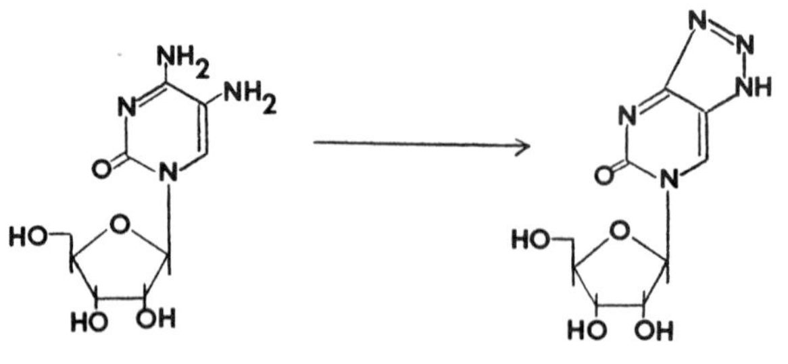

733   **734**

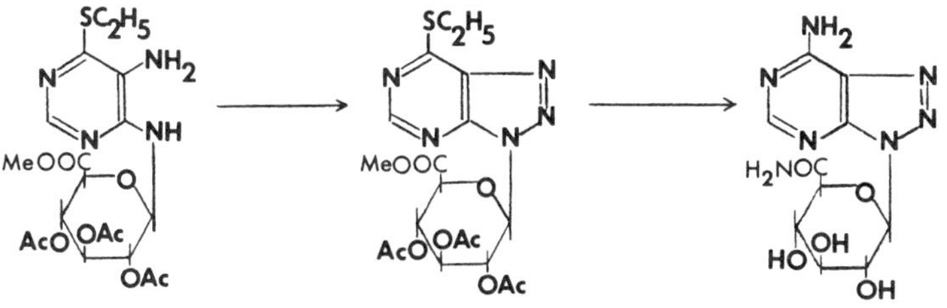

735

nitropyrimidines with sodium azide was described by Fox and co-workers.[864,865] This novel route has been used to prepare a class of nucleosides of potential biochemical interest. The treatment of 5-nitrouridine (**736**)[978] with sodium azide in DMF at room temperature for two days gave an 85% yield of the monohydrate of 4-β-D-ribofuranosyl-*v*-triazolo[4,5-*d*]pyrimidine-5,7-dione[737]. Similarly, 2,3,5-tri-*O*-benzoyl-5-nitrocytidine (**739**)[978] with sodium azide afforded the protected nucleoside **740**, which, after deblocking, gave 7-amino-4-β-D-ribofuranosyl-*v*-triazolo[4,5-*d*]pyrimidin-5-one (**738**) in ~50% yield. Deamination of **738** with nitrous acid furnished **737**. Analogously, 3-β-D-glucopyranosyl-8-azaisoguanine was obtained from 1-(tetra-*O*-acetyl-β-D-glucopyranosyl)-5-nitro-cytosine.[978]

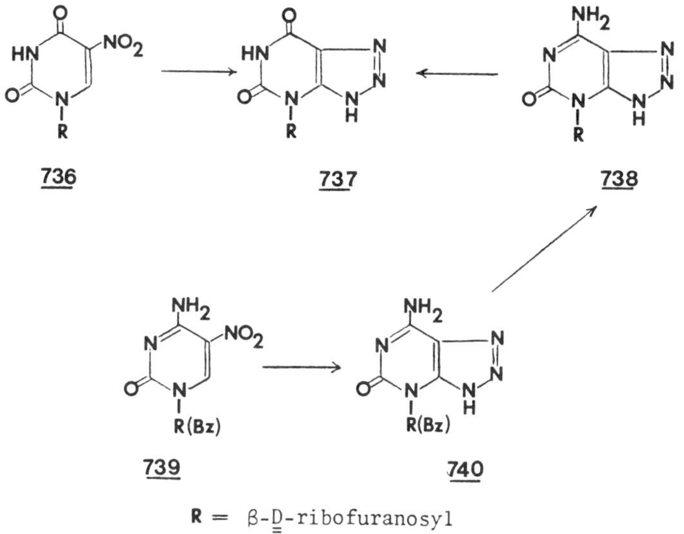

736   **737**   **738**

739   **740**

R = β-D-ribofuranosyl

The proposed[864,865] plausible mechanism for this unexpected cyclization involves the nucleophilic attack by azide ion on C-6 of 5-nitrouridine (**736**) to form **741**. Compound **741** then undergoes an intramolecular nucleophilic attack by C-5 of the *aci*-nitro group on the terminal azide nitrogen to form the 4,5-dihydro-8-azaxanthine derivative **742**, which, by elimination of the elements of nitrous acid, gives the sodium salt of 8-azaxanthosine (**743**). Acidification of **743** thus affords **737**. During the course of the reaction, the liberation of nitrous acid was indeed detected.

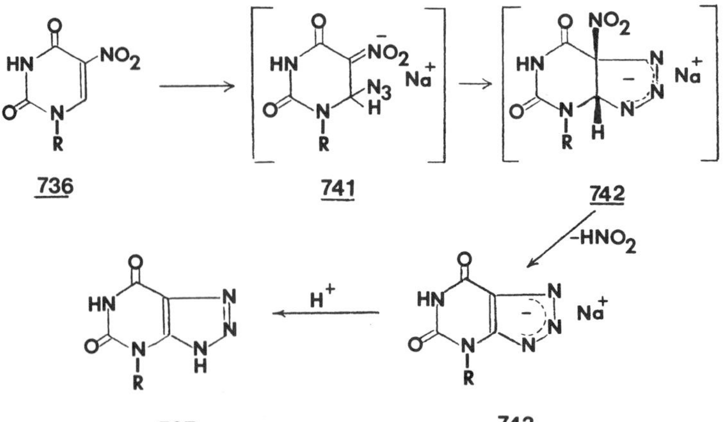

A similar single-step synthesis of 3,5′-cyclo-*v*-triazolo[4,5-*d*]pyrimidine nucleosides was also reported.[979] It involved heating 5′-*O*-benzoyl- or 5′-*O*-tosyl-2′,3′-*O*-isopropylidene-5-bromouridine (**744**, R = OH) with excess sodium azide in DMF at 110°C. Under these conditions, over a 70% yield of 3,5′-cyclo-4-(2,3,*O*-isopropylidene-β-D-ribofuranosyl)-*v*-triazolo[4,5-*d*]pyrimidine-5,7-dione (**745**, R = OH) was obtained, which, on treatment with an acid, gave **746**. Application of this reaction to the 5′-*O*-mesylcytidine derivative (**744**, R = NH₂, R′ = Ms) gave a mixture of two products, from which **745** (R = NH₂) was isolated in an 8% yield. It was shown[979] that this intramolecular thermal cyclization proceeds through a nucleophilic addition of the azide group, followed by elimination of the elements of hydrogen halide.

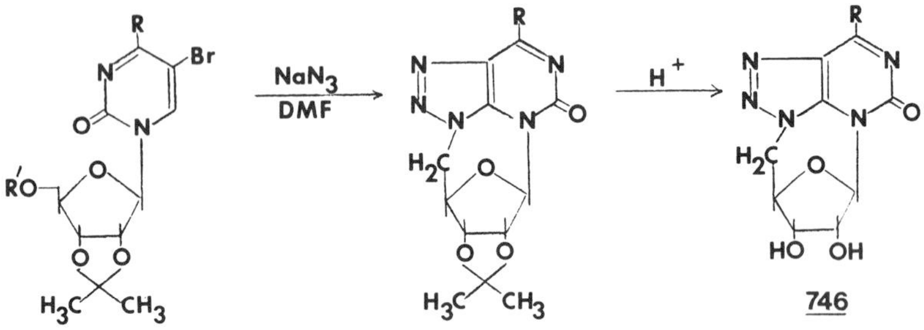

The development of the concept of basic ring opening of suitably substituted purine nucleosides followed by reclosure with a nitrosating agent to give the triazole ring has been reported by Montgomery and Thomas.[912,967,980] Ring opening of 6-chloro-9-(2,3-*O*-isopropylidene-β-D-ribofuranosyl)purine (**747a**) with 0.5 *N* sodium hydroxide solution gave a 67% yield of 4-chloro-5-formamido-6-(2,3-*O*-isopropylidene-β-D-ribofuranosylamino)pyrimidine (**748a**) and a minor amount of 2′,3′-*O*-isopropylideneinosine. Selective hydrolysis of the

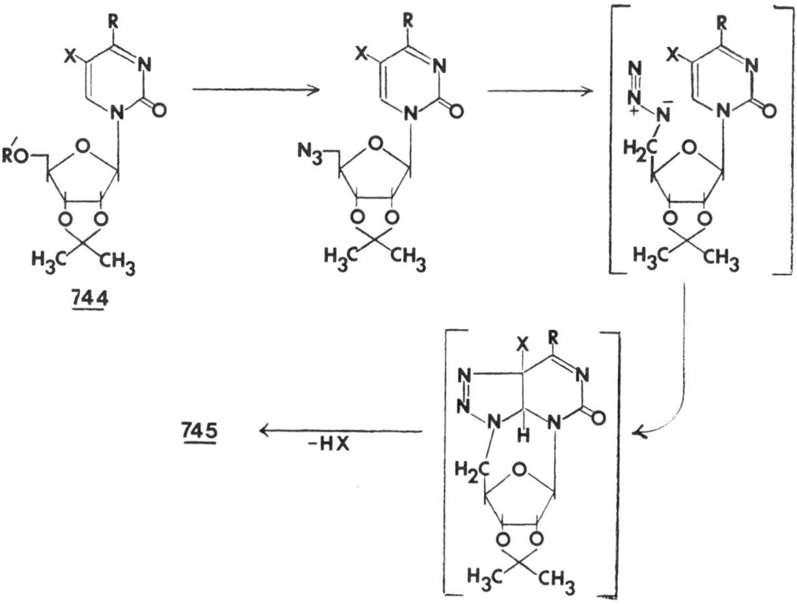

744 → → → 745 ← -HX

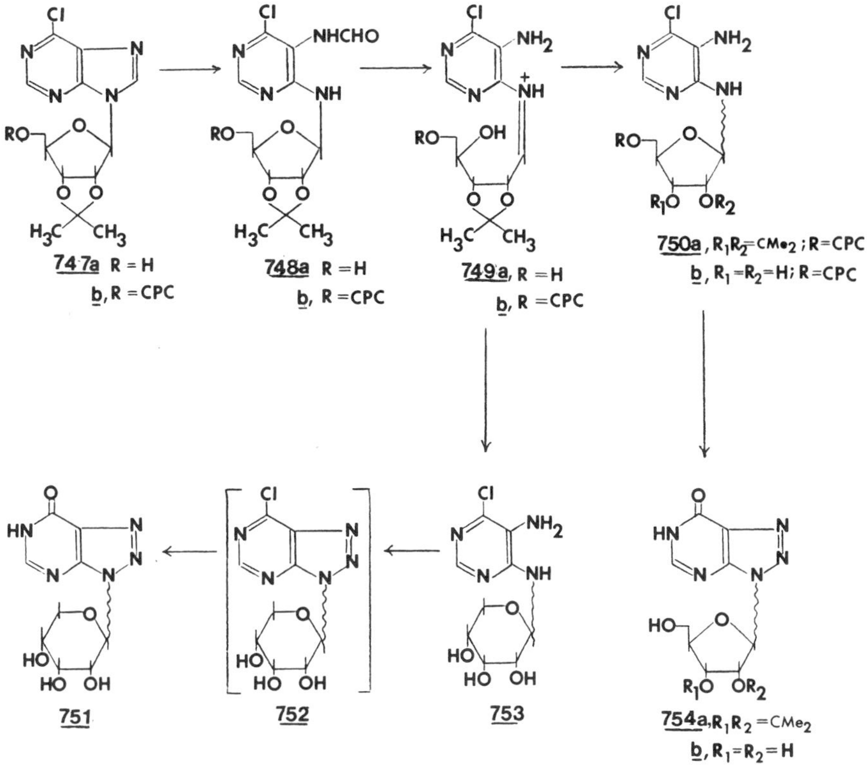

747a R = H
b, R = CPC

748a R = H
b, R = CPC

749a, R = H
b, R = CPC

750a, R_1R_2=CMe$_2$; R=CPC
b, R_1=R_2=H; R=CPC

751

752

753

754a, R_1R_2 = CMe$_2$
b, R_1=R_2=H

formyl group of **748a** by further treatment with a base was not possible; however, the amide linkage was hydrolyzed under mild acidic conditions. The major product was the β-D-ribopyranoside (β-**753**), and a small amount of the α-anomer (α-**753**) was also formed. Treatment of **753** with nitrous acid gave an anomeric mixture of 9-D-ribopyranosyl-8-azahypoxanthine (**751**) via **752** with the chlorine being hydrolyzed under these conditions.

By protecting the 5′-OH group of **747a** with an acid- and base-stable *m*-chlorophenylcarbamoyl group (CPC) (**747b**), the furanose–pyranose isomerization of the sugar moiety was prevented. On treatment with base, **747b** underwent an imidazole ring cleavage to give a 45% yield of **748b**. Acid hydrolysis of the formyl group to give **750a** was followed by ring annulation with nitrous acid to afford **754a**; subsequent removal of the blocking groups gave 8-azainosine (β-**754b**) and its α-anomer (α**754b**) in almost equal amounts.[980] A similar reaction sequence was used to obtain the carbocyclic analogues of 8-azainosine, 8-azaadenosine,[981–983] and puromycin.[984] The syntheses of the carbocyclic analogues of 2′-deoxy- (or arabinosyl-) 8-azaguanosine (**757c** and **757a**) and 2′-deoxy- (or arabinosyl-) 2-amino-8-azaadenosine (**757d** and **757b**) by ring annulation of the appropriate pyrimidine precursors (**755a,b**) have also been reported.[985,986]

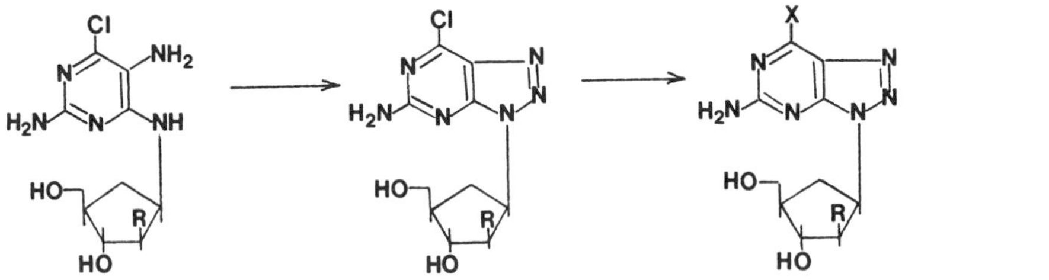

755a, R = OH   756a, R = OH   757a, R = OH; X = OH
b, R = H   b, R = H   b, R = OH; X = NH₂
                                        c, R = H;   X = OH
                                        d, R = H;   X = NH₂

A good deal of effort has been devoted to the preparation of 5,7-disubstituted *v*-triazolo[4,5-*d*]pyrimidine nucleosides by this ring annulation procedure. When 2-acetamido-4-benzylthio-6-chloro-5-nitropyrimidine (**758**) was reacted with stable 2,3-*O*-isopropylidene-β-D-ribofuranosylamine *p*-toluene sulfonate[987] in the presence of sodium bicarbonate (found to be better than triethylamine), the N^4-(2,3-*O*-isopropylidene-β-D-ribofuranosyl)pyrimidine was obtained, which, on catalytic reduction with Pd/C (Raney nickel gave low yield), afforded the ribosylaminopyrimidine (**759**). Nitrosation of **759** furnished the *v*-triazolo[4,5-*d*]pyrimidine derivative **760**. Mild acid hydrolysis (0.5 *N* H_2SO_4) first removed the *N*-acetyl and then the isopropylidene protecting group to give the desired 5-amino-7-benzylthio-3-β-D-ribofuranosyl-*v*-triazolo[4,5-*d*]pyrimidine (**762**). Through a nucleophilic displacement of the benzylthio group, **762** was converted to 8-azaguanosine (**668**), 8-aza-6-thioguanosine (**761**), and other 7-substituted 5-amino-3-β-D-ribofuranosyl-*v*-triazolo[4,5-*d*]pyrimidines.[953,955]

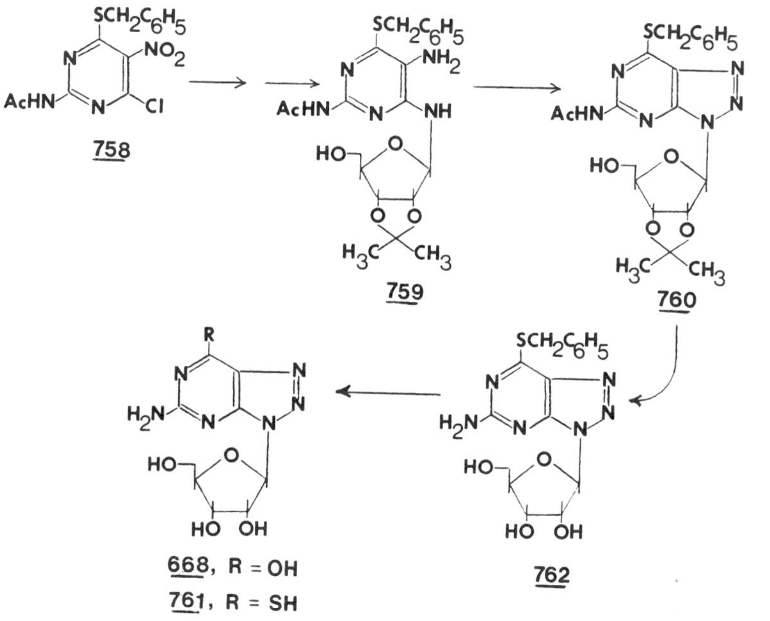

2.4.3.7. *Synthesis Employing the Alkali Metal Salt of a Preformed Aglycon and an O-Methanesulfonyl Sugar.* Schroeder reported[988] in a patent a rather unique method for the preparation of various 3-β-D-psicopyranosyl-*v*-triazolo[4,5-*d*]-pyrimidines. This method involves the condensation of the alkali metal salt, particularly the sodium salt, of an appropriate 8-azapurine with 3-*O*-methane-sulfonyl-D-fructose in an aprotic solvent. It was postulated[988] that the 3-*O*-methanesulfonyl-D-fructose (**763**) undergoes a transient formation of a 2,3-epoxide (**764**), which then reacts with the sodium salt of 8-azaadenine (**765**) with a concomitant opening of the epoxide ring. The 8-azaadenine moiety enters the 2-position of the fructose in the β-position while the OH group, which is re-formed at the 3-position, is below the plane of the ring. This is in an epimeric relation to the OH group in the 3-position of the original fructose ring, and the product furnished is 7-amino-3-β-D-psicopyranosyl-*v*-triazolo[4,5-*d*]pyri-midine (**766**).

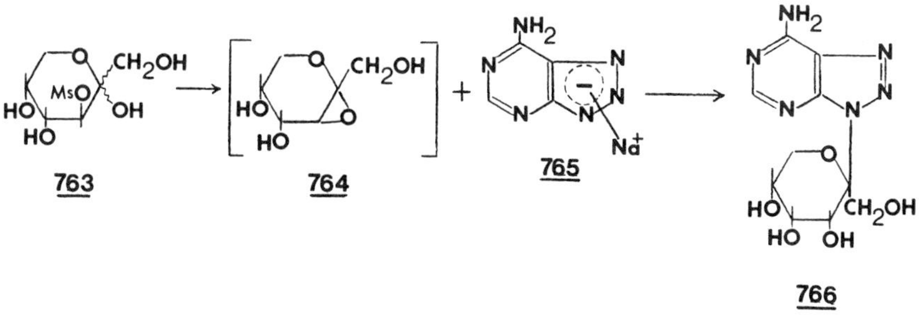

By replacing 8-azaadenine with 8-azahypoxanthine, 8-azaguanine, 7-methylthio-*v*-triazolo[4,5-*d*]pyrimidine, and various other substituted *v*-triazolo-[4,5-*d*]pyrimidines, the corresponding 3-β-D-psicopyranosyl-*v*-triazolo[4,5-*d*]-pyrimidines were prepared.[(988)]

2.4.3.8. Preparation of v-Triazolo[4,5-d]pyrimidine Nucleosides and Nucleotides by an Enzymatic Process. The enzymatic approach[(989)] for the synthesis of guanosine, in the presence of purine nucleoside phosphorylase, was realized by Friedkin[(933)] as early as 1954. This approach was used to prepare 8-azapurine nucleosides. It was shown that both the ribofuranosyl[(933)] and the deoxyribofuranosyl[(933,990)] derivatives of 8-azaguanine could be prepared enzymatically using either ribose 1-phosphate or deoxyribose 1-phosphate in the presence of horse-liver purine nucleoside phosphorylase. On the basis of the specific course of the enzyme reactions, it was assumed that these nucleosides have an N-3 glycosidic linkage. Subsequently, the enzymatic synthesis of 8-azaguanosine phosphates (mono-,[(744,746)] di-,[(991,992)] and triphosphate[(993,994)]) was accomplished by transphosphorylation.

A fermentative conversion of 8-azaguanine to 8-azaguanosine by *Bacillus cereus*[(995,996)] and 8-azahypoxanthine to 8-azainosine by *Flavobacterium aquatile* IFO 3772[(997)] has been described. Several other microorganisms,[(998,999)] particularly those belonging to *Brevibacterium ammoniagenes*,[(1000)] have been employed to obtain various 3-β-D-ribofuranosyl-*v*-triazolo[4,5-*d*]pyrimidine 5'-phosphates.[(1000–1002)] A preparative-scale enzymatic synthesis of 8-azaadenosine 5'-monophosphate using phenyl phosphate as the phosphate donor in the presence of a phosphotransferase from carrots has also been reported.[(513)]

2'-Deoxy-8-azaadenosine (**676**) has been prepared [(937)] by the transfer of 2'-deoxyribose from thymidine or 2'-deoxyinosine to 8-azaadenine with nucleoside deoxyribosyltransferase from *Lactobacillus helveticus*. Deamination of **676** with adenosine deaminase gave 2'-deoxy-8-azainosine (**677**).[(1003)] Similarly, by using the enzyme nucleoside 2'-deoxyribosyltransferase (EC 2.4.2.6) from *Lactobacillus leichmannii* in the presence of 2'-deoxycytidine and 8-azaguanine, 5-amino-3-(2-deoxy-β-D-*erythro*-pentofuranosyl)-*v*-triazolo[4,5-*d*]pyrimidin-7-one (2'-deoxy-8-azaguanosine) was obtained[(764)] in a 21% yield. However, the specific rotation {$[\alpha]_D^{30}$ +79° (*c* 1.0, DMF)} observed for the 2'-deoxy-8-azaguanosine prepared by the enzymatic process does not agree with that of the synthetic product[(939)] {$[\alpha]_D^{28}$ −77.3° (*c* 1.0, DMF)}.

The enzymatic synthesis of 2'-deoxy-8-azaadenosine (**676**),[(1004)] 2-amino-

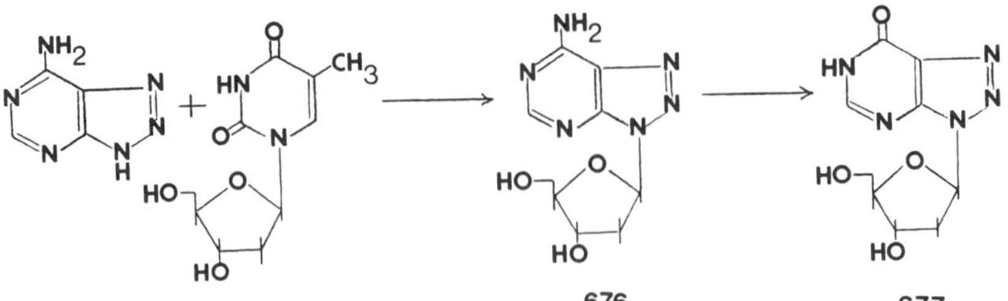

676   **677**

2'-deoxy-8-azaadenosine,[1004,1005] as well as 2-fluoro-2'-deoxy-8-azaadeno-sine[960] using nucleoside deoxyribosyltransferase from *L. leichmannii*[1006,1007] has been reported.

2.4.3.9. Physical Properties. Several reports have appeared dealing with the ^{13}C-NMR spectroscopic study of 8-azaadenosine,[1008] the use of the generalized perturbation theory[1009] to predict the site of protonation in 8-azaadeno-sine,[1010] ORD studies,[1011] and conformation of the 8-azapurine nucleo-sides.[1012–1019] It is of particular interest that the glycosidic bond length in 8-azaadenosine (1.445 Å) is shorter than that in either purine (average value, 1.465 Å) or pyrimidine (1.495 Å) nucleosides. This shortening of the glycosidic bond may be a function of the "high-*anti*" conformation around the glycosidic bond.[1020,1021] This observed "high-*anti*" (intermediate between *syn* and *anti*) conformation of 8-azaadenosine may explain[1012] its weak binding (relative to adenosine) to adenosine deaminase, which is inactive on nucleosides with the *syn* conformation but is active on nucleosides with the *anti* conformation.[1022]

The coding properties of trinucleoside dephosphates containing 8-aza-guanosine have been studies by Grünberger *et al.*,[1023,1024] and it was found that 8-azaguanosine could replace only guanosine.

2.4.3.10. v-Triazolo[4,5-d]pyrimidine Nucleotides. The chemical synthesis of 8-azaguanosine 5'-monophosphate (**769**) was accomplished by Montgomery and Thomas[1025] by the conventional procedure involving the reaction of 2-cyano-ethyl phosphate with the 2',3'-*O*-isopropylidene derivative of 8-azaguanosine (**767**) in pyridine containing DCC, followed by a hydrolytic removal of the blocking groups from **768** to afford **769**. Compound **769** was found to be identical in all respects with an enzymatically prepared sample.[744] The 2'(3')-phosphate and the 2',3'-cyclic phosphate of 8-azaguanosine were also prepared by the same authors.[1026] However, the synthesis of 8-azaguanosine 3',5'-cyclic phosphate has to date not yet been achieved. 8-Azaguanosine 2',3'-cyclic phos-phate was also prepared from 8-azaguanosine 2'(3')-phosphite by the action of hexachloroacetone and DCC.[1027,1028]

Phosphorylation of 8-azaadenosine (**662**) with phosphorus oxychloride in triethyl phosphate gave 8-azaadenosine 5'-monophosphate (**770**),[118,451,454] which was subsequently converted into the 5'-diphosphate[454] and the 5'-tri-phosphate[451,454] by using the conventional procedure. Cyclization of 8-azaade-

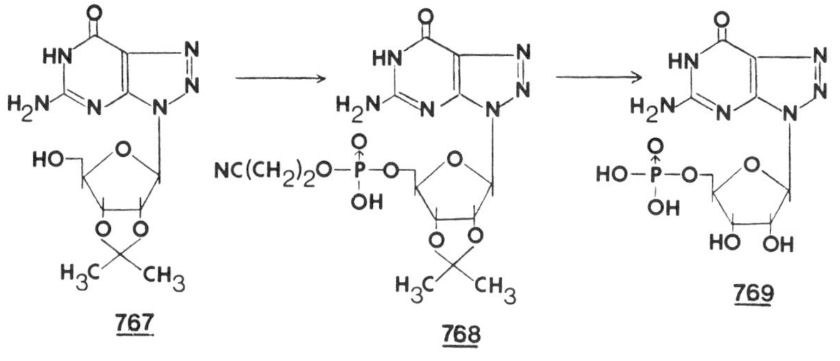

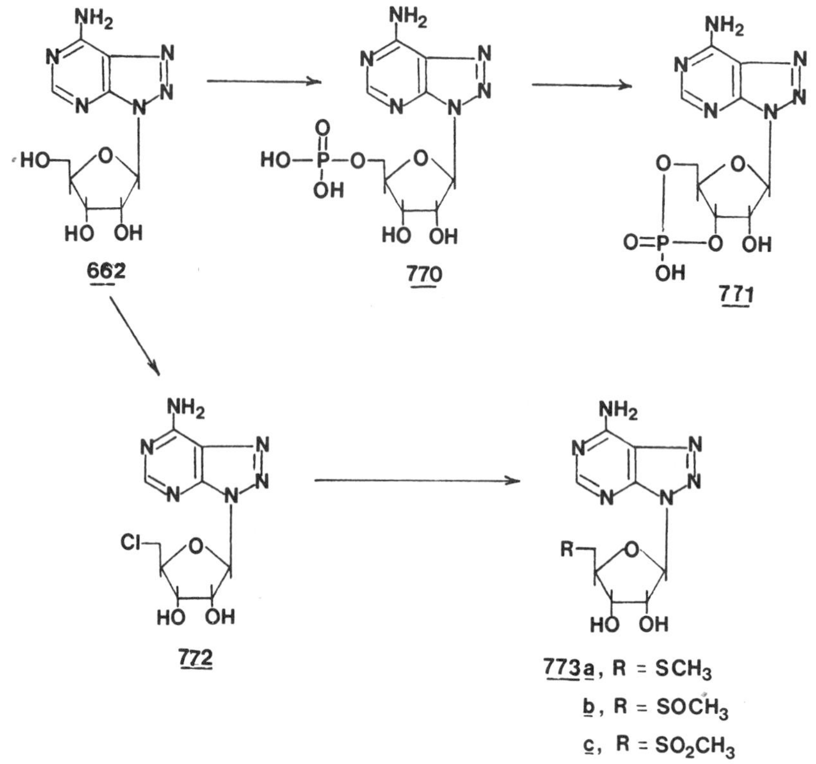

662 → 770 → 771

772 → 773a, R = SCH₃
     b, R = SOCH₃
     c, R = SO₂CH₃

nosine 5'-monophosphate (**770**) with DCC and 4-morpholino-*N,N'*-dicyclo-hexylcarboxamidine in boiling pyridine gave 8-azaadenosine 3',5'-cyclic phosphate (**771**).[118]

5'-Chloro-5'-deoxy-8-azaadenosine (**772**) was found to be an important intermediate for the synthesis of 5'-deoxy-5'-(methylthio)-8-azaadenosine (**773a**). Compound **773a** is an analogue of 5'-deoxy-5'-methylthioadenosine (MTA), which is known to inhibit the transmethylation reactions of *S*-adenosyl-L-methionine (SAM) *in vitro*.[1029,1030] Chlorination of the unprotected 8-azaadenosine with the thionyl chloride–hexamethylphosphoramide reagent[479,480] yielded **772** (58% yield). Compound **772** was converted to 7-amino-3-(5-deoxy-5-methylthio-β-D-ribofuranosyl)-*v*-triazolo[4,5-*d*]pyrimidine (**773a**) by the reaction with methyl mercaptan.[1031] Oxidation of **773a** with 30% hydrogen peroxide gave the corresponding 5'-deoxy-5'-methylsulfinyl derivative (**773b**), and attempts to prepare the sulfone (**773c**) by prolonged oxidation only resulted in extensive glycosyl cleavage. However, **773c** was obtained in high yield[1031] by a rapid oxidation of **773a** with neutral potassium permanganate.

The syntheses of 2',3'-unsaturated[1032] and 2'- and 3'-keto[1033] nucleosides, which are valuable intermediates for the preparation of nucleosides with the 2'- and 3'-inverted configuration of sugar in 8-azapurines, are also described by Moffatt and co-workers in the patent literature.

2.4.3.11. Biochemical Properties of v-Triazolo[4,5-d]pyrimidine Nucleosides and Nucleotides. Several reviews are available[912,953,961] which deal with the biological properties of azapurine nucleosides. 8-Azadenosine (**662**) and 8-azainosine (**663**) are cytotoxic to HEp-2 cells in culture[912] and exhibit *in vivo* activity against L1210 and P388 leukemia.[952,953] 8-Azaadenosine is a substrate for adenosine kinase[781,1034] and inhibits cells lacking adenine phosphoribosyltransferase (APRT). 8-Azaadenosine is also a good substrate for adenosine deaminase. It has a V_{max} (μmol/min per mg) 3½ times that of adenosine, and thus it is rapidly deaminated in cells. Introduction of a fluorine at C-2 of 8-azaadenosine (**706**) did not affect the K_M, but it did reduce the V_{max} to 1/300th that of **662** and decreased the cytotoxicity against HEp-2 cells. The cytotoxicity of 2'-deoxy-8-azaadenosine (**676**) was reduced by the same relative amount by the introduction of fluorine at C-2 (Table III).

9-α-D-Arabinofuranosyl-8-azaadenine (α-ara-8-azaA, α-**689**) is cytotoxic to cultured mammalian cells[1035,1036] and has significant antiviral activity[957,1037] *in vitro*, equal to that of araA against herpes virus type 1. α-Ara-8-azaA is a substrate for adenosine kinase, and its principal metabolite in cell culture is α-ara-8-azaATP.[1035] However, 8-α-D-arabinofuranosyl-8-azaadenine did not show any perceptible growth-inhibitory effects against L1210 leukemia cells in culture.[1038] 8-Aza-6-thioinosine (**694a**) was found to be a substrate for adenosine kinase (EC 2.7.1.20) and was cytotoxic; the 50% inhibitory concentration for HEp-2 cells is 1.8 μM.[1039] 9-β-D-Xylofuranosyl-8-azaadenine (**681**) and 2'-deoxy-8-azaadenosine (**676**) are active as antiviral agents *in vitro*. The 2'-deoxy derivative (**676**) is neither phosphorylated nor catabolized to 8-azahypoxanthine.[937] The α-anomer of 2'-deoxy-8-azaadenosine is only toxic at high levels, and the 5'-substituted derivatives of 8-azaadenosine,[1031] which cannot be phosphorylated, are essentially nontoxic. 8-Aza-GMP inhibits succino-AMP synthetase isolated from promastigotes of *Leishmania donovani*[801] and is a competitive inhibitor of mammalian adenylosuccinate synthetase with respect to GTP.[1040] 8-Azainosine (**663**) is very active against *Leishmania donovani, Trypanosoma cruzi,* and *Trypanosoma gambiense.*[593] It has been shown that 1-(2-deoxy-β-D-*erythro*-pentofuranosyl)-8-azapurin-2-one 5'-monophosphate inhibits the enzyme thmidylate synthetase.[1041]

The antitumor activity and cytotoxicity of 8-azaguanine is presumably due to its conversion to the ribonucleoside triphosphate, which is incorporated into RNA.[924,1042,1043] 8-Azaadenine, 8-azahypoxanthine, and their ribonucleosides do not act as feedback inhibitors or block nucleotide interconversions, and their

Table III. Cytotoxicity of 8-Azaadenine and Derivatives against HEp-2 Cells[a]

Compound	$IC_{50, \mu M}$[b]
8-Azaadenine	20
8-Azaadenosine	0.7
2-Fluoro-8-azaadenosine	3
2'-Deoxy-8-azaadenosine	6
2-Fluoro-2'-deoxy-8-azaadenosine	~30

[a]Reference 961.
[b]The concentration required to inhibit the growth of treated cells to 50% of that of untreated controls.

incorporation into nucleic acids is apparently responsible for their cytotoxicity.[667,1042] This conversion to the 5'-nucleotide seems to be necessary for antitumor activity. The metabolism of 8-azapurines and their ribonucleosides has been studied in great detail[924,930,937,1042,1044–1047] and is illustrated in Fig. 7. A precise interpretation of the activity of these compounds is rather difficult because of their many interconversions; for example, 8-azaadenosine is converted to its triphosphate, which can be incorporated into RNA and may be responsible for cytotoxicity. It is also readily deaminated to afford 8-azainosine, which is converted to its phosphate derivative and then cleaved to furnish 8-azahypoxanthine. 8-Azahypoxanthine is converted directly to 8-azainosinic acid. 8-Azainosinic acid is metabolized to both 8-azaadenylic acid and 8-azaguanylic acid. It is, however, possible that 8-azaguanylic acid is ultimately responsible for the activity of most of the 8-azapurines.

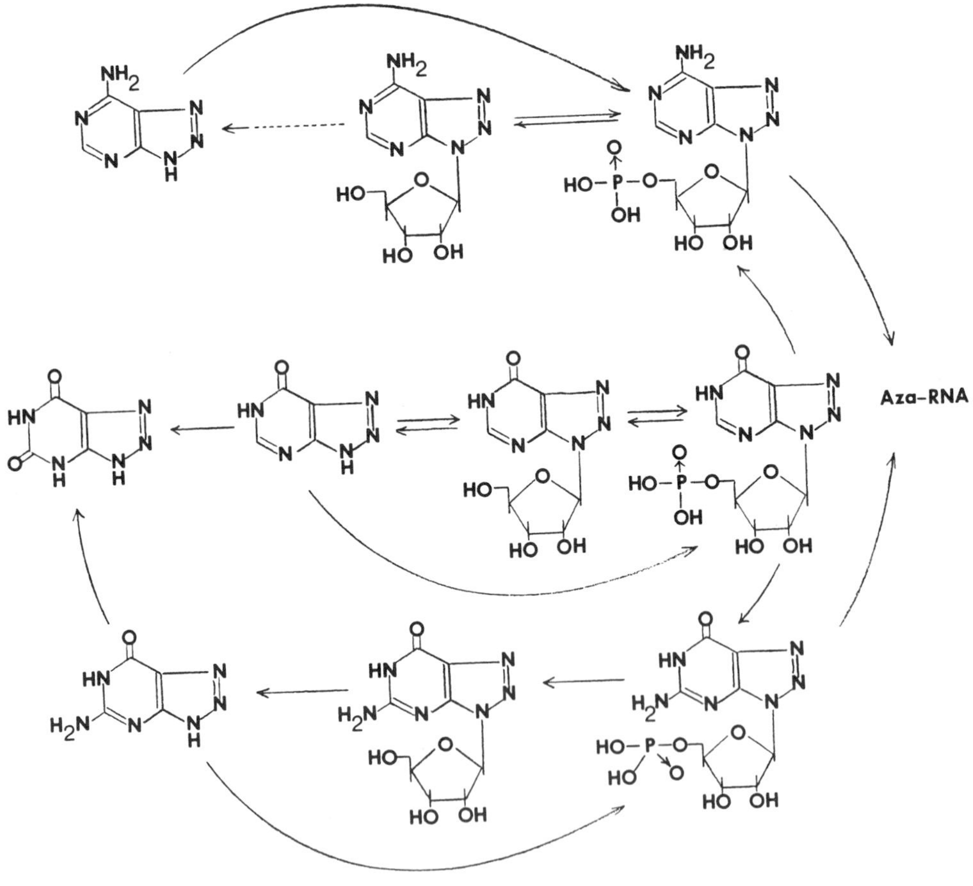

Figure 7. Metabolism of 8-azapurines and their ribonucleosides. From Ref. 953.

2.5. Five-Membered Ring Fused to a Six-Membered Ring with Six Nitrogen Atoms in the Ring System (at Least Three Nitrogen Atoms in Each Ring)

The only nucleoside belonging to the category considered in this section viz., 7-amino-3-β-D-ribofuranosyl-*v*-triazolo[4,5-*d*]-*v*-triazine (2,8-diazaadenosine, **776**), was synthesized by Montgomery and Thomas[884] by the ring-opening and reclosure sequence as described for 2-azapurine nucleosides (Section 2.4.1.1). 6-Benzyloxy-7-amino-3-β-D-ribofuranosyl-*v*-triazolo[4,5-*d*]pyrimidine hydrobromide (**775**) was prepared by the benzylation of the N^6-oxide (**774**) of 7-amino-3-β-D-ribofuranosyl-*v*-triazolo[4,5-*d*]pyrimidine (**662**, 8-azaadenosine)[944] and then ring opened by the action of methanolic ammonia to obtain 5-amino-*N*-benzyloxy-1-β-D-ribofuranosyl-*v*-triazole-4-carboxamidine (**778**) in 54% yield. Raney nickel catalyzed hydrogenolysis of the benzyloxy group of **778** afforded 5-amino-1-β-D-ribofuranosyl-*v*-triazole-4-carboxamidine (**777**). Ring annulation of **777** with aqueous nitrous acid gave 2,8-diazaadenosine (**776**)[884] in a rather low yield (7.5%).

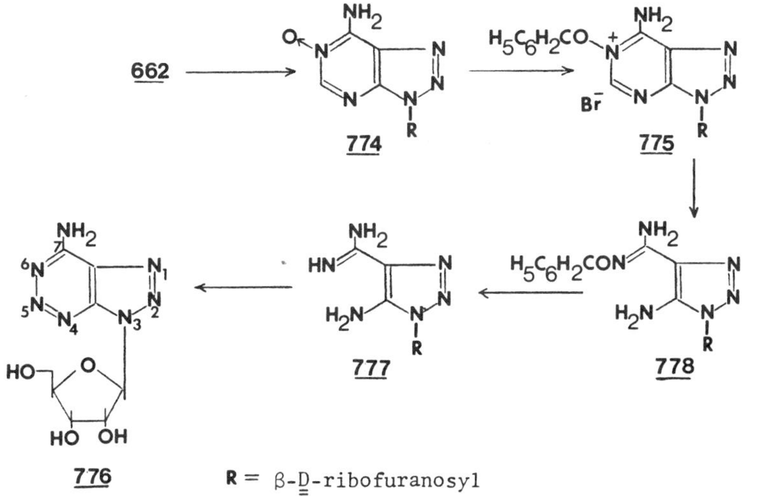

R = β-D-ribofuranosyl

2.6. Purine Analogues with a Bridgehead Nitrogen Atom

The naturally occurring purine nucleosides and nucleotides have received considerable attention in recent years, because of the wide variety of specific biological properties exhibited by them, as well as by a myriad of synthetic analogues. Such activity is largely due to the structural similarities between these nucleosides and the natural enzyme substrates. Therefore, it was of interest to synthesize structural analogues which have the potential either to emulate or to antagonize the functions of the naturally occurring purine nucleosides and nucleotides.

In recent years, such unnatural nucleosides have been prepared,[1048] resembling, at first glance, the natural purine nucleosides, but actually differing in some minor aspect. However, these so-called "counterfeits" have often exhibited considerable biological activity. In some cases, the observed inactivity was correlated with the lack of appropriate binding. Thus, the function of the various nitrogen atoms of purine nucleosides as binding sites for certain important nucleic acid enzymes has become the subject of considerable interest.[1049] The observation that N-3 of the purine-type nucleosides is probably involved in stabilizing the *syn* conformation through intramolecular hydrogen bonding[1050] has stimulated considerable activity toward the synthesis and biological evaluation of nucleosides and nucleotides derived from purine analogues.

A number of *C*-nucleoside analogues structurally related to formycin but containing a bridgehead nitrogen atom have been reported. They include 1,2,4-triazolo[1,5-*a*]pyridine,[1051,1052] 1,2,4-triazolo[4,3-*a*]pyridine,[1051,1052] 1,2,4-triazolo[4,3-*a*]pyrazine,[1053,1054] 1,2,4-triazolo[4,3-*b*]pyridazine,[1055-1057] 1,2,4-triazolo[3,4-*f*]-1,2,4-triazine,[1058] imidazo[5,1-*f*]-1,2,4-triazine,[1059,10660] pyrazolo-[1,5-*a*]-1,3,5-triazine,[1061-1066] and pyrazolo[1,5-*a*]pyrimidine.[1066] However, no attempt has been made to incorporate these nucleosides in this section. Only the syntheses of the nucleosides and nucleotides of purine analogues containing a bridgehead nitrogen atom in which the carbohydrate moiety is attached to a nitrogen atom are discussed. These bridgehead heterocyclic systems are of particular interest since the corresponding *N*-nucleosides (unlike the *C*- nucleosides) lack an N*H* function at position 1 of the purine; therefore, hydrogen bonding of the Watson–Crick type is not possible.

2.6.1. *Imidazo[1,2-a]pyrimidine Nucleosides*

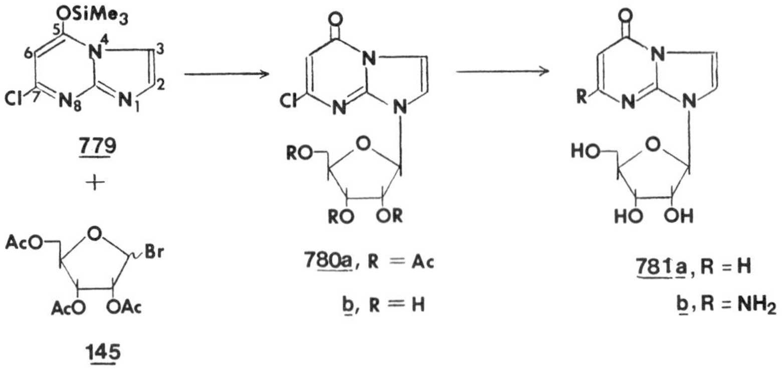

Glycosylation studies of the imidazo[1,2-*a*]pyrimidine ring system, which may be regarded as 1-deazapurine with a bridgehead nitrogen atom, was first reported by Revankar and Robins,[1067] using the trimethylsilyl procedure. Condensation of the trimethylsilyl derivative (**779**) of 7-chloroimidazo[1,2-*a*]-pyrimidin-5-one with 2,3,5-tri-*O*-acetyl-D-ribofuranosyl bromide (**145**) in acetonitrile gave a good yield of 7-chloro-1-(2,3,5-tri-*O*-acetyl-β-D-ribofuranosyl)imidazo[1,2-*a*]pyrimidin-5-one (**780a**). Deacetylation of **780a** with methanolic ammonia at room temperature furnished the free nucleoside **780b**. Dehalogenation of **780b** with Pd/C gave the inosine analogue, 1-β-D-ribofuranosylimidazo[1,2-*a*]pyrimidin-5-one (**781a**). Treatment of **780b** with methanolic ammonia at elevated temperature and pressure furnished 7-amino-1-β-D-ribofuranosylimidazo[1,2-a]pyrimidin-5-one (**781b**), a 1-deazaguanosine analogue possessing a bridgehead nitrogen atom. The anomeric configuration of these nucleosides was assigned as β on the basis of ^{1}H-NMR spectral studies of the 2′,3′-O-isopropylidene derivative of **780b**. The site of ribosylation was determined as N-1 using ^{13}C-NMR spectroscopy.[1068–1070] The nucleosides **780b**, **781a**, and **781b** are devoid of significant antiviral or antitumor activity.[1067]

2.6.2. Imidazo[1,2-c]pyrimidine Nucleosides

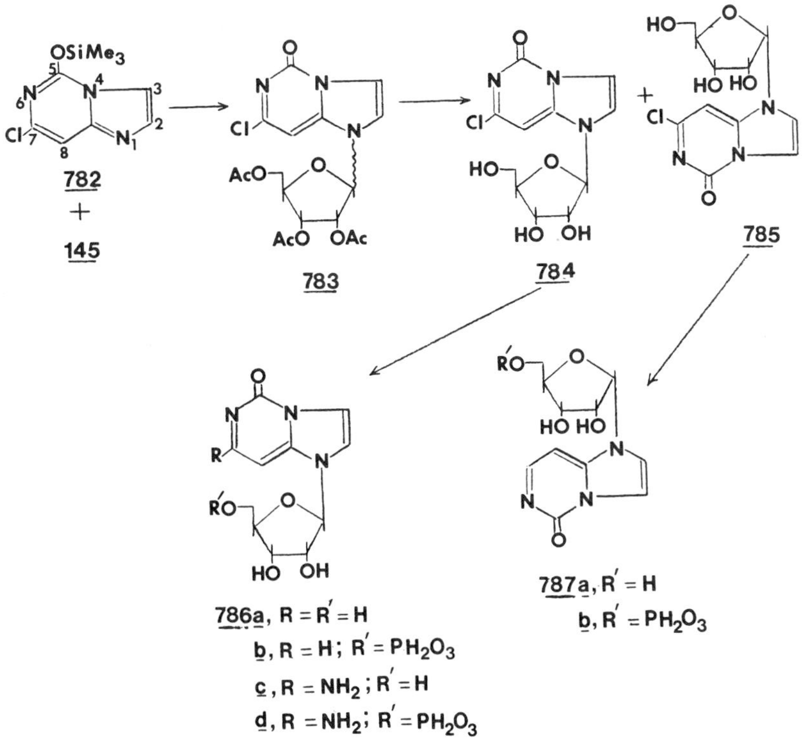

2.6.2.1. Synthesis Employing the Trimethylsilyl Derivative of a Preformed Aglycon and an Acylated Sugar Halide. The nucleoside analogues of imidazo[1,2-c]pyrimidine, which could be regarded as 3-deazapurine with a bridgehead nitrogen atom in which N-7 and C-5 are interchanged, have also been described by Revankar and co-workers.[1071] 7-Chloroimidazo[1,2-c]pyrimidin-5(6H)-one was prepared in excellent yield by the ring annulation of 4-amino-6-chloropyrimidin-2-one with bromoacetaldehyde diethyl acetal in aqueous media. Condensation of the O-trimethylsilyl derivative **782** with 2,3,5-tri-O-acetyl-D-ribofuranosyl bromide (**145**) in acetonitrile at room temperature gave a 93% yield of an anomeric mixture of 7-chloro-1-(2,3,5-tri-O-acetyl-D-ribofuranosyl)imidazo-[1,2-c]pyrimidin-5-one (**783**). The blocked anomers were difficult to separate; however, a better resolution of the anomers was obtained by delaying the separation until after deacetylation. This afforded pure 7-chloro-1-β-D-ribofuranosylimidazo[1,2-*c*]pyrimidin-5-one (**784**) and its α-anomer (**785**), which were readily separated by a combination of fractional crystallization and column chromatography.[1071] The anomeric ratio was nearly 1:1. The anomeric configurations of **784** and **785** were ascertained by ^{1}H-NMR studies[137] of the corresponding 2′,3′-*O*-isopropylidene derivatives, and the site of ribosylation was established as N-1 by a combination of UV and ^{13}C-NMR spectroscopic studies.[1072]

Catalytic dehalogenation of **784** and **785** with Pd/C gave the 3-deazainosine analogue 1-β-D-ribofuranosylimidazo[1,2-*c*]pyramidin-5-one (**786a**) and the corresponding α-anomer (**787a**), respectively. Treatment of **784** with methanolic ammonia at elevated temperature and pressure furnished 7-amino-1-β-D-ribofuranosylimidazo[1,2-*c*]pyrimidin-5-one (**786c**), the 3-deazaguanosine analogue with a bridgehead nitrogen atom. Phosphorylation of the unprotected inosine analogue with phosphorus oxychloride in trimethyl phosphate provided the IMP analogue 1-β-D-ribofuranosylimidazo[1,2-*c*]pyrimidin-5-one 5′-monophosphate (**786b**). Similar phosphorylation of **787a** gave 1-α-D-ribofuranosylimidazo[1,2-*c*]pyrimidin-5-one 5′-monophosphate (**787b**). The GMP analogue 7-amino-1-β-D-ribofuranosylimidazo[1,2-*c*]pyrimidin-5-one 5′-monophosphate (**786d**) was prepared[1071] in a similar fashion.

In view of the interesting biological properties of the nucleosides derived from β-D-arabinofuranose[1073–1081] (for an excellent review, see ref. 1074), the araHx and araG analogues of the imidazo[1,2-*c*]pyrimidine ring system have been prepared.[1082] The synthesis of these analogues was accomplished by a direct glycosylation of the trimethylsilyl derivative **782** with 2,3,5-tri-*O*-benzyl-α-D-arabinofuranosyl chloride[35] in boiling benzene. Under these conditions, a 92% yield of crystalline 7-chloro-1-(2,3,5-tri-*O*-benzyl-β-D-arabinofuranosyl)imidazo[1,2-*c*]pyrimidin-5-one (**788**) was obtained, along with a trace amount of the corresponding α-anomer. Catalytic dehalogenation of **788** gave **789a** in excellent yield, and hydrogenolysis of **789a** furnished 1-β-D-arabino-

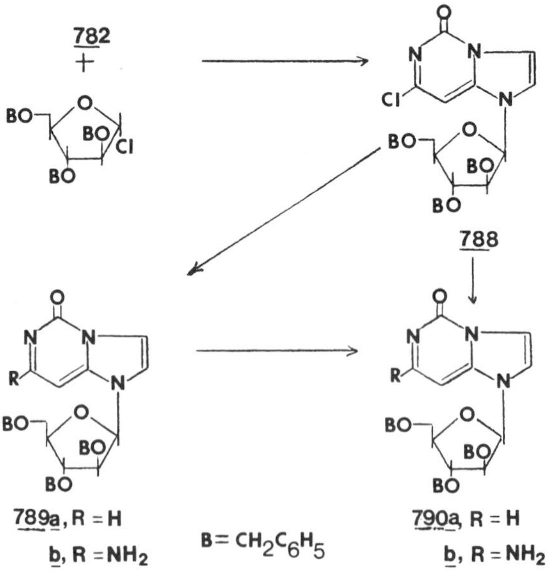

789a, R = H
   b, R = NH$_2$
B= CH$_2$C$_6$H$_5$
790a, R = H
   b, R = NH$_2$

furanosylimidazo[1,2-*c*]pyrimidin-5-one (**790a**). Compound **790a** was also obtained directly from **788** by a reductive hydrogenolysis. Treatment of **788** with methanolic ammonia at 120°C gave a 42% yield of **789b**, which, on subsequent hydrogenolysis, furnished 7-amino-1-β-D-arabinofuranosylimidazo[1,2-*c*]pyrimidin-5-one (**790b**). The site of glycosylation of these arabinosides was assigned as N-1 by UV absorption studies and the anomeric configuration by periodate oxidation followed by a reduction with sodium borohydride.[1082]

2.6.2.2. *Synthesis Employing the Silylated Aglycon and an Acyl Sugar in the Presence of a Lewis Acid.* In an effort to improve on the yield of **784**, Revankar and co-workers[1071] have examined the use of the Lewis acid catalyzed glycosylation procedure.[1083] Treatment of one equivalent of **782** in 1,2-dichloroethane with one equivalent of fully acylated ribofuranose (**51**) in the presence of 1.44 molar equivalent of stannic chloride afforded 7-chloro-1-(2,3,5-tri-*O*-acetyl-β-D-ribofuranosylimidazo[1,2-*c*]pyrimidin-5-one (**791**). Ammonolysis of **791** with methanolic ammonia gave an 81% yield of **784**.

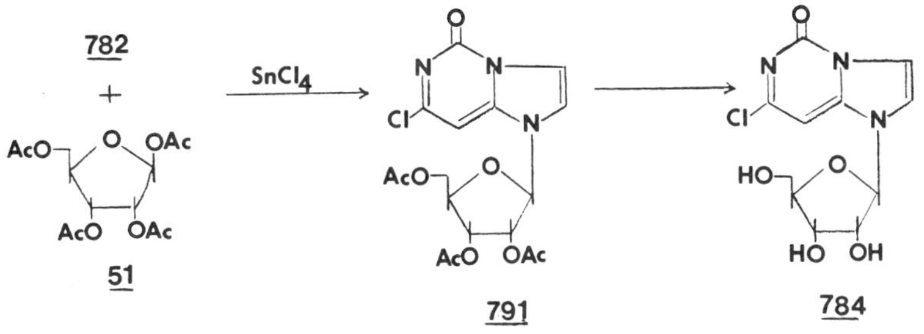

2.6.2.3. *Synthesis Employing the Ring Closure of a Pyrimidine Nucleoside Precursor.* The reaction of chloroacetaldehyde with cytidine under mild acidic conditions (pH range 3.5–4.5) has been reported[1084–1087] to yield a fluorescent 5,6-dihydro-5-oxo-6-β-D-ribofuranosylimidazo[1,2-*c*]pyrimidine (3,N^4-ethenocytidine, ε-cytidine, **792**). A similar reaction using a large excess (about 20-fold) of chloroacetaldehyde and cytosine nucleotides afforded the corresponding 5'-mono-,[1088,1089] di-, and triphosphate derivatives of **792**.[1090] These ε-cytosine nucleotides have fluorescent properties similar to those observed for the corresponding nucleosides. Compound **792** undergoes mercuration readily and completely.[1091] Likewise, nicotinamide 3,N^4-ethenocytosine dinucleotide (ε NCD),[1092] a structural analogue of NAD, and 2'(3')-*O*-isovaleryl-ε-cytidine 5'-diphosphate[1093] have been synthesized. It has been demonstrated that chlo-

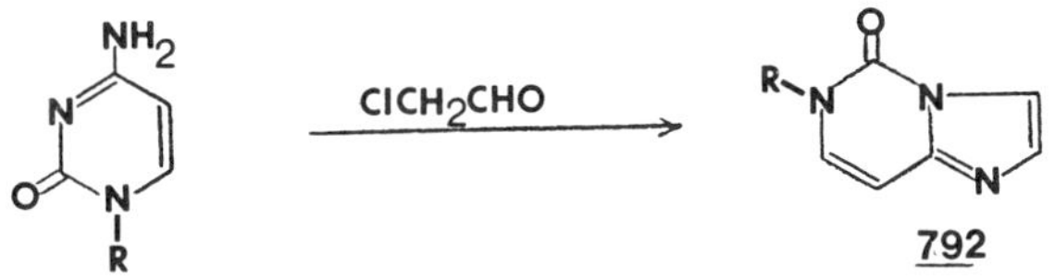

R = β-D-ribofuranosyl

roacetone,[1094] *p*-tosyloxyacetone,[10(4)] phenacyl bromide,[1094] or β-acetylvinyl-triphenylphosphonium bromide[1095] also reacts with cytidine monophosphate analogously to afford 3,N^4-ethenocytidine derivatives, in which the carbohydrate moiety is attached to N-6 of the imidazo[1,2-*c*]pyrimidine ring system.

Several interesting studies describing the dimensions and molecular interactions of **792** and related compounds as revealed by X-ray analysis and other spectroscopic methods have appeared.[1096–1099] None of the imidazo[1,2-*c*]pyrimidine nucleosides exhibited significant antiviral or antitumor activity *in vitro*.[1082]

2.6.3. Imidazo[1,2-a]-s-triazine Nucleosides

2.6.3.1. Synthesis Employing the Silylated Aglycon and an Acyl Sugar in the Presence of a Lewis Acid. The first chemical synthesis of imidazo[1,2-*a*]-*s*-triazine nucleosides was accomplished[1100] by the treatment of the bis-trimethylsilyl derivative of 2-aminoimidazo[1,2-*a*]-*s*-triazin-4-one (**793**) with 1-*O*-acetyl-2,3,5-tri-*O*-benzoyl-β-D-ribofuranose (**57**) in 1,2-dichloroethane in the presence of stannic chloride. This gave a 56% yield of 2-amino-8-(2,3,5-tri-*O*-benzoyl-β-D-ribofuranosyl)imidazo[1,2-*a*]-*s*-triazin-4-one (**794a**). Debenzoylation of **794a** with methanolic sodium methoxide furnished 2-amino-8-β-D-ribofuranosyl-imidazo[1,2-*a*]-*s*-triazin-4-one (5-aza-7-deazaguanosine, **794b**).[1100] Deamination of **794b** in aqueous acetic acid with barium nitrite provided 5-aza-7-deaza-xanthosine (**796**), and the direct phosphorylation of the unprotected **794b** with phosphorus oxychloride in trimethyl phosphate furnished a rather low yield (27.5%) of the *N*-bridgehead GMP analogue 2-amino-8-β-D-ribofuranosylimidazo[1,2-*a*]-*s*-triazin-4-one 5'-monophosphate (**795**). The structure of **794b** was confirmed by ^{13}C-NMR studies[1100] and X-ray analysis.[1101]

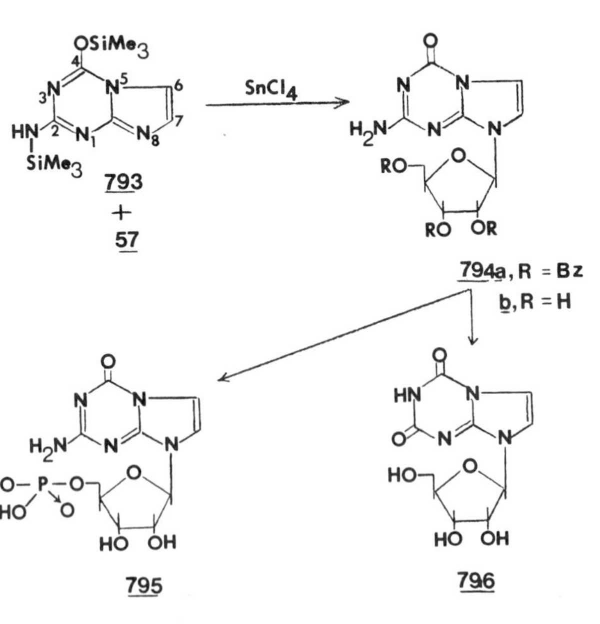

2.6.3.2. Synthesis Employing the Ring Closure of an Imidazole Nucleoside Precursor. Condensation of 1-*O*-acetyl-2,3,5-tri-*O*-benzoyl-D-ribofuranose (**57**) with 2-aminoimidazole has been reported by Moffatt and co-workers[1102] to occur only on the exocyclic amino group to give **797**. Cyclization of **797** with phenoxycarbonyl isocyanate, (phenylthio)carbonyl isothiocyanate, and phenoxythiocarbonyl isothiocyanate gave 1-β-D-ribofuranosylimidazo[1,2-*a*]-*s*-triazines (**798**) bearing oxo and thiono substituents at the 2- and 4- positions. Certain amino-substituted compounds were also prepared via ammonolysis of the thiones or their methylthio derivatives. Similarly, cyclization of 2-amino-1-(2,3,5-tri-*O*-benzoyl-D-ribofuranosyl)imidazole (**799**) with phenoxycarbonyl isocyanate and phenoxycarbonyl isothiocyanate gave an anomeric mixture of 8-(2,3,5-tri-*O*-benzoyl-D-ribofuranosyl)imidazo[1,2-*a*]-*s*-triazines (**800a**) containing oxo and thiono substituents at the 2- and 4- positions.[1103] Further functional group manipulation of **800a** provided the anomeric 8-D-ribofuranosylimidazo[1,2-*a*]-*s*-triazine-2,4(3*H*,8*H*)-dithiones and the corresponding 4-amino-2-oxonucleosides.[1103]

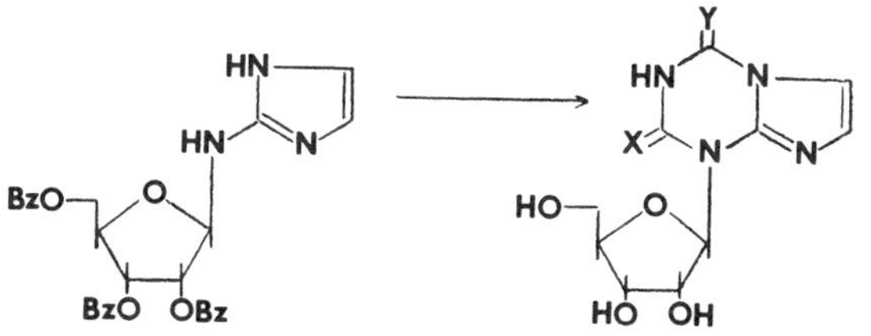

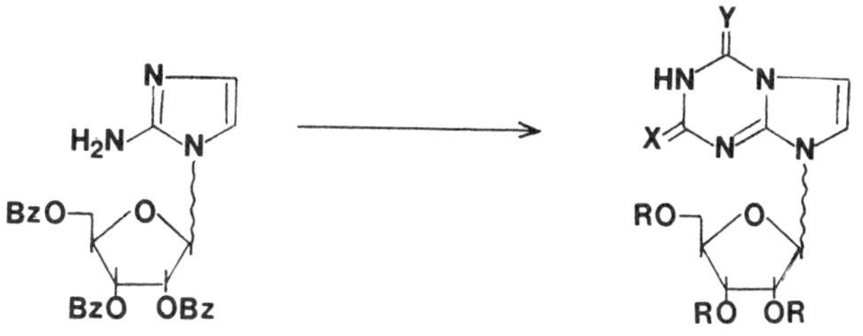

2.6.3.3. Synthesis Employing the Phase-Transfer Glycosylation Procedure. A liquid–liquid phase-transfer glycosylation of **793** with **24** using $CH_2Cl_2/10\%$ aqueous K_2CO_3 solution in the presence of Bu_4NHSO_4 as the catalyst gave an anomeric mixture of the protected nucleosides (**801**), the β:α anomeric ratio being 2:1 (63% total yield).[1104] This was in contrast to earlier phase-transfer glycosylation experiments with pyrrolo[2,3-*d*]pyrimidines, in which, in most cases, only the β-nucleoside was formed.[1105] The formation of α-**801** cannot be avoided due to the unfavorable partition of **793** between the organic and the aqueous phase. This resulted in a prolonged reaction time, during which time

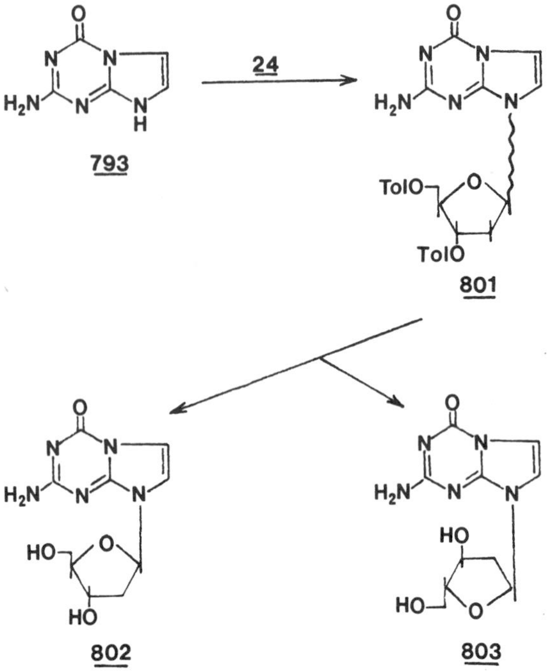

the halogenose **24** anomerized. Deprotection of **801** with MeOH/NH₃ gave
2-amino-8-(2-deoxy-β-D-*erythro*-pentofuranosyl)imidazo[1,2-*a*]-*s*-triazin-4-one
(**802**) and its α-anomer **803**. These anomers were separated on a silica gel column
as crystalline materials.[1104] Both **802** and **803** exhibit the *anti* conformation at
the *N*-glycosidic bond.

5-Aza-7-deazaguanosine (**794b**) and its aglycon moiety (**793**) exhibited
significant antirhinovirus[1100] and antiherpes activity[1101] in cell culture.

2.6.4. Pyrazolo[1,5-a]pyrimidine Nucleosides

Extension of the previous glycosylation studies[1072] to pyrazolo[1,5-*a*]pyri-
midines resulted in the preparation of some interesting compounds. Glycosylation of the trimethylsilyl derivative (**804a**) of 7-hydroxypyrazolo[1,5-*a*] pyrimi-

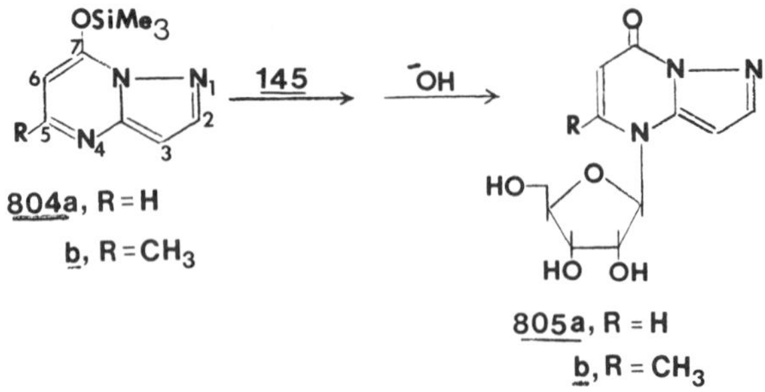

dine[1106] with 2,3,5-tri-*O*-acetyl-D-ribofuranosyl bromide (145) in anhydrous acetonitrile gave a good yield of the blocked nucleoside. Subsequent deacetylation of this compound with methanolic ammonia afforded 4-β-D-ribofuranosylpyrazolo[1,5-*a*]pyrimidin-7-one (805a).[1072] A similar glycosylation of the *O*-trimethylsilyl derivative (804b) of 5-methylpyrazolo[1,5-*a*]pyrimidin-7-one,[1107] followed by deacetylation, gave 5-methyl-4-β-D-ribofuranosylpyrazolo[1,5-*a*]pyrimidin-7-one (805b). Evidence that the glycosylation in this ring system has occurred at the N-4 position was obtained by both ¹H- and ¹³C-NMR studies.[1072]

2.6.5. Pyrazolo[1,5-a]-s-triazine Nucleosides

Substitution of N for CH at the C-6 position in the pyrazolo[1,5-*a*]pyrimidine ring gives the other heterocyclic system, pyrazolo[1,5-*a*]-*s*-triazine. Ribosylation of the trimethylsilyl derivative (806) of 2-methylthiopyrazolo[1,5-*a*]-*s*-triazin-4-one[1108] with 2,3,5-tri-*O*-acetyl-D-ribofuranosyl bromide (145) has been reported[1072] to give an 83% yield of 2-methylthio-3-(2,3,5-tri-*O*-acetyl-β-D-ribofuranosyl)pyrazolo[1,5-*a*]-*s*-triazin-4-one (807). The structure of 807 was assigned on the basis of a UV absorption comparison with the model methyl compound[1108] and ¹H-NMR chemical shift data of the dethiated product, 3-(2,3,5-tri-*O*-acetyl-β-D-ribofuranosyl)pyrazolo[1,5-*a*]-*s*-triazin-4-one (808).[1072] Conclusive evidence for the glycosylation site as N-3 was also obtained by ¹³C-NMR spectroscopy, and a detailed discussion of this evidence has been presented.[1072]

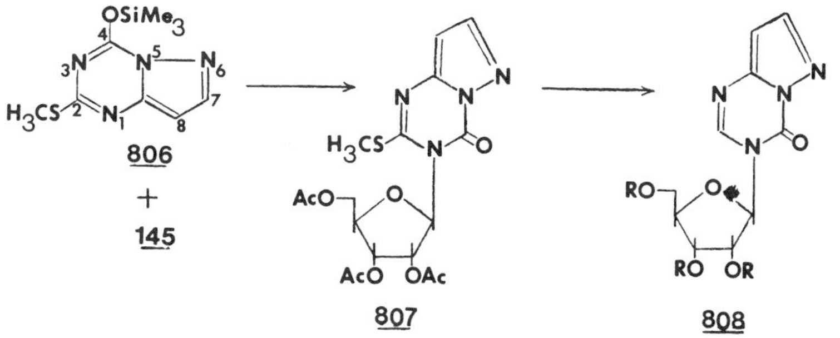

2.6.6. s-Triazolo[1,5-a]pyrimidine Nucleosides

2.6.6.1. Synthesis Employing the Heavy-Metal Salt of a Preformed Aglycon and an Acylated Sugar Halide. Wagner *et al.*[1109] reported the glucosylation studies of 7-hydroxy-[1110,1111] and 7-mercapto-[1112] 5-methyl-*s*-triazolo[1,5-*a*]pyrimidines employing the heavy-metal salt procedure. These glucosyl nucleosides were prepared by the treatment of either the silver or the chloromercuri derivative of 5-methyl-*s*-triazolo[1,5-*a*]pyrimidin-7-one (809) with 2,3,4,6-tetra-*O*-acetyl-α-D-glucopyranosyl bromide (659) in boiling benzene/toluene, which produced 5-methyl-3-(tetra-*O*-acetyl-β-D-glucopyranosyl)-*s*-triazolo[1,5-*a*]pyrimidin-7-one (810). Compound 810 was found to be identical to the product obtained through the condensation of 5-methyl-7-methoxy-*s*-triazolo[1,5-*a*]pyrimidine (811)[1113] with 659, under the Hilbert–Johnson[1114] reaction conditions. Treatment of 810 with P₂S₅ in pyridine gave the 7-thione derivative 813.

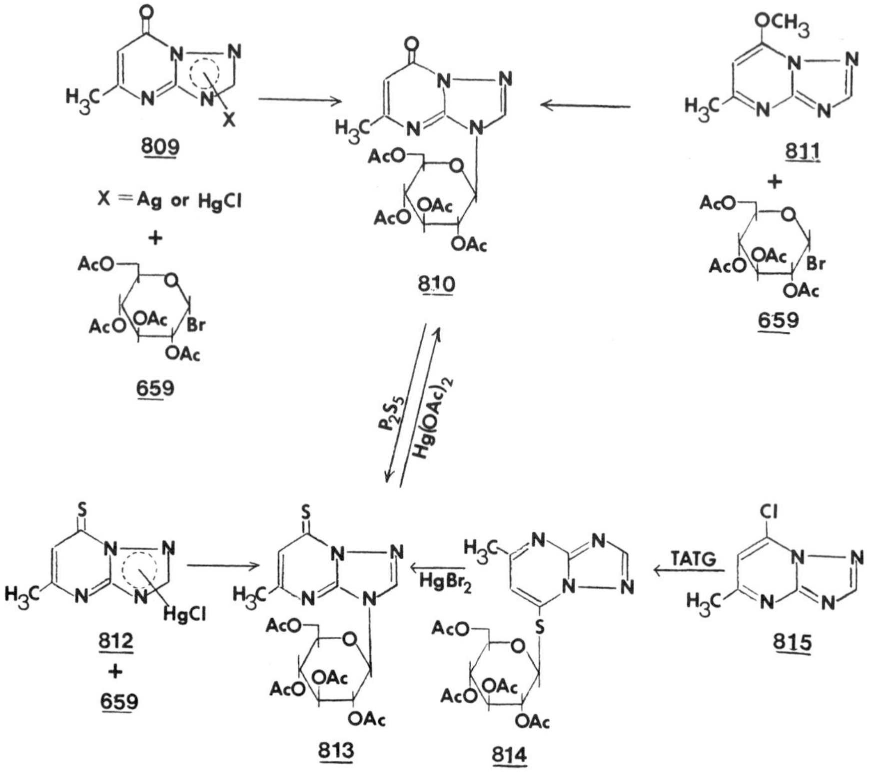

Similarly, a condensation of the chloromercuri derivative of 5-methyl-*s*-triazolo[1,5-*a*]pyrimidine-7-thione (**812**) with **659** in boiling toluene gave crystalline 5-methyl-3-(tetra-*O*-acetyl-β-D-glucopyranosyl)-*s*-triazolo[1,5-*a*]pyrimidine-7-thione (**813**). When this coupling reaction was carried out in aqueous acetone containing sodium hydroxide, a 32% yield of 5-methyl-7-(tetra-*O*-acetyl-β-D-glucopyranosylmercapto)-*s*-triazolo[1,5-*a*]pyrimidine (**814**) was obtained. Compound **814** may also be synthesized in over 92% yield by reacting 7-chloro-5-methyl-*s*-triazolo[1,5-*a*]pyrimidine (**815**) with tetra-*O*-acetyl-1-thioglucose (TATG)[1115] in acetone containing sodium hydroxide. This compound then undergoes a rearrangement when boiled in benzene in the presence of HgBr$_2$ to give **813**. When **813** was heated in dioxane containing mercuric acetate, the 7-thione function was hydrolyzed to give the keto compound **810**. Both **810** and **813** may be deacetylated to obtain the free nucleosides. The site of glucosylation in these nucleosides was established by a comparison of their UV absorption spectra with those of the corresponding model methyl compounds.[1109]

2.6.6.2. Synthesis Employing the Fusion of a Preformed Aglycon and an Acyl Sugar. The iodine-catalyzed fusion of hydroxy-*s*-triazolo[1,5-*a*]pyrimidines with 1,2,3,5-tetra-*O*-acetyl-D-ribofuranose was investigated by Imbach and co-workers[137,1116,1117]

s-Triazolo[1,5-*a*]pyrimidin-5-one (**816**)[1118] was fused at 175°C with 1,2,3,5-

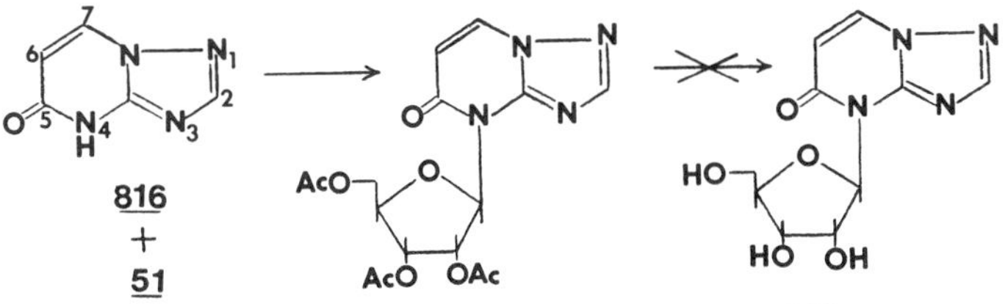

816
+
51

817

818

tetra-*O*-acetyl-β-D-ribofuranose (**51**) in the presence of iodine to obtain crystalline 4-(2,3,5-tri-*O*-acetyl-β-D-ribofuranosyl)-*s*-triazolo[1,5-*a*]pyrimidin-5-one (**817**). Attempts to deacetylate **817** with methanolic ammonia were unsuccessful.[1116] A similar fusion of 7-hydroxy-*s*-triazolo[1,5-*a*]pyrimidine (**819**)[1119] with **51** in the presence of iodine is reported[1117] to yield 3- and 4-(2,3,5-tri-*O*-acetyl-D-ribofuranosyl)-*s*-triazolo[1,5-*a*]pyrimidin-7-ones (**820** and **821**). These blocked nucleosides were successfully deacetylated with methanolic ammonia to obtain the respective free nucleosides. The site of ribosylation was established by UV absorption studies, and the anomeric configuration of these ribonucleosides was determined by ¹H-NMR spectroscopic studies of their corresponding 2′,3′-*O*-isopropylidene derivatives.[1120]

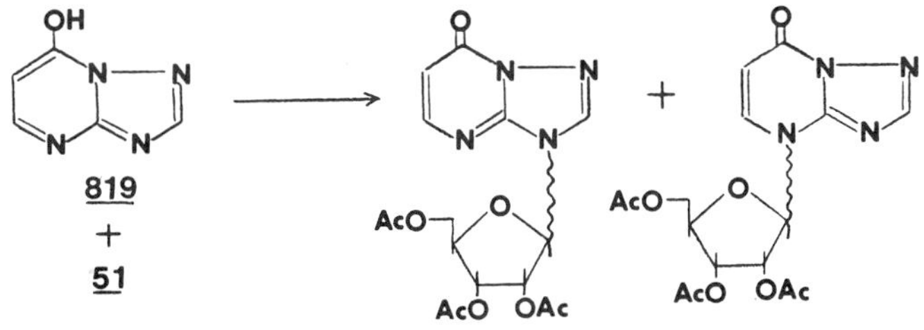

819
+
51

820

821

2.6.6.3. Synthesis Employing the Trimethylsilyl Derivative of a Preformed Aglycon and an Acylated Sugar Halide. The synthesis of the inosine and the guanosine analogue in the *s*-triazolo[1,5-*a*]pyrimidine ring system is of particular interest since these analogues can be viewed as resulting from a simple interchange of N-1 and C-5 of the naturally occurring inosine and guanosine. Synthesis of the inosine analogue was initially accomplished by Robins and co-workers[1121] using the trimethylsilyl procedure. The reaction of the trimethylsilyl derivative (**822**) of *s*-triazolo[1,5-*a*]pyrimidin-7-one with tri-*O*-benzoyl-D-ribofuranosyl bromide (**19**) in acetonitrile led to the formation of two isomeric blocked nucleosides, **823a** and **824a**. Removal of the blocking groups with methanolic ammonia gave 3-β-D-ribofuranosyl-*s*-triazolo[1,5-*a*]pyrimidin-7-one (**823b**) and the N-4 isomer (**824b**) in an overall yield of 45% and 20%, respectively. The site of ribosylation of these nucleosides was originally assigned by UV absorption studies of the model methyl compounds[1121] and was subsequently confirmed by ¹³C-NMR studies.[1072]

Thiation of **823a** with P₂S₅ in pyridine and subsequent deblocking of the

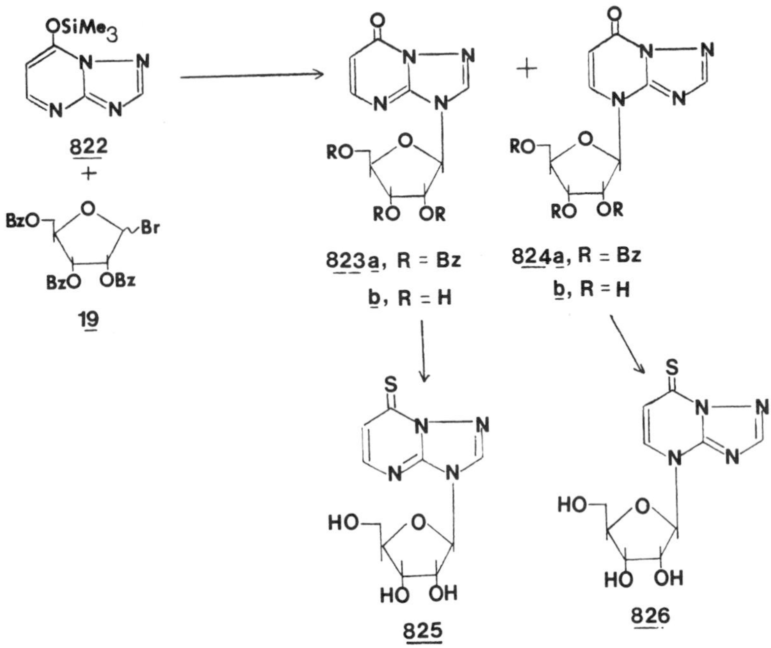

823a, R = Bz   824a, R = Bz
b, R = H   b, R = H

825

826

reaction product with methanolic ammonia gave the thioinosine analogue
825.[1122] Similarly, thiation of **824a** with P_2S_5 in dioxane and deprotection with
sodium methoxide in methanol afforded 4-β-D-ribofuranosyl-*s*-triazolo[1,5-*a*]-
pyrimidin-7-thione (**826**).[1122] The synthesis of the guanosine analogue (**830**)
was accomplished[1123,1124] by employing the trimethylsilyl procedure. Condensation of the trimethylsilyl-derivative (**827**, Y = Cl) of 5-chloro-*s*-triazolo-
[1,5-*a*]pyrimidin-7-one[1125] with **145** in acetonitrile gave the blocked nucleoside

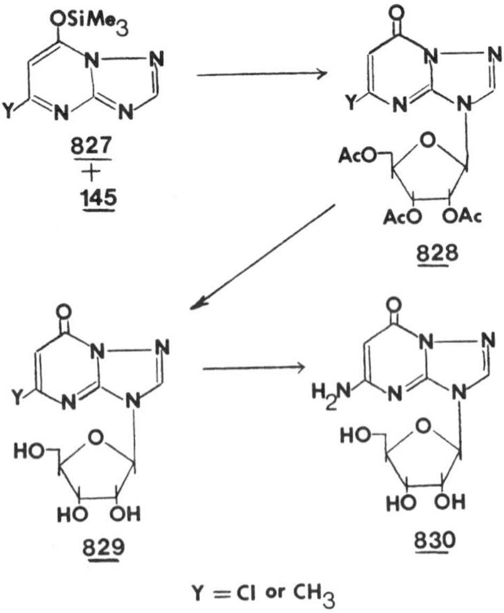

827
+
145

828

829

830

Y = Cl or CH$_3$

(**828**, Y = Cl), which, on deacetylation, afforded 5-chloro-3-β-D-ribofuranosyl-*s*-triazolo[1,5-*a*]pyrimidin-7-one (**829**, Y = Cl). Dehalogenation of **829** (Y = Cl) with Pd/C gave **823b**. A similar glycosylation of the trimethylsilyl-5-methyl-*s*-triazolo[1,5-*a*]pyrimidin-7-one, (**827**, Y = CH$_3$) gave **829** (Y = CH$_3$). Nucleophilic substitution of the 5-chloro group of **829** (Y = Cl) gave some interesting nucleosides, including the guanosine analogue, 5-amino-3-β-D-ribofuranosyl-*s*-triazolo[1,5-*a*]pyrimidin-7-one (**830**).[1123] Compound **830** exhibited moderate antirhinoviral activity *in vitro*.[1067]

2.6.7. s-Triazolo[4,3-a]pyrimidine Nucleosides

2.6.7.1. Synthesis Employing the Fusion of a Preformed Aglycon and an Acyl Sugar.   The chemical syntheses of the nucleosides of the *s*-triazolo[4,3-*a*]pyrimidine ring system have been sparse[1116,1117,1126] and limited only to the synthesis of 5- or 7-hydroxy-*s*-triazolo[4,3-*a*]pyrimidine ribonucleosides. The synthesis of such a nucleoside was accomplished by Imbach and co-workers[1116] by the iodine-catalyzed fusion procedure. The fusion of *s*-triazolo[4,3-*a*]pyrimidin-7-one (**831**) with tetra-*O*-acetyl-**D**-riofuranose (**51**) at 175°C under reduced pressure in the presence of iodine gave a 90% yield of crystalline 8-(2,3,5-tri-*O*-acetyl-β-D-ribofuranosyl)-*s*-triazolo[4,3-*a*]pyrimidin-7-one (**832**).[1116] However, the fusion of *s*-triazolo[4,3-*a*]pyrimidin-5-one (**833**) with **51** using 1% iodine (by weight) as the catalyst in a temperature range of 160–200°C gave only 1-(2,3,5-tri-*O*-acetyl-β-D-ribofuranosyl)-*s*-triazolo[4,3-*a*]pyrimidin-5-one (**834a**).[1117] When the percentage of iodine was raised to 7%, compound **834a** was obtained as the major product along with a minor amount of **821** and traces of **820** (due to a Dimroth rearrangement[1127,1128], as shown in Table IV. Deacetylation of

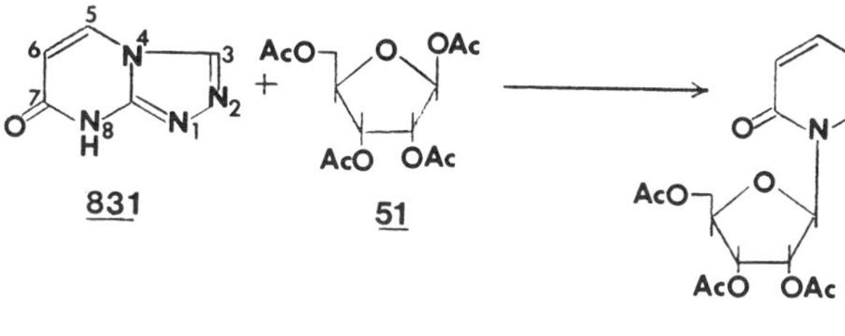

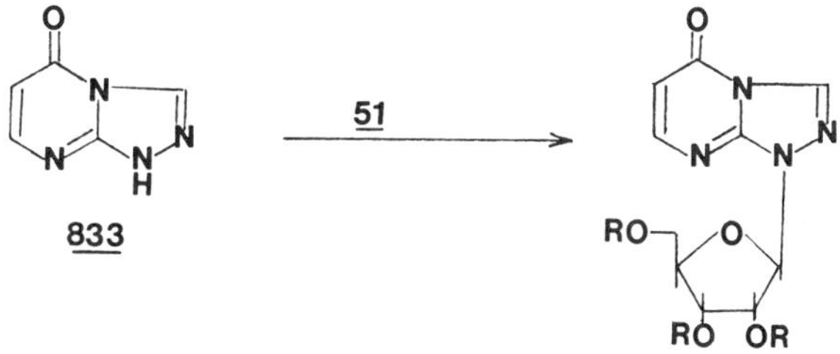

Table IV. Product Distribution in the Fusion
of s-*Triazolo[4,3-a]pyrimidin-5-one*
and *1,2,3,5-Tetra-O-acetyl-D-ribofuranose,*
in Relation to Catalyst and Temperature

Temperature (°C)	Catalyst	Nucleoside derivative		
		834a	**821**	**820**
160	1% I$_2$	100%		
	7% I$_2$	90%	10.%	Traces
180	1% I$_2$	100%		
	7% I$_2$	50%	50%	Traces
200	1% I$_2$	100%		
	7% I$_2$	9%	84%	7%

834a with methanolic ammonia furnished the inosine analogue 1-β-D-ribofura-nosyl-*s*-triazolo[4,3-*a*]pyrimidin-5-one (**834b**).

2.6.7.2. Synthesis Employing the Trimethylsilyl Derivative of a Preformed Aglycon and an Acylated Sugar Halide.   The condensation of the trimethylsilyl derivative of *s*-triazolo[4,3-*a*]pyrimidin-7-one (**835**) with 2,3,5-tri-*O*-acetyl-D-ribofurano-syl chloride in acetonitrile gave a 50% yield of 1-(2,3,5-tri-*O*-acetyl-β-D-ribofura-nosyl)-*s*-triazolo[4,3-*a*]pyrimidin-7-one (**836a**), which readily rearranged to af-ford the more thermodynamically stable **832**, under the above fusion reaction conditions.[1117] Deacetylation of **836a** with methanolic ammonia afforded 1-β-D-ribofuranosyl-*s*-triazolo[4,3-*a*]pyrimidin-7-one (**836b**). The inosine ana-logue **834b** was also prepared[1117] by reacting the trimethylsilyl derivative **837** with the acetylchlorosugar in acetonitrile, followed by deacetylation of the blocked nucleoside (**834a**). The structures of these *s*-triazolo[4,3-*a*]pyrimidine nucleosides were established by the conventional ¹H-NMR (of 2′,3′-*O*-iso-propylidene derivative) and UV absorption (of the model methyl compound) studies.[1117] The reaction of **837** with tetra-*O*-acetyl-D-ribofuranose (**51**) in 1,2-dichloroethane containing stannic chloride afforded an isomeric mixture of **834a** and its N-8 isomer, which were separated by silicic acid column chroma-tography.[1117]

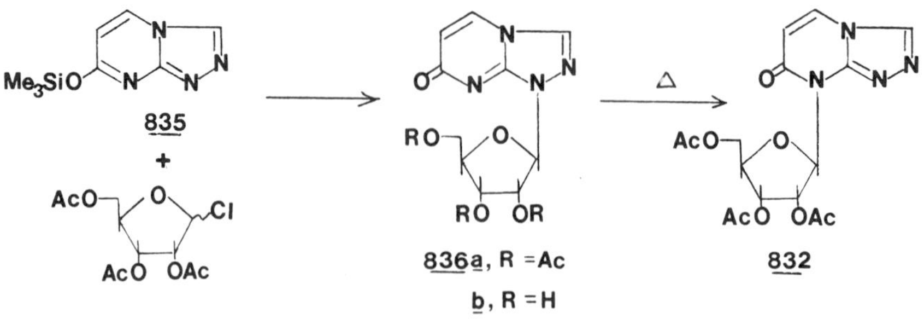

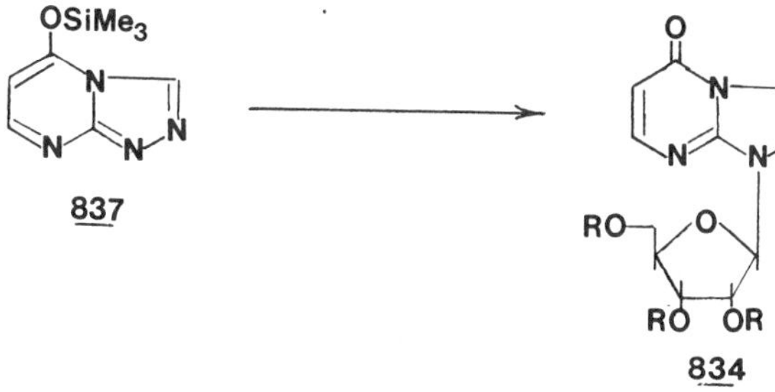

837 → 834

2.6.8. s-*Triazolo[1,5-a]-s-triazine Nucleosides*

The synthesis of the xanthosine analogue (**839b**) in the s-triazolo[1,5-*a*]-*s*-triazine (5-azapurine) ring system has recently been reported[1129] and was accomplished via the ring closure of 3-amino-4-(2,3,5-tri-*O*-benzoyl-β-D-ribofuranosyl)-1,2,4-triazole (**838**). Compound **838** has previously been reported from our laboratory.[1130] Treatment of **838** with chlorocarbonyl isocyanate in anhydrous tetrahydrofuran gave a 98% yield of 1-(2,3,5-tri-*O*-benzoyl-β-D-ribofuranosyl)-*s*-triazolo[1,5-*a*]-*s*-triazine-5,7-dione (**839a**) as the sole product. Debenzoylation of **839a** with sodium methoxide in methanol afforded 5-azaxanthosine (**839b**),[1129] which exhibited significant antiparasitic activity against *Leishmania donovani in vitro*.

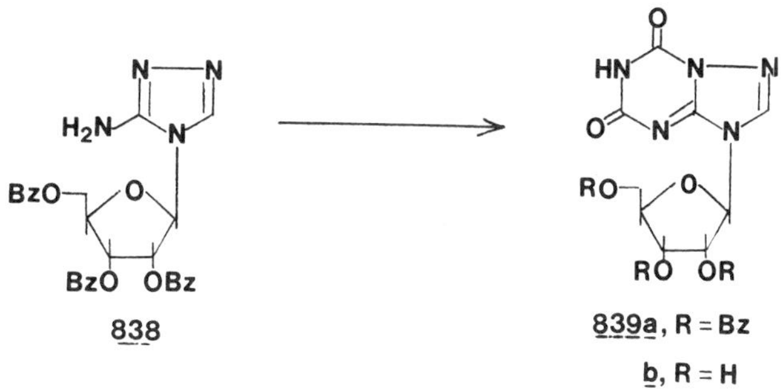

838 → 839a, R = Bz
  b, R = H

2.6.9. *Imidazo[1,2-b]pyrazole Nucleosides*

The first chemical synthesis of imidazo[1,2-*b*]pyrazole nucleosides was reported recently from our laboratory.[1131] The Lewis acid catalyzed trimethylsilyl procedure was employed. Glycosylation of the trimethylsilyl derivative of imidazo[1,2-*b*]pyrazole-7-carbonitrile (**840a**) with 1-*O*-acetyl-2,3,5-tri-*O*-benzoyl-D-ribofuranose (**57**) in the presence of trimethylsilyl trifluoromethanesulfonate gave the isomeric blocked nucleosides **841a** and **842a**. Ammonolysis of this mixture gave 1-β-D-ribofuranosylimidazo[1,2-*b*]pyrazole-7-carbonitrile (**841b**) and the corresponding N-5 glycosyl isomer (**842b**). This is the first known example of a nucleoside in which the glycon moiety is attached to a nitrogen adjacent to a bridgehead nitrogen atom.[1131] The isomeric ratio of **841a** and **842a** was found to be time dependent. Initially, the kinetically controlled isomer

842a was formed almost exclusively. Over a period of time, the thermodynamically controlled isomer **841a** formed, until after six days the ratio of **841a** to **842a** was almost 1:1. This rearrangement of **842a** to **841a** could be accelerated by heating the reaction mixture *per se* or a solution of pure **842a** in acetonitrile in the presence of a Lewis acid catalyst. This conversion is analogous to the rearrangement of 8-glycosyl-8-azapurines to 9-glycosyl-8-azapurines as reported by Montgomery and Elliott.(950) The carbonitrile function of **841b** and **842b** was manipulated to obtain the corresponding carboxamide and carboxamidoxime derivatives.(1131)

840a, R = H
b, R = SCH_3

841a, R = H; R′ = Bz
b, R = R′ = H
c, R = SCH_3; R′ = Bz
d, R = SCH_3; R′ = H

842a, R = H; R′ = Bz
b, R = R′ = H

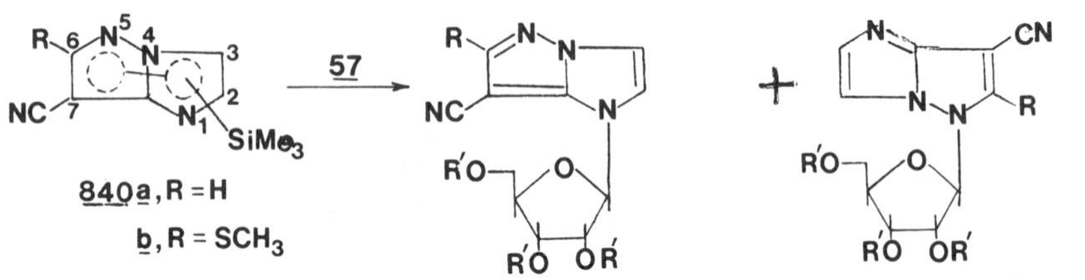

A similar glycosylation of the trimethylsilyl derivative of 6-methylthioimidazo[1,2-*b*]pyrazole-7-carbonitrile (**840b**) with **57**, and subsequent debenzoylation, gave 6-methylthio-1-β-D-ribofuranosylimidazo[1,2-*b*]pyrazole-7-carbonitrile (**841d**). The structural assignments of **841b** and **842b** were made by single-crystal X-ray diffraction analysis.(1131)

2.6.10. s-*Triazolo[5,1-c]*-s-*triazole Nucleosides*

The second fused 5–5-membered azole nucleoside was also reported from our laboratory.(1132) Direct glycosylation of the unsilylated 3-amino-5(7*H*)-*s*-triazolo[5,1-*c*]-*s*-triazole (**843**)(1133) with **57** in the presence of a catalyst, trimethylsilyl trifluoromethanesulfonate, in acetonitrile gave 3-amino-1-(2,3,5-tri-*O*-benzoyl-β-D-ribofuranosyl)-*s*-triazolo[5,1-*c*]-*s*-triazole (**844a**) in an 89%

843 57

844a, R = Bz
b, R = H

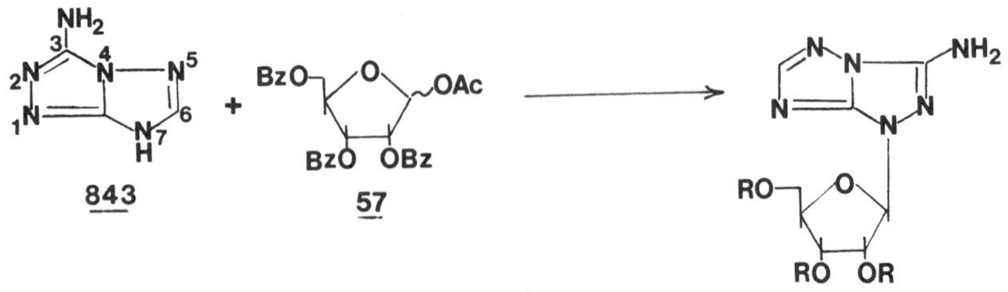

yield. Debenzoylation of **844a** with methanolic ammonia gave 3-amino-1-β-D-ribofuranosyl-*s*-triazolo[5,1-*c*]-*s*-triazole (**844b**).[1132] The absolute structural assignment of **844b** was made by single-crystal X-ray analysis.

Efforts to obtain the desired 7-glycosyl isomer of **843** by the sodium salt glycosylation procedure[30,31] or by the high-temperature glycosylation procedure[696] using **843** and **57** were unsuccessful, and only **844a** was isolated.[1132]

2.6.11. Other Related Purine Analogues with a Bridgehead Nitrogen Atom

2.6.11.1. s-Triazolo[1,5-c]pyrimidine Nucleosides. The synthesis of the 6-β-D-ribofuranosyl-*s*-triazolo[1,5-*c*]pyrimidin-5(6*H*)-one (**846**) by the ring closure of a preformed pyrimidine nucleoside has been described.[1134] *N*-Amination of cytidine with 2,4-dinitrophenoxyamine in dimethylformamide gave 3-amino-cytidine (**845**), which was isolated as the hydrochloride salt. Treatment of **845** with ethyl orthoformate and acetic anhydride at reflux temperature furnished **846** in excellent yield.

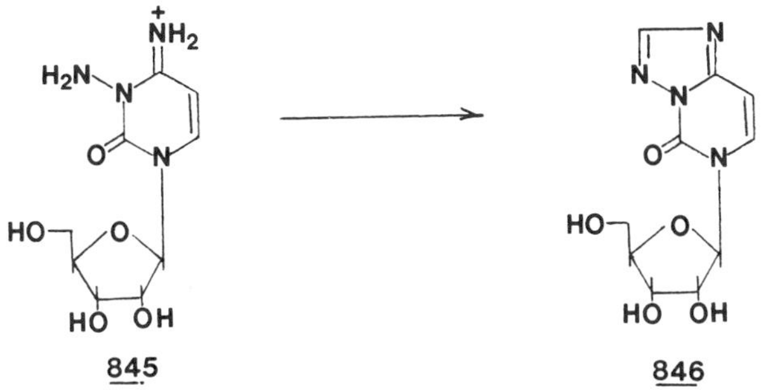

$$\underline{845} \qquad \underline{846}$$

2.6.11.2. s-Triazolo[4,3-c]pyrimidine Nucleosides. *N*⁴-Aminocytidine reacted rapidly with ethyl acetimidate at neutral pH and at room temperature to give 3-methyl-6-β-D-ribofuranosyl-*s*-triazolo[4,3-*c*]pyrimidin-5-one (**848**)[1135] via the intermediate **847**. Since imidate esters are known to react with proteins, the synthesis of **848** is of particular interest for cross-linking between nucleic acid and protein.[1135]

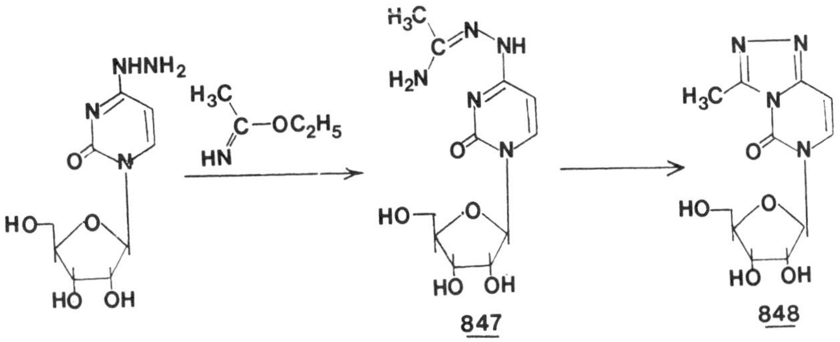

$$\underline{847} \qquad \underline{848}$$

2.6.11.3. *s-Triazolo[4,3-a]pyridine Nucleosides.* s-Triazolo[4,3-a]pyridin-3-one and various C-methyl derivatives of the general structure **849** were converted into the trimethylsilyl derivatives and then subjected to Lewis acid catalyzed glycosylation using 1,2,3,5-tetra-O-acetyl-β-D-ribofuranose.[1136] This procedure provided the protected N-2 glycosyl derivative (**850a**), which, on deprotection with methanolic sodium methoxide, gave the corresponding free nucleosides (**850b**). During the course of these reactions, rearrangements to the isomeric 3-β-D-ribofuranosyl-s-triazolo[1,5-a]pyridin-2-one (**851**) occurred through ring opening and recyclization of the pyridine ring.[1136] The ratio of the rearrangement products was dependent upon the position and number of the methyl substituents.

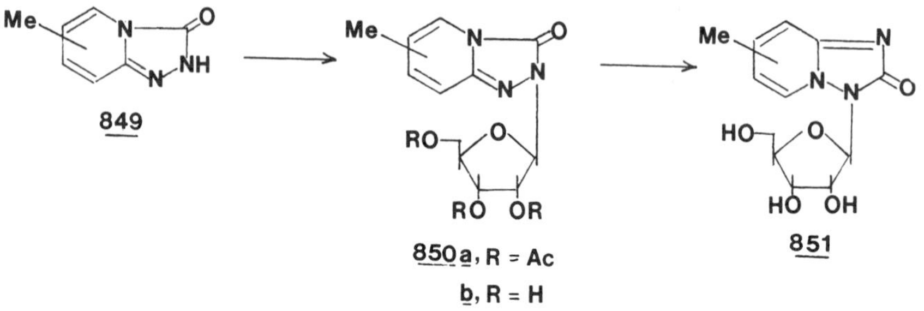

849   →   850a, R = Ac
                b, R = H   →   851

A similar glycosylation of the trimethylsilyl derivative of s-triazolo[4,3-a]-pyridin-3-one (**852**) with 2,3,5-tri-O-benzoyl-D-ribofuranosyl bromide (**19**) in the presence of mercuric bromide gave the expected blocked nucleoside (**854**), along with a major amount of the mesoionic nucleoside **853a**.[1137] Debenzoylation of **853a** with sodium methoxide in methanol readily gave 1-β-D-ribofuranosyl-s-triazolo[4,3-a]pyridine-3-oxide, which was found to be quite stable with a reasonably high melting point.[1137]

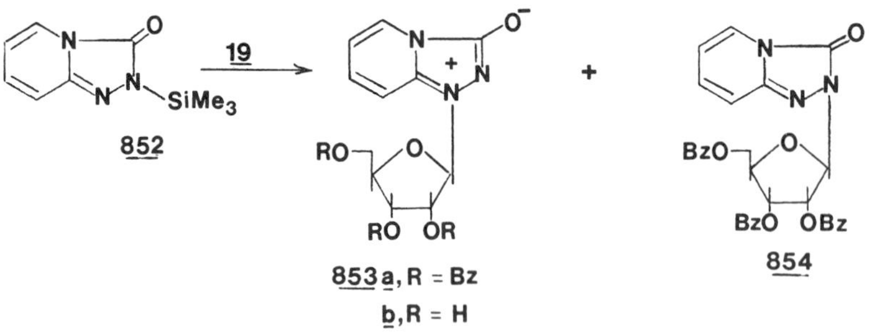

852   →   853a, R = Bz
                b, R = H   +   854

2.7. Miscellaneous Purine Analogues

The characterization[1138–1140] of uric acid ribonucleoside, isolated[1141,1142] from bovine erythrocytes, as 3-D-ribofuranosyluric acid generated early interest in the chemical synthesis of bicyclic heterocyclic nucleosides with the glycosidic linkage on a nitrogen atom in the pyrimidine ring rather than in the five-membered ring portion of the aglycon. 3-β-D-Ribofuranosyladenine (isoadeno-sine) was found[1143] to inhibit the growth of various tumor cell lines, both *in vitro* and *in vivo*, as well as to show significant activity against adeno III virus in culture. This interest was further stimulated by the isolation[703] of 7-β-D-ribofuranosylpyrazolo[3,4-*d*]pyrimidine-4,6-dione (oxoallopurinol riboside, **496**) from the urine of patients treated with allopurinol and the report[1144] that **496**, presumably as the corresponding 5′-monophosphate, inhibits the *de novo* pyrimidine biosynthesis. These findings prompted the synthesis and study of a number of other nucleosides of purine analogues.

2.7.1. Oxazolo[5,4-d]pyrimidine Nucleosides

The synthesis of the first oxazolo[5,4-*d*]pyrimidine nucleoside was reported by Townsend and co-workers.[1145] The trimethylsilyl procedure was employed. The condensation of the trimethylsilyl derivatives (**855**) of various 2-alkyl- or 2-aryl-substituted oxazolo[4,5-*d*]pyrimidin-7-ones[1146] with 2,3,5-tri-*O*-ace-tyl-D-ribofuranosyl bromide (**145**) in benzene at reflux temperature in the presence of mercuric oxide–mercuric bromide furnished good yields of the syrupy blocked nucleosides (**856**). Deacetylation of **856** with methanolic ammo-nia at room temperature gave the corresponding 2-substituted 6-β-D-ribo-furanosyloxazolo[5,4-*d*]pyrimidin-7-ones (**857**).[1147] The site of ribosylation was established as N-6 by a comparison of the UV spectral data obtained for these nucleosides with those of 2,6-dimethyloxazolo[5,4-*d*]pyrimidin-7-one.[1148] The anomeric configuration was assigned as β on the basis of ^{1}H-NMR spectroscopic studies with the corresponding 2′,3′-*O*-isopropylidene derivatives of **857**.

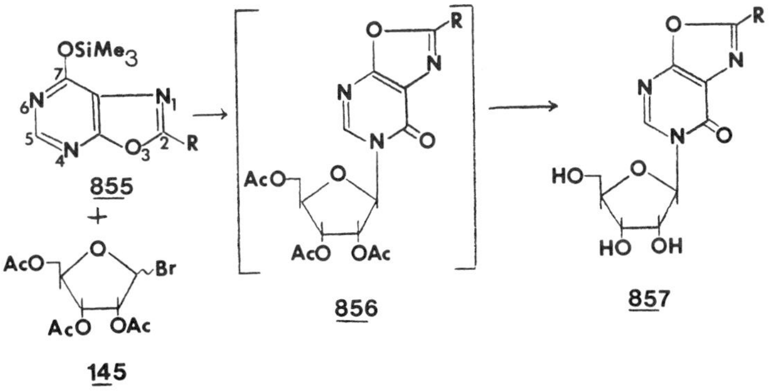

2-Methyl, 2-ethyl, and 2-propyl derivatives of **857** markedly inhibited the *in vitro* growth of leukemia L1210 and *Escherichia coli* cells, the inhibitory concentra-tions ranging from 5×10^{-7} to 8×10^{-4} *M*. Only 2-methyl-6-β-D-ribofuranosyl-oxazolo[5,4-*d*]pyrimidin-7-one is significantly active against leukemia L1210 *in vivo*.[1147] The inhibition of growth by these analogues was prevented in the presence of natural pyrimidines, suggesting that the oxazolo[5,4-*d*]pyrimidine nucleosides interfere with growth by interfering with pyrimidine metabo-lism.[1149]

2.7.2. *Thieno[2,3-d]pyrimidine Nucleosides*

The synthesis of thieno[2,3-*d*]pyrimidine nucleosides was accomplished[1150,1151] by employing the trimethylsilyl procedure. Condensation of the silyl derivative of 4-aminothieno[2,3-*d*]pyrimidin-2-one (**858a**) with either 1,2,3,5-tetra-*O*-acetyl-β-D-ribofuranose (**51**) or 1-*O*-acetyl-2,3,5-tri-*O*-benzoyl-β-D-ribofuranose (**57**) in the presence of stannic chloride and subsequent deacylation provided 4-amino-1-β-D-ribofuranosylthieno[2,3-*d*]pyrimidin-2-one (**859a**).[1150] A similar glycosylation of the silyl derivatives of thieno[2,3-*d*]pyrimidine-2,4-dione (**861a**) and the 5-methyl derivatives of **858a** and **861a** (**858b** and **861b**) gave the corresponding ribonucleosides (**862a**, **859b**, and **862b**), after deprotection of the blocked nucleoside intermediates. Sulfhydrolysis[1152] of **859a** and **859b** with H_2S in pyridine readily gave the corresponding thione compounds **860a** and **860b**, respectively.[1150] Compounds **859a** and **859b** were also converted into the corresponding arabinosyl derivatives (**863a** and **863b**) via the 2,2′-anhydro intermediates.[1150]

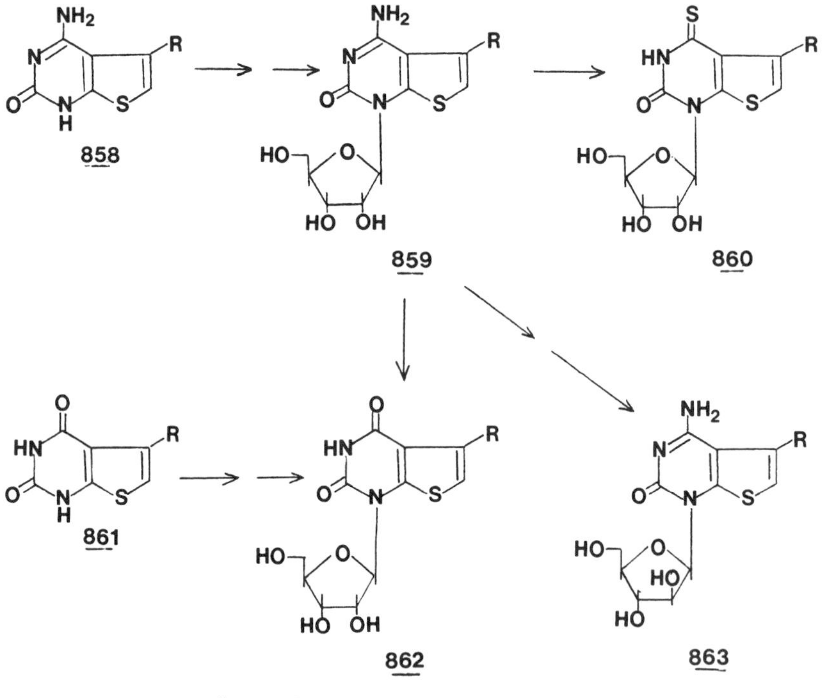

series **a**, R = H
       b, R = CH₃

The Lewis acid catalyzed glycosylation procedure was also used to prepare 3-β-D-ribofuranosylthieno[2,3-*d*]pyrimidin-4-one (**865**).[1151] Thus, condensation of the silylated thieno[2,3-*d*]pyrimidin-4-one (**864**) with **57** in the presence of stannic chloride and subsequent debenzoylation of the reaction product gave **865**. The site of ribosylation and anomeric configuration of **865** were established by ¹H-NMR spectroscopy. Treatment of the tri-*O*-acetyl derivative of **865** with

864 865 866

phosphorus pentasulfide in dioxane furnished 3-(2,3,5-tri-*O*-acetyl-β-D-ribofuranosyl)thieno[2,3-*d*]pyrimidine-4-thione (**866**).(1151) Like the thieno[3,2-*d*]-pyrimidine *C*-nucleoside 7-β-D-ribofuranosylthieno[3,2-*d*]pyrimidin-4(3*H*)-one,(1153) compound **865** inhibited the growth of murine L1210 leukemia cells *in vitro* with an ID_{50} of 3×10^{-5} *M*. The growth inhibition could not be prevented by uridine, cytidine, thymidine, deoxycytidine, cytosine, hypoxanthine, or uridine and hypoxanthine together.(1151)

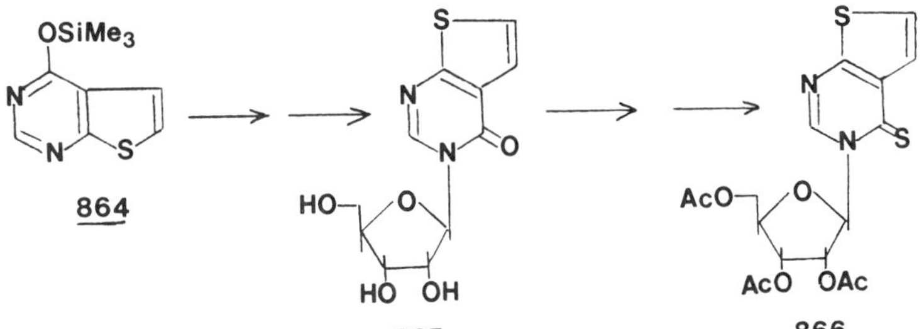

2.7.3. *Thiazolo[5,4-d]pyrimidine Nucleosides*

Like oxazolo[5,4-*d*]pyrimidine nucleosides, the first thiazolo[5,4-*d*]pyrimidine nucleoside was synthesized(54) by the trimethylsilyl procedure. Condensation of the bis-trimethylsilyl derivative (**867**) of thiazolo[5,4-*d*]pyrimidine-5,7-dione(54) with 2,3,5-tri-*O*-benzoyl-D-ribofuranosyl bromide (**19**) in anhydrous DMF gave a mixture of anomeric blocked nucleosides (**868**), which were separated as pure anomers by column chromatography. Removal of the protecting groups from **868** gave 4-β-D-ribofuranosylthiazolo[4,5-*d*]pyrimidine-5,7-dione

867 + 19, 868, 869, 870, 871, 872, 873, 874

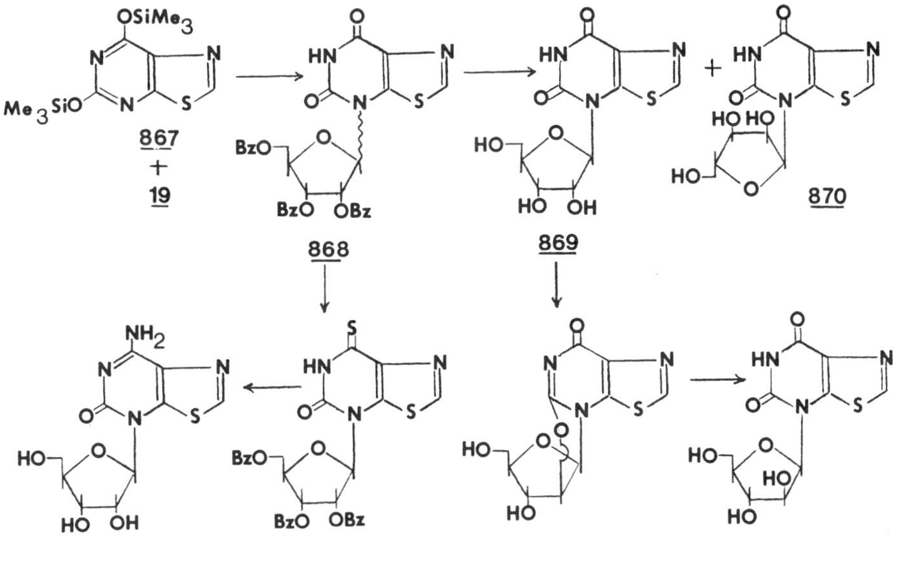

(**869**) and its α-anomer (**870**). Treatment of **869** with diphenyl carbonate in DMF containing sodium hydrogen carbonate gave the intermediate anhydronucleoside **873**, which, in the presence of sodium hydroxide, afforded 4-β-D-arabinofuranosylthiazolo[5,4-*d*]pyrimidine-5,7-dione (**874**). Displacement of the 7-keto group from β-**868** with phosphorus pentasulfide proceeded smoothly to give the 7-thione derivative **872**. Compound **872** was treated with liquid ammonia at room temperature for five days to give the cytidine analogue, 7-amino-4-β-D-ribofuranosylthiazolo[5,4-*d*]pyrimidin-5-one (**871**).[1154] However, the amination using the 7-methylthio derivative of **872** gave a better yield of **871**.

The Lewis acid catalyzed (stannic chloride) condensation of **867** with 1-*O*-acetyl-2,3,5-tri-*O*-benzoyl-β-D-ribofuranose (**57**) in 1,2-dichloroethane gave a good yield of a blocked nucleoside, identified as the N-6 glycosyl derivative (**875**). Subsequent debenzoylation of **875** furnished 6-β-D-ribofuranosylthiazolo[5,4-*d*]pyrimidine-5,7-dione (**876**).[1155] However, a similar glycosylation of the bis-trimethylsilyl derivative of thiazolo[5,4-*d*]pyrimidin-5-one-7-thione (**877**) in the presence of stannic chloride gave a good yield of the 4,6-diribosyl derivative (**878**). The site of ribosylation of these nucleosides was established by UV absorption comparison studies with model methyl compounds, and the anomeric configuration was assigned by a combination of ¹H-NMR, periodate oxidation, and reduction studies.

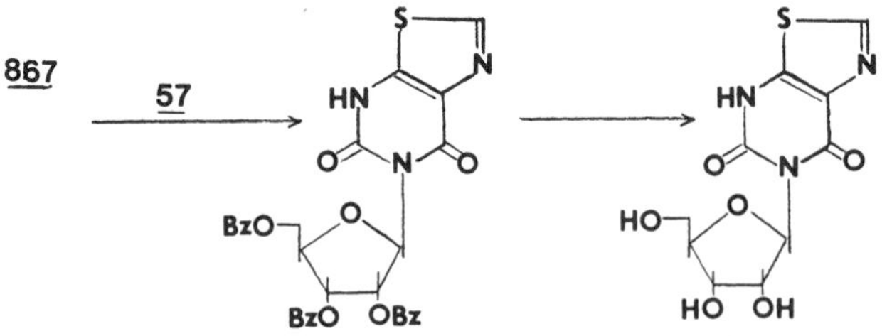

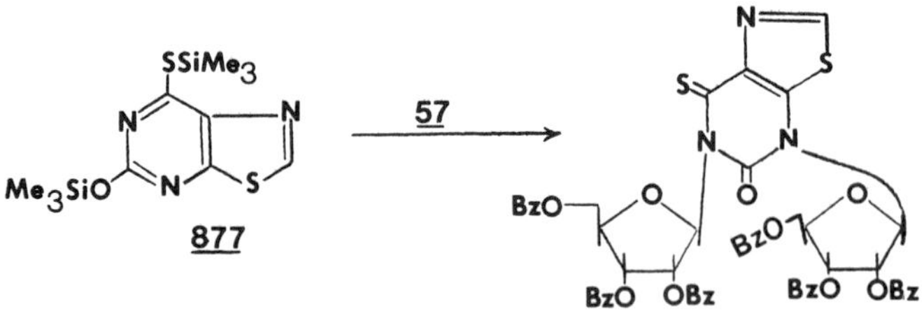

2.7.4. Thiazolo[4,5-d]pyrimidine Nucleosides

2.7.4.1. Synthesis Employing the Trimethylsilyl Derivative of a Preformed Aglycon and an Acyl Sugar. Recently, Robins and co-workers[1156] have reported the synthesis of various thiazolo[4,5-*d*]pyrimidine ribonucleosides as potential immunotherapeutic agents. 5-Aminothiazolo[4,5-*d*]pyrimidine-2,7(3*H*,6*H*)-dione (**879**)[1157] was silylated using hexamethyldisilazane (HMDS), and the resulting trimethylsilyl derivative was treated with **57** in the presence of trimethylsilyl trifluoromethanesulfonate as the catalyst. The major product, 5-amino-3-(2,3,5-tri-*O*-benzoyl-β-D-ribofuranosyl)thiazolo[4,5-*d*]pyrimidine-2,7(3*H*,6*H*)-dione (**880**), was isolated in a 77% yield. Treatment of **880** with sodium methoxide in methanol gave the guanosine analogue, 5-amino-3-β-D-ribofuranosylthiazolo-[4,5-*d*]pyrimidine-2,7(3*H*,6*H*)-dione (**883**), in a 78% yield. When **883** was deaminated with aqueous nitrous acid, the xanthosine analogue **884** was produced.[1156] Replacement of the 5-amino group of **880** by a hydrogen atom was accomplished by the treatment of **880** with *tert*-butyl nitrite in THF to yield 3-(2,3,5-tri-*O*-benzoyl-β-D-ribofuranosyl)thiazolo[4,5-*d*]pyrimidine-2,7(3*H*,-6*H*)-dione (**881a**). Debenzoylation of **881a** using sodium methoxide in methanol or methanolic ammonia provided the inosine analogue **881b**.[1156] Reaction of **880** with P_2S_5 in pyridine at 130–140°C for 29 h gave **882a**. Deprotection of **882a** with MeOH/NH$_3$ gave the 6-thioguanosine analogue 5-amino-7(6*H*)-thioxo- 3-β-D-ribofuranosylthiazolo[4,5-*d*]pyrimidin-2(3*H*)-one (**882b**). The structure of **882b** was confirmed by single-crystal X-ray diffraction analysis.[1156]

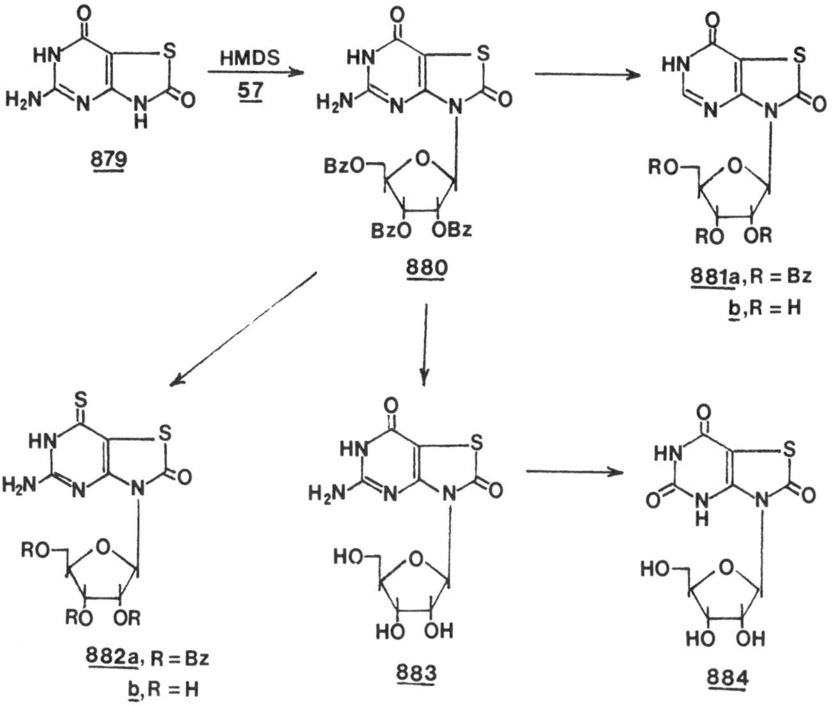

A similar glycosylation of the trimethylsilyl derivative of 5-amino-2-(3*H*)-thioxothiazolo[4,5-*d*]pyrimidin-7(6*H*)-one (**885**)[1156] with **57** resulted in the formation of the protected nucleoside **886**. Treatment of **886** with sodium methoxide in methanol gave the 8-mercaptoguanosine analogue 5-amino-2(3*H*)-thioxo-3-β-D-ribofuranosylthiazolo[4,5-*d*]pyrimidin-7(6*H*)-one (**887**),[1156] in good yield.

The synthesis of the adenosine analogue 7-amino-3-β-D-ribofuranosylthiazolo[4,5-*d*]pyrimidin-2(3*H*)-one (**890b**) was rather cumbersome. Reaction of 7-aminothiazolo[4,5-*d*]pyrimidine–2(3*H*)-one (**889**)[1156] with **57**, under the trimethylsilyl glycosylation conditions, at room temperature, resulted in the formation of the protected 4-ribofuranosyl isomer, 7-amino-4-(2,3,5-tri-*O*-benzoyl-β-D-ribofuranosyl)thiazolo[4,5-*d*]pyrimidin-2-one (**888a**), as the only nucleoside product.[1156] However, when the reaction was carried out at 80°C, the predominant product obtained was the 3-ribofuranosyl isomer **890a**. Both isomers **888a** and **890a** were debenzoylated using sodium methoxide in methanol to obtain the isoadenosine analogue 7-amino-4-β-D-ribofuranosylthiazolo[4,5-*d*]pyrimidin-2-one (**888b**) and the adenosine analogue 7-amino-3-β-D-ribofuranosylthiazolo[4,5-*d*]pyrimidin-2-(3*H*)-one (**890b**),[1156] respectively. Single-crystal X-ray diffraction studies confirmed[1158] the structural assignments of **888b** and **890b**.

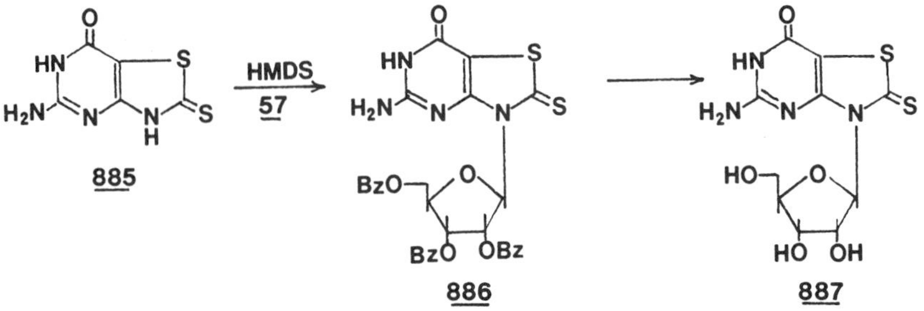

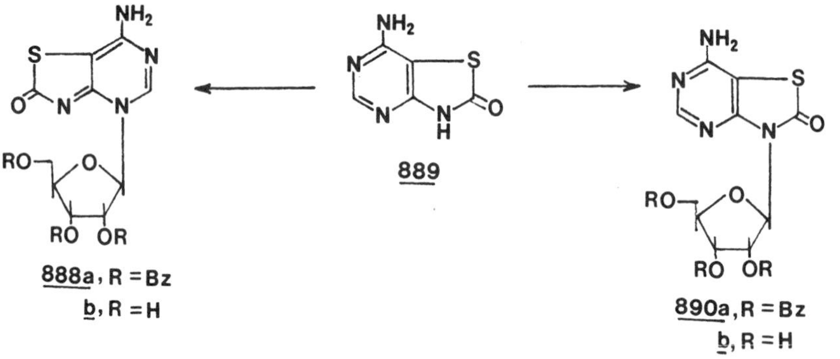

2.7.4.2. Synthesis Employing the Acid-Catalyzed Fusion of a Preformed Aglycon and an Acyl Sugar.   Reaction of 5,7-dichlorothiazolo[4,5-*d*]pyrimidin-2(3*H*)-one (**891**)[1156] with 1,2,3,5-tetra-*O*-acetyl-β-D-ribofuranose (**51**) under fusion (170°C) glycosylation conditions, in the presence of the acid catalyst bis(*p*-nitrophenyl)phosphate, produced an excellent yield of 5,7-dichloro-3-(2,3,5-tri-*O*-acetyl-β-D-ribofuranosyl)thiazolo[4,5-*d*]pyrimidin-2(3*H*)-one (**892**).[1156] Attempts to use **892** for further transformation to obtain the adenosine analogue **890b** were unsuccessful, probably due to the labile nature of the thiazole ring toward nucleophilic ring opening. The nucleophiles used were azide, ammonia, hydrazine, methoxide, and thiourea.[1156]

Nucleosides **883** and **887** exhibited significant immunoactivity in comparison to the known active agents 8-bromoguanosine, 8-mercaptoguanosine, and

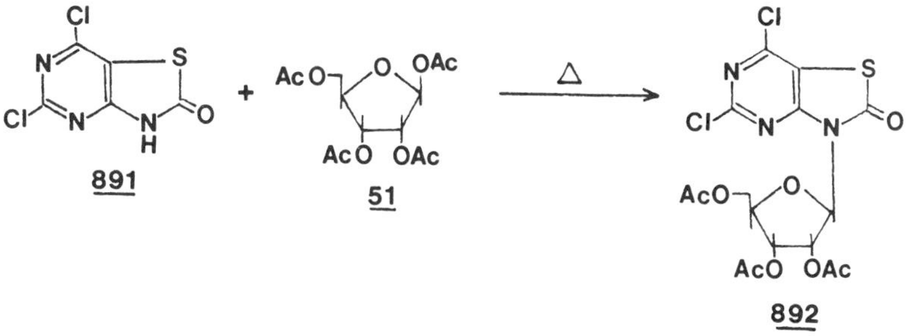

7-methyl-8-oxoguanosine.[1159–1163] Compound **883**, particularly, was twice as potent as 7-methyl-8-oxoguanosine in the murine spleen cell mitogenicity assay. Treatment with **883** produced a fourfold increase in natural killer cell cytotoxicity,[1156] and it induced interferon. Nucleoside **883** was highly active at 50–200 mg/kg in preventing death in mice inoculated i.p. with Semliki Forest, San Angelo, and benzi viruses.[1164] Similarly, **883** was effective against an intranasal challenge of rat coronavirus in suckling rats.[1164]

2.7.5. Oxazolo[4,5-d]pyrimidine Nucleosides

2.7.5.1. Synthesis Employing the Heavy-Metal Salt of a Preformed Aglycon and an Acylated Sugar Halide. Ito *et al.*[1165] reported the synthesis of various nucleosides of 5-amino-7-methyloxazolo[4,5-d]pyrimidin-2-one by the heavy-metal salt procedure. These glycosides were prepared by the treatment of either the silver or the chloromercuri derivative of 5-amino-7-methyloxazolo[4,5-d]pyrimidin-2-one (**893**) with various glycosylacyl bromides in boiling nitromethane, which produced the corresponding blocked nucleosides (**894**). Deprotection of **894** afforded the corresponding 5-amino-7-methyl-3-glycosyloxazolo[4,5-d]pyrimidin-2-ones (**895**).

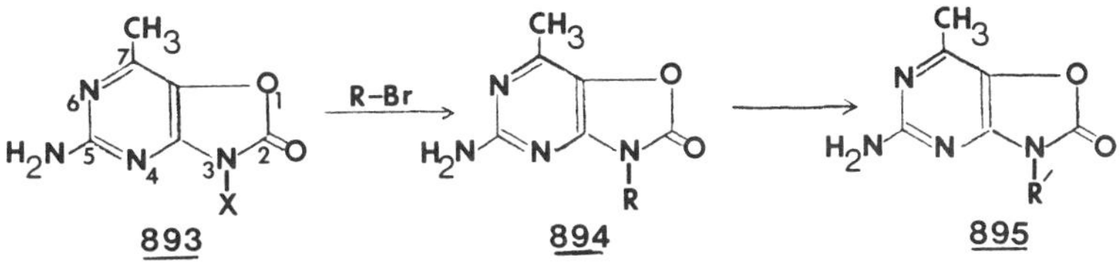

X = Ag or HgCl

R = 2,3,5-tri-*O*-acetyl/benzoyl-D-ribofuranosyl
2,3,4,6-tetra-*O*-acetyl/benzoyl-D-glucopyranosyl
2,3,4-tri-*O*-acetyl/benzoyl-D-xylopyranosyl
2,3,4-tri-*O*-acetyl/benzoyl-D-arabinopyranosyl

R′ = β-D-ribofuranosyl,β-D-glucopyranosyl,β-D-xylopyranosyl,β-D-arabinopyranosyl

2.7.5.2. Synthesis Employing the Acid-Catalyzed Fusion of a Preformed Aglycon and an Acyl Sugar. The fusion of 5-amino-7-methyloxazolo[4,5-d]pyrimidin-2-one with tetra-*O*-acetyl-β-D-ribofuranose (**51**) or 1-*O*-acetyl-2,3,5-tri-*O*-benzoyl-β-D-ribofuranose (**57**) at 160–180°C under reduced pressure in the presence of *p*-toluenesulfonic acid or sodium sulfamate gave the corresponding blocked nucleoside, which, on subsequent deacylation, afforded 5-amino-7-methyl-3-β-D-ribofuranosyloxazolo[4,5-*d*]pyrimidin-2-one. The site of glycosylation of these nucleosides was established as N-3 by UV absorption comparison studies with model methyl compounds,[1166] and the anomeric configuration was assigned as β on the basis of ^{1}H-NMR studies. None of these nucleosides exhibited any antitumor activity.

2.7.6. Imidazo[4,5-d]-1,3-oxazine Nucleosides

Recently, a novel nucleoside antibiotic named oxanosine has been isolated[1167] from *Streptomyces capreolus* MG265-CF3 and identified as 5-amino-3-β-D-ribofuranosylimidazo[4,5-*d*]-1,3-oxazin-7-one (**898**) by X-ray crystallographic studies.[1168] Oxanosine inhibits the growth of *Escherichia coli* K-12 on peptone agar, and this antibacterial activity was antagonized by guanosine.[1167] Oxanosine also inhibited the growth of HeLa cells in culture (IC$_{50}$ 32 μg/ml) and suppressed the growth of L1210 leukemia in mice,[1167] and it was relatively nontoxic. Oxanosine was more effective in inhibiting the growth of rat kidney cells infected with a temperature-sensitive mutant of Rous sarcoma virus at a permissive temperature of 33°C than at a nonpermissive temperature of 39°C.[1169] The modes of antitumor[1170] and antiviral[1171] actions of oxanosine have been studied. It was shown that oxanosine is a competitive inhibitor of GMP synthetase (EC 6.3.5.2) with a K_i value of 7.4 × 10^{-4} M.[1170] The total synthesis of oxanosine has recently been reported.[1172]

Treatment of ethyl 5-amino-1-(2,3-*O*-isopropylidene-β-D-ribofuranosyl)-imidazole-4-carboxylate (**896**) with ethoxycarbonylisothiocyanate gave the thiourea derivative (**897**) as the sole product. Methylation of **897** with methyl iodide gave the *S*-methylisothiourea derivative (**899**). Reaction of **899** with 10 equivalents of 0.2 *N* sodium hydroxide and further treatment with aqueous hydrochloric acid gave oxanosine (**898**),[1172] in rather low yield. Subsequently, 2′-deoxyoxanosine[1173] and 3-deazaoxanosine[1174] have been synthesized. A four-step Barton deoxygenation of oxanosine gave 2′-deoxyoxanosine. 2′-Deoxyoxanosine exhibited stronger antineoplastic activity (IC$_{50}$ of 0.15 μg/liter) than oxanosine (0.53 μg/liter).[1173] However, 3-deazaoxanosine was much less active than oxanosine as an antitumor agent.[1174]

The other imidazo[4,5-*d*]-1,3-oxazine nucleoside reported in the literature[891] was in connection with an unexpected cyclization of an aminoimidazole nucleoside. When 5-amino-1-(2,3,5-tri-*O*-acetyl-β-D-ribofuranosyl)imidazole-4-carboxylic acid (**900**)[1175] was treated with acetic anhydride in pyridine, preferably above 30°C, the 5-amino function of **900** was acetylated, generating the corresponding tetraacetyl derivative (**901**), which immediately ring annulated to furnish 5-methyl-3-(2,3,5-tri-*O*-acetyl-β-D-ribofuranosyl)imidazo[4,5-*d*]-1,3-oxazin-7-one (**902**, R = CH$_3$), in almost quantitative yield.[891] In a similar manner, when acetic anhydride was replaced by propionic anhydride, the corresponding 5-ethyl-3-(2,3,5-tri-*O*-acetyl-β-D-ribofuranosyl)imidazo[4,5-*d*]-1,3-oxazin-7-one (**902**, R = C$_2$H$_5$) was obtained.[891] The structures of these compounds were determined by ^{1}H-NMR studies.

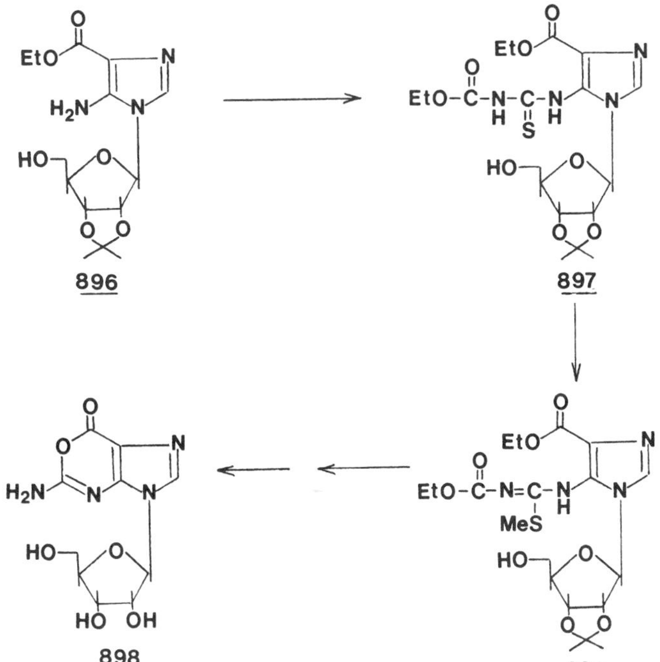

3. Conclusion

In conclusion, the chemistry of heterocyclic analogues of purine nucleosides and nucleotides has developed from its infant to adult stage over the years, and, in certain cases, one can prepare the required anomers or isomers of nucleosides stereospecifically. The biochemistry and molecular biology of certain of these purine nucleoside analogues has opened a new era of understanding of many fundamental biological processes. Use of some of the purine *N*-nucleoside analogues in the treatment of major medical problems, as attested by the use of (to name a few) allopurinol ribonucleoside for antiparasitic chemotherapy; 3-deazaguanosine, 8-azainosine, sangivamycin, and thiosangivamycin for antitumor chemotherapy; and 3-deazaguanosine, 7-deazainosine, and 7-thia-8-oxoguanosine for antiviral chemotherapy, should generate a fruitful area of medicinal chemistry of considerable importance for years to come.

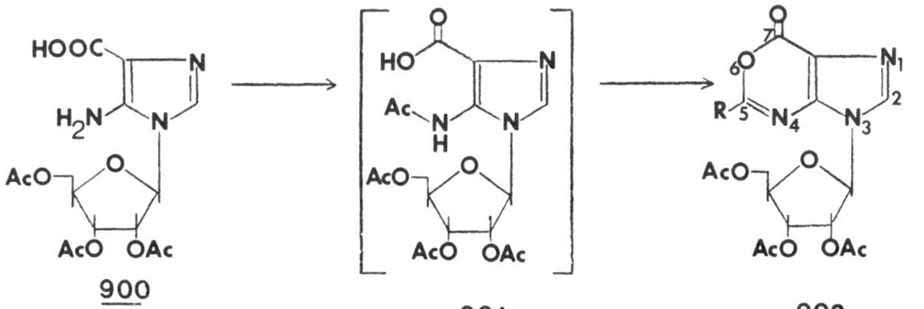

Acknowledgment

We wish to thank Ms. Sandy Young and Mrs. Pauline Elliott for the meticulous and superb preparation of this manuscript. One of us (GRR) also wishes to express his sincerest appreciation to his wife, Ratna, for her valuable assistance and patience.

4. References

1. P. A. Levene and W. A. Jacobs, Nucleic acid from yeast, *Ber.* **42**, 2474–2478 (1909).
2. R. J. Suhadolnik, *Nucleoside Antibiotics*, Wiley Interscience, New York (1970).
3. R. J. Suhadolnik, *Nucleosides as Biological Probes*, Wiley-Interscience, New York (1979).
4. A. M. Michelson, in "The Chemistry of Nucleosides and Nucleotides," Academic Press, New York (1963).
5. W. H. Perkin, Jr. and R. Robinson, Harmine and harmaline. I. *J. Chem. Soc.* **101**, 1775–1787 (1912).
6. O. Kruber, The bases of coal-tar heavy oil. 7-Azaindole in coal tar, *Ber.* **76B**, 128–143 (1943).
7. A. M. Patterson, L. T. Capell, and D. F. Walker, in *The Ring Index*, 2nd ed., p. 166, American Chemical Society, Washington, D.C. (1960).
8. A. E. Chichibabin, Tautomerism in the pyridine series, *Ber.* **60B**, 1607–1617 (1927).
9. M. M. Robison and B. L. Robison, 7-Azaindole—III. Synthesis of 7-aza analogs of some biologically significant indole derivates, *J. Am. Chem. Soc.* **78**, 1247–1251 (1956).
10. M. Hooper, D. A. Patterson, and D. G. Wibberley, Preparation and anti-bacterial activity of isatogens and related compounds, *J. Pharm. Pharmacol.* **17**, 734–741 (1965).
11. T. K. Adler and A. Albert, The biological and physical properties of the azaindoles, *J. Med. Chem.* **6**, 480–483 (1963).
12. M. R. Bell, J. O. Hoppe, H. E. Lape, D. Wood, A. Arnold, and W. H. Selberis, Antihypertensive activity of 7-azaindole-3-acetamidoxime and indole-1-acetamidoxime, *Experientia* **23**, 298–299 (1967).
13. Merck and Co., Inc., Indoles, Neth. Patent 6,510,648 (Feb. 1966) [*CA* **65**, 13711e (1966)].
14. M. H. Fisher, G. Schwartzkopf, Jr., and D. R. Hoff, Azaindole anthelmintic agents, *J. Med. Chem.* **15**, 1168–1171 (1972).
15. L. N. Yakhontov and A. A. Prokopov, Advances in the chemistry of azaindoles, *Russ. Chem. Rev.* **49**, 428–444 (1980).
16. M. N. Preobrazhenskaya, T. D. Miniker, V. S. Martynov, L. N. Yakhontov, N. P. Kostyuchenko, and D. M. Krasnokutskaya, 1-β-D-Glucopyranosides of pyrrolo [2,3-*b*]pyridines (7-azaindoles), *Zh. Org. Khim.* **10**, 745–750 (1974).
17. M. N. Preobrazhenskaya, Synthesis of substituted indoles through indolines, *Usp. Khim.* **36**, 1760–1798 (1967).
18. M. N. Preobrazhenskaya, I. A. Korbukh, V. N. Tolkachev, Ya. V. Dobrynin, and G. I. Vornovitskaya, Studies of purine nucleoside analogues, *INSERM, Nucleosides, Nuclelotides, Biol. Appl.* **81**, 85–116 (1978).
19. L. N. Yakhontov, M. Ya. Uritskaya, and M. V. Rubtsov, Synthesis of 4-methyl-7-azaindole and its 6-chloro-, 6-iodo-, and 6-methoxy derivatives, *Zh. Obshch. Khim.* **34**, 1449–1455 (1964).
20. M. N. Preobrazhenskaya, T. D. Miniker, V. S. Martynov, L. N. Yakhontov, and D. N. Krasnokutskaya, Synthesis of pyrrolo[2,3-*b*]pyridine 1-β-D-ribofuranosides, *Zh. Org. Khim.* **10**, 2449–2452 (1974).
21. (a) M. N. Preobrazhenskaya and T. D. Miniker, 4-methyl-1-β-D-ribofuranosyl-pyrrolo[2,3,-*b*]-pyridine, in: *Nucleic Acid Chemistry, Improved and New Synthetic Procedures, Methods, and Techniques* (L. B. Townsend and R. S. Tipson, eds.), Vol. 2, pp. 749–752, John Wiley and Sons, New York (1978); (b) G. Lupidi, G. Cristalli, M. Marmocchi, F. Riva, and M. Grifantini, Inhibition of adenosine deaminase from several sources by deaza derivatives of adenosine and EHNA, *J. Enzyme Inhibition* **1**, 67–75 (1985).

22. L. V. Ektova, T. D. Miniker, I. V. Yartseva, and M. N. Preobrazhenskaya, Synthesis of 2,3-dioxo-2,3-dihydro-4-methyl-6-chloro-1*H*-pyrrolo [2,3-*b*]pyridine and its 1-α-L-arabino-pyranoside, *Khim. Geterotsikl. Soedin.* **1977**, 1083–1086.

23. V. I. Mukhanov, T. N. Sokolova, T. G. Nikolaeva, Ya. V. Dobrynin, and M. N. Preobrazhen-skaya, Synthesis and study of the biological activity of indole nucleosides. IV. Synthesis of 1-α-L-arabinopyranosides of substituted indoles and 7-azaindoles, *Khim.-Farm. Zh.* **13**, 47–57 (1979).

24. I. Antonini, F. Claudi, G. Cristalli, P. Franchetti, M. Grifantini, and S. Martelli, Synthesis of 4-amino-1-β-D-ribofuranosyl-1*H*-pyrrolo[2,3-*b*]pyridine (1-deazatubercidin) as potential an-titumor agent, *J. Med. Chem.* **25**, 1258–1261 (1982).

25. A. V. Stetsenko and I. P. Kupchevskaya, Glycosyl derivatives of 1*H*-pyrrolo[3,2-*c*]pyridine, *Ukr. Khim. Zh.* (Russ. Ed.) **38**, 503–504 (1972).

26. C. Ducrocq, E. Bisagni, J.-M. Lhoste, J. Mispelter, and J. Defaye, Azaindoles—III. Synthesis of 4-amino-5-azaindole and the corresponding N-5 ribonucleoside (deaza-1-isotubercidin), *Tet-rahedron* **32**, 773–780 (1976).

27. C. Ducrocq, E. Bisagni, J. Defaye, and D. Horton, Synthesis and characterization of 1-deaza-tubercidin, Abstracts of papers, 168th National American Chemical Society meeting, Atlantic City, N.J., Sept. 1974, abstract CARB-68.

28. A. Ducruix, C. Riche, and C. Pascard, Structure and conformation of the nucleoside analogue of deaza-1-isotubercidin, *Tetrahedron Lett.* **1976**, 51–52.

29. A. Ducruix, C. Riche, and C. Pascard, X-ray determination of the molecular structure of an analogue of tubercidin: Deaza-1-isotubercidin, *Acta Crystallogr., Sect. B* **32**, 2467–2471 (1976).

30. Z. Kazimierczuk, G. R. Revankar, and R. K. Robins, Total synthesis of certain 2-, 6- and 2,6-disubstituted-tubercidin derivatives. Synthesis of tubercidin via the sodium salt glycosylation procedure, *Nucleic Acids Res.* **12**, 1179–1192 (1984).

31. Z. Kazimierczuk, H. B. Cottam, G. R. Revankar, and R. K. Robins, Synthesis of 2′-deoxytuber-cidin, 2′-deoxyadenosine, and related 2′-deoxynucleosides via a novel direct stereospecific sodium salt glycosylation procedure, *J. Am. Chem. Soc.* **106**, 6379–6382 (1984).

32. N. S. Girgis, S. B. Larson, R. K. Robins, and H. B. Cottam, The synthesis of 5-azaindoles by substitution-rearrangement of 7-azaindoles upon treatment with certain primary amines, *J. Heterocycl. Chem.* **26**, 317–325 (1989).

33. N. S. Girgis, H. B. Cottam, S. B. Larson, and R. K. Robins, 2-Deoxy-3,7-dideazaguanosine and related compounds. Synthesis of 6-amino-1-(2-deoxy-β-D-*erythro*-pentofuranosyl) and 1-β-D-arbinofuranosyl-1*H*-pyrrolo[3,2-*c*]pyridin-4(5*H*)-one via direct glycosylation of a pyr-role precursor, *Nucleic Acids Res.* **15**, 1217–1226 (1987).

34. M. Hoffer, α-Thymidine, *Chem. Ber.* **93**, 2777–2781 (1960).

35. C. P. J. Glaudemans and H. G. Fletcher, Jr., Syntheses with partially benzylated sugars. III. A simple pathway to a 'cis nucleoside', 9-α-D-ara-binofuranosyladenine (spongoadenosine), *J. Org. Chem.* **28**, 3004–3006 (1963).

36. S. W. Schneller and R. S. Hosmane, Chlorination of 1*H*-pyrrolo [3,2-*c*]pyridin-4,6(5*H*,7*H*)-dione (3,7-dideazaxanthine) and its 5-methyl derivative, *J. Heterocycl. Chem.* **15**, 325–326 (1978).

37. (a) F. Seela and W. Bourgeois, Synthesis of 3,7-dideaza-2′-deoxyadenosine and related pyr-rolo[3,2-*c*]pyridine 2′-deoxyribo- and 2′,3′-dideoxyribonucleosides, *Synthesis* **1988**, 938–943; (b) F. Seela and W. Bourgeois, 2′-Deoxy-3,7-dideazanebularine and 2′-deoxy-3,7-dideazaino-sine: Synthesis of pyrrolo[3,2-*c*]pyridine β-D-2′-deoxy-ribofuranosides by solid–liquid phase-transfer glycosylation, *Heterocycles* **26**, 1755–1760 (1987); (c) G. Cristalli, P. Franchetti, M. Grifantini, G. Nocentini, and S. Vittori, 3,7-Dideazapurine nucleosides. Synthesis and anti-tumor activity of 1-deazatubercidin and 2-chloro-2′-deoxy-3,7-dideazaadenosine, *J. Med. Chem.* **32**, 1463–1466 (1989).

38. S. W. Schneller and R. S. Hosmane, Ring-opening reactions of 1*H*-pyrrolo[3,2-*c*]pyridine-4,6-(5*H*,7*H*)-dione (3,7-dideazaxanthine) and two of its derivatives, *J. Org. Chem.* **43**, 4487–4491 (1978).

39. A. K. Bhattacharya, R. K. Ness, and H. G. Fletcher, Jr., 2-Deoxy-D-*erythro*-pentose. IX. Some relationships among the rotations of acylated aldopentoses, aldopentosyl halides, and an-hydropentitols, *J. Org. Chem.* **28**, 428–435 (1963).

40. B. Frydman, S. J. Reil, J. Boned, and H. Rapoport, Synthesis of substituted 4- and 6-aza-indoles, *J. Org. Chem.* **33**, 3762–3766 (1968).

41. B. A. J. Clark, M. M. S. El-Bakoush, and J. Parrick, Diazaindenes (azaindoles). V. Synthesis, spectra, and tautomerism of 1,5-diazainden-4(5*H*)-one, 1,4- and 1,6-diazainden-2(3*H*)-one and 3-substituted derivatives, *J. Chem. Soc., Perkin Trans. 1* **1974**, 1531–1536.

42. B. A. J. Clark, J. Parrick, P. J. West, and A. H. Kelly, Diazaindenes (azaindoles). III. Reactions of Vilsmeier reagents leading to 3-formyl-1,6-diazaindenes and -1,4-diazabenz[*f*]indene, *J. Chem. Soc. (C)* **1970**, 498–501.

43. W. Herz and D. R. K. Murty, Pyrrolopyridines. III. The madelung cyclization of 3-acyl-amino-4-picolines, *J. Org. Chem.* **25**, 2242–2245 (1960).

44. I. Antonini, G. Cristalli, P. Franchetti, M. Grifantini, S. Martelli, and F. Petrelli, Deaza analogues of adenosine as inhibitors of blood platelet aggregation, *J. Pharm. Sci.* **73**, 366–369 (1984).

45. S. K. Chatterjee, M. M. Dhar, N. Anand, and M. L. Dhar, Potential purine antagonists: Synthesis of N-glycosides of 1-deazapurine, *J. Sci. Ind. Res. India* **19C**, 35–37 (1960).

46. Y. Mizuno, M. Ikehara, T. Itoh, and K. Saito, Studies on condensed systems of aromatic nitrogenous series. XXII. Structural studies of β-D-ribofuranosylimidazopyridines, *J. Org. Chem.* **28**, 1837–1841 (1963).

47. J. Davoll and B. A. Lowy, A new synthesis of purine nucleosides. The synthesis of adenosine, guanosine, and 2,6-diamino-9-β-D-ribofuranosylpurine, *J. Am. Chem. Soc.* **73**, 1650–1655 (1951).

48. H. M. Kissman, C. Pidacks, and B. R. Baker, D-Ribofuranosyl derivatives of 6-dimethylamino-purine, *J. Am. Chem. Soc.* **77**, 18–24 (1955).

49. J. M. Gulland, E. R. Holiday, and T. F. Macrae, Constitution of the purine nucleosides. II. *J. Chem. Soc.* **1934**, 1639–1644.

50. J. Baddiley, Chemistry of nucleosides and nucleotides, in *Nucleic Acids*, Vol. I. (E. Chargaff and J. N. Davidson, eds.), pp. 137–190, Academic Press, New York (1955).

51. P. C. Jain, S. K. Chatterjee, and N. Anand, Potential purine anatagonists: Part IV—Synthesis of *N*-β-D-ribofuranosides of substituted imidazo(*b*)- and imidazo(*c*)pyridines, *Indian J. Chem.* **1**, 30–35 (1963).

52. R. S. Tipson, Action of silver salts of organic acids on bromoacetyl sugars. A new form of tetraacetyl-1-rhamnose, *J. Biol. Chem.* **130**, 55–59 (1939).

53. B. R. Baker, Sterochemistry of nucleoside synthesis, *Ciba Found. Symp.* **1957**, 120–130.

54. C. L. Schmidt, W. J. Rusho, and L. B. Townsend, The synthesis of bicyclic nucleosides related to uridine, 4-β-D-ribofuranosyl)thiazolo[5,4-*d*]-pyrimidines, *J. Chem. Soc., Chem. Commun.* **1971**, 1515–1516.

55. Y. Mizuno, M. Ikehara, T. Itoh, and K. Saito, Synthetic studies of potential antimetabolites, VII. Evidence that the supposed 1-β-D-ribofuranosyl-1*H*-imidazo[4,5-*b*]pyridine is actually 3-β-D-ribofuranosyl-3*H*-imidazo[4,5-*b*]pyridine (1-deazapurine ribofuranoside), *Chem. Pharm. Bull.* **11**, 265–267 (1963).

56. P. C. Jain and N. Anand, Potential purine antagonists. IX. Synthesis of 3-β-D-ribofuranosyl-6-nitroimidazo- and triazolo[4,5-*b*]pyridines, *Indian J. Chem.* **6**, 123–125 (1968).

57. N. S. Miroshnichenko, O. I. Shkrebtii, and A. V. Stetsenko, Preparation of glycoside deriva-tives of 6-bromo-1*H*-imidazo[4,5-*b*]pyridine, *Ukr. Khim. Zh.* (Russ. Ed.) **39**, 277–280 (1973).

58. Y. Mizuno, N. Ikekawa, T. Itoh, and K. Saito, Studies on condensed aromatic nitrogenous compounds. XXV. Product distribution in ribosylation of purines and deazapurines by the mercuri method, *J. Org. Chem.* **30**, 4066–4071 (1965).

59. P. C. Jain, S. K. Chatterjee, and N. Anand, Potential purine antagonists: Part VII—Synthesis of 3-β-D-ribofuranosyl-7-aminoimidazo[4,5-*b*]pyridine (1-deazaadenosine), *Indian J. Chem.* **4**, 403–405 (1966).

60. T. Itoh, S. Kitano, and Y. Mizuno, Synthetic studies of potential antimetabolites. XIII. Synthesis of 7-amino-3-β-D-ribofuranosyl-3*H*-imidazo[4,5-*b*]pyridine (1-deazaadenosine) and related nucleosides, *J. Heterocycl. Chem.* **9**, 465–470 (1972).

61. D. G. Markees and G. W. Kidder, The synthesis of 5-amino-7-hydroxy-1,3,4-imidazopyridine (1-deazaguanine) and related compounds, *J. Am. Chem. Soc.* **78**, 4130–4135 (1956).

62. P. C. Jain and N. Anand, Potential purine antagonists: Part VIII—Synthesis of 6-ribofura-nosyladenosine and 5-ribosylamino-7-chloro-3-β-D-ribofuranosylimidazo[4,5-*b*]pyridine, *Indian J. Chem.* **6**, 616–618 (1968).

63. J. A. Montgomery and K. Hewson, 1-Deaza-6-methylthiopurine ribonucleoside, *J. Med. Chem.* **9**, 354–357 (1966).

64. F. Kögl, G. M. van der Want, and C. A. Salemink, 1-Deazaadenine (7-amino-imidazo[b]pyridine). I. Deazapurine derivatives, *Recl. Trav. Chim. Pays-Bas* **67**, 29–44 (1948).

65. S. Kitano, A. Nomura, Y. Mizuno, T. Okamoto, and Y. Isogai, Synthesis of potential antimetabolites. XVIII. Cytokinin activity of deazakinetin ribofuranosides, *J. Carbohydr. Nucleosides Nucleotides* **2**, 299–307 (1975).

66. K. B. De Roos and C. A. Salemink, Deazapurine derivatives. VIII. Synthesis of 7-substituted 3-β-D-ribofuranosylimidazo[4,5-*b*]pyridines: 1-Deazaadenosine and related compounds, *Recl. Trav. Chim. Pays-Bas* **90**, 654–662 (1971).

67. J. E. Schelling and C. A. Salemink, Deazapurine derivatives. XIII. 5,7-Disubstituted imidazo[4,5-*b*]pyridines. New synthesis of 1-deazaguanine, *Recl. Trav. Chim. Pays-Bas* **93**, 160–162 (1974).

68. K. B. De Roos and C. A. Salemink, Deazapurine derivatives, V. New synthesis of 1- and 3-deazaadenine and related compounds, *Recl. Trav. Chim. Pays-Bas* **88**, 1263–1274 (1969).

69. J. E. Schelling and C. A. Salemink, Deazapurine derivatives, XIV. The synthesis of 1-deazaguanosine, *Recl. Trav. Chim. Pays-Bas* **94**, 153–156 (1975).

70. G. Cristalli, P. Franchetti, M. Grifantini, S. Vittori, K-N. Klotz, and M. J. Lohse, Adenosine receptor agonists: Synthesis and biological evaluation of 1-deaza analogues of adenosine derivatives, *J. Med. Chem.* **31**, 1179–1183 (1988).

71. N. Yamaoka, K. Aso, and K. Matsuda, New synthesis of nucleosides. The syntheses of glycopyranosides of purines, pyrimidine, and benzimidazole, *J. Org. Chem.* **30**, 149–152 (1965).

72. A. V. Stetsenko and E. V. Goshchulyak, Glycosides of halo-substituted imidazo[4,5-*b*]pyridines, *Ukr. Khim. Zh.* (Russ. Ed.) **43**, 165–168 (1977).

73. A. V. Stetsenko, Yu. P. Kovtun, and S. A. Andrianova, Glucosides from substituted nitro-imidazo[4,5-*b*]pyridines, *Ukr. Khim. Zh.* (Russ. Ed.) **47**, 867–870 (1981).

74. S. Okuhara, Heterocyclic glycosides, *Jpn. Patent* 69 25,580 (Oct. 1969) [*CA* **72**, 21927j (1970)].

75. N. J. Leonard and J. A. Deyrup, The chemistry of triacanthine, *J. Am. Chem. Soc.* **84**, 2148–2160 (1962).

76. B. S. Gorton and W. Shive, Synthesis of 1-deazaguanine, *J. Am. Chem. Soc.* **79**, 670–672 (1957).

77. G. W. Kidder and V. C. Dewey, Deazapurines as growth inhibitors, *Arch. Biochem. Biophys.* **66**, 486–492 (1957).

78. H. Kawashima and I. Kumashiro, Studies of purine *N*-oxides. II. The reaction of hypoxanthine 1-*N*-oxide and 2′,3′,5′-tri-*O*-acetylinosine 1-*N*-oxide with phosphoryl chloride, *Bull. Chem. Soc. Jpn.* **40**,639–641 (1967).

79. R. D. Elliott and J. A. Montgomery, Synthesis of 1-deaza-6-thioguanosine and 1-deaza-6-(methylthio)guanosine, *J. Med. Chem.* **21**, 112–114 (1978).

80. T. Itoh, J. Inaba, and Y. Mizuno, Synthesis of 5- and 6-aminoimidazo[4,5-*b*]pyridine ribonucleosides (a new type of adenosine analogs) by coupling reactions, *Heterocycles* **8**, 433–441 (1977).

81. R. L. Letsinger, P. S. Miller, and G. W. Grams, Selective N-debenzoylation of *N,O*-polybenzoylnucleosides, *Tetrahedron Lett.* **1968**, 2621–2624.

82. T. Itoh, T. Sugawara, and Y. Mizuno, A novel synthesis of 1-deazaadenosine, *Heterocycles* **17**, 305–309 (1982).

83. T. Itoh, T. Sugawara, and Y. Mizuno, Studies on the chemical synthesis of potential antimetabolites, 31. A novel synthesis of 1-deazaadenosine and its conversion to 5′-deoxy-5′-isobutylthio-1-deazaadenosine (1-deaza SIBA) and S-(1-deazaadenosyl)homocysteine (1-deaza-SAH), *Nucleosides Nucleotides* **1**, 179–190 (1982).

84. T. Itoh, K. Ono, T. Sugawara, and Y. Mizuno, Studies on the chemical synthesis of potential antimetabolites. 30. Regioselective introduction of a chlorine atom into the imidazo[4,5-*b*]pyridine nucleus, *J. Heterocycl. Chem.* **19**, 513–517 (1982).

85. T. Itoh and Y. Mizuno, Product distribution in the ribosylation reactions of adenine and 1-deazapurine in the presence of stannic chloride, *Heterocycles* **5**, 285–292 (1976).

86. G. Cristalli, P. Franchetti, M. Grifantini, S. Vittori, T. Bordoni, and C. Geroni, Improved synthesis and antitumor activity of 1-deazaadenosine, *J. Med. Chem.* **30**, 1686–1688 (1987).

87. B. L. Cline, R. P. Panzica, and L. B. Townsend, The synthesis of 1-deazaguanosine, *J. Heterocycl. Chem.* **12**, 603–604 (1975).

88. B. L. Cline, R. P. Panzica, and L. B. Townsend, Synthesis of 5-amino-3-(β-D-ribofuranosyl)-imidazo[4,5-*b*]pyridin-7-one (1-deazaguanosine) and related nucleosides, *J. Heterocycl. Chem.* **15**, 839–847 (1978).

89. P. K. Gupta and D. S. Bhakuni, Synthesis of N-glycosides of 1*H*-imidazo[4,5-*b*]pyridine, *Indian J. Chem.* **20B**, 817–819 (1981).

90. H. Inoue, S. Takada, S. Tanigawa, and T. Ueda, A novel purine to 1-deazapurine transformation reaction: Synthesis of 1-deazaadenosine derivatives, *Heterocycles* **15**, 1049–1052 (1981).

91. K. Fukukawa, T. Ueda, and T. Hirano, Synthesis of 2′(*R*)-substituted neplanocin A's (nucleosides and nucleotides, XXXVII), *Chem. Pharm. Bull.* **29**, 597–600 (1981).

92. M. J. Robins and J. S. Wilson, Smooth and efficient deoxygenation of secondary alcohols. A general procedure for the conversion of ribonucleosides to 2′-deoxynucleosides, *J. Am. Chem. Soc.* **103**, 932–933 (1981).

93. K. Pankiewicz, A. Matsuda, and K. A. Watanabe, Nucleosides. 121. Improved and general synthesis of 2′-deoxy C-nucleosides. Synthesis of 5-(2-deoxy-β-D-*erythro*-pentofuranosyl)uracil, -1-methyluracil, -1,3-dimethyluracil, and isocytosine, *J. Org. Chem.* **47**, 485–488 (1982).

94. T. Sugawara, T. Ishikura, T. Itoh, and Y. Mizuno, Studies on the chemical synthesis of potential antimetabolites. 32. Synthesis of β-D-pentofuranosyldeazaadenines as candidate inhibitors for *S*-adenosylhomocysteinases and methyltransferases, *Nucleosides Nucleotides* **1**, 239–251 (1982).

95. K. Kikugawa and M. Ichino, Direct halogenetaion of sugar moiety of nucleosides, *Tetrahedron Lett.* **1971**, 87–90.

96. S. Kitano, Y. Mizuno, M. Ueyama, K. Tori, M. Kamisaku, and K. Ajisaka, Conformation of 1- and 3-deazaadenosines in solution as studied by ^{1}H nuclear magnetic resonance spectroscopy, *Biochem. Biophys. Res. Commun.* **64**, 906–912 (1975).

97. C. Woenckhaus and G. Pfleiderer, Biochemical properties of the coenzyme analogs: nicotinamide-purine dinucleotide, nicotinamide 1-deazapurine dinucleotide, and nicotinamide benzimidazole dinucleotide, *Biochem. Z.* **341**, 495–501 (1965).

98. K. Hoogsteen, The structure of crystals containing a hydrogen-bonded complex of 1-methylthymine and 9-methyladenine, *Acta Crystallogr.* **12**, 833–823 (1959).

99. M. Ikehara, T. Fukui, and S. Uesugi, Polynucleotides XXI. Synthesis and properties of poly 1-deazaadenylic acid and poly 3-deazaadenylic acid, *J. Biochem.* **76**, 107–115 (1974).

100. L. Hagenberg, H. G. Gassen, and H. Matthaei, Synthesis and coding properties of poly(c^{1}A), poly(c^{3}A), poly(c^{7}A) and poly(h^{6}A), *Biochem. Biophys. Res. Commun.* **50**, 1104–1112 (1973).

101. Y. Mizuno, S. Kitano, and A. Nomura, Nucleotides. III. Synthesis of deazaadenosine 3′,5′-cyclic phosphates and related nucleotides of biological interest, *Chem. Pharm. Bull.* **23**, 1664–1670 (1975).

102. M. Smith, G. I. Drummond, and H. G. Khorana, Cyclic phosphates. IV. Ribonucleoside-3′,5′-cyclic phosphates. A general method of synthesis and some properties, *J. Am. Chem. Soc.* **83**, 698–706 (1961).

103. T. Ueda and I. Kawai, A convenient synthesis of ribonucleoside 2′,3′-cyclic phosphates from ribonucleosides and ribonucleotides, *Chem. Pharm. Bull.* **18**, 2303–2308 (1970).

104. Y. Mizuno, S. Kitano, and A. Nomura, Nucleotides. VI. Synthesis and spectral properties of some deazaadenylyldeazaadenosines (dinucleoside monophosphates with unusual CD spectrum) and closely related dinucleoside monophosphates, *Nucleic Acids Res.* **2**, 2193–2207 (1975).

105. E. Ohtsuka, S. Nakamura, M. Yoneda, and M. Ikehara, Polynucleotides. XXIII. A synthesis of ribonucleoside monophosphates using nucleoside 5′-phosphates, *Nucleic Acids Res.* **1**, 323–329 (1974).

106. J. Zemlicka, S. Chladek, A. Holy, and J. Smrt, Oligonucleotidic compounds. XIV. Synthesis of some diribonucleoside phosphates using the *N*-dimethylaminomethylene derivatives of 2′, 3′-*O*-ethoxymethylene ribonucleosides, *Collect Czech. Chem. Commun.* **31**, 3198–3211 (1966).

107. A. Holy, S. Chladek, and J. Zemlicka, Oligonucleotidic compounds. XXIX. Reactions of ribonucleoside 2′(3′)-phosphates with dimethylformamide acetals. *Collect Czech. Chem. Commun.* **34**, 253–271 (1969).

108. K. D. Philips and J. P. Horwitz, Nucleosides. XVII. Benzylation–debenzylation studies on nucleosides, *J. Org. Chem.* **40**, 1856–1859 (1975).

109. O. P. Babbar, Antiviral effect of purine antagonists against Ranikhet disease virus in stationary culture of chorioallantoic membrane of chick embryo, *J. Sci. Ind. Res. India* **20C**, 232–234 (1961).

110. O. P. Babbar and B. L. Chowdhury, Further studies on the antiviral effect of 3-β-D-ribofuranosylimidazo(*b*)pyridine on Ranikhet disease virus, *J. Sci. Ind. Res. India* **21C**, 312–314 (1962).

111. B. M. Gupta, Susceptibility of primary explants of minced chick embryo tissues to vaccinia and Ranikhet disease viruses, and a study of the effect of some newer potential antiviral agents, *Tissue Cult., Proc. Seminar, Baroda (India)*, (C. V. Ramakrishnan, ed.), Dr. W. Junk Publishers, The Hague, Netherlands, pp. 213–216 (1965).

112. U. Agarwal, B. M. Gupta, S. K. Khan, I. Clifford, and K. Chandra, Effect of quinazolones, substituted hydroxyquinazolines, substituted diamines, purines, and nucleoside antagonists on vaccinia virus in chick embryo, *J. Sci. Ind. Res. India* **21C**, 309–312 (1962).

113. L. G. Brantsevich, N. S. Miroshnichenko, A. V. Stetsenko, A. T. Slabospitskaya, and V. V. Checkmacheva, Antimicrobial and antitumor activity of the glycosides of certain condensed imidazole derivatives, *Mikrobiol. Zh. (Kiev)*, **37**, 635–639 (1975).

114. J. H. Rogozinska, C. Kroon, and C. A. Salemink, Influence of alterations in the purine ring on biological activity of cytokinins, *Phytochemistry* **12**, 2087–2092 (1973).

115. M. Ikehara and T. Fukui, Studies of nucleosides and nucleotides. LVIII. Deamination of adenosine naalogs with calf intestine adenosine deaminase, *Biochim. Biophys. Acta* **338**, 512–519 (1974).

116. G. Pfleiderer and C. Woenckhaus, Intramolecular reciprocal effects between heterocycles in coenzymes. II. α-Nicotinamide adenine dinucleotide, *Ann. Chem.* **690**, 170–176 (1965).

117. L. N. Simon, R. J. Bauer, R. L. Tolman, and R. K. Robins, Calf intestine adenosine deaminase. Substrate specificity, *Biochemistry* **9**, 573–577 (1970).

118. J. P. Miller, L. F. Christensen, T. A. Andrea, R. B. Meyer, Jr., S. Kitano, and Y. Mizuno, Interaction of 'AZA' and 'DEAZA' analogs of adenosine cyclic 3′,5′-phosphate with some enzymes of adenosine cyclic 3′,5′-phosphate metabolism: Evidence that the lone pair electrons of N-3 are involved in the binding of adenosine cyclic 3′,5′-phosphate to type II adenosine cyclic 3′,5′-phosphate dependent protein kinase, *J. Cyclic Nucleotide Res.* **4**, 133–144 (1978).

119. M. Ikehara, Deazaadenosine polymers, Jpn. Kokai 74 99,397 (Sept. 1974) [*CA* **82**, 98870z (1975)].

120. Y. Mizuno, T. Itoh, and K. Saito, Studies on condensed systems of aromatic nitrogenous series. XXIII. Synthesis of 1-(β-D-ribofuranosyl)-1*H*-imidazo[1,5-*c*]pyridines, *J. Org. Chem.* **29**, 2611–2615 (1964).

121. C. A. Salemink and G. M. van der Want, Desazapurine derivatives. III. 3-Desazaadenine (4-amino-1*H*-imidazo[c]pyridine) and related bases, *Recl. Trav. Chim. Pays-Bas* **68**, 1013–1029 (1949).

122. S. K. Chakrabarthy and R. Levine, The chemistry of pyrazine and its derivatives. X. The mono- and diacylation of tetramethylpyrazine, *J. Heterocycl. Chem.* **1**, 196–200 (1964).

123. J. A. Montgomery and K. Hewson, 3-Deaza-6-methylthiopurine ribonucleoside, *J. Med. Chem.* **9**, 105–107 (1966).

124. N. S. Miroshnichenko, I. G. Ryabokon, and A. V. Stetsenko, Synthesis of nucleosides from 3*H*-imidazo[4,5-*c*]pyridine derivatives, *Ukr. Khim. Zh.* (Russ. Ed.) **39**, 350–353 (1973).

125. N. S. Miroshnichenko, A. S. Kovalenko, and A. V. Stetsenko, Determination of the structure of 2-methylimidazo[4,5-*c*]pyridine glycosides, *Ukr. Khim. Zh.* (Russ. Ed.) **40**, 258–260 (1974).

126. R. J. Rousseau, L. B. Townsend, and R. K. Robins, The synthesis of 4-amino-β-D-ribofuranosylimidazo[4,5-*c*]pyridine (3-deazaadenosine) and related nucleotides, *Biochemistry* **5**, 756–760 (1966).

127. T. Sato, T. Simadate, and Y. Ishido, Nucleosides and nucleotides (VII). A new method for syntheses of purine ribonucleosides l. *Nippon Kagaku Zasshi* **81**, 1440–1442 (1960).

128. R. T. Borchardt, J. A. Huber, and Y. S. Wu, Potential inhibitors of *S*-adenosylmethionine-dependent methyltransferases. 2. Modification of the base portion of *S*-adenosylhomocysteine, *J. Med. Chem.* **17**, 868–873 (1974).

129. J. A. Montgomery, A. T. Shortnacy, and S. D. Clayton, A comparison of two methods for the preparation of 3-deazapurine ribonucleosides, *J. Heterocycl. Chem.* **14**, 195–197 (1977).

130. P. K. Chiang, G. L. Cantoni, J. P. Bader, W. M. Shannon, H. J. Thomas, and J. A. Montgomery, 5′-Deoxy-5′-(isobutylthio)-3-deazaadenosine and its antiviral effect on Rous sarcoma virus and Gross murine leukemia virus, U.S. Patent. Appl. 937,704 (April 1979) [*CA* **91**, 57424z (1979)].

131. Y. Mizuno, S. Tazawa, and K. Kageura, Synthetic studies of potential antimetabolities. XII. Synthesis of 4-substituted 1-(β-D-ribofuranosyl)-1*H*-imidazo[4,5-*c*]pyridines, *Chem. Pharm. Bull* **16**, 2011–2017 (1968).

132. T. Itoh, T. Yamaguchi, and Y. Mizuno, Studies on the chemical synthesis of potential antimetabolites. 28. Synthesis of 4-amino-1-(2-deoxy-β-D-*erythro*-pentofuranosyl)-1*H*-imidazo[4,5-*c*]pyridine (2′-deoxy-3-deazaadenosine) and its α-anomer, *J. Carbohydr. Nucleosides Nucleotides* **8**, 119–129 (1981).

133. J. A. Montgomery, S. J. Clayton, and P. K. Chiang, 1-β-D-Arabinofuranosyl-1*H*-imidazo[4,5-*c*]pyridine (*ara*-3-deazaadenine), *J. Med. Chem.* **25**, 96–98 (1982).

134. A. V. Stetsenko and N. S. Miroshnichenko, Glycosides derived from 4-chloroimidazo[4,5-*c*]pyridine, *Ukr. Khim. Zh.* (Russ. Ed.) **39**, 703–707 (1973).

135. J. A. May, Jr. and L. B. Townsend, Novel synthesis of 3-deazaadenosine, *J. Chem. Soc., Chem. Commun.* **1973**, 64–65.

136. J. A. May, Jr. and L. B. Townsend, A general synthesis of 4-substituted 1-(β-D-ribofuranosyl)-imidazo[4,5-*c*]pyridines. *J. Chem. Soc., Perkin. Trans. 1* 1975, 125–129.

137. J.-L. Imbach, J.-L. Barascut, B. L. Kam, B. Rayner, C. Tamby, and C. Tapiero, Researches in nucleoside synthesis. III. A new criteria of determination of anomeric configuration of ribonucleosides, *J. Heterocycl. Chem.* **10**, 1069–1070 (1973).

138. J. A. May, Jr. and L. B. Townsend, 3-Deazaadenosine (4-amino-1-β-D-ribofuranosyl-1*H*-imidazo[4,5-*c*]pyridine), in: *Nucleic Acid Chemistry, Improved and New Synthetic Procedures, Methods, and Techniques* (L. B. Townsend and R. S. Tipson, eds.), Vol. 2, pp. 693–699, John Wiley and Sons, New York (1978).

139. G. B. Elion and G. H. Hitchings, Metabolic basis for the actions of analogs of purines and pyrimidines, in: *Advances of Chemotherapy* (A. Goldin, F. Hawking, and R. J. Schnitzer, eds.), Vol. 2, pp. 91–156, Academic Press, New York (1965).

140. A. Goldin, H. B. Wood, Jr., and R. R. Engle, Relation of structure of purine and pyrimidine nucleosides to antitumor activity, *Cancer Chemother. Rep. Suppl.* **1**, 1–272 (1968).

141. R. J. Rousseau, J. A. May, Jr., R. K. Robins, and L. B. Townsend, The synthesis of 3-deaza-6-thioguanine and certain related derivatives, *J. Heterocycl. Chem.* **11**, 233–235 (1974).

142. J. A. May, Jr. and L. G. Townsend, Synthesis of 6-amino-1-(β-D-ribofuranosyl)imidazo[4,5-*c*]pyridin-4-thione (3-deaza-6-thioguanosine), *J. Carbohydr. Nucleosides Nucleotides* **2**, 271–276 (1975).

143. G. R. Revankar and L. B. Townsend, The synthesis of 2-chloro-1-(β-D-ribofuranosyl)benzimidazole and certain related derivatives, *J. Heterocycl. Chem.* **5**, 477–483 (1968).

144. E. Wittenburg, A new synthesis of nucleosides, *Z. Chem.* **4**, 303–304 (1964).

145. J. A. May, Jr. and L. B. Townsend, Synthesis of 6-amino-1-β-D-ribofuranosyl)imidazo[4,5-*c*]pyridine-4-thione (3-deaza-6-thioguanosine) and certain related derivatives, *J. Carbohydr. Nucleosides Nucleotides* **2**, 371–398 (1975).

146. P. D. Cook, R. J. Rousseau, A. M. Mian, R. B. Meyer, Jr., P. Dea, G. Ivanovics, D. G. Streeter, J. T. Witkowski, M. G. Stout, L. N. Simon, R. W. Sidwell, and R. K. Robins, A new class of potent guanine antimetabolites. Synthesis of 3-deazaguanine, 3-deazaguanosine, and 3-deazaguanylic acid by a novel ring closure of imidazole precursors, *J. Am. Chem. Soc.* **97**, 2916–2917 (1975).

147. P. D. Cook, R. J. Rousseau, A. M. Mian, P. Dea, R. B. Meyer, Jr., and R. K. Robins, Synthesis of 3-deazaguanine, 3-deazaguanosine, and 3-deazaguanylic acid by a novel ring closure of imidazole precursors, *J. Am. Chem. Soc.* **98**, 1492–1498 (1976).

148. A. Frankowski, Synthesis of imidazo[4,5-*c*]pyridine and imidazo[4,5-*d*][1,2]diazepine systems and their ribonucleosides, *Tetrahedron* **42**, 1511–1528 (1986).

149. U. Neidballa and H. Vorbrüggen, A general synthesis of N-glycosides. I. Synthesis of pyrimidine nucleosides, *J. Org. Chem.* **39**, 3654–3660 (1974).

150. M. W. Winkley and R. K. Robins, Pyrimidine nucleosides. Part II. The direct glycosidation of 2,6-disubstituted 4-pyrimidones, *J. Chem. Soc.* **1969**, 791–796.
151. A. M. Mian and R. K. Robins, 3-Deazaguanosine and derivatives thereof, U.S. Patent 3,919,193 (1975).
152. R. K. Robins, R. J. Rousseau, and A. M. Mian, 3-Deazaguanine and its derivatives, Ger. Offen. 2,529,533 (Jan. 1977) [*CA* **86**, 171784p (1977)].
153. ICN Pharmaceuticals, Inc., 3-Deazaguanine and 3-Deazaguanosine, Jpn. Kokai 76, 115,495 (1976).
154. P. D. Cook, L. B. Allen, D. G. Streeter, J. H. Huffman, R. W. Sidwell, and R. K. Robins, Synthesis and antiviral and enzymatic studies of certain 3-deazaguanines and their imidazole-carboxamide precursors, *J. Med. Chem.* **21**, 1212–1218 (1978).
155. A. M. Mian and T. A. Khwaja, Synthesis and antitumor activity of 2-deoxyribofuranosides of 3-deazaguanine, *J. Med. Chem.* **26**, 286–291 (1983).
156. G. R. Revankar, P. K. Gupta, A. D. Adams, H. K. Dalley, P. A. McKernan, P. D. Cook, P. G. Canonico, and R. K. Robins, Synthesis and antiviral/antitumor activities of certain 3-deaza-guanine nucleosides and nucleotides, *J. Med. Chem.* **27**, 1389–1396 (1984).
157. M. S. Poonian, W. W. McComas, and M. J. Kramer, Synthesis of arabinofuranosyl derivatives of 3-deazaguanine, *J. Med. Chem.* **22**, 958–962 (1979).
158. P. D. Cook and R. K. Robins, Synthesis of 7- and 9-β-D-ribofuranosides of 3-deaza-6-thioguanine and 3-deaza-2,6-diaminopurine by a novel ring closure of 4(5)-cyano-5(4)-cyanomethylimidazole-β-D-ribofuranosides, *J. Org. Chem.* **43**, 289–293 (1978).
159. P. K. Gupta and D. S. Bhakuni, Synthesis of rhamnopyranosyl derivative of 3-deazaguanine, *Indian J. Chem.* **20B**, 702–703 (1981).
160. H. Tanaka and T. Ueda, Chemical conversion of uridine to 8,5′-*O*-cyclo-3-deazaguanosine, *J. Heterocycl. Chem.* **16**, 411–412 (1979).
161. H. Tanaka, M. Hirayama, A. Matsuda, T. Miyasaka, and T. Ueda, Lithiation of an imidazole nucleoside at the C-5 position. Synthesis of 3-deazaguanosine from uridine, *Chem. Lett.* **1985**, 589–592.
162. H. Tanaka, M. Hirayama, M. Suzuki, T. Miyasaka, A. Matsuda, and T. Ueda, A lithiation route to C-5 substitution of an imidazole nucleoside and its application to the synthesis of 3-deaza-guanosine, *Tetrahedron* **42**, 1971–1980 (1986).
163. J. A. Montgomery, S. J. Clayton, H. J. Thomas, W. M. Shannon, G. Arnett, A. J. Bodner, I.-K. Kion, G. L. Cantoni, and P. K. Chiang, Carbocyclic analogue of 3-deazaadenosine: A novel antiviral agent using *S*-adenosylhomocysteine hydrolase as a pharmacological target, *J. Med. Chem.* **25**, 626–629 (1982).
164. R. J. Rousseau and R. K. Robins, The synthesis of various chloroimidazo[4,5-*c*]pyridines and related derivatives, *J. Heterocycl. Chem.* **2**, 196–201 (1965).
165. P. D. Cook, β-D-Arabinofuranosylimidazo[4,5-*c*]pyridine compounds and methods for their production, U.S. Patent 4,315,000 (Feb. 1982).
166. P. K. Gupta, R. K. Robins, and G. R. Revankar, A new synthesis of certain 7-(β-D-ribofuranosyl) and 7-(2-deoxy-β-D-ribofuranosyl) derivatives of 3-deazaguanine via the sodium salt glycosylation procedure, *Nucleic Acids Res.* **13**, 5341–5352 (1985).
167. (a) R. I. Glazer, M. C. Knode, C. K. H. Tseng, D. R. Haines, and V. E. Marquez, 3-Deazaneplanocin A: A new inhibitor of *S*-adenosylhomocysteine synthesis and its effects in human colon carcinoma cells, *Biochem. Pharmacol.* **35**, 4523–4527 (1986); (b) C. K. H. Tseng, V. E. Marquez, R. W. Fuller, B. M. Goldstein, D. R. Haines, H. McPherson, J. L. Parsons, W. M. Shannon, G. Arnett, and M. Hollingshead, Synthesis of 3-deazaneplanocin A, a powerful inhibitor of *S*-adenosylhomocysteine hydrolase with potent and selective in vitro and in vivo antiviral activities, *J. Med. Chem.* **32**, 1442–1446 (1989).
168. T. Ishikura, S. Oue, T. Itoh, A. Nomura, T. Ueda, and Y. Mizuno, Studies on the chemical synthesis of potential antimetabolites. 37. Synthesis of 3-deazaadenine nucleosides modified at the sugar moiety, *Nucleosides Nucleotides* **3**, 413–422 (1984).
169. P. Serafinowski, Synthesis of some *S*-3′-deoxyadenosyl-L-homocysteine analogues, *Nucleic Acids Res.* **15**, 1121–1137 (1987).
170. T. A. Krenitsky and J. L. Rideout, Deazapurine nucleosides, Eur. Pat. EP 38,568 (Oct. 1981) [*CA* **96**, 120885w (1982)].

171. T. A. Krenitsky and J. L. Rideout, Synthesis of ribosides using bacterial phosphorylase, U.S. Patent 4,347,315 (Aug. 1982) [*CA* **97**, 196824s (1982)].
172. T. A. Krenitsky, J. L. Rideout, G. W. Koszalka, R. B. Inmon, E. Y. Chao, G. B. Elion, V. S. Latter, and R. B. Williams, Pyrazolo[3,4-*d*]pyrimidine ribonucleosides as anticoccidials. 1. Synthesis and activity of some nucleosides of purines and 4-(alkylthio)pyrazolo[3,4-*d*]pyrimidines, *J. Med. Chem.* **25**, 32–35 (1982).
173. J. L. Rideout and T. A. Krenitsky, Deazapurine nucleosides and their formulations, Eur. Pat. Appl. EP 38,569 (Oct. 1981) [*CA* **96**, 91647n (1982).
174. T. A. Krenitsky, J. L. Rideout, E. Y. Chao, G. W. Koszalka, F. Gurney, R. C. Crouch, N. K. Cohn, G. Wolberg, and R. Vinegar, Imidazo[4,5-*c*]pyridines (3-deazapurines) and their nucleosides as immunosuppressive and antiinflammatory agents, *J. Med. Chem.* **29**, 138–143 (1986).
175. P. K. Chiang, G. L. Cantoni, J. P. Bader, W. M. Shannon, H. J. Thomas, and J. A. Montgomery, Adenosylhomocysteine hydrolase inhibitors: Synthesis of 5′-deoxy-5′-(isobutylthio)-3-deazaadenosine and its effect on Rous sarcoma virus and Gross murine leukemia virus, *Biochem. Biophys. Res. Commun.* **82**, 417–423 (1978).
176. R. T. Borchardt, *S*-Adenosyl-L-methionine-dependent macromolecule methyl-transferases: Potential targets for the design of chemotherapeutic agents, *J. Med. Chem.* **23**, 347–357 (1980).
177. R. T. Borchardt, J. A. Huber, and Y. S. Wu, A convenient preparation of *S*-adenosylhomocysteine and related compounds, *J. Org. Chem.* **41**, 565–567 (1976).
178. R. T. Borchardt, Y. S. Wu, J. A. Huber, and A. F. Wycpalek, Potential inhibitors of *S*-adenosylmethionine-dependent methyltransferases. 6. Structural modifications of *S*-adenosylmethionine, *J. Med. Chem.* **19**, 1104–1110 (1976).
179. P. Singh, J. May, Jr., L. B. Townsend, and D. J. Hodgson, Conformational barriers in nucleoside analogs: The crystal structure of 3-deazaadenosine, *J. Am. Chem. Soc.* **98**, 825–830 (1976).
180. D. W. Miles, L. B. Townsend, P. Redington, and H. Eyring, Comparative study of circular dichroism of the conformation of deazapurine nucleosides and that of common purine nucleosides, *Proc. Natl. Acad. Sci. U.S.A.* **73**, 2384–2387 (1976).
181. H.-D. Lüdemann, H. Plach, E. Westhof, and L. B. Townsend, Conformations of 3-deazaadenosine, 3-deaza-8-azaadenosine, and benzimidazole-1-β-D-riboside in solution: HRNMR-, proton-relaxation-time-, and nuclear-Overhauser-effect studies, *Z. Naturforsch., C* **33**, 305–316 (1978).
182. M. Yoshikawa, T. Kato, and T. Takenishi, Studies of phosphorylation. III. Selective phosphorylation of unprotected nucleosides, *Bull. Chem. Soc. Jpn.* **42**, 3505–3508 (1969).
183. J. G. Moffatt and H. G. Khorana, Nucleoside polyphosphates, X. The synthesis and some reactions of nucleoside-5′-phosphoromorpholidates and related compounds. Improved methods for the preparation of nucleoside-5′-polyphosphates, *J. Am. Chem. Soc.* **83**, 649–658 (1961).
184. A. Nakazawa, M. Tokushige, and O. Hayaishi, Studies on the interaction between regulatory enzymes and effectors, *J. Biol. Chem.* **242**, 3868–3872 (1967).
185. C. Woenckhaus and P. Zumpe, Significance of the adenine ring in the coenzyme, nicotinamide adenine dinucleotide. Properties of the coenzyme model nicotinamide 3-deazapurine dinucleotide, *Z. Naturforsch., B* **23**, 484–490 (1968).
186. G. M. Tener, 2-Cyanoethyl phosphate and its use in the synthesis of phosphate esters, *J. Am. Chem. Soc.* **83**, 159–168 (1961).
187. R. E. Borden and M. Smith, Nucleotide synthesis. III. Preparation of nucleoside-3′,5′-cyclic phosphates in strong base, *J. Org. Chem.* **31**, 3247–3253 (1966).
188. S. O. Doskeland, D. Ogreid, R. Ekanger, P. A. Sturm, J. P. Miller, and R. H. Suva, Mapping of the two interchain nucleotide binding sites of adenosine cyclic 3′,5′-phosphate dependent protein kinase I, *Biochemistry* **22**, 1094–1101 (1983).
189. F. Harada, F. Kimura, and S. Nishimura, Primary sequence of tRNA₁^{Val} from *Escherichia coli* B. I. Oligonucleotide sequences of digests of *Escherichia coli* tRNA₁^{Val} with RNase T₁ and pancreatic RNase, *Biochemistry* **10**, 3269–3277 (1971).
190. M. Ikehara and I. Tazawa, Studies of nucleosides and nucleotides. XXIX. Direct synthesis of nucleoside-2′,3′cyclic phosphates, *J. Org. Chem.* **31**, 819–821 (1966).
191. M. Yoshikawa, T. Kato, and T. Takenishi, A novel method for phosphorylation of nucleosides to 5′-nucleotides, *Tetrahedron Lett.* **1967**, 5065–5068.

192. L. B. Allen, J. H. Huffman, R. B. Meyer, Jr., P. D. Cook, J. T. Witkowski, L. N. Simon, R. K. Robins, and R. W. Sidwell, 5th Interscience Congress on Antimicrobial Agents and Chemotherapy, Washington, D.C., Sept. 1975, Abstract no. 245.

193. R. W. Sidwell, L. B. Allen, J. H. Huffman, J. T. Witkowski, P. D. Cook, R. L. Tolman, G. R. Revankar, L. N. Simon, and R. K. Robins, The potential of nucleosides as antiviral agents, *Chemother., Proc. Int. Congr. Chemother.* **6**, 279–294 (1976).

194. L. B. Allen, J. H. Huffman, P. D. Cook, R. B. Meyer, Jr., R. K. Robins, and R. W. Sidwell, Antiviral activity of 3-deazaguanine, 3-deazaguanosine, and 3-deazaguanylic acid, *Antimicrob. Agents Chemother.* **12**, 114–119 (1977).

195. E. De Clercq, R. Bernaerts, D. E. Bergstrom, M.J. Robins, J. A. Montgomery, and A. Holy, Antirhinovirus activity of purine nucleoside analogs, *Antimicrob. Agents Chemother.* **29**, 482–487 (1986).

196. T. A. Khwaja, L. Kigwana, R. B. Meyer, Jr., and R. K. Robins, 3-Deazaguanine, a new purine analog exhibiting antitumor activity, *Proc. Am. Assoc. Cancer Res.* **16**, 162 (1975).

197. T. A. Khwaja and J. Varven, 3-Deazaguanine (3-DG), a potent inhibitor of mammary adenocarcinoma R3230AC and mamaary adenocarcinoma 13762, *Proc. Am. Assoc. Cancer Res.* **17**, 200 (1976).

198. P. Schwartz, D. Hammond, and T. A. Khwaja, Biochemical pharmacology of 3-deazaguanine (3-DG), *Proc. Am. Assoc. Cancer Res.* **18**, 153 (1977).

199. R. S. Rivest, D. Irwin, and H. G. Mandel, Inhibition of macromolecular synthesis by 3-deazaguanine (3DG) in mammalian tumor cells *in vitro*, *Proc. Am. Assoc. Cancer Res.* **21**, 279 (1980).

200. D. G. Streeter and H. H. P. Koyama, Inhibition of purine nucleotide biosynthesis by 3-deazaguanine, its nucleodside and 5′-nucleotide, *Biochem. Pharmacol.* **25**, 2413–2415 (1976).

201. D. L. Lucas, P. K. Chiang, H. K. Webster, R. K. Robins, W. P. Wiesmann, and D. G. Wright, Effects of 3-deazaguanosine and 3-deazaguanine on the growth and maturation of the human promyelocytic leukemia cell line, HL-60, *Adv. Expl. Med. Biol.* **165B**, 321–325 (1984).

202. P. P. Saunders, L.-Y. Chao, T. L. Loo, and R. K. Robins, Actions of 3-deazaguanine and 3-deazaguanosine on variant lines of Chinese hamster ovary cells, *Biochem. Pharmacol.* **30**, 2374–2376 (1981).

203. T. A. Khwaja, L. Momparler, J. C. Varven, and A. M. Mian, Studies on mechanism of antitumor activity of 3-deazagunine (3DG), *Proc. Am. Assoc. Cancer Res.* **20**, 152 (1979).

204. T. R. Matthews, D. W. Yotter, P. D. Cook, R. W. Sidwell, R. K. Robins, and P. F. Dougherty, *In vitro* antibacterial properties of 7-β-D-ribofuranosyl-3-deazaguanine (ICN 4684), 16th Interscience Conference on Antimicrobial Agents and Chemotherapy, Chicago, Oct. 1976, Abstract no. 425.

205. S. Roy-Burman, A. M. Mian, and T. A. Khwaja, 3-Deazaguanine IV: Studies on biological activities and mechanism of action of 3-deazaguanine nucleosides, *Chem. Scri.* **26**, 155–159 (1986).

206. T. R. Matthews, P. E. Dougherty, P. D. Cook, R. W. Sidwell, R. K. Robins, and D. W. Yotter, Therapeutic evaluation of 7-β-D-ribofuranosyl-3-deazaguanine (ICN 4684) in mice, 16th Interscience Conference on Antimicrobial Agents and Chemotherapy, Chicago, Oct. 1976, Abstract no. 426.

207. P. P. Saunders, L-Y. Chao, R. K. Robins, and T. L. Loo, Action of 3-deazaguanine in *Escherichia coli*, *Mol. Pharmacol.* **15**, 691–697 (1979).

208. D. G. Streeter, M. Miller, T. R. Matthews, R. K. Robins, and J. P. Miller, 7-Ribosyl-3-deazaguanine—mechanism of antibacterial action, *Biochem. Pharmacol.* **29**, 1791–1797 (1980).

209. A. M. Mian and S. Furusawa, Antitumor activity and mechanism of action of 6-thio-3-deazaguanine, *Cancer Res.* **47**, 1863–1866 (1987).

210. J. D. Berman, L. S. Lee, R. K. Robins, and G. R. Revankar, Activity of purine analogs against *Leishmania tropica* within human macrophages *in vitro*, *Antimicrob. Agents Chemother.* **24**, 233–236 (1983).

211. P. P. Saunders, M.-T. Tan, C. D. Spindler, R. K. Robins, and W. Plunkett, 3-Deazaguanosine is metabolized to the triphosphate derivative in Chinese hamster cells deficient in hypoxanthine-guanine phosphoribosyltransferase, *J. Biol. Chem.* **261**, 6416–6422 (1986).

212. T. M. Page, S. J. Jacobsen, W. L. Nyhan, J. H. Mangum, and R. K. Robins, The metabolism of

3-deazaguanine and 3-deazaguanosine by human cells in culture, *Int. J. Biochem.* **18**, 957–960 (1986).

213. P. K. Chiang, Conversion of 3T3-L1 fibroblasts to fat cells by an inhibitor of methylation: Effect of 3-deazaadenosine, *Science* **211**, 1164–1166 (1981).

214. A. J. Bodner, G. L. Cantoni, and P. K. Chiang, Antiviral activity of 3-deazaadenosine and 5′-deoxy-5′-isobutylthio-3-deazaadenosine (3-deaza-SIBA), *Biochem. Biophys. Res. Commun.* **98**, 476–481 (1981).

215. J. P. Bader, N. R. Brown, P. K. Chiang, and G. L. Cantoni, 3-Deazaadenosine, an inhibitor of adenosylhomocysteine hydrolase, inhibits reproduction of Rous sarcoma virus and transformation of chick embryo cells, *Virology* **89**, 494–505 (1978).

216. P. K. Chiang, H. H. Richards, and G. L. Cantoni, S-Adenosyl-L-homocysteine hydrolase: Analogs of S-adenosyl-L-homocysteine as potential inhibitors, *Mol. Pharmacol.* **13**, 939–947 (1977).

217. Y. S. Im, P. K. Chiang, and G. L. Cantoni, Guanidoacetate methyltransferase, purification and molecular properties, *J. Biol. Chem.* **254**, 11047–11050 (1979).

218. P. K. Chiang and G. L. Cantoni, Perturbation of biochemical transmethylations by 3-deazaadenosine *in vivo*, *Biochem. Pharmacol.* **28**, 1897–1902 (1979).

219. P. K. Chiang, Y. S. Im, and G. L. Cantoni, Phospholipids biosynthesis by methylations and choline incorporation: Effect of 3-deazaadenosine, *Biochem. Biophys. Res. Commun.* **94**, 174–181 (1980).

220. A. M. Westheimer, S.-Y. Chen, R. T. Borchardt, and Y. Furuichi, S-Adenosylmethionine and its analogs. Structural features correlated with synthesis and methylation of mRNAs of cytoplasmic polyhedrosis virus, *J. Biol. Chem.* **255**, 5924–5930 (1980).

221. P. K. Chiang, A. Guranowski, and J. E. Segall, Irreversible inhibition of S-adenosylhomocysteine hydrolase by nucleoside analogs. *Arch. Biochem. Biophys.* **207**, 175–184 (1981).

222. J.-S. Schanche, T. Schanche, and P. M. Ueland, Inhibition of phospholipid methylation in isolated rat hepatocytes by analogues of adenosine and S-adenosylhomocysteine, *Biochim. Biophys. Acta* **721**, 399–407 (1982).

223. A. Svardal, R. Djurhuus, and P. M. Ueland, Disposition of homocysteine and S-3-deazaadenosylhomocysteine in cells exposed to 3-deazaadenosine, *Mol. Pharmacol.* **30**, 154–158 (1986).

224. M. L. Sherman, T. D. Shafman, D. R. Spriggs, and D. W. Kufe, Inhibition of murine erythroleukemia cell differentiation by 3-deazaadenosine, *Cancer Res.* **45**, 5830–5834 (1985).

225. J. L. Kelley and L. Beauchamp, Antiviral agents, *Annu. Rep. Med. Chem.* **18**, 139–148 (1983).

226. E. De Clercq and J. A. Montgomery, Broad-spectrum antiviral activity of the carbocyclic analog of 3-deazaadenosine, *Antiviral Res.* **3**, 17–24 (1983).

227. D. M. Houston, E. K. Dolence, B. T. Keller, U. Patel-Thombre, and R. T. Borchardt, Potential inhibitors of S-adenosylmethionine-dependent methyltransferases. 8. Molecular dissections of carbocyclic 3-deazaadenosine as inhibitors of S-adenosylhomocysteine hydrolase, *J. Med. Chem.* **28**, 467–471 (1985).

228. D. M. Houston, E. K. Dolence, B. T. Keller, U. Patel-Thombre, and R. T. Borchardt, Potential inhibitors of S-adenosylmethionine-dependent methyltransferases. 9. 2′,3′-Dialdehyde derivatives of carbocylic purine nucleosides as inhibitors of S-adenosylhomocysteine hydrolase, *J. Med. Chem.* **28**, 471–477 (1985).

229. E. De Clercq, Targeted development of new antiviral agents, *Chem. Scr.* **26**, 41–47 (1986).

230. M. Cools and E. De Clercq, Correlation between the antiviral activity of acyclic and carbocyclic adenosine analogues in murine L929 cells and their inhibitory effect on L929 cell S-adenosylhomocysteine hydrolase, *Biochem. Pharmacol.* **38**, 1061–1067 (1989).

231. E. De Clercq and M. Cools, Antiviral potency of adenosine analogues: Correlation with inhibition of S-adenosylhomocysteine hydrolase, *Biochem. Biophys. Res. Commun.* **129**, 306–311 (1985).

232. L. L. Bennett, Jr., R. W. Brockman, P. W. Allan, L. M. Rose, and S. C. Shaddix, Alterations in nucleotide pools induced by 3-deazaadenosine and related compounds, *Biochem. Pharmacol.* **37**, 1233–1244 (1988).

233. T. Okazaki, A. Nakazawa, and O. Hayaishi, Studies on interaction between regulatory enzymes and effectors, *J. Biol. Chem.* **243**, 5266–5271 (1968).

234. C. Woenckhaus and R. Jeck, Fluorometric investigation of the formation of enzyme–coenzyme complexes of pig heart lactate dehydrogenase with coenzyme fragments and coenzyme analogs, *Z. Naturforsch. B* **24**, 1436–1441 (1969).

235. S. F. Velick, Fluorescence spectra and polarization of glyceraldehyde-3-phosphate and lactic dehydrogenase coenzyme complexes, *J Biol. Chem.* **233**, 1455–1467 (1958).

236. T. B. Johnson, Researches on pyrimidines: Synthesis of cytosine-5-acetic acid, *J. Am. Chem. Soc.* **33**, 758–766 (1911).

237. P. S. Ritch and R. I. Glazer, Pyrrolo[2,3-*d*]pyrimidine nucleosides, in: Developments in Cancer Chemotherapy (R. I. Glazer, ed.), pp. 1–33, CRC Press, Boca Raton, Florida (1984).

238. H. Nishimura, K. Katagiri, K. Sato, M. Mayama, and N. Shimaoka, Toyocamycin, a new anti-candida antibiotics, *J. Antibiot. Ser. A* **9**, 60–62 (1956).

239. K. Ohkuma, Chemical structure of toyocamycin, *J. Antiobiot. Ser. A* **13**, 361 (1960).

240. K. Ohkuma, Chemical structure of toyocamycin, *J. Antibiot. Ser. A* **14**, 343–352 (1961).

241. K. Kikuchi, Antibiotics from *Streptomyces* sp. No. E-212. I. Studies on streptomyces antibiotics, XXXV, *J. Antibiot. Ser. A* **8**, 145–147 (1955).

242. K. Katagiri, K. Sato, and S. Nishiyama, New antibiotics from streptomyces No. E-212, *Shionogi Kenkyusho Nempo* **7**, 715–723 (1957).

243. H. Yamamoto, S. Fujii, K. Nakazawa, A. Miyake, H. Hitomi, and M. Imanishi, *Takeda Kenkyusho Nempo* **16**, 26–31 (1957).

244. A. Aszalos, P. Lemanski, R. Robison, S. Davis, and B. Berk, Identification of antibiotic 1037 as toyocamycin, *J. Antibiot. Ser. A* **19**, 285 (1966).

245. F. Arcamone, G. Cassinelli, G. d'Amico, and P. Orezzi, Mannosidohydroxystreptomycin from *Streptomyces sp.*, *Experientia* **24**, 441–442 (1968).

246. J. Davoll, Pyrrolo[2,3-*d*]pyrimidines, *J. Chem. Soc.* **1960**, 131–138.

247. T. Uematsu and R. J. Suhadolnik, 7-Deazaadenine ribonucleosides. The use of periodate oxidation in degradation studies, *J. Org. Chem.* **33**, 726–728 (1968).

248. R. L. Tolman, R. K. Robins, and L. B. Townsend, Pyrrolo[2,3-*d*]pyrimidine nucleoside antibiotics. Total synthesis and structure of toyocamycin, unamycin B, vengicide, antibiotic E-212, and sangivamycin (BA-90912), *J. Am. Chem. Soc.* **90**, 524–526 (1968).

249. R. L. Tolman, R. K. Robins, and L. B. Townsend, Pyrrolopyrimidine nucleosides. III. The total synthesis of toyocamycin, sangivamycin, tubercidin, and related derivatives, *J. Am. Chem. Soc.* **91**, 2102–2108 (1969).

250. M. Matsuoka and H. Umezawa, Unamycin, an antifungal substances produced by *Streptomyces fungicidicus*, *J. Antibiot. Ser. A.* **13**, 114–120 (1960).

251. M. Matsuoka, Biological studies on antifungal substance produced by *Streptomyces fungicidicus*, *J. Antibiot. Ser. A* **13**, 121–124 (1960).

252. N. V. Koninklijke, Vengicide and processes for the preparation of the antibiotics oxy-tetracycline and vengicide, Brit. Patent 764,198 (Dec. 1956) [*CA* **51**, 10009a (1957)].

253. N. V. Koninklijke, Streptomyces antibiotics, Neth. Patent 109,006 (July 1964) [*CA* **62**, 11114g (1965)].

254. K. Anzai and S. Marumo, A new anti-candida antibiotics, *J. Antibiot. Ser. A* **10**, 20–23 (1957).

255. K. Anzai, G. Nakamura, and S. Suzuki, A new antibiotic, tubercidin, *J. Antibiot. Ser. A* **10**, 201–204 (1957).

256. G. Nakamura, Studies on antibiotic actinomycetes. II. On Streptomyces producing a new antibiotic, tubercidin, *J. Antibiot. Ser. A* **14**, 90–93 (1961).

257. S. Suzuki and S. Hakushi, Tubereidin, Jpn. Patent 66 12,671 (1966) [*CA* **66**, 9971 (1967)].

258. J. E. Pike, P. F. Wiley, and L. Slechta, Derivatives of 7-D-ribofuranosylpyrrolopyrimidines, U.S. Patent 3,167,540 (Jan. 1965).

259. W. J. Wechter and A. R. Hanze, 9-β-D-Ribofuranosyl-7-deazapurine 5′-phosphate esters, U.S. Patent 3,336,289 (Aug. 1967).

260. A. R. Hanze, Deazapurine riboside cyclic 3′,5′-phosphates and process therefor, U.S. Patent 3,300,479 (Jan. 1967).

261. E. Higashide, T. Hasegawa, M. Shibata, K. Mizuno, and H. Akaike, Production of sparso-mycin and tubercidin by culture of *Streptomyces cuspidosporus*, *Takeda Kenkyusho Nempo* **25**, 1–14 (1966).

262. E. Higashida, M. Shibata, T. Hasegawa, and K. Mizuno, Tubercidin and sparsomycin production by *Streptomyces cuspidosporus*, Jpn. Patent 71 34,196 (Oct. 1971) [*CA* **76**, 2549d (1972)].
263. S. Shirato, Y. Miyazaki, and I. Suzuki, Antibiotic base analog, tubercidin. II. Fermentation and purification, *Hakko Kogaku Zasshi* **45**, 60–65 (1967).
264. S. Suzuki and S. Shirato, Concentration of tubercidin, Jpn. Patent 69 27,039 (Nov. 1969) [*CA* **72**, 65392k (1970)].
265. S. Suzuki and S. Shirato, Purification of tubercidin, Jpn. Patent 70 26,714 (Sept. 1970) [*CA* **74**, 2681z (1971)].
266. S. Suzuki and S. Marumo, Chemical structure of tubercidin, *J. Antibiot. Ser. A* **14**, 34–38 (1961).
267. S. Suzuki and S. Marumo, Chemical structure of tubercidin, *J. Antibiot. Ser. A* **13**, 360 (1960).
268. Y. Mizuno, M. Ikehara, K. Watanabe, and S. Suzaki, Structural elucidation of tubercidin, *Chem. Pharm. Bull.* **11**, 1091–1094 (1963).
269. Y. Mizuno, M. Ikehara, K. A. Watanabe, S. Suzaki, and T. Itoh, Synthetic studies of potential antimetabolites. IX. The anomeric configuration of tubercidin, *J. Org. Chem.* **28**, 3329–3331 (1963).
270. R. L. Tolman, R. K. Robins, and L. B. Townsend, Pyrrolopyrimidine nucleosides II. The total synthesis of 7-β-D-ribofuranosylpyrrolo[2,3-*d*]pyrimidines related to toyocamycin, *J. Heterocycl. Chem.* **4**, 230–238 (1967).
271. Y. Mizuno, M. Ikehara, K. A. Watanabe, and S. Suzuki, Synthetic studies of potential antimetabolites. X. Synthesis of 4-hydroxy-7-β-D-ribofuranosyl-7*H*-pyrrolo[2,3-*d*]pyrimidine, a tubercidin analog. *J. Org. Chem.* **28**, 3331–3336 (1963).
272. K. V. Rao and D. W. Renn, BA-90912: An antitumor substance, *Antimicrob. Agents Chemother.* 77–79 (1963).
273. K. V. Rao, W. S. Marsh, and D. W. Renn, Sangivamycin and derivatives thereof, U.S. Patent 3,423,398 (1969).
274. T. Kusaka, T. Iwasa, M. Shibata, H. Yamana, and T. Kishi, *Streptomyces purpureofuscus acoagulans* and its product, sangivamycin, *Takeda Kenkyusho Ho* **29**, 406–415 (1970) [*CA* **74**, 29145q (1971)].
275. K. V. Rao, Structure of sangivamycin, *J. Med. Chem.* **11**, 939–941 (1968).
276. S. Nakamura and H. Kondo, A brief review of nucleoside antibiotics, *Heterocycles* **8**, 583–607 (1977).
277. H. M. Goodman, J. Abelson, A. Landy, S. Brenner, and J. D. Smith, Amber suppression: A nucleotide change in the anticodon of a tyrosine transfer RNA, *Nature* **217**, 1019–1024 (1968).
278. B. P. Doctor, J. E. Loebel, M. A. Sodd, and D. B. Winter, Nucleotide sequence of *Escherichia coli* tyrosine transfer ribonucleic acid, *Science* **163**, 693–695 (1969).
279. U. L. RajBhandary, S. H. Chang, H. J. Gross, F. Harada, F. Kimura, and S. Nishimura, *E. coli* tyrosine transfer RNA—primary sequence and direct evidence for base pairing between the terminal sequences, *Fed. Proc., Fed. Am. Soc. Exp. Biol.* **28**, 409 (1969).
280. H. Kasai, Z. Ohashi, F. Harada, S. Nishimura, N. J. Oppenheimer, P. F. Crain, J. G. Liehr, D. L. von Minden, and J. A. McCloskey, Structure of the modified nucleoside Q isolated from *Escherichia coli* transfer ribonucleic acid. 7-(4,5-*cis*-Dihydroxyl-1-cyclopenten-3-ylaminoethyl)-7-deazaguanosine, *Biochemistry* **14**, 4198–4208 (1975).
281. N. Tanaka, R. T. Wu, T. Okabe, H. Yamashita, A. Shimazu, and T. Nishimura, Cadeguomycin, a novel nucleoside analog antibiotic, I. The producing organism, production and isolation of cadeguomycin, *J. Antibiot.* **35**, 272–278 (1982).
282. R. T. Wu, T. Okabe, M. Namikoshi, S. Okuda, T. Nishimura, and N. Tanaka, Cadeguomycin, a novel nucleoside analog antibiotic, II. Improved purification, physiochemical properties and structure assignment, *J. Antibiot.* **35**, 279–284 (1982).
283. T. Kondo, T. Goto, T. Okabe, and N. Tanaka, Synthesis of cadeguomycin (7-deazaguanosine-7-carboxylic acid), *Tetrahedron Lett.* **24**, 3647–3650 (1983).
284. V. G. Beylin, A. M. Kawasaki, C. S. Cheng, and L. B. Townsend, Pyrrolopyrimidine nucleosides 19. A total synthesis of the nucleoside antibiotic cadeguomycin [2-amino-7(β-D-ribofuranosyl)pyrrolo[2,3-*d*]pyrimidin-4-one-5-carboxylic acid], *Tetrahedron Lett.* **24**, 4793–4796 (1983).
285. Dainippon Pharmaceutical Co., Ltd., Antibiotic AB-116, Japan Kokai Tokkyo Koho JP 58 32,893 (Feb. 1983) [*CA* **99**, 37106j (1983)].

286. S. Naruto, H. Uno, A. Tanaka, H. Kotani, and Y. Takase, Kanagawamicin, a new aminonucleoside analog antibiotic from *Actinoplanes kanagawaensis, Heterocycles* **20**, 27–32 (1983).

287. T. Shomura, N. Nishizawa, M. Iwata, J. Yoshida, M. Ito, S. Amano, M. Koyama, M. Kojima, and S. Inouye, Studies on a new nucleoside antibiotic, dapiramicin, I. Producing organism, assay method and fermentation, *J. Antibiot.* **36**, 1300–1304 (1983).

288. N. Nishizawa, Y. Kondo, M. Koyama, S. Omoto, M. Iwata, T. Tsuruoka, and S. Inouye, Studies on a new nucleoside antibiotic, dapiramicin, II. Isolation, physico-chemical and biological characterization, *J. Antibiot.* **37**, 1–5 (1984).

289. H. Seto, N. Otake, M. Koyama, H. Ogino, Y. Kodama, N. Nishizawa, T. Tsuruoka, and S. Inouye, The structure of a novel nucleoside antibiotic, dapiramicin A, *Tetrahedron Lett.* **24**, 495–498 (1983).

290. R. Kazlauskas, P. T. Murphy, R. J. Wells, J. A. Baird-Lambert, and D. D. Jamieson, Halogenated pyrrolo[2,3-*d*]pyrimidine nucleosides from marine organisms, *Aust. J. Chem.* **36**, 165–170 (1983).

291. L. P. Davies, J. A. Baird-Lambert, and J. F. Marwood, Studies on several pyrrolo[2,3-*d*]pyrimidine analogs of adenosine which lack significant agonist activity at A1 and A2 receptors but have potent pharmacological activity in vivo, *Biochem. Pharmacol.* **35**, 3021–3029 (1986).

292. Y. Kato, N. Fusetani, S. Matsunaga, and K. Hashimoto, Bioactive marine metabolites IX. Mycalisines A and B, novel nucleosides which inhibit cell division of fertilized star fish eggs, from the marine sponge *Mycale* sp., *Tetrahedron Lett.* **26**, 3483–3486 (1985).

293. E. A. Meade, S. H. Krawczyk, and L. B. Townsend, A total synthesis of the naturally occurring pyrrolo[2,3-*d*]pyrimidine nucleoside, mycalisine A, *Tetrahedron Lett.* **29**, 4073–4076 (1988).

294. J. E. Pike, L. Slechta, and P. F. Wiley, Tubercidin and related compounds, *J. Heterocycl. Chem.* **1**, 159–161 (1964).

295. M. W. Winkley and R. K. Robins, Pyrimidine nucleosides, Part II. The direct glycosylation of 2,6-disubstituted-4-pyrimidones, *J. Chem. Soc.* **1969**, 791–796.

296. C. W. Noell and R. K. Robins, Aromaticity in heterocyclic systems. II. The application of N.M.R. in a study of the synthesis and structure of certain imidazo[1,2-*c*]pyrimidines and related pyrrolo[2,3-*d*]pyrimidines, *J. Heterocycl. Chem.* **1**, 34–41 (1964).

297. M. W. Winkley, Pyrimidine nucleosides. Part V. Reactions of 3-(β-D-ribofuranosyl)uracils and related compounds, *J. Chem. Soc. (C)* **1970**, 1869–1874.

298. J. A. Montgomery and K. Hewson, Analogs of tubercidin, *J. Med. Chem.* **10**, 665–667 (1967).

299. M. Legraverend, J.-M. Lhoste, J.-M. Bechet, and E. Bisagni, Carbocyclic analogs of 6-methyl-tubericidin, *Eur. J. Med. Chem. (Chim. Ther.)* **18**, 269–272 (1983).

300. M. Legraverend, J.-M. Lhoste, and E. Bisagni, Conformational studies of some carbocyclic nucleoside analogues, *Tetrahedron* **40**, 709–713 (1984).

301. M. Legraverend, E. Bisagni, J.-M. Lhoste, D. Anker, and H. Pacheco, Pseudo-nucleoside derivatives of pyrrolo[2,3-*d*]pyrimidines, synthesis and biological activity, *J. Heterocycl. Chem.* **20**, 925–929 (1983).

302. M. Legraverend, R.-M. N. Ngongo-Tekam, E. Bisagni, and A. Zerial, (±)-2-Amino-3,4-dihydro-7-[2,3-dihydroxy-4-(hydroxymethyl)-1-cyclopentyl]-7*H*-pyrrolo[2,3-*d*]pyrimidine-4-ones: New carbocyclic analogues of 7-deazaguanosine with antiviral activity, *J. Med. Chem.* **28**, 1477–1480 (1985).

303. E. C. Taylor and R. W. Hendess, Synthesis of pyrrolo[2,3-*d*]pyrimidines. The aglycon of toyocamycin, *J. Am. Chem. Soc.* **87**, 1995–2003 (1965).

304. E. C. Taylor and R. W. Hendess, Synthesis of 4-amino-5-cyanopyrrolo[2,3-*d*]pyrimidine, the aglycon of toyocamycin, *J. Am. Chem. Soc.* **86**, 951–952 (1964).

305. M. Bobek, R. L. Whistler, and A. Bloch, Synthesis and biological activity of 4'-thio analogs of the antibiotic toyocamycin, *J. Med. Chem.* **15**, 168–171 (1972).

306. W. Pfleiderer and R. K. Robins, Investigations in the pyrimidine series. XV. Synthesis of pyrimidine nucleosides by means of the fusion condensation method., *Chem. Ber.* **98**, 1511–1513 (1965).

307. H. Iwamura and T. Hashizume, Synthesis in nucleoside antibiotics, Part III. Synthesis of 4-amino-5-cyano-6-methylmercaptopyrrolo[2,3-*d*]pyrimidine-7-β-D-ribosides (toyocamycin analogs), *Agric. Biol. Chem.* **32**, 1010–1015 (1968).

308. Y. Ishido, T. Matsuba, A. Hosono, K. Fujii, T. Sato, S. Isome, A. Maruyama, and Y. Kikuchi,

Non-catalytic fusion reaction of 1,2,3,5-tetra-O-acetyl-β-D-ribofuranose with purine derivatives, *Bull. Chem. Soc.* **40**, 1007–1009 (1967).

309. T. Hashizume and H. Iwamura, A new catalyst for the synthesis of nucleosides, *Tetrahedron Lett.* **1965**, 3095–3102.

310. M. J. Robins and R. K. Robins, Purine nucleosides. XI. The synthesis of 2'-deoxy-9α- and β-D-ribofuranosylpurines and the correlations of their anomeric structure with proton magnetic resonance spectra. *J. Am. Chem. Soc.* **87**, 4934–4940 (1965).

311. L. V. Ektova, V. N. Tolkachev, J. Z. Kornveits, and M. N. Preobrazhenskaya, New total synthesis of the antibiotic tubercidin (7-deazaadenosine), *Bioorg. Khim.* **4**, 1250–1255 (1978).

312. T. Nishimura and I. Iwai, Studies on synthetic nucleosides. II. Novel synthesis of pyrimidine glucosides. *Chem. Pharm. Bull.* **12**, 357–361 (1964).

313. M. G. Stout and R. K. Robins, The synthesis of some quinazoline nucleosides, *J. Org. Chem.* **33**, 1219–1225 (1968).

314. G. R. Revankar and L. B. Townsend, The synthesis of 2-chloro-1-β-D-ribofuranosyl-5,6-dimethylbenzimidazole and certain related derivatives, *J. Heterocycl. Chem.* **5**, 615–620 (1968).

315. R. L. Tolman, G. L. Tolman, R. K. Robins, and L. B. Townsend, Pyrrolo-pyrimidine nucleosides. VI. Synthesis of 1, 3 and 7-β-D-ribofuranosylpyrrolo[2,3-*d*]pyrimidines via silylated intermediates, *J. Heterocycl. Chem.* **7**, 799–806 (1970).

316. E. Wittenburg, Nucleosides and related compounds. III. Synthesis of thymine nucleosides via silylpyrimidine compounds, *Chem. Ber.* **101**, 1095–1114 (1968).

317. J. F. Gerster, B. Carpenter, R. K. Robins, and L. B. Townsend, Pyrrolopyrimidine nucleosides. I. The synthesis of 4-substituted 7-(β-D-ribofuranosyl)pyrrolo[2,3-*d*]pyrimidines from tubercidin, *J. Med. Chem.* **10**, 326–331 (1967).

318. U. Lüpke and F. Seela, Ribosylation of 7H-pyrrolo[2,3-*d*]pyrimidine-4-(3H)-one at N-3, *Chem. Ber.* **112**, 3526–3529 (1979).

319. U. Lüpke and F. Seela, Favored formation of an O-glycoside during ribosylation of 5-methyl-2-methylthio-7H-pyrrolo[2,3-*d*]pyrimidine-4(3H)-one, *Chem. Ber.* **112**, 799–806 (1979).

320. M. Prystas, Nucleic acid components and their analogues. CXIII. Ribosylation of 4- and 5-substituted 6(1H)-pyrimidinones, *Collect. Czech. Chem. Commun.* **33**, 1813–1830 (1968).

321. W. Koenigs and E. Knorr, Über einige derivate des traubenzuckers und der galactose, *Ber* **34**, 957–981 (1901).

322. L. B. Townsend, R. L. Tolman, R. K. Robins, and G. H. Milne, The synthesis of 2-amino-7-(β-D-ribofuranosyl)pyrrolo[2,3-*d*]pyrimidin-4-one (7-deazaguanosine), a nucleoside Q and Q* analog, *J. Heterocycl. Chem.* **13**, 1363–1364 (1976).

323. V. S. Martynov and M. N. Preobrazhenskaya, Glycosylation of 2-mercapto-4-hydroxy-5,6-dihydro-7H-pyrrolo[2,3-*d*]pyrimidine, *Zh. Org. Khim.* **9**, 1980–1982 (1973).

324. D. A. Shuman, A. Bloch, R. K. Robins, and M. J. Robins, Synthesis and biological activity of certain 8-mercaptopurine and 6-mercaptopyrimidine 5-nucleosides, *J. Med. Chem.* **12**, 653–657 (1969).

325. L. V. Ektova, V. N. Tolkachev, N. L. Radyukina, T. P. Ivanova, Ya. V. Dobrynin, and M. N. Preobrazhenskaya, Synthesis and study of nucleosides related to the antibiotic tubercidin, *Bioorg. Khim,* **5**, 1369–1380 (1979).

326. F. Seela, U. Lüpke, and D. Hasselmann, Ribosylation of pyrrolo[2,3-*d*]pyrimidines in the presence of strong bases, *Chem. Ber.* **113**, 2808–2813 (1980).

327. T. Kondo, T. Ohgi, and T. Goto, Synthesis of 5-methyltubercidin and its α-anomer *via* condensation of the anion of 4-methoxy-5-methyl-2-methylthiopyrrolo[2,3-*d*]pyrimidine and 2,3,5-tri-O-benzyl-D-ribofuranosyl bromide, *Agric. Biol. Chem.* **41**, 1501–1507 (1977).

328. G. R. Revankar and R. K. Robins, Use of the sodium salt glycosylation method in nucleoside synthesis, *Nucleosides Nucleotides* **8**, 709–724 (1989).

329. H. B. Cottam, Z. Kazimierczuk, S. Geary, P. A. McKernan, G. R. Revankar, and R. K. Robins, Synthesis and biological activity of certain 6-substituted and 2,6-disubstituted-2'-deoxytubercidins prepared *via* the stereospecific sodium salt glycosylation procedure, *J. Med. Chem.* **28**, 1461–1467 (1985).

330. F. Seela, H.-D. Winkeler, J. Ott, Q.-H. Tran-Tri, D. Hasselmann, D. Franzen, and W. Bussmann, Synthesis of pyrrolo[2,3-*d*]pyrimidine nucleosides by phase transfer glycosylation and their function in polynucleotides, in: *Nucleosides, Nucleotides, and Their Biological Applications*

(J. L. Rideout, D. W. Henry, and L. M. Beacham III, eds.), pp. 181–208, Academic Press, New York (1983).

331. W. J. Middleton, V. A. Engelhardt, and B. S. Fisher, Cyanocarbon chemistry. VIII. Heterocyclic compounds from tetracyanoethylene, *J. Am. Chem. Soc.* **80**, 2822–2829 (1958).

332. F. L. Chung, R. A. Earl, and L. B. Townsend, Reductive debromination of some purine and purine-like nucleosides, *J. Org. Chem.* **45**, 4056–4057 (1980).

333. K. Ramasamy, R. K. Robins, and G. R. Revankar, Total synthesis of 2'-deoxytoyocamycin, 2'-deoxysangivamycin and related 7-β-D-arabinofuranosylpyrrolo[2,3-*d*]pyrimidines *via* ring closure of pyrrole precursors prepared by the sterospecific sodium salt glycosylation procedure, *Tetrahedron* **42**, 5869–5878 (1986).

334. K. Ramasamy, R. K. Robins, and G. R. Revankar, A convenient synthesis of 5-substituted-7-β-D-arabinofuranosylpyrrolo[2,3-*d*]pyrimidines structurally related to the antibiotics toyocamycin and sangivamycin, *J. Heterocycl. Chem.* **25**, 1043–1046 (1988).

335. D. F. Smee, P. A. McKernan, H. A. Alaghamandan, K. B. Frank, K. Ramasamy, G. R. Revankar, and R. K. Robins, Antiviral activities of 2'-deoxyribofuranosyl and arabinofuranosyl analogs of sangivamycin against retro- and DNA viruses, *Antiviral Res.* **10**, 263–278 (1988).

336. T. Matsushita, E. K. Ryu, C. I. Hong, and M. MacCoss, Phospholipid derivatives of nucleoside analogs as prodrugs with enhanced catabolic stability, *Cancer Res.* **41**, 2707–2713 (1981).

337. C. S. Wilcox and R. M. Otoski, Stereoselective preparations of ribofuranosyl chlorides and ribofuranosyl acetates. Solvent effects and stereoselectivity in the reaction of ribofuranosyl acetates with trimethylallylsilane, *Tetrahedron Lett.* **27**, 1011–1014 (1986).

338. K. Ramasamy, N. Imamura, R. K. Robins, and G. R. Revankar, A facile synthesis of tubercidin and related 7-deazapurine nucleosides via the stereospecific sodium salt glycosylation procedure, *Tetrahedron Lett.* **28**, 5107–5110 (1987).

339. K. Ramasamy, N. Imamura, R. K. Robins, and G. R. Revankar, A facile and improved synthesis of tubercidin and certain related pyrrolo[2,3-*d*]pyrimidine nucleosides by the stereospecific sodium salt glycosylation procedure, *J. Heterocycl. Chem.* **25**, 1893–1898 (1988).

340. F. Seela, H. Steker, H. Driller, and U. Bindig, 2-Amino-2'-deoxytubercidin and related pyrrolo[2,3-*d*]pyrimidinyl 2'-deoxyribofuranosides, *Liebigs Ann. Chem.* **1987**, 15–19.

341. B. D. Yuan, R. T. Wu, I. Sato, T. Okabe, H. Suzuki, T. Nishimura, and N. Tanaka, Biological activity of cadeguomycin. Inhibition of tumor growth and metastasis, immunostimulation, and potentiation of 1-β-D-arabinofuranosyl cytosine, *J. Antibiot.* **38**, 642–648 (1985).

342. R. T. Wu, T. Okabe, S. H. Kim, H. Suzuki, and N. Tanaka, Enhancement of pyrimidine nucleoside uptake into K562 and YAC-1 cells by cadeguomycin, *J. Antibiot.* **38**, 1588–1595 (1985).

343. H. Suzuki, S. H. Kim, M. Tahara, K. Okazaki, T. Okabe, R. T. Wu, and N. Tanaka, Potentiation of cytotoxicity of 1-β-D-arabinofuranosylcytosine for K562 human leukemic cells by cadeguomycin, *Cancer Res.* **47**, 713–717 (1987).

344. S. H. Kim, T. Okabe, N. Tanaka, and H. Suzuki, Effects of cadeguomycin on cytotoxicity of cytosine arabinoside and other pyrimidine nucleoside analogs—a comparative study, *J. Antibiot.* **40**, 1776–1777 (1987).

345. T. Kondo, K. Okamoto, M. Yamamoto, and T. Goto, A total synthesis of cadeguomycin, a nucleoside antibiotic produced by *Streptomyces hygroscopicus*, *Tetrahedron* **42**, 199–205 (1986).

346. K. Ramasamy, R. V. Joshi, R. K. Robins, and G. R. Revankar, Total and stereospecific synthesis of cadeguomycin, 2'-deoxycadeguomycin, *ara*-cadeguomycin and certain related nucleosides, *J. Chem. Soc., Perkin Trans 1*, 2375–2384 (1989).

347. K. Okamoto, T. Kondo, and T. Goto, Synthesis of *ara*-cadeguomycin. 2-Amino-3,4-dihydro-4-oxo-7-(β-D-arabinofuranosyl)-7*H*-pyrrolo[2,3-*d*]pyrimidine-5-carboxylic acid, *Bull. Chem. Soc. Jpn.* **59**, 1915–1919 (1986).

348. K. Ramasamy, R. K. Robins, and G. R. Revankar, Total and stereospecific synthesis of 2'-deoxycadeguomycin, *J. Chem. Soc., Chem. Commun.* **1989**, 560–562.

349. P. D. Cook, 7-β-D-Arabinofuranosyl-7*H*-pyrrolo[2,3-*d*]pyrimidine compounds and methods for their production, U.S. Patent 4,439,604 (March 1984).

350. A. J. Cocuzza, Total synthesis of 7-iodo-2',3'-dideoxy-7-deazapurine nucleosides, key intermediates in the preparation of reagents for the automated sequencing of DNA, *Tetrahedron Lett.* **29**, 4061–4064 (1988).

351. N. K. Saxena, B. M. Hagenow, G. Genzlinger, S. R. Turk, J. C. Drach, and L. B. Townsend, Synthesis and antiviral activity of certain 4-substituted and 2,4-disubstituted 7-[(2-hydroxy-ethoxy)methyl]pyrrolo[2,3-*d*]pyrimidines, *J. Med. Chem.* **31**, 1501–1506 (1988).

352. (a) P. K. Gupta, S. Daunert, M. R. Nassiri, L. L. Wotring, J. C. Drach, and L. B. Townsend, Synthesis, cytotoxicity, and antiviral activity of some acyclic analogues of the pyrrolo[2,3-*d*]pyrimidine nucleoside antibiotics tubercidin, toyocamycin, and sangivamycin, *J. Med. Chem.* **32**, 404–408 (1989); (b) P. K. Gupta, M. R. Nassiri, L. A. Coleman, L. L. Wotring, J. C. Drach, and L. B. Townsend, Synthesis, cytotoxicity, and antiviral activity of certain 7-[(2-hydroxy-ethoxy)methyl]pyrrolo[2,3-*d*]pyrimidine nucleosides related to toyocamycin and sangivamycin, *J. Med. Chem.* **32**, 1420–1425 (1989).

353. W. P. Weber and G. W. Gokel, *Phase Transfer Catalysis in Organic Synthesis*, Springer-Verlag, Berlin (1977).

354. C. M. Starks and C. Liotta, *Phase Transfer Catalysis—Principles and Techniques*, Academic Press, New York (1978).

355. E. V. Dehmlow and S. S. Dehmlow, *Phase Transfer Catalysis*, Verlag Chemie, Weinheim, FRG (1983).

356. F. Seela, U. Bindig, H. Driller, W. Herdering, K. Kaiser, A. Kehne, H. Rosemeyer, and H. Steker, Synthesis and application of isosteric purine 2'-deoxyribofuranosides, *Nucleosides Nucleotides* **6**, 1–23 (1987).

357. R. Barker and H. G. Fletcher, Jr., 2,3,5-Tri-*O*-ribosyl and -L-arabinosyl bromides, *J. Org. Chem.* **26**, 4605–4609 (1961).

358. F. Seela and D. Hasselmann, Synthesis of 7-deazainosine by phase transfer glycosylation, *Chem. Ber.* **113**, 3389–3393 (1980).

359. F. Seela and S. Menkhoff, 2-Methylthio-7-deazainosine and its α-anomerdebenzylation of thionucleosides with boron trichloride, *Liebigs Ann. Chem.* **1982**, 813–816.

360. F. Seela and D. Hasselman, Synthesis of 2-amino-3,7-dihydro-7-(β-D-ribofuranosyl)-4*H*-pyrrolo[2,3-*d*]pyrimidin-4-one(7-deazaguanosine), the parent compound of the nucleoside Q, *Chem. Ber.* **114**, 3395–3402 (1981).

361. F. Seela, D. Hasselmann, and H.-D. Winkeler, Synthesis of α-7-deazaguanosine and influence of base concentration on phase-transfer glycosylation, *Liebigs Ann. Chem.* **1982**, 499–506.

362. F. Seela and W. Bussman, N⁴-(3-Methyl-2-butenyl)-2-(methylthio)tubercidin and its α-ano-mer-7-deazapurine nucleosides with potential anticytokinin activity, *Nucleosides Nucleotides* **1**, 253–261 (1982).

363. F. Seela and W. Bussman, Anomers of tubercidin derivatives with 2-methylthio and N⁴-(Δ²-isopentenyl)residues, *Liebigs Ann. Chem.* **1984**, 1972–1980.

364. F. Seela, A. Kehne, and H.-D. Winkeler, Synthesis of acyclo-7-deazaguanosine by regiospecific phase-transfer alkylation of 2-amino-4-methoxy-7*H*-pyrrolo[2,3-*d*]pyrimidine, *Liebigs Ann. Chem.* **1983**, 137–146.

365. H.-D. Winkeler and F. Seela, Synthesis of 2-amino-7-(2'-deoxy-β-D-*erythro*-pentofurano-syl)-3,7-dihydro-4*H*-pyrrolo[2,3-*d*]pyrimidine-4-one, a new isostere of 2'-deoxyguanosine, *J. Org. Chem.* **48**, 3119–3122 (1983).

366. F. Seela and H.-P. Muth, Synthesis of 7-deaza-2',3'-dideoxyguanosine by deoxygenation of its 2'-deoxy-β-D-ribofuranoside, *Liebigs Ann. Chem.* **1988**, 215–219.

367. F. Seela, H.-D. Winkeler, H. Driller, and S. Menkhoff, Synthesis of 7-deaza-2'-deoxyguanosine by phase transfer glycosylation and preparation of suitable derivatives for oligonucleotide synthesis, *Nucleic Acids Res. Symp. Ser.* **14**, s245–246 (1984).

368. F. Seela, B. Westermann, and U. Bindig, Liquid–liquid and solid–liquid phase-transfer glycosylation of pyrrolo[2,3-*d*]pyrimidines: Stereospecific synthesis of 2-deoxy-β-D-ribofura-nosides related to 2'-deoxy-7-carbaguanosine, *J. Chem. Soc., Perkin Trans. 1* **1988**, 697–701.

369. F. Seela, H. Steker, H. Driller, and U. Bindig, 2-Amino-2'-deoxytubercidin and related pyrrolo[2,3-*d*]pyrimidinyl 2'-deoxyribofuranosides, *Liebigs Ann. Chem.* **1987**, 15–19.

370. F. Seela and A. Kehne, 2'-Deoxytubercidin—synthesis of a 2'-deoxyadenosine isostere by phase-transfer glycosylation, *Liebigs Ann. Chem.* **1983**, 876–884.

371. F. Seela, H.-P. Muth, and U. Bindig, Synthesis of 6-substituted 7-carbapurine 2',3'-dideoxy-nucleosides: Solid–liquid phase-transfer glycosylation of 4-chloropyrrolo[2,3-*d*]pyrimidine and deoxygenation of its 2'-deoxyribofuranoside, *Synthesis* **1988**, 670–674.

372. H. Rosemeyer and F. Seela, Stereoselective synthesis of pyrrolo [2,3-*d*]pyrimidine *α*- and *β*-D-ribonucleosides from anomerically pure D-ribofuranosyl chlorides: Solid–liquid phase-transfer glycosylation and ¹⁵N-NMR spectra, *Helv. Chim. Acta* **71**, 1573–1585 (1988).

373. F. Seela, H. Rosemeyer, A. Biesewig, and T. Jürgens, 7-Carbapurine ribofuranosides: Synthesis by solid–liquid phase-transfer glycosylation and ¹⁵N-NMR spectra. *Nucleosides Nucleotides* **7**, 581–584 (1988).

374. H.-D. Winkeler and F. Seela, 4-Amino-7-(*β*-D-arabinofuranosyl)pyrrolo[2,3-*d*]pyrimidine—the synthesis of *ara*-tubercidin by phase-transfer catalysis, *Chem. Ber.* **113**, 2069–2080 (1980).

375. F. Seela and H.-D. Winkeler, Preferred *β*-glycoside formation by phase transfer catalysis in the synthesis of D-arabinofuranosyl-7-deazapurine nucleosides, *Angew. Chem. Int. Ed. Engl.* **18**, 536 (1979).

376. F. Seela and H.-D. Winkeler, Synthesis of *ara*-tubercidin and its *α*-anomer by phase-transfer glycosylation of pyrrolo[2,3-*d*]pyrimidines, *Nucleic Acids Res. Symp. Ser.* **9**, s107–110 (1981).

377. F. Seela, Q.-H. Tran-Thi, and H.-D. Winkeler, Synthesis of *ara*-tubercidin and its 5'-phosphates *via* phase transfer glycosylation of 4-amino-2-methylthio-7*H*-pyrrolo[2,3-*d*]pyrimidine, *Chem. Ber.* **114**, 1217–1225 (1981).

378. F. Seela and H.-D. Winkeler, Selective 7-glycosylation of 4-amino-7-*H*-pyrrolo[2,3-*d*]pyrimidine to *ara*-tubercidin and its *α*-anomer, *Angew. Chem. Int. Ed. Engl.* **20**, 97–98 (1981).

379. U. Lupke and F. Seela, 7-(*β*-D-Arabinofuranosyl)pyrrolo[2,3-*d*]pyrimidin-4-(3*H*)-one—the 7-deaza derivative of the antiviral nucleoside ara-H, *Chem. Ber.* **112**, 3423–3440 (1979).

380. F. Seela and H.-D. Winkeler, 2-Amino-7-(*β*-D-arabinofuranosyl)pyrrolo[2,3-*d*]pyrimidin-4-(3*H*)-one. Synthesis of *ara*-7-deazaguanosine *via* phase-transfer glycosylation, *J. Org. Chem.* **47**, 226–230 (1982).

381. F. Seela and H.-D. Winkeler, Synthesis of the *α*-anomer of *ara*-7-deazaguanosine via phase-transfer glycosylation—control of anomer formation by the quaternary ammonium salt, *Liebigs Ann. Chem.* **1982**, 1634–1642.

382. F. Seela and H.-D. Winkeler, 2-Amino-7-*β*-D-arabinofuranosyl-4-methoxy-7*H*-pyrrolo[2,3-*d*]-pyrimidine: A facile preparation and anomerization of a 7-deazapurine nucleoside, *Carbohydr. Res.* **118**, 29–35 (1983).

383. F. Seela and H. Driller, 7-(*β*-D-Arabinofuranosyl)-2,4-dichloro-7*H*-pyrrolo[2,3-*d*]pyrimidine—synthesis, selective displacement of halogen, and influence of glyconic protecting groups on the reactivity of the aglycon, *Liebigs Ann. Chem.* **1984**, 722–733.

384. F. Seela and U. Liman, *ara*-7-Deazaxanthosine—a xanthine nucleoside with a stable N-glycosidic bond. *Liebigs Ann. Chem.* **1984**, 273–282.

385. J. J. Fox, K. A. Watanabe, and A. Bloch, Nucleoside antibiotics, *Prog. Nucleic Acid Res. Mol. Biol.* **5**, 251–313 (1966).

386. G. Acs, E. Reich, and M. Mori, Biological and biochemical properties of the analogue antibiotic tubercidin, *Proc. Natl. Acad. Sci. U.S.A.* **52**, 493–501 (1964).

387. A. Bloch, M. T. Hakala, E. Mihich, and C. A. Nichol, Studies on the biological activity and mode of action of 7-deazainosine, *Proc. Am. Assoc. Cancer Res.* **5**, 6 (1964).

388. S. P. Owen and C. C. Smith, Cytotoxicity and antitumor properties of the abnormal nucleoside tubercidin (NSC-56408), *Cancer Chemother. Rep.* **36**, 19–22 (1964).

389. M. Saneyoshi, R. Tokuzen, and F. Fukuoka, Antitumor activities and structural relationship of tubercidin, toyocamycin, and their derivatives, *Gann* **56**, 219–222 (1965).

390. J. A. Cavins, Initial toxicity study of sangivamycin, *Proc. Am. Assoc. Cancer Res.* **7**, 12 (1966).

391. R. K. Robins and G. R. Revankar, Purine analogs and related nucleosides and nucleotides as antitumor agents, *Med. Res. Rev.* **5**, 273–296 (1985).

392. Upjohn Co., *N*⁶-Methyltubercidin, Neth. Patent Appl. 6,513,128 (April 1966) [*CA* **65**, 15492c (1966)].

393. P. F. Wiley, Process for the production of *N*⁶-methyltubercidin and intermediate, U.S. Patent 3,364,198 (Jan. 1968).

394. B. C. Hinshaw, R. K. Robins, and L. B. Townsend, The synthesis of 4-hydroxylamino-7-(*β*-D-ribofuranosyl)pyrrolo[2,3-*d*]pyrimidine (7-deaza-HAPR), *J. Heterocycl. Chem.* **5**, 885 (1968).

395. L. V. Ektova, V. N. Tolkachev, N. L. Radyukina, T. P. Ivanova, Ya. V. Dobrynin, and M. N. Preobrazhenskaya, Synthesis and investigation of nucleosides related to tubercidin antibiotic. *Bioorg. Khim* **5**, 1369–1380 (1979).

396. T. Sheradsky, *O*-(2,4-Dinitrophenyl)oximes. Synthesis and cyclization to 5,7-dinitrobenzofurans, *J. Heterocycl. Chem.* **4**, 413–414 (1967).

397. G-F. Huang, T. Okamoto, M. Maeda, and Y. Kawazoe, Studies on chemical alterations of nucleic acids and their components. VI. N-Amination of some nucleic acid bases containing basic nitrogen with hydroxylamine-*O*-esters, *Tetrahedron Lett.* **1973**, 4541–4544.

398. Upjohn Co., 4-(2-Hydroxyethyl)tubercidin, Neth. Patent Appl. 6,514,129; (April, 1966) [*CA* **65**, 20206 (1966)].

399. T. Uematsu and R. J. Suhadolnik, 5-Hydroxymethyltubercidin. Synthesis, biological activity, and role in pyrrolopyrimidine biosynthesis, *J. Med. Chem.* **16**, 1405–1407 (1973).

400. D. E. Bergstrom, A. J. Brattesani, M. J. Ogawa, and M. J. Schweickert, Pyrrolo[2,3-*d*]pyrimidine nucleoside antibiotic analogues. Synthesis *via* organopalladium intermediates derived from 5-mercuritubercidin, *J. Org. Chem.* **46**, 1423–1431 (1981).

401. J. D. Westover, G. R. Revankar, R. K. Robins, R. D. Madsen, J. R. Ogden, J. A. North, R. W. Mancuso, R. J. Rousseau, and E. L. Stephen, Synthesis and antiviral activity of certain 9-β-D-ribofuranosylpurine-6-carboxamides, *J. Med. Chem.* **24**, 941–946 (1981).

402. R. J. Goebel, A. D. Adams, P. A. McKernan, B. K. Murray, R. K. Robins, G. R. Revankar, and P. G. Canonico, Synthesis and antiviral activity of certain carbamoylpyrrolopyrimidine and pyrazolopyrimidine nucleosides, *J. Med. Chem.* **25**, 1334–1338 (1982).

403. B. C. Hinshaw, J. F. Gerster, R. K. Robins, and L. B. Townsend, Pyrrolopyrimidine nucleosides. IV. The synthesis of certain 4,5-disubstituted-7-(β-D-ribofuranosyl)pyrrolo[2,3-*d*]pyrimidines related to the pyrrolo[2,3-*d*]pyrimidine nucleoside antibiotics, *J. Heterocycl. Chem.* **6**, 215–221 (1969).

404. E. Reich, Biochemical studies and applications of nucleoside analogs, 156th National Meeting of the American Chemical Society, Atlantic City, N.J., 1968, Abstract MEDI-30.

405. A. F. Cook and M. J. Holman, Synthesis of the natural product 5′-deoxy-5-iodotubercidin and related halogenated analogs, *Nucleosides Nucleotides* **3**, 401–411 (1984).

406. D. E. Bergstrom and A. Brattesani, Halogenation of tubercidin by *N*-halosuccinimides. A direct route to 5-bromotubercidin, a reversible inhibitor of RNA synthesis in eukaryotic cells, *Nucleic Acids Res.* **8**, 6213–6219 (1980).

407. B. Brdar, D. B. Rifkin, and E. Reich, Studies of Rous sarcoma virus. Effects of nucleoside analogues on virus synthesis, *J. Biol. Chem.* **248**, 2397–2408 (1973).

408. L. Ossowski and E. Reich, Effects of nucleoside analogs on transcription of simian virus 40, *Virology* **50**, 630–639 (1972).

409. T. G. Easton and E. Reich, Muscle differentiation in cell culture. Effects of nucleoside inhibitors and Rous sarcoma virus, *J. Biol. Chem.* **247**, 6420–6431 (1972).

410. K. H. Schram and L. B. Townsend, Pyrrolopyrimidine nucleosides. XII. Halogenation of 7-(β-D-ribofuranosyl)pyrrolo[2,3-*d*]pyrimidine, *J. Carbohydr. Nucleosides Nucleotides* **2**, 177–184 (1975).

411. R. L. Tolman, R. K. Robins, and L. B. Townsend, Pyrrolopyrimidine nucleosides VII. A study on electrophilic and nucleophilic substitution at position six of certain pyrrolo[2,3-*d*]pyrimidine nucleosides, *J. Heterocycl. Chem.* **8**, 703–706 (1971).

412. S. M. Greenberg, L. O. Ross, and R. K. Robins, Potential purine antagonists XXI. Preparation of some 9-phenyl-6-substituted purines, *J. Org. Chem.* **24**, 1314–1317 (1959).

413. D. Martin and S. Rackow, Cyanic acid esters. VI. Nitrile group transfer by aryl cyanates, *Chem. Ber.* **98**, 3662–3671 (1965).

414. C. Temple, Jr., C. L. Kussner, and J. A. Montgomery, Pyrimido[5,4-*e*]-*as*-triazines. IV. The preparation and some reactions of pyrimido[5,4-*e*]-*as*-triazine-5(6*H*)-thiones, *J. Org. Chem.* **34**, 3161–3165 (1969).

415. R. L. Tolman and L. B. Townsend, The synthesis of 4,5-diamino-8-β-D-ribofuranosylpyrazolo [3′,4′-5,4]pyrrolo[2,3-*d*]pyrimidine, a novel tricyclic nucleoside, *Tetrahedron Lett.* **1968**, 4815–4818.

416. C. L. Liotta and H. P. Harris, The chemistry of "naked" anions. I. Reactions of the 18-crown-6-complex of potassium fluoride with organic substrates in aprotic organic solvents, *J. Am. Chem. Soc.* **96**, 2250–2252 (1974).

417. S. Watanabe and T. Ueda, Thiocyanation of tubercidin and its derivatization to 6-propyl- and 6-cyano derivatives (nucleosides and nucleotides. 41), *Nucleosides Nucleotides* **1**, 191–203 (1982).

418. S. Watanabe and T. Ueda, Introduction of substituents to the 7(8)-position of 7-deazaadeno-sine (tubercidin): Conversion to toyocamycin, *Nucleic Acids Res. Symp. Ser.* **8**, s21–24 (1980).

419. S. Watanabe and T. Ueda, Conversion of tubercidin to toyocamycin: Some properties of tubercidin derivatives (nucleosides and nucleotides. 47), *Nucleosides Nucleotides* **2**, 113–125 (1983).

420. B. C. Hinshaw, J. F. Gerster, R. K. Robins, and L. B. Townsend, Pyrrolopyrimidine nucleosides. V. A study on the relative chemical reactivity of the 5-cyano group of the nucleoside antibiotic toyocamycin and desaminotoyocamycin. The synthesis of analogs of sangivamycin, *J. Org. Chem.* **35**, 236–241 (1970).

421. B. C. Hinshaw, O. Leonoudakis, K. H. Schram, and L. B. Townsend, Pyrrolopyrimidine nucleosides. Part X. Synthesis of certain 4,5-disubstituted 7-(β-D-ribofuranosyl)pyrrolo-[2,3-*d*]pyrimidines related to toyocamycin and sangivamycin, *J. Chem. Soc., Perkin Trans. 1* **1975**, 1248–1253.

422. G. H. Milne and L. B. Townsend, Pyrrolopyrimidine nucleosides XIII. Synthesis and chemical reactivity of certain selenopyrrolo[2,3-*d*]pyrimidine nucleosides, *J. Heterocyl. Chem.* **13**, 745–748 (1976).

423. L. B. Townsend and G. H. Milne, Synthesis, chemical reactivity, and chemotherapeutic activity of certain selenonucleosides and nucleosides related to the pyrrolo[2,3-*d*]pyrimidine nucleo-side antibiotics, *Ann. N. Y. Acad. Sci.* **255**, 91–103 (1975).

424. G. H. Milne and L. B. Townsend, Pyrrolopyrimidine nucleosides XIV. The synthesis of 7-(β-D-ribofuranosyl)pyrrolo[2,3-*d*]pyrimidin-4-selone and certain related derivatives, *J. Carbohydr. Nucleosides Nucleotides* **3**, 177–183 (1976).

425. K-Y. Zee-Cheng and C. C. Cheng, Common receptor-complement feature among some antileukemic compounds, *J. Pharm. Sci.* **59**, 1630–1633 (1970).

426. K. H. Schram and L. B. Townsend, Pyrrolopyrimidine nucleosides VIII. Synthesis of san-givamycin derivatives possessing exocyclic heterocycles at C5, *J. Carbohydr. Nucleosides Nucleo-tides* **1**, 39–54 (1974).

427. D. G. Nielson, R. Roger, J. W. M. Heatlie, and L. R. Newlands, The chemistry of amidrazones, *Chem. Rev.* **70**, 151–170 (1970).

428. K. H. Schram and L. B. Townsend, Pyrrolopyrimidine nucleosides. Part XI. Influence of amino-groups at C-4 and C-6 or an amino-group at C-6 on the reactivity of a 5-cyano-group in pyrrolo[2,3-*d*]pyrimidine nucleosides, *J. Chem. Soc., Perkin Trans. 1* **1975**, 1253–1257.

429. C. S. Cheng, B. C. Hinshaw, R. P. Panzica, and L. B. Townsend, Synthesis of 2-amino-5-cyano-7-(β-D-ribofuranosyl)pyrrolo[2,3-*d*]pyrimidin-4-one. An important precursor for the synthesis of nucleoside Q and Q*, *J. Am. Chem. Soc.* **98**, 7870–7872 (1976).

430. K. H. Schram and L. B. Townsend, The synthesis of 6-amino-4-methyl-8-(β-D-ribofurano-syl)(4-H,8-H)pyrrolo[4,3-2-*de*]pyrimido[4,3-*c*]pyridazine, a new tricyclic nucleoside, *Tetra-hedron Lett.* **1971**, 4757–4760.

431. K. H. Schram and L. B. Townsend, Fluorescent nucleoside derivatives of imidazo[1,2-*c*]-pyrrolo[2,3-*d*]pyrimidine, a new and novel heterocyclic ring system, *Tetrahedron Lett.* **1974**, 1345–1348.

432. K. H. Schram, S. J. Manning, and L. B. Townsend, Tricyclic nucleosides I. Synthesis of the new tricyclic ring system tetrazolo[1,5-*c*]pyrrolo[2,3-*d*]pyrimidine and certain tetrazolo[1,5-*c*]pyr-rolo[2,3-*d*]pyrimidine ribonucleosides, *J. Heterocycl. Chem.* **12**, 1021–1023 (1975).

433. L. B. Townsend, G. A. Bhat, F.-L. Chung, K. H. Schram, R. P. Panzica, and L. L. Wotring, Synthesis, chemical reactivity, biological and chemotherapeutic activity of certain tricyclic nucleosides, in: *Nucleosides, Nucleotides and Their Biological Applications* (J.-L. Barascut and J.-L. Imbach, eds.), pp. 37–81, INSERM, Paris (1979).

434. G. A. Bhat, K. H. Schram, and L. B. Townsend, The synthesis of certain fluorescent imid-azo[1,2-*c*]pyrrolo[3,2-*e*]pyrimidine nucleoside derivatives related to ϵ-adenosine, *J. Carbohydr. Nucleosides Nucleotides* **7**, 333–345 (1980).

435. F.-L. Chung, K. H. Schram, R. P. Panzica, R. A. Earl, L. L. Wotring, and L. B. Townsend, Synthesis of certain [6:5:6]linear tricyclic nucleosides as potential antitumor agents, *J. Med. Chem.* **23**, 1158–1166 (1980).

436. F.-L. Chung, R. A. Earl, and L. B. Townsend, The novel synthesis of a [6:5:6]linear iso-guanosine type tricyclic nucleoside using carbonyl sulfide, *Tetrahedron Lett.* **21**, 1599–1602 (1980).

437. F.-L. Chung, R. A. Earl, and L. B. Townsend, 5-Amino-9-(β-D-ribofuranosyl)-*v*-triazino-[4,5-*b*]pyrimido[4,5-*d*]pyrrol-4-one and unusual ring opening of this new ring system with the Vilsmeier–Haack reagent, *J. Org. Chem.* **45**, 2532–2535 (1980).

438. G. A. Bhat, K. H. Schram, and L. B. Townsend, Synthesis of 7-(β-D-ribofuranosyl)pyrrolo[3,2-*e*]-1,2,4-triazolo[1,5-*c*]pyrimidines and 7-(β-D-ribofuranosyl)pyrrolo[3,2-*e*]-1,2,4-triazolo[4,3-*c*]pyrimidine, *J. Carbohydr. Nucleosides Nucleotides* **8**, 145–160 (1981).

439. M. A. Hernandez, F.-L. Chung, R. A. Earl, and L. B. Townsend, A general route for the facile synthesis of 4-thioxopyrimidin-2-one derivatives *via* the annulation of cyclic *O*-aminonitriles using carbonyl sulfide, *J. Org. Chem.* **46**, 3941–3945 (1981).

440. M. H. Fleysher, A. Bloch, M. T. Hakala, and C. A. Nichol, Synthesis and biological activity of some new N^6-substituted purine nucleosides, *J. Med. Chem.* **12**, 1056–1061 (1969).

441. J. W. Jones and R. K. Robins, Purine nucleosides, III. Methylation studies of certain naturally occurring purine nucleosides, *J. Am. Chem. Soc.* **85**, 193–201 (1963).

442. A. F. Lewis and L. B. Townsend, Derivatives of the nucleoside antibiotics, toyocamycin and sangivamycin. Analogs of N^6-(Δ^2-isopentenyl)-adenosine, *J. Heterocycl. Chem.* **11**, 71–72 (1974).

443. S. M. Hecht, Mass spectra of some heterocycles related to N^6-(Δ^2-isopentyl)adenosine, *Anal. Biochem.* **44**, 262–275 (1971).

444. H. Iwamura, T. Ito, Z. Kumazawa, and Y. Ogawa, Anticytokinin activity of 4-furfurylamino-7-(β-D-ribofuranosyl)pyrrolo[2,3-*d*]pyrimidine, *Biochem. Biophys. Res. Commun.* **57**, 412–416 (1974).

445. H. Iwamura, T. Ito, Z. Kumazawa, and Y. Ogawa, Synthesis and anticytokinin activity of 4-substituted-7-(β-D-ribofuranosyl)pyrrolo[2,3-*d*]pyrimidines, *Phytochemistry* **14**, 2317–2321 (1975).

446. H. Iwamura, G. Kumazawa, and Y. Ogawa, 4-Methylamino-7-(β-D-ribofuranosyl)pyrrolo-[2,3-*d*]pyrimidine derivatives as plant growth regulators, Jpn. Kokai 75,125,033 (1975).

447. F. Skoog, R. Y. Schmitz, S. M. Hecht, and R. B. Frye, Anticytokinin activity of substituted pyrrolo[2,3-*d*]pyrimidines, *Proc. Natl. Acad. Sci. U.S.A.* **72**, 3508–3512 (1975).

448. A. R. Hanze, Nucleic acids. V. Nucleoside derivatives of tubercidin (7-deazaadenosine), *Biochemistry* **7**, 932–939 (1968).

449. I. Goia, M. Kezdi, and O. Birzu, Tubercidin 5'-monophosphate, Rom. RO 76,852 (July 1981) [*CA* **99**, 158786q (1983)].

450. F. Seela, Q.-H. Tran-Thi, H. Mentzel, and V. A. Erdmann, Favored incorporation of tubercidin in poly(adenylic,7-deazaadenylic acids) and their function as messenger ribonucleic acids in protein synthesis, *Biochemistry* **20**, 2559–2564 (1981).

451. W. Friest and F. Cramer, Synthesis of AMP and ATP analogs, in: *Nucleic Acid Chemistry, Improved and New Synthetic Procedures, Methods, and Techniques* (L. B. Townsend and R. S. Tipson, eds.), Vol. 2, pp. 827–836, John Wiley and Sons, New York (1978).

452. Y. Tonomura, K. Imamura, M. Ikehara, H. Uno, and F. Harada, Interaction between synthetic ATP analogues and actomyosin systems. IV, *J. Biol. Chem.* **61**, 460–472 (1967).

453. J. P. Miller, K. H. Boswell, K. Muneyama, R. L. Tolman, M. B. Scholten, R. K. Robins, L. N. Simon, and D. A. Shuman, activity of tubercidin-, toyocamycin- and sangivamycin-3',5'-cyclic phosphates and related compounds with some enzymes of adenosine-3',5'-cyclic phosphate metabolism. *Biochem. Biophys. Res. Commun.* **55**, 843–849 (1973).

454. B. G. Hughes, P. C. Srivastava, D. D. Muse, and R. K. Robins, 2',5'-Oligoadenylates and related 2',5'-oligonucleotide analogues. 1. Substrate specificity of the interferon-induced murine 2',5'-oligoadenylate synthetase in enzymatic synthesis of oligomers, *Biochemistry* **22**, 2116–2126 (1983).

455. R. J. Suhadolnik, T. Uematsu, H. Uematsu, and R. G. Wilson, The incorporation of sangivamycin 5'-triphosphate into polyribonucleotide by ribonucleic acid polymerase from *Micrococcus lysodeikticus, J. Biol, Chem.* **243**, 2761–2766 (1968).

456. F. Seela, H. Driller, A. Kehne, and K. Kaiser, Self-complementary oligomers containing 7-deaza-2'-deoxyguanosine or 2'-deoxytubercidin, *Chem. Scr.* **26**, 173–178 (1986).

457. P. J. Barr, R. M. Thayer, P. Laybourn, R. C. Najarian, F. Seela, and D. R. Tolan, 7-Deaza-2'-deoxyguanosine 5'-triphosphate: Enhanced resolution in M13 dideoxy sequencing *Bio Techniques* **4**, 428–432 (1986).

458. J. Hashimoto, T. Uchida, and F. Egami, Action of ribonucleases T, T$_2$, and U$_2$ on dinucleoside monophosphates containing 7-deazapurine base, *Biochim. Biophys. Acta.* **199**, 535–536 (1970).

459. M. Ikehara and I. Tazawa, Studies on nucleosides and nucleotides. XXIX. Direct synthesis of nucleoside-2′,3′-cyclic phosphates, *J. Org. Chem.* **31**, 819–821 (1966).

460. J. Moravek and J. Skoda, Preparation of nucleotides and oligonucleotides by thermal reaction, Czech. Patent 133,464 (1969).

461. N. K. Kochetkov, E. I. Budowsky, and V. N. Shibaev, Isocytidine 5′-phosphate, in *Synthetic Procedures in Nucleic Acid Chemistry* (W. W. Zorbach and R. S. Tipson, eds.), pp. 477–481, Interscience, New York (1968).

462. P. F. Torrence, E. De Clercq, J. A. Waters, and B. Witkop, A potent interferon inducer derived from poly(7-deazainosinic acid), *Biochemistry* **13**, 4400–4408 (1974).

463. J. G. Moffatt and H. G. Khorana, Nucleoside polyphosphates. X. The synthesis and some reactions of nucleoside-5′-phosphoromorpholidates and related compounds. Improved methods for the preparation of nucleoside-5′-polyphosphates, *J. Am. Chem. Soc.* **83**, 649–658 (1961).

464. M. Ikehara, F. Harada, and E. Ohtsuka, Polynucleotides. III. Synthesis of four trinucleoside diphosphates containing tubercidin (7-deazaadenosine) and N^6-dimethyladenosine. *Chem. Pharm. Bull.* **14**, 1338–1346 (1966).

465. M. Smith, D. H. Rammler, I. H. Goldberg, and H. G. Khorana, Studies on polynucleotides. XIV. Specific synthesis of the C$_{3'}$-C$_{5'}$ interribonucleotide linkage. Syntheses of uridylyl-(3′→5′)-uridine and uridylyl-(3′→5′)-adenosine, *J. Am. Chem. Soc.* **84**, 430–440 (1962).

466. M. Ikehara and E. Ohtsuka, Studies on coenzyme analogs. XV. A novel phosphorylating agent, *P*-diphenyl-*P*′-morpholino pyrophosphorochloridate, *Chem. Pharm. Bull.* **11**, 961–967 (1963).

467. Q.-H. Tran-Thi, D. Franzen, and F. Seela, 7-Deazaguanosine-3′,5′-phosphate: An isosteric cGMP analogue with high affinity for cyclonucleotide phosphodiesterase, *Angew. Chem. Int. Ed. Engl.* **21**, 367–368 (1982).

468. F. Seela, Q.-H. Tran-Thi, and D. Franzen, Poly(7-deazaguanylic acid), the homopolynucleotide of the parent nucleoside of quenosine, *Biochemistry* **21**, 4338–4343 (1982).

469. P. F. Wiley, J. H. Johnson, and R. R. Hanze, Alkyl and acyl derivatives of tubercidin, *J. Antibiot.* **24**, 720–727 (1976).

470. J. K. Coward, D. L. Bussolotti, and C-D. Chang, Analogs of S-adenosylhomocysteine as potential inhibitors of biological transmethylation. Inhibition of several methylases by S-tubercidinylhomocysteine, *J. Med. Chem.* **17**, 1286–1289 (1974).

471. R. T. Borchardt, J. A. Huber, and Y. S. Wu, Potential inhibitors of S-adenosylmethionine-dependent methyltransferases. 4. Further modifications of the amino acid and base portions of S-adenosyl-L-homocysteine, *J. Med. Chem.* **19**, 1094–1099 (1976).

472. J. K. Coward, N. C. Motola, and J. D. Moyer, Polyaminebiosynthesis in rat prostate. Substrate and inhibitor properties of 7-deaza analogues of decarboxylated S-adenosylmethionine and 5′-methylthioadenosine, *J. Med. Chem.* **20**, 500–505 (1977).

473. M. J. Robins, S. R. Naik, and A. S. K. Lee, Nucleic acid related compounds. 12. The facile and high yield stannous chloride catalyzed monomethylation of the cis-glycol system of nucleosides by diazomethane, *J. Org. Chem.* **39**, 1891–1899 (1974).

474. M. J. Robins, A. S. K. Lee, and S. R. Naik, 2′- and 3′-O-Methyl-nucleosides, in: *Nucleic Acid Chemistry, Improved and New Synthetic Procedures, Methods, and Techniques* (L. B. Townsend and R. S. Tipson, eds.), Vol. 2, pp. 759–764, John Wiley and Sons, New York (1978).

475. C. A. Dekker, Separation of nucleoside mixtures on Dowex-1 (OH$^-$), *J. Am. Chem. Soc.* **87**, 4027–4029 (1965).

476. K. Anzai and M. Matsui, Tubercidin. Its conversion into 5′-deoxytubercidin, *Bull. Chem. Soc. Jpn.* **46**, 618–623 (1973).

477. K. Anzai and M. Matsui, Compounds relating to dibenzoyladenine riboside. The choice between the N^6,1-dibenzoyl and N^6,N^6-dibenzoyl structures, *Bull. Chem. Soc. Jpn.* **46**, 3228–3232 (1973).

478. Y. Wang, H. P. C. Hogencamp, R. A. Long, G. R. Revankar, and R. K. Robins, A convenient synthesis of 5′-deoxyribonucleosides, *Carbohydr. Res.* **59**, 449–457 (1977).

479. K. Kikugawa, K. Iizuka, Y. Higuchi, H. Hirayama, and M. Ichino, Platelet aggregation

inhibitors. 2. Inhibition of platelet aggregation by 5'-, 2-, 6-, and 8- substituted adenosines, *J. Med. Chem.* **15**, 387–390 (1972).

480. H. G. Kuivila, Organotin hydrides and organic free radicals, *Acc. Chem. Res.* **1**, 299–305 (1968).

481. K. Anzai and M. Matsui, Cyclonucleosides related to adenosine and tubercidin, *Agric. Biol. Chem.* **37**, 301–305 (1973).

482. K. Anzai and M. Matsui, Reinvestigation on cyclonucleosides related to adenosine and tubercidin, *Agric. Biol. Chem.* **37**, 2431–2432 (1973).

483. K. Anzai and M. Matsui, Tubercidin: Its conversion to 5'-amino-5'-deoxytubercidin, *Agric. Biol. Chem.* **37**, 921–923 (1973).

484. T. A. Glassman, C. Cooper, L. W. Harrison, and T. J. Swift, A proton magnetic resonance study of metal ion–adenine ring interactions in metal ion complexes with adenosine triphosphate, *Biochemistry* **10**, 843–851 (1971).

485. M. J. Robins, J. R. McCarthy, Jr., R. A. Jones, and R. Mengel, Nucleic acid related compounds. 5. The transformation of formycin and tubercidin into 2'- and 3'- deoxynucleosides, *Can. J. Chem.* **51**, 1313–1321 (1973).

486. T. C. Jain, A. F. Russell, and J. G. Moffatt, Reactions of 2-acyloxyisobutyryl halides with nucleosides. III. Reactions of tubercidin and formycin, *J. Org. Chem.* **38**, 3179–3186 (1973).

487. E. M. Filachione, J. H. Lengel, and C. H. Fisher, Pyrolytic preparation of α-carbalkoxyalkyl acrylates and methacrylates, *J. Am. Chem. Soc.* **68**, 330–333 (1946).

488. S. Greenberg and J. G. Moffatt, Reactions of 2-acyloxyisobutyryl halides with nucleosides. I. Reactions of model diols and of uridine, *J. Am. Chem. Soc.* **95**, 4016–4025 (1973).

489. A. F. Russell, S. Greenberg, and J. G. Moffatt, Reactions of 2-acyloxyisobutyryl halides with nucleosides. II. Reactions of adenosine, *J. Am. Chem. Soc.* **95**, 4025–4030 (1973).

490. M. J. Robins, R. Mengel, and R. A. Jones, Nucleic acid related compounds. VII. Conversion of ribonucleoside 2',3'-ortho esters into deoxy, epoxy, and unsaturated nucleosides, *J. Am. Chem. Soc.* **95**, 4074–4076 (1973).

491. M. J. Robins and W. H. Muhs, Synthesis of 2'-deoxytubercidin {4-amino-7-(2-deoxy-β-D-*erythro*-pentofuranosyl)pyrrolo[2,3-*d*]pyrimidine} from the parent antibiotic, *J. Chem. Soc., Chem. Commun.* 1976, 269.

492. M. J. Robins, J. S. Wilson, and F. Hansske, Nucleic acid related compounds. 42. A general procedure for the efficient deoxygenation of secondary alcohols. Regiospecific and stereoselective conversion of ribonucleosides to 2'-deoxynucleosides, *J. Am. Chem. Soc.* **105**, 4059–4065 (1983).

493. R. A. Lessor and N. J. Leonard, Synthesis of 2'-deoxynucleosides by deoxygenation of ribonucleosides, *J. Org. Chem.* **46**, 4300–4301 (1981).

494. K. Fukukawa, T. Ueda, and T. Hirano, Nucleosides and nucleotides. XXXXV. Facile deoxygenation of neplanocin A and nucleosides by the use of tri-*n*-butyltin hydride, *Chem. Pharm. Bull.* **31**, 1842–1847 (1983).

495. (a) T. Maruyama, L. L. Wotring, and L. B. Townsend, Pyrrolopyrimidine nucleosides. 18. Synthesis and chemotherapeutic activity of 4-amino-7-(3-deoxy-β-D-ribofuranosyl)pyrrolo[2,3-*d*]pyrimidine-5-carboxamide (3'-deoxysangivamycin) and 4-amino-7-(2-deoxy-β-D-ribofuranosyl)pyrrolo[2,3*d*]pyrimidine-5-carboxamide (2'-deoxysangivamycin), *J. Med. Chem.* **26**, 25–29 (1983); (b) S. H. Krawczyk and L. B. Townsend, 2'-3'-Dideoxyadenosine analogs of the nucleoside antibiotics tubercidin, toyocamycin and sangivamycin, *Nucleosides Nucleotides* **8**, 97–115 (1989).

496. S. A. Brinkley, A. Lewis, W. J. Critz, L. L. Witt, L. B. Townsend, and R. L. Blakley, Enzymatic preparation of the 5'-triphosphates of 2'-deoxytubercidin, 2'-deoxytoyocamycin, and 2'-deoxyformycin and the allosteric effects of these nucleotides on ribonucleotide reductase, *Biochemistry* **17**, 2350–2356 (1978).

497. C. U. Pittman, Jr., S. P. McManus, and J. W. Larsen, 1,3-Dioxolan-2-ylium and related heterocyclic cations, *Chem. Rev.* **73**, 357–438 (1972).

498. M. J. Robins, Chemical transformations of naturally occurring nucleosides to otherwise difficultly accessible structures, *Ann. N. Y. Acad. Sci.* **255**, 104–120 (1975).

499. J. Farkas and F. Sorm, Nucleic acids components and their analogues. XCIV. Synthesis of 6-amino-9(1-deoxy-β-D-psicofuranosyl)purine, *Collect. Czech. Chem. Commun.* **32**, 2663–2667 (1967).

500. M. J. Robins, Y. Fouron, and W. H. Muhs, Nucleic acid related compounds. 25. Syntheses of arabino, xylo, and lyxo-anhydro sugar nucleosides from tubercidin ribo-epoxide, *Can. J. Chem.* **55**, 1260–1267 (1977).

501. K. Anzai and M. Matsui, Tubercidin. Its conversion to 2′,3′-didehydro-2′,3′-dideoxy derivatives, *Agric. Biol. Chem.* **37**, 345–348 (1973).

502. E. J. Corey and R. A. E. Winter, A new, stereospecific olefin synthesis from 1,2-diols, *J. Am. Chem. Soc.* **85**, 2677–2678 (1963).

503. K. Anzai and M. Matsui, 8,5′-*O*-Cyclonucleosides of tubercidin, *Bull. Chem. Soc. Jpn.* **47**, 417–420 (1974).

504. A. R. Hanze, Upjohn Co., Arabinofuranosylcytosines, Neth. Patent Appl. 6,606,669 (Dec. 1966) [*CA* **67**, 82375w (1967)].

505. A. R. Hanze, 7-Deazaadenine 2′,5′- and 3′,5′-dinucleoside phosphate and process therefor, U.S. Patent 3,309,358 (March 1967).

506. A. R. Hanze, Dinucleoside 3′,5′- and 2′,5′-phosphates containing one 7-deazapurine riboside moiety, U.S. Patent 3,337,530 (August 1967).

507. F. Seela, J. Ott, and E. Hissmann, (2′→5′)- and (3′→5′)- Tubercidylyltubercidins—Synthesis *via* phosphite triester and investigation of secondary structure, *Liebigs Ann. Chem.* **1984**, 692–707.

508. F. Seela, Q.-H. Tran-Thi, H. Mentzel, and V. A. Erdmann, Favored incorporation of tubercidin in poly(adenylic, 7-deazaadenylic acids) and their function as messenger ribonucleic acids in protein synthesis, *Biochemistry* **20**, 2559–2564 (1981).

509. T. M. Jacob and H. G. Khorana, Studies on polynucleotides. XXX. A comparative study of reagents for the synthesis of the C$_3$,–C$_5$, internucleotidic linkage, *J. Am. Chem. Soc.* **86**, 1630–1635 (1964).

510. E. De Clercq, P. F. Torrence, T. Fukui, and M. Ikehara, Role of purine N-3 in the biologic activities of poly(A) and poly(I), *Nucleic Acids Res.* **3**, 1591–1601 (1976).

511. A. Ono, M. Sato, Y. Ohtani, and T. Ueda, Synthesis of deoxyoligonucleotides containing 7-deazaadenine: Recognition and cleavage by restriction endonuclease *Bgl* II and *Sau* 3 AI (nucleosides and nucleotides, part 55), *Nucleic Acids Res.* **12**, 8939–8949 (1984).

512. S. Shirato, K. Yoshida, and Y. Miyazaki, Antibiotic base analog, tubercidin. III. Biochemical synthesis of tubercidin 5′-monophosphate, *Hakko Kogaku Zasshi* **46**, 233–240 (1968) [*CA* **69**, 17001y (1968)].

513. R. Marutzky, H. Peterssen-Borstel, J. Flossdorf, and M-R. Kula, Large scale enzymatic synthesis of nucleoside 5′-monophosphates using a phosphotransferase from carrots, *Biotechnol. Bioeng.* **16**, 1449–1458 (1974).

514. C. T. Hardesty, N. A. Chaney, V. S. Waravdekar, and J. A. R. Mead, Enzymatic phosphorylation of sangivamycin, *Biochim. Biophys. Acta.* **195**, 581–583 (1969).

515. R. M. Stroud, The crystal and molecular structure of tubercidin, C$_{11}$H$_{14}$N$_4$O$_4$ *Acta Crystallogr., Sect. B* **29**, 690–696 (1973).

516. J. Abola and M. Sundaralingam, Refinement of the crystal structure of tubercidin, *Acta Crystallogr., Sect. B* **29**, 697–703 (1973).

517. A. Lo, E. Shefter, and T. G. Cochran, Analysis of *N*-glycosyl bond length in crystal structures of nucleosides and nucleotides, *J. Pharm. Sci.* **64**, 1701–1710 (1975).

518. V. Zabel, W. Saenger, and F. Seela, Structures of 2′-deoxytubercidin (I) and 2′-deoxytubercidin dihydrate (II, *Acta Crystallogr, Sect. C* **43**, 131–134 (1987).

519. C. Altona and M. Sundaralingam, Conformational analysis of the sugar ring in nucleosides and nucleotides. A new description using the concept of pseudo-rotation, *J. Am. Chem. Soc.* **94**, 8205–8212 (1972).

520. E. Westhof, O. Röder, I. Croneiss, and H.-D.Lüdermann, Ribose conformations in the common purine (β) ribosides, in some antibiotic nucleosides, and in some isopropylidene derivatives: A comparison, *Z. Naturforsch., C* **30**, 131–140 (1975).

521. D. L. Von Minden, J. G. Liehr, M. H. Wilson, and J. A. McCloskey, Mechanism of electron impact induced elimination of methylenimine from dimethylamino heteroaromatic compounds, *J. Org. Chem.* **39**, 285–289 (1974).

522. Y. Tondeur, M. Shorter, M. E. Gustafson, and R. C. Pandey, Fast atom bombardment mass spectometry and tandem mass spectrometry in antibiotics: Identification of nucleoside antitumor antibiotic toyocamycin in fermentation broth, *Biomed. Mass. Spectrom.* **11**, 622–628 (1984).

523. P. Prusiner and M. Sundaralingam, The crystal and molecular structure of toyocamycin monohydrate, a nucleoside antibiotic, *Acta Crystallog., Sect. B* **34**, 517–523 (1978).

524. J. G. Liehr, C. L. Weise, P. F. Crain, C. H. Milne, D. S. Wise, L. B. Townsend, and J. A. McCloskey, Mass spectrometry of nucleoside derivatives. Seleno analogs of purine and pyrimidine bases and nucleosides, *J. Heterocycl. Chem.* **16**, 1263–1272 (1979).

525. M.-T. Chenon, R. J. Pugmire, D. M. Grant, R. P. Panzica, and L. B. Townsend, Carbon-13 magnetic resonance. XXV. A basic set of parameters for the investigation of tautomerism in purines established from carbon-13 magnetic resonance studies using certain purines and pyrrolo[2,3-*d*]pyrimidines, *J. Am. Chem. Soc.* **97**, 4627–4636 (1975).

526. M. Ikehara and T. Fukui, Some physical properties of poly 7-deazaadencyclic acid (poly tubercidin phosphoric acid), *J. Mol. Biol.* **38**, 437–441 (1968).

527. M. Ikehara and E. Ohtsuka, Stimulation of the binding of aminoacyl-*s*-RNA to ribosomes by tubercidin (7-deazaadenosine) and N^6-dimethyladenosine containing trinucleoside diphosphate analogs, *Biochem. Biophys. Res. Commun.* **21**, 257–264 (1965).

528. S. C. Uretsky, G. Acs, E. Reich, M. Mori and L. Altwerger, Pyrrolopyrimidine nucleotides and protein synthesis, *J. Biol. Chem.* **243**, 306–312 (1968).

529. T. Ongi, T. Goto, H. Kasai, and S. Nishimura, Stereochemistry of the cyclopentene side chain in the nucleoside Q obtained from *Escherichia coli* tRNA, *Tetrahedron Lett.* **1976**, 367–370.

530. J. A. McCloskey and S. Nishimura, Modified nucleosides in transfer RNA *Acc. Chem. Res.* **10**, 403–410 (1977).

531. F. Harada and S. Nishimura, Possible anticodon sequences of tRNA[His], tRNA[Asn], and tRNA[Asp] from *Escherichia coli* B. Universal presence of nucleoside Q in the first position of the anticodons of these transfer ribonucleic acids, *Biochemistry* **11**, 301–308 (1972).

532. H. Kasai, Y. Kuchino, K. Nihei, and S. Nishimura, Distribution of the modified nucleoside Q and its derivatives in animal and plant transfer RNA's, *Nucleic Acids Res.* **2**, 1931–1939 (1975).

533. B. N. White, G. M. Tener, J. Holden, and D. T. Suzuki, Activity of a transfer RNA modifying enzyme during the development of *Drosophila* and its relationship to the *su(s)* locus, *J. Mol. Biol.* **74**, 635–651 (1973).

534. P. F. Crain, S. K. Sethi, J. R. Katze, and J. A. McCloskey, Structure of an amniotic fluid component, 7-(4,5-*cis*-dihydroxyl-1-cyclopenten-3-ylaminomethyl)-7-deazaguanine (quenine), a substrate for tRNA:guanine transglycosylase, *J. Biol. Chem.* **255**, 8405–8407 (1980).

535. T. Ohgi, T. Kondo, and T. Goto, Synthetic studies on nucleoside Q(I). Synthesis of Q-base, *Nucleic Acids Res. Spec. Publ.* **2**, s83–85 (1976).

536. S. Nishimura, Characterization, function, and biosynthesis of modified nucleotides in tRNA, *J. Biochem.* **79**, 35p (1976).

537. Y. Kuchino, H. Kasai, K. Nihei, and S. Nishimura, Biosynthesis of the modified nucleoside Q in transfer RNA, *Nucleic Acids Res.* **3**, 393–398 (1976).

538. R. J. Suhadolnik and T. Uematsu, Biosynthesis of the pyrrolopyrimidine nucleoside antibiotic, toyocamycin, *J. Biol. Chem.* **245**, 4365–4371 (1970).

539. H. Kasai, K. Nakanishi, R. D. Macfarlane, D. F. Torgerson, Z. Ohashi, J. A. McCloskey, H. J. Gross, and S. Nishimura, The structure of Q* nucleoside isolated from rabbit liver transfer ribonucleic acid, *J. Am. Chem. Soc.* **98**, 5044–5046 (1976).

540. I. R. Lehman and E. A. Pratt, On the structure of the glucosylated hydroxymethylcytosine nucleosides of coliphages T2, T4, and T6, *J. Biol. Chem.* **235**, 3254–3259 (1960).

541. T. Ohgi, T. Kondo, and T. Goto, Total synthesis of nucleoside Q, *Tetrahedron Lett.* **1977**, 4051–4054.

542. S. Noguchi, Z. Yamaizumi, T. Ohgi, T. Goto, Y. Nishimura, Y. Hirota, and S. Nishimura, Isolation of Q nucleoside precursor present in tRNA of an *E. coli* mutant and its characterization as 7-cyano-7-deazaguanosine, *Nucleic Acids Res.* **5**, 4215–4223 (1978).

543. T. Ohgi, T. Kondo, and T. Goto, Synthetic studies on possible biosynthetic intermediates of nucleoside Q, *Nucleic Acids Res. Spec. Publ.* **3**, s97–99 (1977).

544. N. Okada, S. Noguchi, S. Nishimura, T. Ohgi, T. Goto, P. F. Crain, and J. A. McCloskey, Structure determination of a nucleoside Q precursor isolated from *E. coli* tRNA: 7-(Aminomethyl)-7-deazaguanosine, *Nucleic Acids Res.* **5**, 2289–2296 (1978).

545. T. Ohgi, T. Kondo, and T. Goto, Total synthesis of optically pure nucleoside Q. Determination of absolute configuration of natural nucleoside Q, *J. Am. Chem. Soc.* **101**, 3629–3633 (1979).

546. N. Okada and S. Nishimura, Enzymatic synthesis of Q nucleoside containing mannose in the anticodon of tRNA: Isolation of a novel mannosyltransferase from a cell-free extract of rat liver, *Nucleic Acids Res.* **4**, 2931–2937 (1977).

547. S. Yokoyama, T. Miyazawa, Z. Yamaizumi, and S. Nishimura, The conformational properties of Q nucleotide in aqueous solution, *Nucleic Acids Res. Spec. Publ.* **5**, s363–365 (1979).

548. S. Yokoyama, T. Miyazawa, Y. Iitaka, Z. Yamaizumi, H. Kazai, and S. Nishimura, Three dimensional structure of hyper-modified nucleoside Q located in the wobbling position of tRNA, *Nature* **282** 107–109 (1979).

549. S. Yokoyama, T. Miyazawa, Y. Iitaka, Z. Yamaizumi, H. Kasai, and S. Nishimura, Molecular structure of Q nucleotide, *Nucleic Acids Res. Symp. Ser.* **6**, s75–76 (1979).

550. R. J. Suhadolnik, Cordycepin, psicofuranine, decoyinine, tubercidin and toyocamycin, *Antibiotics* **2**, 400–408 (1967).

551. G. Acs and E. Reich, Tubercidin and related pyrrolopyrimidin antibiotics, *Antibiotics* **1**, 494–498 (1967).

552. C. A. Nichol, Antibiotics resembling adenosine: Tubercidin, toyocamycin, sangivamycin, formycin, psicofuranine, and decoyinine, *Handbook Exp. Pharmacol.* **38(2)**, 434–457 (1975).

553. C. G. Smith, L. M. Reineke, and H. Harpootlian, Uptake of tubercidin and 7-deazainosine by human blood cells, *Proc. Am. Assoc. Cancer Res.* **7**, 66 (1966).

554. C. G. Smith, G. D. Gray, R. G. Carlson, and A. R. Hanze, Biochemical and biological studies with tubercidin (7-deazaadenosine), 7-deazainosine and certain nucleotide derivatives of tubercidin, *Adv. Enzyme Reg.* **5**, 121–151 (1967).

555. C. G. Smith, L. M. Reineke, M. R. Burch, A. M. Shefner, and E. E. Muirhead, Studies on the uptake of tubercidin (7-deazaadenosine) by blood cells and its distribution in whole animals, *Cancer Res.* **30**, 69–75 (1970).

556. R. J. Suhadolnik, T. Uematsu, H. Uematsu, and R. G. Wilson, The incorporation of san-givamycin 5′-triphosphate into polyribonucleotide by ribonucleic acid polymerase from *Micrococcus lysodeikticus*, *J. Biol. Chem.* **243**, 2761–2766 (1968).

557. R. J. Suhadolnik, T. Uematsu, and H. Uematsu, Toyocamycin: Phosphorylation and incorporation into RNA and DNA, and the biochemical properties of the triphosphate, *Biochim. Biophys. Acta* **149**, 41–49 (1967).

558. B. Brdar and E. Reich, 7-Deazanebularin. Metabolism in cultures of mouse fibroblasts and incorporation into cellular and viral nucleic acids, *J. Biol. Chem.* **247**, 725–730 (1972).

559. C. T. Hardesty, N. A. Chaney, V. S. Waravdekar, and J. A. R. Mead, The disposition of the antitumor agent, sangivamycin, in mice, *Cancer Res.* **34**, 1005–1009 (1974).

560. C. E. Cass, M. Selner, T. H. Tan, W. H. Muhs, and M. J. Robins, Comparison of the effects on cultured L1210 leukemia cells of the ribosyl, 2′-deoxyribosyl, and xylosyl homologs of tubercidin and adenosine alone or in combination with 2′-deoxycoformycin, *Cancer Treat. Rep.* **66**, 317–326 (1982).

561. P. S. Ritch, R. I. Glazer, R. E. Cunningham, and S. E. Shackney, Kinetic effects of sangivamycin in sarcoma 180 *in vitro*, *Cancer Res.* **41**, 1784–1788 (1981).

562. R. I. Glazer and K. D. Hartman, Cytokinetic and biochemical effects of sangivamycin in human colon carcinoma cells in culture, *Mol. Pharmacol.* **20**, 657–661 (1981).

563. C. A. Nichol, A. Bloch, and E. Mihich, Biological and biochemical activities of tubercidin and related nucleosides, *Cancer Chemother., Proc. Takeda Int. Conf., Osaka*, 185–195 (1966).

564. P. R. Ritch and M. Helmsworth, Pyrrolopyrimidine lethality in relation to ribonucleic acid synthesis in sarcoma 180 cells *in vitro*, *Biochem. Pharmacol.* **31**, 2686–2688 (1982).

565. A. I. Kravchenko, V. A. Chernov, L. I. Sncherbakova, L. N. Filitis, G. N. Pershin, and V. N. Sokolova, Pyrrolo[3,2-*d*]pyrimidines as potential antitumor agents, *Farmakol. Toksikol. (Moscow)* **42**, 659–665 (1979).

566. R. I. Glazer, K. D. Hartman, and O. J. Cohen, Effect of sangivamycin and xylosyladenine on the synthesis and methylation of polysomal ribonucleic acid in Ehrlich ascites cells *in vitro*, *Biochem. Pharmacol.* **30**, 2697–2701 (1981).

567. L. J. Stekol'nikov, Anticancerous antibiotic, *Priroda (Moscow).* **1970**(5), 109.

568. J. A. Cavins, T. C. Hall, K. B. Olson, C. L. Khung, J. Horton, J. Colsky, and R. K. Shadduck, Initial toxicity study of sangivamycin (NSC-65346), *Cancer Chemother. Rep.* **51**, 197–200 (1967).

569. C. G. Zubrod, S. Schepartz, J. Leiter, K. M. Endicott, L. M. Carrese, and C. G. Baker, The chemotherapy program of the National Cancer Institute: History, analysis, and plans, *Cancer Chemother. Rep.* **50**, 349–381 (1966).

570. R. W. Brockman, R. W. Rundles, R. E. Parks, T. A. Khwaja, and H. G. Mandel, Purines, *Med. Pediatr. Oncol.* **11**, A388–392 (1983).

571. J. W. DeJong, Partial purification and properties of rat-heart adenosine kinase, *Arch. Int. Physiol. Biochim.* **85**, 557–569 (1977).

572. T. D. Palella, C. M. Andres, and I. H. Fox, Human placental adenosine kinase. Kinetic mechanism and inhibition, *J. Biol. Chem.* **255**, 5264–5269 (1980).

573. A. C. Newby, The interaction of inhibitors with adenosine metabolising enzymes in intact isolated cells, *Biochem. Pharmacol.* **30**, 2611–2615 (1981).

574. S. Shirato, Y. Miyazaki, and M. Yamamoto, The antibiotic base analog, tubercidin. I. Microbiological assay, *Hakko Kogaku Zasshi* **44**, 133–138 (1966).

575. A. Bloch and C. A. Nichol, Nucleoside antibiotics related to adenosine, *Antimicrob. Agents Chemother.* **1964**, 530–539.

576. A. Bloch, R. J. Leonard, and C. A. Nichol, Oh the mode of action of 7-deazaadenosine (tubercidin), *Biochim. Biophys. Acta* **138**, 10–25 (1967).

577. J. J. Jaffe, E. Meymarian, and H. M. Doremus, Antischistosomal action of tubercidin administered after absorption into red cells, *Nature* **230**, 408–409 (1971).

578. A. F. Ross and J. J. Jaffe, Effects of tubercidin and its ribonucleotides on various metabolic pathways in *Schistosoma mansoni, Biochem. Pharmacol.* **21**, 3059–3069 (1972).

579. J. J. Jaffe, H. M. Doremus, H. A. Dunsford, W. S. Kammerer, and E. Meymarian, Antischistosomal activity of tubercidin in monkeys, *Am. J. Trop. Med. Hygiene* **22**, 62–72 (1973).

580. J. J. Jaffe, H. M. Doremus, H. A. Dunsford, and E. Meymarian, Comparative efficacy of one and two treatments with tubercidin against *Schistosomiasis japonica* in monkeys, *Am. J. Trop. Med. Hygiene* **23**, 65–70 (1974).

581. R. J. Stegman, A. W. Senft, P. R. Brown, and R. E. Parks, Jr., Pathways of nucleotide metabolism in *Schistosoma mansoni*—IV. Incorporation of adenosine analogs *in vitro, Biochem. Pharmacol.* **22**, 459–468 (1973).

582. L. R. Duvall, Tubercidin, *Cancer Chemother. Rep.* **30**, 61–62 (1963).

583. G. J. McCormick, C. J. Canfield, and G. P. Willet, *In vitro* antimalarial activity of nucleic acid precursor analogues in the simian malaria *Plasmodium knowlesi, Antimicrob. Agents Chemother.* **6**, 16–21 (1974).

584. R. E. Kohls, A. J. Lemin, and P. W. O'Connell, New chemosterilants against the house fly, *J. Econ. Entomol.* **59**, 745–746 (1966).

585. Upjohn Company, Delaware, Improvements in or relating to insecticidal compositions, Brit. Patent 1,102,029 (Feb. 1968).

586. H. Iwamura, T. Ito, Z. Kumazawa, J. Eguchi, M. Mogami, and S. Okuda, Antifungal activity of substituted 7-(β-D-ribofuranosyl)pyrrolo[2,3-*d*]pyrimidines, *Agric. Biol. Chem.* **40**, 1431–1433 (1976).

587. L. Ehrman, Antibiotics and infectious hybrid sterility in *Drosophila paulistorum, Mol. Gen. Genet.* **103**, 218–222 (1968).

588. O. Yamada, Y. Kaise, F. Futatsuya, S. Ishida, K. Ito, H. Yamamoto, and K. Munakata, Studies on plant growth-regulating activities of anisomycin and toyocamycin, *Agric. Biol. Chem.* **36**, 2013–2015 (1972).

589. H. Iwamura, Z. Kumazawa, S. Nagato, and S. Okuda, Plant growth retarding activity of the 4-substituted-7-(β-D-ribofuranosyl)pyrrolo[2,3-*d*]pyrimidine anticytokinins, *Agric. Biol. Chem.* **40**, 1653–1654 (1976).

590. H. Iwamura, T. Ito, N. Masuda, A. Mizuno, K. Koshimizu, Z. Kumazawa, S. Nagato, and S. Okuda, Effects of substituted pyrrolo[2,3-*d*]pyrimidine derivatives on plant growth, *Nippon Noyaku Gakkaishi* **6**, 9–15 (1981).

591. E. Mihich, C. L. Simpson, and A. I. Mulhern, Comparative study of the toxicologic effects of 7-deazaadenosine (tubercidin) and 7-deazainosine, *Cancer Res.* **29**, 116–123 (1969).

592. A. Bloch, E. Mihich, R. J. Leonard, and C. A. Nichol, Studies on the biologic activity and mode of action of 7-deazainosine, *Cancer Res.* **29**, 110–115 (1969).

593. J. J. Marr, R. L. Berens, N. K. Cohn, D. J. Nelson, and R. S. Klein, Biological action of inosine analogs in *Leishmania* and *Trypanosoma* spp., *Antimicrob. Agents Chemother.* **25**, 292–295 (1984).

594. M. H. El Kouni, D. Diop. P. O'Shea, R. Carlisle, and J-P. Sommadossi, Prevention of tubercidin host toxicity by nitrobenzylthioinosine 5'-monophosphate for the treatment of schistosomiasis, *Antimicrob. Agents. Chemother.* **33**, 824–827 (1989).

595. A. M. Kapuler, D. C. Ward, N. Mendelsohn, H. Klett, and G. Acs, Utilization of substrate analogs by mengovirus induced RNA polymerase, *Virology* **37**, 701–706 (1969).

596. L. Ossowski and E. Reich, Effects of nucleoside analogs on transcription of simian virus 40, *Virology* **50**, 630–639 (1972).

597. B. Brdar, D. B. Rifkin, and E. Reich, Studies on Rous sarcoma virus. Effects of nucleoside analogues on virus synthesis, *J. Biol. Chem.* **248**, 2397–2408 (1973).

598. E. De Clercq, A. Billiau, P. F. Torrence, J. A. Waters, and B. Witkop, Antiviral and antimetabolic activities of poly(7-deazaadenylic acid) and poly(7-deazainosinic acid), *Biochem. Pharmacol.* **24**, 2233–2238 (1975).

599. S. A. Moyer and K. S. Holmes, The specific inhibition of vesicular stomatitis virus replication by toyocamycin, *Virology* **98**, 99–107 (1979).

600. M. Mauchauffe, R. Hamelin, A. Tavitian, M. L. Michel, and C. J. Larsen, Effects of toyocamycin on the biological activity of a murine oncornavirus produced by a chronically infected cell line, *Biomed. Express* **31**, 17–20 (1979).

601. D. E. Bergstrom, A. J. Brattesani, M. K. Ogawa, P. A. Reddy, M. J. Schweickert, J. Balzarini, and E. De Clercq, Antiviral activity of C-5 substituted tubercidin analogues, *J. Med. Chem.* **27**, 285–292 (1984).

602. E. De Clercq, D. E. Bergstrom, A. Holy, and J. A. Montgomery, Broad-spectrum antiviral activity of adenosine analogues, *Antiviral Res.* **4**, 119–133 (1984).

603. S. Nishimura, F. Harada, and M. Ikehara, The selective utilization of tubercidin triphosphate as an ATP analog in the DNA-dependent RNA polymerase system, *Biochim. Biophys. Acta.* **129**, 301–309 (1966).

604. W. H. Wolberg, Effect of tubercidin on nucleoside incorporation in human tumors, *Biochem. Pharmacol.* **14**, 1921–1925 (1965).

605. E. De Clercq, J. Balzarini, D. Madej, F. Hansske, and M. J. Robins, Nucleic acid related compounds. 51. Synthesis and biological properties of sugar-modified analogues of the nucleoside antibiotics tubercidin, toyocamycin, sangivamycin, and formycin, *J. Med. Chem.* **30**, 481–486 (1987).

606. E. De Clercq and M. J. Robins, Xylotubercidin against herpes simplex virus type 2 in mice, *Antimicrob. Agents Chemother.* **30**, 719–724 (1986).

607. A. Tavitian, S. C. Uretsky, and G. Acs, Selective inhibition of ribosomal RNA synthesis in mammalian cells, *Biochim. Biophys. Acta* **157**, 33–42 (1968).

608. A. Tavitian, S. C. Uretsky, and G. Acs, The effect of toyocamycin on cellular RNA synthesis, *Biochim. Biophys. Acta* **179**, 50–57 (1969).

609. L. Sverak, R. A. Bonar, A. J. Langlois, and J. W. Beard, Inhibition by toyocamycin of RNA synthesis in mammalian cells and in normal and avian tumor virus-infected chick embryo cells, *Biochim. Biophys. Acta* **224**, 441–450 (1970).

610. R. A. Bonar, J. F. Chabot, A. J. Langlois, L. Sverak, L. Veprek, and J. W. Beard, Influence of toyocamycin on avian leukemia myeloblasts: Cell growth, ultrastructure, RNA synthesis, and elaboration of BAI strain A virus, *Cancer Res.* **30**, 753–762 (1970).

611. J. W. Weiss and H. C. Pitot, Inhibition of ribosomal RNA maturation in Novikoff hepatoma cells by toyocamycin, tubercidin, and 6-thioguanosine, *Cancer Res.* **34**, 581–587 (1974).

612. L. L. Bennett, Jr. and D. Smithers, Feedback inhibition of purine biosynthesis in H. Ep. # 2 cells by adenine analogs, *Biochem. Pharmacol.* **13**, 1331–1339 (1964).

613. J. F. Henderson and M. K. Y. Khoo, On the mechanism of feedback inhibition of purine biosynthesis *de novo* in Ehrlich ascites tumor cells *in vitro*, *J. Biol. Chem.* **240**, 3104–3109 (1965).

614. L. L. Bennett, Jr., M. H. Vail, S. Chumley, and J. A. Montgomery, Activity of adenosine analogs against a cell culture line resistant to 2-fluoroadenine, *Biochem. Pharmacol.* **15**, 1719–1728 (1966).

615. L. L. Bennett, Jr., H. P. Schnebli, M. H. Vail, P. W. Allan, and J. A. Montgomery, Purine ribonucleoside kinase activity and resistance to some analogs of adenosine, *Mol. Pharmacol.* **2**, 432–443 (1966).

616. A. Zerial, M. Zerial, M. Legraverend, and E. Bisagni, Antiviral activities of 54,247-RP, a carbocyclic analog of 7-deazaguanosine in cell cultures and animals, *Ann. Inst. Pasteur/Virol.* **137E**, 317–325 (1986).

617. M.-I. Lim, R. S. Klein, and J. J. Fox, Synthesis of the pyrrolo[3,2-d]pyrimidine C-nucleoside isostere of inosine, *Tetrahedron Lett.* **21**, 1013–1016 (1980).

618. M.-I. Lim and R. S. Klein, Synthesis of "9-deazaadenosine"; a new cytotoxic C-nucleoside isostere of adenosine, *Tetrahedron Lett.* **22**, 25–28 (1981).

619. R. S. Klein, M.-I. Lim, W.-Y. Ren, and J. H. Burchenal, Antileukemic β-glycosyl C-nucleosides, *Eur. Pat. Appl.* EP 71,227 (Feb. 1983).

620. M.-I. Lim, W.-Y. Ren, B. A. Otter, and R. S. Klein, Synthesis of "9-deazaguanosine" and other new pyrrolo[3,4-d]pyrimidine C-nucleosides, *J. Org. Chem.* **48**, 780–788 (1983).

621. N. S. Girgis, H. B. Cottam, S. B. Larson, and R. K. Robins, 9-Deazapurine nucleosides. The synthesis of certain N-5-2'-deoxy-β-*erythro*-pentofuranosyl and N-5-β-D-arabinofuranosyl-pyrrolo[3,2-d]pyrimidines, *J. Heterocycl. Chem.* **24**, 821–827 (1987).

622. N. S. Girgis, R. K. Robins, and H. B. Cottam, 9-Deazapurine nucleosides. The synthesis of certain N-5-β-D-ribofuranosylpyrrolo[3,2-d]pyrimidines, *J. Heterocycl. Chem.* **27**, 171–175 (1990).

623. K. Imai, Studies on nucleic acid antagonists. VII. Synthesis and characterization of 1,4,6-triazaindenes (5H-pyrrolo[3,2-d]pyrimidines), *Chem. Pharm. Bull.* **12**, 1030–1042 (1964).

624. R. I. Glazer, K. D. Hartman, and M. C. Knode, 9-Deazaadenosine. Cytocidal activity and effects on nucleic acids and protein synthesis in human colon carcinoma cells in culture, *Mol. Pharmacol.* **24**, 309–315 (1983).

625. M. Y. Chu, L. B. Landry, R. S. Klein, M.-I. Lim, A. E. Bogden, and G. W. Crabtree, 9-Deazaadenosine, a new potent antitumor agent, *Proc. Am. Assoc. Cancer Res.* **23**, 220 (1982).

626. M. Y. Chu, L. B. Zukerman, S. Sato, G. W. Crabtree, R. E. Bogden, M.-I. Lim, and R. S. Klein, 9-Deazaadenosine—a new potent antitumor agent, *Biochem. Pharmacol.* **33**, 1229–1234 (1984).

627. T. P. Zimmerman, R. D. Deeprose, G. Wolberg, C. R. Stopford, G. S. Duncan, W. H. Miller, R. L. Miller, M.-I. Lim, W.-Y. Ren, and R. S. Klein, Inhibition of lymphocyte function by 9-deazaadenosine, *Biochem. Pharmacol.* **32**, 1211–1217 (1983).

628. R. L. Miller, D. L. Adamczyk, J. L. Rideout, and T. A. Krenitsky, Purification, characterization, substrate and inhibitor specificity of adenosine kinase from several *Eimera* species, *Mol. Biochem. Parasitol.* **6**, 209–223 (1982).

629. I. A. Korbukh, F. F. Blanko, M. N. Preobrazhenskaya, and H. Dorn, Synthesis of 4-chloropyrazolo[3,4-b]pyridine riboside, *Zh. Org. Khim.* **7**, 2633 (1971).

630. I. A. Korbukh, F. F. Blanko, M. N. Preobrazhenskaya, H. Dorn, N. G. Kondakova, T. I. Sukhova, and N. P. Kostyuchenko, Glycosides of pyrazole and condensed pyrazole heterocycles. II. Synthesis of pyrazolo[3,4-b]pyridine glycosides, *Zh. Org. Khim.* **9**, 1266–1272 (1973).

631. H. Reimlinger, M. A. Peiren, and M. Robert, Synthesis with heterocyclic amines. I. Reactions of 3(5)-aminopyrazole with α,β-unsaturated esters. Preparation and characterization of isomeric oxodihydropyrazolopyrimidines, *Chem. Ber.* **103**, 3252–3265 (1970).

632. J. D. Ratajczyk and L. R. Swett, The cyclocondensation of 5-amino-1,3-dimethylpyrazole with ethyl acetoacetate. Synthesis of isomeric pyrazolopyridones, *J. Heterocycl. Chem.* **12**, 517–522 (1975).

633. I. A. Korbukh, F. F. Blanko, Kh. Dorn, and M. N. Preobrazhenskaya, Structure of hydroxypyrazolo[3,4-b]pyridine and nucleosides synthesized from it, *Zh. Org. Khim.* **12**, 2043–2044 (1976).

634. M. N. Preobrazhenskaya, I. A. Korbukh, and F. F. Blanko, The structure of pyrazole and fused pyrazole trimethylsilyl derivatives and the site of their glycosylation, *J. Carbohydr. Nucleosides Nucleotides* **2**, 73–78 (1975).

635. Y. S. Sanghvi, S. B. Larson, R. C. Willis, R. K. Robins, and G. R. Revankar, Synthesis and biological evaluation of certain C-4 substituted pyrazolo[3,4-b]pyridine nucleosides, *J. Med. Chem.* **32**, 945–951 (1989).

636. H. Dorn and R. Ozegowski, Unambiguous synthesis of 4,7-dihydro-4-oxo-1H-pyrazolo-[3,4-b]pyridine—Further comments on the "(N-C)-rearrangement" of (2-alkoxycarbonyl-vinylamino—pyrazoles, *J. Prakt. Chem.* **324**, 557–562 (1982).

637. S. B. Larson, Y. S. Sanghvi, G. R. Revankar, and R. K. Robins, Crystal structures of four pyrazolo[3,4-b]pyridine nucleosides, *Acta Crystallogr.* **C46**, 791–797 (1990).

638. C. M. Smith, G. Zomber, and J. F. Henderson, Inhibitors of hypoxanthine metabolism in Ehrlich ascites tumor cells *in vitro*, *Cancer Treat. Rep.* **60**, 1567–1584 (1976).

639. P. K. Gupta, N. K. Dalley, R. K. Robins, and G. R. Revankar, Synthesis of β-D-ribo- and 2'-deoxy-β-D-ribofuranosyl derivatives of 6-aminopyrazolo[4,3-*c*]pyridin-4(5*H*)-one by a ring closure of pyrazole nucleoside precursors, *J. Heterocyl. Chem.* **23**, 59–64 (1986).

640. K. W. Ehler, R. K. Robins, and R. B. Meyer, Jr., 6-Aminopyrazolo[4,3-*c*]pyridin-4(5*H*)-one, a novel analogue of guanine, *J. Med. Chem.* **20**, 317–318 (1977).

641. R. J. Rousseau, R. K. Robins, and L. B. Townsend, Purine nucleosides. XX. The synthesis of 7-β-D-ribofuranosylpurines from imidazole nucleoside derivatives, *J. Am. Chem. Soc.* **90**, 2661–2668 (1968).

642. R. J. Rousseau, R. P. Panzica, S. M. Reddick, R. K. Robins, and L. B. Townsend, Purine nucleosides. XXVI. A general synthesis of 6-substituted 7-(β-D-ribofuranosyl)purines. A reinvestigation and corroboration of the position of glycosylation of 6-dimethylamino-7-(β-D-ribofuranosyl)purine, *J. Org. Chem.* **35**, 631–635 (1970).

643. N. J. Leonard and R. A. Laursen, Synthesis of 3-β-D-ribofuranosyladenine and (3-β-D-ribofuranosyladenine)-5'-phosphate, *Biochemistry* **4**, 354–365 (1965).

644. E. J. Reist, A. Benitez, L. Goodman, B. R. Baker, and W. W. Lee, Potential anticancer agents. LXXVI. Synthesis of purine nucleosides of β-D-arabinofuranose, *J. Org. Chem.* **27**, 3274–3279 (1962).

645. E. Schipper and A. R. Day, Studies in imidazoles. II. Imidazo[*b*]pyrazines, *J. Am. Chem. Soc.* **74**, 350–353 (1952).

646. F. L. Muehlmann and A. R. Day, Metabolite analogs. V. Preparation of some substituted pyrazines and imidazo[*b*]pyrazines, *J. Am. Chem. Soc.* **78**, 242–244 (1956).

647. G. Palamidessi and F. Luini, Pyrazine derivatives. XI. Tetrachloropyrazine, *Farmaco. Ed. Sci.* **21**, 811–817 (1966).

648. R. P. Panzica and L. B. Townsend, Synthesis of 5,6-dimethyl-1-(β-D-ribofuranosyl)imidazo-[4,5-*b*]pyrazine by ring closure of an imidazole nucleoside, a new bicyclic nucleoside, *Tetrahedron Lett.* **1970**, 1013–1015.

649. R. P. Panzica and L. B. Townsend, Synthesis of imidazo[4,5-*b*]pyrazine nucleosides, *J. Chem. Soc., Perkin Trans. 1* **1973**, 244–248.

650. R. A. Sharma, M. Bobek, F. E. Cole, and A. Bloch, Synthesis and biological activity of some imidazo[4,5-*b*]pyrazines and their ribonucleosides as purine analogs, *J. Med. Chem.* **16**, 643–647 (1973).

651. C. M. Smith, L. J. Fontenelle, H. Muzik, A. R. P. Paterson, H. Unger, L. W. Brox, and J. F. Henderson, Inhibitors of inosinate dehydrogenase activity in Ehrlich ascites tumor cells *in vitro*, *Biochem. Pharmacol.* **23**, 2727–2735 (1974).

652. T. Kuraishi and R. N. Castle, The synthesis of imidazo[4,5-*e*]-, *s*-triazolo[4,3-*b*]- and tetrazolo-[1,5-*b*]pyridazines, *J. Heterocyl. Chem.* **1**, 42–47 (1964).

653. D. M. Halverson and R. N. Castle, The synthesis of 3-(β-D-ribofuranosyl)-imidazo[4,5-*c*]pyridazines, *J. Heterocyl. Chem.* **11**, 39–42 (1974).

654. H. Murakami and R. N. Castle, The synthesis of imidazo[4,5-*c*]- and *v*-triazolo[4,5-*c*]pyridazines, *J. Heterocyl. Chem.* **4**, 555–563 (1967).

655. M. Tisler and B. Stanovnik, Azolo- and azinopyridazines and some oxa and thia analogs, in: *The Chemistry of Heterocyclic Compounds* (R. N. Castle, ed.), Vol. 27, pp. 761–1056, John Wiley and Sons, New York (1973).

656. J. A. Carbon, Synthesis of some imidazo[4,5-*d*]pyridazines and imidazo-[4,5-*d*]triazolo-[4,3-*b*]pyridazines, *J. Org. Chem.* **25**, 579–582 (1960).

657. J. A. Carbon, The preparation of several 4-substituted imidazo [4,5-*d*]pyridazines as possible purine antimetabolites, *J. Am. Chem. Soc.* **80**, 6083–6088 (1958).

658. R. P. Gagnier, M. J. Halat, and B. A. Otter, Synthesis and NMR studies of some imidazo[4,5-*d*]pyridazine nucleosides, *J. Heterocyl. Chem.* **21**, 481–489 (1984).

659. C. Tapiero, J.-L. Imbach, R. P. Panzica, and L. B. Townsend, The synthesis of 1-(β-D-ribofuranosyl)imidazo[4,5-*d*]pyridazine-4,7-dione *via* ring closure of an imidazole nucleoside, *J. Carbohydr. Nucleosides Nucleotides* **3**, 191–195 (1976).

660. P. D. Cook, P. Dea, and R. K. Robins, Synthesis of imidazo[4,5-*d*]pyridazine nucleosides related to inosine, *J. Heterocyl. Chem.* **15**, 1–8 (1978).

661. R. G. Jones, Reactions of hydrazine with heterocyclic 1,2-dicarboxylic acids and esters, *J. Am. Chem. Soc.* **78**, 159–163 (1956).

662. E. M. Acton, K. J. Ryan, and L. Goodman, Synthesis of C-nucleoside analogues by 1,3-dipolar addition of a 1-diazo-sugar to acetylenes, *J. Chem. Soc., Chem. Commun.* **1970**, 313–314.

663. N. R. Patel, W. M. Rich, and R. N. Castle, The synthesis of ω-dialkylaminoalkylaminopyrazino[2,3-d]-, pyrido[2,3-d]-, imidazo[4,5-c]- and imidazo[4,5-d]pyridazines, *J. Heterocycl. Chem.* **5**, 13–24 (1968).

664. R. P. Gagnier, M. J. Halat, and B. A. Otter, 1-(β-D-ribofuranosyl)imidazo[4,5-d]pyridazin-4(5H)-one: A new analogue of inosine, *J. Heterocyl. Chem.* **19**, 221–223 (1982).

665. J. A. Montgomery and H. J. Thomas, The use of the allyl group as a blocking group for the synthesis of N-substituted purines, *J. Org. Chem.* **30**, 3235–3236 (1965).

666. A. Guranowski, J. A. Montgomery, G. L. Cantoni, and P. K. Chiang, Adenosine analogues as substrates and inhibitors of 2-adenosylhomocysteine hydrolase, *Biochemistry* **20**, 110–115 (1981).

667. L. L. Bennett, Jr. and J. A. Montgomery, Design of anticancer agents. Problems and approaches, in: *Methods in Cancer Research* (H. Busch, ed.), Vol. 3, pp. 549–631, Academic Press, New York (1967).

668. H. Dorn and H. Dilcher, Potential cytostatic agents. XV. Synthesis and methylation of 1H-pyrazolo[3,4-b]pyrazines, *Justus Liebigs Ann. Chem.* **717**, 118–123 (1968).

669. I. A. Korbukh, M. N. Preobrazhenskaya, Kh. Dorn, N. G. Kondakova, and N. P. Kostyuchenko, Glycosides of pyrazole and condensed pyrazole systems. VI. Glycosides of pyrazolo[3,4-b]pyrazine, *Zh. Org. Khim.* **10**, 1095–1101 (1974).

670. L. Cecchi, A. Costanzo, L. P. Vettori, G. Auzzi, F. Bruni, and F. DeSio, Synthesis of 1-N-glycosides of 3-phenylpyrazolo[3,4-b]pyrazine, *Farmaco. Ed. Sci.* **38**, 24–28 (1983).

671. I. A. Korbukh, N. G. Yakunina, H. Dorn, and M. N. Preobrazhenskaya, Glycosides of pyrazole and condensed pyrazole heterocyclic systems, XIV. Synthesis of N-oxides of pyrazolo-[3,4-b]pyrazine and its riboside, *Zh. Org. Khim.* **12**, 900–903 (1976).

672. R. K. Robins, The purines and related ring systems, in: *Heterocyclic Compounds* (R. C. Elderfield, ed.), Vol. 8, pp. 162–442, John Wiley and Sons, New York (1967).

673. W. H. Prusoff and A. D. Welch, Studies on the mechanism of action of 6-azathymine, II. Azathymine deoxyriboside, a microbial inhibitor, *J. Biol. Chem.* **218**, 929–939 (1956).

674. W. H. Prusoff, Studies on the mechanism of action of azathymine, III. Relationship between incorporation into deoxypentose nucleic acid and inhibition, *J. Biol. Chem.* **226**, 901–910 (1957).

675. J. Davoll and K. A. Kerridge, The preparation of some 1- and 2-β-D-ribofuranosylpyrazolo-[3,4-d]pyrimidines, *J. Chem. Soc.* **1961**, 2589–2591.

676. C. C. Cheng and R. K. Robins, Potential purine antagonists. VI. Synthesis of 1-alkyl- and 1-aryl-4-substituted pyrazolo[3,4-d]pyrimidines, *J. Org. Chem.* **21**, 1240–1256 (1956).

677. P. Schmidt, K. Eichenberger, M. Wilhelm, and J. Druey, Chemotherapeutic studies in the heterocyclic series. XXVI. Pyrazolopyrimidines. 4. Aminopyrazolo[3,4-d]pyrimidines, *Helv. Chim. Acta* **42**, 763–772 (1959).

678. C. C. Cheng and R. K. Robins, Potential purine antagonists. XII. Synthesis of 1-alkyl-(aryl)-4,6-disubstituted pyrazolo[3,4-d]pyrimidines, *J. Org. Chem.* **23**, 852–861 (1958).

679. J. A. Montgomery, S. J. Clayton, and W. E. Fitzgibbon, Jr., The ribonucleosides of 4-aminopyrazolo[3,4-d]pyrimidine, *J. Heterocycl. Chem.* **1**, 215–216 (1964).

680. R. A. Earl and L. B. Townsend, Pyrazolopyrimidine nucleosides. Part VII. The synthesis of certain pyrazolo[3,4-d]pyrimidine nucleosides related to the nucleoside antibiotics toyocamycin and sangivamycin, *J. Heterocycl. Chem.* **11**, 1033–1039 (1974).

681. S. M. Hecht, R. B. Frye, D. Werner, T. Fukui, and S. D. Hawrelak, The synthesis and biological activity of pyrazolo[3,4-d]pyrimidine nucleosides and nucleotides related to tubercidin, toyocamycin, and sangivamycin, *Biochemistry* **15**, 1005–1015 (1976).

682. F. F. Blanko, I. A. Korbukh, and M. N. Preobrazhenskaya, Ribosylation of 4-methylthiopyrazolo[3,4-d]pyrimidine, *Zh. Org. Khim.* **12**, 1132–1133 (1976).

683. R. K. Robins, Potential purine antagonists. I. Synthesis of some 4,6-substituted pyrazolo[3,4-d]pyrimidines, *J. Am. Chem. Soc.* **78**, 784–790 (1956).

684. I. A. Korbukh, N. G. Yakunina, and M. N. Preobrazhenskaya, Synthesis and reactions of the nucleosides of 4- and 4,6-substituted pyrazolo[3,4-d]pyrimidines, *Bioorg. Khim.* **6**, 1632–1638 (1980).

685. I. A. Korbukh, N. G. Yakunina, and M. N. Preobrazhenskaya, Synthesis of 1-β-D-xylofurano-sides and 1-α-D-arabinofuranosides of 4-substituted pyrazolo[3,4-d]pyrimidines, *Bioorg. Khim,* **11**, 1656–1660 (1985).

686. J. E. Rideout, T. A. Krenitsky, and G. B. Elion, 4-Substituted thio-1-β-D-ribofuranosylpyr-azolo[3,4-d]pyrimidines, intermediates, pharmaceutical formulations and medical uses. Eur. Patent Appl. 21,293 (Jan. 1981).

687. J.-L. G. Montero, G. A. Bhat, R. P. Panzica, and L. B. Townsend, Pyrazolopyrimidine nucleo-sides, Part VIII. The synthesis of certain 4-substituted pyrazolo[3,4-d]pyrimidine nucleo-sides, *J. Heterocycl. Chem.* **14**, 483–487 (1977).

688. J. E. Rideout, T. A. Krenitsky, and G. B. Elion, 4-Substituted thio-1-β-D-ribofuranosyl pyr-azolo[3,4-d]pyrimidines, intermediates, pharmaceutical formulations and medical uses, *Eur. Patent Appl.* 21,193 (1981).

689. I. A. Korbukh, Yu. N. Bulychev, and M. N. Preobrazhenskaya, Synthesis of 3-cyano-4,6-dimethylthiopyrazolo[3,4-d]pyrimidine 1-riboside, *Khim. Geterotsikl. Soedin.* **1979**, 1687–1692.

690. Yu. N. Bulychev, I. A. Korbukh, and M. N. Preobrazhenskaya, Chemical reactions in a series of trisubstituted pyrazolo[3,4-d]pyrimidines and their 1-ribosides, *Khim. Geterotsikl. Soedin.* **1980**, 243–250.

691. Yu. N. Bulychev, I. A. Korbukh, and M. N. Preobrazhenskaya, Synthesis of derivatives of pyrazolo[3,4-d]pyrimidin-3-ylacetic acid and their nucleosides, *Khim. Geterotsikl. Soedin.* **1981**, 536–545.

692. I. A. Korbukh, Y. N. Bulychev, N. G. Yakunina, and M. N. Preobrazhenskaya, The nucleosides of substituted pyrazolo[3,4-d]pyrimidines, *Nucleic Acids Res. Spec. Publ.* **9**, s73–75 (1981).

693. Yu. N. Bulychev, I. A. Korbukh, and M. N. Preobrazhenskaya, Synthesis of 3-substituted 4-methylthio and 4-aminopyrazolo[3,4-d]pyrimidines and their ribosides, *Khim. Geterotsikl. Soedin.* **1984**, 253–258.

694. P. Finlander and E. B. Pedersen, Phosphorus pentoxide in organic synthesis. VI. Phosphorus pentoxide and amine hydrochlorides in the synthesis of 1,5-dihydro-6-methyl-4H-pyr-azolo[3,4-d]pyrimidin-4-ones which are applicable for nucleoside syntheses, *Chem. Scr.* **23**, 23–28 (1984).

695. R. A. Earl and L. B. Townsend, Crystalline 2,3,5-tri-O-acetyl-β-D-ribofuranosyl chloride, *J. Carbohydr. Nucleosides Nucleotides* **1**, 177–182 (1974).

696. H. B. Cottam, C. R. Petrie, P. A. McKernan, R. J. Goebel, N. K. Dalley, R. B. Davidson, R. K. Robins, and G. R. Revankar, Synthesis and biological activity of certain 3,4-disubstituted pyrazolo[3,4-d]pyrimidine nucleosides, *J. Med. Chem.* **27**, 1119–1127 (1984).

697. C. R. Petrie, H. B. Cottam, P. A. McKernan, R. K. Robins, and G. R. Revankar, Synthesis and biological activity of 6-azacadeguomycin and certain 3,4,6-trisubstituted pyrazolo[3,4-d]py-rimidine ribonucleosides, *J. Med. Chem.* **28**, 1010–1016 (1985).

698. J. D. Anderson, N. K. Dalley, G. R. Revankar, and R. K. Robins, Synthesis of certain 3-alk-oxy-1-β-D-ribofuranosylpyrazolo[3,4-d]pyrimidines structurally related to adenosine, inosine and guanosine, *J. Heterocycl. Chem.* **23**, 1869–1878 (1986).

699. G. R. Revankar and L. B. Townsend, Pyrazolopyrimidine nucleosides. Part I. Synthesis of pyrazolopyrimidine nucleosides related to the nucleoside antibiotic tubercidin, *J. Chem. Soc. (C)* **1971**, 2240–2442.

700. H. Steinmaus, Antihyperuricemic 7-β-D-ribofuranosyl-4,6-dihydroxypyrazolo[3,4-d]pyrimi-dine, Ger. Offen. 2,224,379 (Nov. 1973) [*CA* **80**, 48338v (1974)].

701. H. Steinmaus, Antihyperuricemic 7-β-D-ribofuranosyl-4,6-dihydroxypyrazolo[3,4-d]pyrimi-dine, French Patent 2,184,957 (1973).

702. H. Steinmaus, Allopurinol β-riboside, Ger. Offen. 2,226,673 (Dec. 1973) [*CA* **80**, 60154q (1974)].

703. T. A. Krenitsky, G. B. Elion, R. A. Strelitz, and G. H. Hitchings, Ribonucleosides of allopurinol and oxoallopurinol. Isolation from human urine, enzymatic synthesis, and characterization, *J. Biol. Chem.* **242**, 2675–2682 (1967).

704. H. W. Hamilton and J. A. Bristol, C_4-Substituted 1-β-D-ribofuranosylpyrazolo[3,4-d]pyrimi-dines as adenosine agonist analogues, *J. Med. Chem.* **26**, 1601–1606 (1983).

705. F. W. Lichtenthaler, P. Voss, and A. Heerd, Nucleosides, XX. Stannic chloride catalyzed

glycosylations of silylated purines with fully acylated sugars, *Tetrahedron Lett.* **1974**, 2141–2144.

706. F. W. Lichtenthaler and E. Cuny, Nucleosides, 38. The ribonucleosides of allopurinol, *Chem. Ber.* **114**, 1610–1623 (1981).

707. E. Cuny and F. W. Lichtenthaler, Synthesis of allopurinol-ribosides, *Nucleic Acids Res. Spec. Publ.* **1**, s25–28 (1975).

708. H. C. Marshmann and H. G. Horn, Silicon-29 chemical shifts of trimethylsilyl esters of inorganic and organic acids, *Z. Naturforsch., B* **27**, 1448–1451 (1972).

709. H. B. Cottam, G. R. Revankar, and R. K. Robins, A convenient synthesis of 6-amino-1-β-D-ribofuranosylpyrazolo[3,4-*d*]pyrimidin-4-one and related 4,6-disubstituted pyrazolopyrimidine nucleosides, *Nucleic Acids Res.* **11**, 871–882 (1983).

710. B. G. Ugarkar, H. B. Cottam, P. A. McKernan, R. K. Robins, and G. R. Revankar, Synthesis and antiviral/antitumor activities of certain pyrazolo[3,4-*d*]pyrimidine-4(5*H*)-selone nucleosides and related compounds, *J. Med. Chem.* **27**, 1026–1030 (1984).

711. R. P. Panzica, G. A. Bhat, R. A. Earl, J.-L. G. Montero, L. W. Roti Roti, and L. B. Townsend, The chemistry and biological activity of certain 4-substituted and 3,4-disubstituted pyrazolo[3,4-*d*]pyrimidine nucleosides, in: *Chemistry and Biology of Nucleosides and Nucleotides* (R. E. Harmon, R. K. Robins, and L. B. Townsend, eds.), pp. 121–134, Academic Press, New York (1978).

712. G. A. Bhat and L. B. Townsend, Synthesis of certain fluorescent tricyclic nucleosides derived from pyrazolo[3,4-*d*]pyrimidine nucleosides, *J. Chem. Soc., Perkin Trans. 1* **1981**, 2387–2393.

713. G. A. Bhat, J.-L. G. Montero, R. P. Panzica, L. L. Wotring, and L. B. Townsend, Pyrazolopyrimidine nucleosides. 12. Synthesis and biological activity of certain pyrazolo[3,4-*d*]pyrimidine nucleosides related to adenosine, *J. Med. Chem.* **24**, 1165–1172 (1981).

714. J. L. Rideout, T. A. Krenitsky, G. W. Koszalka, N. K. Cohn, E. Y. Chao, G. B. Elion, V. S. Latter, and R. B. Williams, Pyrazolo[3,4-*d*]pyrimidine ribonucleosides as anticoccidials. 2. Synthesis and activity of some nucleosides of 4-(alkylamino)-1*H*-pyrazolo[3,4-*d*]pyrimidines, *J. Med. Chem.* **25**, 1041–1044 (1982).

715. R. A. Earl, R. P. Panzica, and L. B. Townsend, Pyrazolopyrimidine nucleosides. Part III. Synthesis of 1- and 2-(β-D-ribofuranosyl)-pyrazolo[3,4-*d*]pyrimidines from pyrazole nucleoside derivatives, *J. Chem. Soc., Perkin Trans. 1* **1972**, 2672–2677.

716. C. C. Cheng, Pyrazoles. II. Reactions of 1-methyl-5-amino-4-pyrazolecarboxamide and nitrous acid. Introduction of a nitro group at position 5 in the pyrazole ring, *J. Heterocycl. Chem.* **5**, 195–197 (1968).

717. C. W. Noell and C. C. Cheng, Pyrazoles, III. Antileukemic activity of 3,(3,3-dimethyl-1-triazeno)pyrazole-4-carboxamide, *J. Med. Chem.* **12**, 545–546 (1969).

718. H. Tanaka, T. Hayashi, and K. Nakayama, Riboside and ribotide of 5 (or 3)-aminopyrazole-4-carboxamide. The chemical preparation from the corresponding 4-hydroxypyrazolo-[3,4-*d*]pyrimidine derivatives by ring opening reaction. *Agric. Biol. Chem.* **37**, 1731–1736 (1973).

719. R. R. Schmidt, J. Karg, and W. Guilliard, Hydrazinoribose—an intermediate for the specific formation of natural nucleosides and their derivatives, *Angew. Chem.* **87**, 69 (1975).

720. R. Schmidt and K. Klemm, Hydrazinosugar derivative, Ger. Offen. 2,426,279 (1975).

721. R. R. Schmidt, J. Karg, and W. Guilliard, Glycosylhydrazines, 2. Synthesis of pyrazole nucleosides via ribosylhydrazines, *Chem. Ber.* **110**, 2433–2444 (1977).

722. R. R. Schmidt, W. Guilliard, and J. Karg, Glycosylhydrazines, 3. Pyrazolo[3,4-*d*]pyrimidine nucleosides, *Chem. Ber.* **110**, 2445–2455 (1977).

723. G. Shaw, Purines, pyrimidines and glyoxalines, Part I. New synthesis of glyoxalines and pyrimidines, *J. Chem. Soc.* **1955**, 1834–1840.

724. K.-H. Jung, R. R. Schmidt, and D. Heermann, Glycosylhydrazines, 5. Synthesis of pyrazole-, pyrazolo[3,4-*d*]pyrimidine-, and 1-*H*-1,2,4-triazole nucleoside-5′-amide derivatives via riburonamide hydrazones, *Chem. Ber.* **114**, 2834–2843 (1981).

725. R. R. Schmidt, W. Guilliard, and D. Heermann, Glycohydrazines, 6. Synthesis of pyrazole, pyrazolo[3,4-*d*]pyrimidine, and 1-*H*-1,2,4-triazolegluconucleosides from glucose hydrazones, *Liebigs. Ann. Chem.* **1981**, 2309–2317.

726. R. R. Schmidt, W. Guilliard, D. Heermann, and M. Hoffmann, Synthesis of pyrazole, pyrazolo[3,4-*d*]pyrimidine, and 1*H*-1,2,4-triazole arabinonucleosides from 2,3,5-tri-*O*-benzyl-D-arabinose hydrazone, *J. Heterocycl. Chem.* **20**, 1447–1451 (1983).

727. S. Oshiro, Y. Urabe, T. Wakabayashi, and K. Okumura, 4-(4-Amino-1*H* or 2*H*-pyrazolo-[3,4-*d*]pyrimidin-1 or 2-yl)-2,3-dihydroxybutyric acid, Jpn. Patent 73, 30,075 (Sept. 1973 [*CA* **80**, 121279h (1974)].

728. S. Oshiro, Y. Urabe, T. Wakabayashi, and K. Okumura, 5-(4-Amino-1*H* or 2*H*-pyrazolo-[3,4-*d*]pyrimidin-1 or 2-yl)-5-deoxypentose derivatives. Jpn. Patent 73, 30,074 (Sept. 1973) [*CA* **80**, 121284f (1974)].

729. H. Takahashi, N. Nimura, and H. Ogura, A novel one-step synthesis of thioquinazoline glycosides and pyrazolopyrimidine glycoside analogs, *Chem. Pharm. Bull.* **27**, 1143–1146 (1979).

730. H. Ogura, H. Takahashi, and E. Kudo, C-glycosyl nucleosides. X. Synthesis of glycosylamino-pyrimido[4,5-*e*]-1,3,4-thiadiazines and their desulfurization, *J. Carbohydr. Nucleosides Nucleotides* **5**, 329–341 (1978).

731. H. Ogura, H. Takahashi, and M. Sakaguchi, Synthesis of nucleoside analogs from hydrazine derivatives, *Nucleic Acids Res. Spec. Publ.* **5**, s251–254 (1978).

732. F. Seela and H. Steker, Facile synthesis of 2'-deoxyribofuranosides of allopurinol and 4-amino-1*H*-pyrazolo[3,4-*d*]pyrimidine via phase-transfer glycosylation, *Helv. Chim. Acta* **68**, 563–570 (1985).

733. F. Seela and H. Steker, Synthesis of the β-D-deoxyribofuranoside of 6-amino-1*H*-pyrazolo[3,4-*d*]pyrimidin-4(5*H*)-one—a new isoster of 2'-deoxyguanosine, *Heterocycles* **23**, 2521–2524 (1985).

734. F. Seela and H. Steker, Synthesis of 2'-deoxyribofuranosides of 8-aza-7-deazaguanine and related pyrazolo[3,4-*d*]pyrimidines, *Helv. Chim. Acta* **69**, 1602–1613 (1986).

735. F. Seela and H. Driller, 8-Aza-7-deaza-2',3'-dideoxyguanosine: Deoxygenation of its 2'-deoxy-β-D-ribofuranoside, *Helv. Chim. Acta.* **71**, 757–761 (1988).

736. F. Seela and H. Steker, Synthesis and hydrolytic stability of 4-substituted pyrazolo[3,4-*d*]pyrimidine 2'-deoxyribofuranosides, *J. Chem. Soc., Perkin Trans. 1* **1985**, 2573–2576.

737. F. Seela and K. Kaiser, 8-Aza-7-deaza-2',3'-dideoxyadenosine: Synthesis and conversion into allopurinol 2',3'-dideoxyribofuranoside, *Chem. Pharm. Bull.* **36**, 4153–4156 (1988).

738. A. Kornberg, I. Liebermann, and E. S. Simms, Enzymatic synthesis of purine nucleotides, *J. Biol. Chem.* **215**, 417–427 (1955).

739. C. N. Remy, W. T. Remy, and J. M. Buchanan, Biosynthesis of the purines, VIII. Enzymatic synthesis and utilization of α-5-phosphoribosylpyrophosphate, *J. Biol. Chem.* **217**, 885–895 (1955).

740. I. Liebermann, A. Kornberg, and E. S. Simms, Enzymatic synthesis of pyrimidine nucleotides. Orotidine 5'-phosphate and uridine 5'-phosphate, *J. Biol. Chem.* **215**, 403–415 (1955).

741. J. Preiss and P. Handler, Enzymatic synthesis of nicotinamide mononucleotide, *J. Biol. Chem.* **225**, 759–770 (1957).

742. S. C. Hartman, B. Levenberg, and J. M. Buchanan, Biosynthesis of purines, XI. Structure, enzymatic synthesis, and metabolism of glycinamide ribotide and (α-*N*-formyl)-glycinamide ribotide, *J. Biol. Chem.* **221**, 1057–1070 (1956).

743. D. A. Goldthwait, R. A. Peabody, and G. R. Greenberg, Glycine ribotide intermediates in the *de novo* synthesis of inosinic acid, *J. Am. Chem. Soc.* **76**, 5258–5259 (1954).

744. J. L. Way and R. E. Parks, Jr., Enzymatic synthesis of 5'-phosphate nucleotides of purine analogues, *J. Biol. Chem.* **231**, 467–480 (1958).

745. C. Auscher, N. Mercier, C. Pasquier, and F. Delbarre, Allopurinol and thiopurinol: Effect *in vivo* on urinary oxypurine excretion and rate of synthesis of their ribonucleotides in different enzymatic deficiencies, *Adv. Exp. Med. Biol.* **41B**, 657–662 (1974).

746. R. J. McCollister, W. R. Gilbert, Jr., D. M. Ashton, and J. B. Wyngaarden, Pseudofeedback inhibition of purine synthesis by 6-mercaptopurine ribonucleotide and other purine analogues, *J. Biol. Chem.* **239**, 1560–1563 (1964).

747. E. D. Korn, C. N. Remy, H. C. Wasilejko, and J. M. Buchanan, Biosynthesis of the purines, VII. Synthesis of nucleotides from bases by partially purified enzymes, *J. Biol. Chem.* **217**, 875–883 (1955).

748. T. A. Krenitsky, G. W. Koszalka, and J. V. Tuttle, Purine nucleoside synthesis, an efficient method employing nucleoside phosphorylases, *Biochemistry* **20**, 3615–3621 (1981).

749. Ajinomoto Co., Inc., Ribofuranosylpurine or deoxyribofuranosylpurine derivatives, Japan Kokai Tokkyo Koho JP 83 63,393 (April 1983).

750. J. A. Fyfe, R. L. Miller, and T. A. Krenitsky, Kinetic properties and inhibition of orotidine 5'-phosphate decarboxylase. Effect of some allopurinol metabolites on the enzyme, *J. Biol. Chem.* **248**, 3801–3809 (1973).

751. T. Spector, R. L. Miller, J. A. Fyfe, and T. A. Krenitsky, GMP synthetase from *Escherichia coli* B-96 interactions with substrate analogs, *Biochim. Biophys. Acta* **370**, 585–591 (1974).

752. T. A. Krenitsky, R. Papaioannou, and G. B. Elion, Human hypoxanthine phosphoribosyltransferase, I. Purification, properties, and specificity, *J. Biol. Chem.* **244**, 1263–1270 (1969).

753. Kyowa Hakko Kogyo Co., Ltd., Japan, Process for producing 1-β-D-ribofuranoside 5'-phosphoric and pyrophosphoric acid esters of 1-*H*-pyrazolo[3,4-*d*]pyrimidines, Brit. Patent 1,169,624 (Nov. 1969).

754. Kyowa Ferm. Ind., Preparation of 4-hydroxy-1-β-D-ribofuranosylpyrazolo[3,4-*d*]pyrimidine, French Patent 2,014,141 (April 1970).

755. K. Nakayama, Microbial conversion of pyrazolopyrimidines to their ribotides, Jpn. Patent 71 20,756 (June 1971) [*CA* **75**, 97268c (1971)].

756. H. Tanaka and K. Nakayama, Production of ribotides of 4-hydroxy- and 4-aminopyrazolo[3,4-*d*]pyrimidine by *Brevibacterium ammoniagenes*, *Agric. Biol. Chem.* **36**, 1405–1412 (1972).

757. K. Nakayama, A. Furuya, and F. Kato, Ribosides of purine and pyrimidine analogs, Ger. Offen. 2,209,078 (Nov. 1972); [*CA* **78**, 27908z (1973)].

758. P. J. Curtis and D. R. Thomas, The metabolism of exogenous adenine and purine analogues by fungi, *Biochem. J.* **82**, 381–384 (1962).

759. O. Kanamitsu, Microbial production of ribosides of pyrazolo[3,4-*d*]pyrimidines, Jpn. Kokai 73 91,284 (Nov. 1973) [*CA* **80**, 94260m (1974)].

760. K. Nakayama and H. Tanaka, 4-Hydroxy-1-β-D-ribofuranosylpyrazolo[3,4-*d*]pyrimidine as a xanthine oxidase inhibitor, Ger. Offen. 1,927,136 (Dec. 1969) [*CA* **72**, 133149b (1970)].

761. K. Nakayama and H. Tanaka, 4-Hydroxyl-1-β-D-ribofuranosylpyrazolo[3,4-*d*]pyrimidine, Jpn. Patent 71 07,699 (Feb. 1971) [*CA* **75**, 36107c (1971)].

762. T. Sakai, K. Ushio, I. Ichimoto, and S. Omata, Ribosylation of 4-hydroxypyrazolo[3,4-*d*]-pyrimidine by *Erwinia carotovora*, *Agric. Biol. Chem.* **38**, 433–438 (1974).

763. M.-T. Chenon, R. J. Pugmire, D. M. Grant, R. P. Panzica, and L. B. Townsend, Carbon-13 NMR spectra of *C*-nucleosides. II. A study on the tautomerism of formycin and formycin B by the use of CMR spectroscopy, *J. Heterocycl. Chem.* **10**, 431–433 (1973).

764. M. G. Stout, D. E. Hoard, M. J. Holman, E. S. Wu, and J. M. Siegel, Preparation of 2'-deoxyribonucleosides via nucleoside deoxyribosyl transferase, *Methods Carbohydr. Chem.* **7**, 19–24 (1976).

765. R. Cardinaud and J. Holguin, Thin-layer chromatography of purine and pyrimidine bases and deoxyribonucleoside analogues. II. *J. Chromatogr.* **115**, 673–677 (1975).

766. R. Cardinaud, Thin-layer chromatography of purine bases an deoxyribonucleoside analogues. IV. *J. Chromatogr.* **154**, 345–348 (1978).

767. I. A. Korbukh, O. V. Goryunova, Yu. V. Stukalov, T. P. Ivanova, Ya. V. Dobrynin, and M. N. Preobrazhenskaya, Biotransformation of 1-β-D-ribofuranosyl-4-methylthiopyrazolo[3,4-*d*]-pyrimidine and its 5'-monophosphate, *Bioorg. Khim.* (Russ.) **10**, 963–969 (1984).

768. J. D. Anderson, R. K. Robins, and G. R. Revankar, Synthesis of pyrazolo[3,4-*d*]pyrimidine ribonucleoside 3',5'-cyclic phosphates related to *c*AMP, *c*IMP and *c*GMP, *Nucleosides Nucleotides* **6**, 853–863 (1987).

769. F. Seela and K. Kaiser, Phosphoramidites of base-modified 2'-deoxyinosine isosteres and solid-phase synthesis of d(GCI*CGC) oligomers containing an ambiguous base, *Nucleic Acids Res.* **14**, 1825–1844 (1986).

770. F. Seela and K. Kaiser, 8-Aza-7-deazaadenine N^8- and N^9-(β-D-2'-deoxyribofuranosides): Building blocks for automated DNA synthesis and properties of oligodeoxyribonucleotides, *Helv. Chim. Acta.* **71**, 1813–1823 (1988).

771. F. Seela and H. Driller, 8-Aza-7-deaza-2'-deoxyguanosine: Phosphoramidite synthesis and properties of octanucleotides, *Helv. Chim. Acta* **71**, 1191–1198 (1988).

772. S. Sprang, R. Scheller, D. Rohrer, and M. Sundaralingam, Conformational analysis of 8-aza-nucleosides. Crystal and molecular structure of 8-azatubercidin monohydrate, a nucleoside analogue exhibiting the "high anti" conformation, *J. Am. Chem. Soc.* **100**, 2867–2872 (1978).

773. S. B. Larson, J. D. Anderson, G. R. Revankar, and R. K. Robins, Structure of a pyrazolo-pyridine nucleoside 5'-phosphate, *Acta Crystallogr., Sect. C* **44**, 191–193 (1988).

774. T. M. Savarese, D. L. Dexter, R. E. Parks, Jr., and J. A. Montgomery, 5'-Deoxy-5'-methylthio-adenosine phosphorylase-II. Role of the enzyme in the metabolism and antineoplastic action of adenine-substituted analogs of 5'-deoxy-5'-methylthioadenosine, *Biochem. Pharmacol.* **32**, 1907–1916 (1983).

775. T. S. Chan, K. Ishii, C. Iong, and H. Green, Purine excretion by mammalian cells deficient in adenosine kinase, *J. Cell. Physiol.* **81**, 315–322 (1973).

776. L. L. Wotring and L. B. Townsend, Study of the cytotoxicity and metabolism of 4-amino-3-carboxamido-1-(β-D-ribofuranosyl)pyrazolo[3,4-*d*]pyrimidine using inhibitors of adenosine kinase and adenosine deaminase, *Cancer Res.* **39**, 3018–3023 (1979).

777. R. Kojima, Anticancer activity of allopurinol ribotide, *Igaku to Seibutsugaku* **97**, 107–109 (1978) [*CA* **91**, 117296w (1979)].

778. K. C. Agarwal, R. E. Parks, Jr., and L. B. Townsend, Adenosine analogs and human platelets—II. Inhibition of ADP-induced aggregation by carbocyclic adenosine and imidazole-ring modified analogs. Significance of alterations in the nucleotide pools, *Biochem. Pharmacol.* **28**, 501–510 (1979).

779. L. L. Bennett, Jr., P. W. Allan, D. Smithers, and M. H. Vail, Resistance to 4-aminopyr-azolo[3,4-*d*]pyrimidine, *Biochem. Pharmacol.* **18**, 725–740 (1969).

780. J. Wierzchowski and D. Shugar, Sensitive fluorimetric assay for adenosine deaminase with formycin as substrate; and substrate and inhibitor properties of some pyrazolopyrimidine and related analogues, *Z. Naturforsch., C* **38**, 67–73 (1983).

781. H. P. Schnebli, D. L. Hill, and L. L. Bennett, Jr., Purification and properties of adenosine kinase from human tumor cells of type H.Ep. No. 2, *J. Biol. Chem.* **242**, 1997–2004 (1967).

782. R. L. Miller, D. L. Adamczyk, W. H. Miller, G. W. Koszalka, J. L. Rideout, L. M. Beacham III, E. Y. Chao, J. J. Haggerty, T. A. Krenitsky, and G. B. Elion, Adenosine kinase from rabbit liver, II. Substrate and inhibitor specificity, *J. Biol. Chem.* **254**, 2346–2352 (1979).

783. D. F. Smee, P. A. McKernan, L. D. Nord, R. C. Willis, C. R. Petrie, T. M. Riley, G. R. Revankar, R. K. Robins, and R. A. Smith, Novel pyrazolo[3,4-*d*]pyrimidine nucleoside analog with broad-spectrum antiviral activity, *Antimicrob. Agents Chemother.* **31**, 1535–1541 (1987).

784. T. A. Bektemirov, E. V. Chekunova, I. A. Korbukh, Yu. N. Bulychev, N. G. Yakunina, and M. N. Preobrazhenskaya, Antiviral activity of substituted 6-methylthiopyrazolo[3,4-*d*]pyrimi-dines and their ribosides, *Acta Virol.* **25**, 326–329 (1981).

785. Ya. V. Dobrynin, T. A. Bektemirov, T. P. Ivanova, E. V. Chekunova, O. G. Andzhaparidze, I. A. Korbukh, Yu. N. Bulychev, N. G. Yakunina, and M. N. Preobrazhenskaya, Cytotoxic and antiviral activity of 4- and 3,4-disubstituted 6-(methylthio)pyrazolo[3,4-*d*]pyrimidines and their ribosides, *Khim.-Farm. Zh.* **14**, 10–15 (1980).

786. J. L. Avila, M. A. Polegre, A. Avila, and R. K. Robins, Action of pyrazolopyrimidine derivatives on American *Leishmania* spp. promastigotes, *Comp. Biochem. Physiol.* **83C**, 285–289 (1986).

787. J. L. Avila, M. A. Polegre, and R. K. Robins, Action of pyrazolopyrimidine derivatives on *Trypanosoma rangeli* culture forms, *Comp. Biochem. Physiol.* **83C**, 291–294 (1986).

788. W. Trager, Nutrition and biosynthetic capabilities of flagellates. Problems of *in vitro* cultivation and differentiation, *Ciba Found. Symp.* **1974**, 225–254.

789. J. J. Marr, R. L. Berens, and D. J. Nelson, Purine metabolism in *Leishmania donovani* and *Leishmania braziliensis*, *Biochim. Biophys. Acta* **544**, 360–371 (1978).

790. W. E. Gutteridge and M. Gaborak, A. re-examination of purine and pyrimidine synthesis in the three main forms of *Trypanosoma cruzi*, *Int. J. Biohem.* **10**, 415–422 (1979).

791. C. R. Ceron, R. A. Caldas, C. R. Felix, M. H. Mundim, and I. Roitman, Purine metabolism in trypanosomatides, *J. Protozool.* **26**, 479–483 (1979).

792. R. L. Berens, J. J. Marr, S. W. LaFon, and D. J. Nelson, Purine metabolism in *Trypanosoma cruzi*, *Mol. Biochem. Parasitol.* **3**, 187–196 (1981).

793. M. A. Pfaller and J. J. Marr, Antileishmanial effect of allopurinol, *Antimicrob. Agents. Chemother.* **5**, 469–472 (1974).

794. J. J. Marr and R. L. Berens, Antileishmanial effect of allopurinol. II. Relationship of adenine

metabolism in *Leishmania* species to the action of allopurinol, *J. Infect. Dis.* **136**, 724–732 (1977).

795. D. J. Nelson, C. J. L. Bugge, G. B. Elion, R. L. Berens, and J. J. Marr, Metabolism of pyrazolo[3,4-*d*]pyrimidines in *Leishmania braziliensis* and *Leishmania donovani*, allopurinol, oxipurinol, and 4-aminopyrazolo[3,4-*d*]pyrimidine, *J. Biol. Chem.* **254**, 3959–3964 (1979).

796. J. J. Marr and R. L. Berens, Pyrazolopyrimidine metabolism in the pathogenic Trypanosomatidae, *Mol. Biochem. Parasitol.* **7**, 339–356 (1983).

797. B. C. Walton, J. Harper III, and R. A. Neal, Effectiveness of allopurinol against *Leishmania braziliensis panamensis* in *aotus trivirgatus*, *Am. J. Trop. Med. Hyg.* **32**, 46–50 (1983).

798. R. L. Berens, J. J. Marr, F. Steela Da Cruz, and D. J. Nelson, Effect of allopurinol on *Trypanosoma cruzi*: Metabolism and biological activity in intracellular and bloodstream forms, *Antimicrob. Agents. Chemother.* **22**, 657–661 (1982).

799. D. J. Nelson, S. W. LaFon, G. B. Elion, J. J. Marr, and R. L. Berens, Comparative metabolism of a new antileishmanial agent, allopurinol riboside, in the parasite and the host cell, *Adv. Exp. Med. Biol.* **122B**, 7–12 (1980).

800. J. J. Marr, R. L. Berens, and D. J. Nelson, Antitrypanosomal effect of allopurinol: Conversion *in vivo* to aminopyrazolopyrimidine nucleotides by *Trypanosoma cruzi*, *Science* **201**, 1018–1020 (1978).

801. T. Spector, T. E. Jones, and G. B. Elion, Specificity of adenylosuccinate synthetase and adenylosuccinate lyase from *Leishmania donovani*, selective amination of an antiprotozoal agent, *J. Biol. Chem.* **254**, 8422–8426 (1979).

802. T. Spector, R. L. Berens, and J. J. Marr, Adenylosuccinate synthetase and adenylosuccinate lyase from *Trypanosoma cruzi*, *Biochem. Pharmacol.* **31**, 225–229 (1982).

803. J. L. Avila, A. Avila, and M. A. Casanova, Differential metabolism of allopurinol and derivatives in *Trypanosoma rangeli* and *T. cruzi* culture forms, *Mol. Biochem. Parasitol.* **4**, 265–272 (1981).

804. S. W. LaFon, D. J. Nelson, R. L. Berens, and J. J. Marr, Purine and pyrimidine salvage pathways in *Leishmania donovani*, *Biochem. Pharmacol.* **31**, 231–238 (1982).

805. D. J. Nelson, S. W. LaFon, J. V. Tuttle, W. H. Miller, R. L. Miller, T. A. Krenitsky, G. B. Elion, R. L. Berens, and J. J. Marr, Allopurinol ribonucleoside as an antileishmanial agent; biological effects, metabolism and enzymatic phosphorylation, *J. Biol. Chem.* **254**, 11544–11549 (1979).

806. J. L. Avila, A. Avila, E. Munoz, and H. Monzon, *Trypanosoma cruzi*: 4-Aminopyrazolopyrimidine in the treatment of experimental Chagas' disease, *Exp. Parasitol.* **56**, 236–240 (1983).

807. J. L. Avila, New rational approaches to Chagas' disease chemotherapy, *Interciencia* **8**, 405–417 (1983).

808. J. L. Avila and M. A. Casanova, Comparative effects of 4-aminopyrazolopyrimidine, its 2'-deoxyriboside derivative, and allopurinol on *in vitro* growth of American *Leishmania* species, *Antimicrob. Agents Chemother.* **22**, 380–385 (1982).

809. J. J. Marr and R. L. Berens, Antileishmanial action of 4-thiopyrazolo[3,4-*d*]pyrimidine and its ribonucleoside, *Biochem. Pharmacol.* **31**, 143–148 (1982).

810. T. Spector and T. E. Jones, Guanosine 5'-monophosphate reductase from *Leishmania donovani*, a possible chemotherapeutic target, *Biochem. Pharmacol.* **31**, 3891–3897 (1982).

811. J. L. Rideout, T. A. Krenitsky, E. Y. Chao, G. B. Elion, R. B. Williams, and V. S. Latter, Pyrazolo[3,4-*d*]pyrimidine ribonucleosides as anticoccidials. 3. Synthesis and activity of some nucleosides of 4-[(aryl-alkenyl)thio]pyrazolo[3,4-*d*]pyrimidines, *J. Med. Chem.* **26**, 1489–1494 (1983).

812. J. F. Henderson and I. G. Junga, The metabolism of 4-aminopyrazolo[3,4-*d*]pyrimidine in normal and neoplastic tissues, *Cancer Res.* **21**, 118–129 (1961).

813. J. F. Henderson and I. G. Junga, Effects of 4-aminopyrazolo[3,4-*d*]pyrimidine and 1-methyl-4-aminopyrazolo[3,4-*d*]pyrimidine on nucleic acid purine metabolism in ascites tumors, *Cancer Res.* **21**, 173–177 (1961).

814. R. E. A. Gadd and J. F. Henderson, Studies of the binding of adenine to adenine phosphoribosyltransferase, *Can. J. Biochem.* **48**, 295–301 (1970).

815. H. A. Simmonds, P. J. Hatfield, J. S. Cameron, A. S. Jones, and A. Cadenhead, Metabolic studies of purine metabolism in the pig during the oral administration of guanine and allopurinol, *Biochem. Pharmacol.* **22**, 2537–2551 (1973).

816. D. O. Schachtschabel and J. J. Ferro, The effect of 4-aminopyrazolo[3,4-*d*]pyrimidine [APP] and guanine on Ehrlich ascites cells in culture, *Exp. Cell. Res.* **48**, 319–326 (1967).

817. S. Reiter, H. A. Simmonds, D. R. Webster, and A. R. Watson, On the metabolism of allopurinol: Formation of allopurinol-1-riboside in purine nucleoside phosphorylase deficiency, *Biochem. Pharmacol.* **32**, 2167–2174 (1983).

818. D. J. Nelson and G. B. Elion, Metabolic studies of high doses of allopurinol in humans, *Adv. Exp. Med. Biol.* **165A**, 167–170 (1984).

819. D. J. Nelson and G. B. Elion, Metabolism of allopurinol-6-^{14}C. Lack of incorporation of allopurinol into nucleic acids, *Biochem. Pharmacol.* **24**, 1235–1237 (1975).

820. I. H. Fox, J. B. Wyngaarden, and W. N. Kelley, Depletion of erythrocyte phosphoribosylpyrophosphate in man. A newly observed effect of allopurinol, *New Engl. J. Med.* **283**, 1177–1182 (1970).

821. E. W. Holmes, M. D. Pehlke, and W. N. Kelley, Human IMP dehydrogenase, kinetics and regulatory properties, *Biochem. Biophys. Acta* **364**, 209–217 (1974).

822. W. N. Kelley and J. B. Wyngaarden, Effects of allopurinol and oxipurinol on purine synthesis in cultured human cells, *J. Clin. Invest.* **49**, 602–609 (1970).

823. W. Kaiser and K. Stocker, Purine and pyrimidine biosynthesis in *Neurospora crassa* and human skin fibroblasts. Alteration by ribosides and ribotides of allopurinol and oxipurinol, *Adv. Exp. Med. Biol.* **41B**, 629–635 (1974).

824. D. J. Nelson, C. J. L. Bugge, H. C. Krasny, and G. B. Elion, Formation of nucleotides of [6-^{14}C]allopurinol and [6-^{14}C]oxipurinol in rat tissues and effects on uridine nucleotide pools, *Biochem. Pharmacol.* **22**, 2003–2022 (1973).

825. G. B. Elion and D. J. Nelson, Ribonucleotides of allopurinol and oxipurinol in rat tissues and their significance in purine metabolism, *Adv. Exp. Med. Biol.* **41B**, 639–652 (1974).

826. L. Sweetman, Urinary and cerebrospinal fluid oxypurine levels and allopurinol metabolism in the Lesch–Nyhan syndrome, *Fed. Proc.* **27**, 1055–1059 (1968).

827. M. Hori, E. Ito, T. Takita, G. Koyama, T. Takeuchi, and H. Umezawa, A new antibiotic, formycin, *J. Antibiot. Ser. A* **17**, 96–99 (1964).

828. G. Koyama and H. Umezawa, Formycin B and its relation to formycin, *J. Antibiot. Ser. A* **18**, 175–177 (1965).

829. S. Aizawa, T. Hidaka, N. Otake, H. Yonehara, K. Isono, N. Igarashi, and S. Suzuki, Studies on a new antibiotic, laurusin, *Agric. Biol. Chem.* **29**, 375–376 (1965).

830. N. Ishida, A. Izawa, M. Homma, K. Kumagai and Y. Shimizu, Anti-myxovirus activity of formycin B, *J. Antibiot. Ser. A* **20**, 129–131 (1967).

831. T. Sawa, Y. Fukagawa, I. Homma, T. Wakashiro, T. Takeuchi, M. Hori, and T. Komai, Metabolic conversion of formycin B to formycin A and to oxoformycin B in *Nocardia interforma*, *J. Antibiot.* **21**, 334–339 (1968).

832. R. E. Parks, Jr., J. D. Stoeckler, C. Cambor, T. M. Savarese, G. W. Crabtree, and S.-H. Chu, in: *Molecular Actions and Targets for Cancer Chemotherapeutic Agents* (A. Sartorelli, J. S. Lazo, and J. R. Bertino, eds.), pp. 229–252, Academic Press, New York (1981).

833. M. R. Sheen, B. K. Kim, and R. E. Parks, Jr., Purine nucleoside phosphorylase from human erythrocytes. III. Inhibition by the inosine analog formycin B of the isolated enzyme and of nucleoside metabolism in intact erythrocytes and sarcoma 180 cells, *Mol. Pharmacol.* **4**, 293–299 (1968).

834. L. B. Townsend, in: *Handbook of Biochemistry and Molecular Biology, Nucleic Acids*, 3rd ed. (D. G. Fasman, Ed.), Vol. 1, pp. 271–401, CRC Press, Cleveland, Ohio (1975).

835. R. A. Long, A. F. Lewis, R. K. Robins, and L. B. Townsend, Pyrazolopyrimidine nucleosides, Part II. 7-Substituted 3-β-D-ribofuranosylpyrazolo[4,3-*d*]pyrimidines related to and derived from the nucleoside antibiotics formycin and formycin B, *J. Chem. Soc. (C)* **1971**, 2441–2446.

836. G. H. Milne and L. B. Townsend, Pyrazolopyrimidine nucleosides. Part IV. Synthesis and chemical reactivity of the C-nucleoside selenoformycin B and derivatives, *J. Chem. Soc., Perkin Trans 1*, **1972**, 2677–2681.

837. M. J. Robins and E. M. Trip, Sugar-modified N^6-(3-methyl-2-butenyl)-adenosine derivatives, N^6-benzyl analogs, and cytokinin-related nucleosides containing sulfur or formycin, *Biochemistry* **12**, 2179–2187 (1973).

838. L. B. Townsend, R. A. Long, J. P. McGraw, D. W. Miles, R. K. Robins, and H. Eyring,

Pyrazolopyrimidine nucleosides. V. Methylation of the C-nucleoside antibiotic formycin and structural elucidation of products by magnetic circular dichroism spectroscopy, *J. Org. Chem.* **39**, 2023–2027 (1974).

839. A. F. Lewis and L. B. Townsend, Pyrazolo[4,3-*d*]pyrimidine nucleosides. 9. Studies on the isomeric *N*-methylformycins, *J. Am. Chem. Soc.* **102**, 2817–2822 (1980).

840. A. F. Lewis and L. B. Townsend, Pyrazolopyrimidine nucleosides. 13. Synthesis of the novel C-nucleoside 5-amino-3-(β-D-ribofuranosyl)pyrazolo[4,3-*d*]pyrimidin-7-one, a guanosine analogue related to the nucleoside antibiotic formycin B, *J. Am. Chem. Soc.* **104**, 1073–1077 (1982).

841. E. M. Acton and K. J. Ryan, Synthesis of 3-glycofuranosyl-5-aminopyrazolo[4,3-*d*]pyrimidine-7-thiones: Thioguanosine-type C-nucleosides, *J. Org. Chem.* **49**, 528–536 (1984).

842. B. G. Ugarkar, R. K. Robins, and G. R. Revankar, Synthesis of 1-methyl-3-β-D-ribofuranosyl-pyrazolo[4,3-*d*]pyrimidin-7(6*H*)-selone and certain related nucleosides and nucleotides, *Nucleosides Nucleotides* **3**, 233–244 (1984).

843. B. G. Ugarkar, G. R. Revankar, and R. K. Robins, A simple oxidation of formycin to oxoformycin and oxoformycin B. Synthesis of 6-methyloxoformycin, a C-nucleoside analog of doridosine, *J. Heterocycl. Chem.* **21**, 1865–1870 (1984).

844. A. Rosowsky, V. C. Solan, and L. J. Gudas, Improved synthesis of 2′-deoxyformycin A and studies of its *in vitro* activity against mouse lymphoma of T-cell origin, *J. Med. Chem.* **28**, 1096–1099 (1985).

845. M. J. Robins, J. R. McCarthy, Jr., R. A. Jones, and R. Mengel, Nucleic acid related compounds. 5. The transformation of formycin and tubercidin into 2′- and 3′-deoxynucleosides, *Can. J. Chem.* **51**, 1313–1321 (1973).

846. T. C. Jain, A. F. Russell, and J. C. Moffatt, Reactions of 2-acyloxyisobutyryl halides with nucleosides. III. Reactions of tubercidin and formycin, *J. Org. Chem.* **38**, 3179–3186 (1973).

847. J. G. Buchanan, D. Smith, and R. H. Wightman, C-Nucleoside studies-15. Synthesis of 3-β-D-arabinofuranosylpyrazoles and the D-arabinofuranosyl analogue of formycin, *Tetrahedron* **40**, 119–123 (1984).

848. J. G. Buchanan, D. Smith, and R. H. Wightman, C-nucleoside studies. Part 19. The synthesis of the β-D-xylofuranosyl analogue of formycin, *J. Chem. Soc., Perkin Trans. 1* **1986**, 1267–1271.

849. S. H. Chu, L. Ho, E. Chu, T. Savarese, Z. H. Chen, E. C. Rowe, and M. Y. W. Chu, 5′-Halogenated formycins as inhibitors of 5′-deoxy-5′-methylthioadenosine phosphorylase: Protection of cells against the growth-inhibitory activity of 5′-halogenated adenosines, *Nucleosides Nucleotides* **5**, 185–200 (1986).

850. P. Rainey and D. V. Santi, Metabolism and mechanism of action of formycin B in *Leishmania*, *Proc. Natl. Acad. Sci. U.S.A.* **80**, 288–292 (1983).

851. O. V. Budanova, I. A. Korbukh, and M. N. Preobrazhenskaya, Ribosylation of 7-methylthio-pyrazolo[3,4-*d*]pyrimidine, *Zh. Org. Khim.* **12** 1131 (1976).

852. O. V. Goryunova, I. A. Korbukh, and M. N. Preobrazhenskaya, Glycosides of pyrazole and condensed pyrazole heterocyclic systems. XXI. Nucleosides of 7-substituted pyrazolo-[4,3-*d*]pyrimidines, *Zh. Org. Khim.* **14**, 651–660 (1978).

853. O. V. Goryunova, I. A. Korbukh, M. N. Preobrazhenskaya, and A. I. Chernyshev, Nucleosides of 5-methylthio-6*H*-pyrazolo[4,3-*d*]pyrimidin-7-one, *Bioorg. Khim.* **5**, 1361–1368 (1979).

854. P. G. Baraldi, D. Simoni, V. Periotto, S. Manfredini, M. Guarneri, R. Manservigi, E. Cassai, and V. Bertolasi, Pyrazolo[4,3-*d*]pyrimidine nucleosides. Synthesis and antiviral activity of 1-β-D-ribofuranosyl-3-methyl-6-substituted-7*H*-pyrazolo[4,3-*d*]pyrimidin-7-ones, *J. Med. Chem.* **27**, 986–990 (1984).

855. R. K. Robins, F. W. Furcht, A. D. Grauer, and J. W. Jones, Potential purine antagonists. II. Synthesis of some 7- and 5,7-substituted pyrazolo[4,3-*d*]pyrimidines, *J. Am. Chem. Soc.* **78**, 2418–2422 (1956).

856. P. C. Jain, S. K. Chatterjee, and N. Anand, Potential purine antagonists: Part V—Synthesis of *N*-β-D-ribofuranosides of triazolo[4,5-*b*]-and triazolo[4,5-*c*]pyridines, *Indian J. Chem.* **3**, 84–85 (1965).

857. A. E. Chichibabin and A. V. Kirsanov, α,β′-Diaminopyridine and α,β-diaminopyridine, *Ber.* **60B**, 766–776 (1927).

858. K. B. De Roos and C. A. Salemink, Deazapurine derivatives. IX. Synthesis of (7-substi-

tuted)-1-, 2- and 3-β-D-ribofuranosyl-*vic*-triazolo[4,5-*b*]pyridines. 1-Deaza-8-azaadenosine and related compounds, *Recl. Trav. Chim. Pays-Bas.* **90**, 1181–1196 (1971).

859. K. B. De Roos and C. A. Salemink, Deazapurine derivatives. VII. Synthesis of substituted imidazo- and triazolopyridines, *Recl. Trav. Chim. Pays-Bas.* **90**, 1166–1180 (1971).

860. D. E. Burton, A. J. Lambie, D. W. J. Lane, G. T. Newbold, and A. Percival, Halogeno-*o*-phenylenediamines and derived heterocycles. Part I. Reductive fission of benzotriazoles to *o*-phenylenediamines, *J. Chem. Soc. (C)* **1968**, 1268–1273.

861. B. L. Cline, R. P. Panzica, and L. B. Townsend, The synthesis of 5-amino-*v*-triazolo[4,5-*b*]pyridin-7-one (1-deaza-8-azaguanine) and 5-amino-3-(β-D-ribofuranosyl)-*v*-triazolo[4,5-*b*]pyridin-7-one (1-deaza-8-azaguanosine), *J. Heterocyl. Chem.* **13**, 1365–1367 (1976).

862. B. L. Cline, R. P. Panzica, and L. B. Townsend, 5-Amino-3-(β-D-ribofuranosyl)-*v*-triazolo[4,5-*b*]pyridin-7-one (1-deaza-8-azaguanosine) and certain related derivatives, *J. Org. Chem.* **43**, 4910–4915 (1978).

863. B. L. Cline, P. E. Fagerness R. P. Panzica, and L. B. Townsend, The use of carbon-13 magnetic resonance chemical shifts and long-range ^{13}C–^{1}H coupling constants for assigning the site of glycosylation on nitrogen heterocycles, *J. Chem. Soc., Perkin Trans 2* **1980**, 1586–1591.

864. H. U. Blank and J. J. Fox, Pyrimidines. VII. A simple conversion of 2-oxo-5-nitro heterocycles to *v*-triazolo derivatives by sodium azide, *J. Am. Chem. Soc.* **90**, 7175–7176 (1968).

865. H. U. Blank, I. Wempen, and J. J. Fox, Pyrimidines. IX. A new synthesis of 8-azapurines and *v*-triazolo[4,5-*b*]pyridines, *J. Org. Chem.* **35**, 1131–1138 (1970).

866. B. M. Lynch and S. C. Sharma, Syntheses, structural and conformational assignments, and conversions of pyridine and triazolopyridine nucleosides, *Can. J. Chem.* **54**, 1029–1038 (1976).

867. O. von Bremer, Reactivity of the methoxy derivatives of 3-nitropyridine and new derivatives of 3,4-pyridinopyrazine, *Ann. Chem.* **529**, 290–298 (1937).

868. J. A. May, Jr. and L. B. Townsend, Synthesis of *v*-triazolo[4,5-*c*]pyridine nucleosides and 4-(β-D-ribofuranosyl)amino-1,2,3-thiadiazolo[5,4-*b*]pyridine via a rearrangement, *J. Org. Chem.* **41**, 1449–1456 (1976).

869. Z. Talik and E. Plazek, The nitration of 2-chloro-4-aminopyridine, *Roczniki Chem.* **30**, 1139–1149 (1956).

870. R. B. Meyer, Jr., G. R. Revankar, P. D. Cook, K. W. Ehler, M. P. Schweizer, and R. K. Robins, Synthesis of 6-amino-1,2,3-triazolo[4,5-*c*]pyridin-4(5*H*)-one (8-aza-3-deazaguanine) and 6-amino-1-(β-D-ribofuranosyl)-1,2-3-triazolo[4,5-*c*]pyridin-4(5*H*)-one (8-aza-3-deazaguanosine) via novel ring closure procedures, *J. Heterocyl. Chem.* **17**, 159–169 (1980).

871. R. A. Earl and L. B. Townsend, The synthesis of 8-aza-3-deazaguanosine [6-amino-1-(β-D-ribofuranosyl)-*v*-triazolo[4,5-*c*]pyridin-4-one] via a novel 1,3-dipolar cycloaddition reaction, *Can. J. Chem.* **58**, 2550–2561 (1980).

872. R. O. Roblin, Jr., J. O. Lampen, J. P. English, Q. P. Cole, and J. R. Vaughan, Jr., Studies in chemotherapy. VIII. Methionine and purine antagonists and their relation to the sulfonamides, *J. Am. Chem. Soc.* **67**, 290–294 (1945).

873. D. W. Woolley and E. Shaw, Some imidazo-1,2,3-triazines and their biological relationship to the purines, *J. Biol. Chem.* **189**, 401–410 (1951).

874. E. Shaw and D. W. Wooley, Imidazo-1,2,3-triazines as substrates and inhibitors for xanthine oxidase, *J. Biol. Chem.* **194**, 641–654 (1952).

875. J. J. Biesele, Purine antagonism and differential toxicity of some imidazo-1,2,3-triazines in mouse-tumor tissue cultures, *Cancer* **5**, 787–791 (1952).

876. A. Fjelde, Effect of two purine analogs on a human tumor in tissue culture, *Z. Krebsforsch.* **61**, 364–367 (1956).

877. R. Guthrie and W. N. Lu, Antagonism between purines in purine requiring *Bacillus subtilis* mutants, *Arch. Biochem. Biophys.* **108**, 398–402 (1964).

878. J. G. Cappuccino, M. George, P. C. Merker, and G. S. Tarnowski, Growth inhibition of *Clostridium feseri* by carcinostatic purine and pyrimidine analogs. I. Effect of medium on growth inhibition, *Cancer Res.* **24**, 1243–1248 (1964).

879. M. A. Stevens, H. W. Smith, and G. B. Brown, Purine *N*-oxides. VIII. *N*-Oxides of azapurines, *J. Am. Chem. Soc.* **82**, 3189–3193 (1960).

880. M. A. Stevens, H. W. Smith, and G. B. Brown, Purine *N*-oxides. V. Oxides of adenine nucleotides, *J. Am. Chem. Soc.* **81**, 1734–1738 (1959).

881. M. A. Stevens, D. I. Magrath, H. W. Smith, and G. B. Brown, Purine *N*-oxides. I. Mono-oxides of aminopurines, *J. Am. Chem. Soc. 80*, 2755–2758 (1958).

882. C. M. Baugh and E. N. Shaw, The reaction of β-propiolactone with inosine 5′-phosphate. The preparation of *N*-(5-amino-1-β-D-ribofuranosylimidazole 4-carbonyl)-β-alanine 5′-phosphate, *Biochim. Biophys. Acta* **114**, 213–221 (1966).

883. J. A. Montgomery and H. J. Thomas, 2-Azaadenosine (4-amino-7-β-D-ribofuranosyl-7*H*-imidazo[4,5-*d*]-*v*-triazine), *J. Chem. Soc., Chem. Commun.* **1969**, 458.

884. J. A. Montgomery and H. J. Thomas, Nucleosides of 2-azapurines and certain ring analogs, *J. Med. Chem.* **15**, 182–187 (1972).

885. J. A. Montgomery and H. J. Thomas, 4-Amino-7-β-D-ribofuranosyl-7*H*-imidazo[4,5-*d*]-*v*-triazine (2-azaadenosine), in: *Nucleic Acid Chemistry, Improved and New Synthetic Procedures, Methods, and Techniques* (L. B. Townsend and R. S. Tipson, eds.), Vol. 2, pp. 681–685, John Wiley and Sons, New York (1978).

886. J. A. Montgomery, A. G. Laseter, A. T. Shortnacy, S. J. Clayton, and H. J. Thomas, Nucleosides of 2-azapurines. 7*H*-Imidazo[4,5-*d*]-1,2,3-triazines. 2. *J. Med. Chem.* **18**, 564–567 (1975).

887. M. R. Grimmett, Advances in imidazole chemistry, *Adv. Heterocycl. Chem.* **12**, 103–183 (1970).

888. R. P. Panzica, R. J. Rousseau, R. K. Robins, and L. B. Townsend, A study on the relative stability and a quantitative approach to the reaction mechanism of the acid-catalyzed hydrolysis of certain 7- and 9-β-D-ribofuranosyl purines, *J. Am. Chem. Soc.* **94**, 4708–4714 (1972).

889. M. Kawana, G. A. Ivanovics, R. J. Rousseau, and R. K. Robins, Azapurine nucleosides. 3. Synthesis of 7-(β-D-ribofuranosyl)imidazo[4,5-*d*]-*v*-triazin-4-one (2-azainosine) and related derivatives, *J. Med. Chem.* **15**, 841–843 (1972).

890. R. P. Panzica and L. B. Townsend, The synthesis of 2-azainosine and related derivatives by ring annulation of imidazole nucleosides, *J. Heterocycl. Chem.* **9**, 623–628 (1972).

891. P. C. Strivastava, G. A. Ivanovics, R. J. Rousseau, and R. K. Robins, Nucleosides of 4-substituted imidazoles, *J. Org. Chem.* **40**, 2920–2924 (1975).

892. B. Rayner, C. Tapiero, and J.-L. Imbach, Synthesis of nucleosides. II. α-Anomers in the purine series, *J. Heterocycl. Chem.* **10**, 417–418 (1973).

893. B. Rayner, C. Tapiero, and J.-L. Imbach, Synthesis of nucleosides XXI. Stereospecific preparation of α-ribonucleosides, *J. Heterocycl. Chem.* **19**, 593–596 (1982).

894. J. I. Andrés, R. Herranz, and M. T. G. Lopez, Synthesis of 8-methyl-2-azainosine and related nucleosides, *J. Heterocycl. Chem.* **21**, 1221–1224 (1984).

895. R. J. Rousseau and G. A. Ivanovics, 2-Azainosine, U.S. Patent 3,803,126 (April 1974).

896. R. B. Meyer, Jr. and D. A. Shuman, 2-Substituted cyclic-AMP derivatives, U.S. Patent 3,917,583 (Nov. 1975).

897. R. B. Meyer, Jr., D. A. Shuman, and R. K. Robins, A new purine ring closure and the synthesis of 2-substituted derivatives of adenosine cyclic 3′,5′-phosphate, *J. Am. Chem. Soc.* **96**, 4962–4966 (1974).

898. N. Yamaji and M. Kata, The synthesis of 2-azaadenosine-3′,5′-cyclic-phosphate via 1, *N*⁶-ethenoadenosine-3′,5′-cyclic phosphate, *Chem. Lett.* **1975**, 311–314.

899. N. Yamaji, M. Kato, J. Ishiyama, and F. Yoshida, 2-Azapurines, *Jpn., Kokai* 76 70,793 (1976).

900. G. H. Jones, D. V. Murthy, D. Tegg, R. Golling, and J. G. Moffatt, Analogs of adenosine 3′,5′-cyclic phosphate II. Synthesis and enzymatic activity of derivatives of 1, *N*⁶-ethenoadenosine 3′,5′-cyclic phosphate, *Biochem. Biophys. Res. Commun.* **53**, 1338–1343 (1973).

901. K. F. Yip and K. C. Tsou, Synthesis of fluorescent adenosine derivatives, *Tetrahedron Lett.* **1973**, 3087–3090.

902. G. A. Ivanovics, R. J. Rousseau, M. Kawana, P. C. Srivastava, and R. K. Robins, The synthesis of 2-substituted derivatives of 5-amino-1-β-D-ribofuranosylimidazole-4-carboxamide. Ring opening reactions of 2-azapurine nucleosides, *J. Org. Chem.* **39**, 3651–3654 (1974).

903. P. C. Srivastava, A. R. Newman, and R. K. Robins, The synthesis of 5′-deoxy-2-azainosine from AICA riboside, *J. Carbohydr. Nucleosides Nucleotides* **3**, 327–333 (1975).

904. V. I. Ofitserov, V. C. Mokrushin, I. A. Korbukh, M. N. Preobrazhenskaya, and Z. V. Pushkareva, Synthesis of 4-methylthio-7-(β-ribofuranosyl)-imidazo[4,5-*d*][1,2,3]triazine, *Zh. Org. Khim.* **11**, 909–910 (1975).

905. V. A. Bakulev, V. S. Mokrushin, V. I. Ofitserov, Z. V. Pushkareva, and A. N. Grishakov, Synthesis of analogs of 5(4)-aminoimidazole-4-(methylthio)imidazo[4,5-*d*]-1,2,3-triazine, *Khim. Geterotsikl. Soedin.* **1979**, 836–838.

906. V. I. Ofitserov, Z. V. Pushkareva, V. S. Mokrushin, and K. V. Aglitskaya, Intermolecular cyclization of 5(4)-diazoimidazole-4(5)-thio carboxamide, *Khim. Geterotsikl. Soedin.* **1973**, 1292.
907. H.-D. Lüdemann and E. Westhof, Conformations of the nucleoside analogs formycin, 2-aza-adenosine, and nebularine in solution, *Z. Naturforsch., C* **32**, 528–538 (1977).
908. M. Tatibana and H. Yoshikawa, Formation of 2-azaadenine and 2,6-diaminopurine analogues of adenosine triphosphate in human erythrocytes, *Biochim. Biophys. Acta* **57**, 613–615 (1962).
909. M. Tatibana, T. Hashimoto, and H. Yoshikawa, 2-Azaadenosine triphosphate as a substitute for adenosine triphosphate in active transport of potassium across the erythrocycte membrane, *Biochim. Biophys. Acta.* **71**, 464–465 (1963).
910. M. Taitbana, T. Hashimoto, and H. Yoshikawa, 2-Azaadenine and 2,6-diaminopurine analogues of adenosine triphosphate in active transport of potassium across the human erythrocyte membrane, *J. Biochem. (Tokyo)* **53**, 214–218 (1963).
911. L. L. Bennett, Jr., P. W. Allan, J. W. Carpenter, and D. L. Hill, Nucleosides of 2-azapurines—cytotoxicities and activities as substrates for enzymes metabolizing purine nucleosides, *Biochem. Pharmacol.* **25**, 517–521 (1976).
912. J. A. Montgomery and H. J. Thomas, The synthesis and biologic evaluation of azapurine nucleosides, *Jerusalem Symp. Quantum Chem. Biochem.* **4**, 446–453 (1972).
913. G. W. Kidder, V. C. Dewey, R. E. Parks, Jr., and G. L. Woodside, Purine metabolism in *tetrahymena* and its relation to malignant cells in mice, *Science* **109**, 511–514 (1949).
914. G. W. Kidder and V. C. Dewey, The biological activity of substituted purines, *J. Biol. Chem.* **179**, 181–187 (1949).
915. H.-D. Lüdemann, E. Westhof, and I. Cuno, Ribose conformation of 8-azapurine nucleosides in solution, *Z. Naturforsch, C* **31**, 135–140 (1976).
916. K. Anzai and S. Suzuki, Chemical structure of pathocidin, *J. Antibiot. Ser. A* **14**, 253 (1961).
917. K. Anzai, J. Nagatsu, and S. Suzuki, Pathocidin, a new antifungal antibiotic. I. Isolation, physical and chemical properties, and biological activities, *J. Antibiot. Ser. A* **14**, 340–342 (1961).
918. R. E. F. Matthews, Incorporation of 8-azaguanine into nucleic acid of tobacco mosaic virus, *Nature* **171**, 1065–1066 (1953).
919. I. Lasnitzki, R. E. P. Matthews, and J. D. Smith, Incorporation of 8-azaguanine into nucleic aids, *Nature* **173**, 346–348 (1954).
920. H. G. Mandel and P.-E. Carlo, The incorporation of guanine into nucleic acids of tumor-bearing mice, *J. Biol. Chem.* **201**, 335–341 (1953).
921. R. W. Brockman, C. Sparks, D. J. Hutchison, and H. E. Skipper, A mechanism of resistance to 8-azaguanine I. Microbiological studies on the metabolism of purines and 8-azapurines, *Cancer Res.* **19**, 177–188 (1959).
922. R. W. Brockman, L. L. Bennett, Jr., M. S. Simpson, A. R. Wilson, J. R. Thompson, and H. E. Skipper, A mechanism of resistance to 8-azaguanine, II. Studies with experimental neoplasms, *Cancer Res.* **19**, 856–869 (1959).
923. J. Gut, Aza analogs of pyrimidine and purine bases of nucleic acids, in: *Advances in Heterocyclic Chemistry*, (A. R. Katritzky, ed.), Vol. 1, pp. 189–251, Academic Press, New York (1963).
924. R. E. F. Mathews, Biosynthetic incorporation of metabolic analogs, *Pharmacol. Rev.* **10**, 359–406 (1958).
925. H. G. Mandel, The physiological disposition of some anticancer agents, *Pharmacol. Rev.* **11**, 743–838 (1959).
926. P. Roy-Burman, Analogues of Nucleic Acid Components, *Recent Results in Cancer Research*, (P. Rentchnick, ed.), pp. 28, Springer, New York (1970).
927. R. E. Parks, Jr. and K. C. Agarwal, 8-Azaguanine, *Handbook Exp. Pharmacol.* **38(2)**, 458–467 (1975).
928. R. K. Robins, Antitumor activity and structural relationships of purine derivatives and related compounds against neoplasms in experimental animals, *J. Med. Chem.* **7**, 186–199 (1964).
929. P. W. Allan and L. L. Bennett, Jr., Metabolism of 8-azainosine, *Proc. Am. Assoc. Cancer Res.* **11**, 2 (1970).
930. L. L. Bennett, Jr., M. H. Vail, P. W. Allan, and W. R. Laster, Jr., Studies with 8-azainosine, a cytotoxic nucleoside with antitumor activity, *Cancer Res.* **33**, 465–471 (1973).
931. J. Davoll, The synthesis of the *v*-triazolo[*d*]pyrimidine analogues of adenosine, inosine, guanosine, and xanthosine, and a new synthesis of guanosine, *J. Chem. Soc.* **1958**, 1593–1599.

932. D. S. Bhakuni, S. Roy, P. K. Gupta, and B. L. Chowdhury, 2-Alkylthio-8-azaadenines and their corresponding ribosides and rhamnosides as antiviral agents, *Indian J. Chem.* **23B**, 369–371 (1984).

933. M. Friedkin, Enzymatic synthesis of azaguanine riboside and azaguanine desoxyriboside, *J. Biol. Chem.* **209**, 295–301 (1954).

934. K. J. M. Andrews and W. E. Barber, The synthesis of 6-dimethylamino-9-β-D-ribofuranosyl-purine-5+-phosphate, *J. Chem. Soc.* **1958**, 2768–2771.

935. R. B. Angier and J. W. Marsico, 1-, 2- and 3-Monoalkyl- and 2-(β-D-ribofuranosyl) derivatives of 7-dimethylamino-*v*-triazolo(*d*)pyrimidine and related compounds, *J. Org. Chem.* **25**, 759–765 (1960).

936. G. L. Tong, W. W. Lee, L. Goodman, and S. Frederiksen, Synthesis of some 2'-deoxyriboses of 8-azaadenine, *Arch. Biochem. Biophys.* **112**, 76–81 (1965).

937. S. Frederiksen, Deoxyribosyl-8-azaadenine. Its deamination to deoxyribosyl-8-azahypoxanthine and effect on nucleic acid synthesis in Ehrlich ascites cells *in vitro*, *Biochim. Biophys. Acta* **87**, 574–582 (1964).

938. W. W. Lee, A. P. Martinez, G. L. Tong, and L. Goodman, Simultaneous formation of both α- and β- nucleosides by the fusion method, *Chem. Ind. (London)* **52**, 2007–2008 (1963).

939. W. Hutzenlaub, R. L. Tolman, and R. K. Robins, Azapurine nucleosides. 1. Synthesis and antitumor activity of certain 3-β-D-ribofuranosyl- and 2'-deoxy-D-ribofuranosyl-*v*-triazolo[4,5-*d*]pyrimidines, *J. Med. Chem.* **15**, 879–883 (1972).

940. R. D. Guthrie and S. C. Smith, An improved preparation of 1,2,3,5-tetra-*O*-acetyl-β-D-ribofuranose, *Chem. Ind. (London)* **1968**, 547–548.

941. R. L. Tolman and D. A. Baker, 1-*O*-Acetyl-2,3,5-tri-*O*-benzoyl-D-arabinofuranose and its use in glycosylation by fusion, *Methods Carbohydr. Chem.* **7**, 59–62 (1976).

942. D. A. Baker, R. A. Harder, Jr., and R. L. Tolman, Stereospecific synthesis of nucleosides by the fusion reaction, *J. Chem. Soc., Chem. Commun.* **1974**, 167–168.

943. R. U. Lemieux, Some implications in carbohydrate chemistry of theories relating to the mechanisms of replacement reactions, *Adv. Carbohydr. Chem.* **9**, 32–57 (1954).

944. J. A. Montgomery, H. J. Thomas, and S. J. Clayton, A convenient synthesis of 8-azaadenosine, *J. Heterocycl. Chem.* **7**, 215–217 (1970).

945. J. A. Montgomery and K. Hewson, Nucleosides of 2-fluoroadenine, *J. Med. Chem.* **12**, 498–504 (1969).

946. J. A. Montgomery and H. J. Thomas, Nucleosides of 8-azapurines (*v*-triazolo[4,5-*d*]pyrimidines), *J. Med. Chem.* **15**, 305–307 (1972).

947. K. R. Darnall and L. B. Townsend, 3-β-D-Arabinofuranosyladenine, *J. Heterocycl. Chem.* **3**, 371–373 (1966).

948. J. A. Montgomery and H. J. Thomas, 9-α-D-Arabinofuranosyl-8-azaadenine-2-^{14}C, *J. Labelled Compd. Radiopharm.* **15**, 727–730 (1978).

949. R. Weiss, R. K. Robins, and C. W. Noell, Potential purine antagonists. XXIII. Synthesis of some 7-substituted amino-*v*-triazolo(*d*)pyrimidines, *J. Org. Chem.* **25**, 765–770 (1960).

950. J. A. Montgomery and R. D. Elliott, Molecular-sieve catalyzed *N*→*N* glycosyl migration in the *v*-triazolo[4,5-*d*]pyrimidine ring system, *J. Chem. Soc., Chem. Commun.* **1972**, 1279–1280.

951. J. A. Montgomery and R. D. Elliott, 7-Amino-3-β-D-ribofuranosyl-3*H*-1,2,3-triazolo[4,5-*d*]pyrimidine (8-azaadenosine), in: *Nucleic Acid Chemistry, Improved and New Synthetic Procedures, Methods, and Techniques* (L. B. Townsend and R. S. Tipson, eds.), Vol. 2, pp. 687–691, John Wiley and Sons, New York (1978).

952. R. D. Elliott and J. A. Montgomery, Analogues of 8-azainosine, *J. Med. Chem.* **20**, 116–120 (1977).

953. J. A. Montgomery, R. D. Elliott, and H. J. Thomas, The synthesis and evaluation of azapurine nucleosides as cytotoxic agents, *Ann. N. Y. Acad. Sci.* **255**, 292–304 (1975).

954. J. A. Montgomery, H. J. Thomas, R. D. Elliott, and A. T. Shortnacy, The use of microparticulate reversed-phase packing in high-pressure liquid chromatography of nucleosides, in: *Chemistry and Biology of Nucleosides and Nucleotides* (R. E. Harmon, R. K. Robins, and L. B. Townsend, eds.), pp. 37–53, Academic Press, New York (1978).

955. R. D. Elliott and J. A. Montgomery, Analogues of 8-azaguanosine, *J. Med. Chem.* **19**, 1186–1191 (1976).

956. J. A. Montgomery, A. T. Shortnacy, G. Arnett, and W. M. Shannon, 2-Substituted derivatives of 9-α-D-arabinofuranosyladenine and 9-α-D-arabinofuranosyl-8-azaadenine, *J. Med. Chem.* **20**, 401–404 (1977).

957. C. W. Smith, R. W. Sidwell, R. K. Robins, and R. L. Tolman, Azapurine nucleosides. 2. Synthesis and antiviral activity of 7-amino-3-α-D-arabinofuranosyl-v-triazolo[4,5-d]pyrimidine and related nucleosides, *J. Med. Chem.* **15**, 883–887 (1972).

958. R. L. Tolman and C. W. Smith, Arabinofuranosyl-8-azaadenines, U.S. Patent 3,826,803 (1974).

959. P. Bitterli and H. Erlenmeyer, Some derivatives of triazolopyrimidine, *Helv. Chim. Acta* **34**, 835–840 (1951).

960. J. A. Montgomery, A. T. Shortnacy, and J. A. Secrist III, Synthesis and biological evaluation of 2-fluoro-8-azaadenosine and related compounds, *J. Med. Chem.* **26**, 1483–1489 (1983).

961. J. A. Montgomery, The chemistry and biology of nucleosides of purines and ring analogs, in: *Nucleosides, Nucleotides and Their Biological Applications* (J. L. Rideout, D. W. Henry, and L. M. Beacham III, eds.), pp. 19–46, Academic Press, New York (1983).

962. J. Baddiley, J. G. Buchanan, and G. O. Osborne, The preparation of 7- and 9-glucopyranosyl and -xylopyranosyl derivatives of 8-azaxanthine (5,7-dihydroxy-v-triazolo[d]pyrimidine), *J. Chem. Soc.* **1958**, 1651–1657.

963. J. Baddiley, J. G. Buchanan, and G. O. Osborne, The preparation of 7- and 9-ribofuranosyl derivatives of 8-azaxanthine. A note on the preparation of 9-glycopyranosylxanthine, *J. Chem. Soc.* **1958**, 3606–3610.

964. R. A. Baxter and F. S. Spring, Application of the Hofmann reaction to the synthesis of heterocyclic compounds. Part IV. The synthesis of 9-d-xylopyranosidoxanthine, *J. Chem. Soc.* **1947**, 378–381.

965. G. A. Howard, A. C. McLean, G. T. Newbold, F. S. Spring, and A. R. Todd, Application of the Hofmann reaction to the synthesis of heterocyclic compounds. Part VI. Experiments on the synthesis of purine nucleosides. Part XXI. A synthesis of xanthosine, *J. Chem. Soc.* **1949**, 232–234.

966. R. Carrington, G. Shaw, and D. V. Wilson, Purines, pyrimidines, and imidazoles, Part XXIII. The use of 5-phospo-β-D-ribosyl azide in a new direct synthesis of nucleotides, *J. Chem. Soc.* **1965**, 6864–6870.

967. J. A. Montgomery and H. J. Thomas, A new approach to the synthesis of nucleosides of 8-azapurines (3-glycosyl-v-triazolo[4,5-d]pyrimidines), *J. Org. Chem.* **36**, 1962–1967 (1971).

968. V. Chretien and B. Gross, Synthesis of 8-azapurine glycosides starting from 1-azidoglycosides, *Tetrahedron* **38**, 103–112 (1982).

969. R. L. Tolman, C. W. Smith, and R. K. Robins, Anomerization of glycosyl azides in a two-step 1,3-dipolar cycloaddition reaction, *J. Am. Chem. Soc.* **94**, 2530–2532 (1972).

970. W. Traube, Der synthetische aufbau der harnsäure, des xanthins, theobromins, theophyllins und caffeins aus der cyanessigsäure, *Ber. Chem.* **33**, 3035–3056 (1900).

971. S. Gabriel and J. Colman, Synthesen in der purinreihe, *Ber.* **34**, 1234–1257 (1901).

972. J. Davoll and D. D. Evans, The synthesis of 9-glycitylpurines, 3-glycityl[1,2,3]triazolo[d]pyrimidines, 8-glycitylpteridines, and 10-glycitylbenzo[g]pteridines, including riboflavin and riboflavin 2-imine, *J. Chem. Soc.* **1960**, 5041–5049.

973. D. L. Ross, C. G. Skinner, and W. Shive, 9-Ribityl derivatives of guanine and 8-azaguanine, *J. Org. Chem.* **26**, 3582–3583 (1961).

974. W. Pfleiderer and E. Bühler, Pteridines, XXXII. New synthesis of pteridine-N-8-, purine-N-9- and triazolo[4,5-d]pyrimidin-N-3-glycosides, *Chem. Ber.* **99**, 3022–3039 (1966).

975. H. Rokos and W. Pfleiderer, Studies in the pyrimidine series, XXVI. A new synthesis of 4-(glycosylamino)-5-nitropyrimidines, *Chem. Ber.* **104**, 748–769 (1971).

976. J. J. Fox and D. V. Praag, Nucleosides. VIII. Synthesis of 5-nitrocytidine and related nucleosides, *J. Org. Chem.* **26**, 526–532 (1961).

977. T. Kishikawa and H. Yuki, Studies on chemotherapeutic agents. II. A synthesis of purine nucleosides of D-glucuronic acid, *Chem. Pharm. Bull.* **14**, 1360–1364 (1966).

978. I. Wempen, I. L. Doerr, L. Kaplan, and J. J. Fox, Pyrimidine nucleosides. VI. Nitration of nucleosides, *J. Am. Chem. Soc.* **82**, 1624–1629 (1960).

979. T. Sasaki, K. Minamoto, M. Kino, and T. Mizuno, Reactions of the derivatives of 5-bromopyrimidine nucleosides with sodium azide, *J. Org. Chem.* **41**, 1100–1104 (1976).

980. J. A. Montgomery and H. J. Thomas, Furanose-pyranose isomerization in a synthesis of 8-azapurine nucleosides, *J. Chem. Soc., Chem. Commun.* **1970**, 265.

981. Y. F. Shealy, J. D. Clayton, and C. A. O'Dell, Cyclopentyl derivatives of 8-azahypoxanthine and 8-azaadenine. Carbocyclic analogs of 8-azainosine and 8-azaadenosine, *J. Heterocycl. Chem.* **10**, 601–605 (1973).

982. M. S. Zedeck, A. C. Sartorelli, P. K. Chang, K. Raska, Jr., R. K. Robins, and A. D. Welch, Inhibition of the steroid-induced synthesis of Δ^5-3-ketosteroid isomerase in *Pseudomonas testosteroni* by a new purine deoxyribonucleoside analog, 6-chloro-8-aza-9-cyclopentylpurine, *Mol. Pharmacol.* **3**, 386–395 (1967).

983. P. K. Chang, L. J. Sciarini, A. C. Sartorelli, and M. S. Zedeck, Synthesis and biological activity of some 8-aza-9-cyclopentylpurines, *J. Med. Chem.* **11**, 513–515 (1968).

984. P. H. Duquett, C. L. Ritter, and R. Vince, Puromycin analogs. Ribosomal binding and peptidyl transferase substrate activity of a carbocyclic analog of 8-azapuromycin, *Biochemistry* **13**, 4855–4859 (1974).

985. Y. F. Shealy, C. A. O'Dell, W. M. Shannon, and G. Arnett, Synthesis and antiviral activity of carbocyclic analogues of 2'-deoxyribonucleosides of 2-amino-6-substituted-purines and of 2-amino-6-substituted-8-azapurines, *J. Med. Chem.* **27**, 1416–1421 (1984).

986. H. Lee and R. Vince, Carbocyclic analogs of arabinosylpurine nucleosides, *J. Pharm. Sci.* **69**, 1019–1021 (1980).

987. N. J. Cusack, B. J. Hildick, D. H. Robinson, P. W. Rugg, and G. Shaw, Purines, pyrimidines, and imidazoles. Part XL. A new synthesis of a D-ribofuranosylamine derivative and its use in the synthesis of pyrimidine and imidazole nucleosides, *J. Chem. Soc., Perkin Trans. 1*, **1973**, 1720–1731.

988. W. Schroeder, N-Glycosylation of aldoses and ketoses, U.S. Patent 2,993,039 (1961).

989. H. M. Kalekar, The enzymatic synthesis of purine ribosides, *J. Biol. Chem.* **167**, 477–486 (1947).

990. M: Kanda and Y. Takagi, Purification and properties of a bacterial deoxyribose transferase, *J. Biochem.* **46**, 725–732 (1959).

991. W. W. Kielley, H. M. Kalekar, and L. B. Bradley, The hydrolysis of purine and pyrimidine nucleoside triphosphates by myosin, *J. Biol. Chem.* **219**, 95–101 (1956).

992. D. H. Levin, The polymerization of 8-azaguanosine 5'-diphosphate by polynucleotide phosphorylase, *Biochim. Biophys. Acta.* **61**, 75–81 (1962).

993. J. L. Way, J. L. Dahl, and R. E. Parks, Jr., Polyphosphate nucleosides of purine analogues, *J. Biol. Chem.* **234**, 1241–1243 (1959).

994. G. Marbaix, Synthesis of azaguanine triphosphate from azaguanine by rabbit reticulocytes, *Arch. Int. Physiol. Biochim.* **72**, 332–334 (1964).

995. J. Kara, J. Skoda, and F. Sorm, Fermentative preparation of 8-azaguanosine, *Collect. Czech. Chem. Commun.* **26**, 1386–1392 (1961).

996. J. Skoda and J. Kara, Fermentation production of 8-azaguanosine, Czech. Patent 101,896 (Dec. 1961) [*CA* **58**, 13096a (1963)].

997. O. Kanamitsu, Ribosides of nucleic acid base derivatives and analogs, Jpn. Kokai 74 07,494 (Jan. 1974) [*CA* **81**, 48570k (1974)].

998. K. Mitsugi, S. Okumura, N. Katsuya, and A. Uemura, Preparation of nucleotides from nucleosides by bacteria, Jpn. Patent 68 28,960 (Dec. 1968) [*CA* **70**, 95459 (1969)].

999. O. Kanemitsu, Fermentative manufacture of 8-azaguanosine 5'-phosphate and 8-azaguanosine, Jpn. Patent 72 37,035 (Sept. 1972) [*CA* **78**, 2742c (1973)].

1000. H. Tanaka and K. Nakayama, Production of nucleic acid-related substances by fermentative processes, Part XXXX. Production of ribotides of purine and 8-azapurine derivatives by salvage synthesis with *Brevibacterium ammoniagenes*, *Agric. Biol. Chem.* **36**, 464–471 (1972).

1001. S. Kinoshita, K. Nakayama, Z. Sato, and H. Tanaka (Kyowa Fermentation Industry Co., Ltd), Manufacture of nucleotides by fermentation, Jpn. Patent 66 17,636 (1966) [*CA* **66**, 36618j (1967)].

1002. Kyowa Hakko Kogyo Co., Japan, Process for the production of ribosylphosphates of 8-azapurine derivatives by fermentation, Brit. Patent 1,080,426 (August 1967).

1003. S. Fredericksen, Specificity of adenosine deaminase toward adenosine and 2'-deoxyadenosine analogues, *Arch. Biochem. Biophys,* **113**, 383–388 (1966).

1004. Takeda Chemical Industries, Ltd., Purine nucleosides, French Patent 1,402,909 (June 1965) [*CA* **63**, 17100e (1965)].

1005. M.-C. Huang, K. Hatfield, A. W. Roetker, J. A. Montgomery, and R. L. Blakley, Analogs of 2'-deoxyadenosine: Facile enzymatic preparation and growth inhibitory effects on human cell lines, *Biochim. Pharmacol.* **30**, 2663–2671 (1981).

1006. J. Holguin, R. Cardinaud, and C. A. Salemink, Trans-*N*-deoxyribosylase: Substrate specificity studies—purine bases as acceptors, *Eur. J. Biochem.* **54**, 515–520 (1975).

1007. W. S. Beck and M. Levin, Purification, kinetics, and repression control of bacterial trans-*N*-deoxyribosylase, *J. Biol. Chem.* **238**, 702–709 (1963).

1008. T. R. Krugh, Tautomerism of the nucleoside antibiotic formycin, as studied by carbon-13 nuclear magnetic resonance, *J. Am. Chem. Soc.* **95**, 4761–4762 (1973).

1009. G. Klopman, Chemical reactivity and the concept of charge- and frontier-controlled reactions, *J. Am. Chem. Soc.* **90**, 223–234 (1968).

1010. T. A. Glassman, G. Klopman, and C. Cooper, Use of the generalized perturbation theory to predict the interaction of purine nucleotides with metal ions, *Biochemistry* **12**, 5013–5019 (1973).

1011. T. R. Emerson, R. J. Swan, and T. L. V. Ulbricht, The optical rotatory dispersion of purine nucleosides, *Biochem. Biophys. Res. Commun.* **22**, 505–510 (1966).

1012. P. Singh and D. J. Hodgson, 8-Azaadenosine. Crystallographic evidence for a "high-anti" conformation around a shortened glycosidic linkage, *J. Am. Chem. Soc.* **96**, 5276–5278 (1974).

1013. P. Singh and D. J. Hodgson, 8-Azaadenosine. Crystal structure of its monohydrate and conformational analysis for rotation around the glycosyl bond, *J. Am. Chem. Soc.* **99**, 4807–4815 (1977).

1014. C.-H. Lee, F. E. Evans, and R. H. Sarma, Interrelation between glycosidic torsion, sugar pucker, and backbone conformation in 5'-β-nucleotides. A ^{1}H and ^{31}P fast Fourier tranform nuclear magnetic resonance investigation of the confornmation of 8-aza-5'-β-adenosine monophosphate, *J. Biol. Chem.* **250**, 1290–1296 (1975).

1015. H.-D. Lüdemann, E. Westhof, and I. Cuno, Ribose conformations of 8-azapurine nucleosides in solution, *Z. Naturforsch. C* **31**, 135–140 (1976).

1016. D. C. Ward and E. Reich, Relationship between nucleoside conformation and biological activity, *Ann Rep. Med. Chem.* **5**, 274–284 (1969).

1017. A. Saran, C. Mitra, and B. Pullman, Molecular orbital studies on the structure of nucleoside analogs, *Biochim. Biophys. Acta* **517**, 255–264 (1978).

1018. D. L. Miles, D. W. Miles, and H. Eyring, Conformational component of structure–activity relationships in nucleosides, in: *Physiological and Regulatory Functions of Adenosine and Adenine Nucleotides* (H. P. Baer and G. I. Drummond, eds.), pp. 283–294, Raven Press, New York (1979).

1019. I. Ekiel, E. Darzynkiewicz, and D. Shugar, Conformational parameters of the carbohydrate moieties of α-arabinonucleosides, *Carbohydr. Res.* **92**, 21–36 (1981).

1020. M. Sundaralingam, Stereochemistry of nucleic acid constituents. III. Crystal and molecular structure of adenosine 3'-phosphate dihydrate (adenylic acid b), *Acta Crystallogr.* **21**, 495–506 (1966).

1021. G. H.-Y. Lin, M. Sundaralingam, and S. K. Arora, Sterochemistry of nucleic acids and their constituents. XV. Crystal and molecular structure of 2-thiocytidine dihydrate, a minor constituent of transfer ribonucleic acid, *J. Am. Chem. Soc.* **93**, 1235–1241 (1971).

1022. K. K. Ogilvie, L. Slotin, and P. Rheault, Novel substrate of adenosine deaminase, *Biochem. Biophys. Res. Commun.* **45**, 297–300 (1971).

1023. D. Grünberger, A. Holy, and F. Sorm, Synthesis and coding properties of 8-azaguanosine containing triribonucleoside diphosphates, *Biochim. Biophys. Acta* **161**, 147–155 (1968).

1024. D. Grünberger, A. Holy, and F. Sorm, Coding properties of trinucleoside diphosphates containing anomalous bases, *Abh. Dtsch. Akad. Wiss. Berlin Kl. Med.* **1968**, 487–498.

1025. J. A. Montgomery and H. J. Thomas, Synthesis of potential anticancer agents. XXVII. The ribonucleotides of 6-mercaptopurine and 8-azaguanine, *J. Org. Chem.* **26**, 1926–1929 (1961).

1026. H. J. Thomas, K. Hewson, and J. A. Montgomery, The 2'(3')-phosphates of 6-mercaptopurine ribonucleoside and 8-azaguanosine, *J. Org. Chem.* **27**, 192–194 (1962).

1027. A. Holy and J. Smrt, Oligonucleotide compounds. XI. Synthesis of ribonucleoside-2',3'-phosphites, *Collect. Czech. Chem. Commun.* **31**, 1528–1534 (1966).

1028. D. Grünberger, L. Meissner, A. Holy, and F. Sorm, The coding properties of polymers and trinucleoside diphosphates containing 8-azaguanosine, *Collect. Czech. Chem. Commun.* **32**, 2625–2633 (1967).

1029. J. K. Coward and E. P. Slisz, Analogs of S-adenoxylhomocysteine as potential inhibitors of biological transmethylation. Specificity of the S-adenosylhomocysteine binding site, *J. Med. Chem.* **16**, 460–463 (1973).

1030. J. Hildesheim, R. Hildesheim, P. Blanchard, G. Farrugia, and R. Michelot, Synthetic inhibitors of tRNA methyltransferases. Analogs of S-adenosylhomocysteine, *Biochimie* **55**, 541–546 (1973).

1031. J. A. Montgomery, A. T. Shortnacy, and H. J. Thomas, Analogs of 5'-deoxy-5'-(methylthio)-adenosine, *J. Med. Chem.* **17**, 1197–1207 (1974).

1032. J. P. Verheyden and J. G. Moffatt, 2',3'-Unsaturated nucleosides, U.S. Patent 3,817,982 (1974) [*CA* **81**, 63942b (1974)].

1033. A. F. Cook and J. G. Moffatt, Antimetabolic 2'- and 3'-oxonucleosides, U.S. Patent 3,491,085 (Jan. 1970) [*CA* **72**, 111785r (1970)].

1034. L. L. Bennett, Jr. and D. L. Hill, Structural requirements for activity of nucleosides as substrates for adenosine kinase: Orientation of substituents on the pentofuranosyl ring, *Mol. Pharmacol.* **11**, 803–808 (1975).

1035. L. L. Bennett, Jr., P. W. Allan, D. L. Hill, H. J. Thomas, and J. W. Carpenter, Metabolic studies with an α-nucleoside, 9-α-D-arabinofuranosyl-8-azaadenine, *Mol. Pharmacol.* **12**, 242–250 (1976).

1036. L. L. Bennett, Jr., P. W. Allan, S. C. Shaddix, W. M. Shannon, G. Arnett, L. Westbrook, J. C. Drach, and C. M. Reinke, Biological activities and modes of action of 9-α-D-arabinofuranosyl-adenine and 9-α-D-arabinofuranosyl-8-azaadenine, *Biochem. Pharmacol.* **30**, 2325–2332 (1981).

1037. L. L. Bennett, Jr., W. M. Shannon, P. W. Allan, and G. Arnett, Studies on biochemical basis for the antiviral activities of some nucleoside analogs, *Ann. N. Y. Acad. Sci.* **255**, 342–348 (1975).

1038. C. E. Cass, T. H. Tan, and M. Selner, Antiproliferative effects of 9-β-D-arabinofuranosyl-adenine 5'-monophosphate and related compounds in combination with adenosine deaminase inhibitors against mouse leukemia L1210/C2 cells in culture, *Cancer Res.* **39**, 1563–1569 (1979).

1039. L. L. Bennett, Jr., L. M. Rose, P. W. Allan, D. Smithers, D. J. Adamson, R. D. Elliott, and J. A. Montgomery, Metabolism and metabolic effects of 8-aza-6-thioinosine and its rearrangement product, N-β-D-ribofuranosyl[1,2,3]thiadiazolo[5,4,*d*]pyrimidin-7-amine, *Mol. Pharmacol.* **16**, 981–996 (1979).

1040. T. Spector and R. L. Miller, Mammalian adenylosuccinate synthetase nucleotide monophosphate substrates and inhibitors, *Biochim. Biophys. Acta* **445**, 509–517 (1976).

1041. T. I. Kalman and D. Goldman, Inactivation of thymidylate synthetase by a novel mechanism-based enzyme inhibitor: 1-(β-D-2'-Deoxyribofuranosyl)-8-azapurin-2-one 5'-monophosphate, *Biochem. Biophys. Res. Commun.* **102**, 682–689 (1981).

1042. J. A. Montgomery, Biochemical basis for the drug actions of purines, *Prog. Med. Chem.* **7**, 69–123 (1970).

1043. M. E. Balis, Incorporation of analogs, in: *Frontiers of Biology—Antagonists and Nucleic Acids* (A. Neuberger and E. L. Tatum, eds.), Vol. 10, pp. 162–176, North-Holland, Amsterdam (1968).

1044. E. Lodemann and A. Wacker, Specificity of nucleosidephosphorylases. Enzymatic determination of nucleoside configuration, *Z. Naturforsch., B* **22**, 42–44 (1967).

1045. A. W. Paula and L. L. Bennett, Jr., Cytotoxicity and metabolism of some nucleosides of adenine (A), 2-AZAA, and 8-AZAA, *Proc. Am. Assoc. Cancer Res.* **14**, 16 (1973).

1046. L. L. Bennett, Jr. and P. W. Allan, Metabolism and metabolic effects of 8-azainosine and 8-azaadenosine, *Cancer Res.* **36**, 3917–3923 (1976).

1047. E. N. Spremulli, G. W. Crabtree, D. L. Dexter, S. H. Chu, D. M. Farineau, L. Y. Ghoda, D. L. McGowan, I. Diamond, R. E. Parks, Jr., and P. Calabresi, Biochemical pharmacology and toxicology of 8-azaadenosine alone and in combination with 2'-deoxycoformycin (pentostatin), *Biochem. Pharmacol.* **31**, 2415–2421 (1982).

1048. J. J. Fox, K. A. Watanabe, R. S. Klein, C. K. Chu, S. Y.-K. Tam. U. Reichman, H. Hirota, J.-S. Hwang, F. G. De Las Heras, and I. Wempen, New C-nucleoside isosteres of some nucleoside antibiotics, in: *Chemistry and Biology of Nucleosides and Nucleotides* (R. E. Harmon, R. K. Robins, and L. B. Townsend, eds.), pp. 415–439, Academic Press, New York (1978).

1049. P. C. Zamecnik, Unsettled questions in the field of protein synthesis, *Biochem. J* **85**, 257–264 (1962).

1050. W. Saenger, Structure and function of nucleosides and nucleotides, *Angew. Chem. Int. Ed. Engl.* **12**, 591–601 (1973).

1051. T. Huynh-Dinh, J. Igolen, J.-P. Marquet, E. Bisagni, and J.-M. Lhoste, Synthesis of C-nucleosides, 13. *s*-Triazolo[4,3-*a*]- and [1,5-*a*]pyridine derivatives, *J. Org. Chem.* **41**, 3124–3128 (1976).

1052. J. Kobe, B. Brdar, and J. Soric, Formycin analogs II. Antiviral and cytotoxic *s*-triazolo[4,3-*a*]- and [1,5-*a*]pyridine derivatives, *Nucleosides Nucleotides* **5**, 135–151 (1986).

1053. T. Huynh-Dinh, R. S. Sarfati, C. Gouyette, and J. Igolen, Synthesis of C-nucleosides. 17. *s*-Triazolo[4,3-*a*]pyrazines, *J. Org. Chem.* **44**, 1028–1035 (1979).

1054. S. W. Schneller, R. D. Thompson, J. G. Cory, R. A. Olsson, E. De Clercq, I.-K. Kim, and P. K. Chiang, Biological activity and a modified synthesis of 8-amino-3-β-D-ribofuranosyl-1,2,4-triazolo[4,3,-*a*]pyrazine, an isomer of formycin, *J. Med. Chem.* **27**, 924–928 (1984).

1055. M. Legraverend, E. Bisagni, and J.-M. Lhoste, Synthesis of *s*-triazolo[4,3-*b*]pyridazine C-nucleosides, *J. Heterocycl. Chem.* **18**, 893–898 (1981).

1056. Y. Kang, S. B. Larson, R. K. Robins, and G. R. Revankar, Synthesis and biological evaluation of certain 3-β-D-ribofuranosyl-1,2,4-triazolo[4,3-*b*]pyridazines related to formycin prepared via ring closure of pyridazine precursors, *J. Med. Chem.* **32**, 1547–1551 (1989).

1057. S. B. Larson, Y. Kang, G. R. Revankar, and R. K. Robins, Structure of 8-amino-3-β-D-ribofuranosyl-1,2,4-triazolo[4,3,*b*]pyridazine, an analog of formycin A, *Acta Crystallogr., Sect. C* **45**, 1093–1095 (1989).

1058. K. Ramasamy, B. G. Ugarkar, P. A. McKernan, R. K. Robins, and G. R. Revankar, Synthesis and antitumor activity of certain 3-β-D-ribofuranosyl-1,2,4-triazolo[3,4-*f*]-1,2,4-triazines related to formycin prepared via ring closure of a 1,2,4-triazine preursor, *J. Med. Chem.* **29**, 2231–2235 (1986).

1059. L. J. S. Knutsen, B. D. Judkins, R. F. Newton, D. I. C. Scopes, and G. Klinkert, Synthesis of imidazo-fused bridgehead-nitrogen 2'-deoxyribo-C-nucleosides: Coupling–elimination reactions of 2,5-anhydro-3,4,6-tri-*O*-benzoyl-D-allonic acid, *J. Chem. Soc., Perkin Trans. 1* **1985**, 621–630.

1060. W. L. Mitchell, M. L. Hill, R. F. Newton, P. Ravenscroft, and D. I. C. Scopes, Synthesis of C-nucleoside isosteres of 9-(2-hydroxyethoxymethyl)guanine (acyclovir), *J. Heterocycl. Chem.* **21**, 697–699 (1984).

1061. F. G. De Las Heras, C. K. Chu, S. Y.-K. Tam, R. S. Klein, K. A. Watanabe, and J. J. Fox, Nucleosides XCVII. Synthesis of an 8-(D-ribofuranosyl)pyrazolo[1,5-*a*]-1,3,5-triazine. A new type of C-nucleoside, *J. Heterocycl. Chem.* **13**, 175–177 (1976).

1062. S. Y.-K. Tam, J.-S. Hwang, F. C. De Las Heras, R. S. Klein, and J. J. Fox, Nucleosides CV. Synthesis of the 8-(β-D-ribofuranosyl)pyrazolo[1,5-*a*]1,3,5-triazine isosteres of adenosine and inosine, *J. Heterocycl. Chem.* **13**, 1305–1308 (1976).

1063. S. Y.-K. Tam. R. S. Klein, I. Wempen, and J. J. Fox, Nucleosides. 112. Synthesis of some new pyrazolo[1,5-*a*]-1,3,5-triazines and their C-nucleosides, *J. Org. Chem.* **44**, 4547–4553 (1979).

1064. C. K. Chu, K. A. Watanabe and J. J. Fox, Nucleosides. 117. Synthesis of 4-oxo-8-(β-D-ribofuranosyl)-3*H*-pyrazolo[1,5-*a*]-1,3,5-triazine (OPTR) via 3-amino-2*N*-carbamoyl-4-(β-D-ribofuranosyl)pyrazole (ACPR) derivatives, *J. Heterocycl. Chem.* **17**, 1435–1439 (1980).

1065. C. K. Chu, A facile synthesis of 8-(β-D-ribofuranosyl)-4-thioxo-3*H*-pyrazolo[1,5-*a*]-1,3,5-triazine and its α-anomer, *J. Heterocycl. Chem.* **21**, 389–392 (1984).

1066. C. K. Chu, J. J. Suh, M. Mesbah, and S. J. Cutler, Ring transformation reactions of C-nucleosides: Facile synthesis of pyrazolo[1,5-*a*]pyrimidine and pyrazolo[1,5-*a*]triazine C-nucleosides, *J. Heterocycl. Chem.* **23**, 349–352 (1986).

1067. G. R. Revankar and R. K. Robins, Synthesis and biological activity of some nucleosides resembling guanosine: Imidazo[1,2-*a*]pyrimidine nucleosides, *Ann. N. Y. Acad. Sci.* **255**, 166–176 (1975).

1068. R. J. Pugmire and D. M. Grant, Carbon-13 magnetic resonance. X. The six-membered nitrogen heterocycles and their cations, *J. Am. Chem. Soc.* **90**, 697–706 (1968).

1069. R. J. Pugmire and D. M. Grant, Carbon-13 magnetic resonance. XIX. Benzimidazole, purine, and their anionic and cationic species, *J. Am. Chem. Soc.* **93**, 1880–1887 (1971).

1070. R. J. Pugmire, D. M. Grant, L. B. Townsend, and R. K. Robins, Carbon-13 magnetic resonance. XXII. The N-methylpurines, *J. Am. Chem. Soc.* **95**, 2791–2796 (1973).

1071. D. G. Bartholomew, P. Dea, R. K. Robins, and G. R. Revankar, Imidazo[1,2,c]pyrimidine nucleosides. Synthesis of N-bridgehead inosine monophosphate and guanosine monophosphate analogues related to 3-deazapurines, *J. Org. Chem.* **40**, 3708–3713 (1975).

1072. P. Dea, G. R. Revankar, R. L. Tolman, R. K. Robins, and M. P. Schweizer, Use of carbon-13 and proton magnetic resonance studies for the determination of glycosylation site in nucleosides of fused nitrogen heterocycles, *J. Org. Chem.* **39**, 3226–3231 (1974).

1073. F. M. Schabel, Jr., Antiviral activity of 9-β-D-arabinofuranosyladenine, *Chemotherapy* **13**, 321–338 (1968).

1074. J. Kangilaski, A review of preliminary results of adenine arabinoside studies, *J. Am. Med. Assoc.* **230**, 189–201 (1974).

1075. A. M. Mian, R. Harris, R. W. Sidwell, R. K. Robins, and T. A. Khwaja, Synthesis and biological activity of 9-β-D-arabinofuranosyladenine cyclic 3′,5′-phosphate and 9-β-D-arabinofuranosylguanine cyclic 3′,5′-phosphate, *J. Med. Chem.* **17**, 259–263 (1974).

1076. R. W. Sidwell, L. B. Allen, J. F. Huffman, T. A. Khwaja, R. L. Tolman, and R. K. Robins, Anti DNA virus activity of the ′-nucleotide and 3′,5′-cyclic nucleotide of 9-β-D-arabinofuranosyladenine, *Chemotherapy* **19**, 325–340 (1973).

1077. G. R. Revankar, J. H. Huffman, L. B. Allen, R. W. Sidwell, R. K. Robins, and R. L. Tolman, Synthesis and antiviral activity of certain 5′-monophosphates of 9-D-arabinofuranosyladenine and 9-D-arabinofuranosylhypoxanthine, *J. Med. Chem.* **18**, 721–726 (1975).

1078. L. B. Allen, J. M. Thompson, J. H. Huffman, G. R. Revankar, R. L. Tolman, L. N. Simon, R. K. Robins, and R. W. Sidwell, Inhibition of experimental deoxyribonucleic acid virus-induced encephalitis by 9-β-D-arabinofuranosylhypoxanthine 5′-monophosphate, *Antimicrob. Agents Chemother.* **8**, 468–473 (1975).

1079. L. B. Allen, J. H. Huffman, G. R. Revankar, R. L. Tolman, L. N. Simon, R. K. Robins, and R. W. Sidwell, Efficacy of 9-β-D-arabinofuranosylhypoxanthine 5′-monophosphate in therapy of equine abortion virus-induced hepatitis in hamsters, *Antimicrob. Agents Chemother.* **8**, 474–478 (1975).

1080. R. W. Sidwell, L. B. Allen, J. H. Huffman, G. R. Revankar, R. K. Robins, and R. L. Tolman, Viral keratitis-inhibitory effect of 9-β-D-arabinofuranosylhypoxanthine 5′-monophosphate, *Antimicrob. Agents Chemother.* **8**, 463–467 (1975).

1081. L. B. Allen, C. J. Hintz, S. M. Wolf, H. J. Huffman, L. N. Simon, R. K. Robins, and R. W. Sidwell, Effect of 9-β-D-arabinofuranosylhypoxanthine 5′-monophosphate on genital lesions and encephalitis induced by *herpesvirus hominis* type 2 in female mice, *J. Infect. Dis.* **133**, A178–183 (1976).

1082. D. G. Bartholomew, J. H. Huffman, T. R. Matthews, R. K. Robins, and G. R. Revankar, Imidazo[1,2-c]pyrimidine nucleosides. Synthesis and biological evaluation of certain 1-(β-D-arabinofuranosyl)imidazo[1,2-c]pyrimidines, *J. Med. Chem.* **19**, 814–816 (1976).

1083. U. Niedballa and H. Vorbrüggen, A general synthesis of pyrimidine nucleosides, *Angew. Chem. Int. Ed. Engl.* **9**, 461–462 (1970).

1084. N. K. Kochetkov, V. N. Shibaev, and A. A. Kost, New reaction of adenine and cytosine derivatives, potentially useful for nucleic acids modification, *Tetrahedron Lett.* **1971**, 1993–1996.

1085. N. K. Kochetkov, V. N. Shibaev, and A. A. Kost, Modification of nucleic acid components with α-haloaldehydes, *Dokl. Akad. Nauk. SSSR* **205**, 100–103 (1972).

1086. J. R. Barrio, J. A. Secrist III, and N. J. Leonard, Fluorescent adenosine and cytidine derivatives, *Biochem. Biophys. Res. Commun.* **46**, 597–604 (1972).

1087. J. A. Secrist III, J. R. Barrio, N. J. Leonard, and G. Weber, Fluorescent modification of adenosine-containing coenzymes. Biological activities and spectroscopic properties, *Biochemistry* **11**, 3499–3506 (1972).

1088. B. S. Kramer, C. C. Fenselau, and D. B. Ludlum, Reaction of BCNU [1,3-bis(2-chloroethyl)-1-nitrosourea] with polycytidylic acid. Substitution of the cytosine ring. *Biochem. Biophys. Res. Commun.* **56**, 783–788 (1974).

1089. D. B. Ludlum, B. S. Kramer, J. Wang, and C. C. Fenselau, Reaction of 1,3-bis(2-chloroethyl)-1-nitrosourea with synthetic polynucleotides, *Biochemistry* **14**, 5480–5485 (1975).

1090. J. R. Barrio, L. G. Dammann, L. H. Kirkegaard, R. L. Switzer, and N. H. Leonard, Enzymatic activity of γ-^{32}P labelling of fluorescent derivatives of cytidine triphosphate and adenosine triphosphate, *J. Am. Chem. Soc.* **95**, 961–962 (1973).

1091. S. D. Rose, Mercuration of modified nucleotides: Chemical methods toward nucleic acid sequencing by electron microscopy, *Biochim. Biophys. Acta* **361**, 231–235 (1974).

1092. J. C. Greenfield, N. J. Leonard, and R. I. Gumport, Nicotinamide 3,N^4-ethenocytosine dinucleotide, an analog of nicotinamide adenine dinucleotide. Synthesis and enzyme studies, *Biochemistry* **14**, 698–706 (1975).

1093. G. C. Walker and O. C. Uhlenbeck, Stepwise enzymatic oligoribonucleotide synthesis including modified nucleotides, *Biochemistry* **14**, 817–824 (1975).

1094. N. K. Kochetkov, V. N. Shibaev, A. A. Kost, A. P. Razjivin, and A. Yu. Borisov, New fluorescent cytidine 5′-phosphate derivatives, *Nucleic Acids Res.* **3**, 1341–1349 (1976).

1095. C. Ivancsics and E. Zbiral, Reactions with organophosphorous compounds. Structural transformation on nucleotides, nucleosides, and nucleoside bases using β-acylvinylphosphonium salts, *Monatsh. Chem.* **106**, 417–428 (1975).

1096. A. H.-J. Wang, J. R. Barrio, and I. C. Paul, Crystal and molecular structure of 3,N^4-ethenocytidine hydrochloride. A study of the dimensions and molecular interactions of the fluorescent ϵ-cytidine system, *J. Am. Chem. Soc.* **98**, 7401–7408 (1976).

1097. J. R. Barrio, P. D. Sattsangi, B. A. Gruber, L. G. Dammann, and N. J. Leonard, Species responsible for the fluorescence of 3,N^4-ethenocytidine, *J. Am. Chem. Soc.* **98**, 7408–7414 (1976).

1098. M. Jaskolski, W. Krzyzosiak, H. Sierzputowska-Gracz, and M. Wiewiorowski, Comparative structural analysis of cytidine, ethenocytidine, and their protonated salts. I. Crystal and molecular structure of ethenocytidine, *Nucleic Acids Res.* **9**, 5423–5442 (1981).

1099. W. Krzyzosiak, M. Jaskolski, H. Sierzputowska-Gracz, and M. Wiewiorowski, Comparative structural analysis of cytidine, ethenocytidine, and their protonated salts. II. IR spectral studies, *Nucleic Acids Res.* **10**, 2741–2753 (1982).

1100. S.-H. Kim, D. G. Bartholomew, L. B. Allen, R. K. Robins, G. R. Revankar, and P. Dea, Imidazo[1,2-*a*]-*s*-triazine nucleosides. Synthesis and antiviral activity of the *N*-bridgehead guanine, guanosine, and guanosine monophosphate analogues of imidazo[1,2-*a*]-*s*-triazine, *J. Med. Chem.* **21**, 883–889 (1978).

1101. B. Kojic-Prodic, Z. Ruzic-Toros, L. Golic, B. Brdar, and J. Kobe, Conformation and structure of 2-amino-8-(β-D-ribofuranosyl-imidazo[1,2-*a*]-*s*-triazin-4-one (5-aza-7-deazaguanosine), a potent antiviral nucleoside, *Biochim. Biophys. Acta* **698**, 105–110 (1982).

1102. E. J. Prisbe, J. P. H. Verheyden, and J. G. Moffatt, 5-Aza-7-deazapurine nucleosides. 1. Synthesis of some 1-(β-D-ribofuranosyl)imidazo[1,2-*a*]-1,3,5-triazines, *J. Org. Chem.* **43**. 4774–4784 (1978).

1103. E. J. Prisbe, J. P. H. Verheyden, and J. G. Moffatt, 5-Aza-7-deazapurine nucleosides. 2. Synthesis of some 8-(D-ribofuranosyl)imidazo[1,2-*a*]-1,3,5-triazine derivatives, *J. Org. Chem.* **43**, 4784–4794 (1978).

1104. H. Rosemeyer and F. Seela, 5-Aza-7-deaza-2′-deoxyguanosine: Studies on the glycosylation of weakly nucleophilic imidazo[1,2-*a*]-*s*-triazinyl anions, *J. Org. Chem.* **52**, 5136–5143 (1987).

1105. F. Seela, U. Bindig, H. Driller, W. Herdering, K. Kaiser, A. Kehne, H. Rosemeyer, and H. Steker, Synthesis and application of isoteric purine 2′-deoxyribofuranosides, *Nucleosides Nucleotides* **6**, 11–23 (1987).

1106. H. Reimlinger, M. A. Peiren, and M. Robert, Synthesis with heterocyclic amines. I. Reactions of 3(5)-aminopyrazole with α-β-unsaturated esters. Preparation and characterization of isomeric oxodihydropyrazolopyrimidines, *Chem. Ber.* **103**, 3252–3265 (1970).

1107. Y. Makisumi, Studies on the azaindolizine compounds. X. Synthesis of 5,7-disubstituted pyrazolo[1,5-*a*]pyrimidines, *Chem. Pharm. Bull.* **10**, 612–620 (1962).

1108. J. Kobe, R. K. Robins, and D. E. O'Brien, The synthesis and chemical reactions of certain pyrazolo[1,5-*a*]-1,3,5-triazines, *J. Heterocycl. Chem.* **11**, 199–204 (1974).

1109. G. Wagner, G. Valz, B. Dietzsch, and G. Fischer, Glycosides of heterocycles. 53. Glycosides of 7-hydroxy- and 7-mercapto-*s*-triazolo[1,5-*a*]pyrimidines, *Pharmazie* **30**, 134–141 (1975).

1110. C. Bulow and K. Haas, Synthetic attempts to prepare derivatives of heterocondensed heterocyclic 1,3-triazo-7,O′-pyrimidine, *Ber.* **42**, 4638–4644 (1909).

1111. E. L. Birr and W. Walther, The constitution of the triazaindolizines, *Chem. Ber.* **86**, 1401–1403 (1953).

1112. C. F. H. Allen, H. R. Beilfuss, D. M. Burness, G. A. Reynolds, J. F. Tinker, and J. A. Van Allan, The structure of certain polyazaindenes. II. The product from ethyl acetoacetate and 3-amino-1,2,4-triazole, *J. Org. Chem.* **24**, 787–796 (1959).

1113. Y. Makisumi and H. Kano, Studies on the azaindolizine compounds. XII. Alkyl rearrangement of 5-methyl-7-alkoxy-*s*-triazolo[1,5-*a*]pyrimidines, *Chem. Pharm. Bull.* **11**, 67–75 (1963).

1114. G. E. Hilbert and T. B. Johnson, Researches on pyrimidines. CXVII. A method for the synthesis of nucleosides, *J. Am. Chem. Soc.* **52**, 4489–4494 (1930).

1115. M. Cerny, J. Vrkoc, and J. Stanek, Preparation of acylated glycopyranosyl mercaptans, *Collect. Czeh. Chem. Commun.* **24**, 64–69 (1959).

1116. J.-L. Barascut and J.-L. Imbach, Synthetic nucleosides. VIII. Preparation of ribofuranosides of *s*-triazolopyrimidinones via silylderivatives and by fusion reactions, *Bull. Soc. Chim. Fr., Pt. 2* **1975**, 2561–2566.

1117. J.-L. Barascut, C. O. De Marichard, and J.-L. Imbach, Synthetic nucleosides. XII. Study of the ribosylation of triazolopyrimidines, *J. Carbohydr. Nucleosides Nucleotides* **3**, 281–305 (1976).

1118. H. Reimlinger and M. A. Peiren, Syntheses with heterocyclic amines. II. Reactions of 3-amino-1-2,4-triazole with methyl propiolate. Preparation and characterization of four isomeric oxodihydro-*s*-triazolopyrimidines, *Chem. Ber.* **103**, 3266–3277 (1970).

1119. Y. Makisumi and H. Kano, Synthesis of potential anticancer agents. III. 7-Substituted *s*-triazolo[2,3-*a*]pyrimidines and their halogenated compounds, *Chem. Pharm. Bull.* **7**, 907–911 (1959).

1120. B. Rayner, C. Tapiero, and J.-L. Imbach, Studies on Δδ criterion for determining the anomeric configuration of ribofuranosyl nucleosides, *Carbohydr. Res.* **47**, 195–202 (1976).

1121. M. W. Winkley, G. F. Judd, and R. K. Robins, Synthesis of 3- and 4-(β-D-ribofuranosyl)-*s*-triazolo[2,3-*a*]pyrimid-7-one, isomers of inosine containing a bridgehead nitrogen atom, *J. Heterocycl. Chem.* **8**, 237–240 (1971).

1122. J.-L. Barascut, C. O. De Marichard, and J.-L. Imbach, Synthesis of nucleosides XV. Synthesis of isosteres of 6-mercaptopurine nucleosides, *J. Carbohydr. Nucleosides Nucleotides* **5**, 149–162 (1978).

1123. G. R. Revankar, R. K. Robins, and R. L. Tolman, *s*-Triazolo[1,5-*a*]pyrimidine nucleosides. Site of *N*-glycosylation studies and the synthesis of an *N*-bridgehead guanosine analog, *J. Org. Chem.* **39**, 1256–1262 (1974).

1124. R. L. Tolman and G. R. Revankar, *s*-Triazolo[2,3-*a*]pyrimidine nucleosides, U.S. Patent 3,868,361 (Feb. 1975).

1125. Y. Makisumi, Synthesis of potential anticancer agents, IV. 5,7-Disubstituted *s*-triazolo-[2,3-*a*]pyrimidines, *Chem. Pharm. Bull.* **9**, 801–808 (1961).

1126. G. R. Revankar and R. K. Robins, The synthesis of heterocyclic analogs of purine nucleosides and nucleotides containing a bridgehead nitrogen atom, in: *Chemistry and Biology of Nucleosides and Nucleotides* (R. E. Harmon, R. K. Robins, and L. B. Townsend, eds.), pp. 287–299, Academic Press, New York (1978).

1127. B. Rathke, Über monophenylisocyanursaure; über ein viertes triphenylmelamin und seine umwandlung in das normale, *Ber.* **21**, 867–873 (1988).

1128. O. Dimroth, Intramolecular rearrangements, *Liebigs Ann. Chem.* **364**, 183–226 (1909).

1129. B. G. Ugarkar, K. G. Upadhya, R. K. Robins, and G. R. Revankar, Synthesis of 5-azaxanthosine and related compounds, *Nucleosides Nucleotides*, in press.

1130. M. Fuertes, R. K. Robins, and J. T. Witkowski, Synthesis of 3-amino-1,2,4-triazole ribonucleosides, *J. Carbohydr. Nucleosides Nucleotides* **3**, 169–175 (1976).

1131. S. G. Wood, N. K. Dalley, R. D. George, R. K. Robins, and G. R. Revankar, Synthesis and structural studies of certain novel imidazo[1,2-*b*]pyrazole nucleosides, *J. Org. Chem.* **49**, 3534–3540 (1984).

1132. S. G. Wood, N. K. Dalley, R. D. George, R. K. Robins, and G. R. Revankar, Synthesis and X-ray crystal structure of 3-amino-1-β-D-ribofuranosyl-*s*-triazolo[5,1-*c*]-*s*-triazole, *Nucleosides Nucleotides* **3**, 187–194 (1984).

1133. K. T. Potts and C. Hirsch, 1,2,4-Triazoles. XVIII. The synthesis of 5-*H*-*s*-triazolo[5,1-*c*]-*s*-triazole and its derivatives, *J. Org. Chem.* **33**, 143–150 (1968).

1134. M. Maeda and Y. Kawazoe, Studies on chemical alterations of nucleic acids and their components. X. Syntheses and reactivities of 3-aminopyrimidine nucleosides, *Chem. Pharm. Bull.* **23**, 844–852 (1975).

1135. H. Hayatsu, A. Kitajo, K. Sugihara, N. Nitta, and K. Negishi, The reaction between 4-aminocytidine and imidate esters, *Nucleic Acids Res. Spec. Publ.* **5**, s315–318 (1978).

1136. B. M. Lynch and S. C. Sharma, Triazolopyridine nucleosides. II. Glycosylation of 3-oxo-*s*-triazolo[4,3-*a*]pyridines with accompanying conversion into 2-oxo-*s*-triazolo[1,5-*a*]pyridine nucleosides, *Can J. Chem.* **55**, 831–840 (1977).

1137. A. Saito and B. Shimizu, Synthesis of mesoionic triazolopyridine. I. N-Alkylation of 1,2,4-triazolo[4,3-*a*]pyridin-3(2*H*)-one, *Bull. Chem. Soc. Jpn.* **50**, 1596–1599 (1977).

1138. A. R. Davis, E. B. Newton, and S. R. Benedict, The combined uric acid in beef blood, *J. Biol. Chem.* **54**, 595–599 (1922).

1139. R. Falconer and J. M. Gulland, The constitution of the purine nucleosides. Part VIII. Uric acid riboside, *J. Chem. Soc.* **1939**, 1369–1371.

1140. M. M. Jezewska, B. Gorzkowski, and T. Sawicka, The structure of uric acid riboside and changes in its content in moths, *Acta Biochim. Pol.* **14**, 71–75 (1967).

1141. H. S. Forrest, D. Hatfield, and J. M. Lagowski, Uric acid riboside. Part I. Isolation and reinvestigation of the structure, *J. Chem. Soc.* **1961**, 963–968.

1142. R. Lohrmann, J. M. Lagowski, and H. S. Forrest, 3-Ribosyluric acid. Part III. Unambiguous syntheses of 3-ribosyluric acid and related compounds, *J. Chem. Soc.* **1964**, 451–459.

1143. K. Gerzon, I. S. Johnson, G. B. Boder, J. C. Cline, P. J. Simpson, C. Speth, N. J. Leonard, and R. A. Laursen, Biological activities of 3-isoadenosine, *Biochim. Biophys. Acta* **119**, 445–461 (1966).

1144. W. N. Kelley and T. D. Beardmore, Allopurinol: Alteration in pyrimidine metabolism in man, *Science* **169**, 388–390 (1970).

1145. V. D. Patil, D. S. Wise, and L. B. Townsend, The synthesis of 2-ethyl-6-(β-D-ribofuranosyl)-oxazolo[5,4-*d*]pyrimidin-7-one, *J. Heterocycl. Chem.* **10**, 277–278 (1973).

1146. V. D. Patil and L. B. Townsend, The synthesis of certain 2,7-disubstituted oxazolo[5,4-*d*]-pyrimidines, *J. Heterocycl. Chem.* **8**, 503–505 (1971).

1147. V. D. Patil, D. S. Wise, L. B. Townsend, and A. Bloch, Synthesis and biological activity of selected 2-substituted 6-(β-D-ribofuranosyl)-oxazolo[5,4-*d*]pyrimidin-7-ones, *J. Med. Chem.* **17**, 1282–1285 (1974).

1148. M. Sekiya and J. Suzuki, Azole series. III. Reactions of 2-acylamino-2-cyanoacetamides leading to 5-aminooxazole-4-carboxamides and to oxazolo[5,4-*d*]pyrimidines, *Chem. Pharm. Bull.* **18**, 2242–2246 (1970).

1149. L. B. Townsend, V. D. Patil, and A. Bloch, Synthesis and biological activity of ribofuranosyl derivatives of oxazolo[5,4-*d*]pyrimidines, *Proc. Am. Assoc. Cancer Res.* **12**, 74 (1971).

1150. V. D. Patil, D. S. Wise, and L. B. Townsend, The synthesis of thieno[2,3-*d*]pyrimidine nucleosides related to the naturally occurring nucleosides cytidine and uridine, *J. Chem. Soc., Perkin Trans 1*, **1980**, 1853–1858.

1151. V. D. Patil, D. S. Wise, L. L. Wotring, L. C. Bloomer, and L. B. Townsend, Synthesis and biological activity of a novel adenosine analogue, 3-β-D-ribofuranosylthieno[2,3-*d*]pyrimidin-4-one, *J. Med. Chem.* **28**, 423–427 (1985).

1152. T. Ueda, K. Miura, M. Imazawa, and K. Odajima, Nucleosides and nucleotides. X. Synthesis of 4-thiouracil nucleosides and nucleotides by the solvolysis of cytidine and its phosphates with hydrogen sulfide, *Chem. Pharm. Bull.* **22**, 2377–2382 (1974).

1153. W.-Y. Ren, M.-I. Lim, B. A. Otter, and R. S. Klein, Synthetic studies of the thieno[3,2-*d*]pyrimidine C-nucleoside isostere of inosine, *J. Org. Chem.* **47**, 4633–4637 (1982).

1154. S. J. Childress and R. L. McKee, Thiazolopyrimidines, *J. Am. Chem. Soc.*, **73**, 3862–3864 (1951).

1155. C. L. Schmidt and L. B. Townsend, Bicyclic nucleosides related to pyrimidine nucleosides. IV. Synthesis of 4- and 6-ribofuranosylthiazolo[5,4-*d*]pyrimidines and 4-arabinofuranosylthiazolo[5,4-*d*]pyrimidines, *J. Org. Chem.* **40**, 2476–2481 (1975).

1156. K. Nagahara, J. D. Anderson, G. D. Kini, N. K. Dalley, S. B. Larson, D. F. Smee, Ai Jin, B. S. Sharma, W. B. Jolley, R. K. Robins, and H. B. Cottam, Thiazolo[4,5-*d*]pyrimidine nucleosides.

The synthesis of certain 3-β-D-ribofuranosylthiazolo[4,5-*d*]pyrimidines as potential immuno-therapeutic agents, *J. Med. Chem.* **32**, 407–415 (1990).

1157. J. A. Baker and P. V. Chatfield, Synthesis of derivatives of thiazolo[4,5-*d*]pyrimidine. Part II, *J. Chem. Soc. (C)* **1970**, 2478–2484.

1158. S. B. Larson, J. D. Anderson, H. B. Cottam, and R. K. Robins, Structure of the sodium salt of a thiazolopyrimidine, a guanine analog, *Acta Crystallogr.,* **C45**, 1073–1076 (1989).

1159. M. G. Goodman and W. O. Weigle, Intracellular lymphocyte activation and carrier-mediated transport of C8-substituted guanine ribonucleosides, *Proc. Natl. Acad. Sci. U.S.A.* **81**, 862–866 (1984).

1160. A. Ahmad and J. J. Mond, Restoration of *in vitro* responsiveness of XID B cells to TNP-FICOLL by 8-mercaptoguanosine, *J. Immunol.* **136**, 1223–1226 (1986).

1161. T. L. Feldbush and Z. K. Ballas, Lymphokine-like activity of 8-mercaptoguanosine: Induction of T and B cell differentiation, *J. Immunol.* **134**, 3204–3211 (1985).

1162. G. C. Koo, M. E. Jewell, C. L. Manyak, N. H. Sigal, and L. S. Wicker, Activation of murine natural killer cells and macrophages by 8-bromoguanosine, *J. Immunol.* **140**, 3249–3252 (1988).

1163. M. G. Goodman and W. J. Hennen, Distinct effects of dual substitution on inductive and differentiative activities of C8-substituted guanine ribonucleosides, *Cell. Immunol.* **102**, 395–402 (1986).

1164. D. F. Smee, H. A. Alaghamandan, H. B. Cottom, B. S. Sharma, W. B. Jolley, and R. K. Robins, Broad-spectrum *in vivo* antiviral activity of 7-thia-8-oxoguanosine, a novel immunopotentiating agent related to guanosine, *Antimicrob. Agents Chemother.* **33**, 1487–1492 (1989).

1165. I. Ito, N. Oda, T. Kato, and K. Ota, Synthesis of compounds related to antitumor agents. II. Studies on the synthesis of oxazolo[4,5-*d*]pyrimidine nucleoside derivatives, *Chem. Pharm. Bull.* **23**, 2104–2108 (1975).

1166. N. Oda, Y. Kanie, and I. Ito, Synthesis of compounds related to antitumor agents. I. Studies on the synthesis of oxazolo[4,5-*d*]pyrimidine and pyrimido[5,4-*b*][1,4]oxazine derivatives, *Yakugaku Zasshi* **93**, 817–821 (1973).

1167. N. Shimada, N. Yagisawa, H. Naganawa, T. Takita, M. Hamada, T. Takeuchi, and H. Umezawa, Oxanosine, a novel nucleoside from actinomycetes, *J. Antibiot.* **34**, 1216–1218 (1981).

1168. H. Nakamura, N. Yagisawa, N. Shimada, T. Takita, H. Umezawa, and Y. Iitaka, The X-ray structure determination of oxanosine, *J. Antibiot.* **34**, 1219–1221 (1981).

1169. Y. Uehara, M. Hasegawa, M. Hori, and H. Umezawa, Differential sensitivity of RSV[ts] (temperature-sensitive Rous-sarcoma virus)-infected rat kidney cells to nucleoside antibiotics at permissive and nonpermissive temperatures, *Biochem. J.* **232**, 825–831 (1985).

1170. N. Yagisawa, N. Shimada, T. Takita, M. Ishizuka, T. Takeuchi, and H. Umezawa, Mode of action of oxanosine, a novel nucleoside antibiotic, *J. Antibiot.* **35**, 755–759 (1982).

1171. Y. Uehara, M. Hasegawa, M. Hori, and H. Umezawa, Increased sensitivity to oxanosine, a novel nucleoside antibiotic, of rat kidney cells upon expression of the integrated viral src gene, *Cancer Res.* **45**, 5230–5234 (1985).

1172. N. Yagisawa, T. Takita, H. Umezawa, K. Kato, and N. Shimada, A facile total synthesis of oxanosine, a novel nucleoside antibiotic, *Tetrahedron Lett.* **24**, 931–932 (1983).

1173. K. Kato, N. Yagisawa, N. Shimada, M. Hamada, T. Takita, K. Maeda, and H. Umezawa, Chemical modification of oxanosine. I. Synthesis and biological properties of 2′-deoxyoxanosine, *J. Antibiot.* **37**, 941–942 (1984).

1174. S. Niitsuma, K. Kato, T. Takita, and H. Umezawa, Synthesis of 3-deazaoxanosines, C-nucleoside isosteres of oxanosine, *Tetrahedron Lett.* **26**, 5785–5786 (1985).

1175. P. C. Srivastava, R. W. Mancuso, R. J. Rousseau, and R. K. Robins, Nucleoside peptides. 6. Synthesis of certain N-[5-amino-1-(β-D-ribofuranosyl)imidazole-4-carbonyl]amino acids related to naturally occurring intermediates in the purine biosynthetic pathway, *J. Med. Chem.* **17**, 1207–1211 (1974).

Chapter 5

Substrate Binding of Adenine Nucleosides and Nucleotides to Certain Enzymes

Alexander Hampton

1. Introduction

Adenine nucleosides and nucleotides are utilized as enzyme substrates in a wide variety of metabolic conversions. The mode of substrate binding of adenine nucleosides and nucleotides has presumably evolved from interactions that occurred in prebiotic times between primitive precursors of nucleotides and polypeptides. Knowledge of the mode of binding of these substrates is clearly essential for understanding mechanisms of enzyme-catalyzed transformations of adenine nucleosides and nucleotides. Furthermore, such knowledge can afford insights into evolutionary relationships between certain classes of enzymes as well as insights into principles involved in the organization and control of important areas of cellular metabolism. Additionally, information on the mode of binding of any type of substrate constitutes an important basis for the development of alternate substrates or of powerful and specific enzyme inhibitors that might serve as useful chemotherapeutic agents or as tools for studies of metabolism or enzyme mechanism. This chapter is devoted to aspects of the substrate binding of adenosine, adenosine 5'-mono-, di-, and triphosphates, and the adenine nucleotide portions of nicotinamide adenine dinucleotide and coenzyme A. A comprehensive survey of the very extensive literature has not been attempted; instead, an account has been given of major developments and contributions in the field by the various chemical and physical approaches so far found most useful.

Much valuable information on adenine nucleoside and nucleotide substrate binding has been obtained through studies of the substrate and inhibitor properties of analogues derived from the normal substrate by atom deletions or substitutions. In the present treatment of this subject, the kinetic data discussed

---

Alexander Hampton • Institute for Cancer Research, Fox Chase Cancer Center, Philadelphia, Pennsylvania 19111.

are, in the interests of simplicity and ease of interpretation, restricted in the main to those of compounds derived from the substrate or a substrate analogue by either single or sequential structural changes. Other useful probes of substrate binding discussed here are (a) those obtained by the attachment of substituents to the substrate, (b) those which appear to act as affinity labels for the substrate binding site, and (c) those which are conformationally restricted, either by intramolecular covalent bonds or by the steric effects of substituents. It is well to bear in mind that an interpretation of the effects of structural changes in a purine ring system on affinity for an enzyme is sometimes difficult because of complex electronic effects in the purine ring that can result from atom substitutions, from variations in existing groups, or from attachment of additional atoms or groups. Such electronic effects include, for example, changes in tautomeric equilibria, changes in basicity of exocyclic or ring nitrogens, shifts in the most basic center from one nitrogen to another, alterations in the electrophilicity of ring carbons, and alterations in the tendencies of the ring system to participate in charge-transfer interactions of various types.

For the majority of the enzymes discussed here, a detailed picture is emerging of the conformation of enzyme-bound adenosine or adenine nucleotides and of the structural requirements for binding of these substrates. X-ray crystallographic studies have, in several instances, been performed at sufficiently high resolution to allow identification of enzymatic amino acid residues that might interact with functional groups or individual atoms of the substrates. Future work at higher resolution can be expected to provide a more detailed delineation of interactions between various atoms of these substrates and atoms of their binding sites. A valuable approach directed to this end is to crystallize separately a free enzyme and its complex with a substrate, as illustrated by studies with hexokinase and the hexokinase–glucose complex.[1] A frequent difficulty with nucleotidic substrates, exemplified in studies with muscle adenylate kinase,[2] is to achieve significant binding of the nucleotide to enzyme crystals due to competition with sulfate anions arising from the relatively high concentrations of ammonium sulfate which are usually present. There is some promise that this difficulty will be surmountable through the use of affinity labels or of high-affinity nucleotide derivatives such as the powerful bisubstrate inhibitor P^1, P^5-di(adenosine-5′)pentapolyphosphate, which was reported recently to form a crystalline complex with an adenylate kinase.[2] Finally, with regard to enzyme–substrate complexes as they exist in solution, nuclear magnetic resonance techniques are proving increasingly useful in the delineation of conformations, arrangements, and distances from specific enzyme protons of enzyme-bound substrates. Examples of the application of these techniques to nucleotidic substrates are given in Sections 4 and 5.

2. Substrate Binding of Adenosine

2.1. Adenosine Kinase

Adenosine kinase (EC 2.7.1.20) catalyzes the transfer of phosphate from a nucleoside 5′-triphosphate to adenosine (**1**) to give adenosine 5′-phosphate

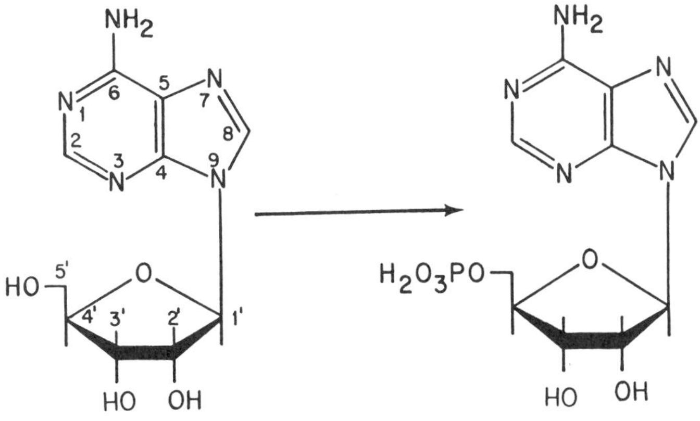

1   2

(AMP) (**2**). Preparations of the enzyme, some apparently homogeneous, have been obtained from yeast and mammalian tissues.[3,4] Cytotoxic or antibiotic properties are exhibited by a variety of purine nucleosides and appear in most cases to be associated with intracellular formation of the corresponding nucleoside 5′-phosphates catalyzed by adenosine kinase.[5,6] This has prompted investigations of the substrate and inhibitor properties of adenosine analogues and derivatives with preparations of the enzyme from several mammalian sources, most extensively from established lines of human and mouse neoplastic cells and from rabbit liver, as described below. Kinetic analysis has suggested that with a human placental enzyme preparation the preferred kinetic sequence was E→E–AR→E–AR–MgATP→E–AMP–MgADP→E–AMP (where AR = adenosine).[7] In contrast, kinetic, substrate binding, and isotope exchange studies with a preparation from L1210 mouse leukemia cells led to the conclusion that the most probable sequence involved a phosphorylated enzyme intermediate and was E→MgATP–E–AR (random order of addition) → P$_i$–E–AR + MgADP→E–AMP→E–AMP; in addition, studies with the L1210 enzyme and labeled ATP indicated a ping-pong kinetic mechanism in which E→MgATP→P$_i$–E and P$_i$–E + AR→P$_i$–E–AR→E–AMP.[4]

2.1.1. Binding of the Adenine Moiety

Bennett and his associates[8] purified adenosine kinase 175-fold from human tumor cells of type HEp-2 and obtained a preparation free of adenosine deaminase and adenylate kinase activities. Purine nucleoside analogues and derivatives differing from adenosine in the purine moiety were tested as substrates at concentrations 500-fold higher than the K_M value of adenosine itself (Table I). Purine ribonucleoside was a better phosphate acceptor than adenosine, indicating that the 6-amino group of adenosine plays little, if any, role in catalytic events. Replacement of the 6-amino group by the small electron-donating or electron-accepting groups —CH$_3$, —SCH$_3$, —OCH$_3$, —NHCH$_3$, —Cl, —NHNH$_2$, or —N(CH$_3$)$_2$ permitted high substrate activity. Replacement of the 1- or 3-nitro-

Table I. Phosphorylation of Nucleosides by Adenosine Kinase of HEp-2 Cells[a]

Substrate (1 mM)	Nucleotide formed (nmol/min per mg protein)
Adenosine	148
Guanosine	<10
Inosine	<10
Purine ribonucleoside	510
1-Methyladenosine	158
Adenosine N^1-oxide	597
2-Methyladenosine	<10
2-Methoxyadenosine	<10
2-Aminoadenosine	54
2-Hydrazinoadenosine	68
2-Dimethylaminoadenosine	<10
2-Fluoroadenosine	349
2-Chloroadenosine	<10
2-Bromoadenosine	<10
6-Methylpurine ribonucleoside	468
6-Chloropurine ribonucleoside	424
6-Methoxypurine ribonucleoside	309
6-Methylthiopurine ribonucleoside	487
6-Hydrazinopurine ribonucleoside	216
N^6-Methyladenosine	571
N^6,N^6-Dimethyladenosine	235
N^6-Allyladenosine	55
1-Deaza-6-methylthiopurine ribonucleoside	<20
3-Deaza-6-methylthiopurine ribonucleoside	<10
7-Deazaadenosine	643
8-Azaadenosine	279
8-Aza-9-deazaadenosine	149

[a]Data from ref. 8.

gen of 6-methylthiopurine ribonucleoside by carbon abolished substrate activity, indicating possible electron donation by these nitrogens to atoms of the enzyme. This view is consistent with the lack of activity of inosine and guanosine, in which N-1 is weakly acidic rather than weakly basic. In this regard, the high activity of adenosine N^1-oxide, despite its relatively weak basicity, might be associated with electron donation to the enzyme from O^1 rather than from N-1. Substrate activity of 2-substituted adenosines is seen from Table I to vary widely with no obvious correlation with the size or electronic nature of the substituents. Binding of the imidazole portion of the purine ring system appears to be less specific than binding of the pyrimidine portion because neither N-7, C-8, nor N-9 is essential for activity, and 8-aza-9-deazaadenosine (formycin) is as active as adenosine itself under the conditions of the velocity data of Table I.

A partially purified preparation of adenosine kinase from rabbit liver resembled the HEp.-2 human enzyme in that replacement of the 6-amino group of adenosine by —H, —Cl, or —NHCH$_3$ or insertion of —F or —NH$_2$ at the 2-position permitted good substrate activity.[9] However, 1-methyladenosine showed no substrate activity with the rabbit liver preparation yet was about as effective as adenosine as a substrate of the HEp-2 preparation.[8] More recent studies utilizing extensively purified rabbit liver enzyme have confirmed this finding.[10] It is not clear whether 1-methyladenosine was phosphorylated by a kinase other than adenosine kinase in the HEp-2 preparation or whether the apparent differential substrate effects arise from a structural difference between rabbit liver and HEp-2 adenosine kinases. It is noteworthy that no selectivity was manifested by adenosine N^1-oxide, which exhibited high activity with both of these kinases. The rabbit liver enzyme further resembled that from HEp-2 cells in that replacement of N-1 or N-3 by carbon abolished or greatly reduced substrate activity.[10] In addition, inosine and guanosine were, at best, feeble substrates for both enzymes, whereas 7-deazaadenosine (tubercidin), 8-aza-adenosine, and 8-aza-9-deazaadenosine (formycin) were good substrates.[8–10]

Substituent size at C-6 of adenosine has been shown to be a major determinant of substrate activity with human and mouse adenosine kinases. Thus, with the HEp-2 enzyme, N^6-allyladenosine was a much less effective substrate than N^6-methyladenosine, while 6-allylthio- and 6-benzylthiopurine ribonucleosides, in contrast to the 6-methylthio derivative, showed no substrate activity.[8] N^6-Allyl- and N^6-benzyladenosine were phosphorylated about one-third as rapidly as adenosine by a partially purified preparation of adenosine kinase from mouse sarcoma -180 cells; N^6-pentyl- and N^6-phenyladenosine were not significantly phosphorylated but were potent inhibitors.[11] These properties of N^6-substituted adenosines, together with studies of competitive inhibition of 6-ureidopurine ribonucleosides,[12] have indicated that adenosine kinase of sarcoma-180 cells possesses a hydrophobic area adjacent to N^6 of enzyme-bound adenosine. In the case of highly purified adenosine kinase from rabbit liver, attachment of alkyl or aryl groups to N^6 produced similar effects on substrate and inhibitor properties.[10]

2.1.2. Binding of the Ribofuranose Ring

Nucleosides of adenine and 8-azaadenine in which the 1'-, 2'-, 3'-, and 4'-substituents were either *cis* or *trans* to the purine ring, as well as 2'- and 3'-dexoyadenosines, have been examined as substrates of a partially purified preparation of adenosine kinase from HEp-2 cells.[13] The 9-glycosyl groups studied included α- and β-D-ribofuranosyl, α- and β-D-arabinofuranosyl, β-D-xylofuranosyl, and α-L-xylofuranosyl. The results indicated that three conditions necessary for substrate activity are the presence of a 2'-hydroxyl group, a *trans* relationship of this hydroxyl to the purinyl substituent, and an ability to assume a favorable torsional angle about the glycosidic bond. Thus, the observed inactivity of 1'-hydroxymethyladenosine and of those nucleosides with a 2'-hydroxyl group *cis* to the purine ring is ascribable, at least in part, to insufficient freedom of rotation about this bond. The studies revealed further that the 3'-hydroxyl and 4'-hydroxymethyl groups can be either *cis* or *trans* to the purine ring, although departures from the orientations found in adenosine tended to decrease substrate activity. In addition, the furanose oxygen is not essential, inasmuch as replacement of it by a methylene group decreased the rate of enzyme-catalyzed phosphorylation by only 60%.

Adenosine kinase purified partially[9] or to apparent homogeneity[10] from rabbit liver resembled the preparation from HEp-2 cells in that 1'-hydroxy-methyladenosine (psicofuranine) was not a substrate. In addition, replacement of the 2'-hydroxyl group of adenosine by hydrogen or its inversion to give arabinofuranosyladenine reduced substrate activity much more drastically than the same operations at C-3'. Adenosine kinase purified to apparent homogeneity from L1210 mouse leukemia cells did not appear to phosphorylate 2'-deoxy-adenosine or arabinofuranosyladenine.[14] The rate of phosphorylation of 2'-deoxyadenosine by apparently homogeneous preparations of the enzyme from rat liver[15] and rat brain[16] was 1.5 and 20%, respectively, that of adenosine. It should be noted that mammalian 2'-deoxyadenosine kinase is an enzyme distinct from adenosine kinase.[17] With the rabbit liver enzyme, replacement of the 2'- or 3'- hydroxyl groups of adenosine by an amino group permitted appreciable substrate activity to persist, whereas introduction of an amino group at the 5'-position apparently prevented substrate activity but produced an effective reversible inhibitor, $K_i = 0.02$ μM (K_M of adenosine, 0.4 μM).[10] The inhibition constant of 2', 5'-dideoxyadenosine as a competitive inhibitor of the rabbit enzyme was 100-fold higher than that of 5'-deoxyadenosine,[10] indicating that the 2'-hydroxyl group probably makes an important contribution to the binding of adenosine to its site. This conclusion is supported by the relatively high K_M value (575 μM) of 2'-deoxyadenosine with rabbit liver adenosine kinase.[10]

2.2. Adenosine Aminohydrolase

Adenosine aminohydrolase (EC 3.5.4.4), which catalyzes conversion of adenosine (**1**) to inosine (**3**), is widely distributed among living organisms.[18] Abnormally low adenosine aminohydrolase activity is implicated as a causative factor in certain types of combined immunodeficiency disease.[19,20] On the

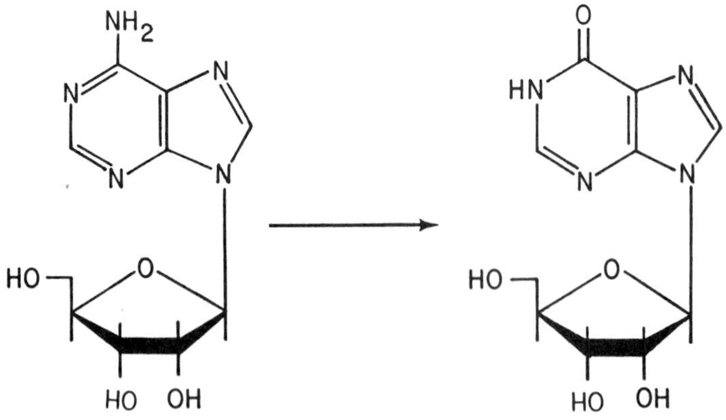

1   3

other hand, the enzyme inactivates a number of adenosine analogues which possess chemotherapeutic properties, and, in consequence, considerable effort has been directed toward studies of structural requirements for substrate activity and for binding to the substrate site, with the object of assisting the design of strong inhibitors of the enzyme as well as the design of adenosine analogues resistant to enzymatic deamination. Most of this work has utilized the commercially available, partially purified adenosine aminohydrolase of calf duodenum, and the present discussion of substrate binding refers to results obtained with enzyme from that source.

2.2.1. Binding of the Adenine Moiety

Table II lists kinetic constants of some purine-modified analogues and derivatives of adenosine which have been examined as substrates and/or inhibitors of calf duodenal adenosine aminohydrolase. All inhibitions listed are competitive in nature with respect to adenosine. Table II shows that K_M and K_i values for a given compound do not differ more than twofold. On this basis in the present discussion it is assumed that when a K_i value is not available, the K_M value serves as an approximate measure of the relative affinity of a compound for the enzyme under the test conditions of Table II. Removal of the 6-amino group of

Table II. Kinetic Constants with Calf Duodenal Adenosine Aminohydrolase for Compounds in Which the Purine Moiety of Adenosine Is Modified

Compound	Rel V_{max}	K_M (μM)	K_i (μM)	Reference
Adenosine	100	35		21, 22
9-β-D-Ribofuranosylpurine			9.3	23
Adenine	0.0014	150	280	23, 24
1-Methyladenosine	0.0014	96		25
	0.17	96		23
2-Aminoadenosine	25	34		26
2-Fluoroadenosine	—^{a}			
6-Chloropurine ribonucleoside	40	667		21
6-Methoxypurine ribonucleoside	0.45	32	25	26, 27
N^6-Hydroxyadenosine	38	45		27
N^6-Methyladenosine	0.16	11		27
N^6-Ethyladenosine	0.14	120	210	27
N^6,N^6-Dimethyladenosine	0		110	27
6-Mercaptopurine ribonucleoside	0		370	28
8-Aminoadenosine	3.7	137	89	25
8-Oxoadenosine	10	63	123	25
8-Azaadenosine	217	96	145	25
8-Fluoroadenosine	1	220		29
8-Bromoadenosine	0			25
8-Methylthioadenosine	0			29
1-Deazaadenosine	0		2	29
3-Deazaadenosine	0			27, 29
7-Deazaadenosine	0			29

^{a}Reported to be a substrate in ref. 21; substrate constants were not obtained.

adenosine is seen not to diminish affinity whereas removal of the 9-(β-D-ribofuranosyl) group decreases affinity about 10-fold. The diminution in binding upon removal of the ribofuranosyl group varies considerably among 6-substituted purine ribonucleosides: when the substituent is hydrogen, chloro, or methylamino, binding is diminished by factors of 130, 2, and 75, respectively.[24] The affinity of 2-aminoadenine is 8.5 times less than that of 2-aminoadenosine.[24] It has been shown[30] that a removal of the ribofuranosyl group from a 6-substituted ribonucleoside has only a 2.5-fold effect on the ease of non-enzymatic nucleophilic displacement of the 6-substituent, whereas with adenosine it brings about a more than 10^4-fold reduction in V_{max} of the nucleophilic displacement catalyzed by adenosine aminohydrolase (Table II). This prompted Wolfenden *et al.*[30] to propose that the ribose moiety of adenosine may stabilize the transition state for enzymatic deamination. In accord with this view, removal of ribose from the transition state analogue 1,6-dihydro-6-hydroxymethylpurine ribonucleoside was found to lower affinity for the enzyme by several orders of magnitude.[30]

The low V_{max} of 1-methyladenosine (Table II) might result from steric interference by the methyl group with the catalyzed reaction or from decreased electrophilicity at C-6 associated with the 10^4-fold increase in basicity occasioned by introduction of the 1-methyl group, or from both factors.[31] Small substituents on C-2 permit substrate activity, as do a variety of leaving groups (chloro, methoxy, hydroxylamino, or methylamino) attached to C-6 in place of the 6-amino group of adenosine. N^6,N^6-Dialkyladenosines have lower V_{max} values and lower enzyme affinities than the corresponding N^6-monoalkyladenosines; for example, N^6,N^6-diethyladenosine, in contrast to N^6-ethyladenosine (Table II), showed no substrate or inhibitor properties.[27] With increasing size of N^6-monoalkyl substituents, there is a sharp reduction both in substrate activity and in affinity for the enzyme: N^6-*n*-propyl-, N^6-isopropyl, and N^6-benzyladenosines, for example, are neither substrates nor inhibitors.[27]

It has been proposed from kinetic evidence that nucleophilic attack at C-6 of adenosine can be the rate-limiting event in catalysis by this enzyme.[32] On the basis of this view, suitably positioned electron-withdrawing groups might be able to enhance V_{max}, and it is of interest to note that replacement of C-8 of adenosine by nitrogen doubles V_{max} (Table II) and that, in the case of adenine as substrate, replacement of C-8 by oxygen or sulfur serves to increase V_{max} by several orders of magnitude.[23] Small groups (fluoro, oxo, amino) introduced at the 8-position of adenosine permit some substrate activity but reduce affinity for the enzyme, as judged by K_M and K_i values, whereas larger groups such as bromo,[25] iodo,[29] thio,[25] methylthio,[25,29] methoxy,[25] ethylamino,[25] or azido[29] prevent substrate activity and give rise to poor inhibitors. These effects could arise from lack of substituent tolerance at the 8-position in the enzyme–adenosine complex and/or from restricted rotation about the 9,1' (glycosidic) bond brought about by the 8-substituent. That the latter effect could be operative in the case of 8-thioadenosine is indicated by the finding that 8,2'-anhydro-8-mercapto-9-β-D-arabinofuranosylpurine (**4**), a fixed 9,1'-rotamer with partial *anti* character, is a substrate (rel. V_{max}, 7%; K_M, 200 μM) of adenosine aminohydrolase.[33] Circular dichroism studies[34,35] suggested that 8-substituted adenosines, as expected

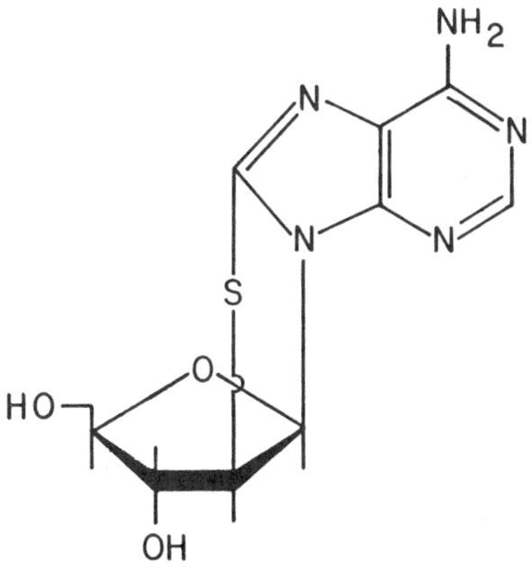

4

from inspection of space-filling models, exist in aqueous solution predominantly as populations of *syn* conformers. Adenosine itself, on the other hand, appears to be present in aqueous solution mainly as *anti* conformers, as indicated by a variety of physical properties, notably by ^{1}H-NMR spectra.[34,36–38] The question of the conformation about the 9,1′ bond of enzyme-bound adenosine is discussed in Section 2.2.3.

Replacement of N-1, N-3, or N-7 of adenosine by methine appears to abolish substrate activity (Table II). 1-Deazaadenosine was a powerful competitive inhibitor, whereas 3- and 7-deazaadenosines exhibited no inhibitory properties, from which it was concluded[29] that N-1 is required for the enzyme-catalyzed hydrolysis whereas N-3 and N-7 play roles in complexation of adenosine to the enzyme.

8-Aminoimidazo[4,5-*g*]quinazoline (**5**) is a good substrate with a V_{max} 85% that of adenosine and a K_M only 2.5 times higher than that of adenosine.[39] The corresponding adenosine analogue **6** possesses essentially the same substrate constants as **5**, showing that a linear extension of adenine by incorporation of the fused benzene ring of **5** brings about a marked enhancement of catalysis that is similar to that effected by the ribofuranosyl group of adenosine. As discussed earlier in this section, it has been proposed[30] that a function of the ribofuranosyl group of adenosine may be to help stabilize the transition state for the hydrolytic deamination.

Wolfenden and co-workers[23,32] have suggested that the transition state of the catalyzed reaction may be reached during attack by the enzyme or by water on C-6 of adenosine to produce a tetrahedral intermediate **7** (R^1 = H or enzyme, R^2 = β-D-ribofuranosyl). The two 6-epimers of 1,6-dihydro-6-hydroxymethyl-purine ribonucleoside (**8**) (R = β-D-ribofuranosyl) were prepared as potential transition state analogues which, in appropriate conformation, bear a steric resemblance to the tetrahedral intermediate **7** for water attack on adenosine. One of the epimers was found to be a powerful competitive inhibitor of ade-

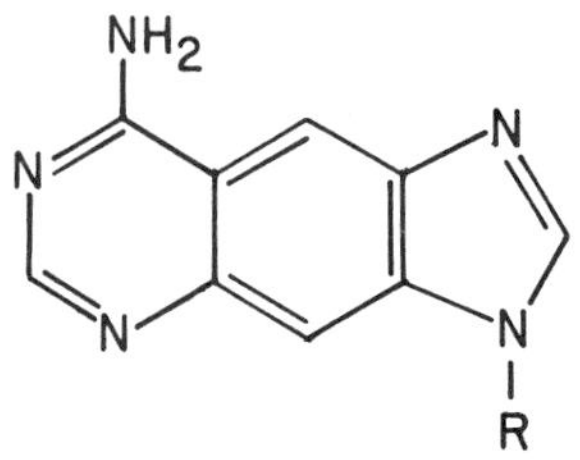

5 ,   R = H

6 ,   R = β-D-ribofuranosyl

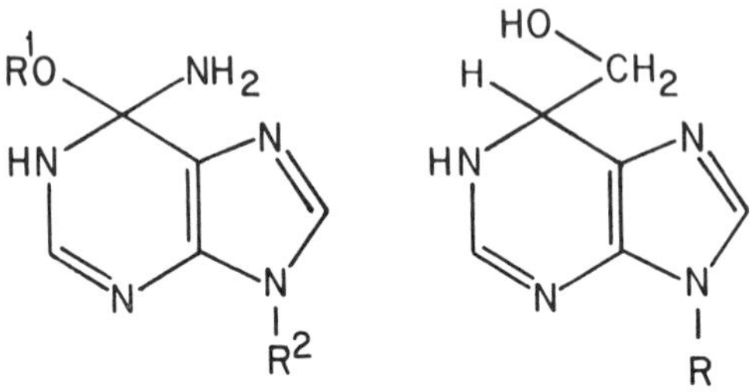

7   8

nosine aminohydrolase with a K_i value about 40 times lower than the K_M value for adenosine.[40] The deoxyadenosine analogue deoxycoformycin (**9**), in which C-8 (equivalent to C-6 of adenosine) is in the R configuration, inhibits adenosine aminohydrolase approximately 100 times more strongly than the above epimer of **8**,[41] suggesting that the postulated type of tetrahedral intermediate formed from enzyme-bound adenosine would likewise have the C-6 in the (R) configuration, as illustrated in structure **10** (R^1 = H or enzyme).

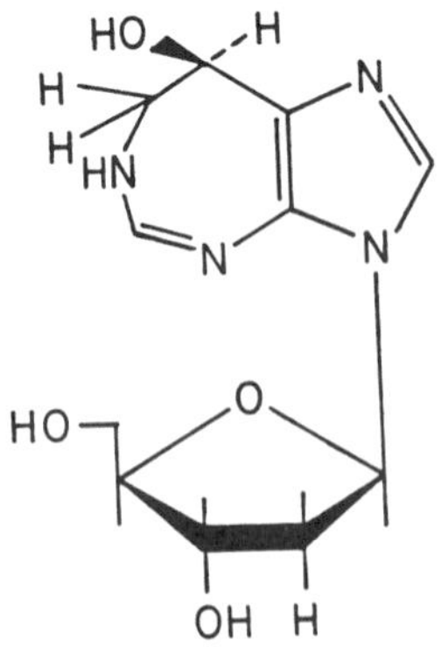

9

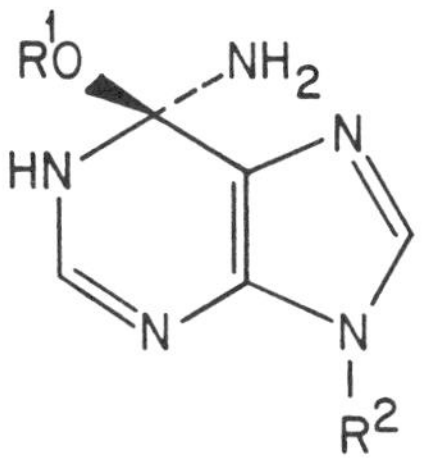

10

2.2.2. Binding of the Ribofuranose Ring

Several laboratories have studied the effects of structural changes in the carbohydrate moiety of adenosine on substrate activity with adenosine aminohydrolase. Some key findings are given in Table III. Replacement of the 2'- or 3'-hydroxyl groups of adenosine by hydrogen has little effect on either the K_M or V_{max}, whereas replacement of the 5'- hydroxyl group abolishes substrate activity. Thus, 2', 3'-dideoxyadenosine is almost as effective a substrate as adenosine, but 2',3',5'-trideoxyadenosine is not a substrate under the same conditions. The requirement for a 5'-hydroxyl group is not absolute, however, as shown by the ability of 2,5'-anhydroformycin (Section 2.2.3) and 9-(5-deoxy-β-D-*erythro*-pent-4-enofuranosyl)adenine[43] (**11**) to act as weak substrates. It is not yet clear whether one or both atoms of the 5'-hydroxyl group participate in catalysis. Thus, 5'-O-acetyl- and 5'-O-benzyladenosines are not substrates,[42] indicating that the 5'-hydroxylic hydrogen could be involved in catalytic events or that an acetyl or benzyl group may be too large to fit into the enzyme–substrate complex. Replacement of the 5'-hydroxyl group of adenosine with an amino group (giving **12**) did permit enzymatic deamination, although the V_{max} was only

Table III. Kinetic Constants with Calf Duodenal Adenosine Aminohydrolase for Compounds in Which the Ribofuranosyl Moiety of Adenosine Is Modified

Compound	Rel. V_{max}	K_M (μM)	Reference
Adenosine	100	35	
2'-Deoxyadenosine	93	22	21
3'-Deoxyadenosine	55	52	22,27
5'-Deoxyadenosine	0		42
2',3'-Dideoxyadenosine	82–100	77–87	21, 22
2',3',5'-Trideoxyadenosine	0		42
4'-Thio-4'-deoxyadenosine	37	33	21
2',3'-O-Isopropylideneadenosine	49	48	27
2'-O-Methyladenosine	53	22	27
9-Arabinosyladenine	25	140	22
9-Xylosyladenine	57	100	22
9-Lyxosyladenine	1.5	170	22
5'-Deoxyxylosyladenine	6.5	80	22

11

12   13

approximately 0.4% of that given by adenosine itself.[42] A 3'-hydroxyl group *cis* to the adenine moiety was better able to take over the function of the 5'-hydroxyl group in catalysis, as shown by the relative V_{max} (6.5%) of 5'-deoxyxylofuranosyladenine (13). It has been pointed out[44] that, given a suitable conformation of the sugar ring, the 3'-hydroxylic hydrogen of 13 can occupy some of the same spatial positions relative to that ring as does the 5'-hydroxylic hydrogen of adenosine. On the other hand, 5'-deoxyarabinofuranosyladenine (14), unlike arabinofuranosyladenine itself (Table III), is not a substrate,[42] presumably because the 2'-hydroxylic hydrogen is too far removed from enzymic groups which interact with the 3'- or 5'-hydroxylic hydrogens of 13 and adenosine, respectively.

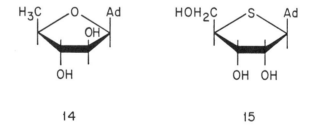

14   15

Replacement of the 4' furanose oxygen by sulfur (15) apparently does not influence binding, as judged by the K_M value, and decreases V_{max} by about 60%, an effect perhaps associated with the larger size of decreased hydrogen bonding capacity of sulfur in comparison with oxygen. Consistent with this view, when the furanose oxygen of 2',3'-dideoxyadenosine is replaced by a methylene group[45] (compound 16), all substrate activity is lost and the K_i value indicates a sixfold decrease in binding to the enzyme.

The introduction of a hydroxymethyl group at the 1'-position of adenosine (psicofuranine, 17) prevents substrate activity.[21] It is not yet clear to what extent this may reflect limited bulk tolerance at the 1'-position and to what extent it would result from the marked steric constraint by the 1'-hydroxymethyl group on rotation about the 9,1' bond. This restricted rotation is evident from an inspection of space-filling models, and it has been concluded to be an important factor for substrate activity with adenosine kinase (Section 2.1.2).

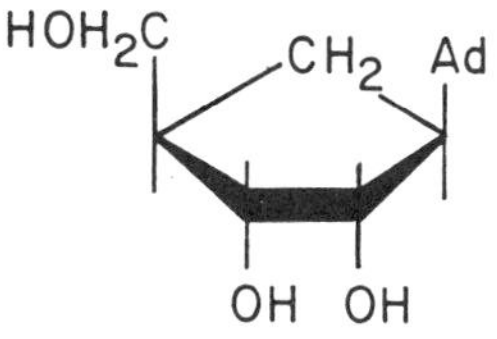

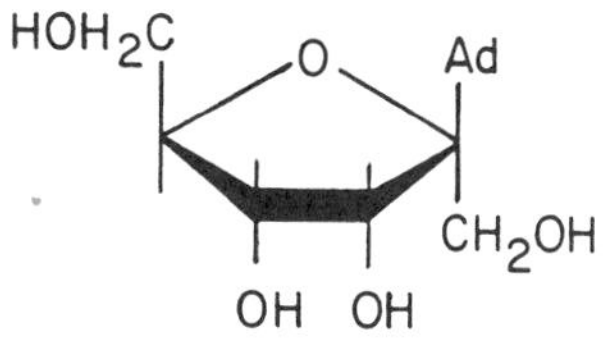

In view of the evidence for nonparticipation of the 2'- and 3'-hydroxyls in the enzymatic deamination, it is not surprising that substrate activity persists when groups are attached to O-2' and/or 0-3'. Thus, 2'-*O*-methyl, 2',3'-*O*-isopropylidene, 3'-*O*-acetyl, and 3'-*O*-tosyl groups can be attached to adenosine or 2'-deoxyadenosine without preventing substrate activity, although activity tends to decrease with an increase in the size of the substituent.[42] The 2'-*O*-methyl or 2',3'-*O*-isopropylidene substituents have no marked effect on K_M.[27] Alkylation of the methyl group of 2',3'-*O*-isopropylideneadenosine that is *endo* to the ribofuranose ring has little effect on the K_M, whereas alkylation of the *exo* methyl increases the K_M 10- to 20-fold.[46] An intact five-membered ring attached to N-9 of adenine is not essential for substrate activity, inasmuch as 9-(2-hydroxyethoxymethyl)adenine (**18**) was a substrate (V_{max} 1.4% that of adenosine) and was bound to the enzyme only about four times less strongly than adenosine,[47] whereas 9-(ethoxymethyl)adenine (**19**), analogously to 2',3',5'-trideoxyadenosine,[42] was devoid of substrate activity.[47]

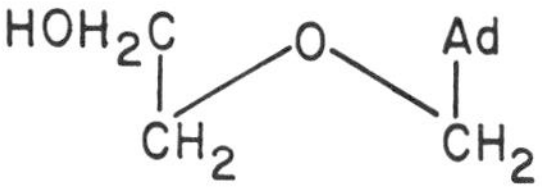

2.2.3. *Conformation of Enzyme-Bound Adenosine*

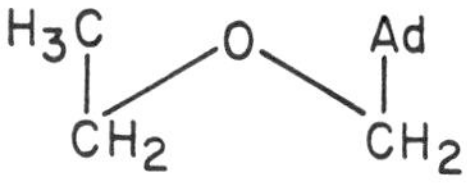

The major possible conformational variations in the adenosine molecule are those associated with modes of puckering of the ribofuranose ring and with rotation around the 9,1' (glycosidic) and 4',5' bonds. In regard to the glycosidic bond, a variety of physical analyses, mostly spectroscopic in nature, have been carried out with aqueous solutions of adenosine (see Section 2.2.1) and have for the most part indicated that rotational barriers are low and that *anti* rotamers probably predominate over *syn* rotamers. A recent study showed a quantitative agreement between calculations of glycosidic torsional angles based on H-2' and C-2' chemical shifts of various nucleosides and indicated that, at pD 8, 52 ± 5% of adenosine is present in the *anti* form.[48] The *syn* and *anti* ranges are centered at ϕ_{CN} = +150° and −30°, respectively, each encompasses an angle of approximately 90° [ϕ_{CN} represents the angle between the C(1')—O(4') bond and the plane of the purine ring as viewed from that ring].[49] 8,2'-Anhydro-8-mercapto-9-β-D-arabinofuranosyladenine (**4**) (V_{max} 7–29% relative to adenosine[29,33]), its 8-hydroxy analogue, **20** (rel. V_{max}, 2–4%[29,50]), and its 8-amino analogue (rel. V_{max}, 3%[29]), are substrates of adenosine deaminase whereas 8,3'-anhydro-8-hydroxy-9-β-D-xylofuranosyladenine (**21**)[29,50] and its 8-mercapto analogue[29] are not. The (*S*)-epimer of 8,5'-cycloadenosine (**22**) and the corresponding (*R*)-epimer were likewise not substrates, and **22** was a weak inhibitor (K_i = 0.3 m*M*).[51] These findings, together with evidence (Section 2.2.1) for little or no substrate activity among adenosine derivatives constrained in *syn* conformations by bulky 8-substituents, suggest an *anti* conformation for enzyme-bound adenosine. The degree of *anti* character is uncertain because of the role of the 5'-hydroxyl group of adenosine in binding and catalysis, and the possibility that the 5'-hydroxyl groups of certain 8,3'- and 8,5'-cyclo structures, for example, **21** and **22**, may be unable to orient correctly. It should be noted that ϕ_{CN} values as high as −122° (that of the 3',5'-cyclic phosphate of **4**) are compatible with substrate activity.[52] An *anti* conformation for enzyme-bound adenosine is consistent also with the observation that 2,5'-anhydroformycin (**23**) undergoes slow aminolysis, albeit with relatively high levels of enzyme, under conditions in which 4,5'-anhydroformycin (**24**) does not.[53]

2.2.4. Active-Site-Directed Irreversible Inhibition

A series of 9-(substituted aralkyl)adenines (**25–31**) have been synthesized by Schaeffer *et al.* and studied as reversible and irreversible inhibitors of the

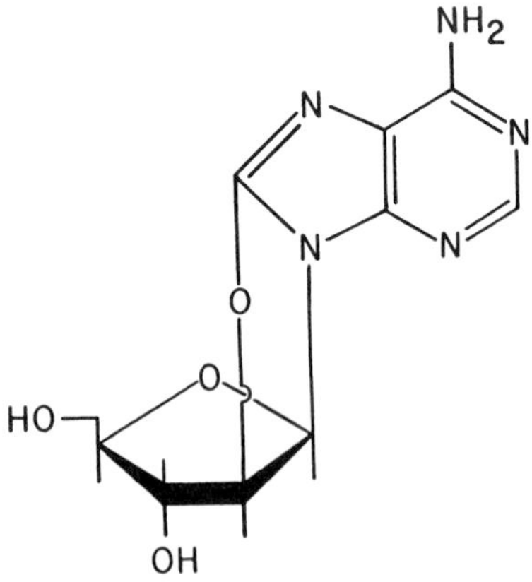

20

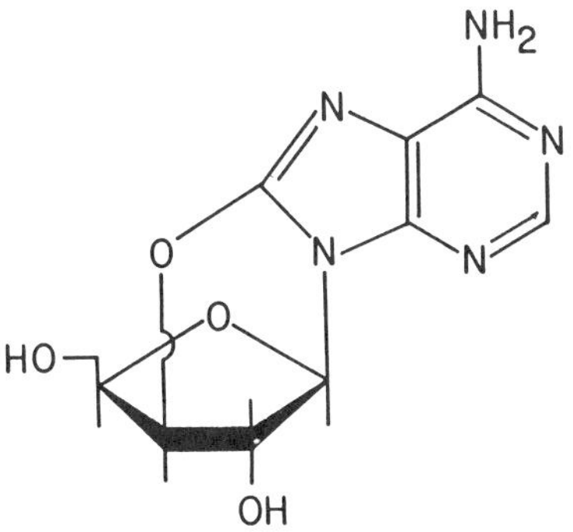

21

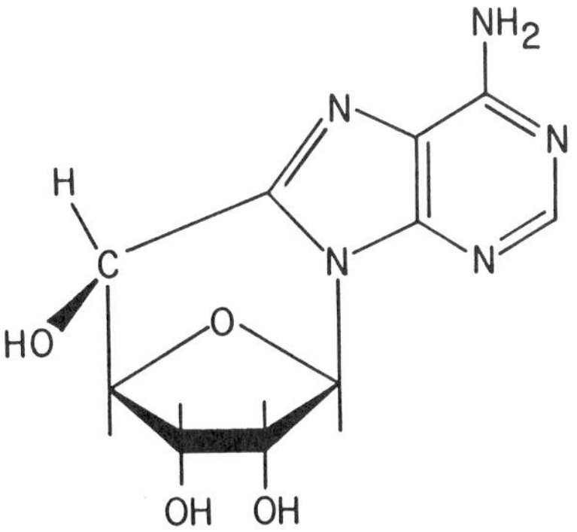

22

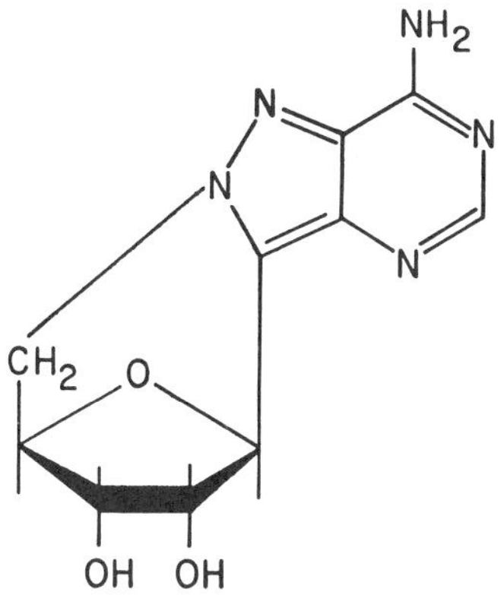

23

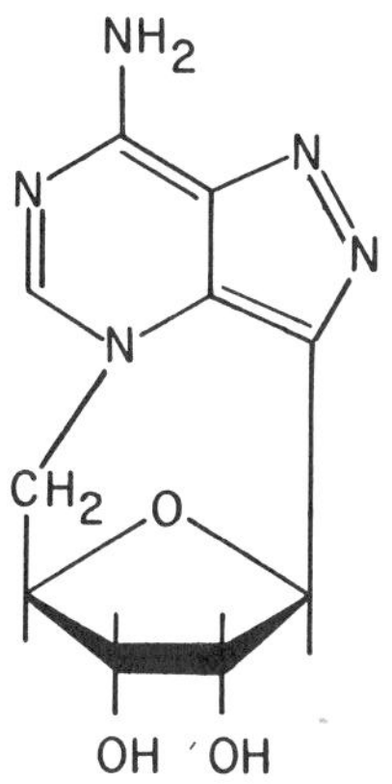

24

enzyme.[54] Of these, 9-(*o*-, *m*-, and *p*-bromoacetamido-benzyl)adenines (**25–27**) were good reversible inhibitors; in addition, the *o*- and *p*-isomers were good irreversible inhibitors and the *m*-isomer was a weaker irreversible inhibitor. Iodoacetamide at relatively high concentration did not cause inactivation, indicating that inactivation by these compounds involves a reversible formation of an enzyme–inhibitor complex prior to alkylation of the enzyme; furthermore, the inactivations followed pseudo-first-order kinetics and exhibited a saturation effect. Kinetic analysis of the inactivation process yielded values of the enzyme–inhibitor dissociation constants that were in agreement with the same constants evaluated from the kinetic behavior of the compounds as reversible competitive

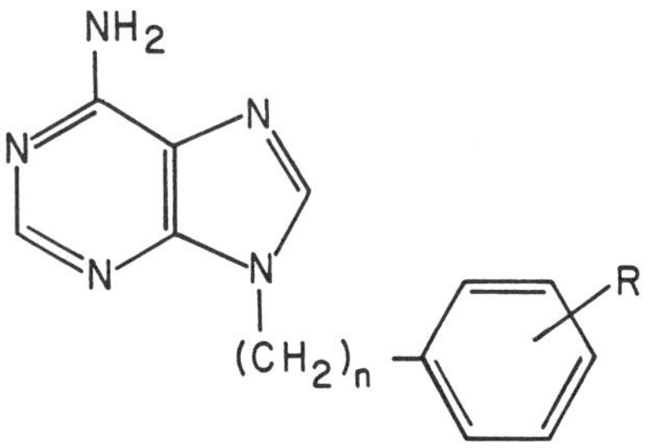

(25) n = 1, R = o –NHCOCH$_2$Br

(26) n = 1, R = m –NHCOCH$_2$Br

(27) n = 1, R = p –NHCOCH$_2$Br

(28) n = 2, R = m –NHCOCH$_2$Br

(29) n = 2, R = p –NHCOCH$_2$Br

(30) n = 3, R = o –NHCOCH$_2$Br

(31) n = 3, R = p –NHCOCH$_2$Br

inhibitors, thus providing strong additional evidence that the inactivations are substrate site directed.[55]

Of the remaining 9-substituted adenines, the phenethyl derivative **28** caused rapid irreversible enzyme inactivation whereas its p-isomer **29** and the two phenylpropyl derivatives **30** and **31** did not bring about inactivation.[54] One mole of compounds **25** or **27** became bound per mole of inactivated adenosine aminohydrolase. The alkylated amino acid is lysine with **27**[56] and a mixture of lysine and (predominantly) histidine with **25**.[57]

2.2.5. Reversible Binding by 9-Substituted Adenines

Schaeffer and his co-workers[58–62] carried out a series of systematic studies on the inhibition of calf intestinal mucosal adenosine aminohydrolase by 9-substituted adenines of the general structure **32** (R^1,R^2 = H or alkyl). The enzyme was found to possess (a) a large hydrophobic area which interacted with the alkyl groups of 9-alkyladenines,[58] (b) a hydroxyl binding site accessible to 9-substituents,[59] and (c) a specific methyl binding site for 9-(2-hydroxypropyl)adenine (**32**, R^1 = H, R^2 = CH$_3$).[60] Studies with 9-(1-hydroxy-2-alkyl)adenines (**32**, R^1 = alkyl, R^2 = H) showed that stronger binding to the enzyme occurs when the chiral center proximal to N-9 has the R configuration.[61] In addition, studies with 9-(2-hydroxypropyl)adenine showed that binding is promoted when the secondary alcoholic chiral center of **32** has the S configuration.[60] When the steric and structural features which promote binding were combined, enhanced inhibition was observed, implying that the adenine moieties of compounds **32**

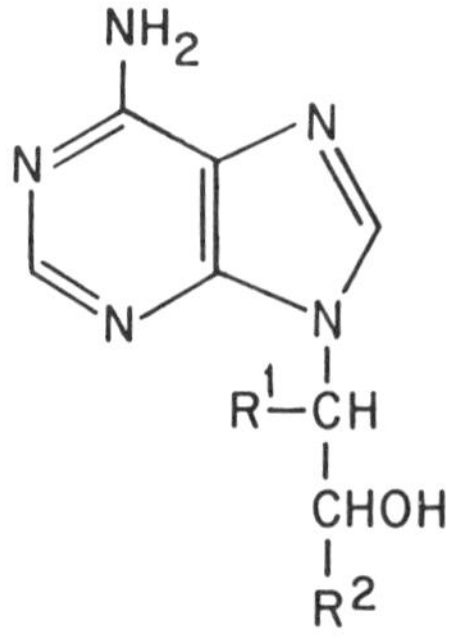

$$NH_2$$

32

$(R^1 = H$ or alkyl, $R^2 = H$ or $CH_3)$ bind to the same site of the enzyme and that adjacent to this site are sites that interact with the hydroxyl, methyl, and hydrophobic sections of the 9-substituents.[62] This systematic approach has resulted in the design of powerful reversible inhibitors of adenosine aminohydrolase. For example, in the presence of 66 μM adenosine, 50% inhibition of the enzymatic reaction is brought about by 10 nM *erythro*-9-(2-hydroxy-3-nonyl)adenine (**32**, R^1 = C_6H_{13}, R^2 = CH_3).[62] Replacement of the β-D-ribofuranosyl group in the transition state analogue **8** by an *erythro*-(2-hydroxyl-3-nonyl) group caused the inhibitory potency toward the enzyme to decrease rather than to increase,[63] suggesting that this 9-substituent may hinder a change in the conformation of the enzyme that is required to enhance its affinity for the transition state involving the N(1)—C(6) system of adenosine.

3. Substrate Binding of Adenosine 5'-Phosphate

3.1. Adenylate Kinase

Adenylate kinase (ATP:AMP phosphotransferase, EC 2.7.4.3) catalyzes the reversible reaction ATP + AMP$\rightleftharpoons$2 ADP. A divalent metallic ion such as magnesium or manganese is required. The enzyme occurs widely among microorganisms and plant and animal tissues.[64] It constitutes a major biosynthetic means for the conversion of AMP to ADP and appears to play an important role in the control of energy metabolism in living systems by regulating the relative concentrations of AMP, ADP, and ATP.[64] In mammals, as exemplified by humans[64,65] and the rat,[64,66–68] the enzyme occurs in at least three isozymic forms. Most of the information regarding substrate binding to adenylate kinase has been obtained in studies of muscle isozymes, principally those of the rabbit and pig.

That adenylate kinases possess distinct, nonidentical binding sites for AMP and ATP is indicated by several lines of evidence. Thus, with regard to the effects of substrate modification, ITP and GTP are substrates[69] whereas IMP and GMP are not[70]; a 1,N^6-etheno group permits substrate activity with ATP but none with AMP[71]; N^6-benzoyl-ATP is a good substrate (V_{max} ca. 25% that of ATP) of the rabbit, pig, and carp muscle enzymes whereas N^6-benzoyl-AMP shows no substrate activity and behaves as a noncompetitive inhibitor with respect to AMP[72]; 2′,3′-O-isopropylidene-ATP has a V_{max} value ca. 50% that of ATP, but the corresponding AMP derivative is not a substrate of rabbit, pig, or carp muscle adenylate kinases.[73] Nonidentical substrate binding sites are further indicated by the finding that the NMR signals of H-2 and H-8 of adenine exhibit little shift upon binding of AMP to the pig muscle enzyme but do become shifted when ATP binds.[74] An X-ray diffraction analysis of the pig muscle enzyme has led to assignments of the binding sites for AMP and ATP which are consistent with the relatively limited bulk tolerance near the adenosine moiety of AMP and with the ^{1}H-NMR shifts of H-2 and H-8 upon binding of ATP, the latter being attributable to an interaction of the adenine of ATP with tyrosine-95 of the enzyme.[2] Electron paramagnetic resonance studies with MnATP[70] and studies of ^{31}P-NMR signals of free and enzyme-bound adenine nucleotides[75] support the conclusion that the AMP site also binds ADP or ATP, provided these are uncomplexed to metal ion, whereas the ATP site binds ADP or ATP with or without a coordinated metal ion. Properties of the AMP site are discussed in the sections which follow. Properties of the ATP site are considered separately in Section 4.1.

3.1.1. Binding of the Adenine Moiety

With the rabbit muscle enzyme, CMP can substitute for AMP as a substrate[76] whereas IMP and GMP cannot.[70] The major conformational features of IMP and GMP in aqueous solution are similar to those of AMP,[77] and the above substrate specificity suggests that one or both nitrogens of the amidine system of AMP [N(1)—C(6)—N^6] or of CMP [N(3)—C(4)—N^4] may play important roles in the catalytic cycle. Consistent with this view is the lack of substrate activity of 1,N^6-etheno-AMP with the rabbit muscle enzyme[71] and of 1-methyl-AMP and N^6-methyl-AMP with the rat muscle enzyme and with the two major isozymes of rat liver.[78] Additionally, 1-methyl-AMP shows little tendency to inhibit these three rat isozymes, while N^6-methyl-AMP behaves as a weak noncompetitive inhibitor with respect to AMP.[78] Similarly, N^6-benzoyl-AMP is not a substrate of rabbit, pig, or carp muscle AMP kinase and is a noncompetitive inhibitor.[72] An analysis of proton relaxation times has indicated that the conformation of 1-methyl-AMP in aqueous solution is similar overall to that of AMP itself.[79] The drastic reduction in affinity for the AMP site associated with the attachment of a methyl group at N-1 or N^6 can thus be interpreted most simply as steric in origin, although electronic effects resulting from methylation at N-1 could also play an important role.

Introduction of a chloro group at C-2 of AMP permits substrate activity to persist with the rabbit muscle enzyme.[80] With the three rat isozymes mentioned above, a 2-chloro group also permits substrate activity, but the more bulky 2-methylthio group drastically reduces substrate activity and produces a weak inhibitor noncompetitive with AMP.[78] Since the 2-methylthio group produces little perturbation in the conformation of AMP in aqueous solution, as judged by ^{1}H-NMR studies,[81] the poor substrate and inhibitor properties of 2-methylthio-AMP indicate that C-2 of enzyme-bound AMP may be in close proximity to one or more atoms of the enzyme. This is in accord with the foregoing evidence that N-1 and/or N[6], which are nearby, may directly participate in AMP binding and possibly in other phases of catalysis. An X-ray diffraction analysis of pig muscle AMP kinase at 0.6 nm[2] has permitted an assignment of the AMP binding site to a relatively small and almost totally enclosed pocket in which the 6-amino group of AMP could be hydrogen bonded to Cys-25 or Asp-93. That N-7 of AMP is not essential for substrate activity is revealed by the ability of CMP[76] and of 7-deaza-AMP (tubercidin 5'-phosphate)[82] to act as phosphate acceptors.

8-Bromo-AMP (**33**) is not a substrate of rabbit muscle adenylate kinase[73,83] nor of the isozymes from rat muscle and liver[78] and behaves as a weak noncompetitive inhibitor with respect to AMP[78] Proton, ^{31}p and ^{13}C magnetic resonance studies[84–86] indicate that in aqueous solution 8-bromo-AMP exists principally as *syn*-type rotamers about the glycosidic bond. On the other hand, AMP itself is present in aqueous solution mainly as *anti* conformers.[48,77]

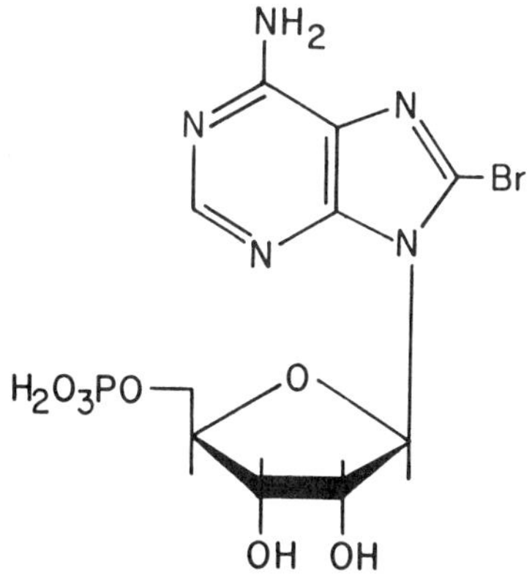

33

3.1.2. Binding of the Ribofuranose Ring

An X-ray diffraction analysis of pig muscle AMP kinase[2] has indicated that the 2'-OH of enzyme-bound AMP might form a hydrogen bond with the carbonyl of Ser-19 and that the 3'-OH might be bonded to the sidechain of Gln-24. It is possible that such interactions may modulate binding and/or catalysis, because replacement of the 2'- or 3'-hydroxyls groups of AMP by hydrogen produces maximal velocities that are 32% and 40%, respectively, that of AMP with the pig muscle enzyme and 20% that of AMP with the rabbit muscle enzyme,[87] whereas the K_M of 2'-deoxy-AMP with the above three rat isozymes is 9–11 times higher than the K_M of AMP, and its V_{max} is 70–95% less.[78] In addition 2'- and 3'-*O*-methyl-AMP are weak substrates and weak competitive inhibitors of the rat isozymes, and AMP derivatives with 2',3'-*O*-acetal groups are similarly weak inhibitors but exhibit no substrate activity.[78] No substrate activity with pig or rabbit muscle AMP kinase is shown by 2′,3′-di-*O*-acetyl-AMP, 2′,3′-*O*-ethylidene-AMP, or 2′,3′-*O*-isopropylidene-AMP, although these derivatives are competitive inhibitors with affinities approximately 2–4 times less than that of AMP itself.[73] These indications of limited bulk tolerance near O-2′ and O-3′ of enzyme-bound AMP are in accord with the X-ray diffraction evidence for an AMP binding pocket of restricted dimensions in pig muscle AMP kinase.[2]

3.1.3. The Ribose–Phosphate Bridge

In the case of rabbit muscle AMP kinase, replacement of the CH_2—O—P system of AMP by the phosphonate system CH_2—CH_2-P reduces the V_{max} to 6% of that observed with AMP as substrate[88] but does not significantly alter the enzyme–substrate dissociation constant.[89] When a cyano or hydroxyl group is introduced α to phosphorus, substrate activity with the pig and rabbit muscle enzymes persists,[89,90] showing that the respective enzyme–AMP complexes can accommodate a cyano or hydroxyl group in the O-5′ region. In a series of α-acylaminomethyl phosphonate analogues of AMP in which the system $CH_2CH(CH_2NHCOR)PO_3H_2$ replaced $CH_2OPO_3H_2$,[73] substrate activity and competitive inhibition with respect to AMP were observed for R = CH_3. With longer aliphatic chains (R = C_2H_5, OCH_3, OC_2H_5, OC_3H_7), substrate activity was not detectable but the inhibitions remained competitive, suggesting that the longer chains do not prevent binding to the AMP site, but do block the catalytic cycle. Additional methylenes promoted binding in the series, indicating that the α substituents interact with a lipophilic area of the enzyme in the vicinity of O-5′ in the enzyme–AMP complex.

Substrate binding of AMP has been studied also by means of single substituents attached to C-5′ in either the 6′-deoxy-β-D-allofuranosyl (**34**) or the 6′-deoxy-α-L-talofuranosyl (**35**) configuration. *allo*-5′-Methyl-AMP and *talo*-5′-methyl-AMP (**34, 35**; R = H) were substrates of pig and rabbit AMP kinases.[91] Both enzymes exhibited a marked preference for the *talo* configuration. A group of 5′-*C*-acylaminomethyl AMP derivatives of the *allo* configuration (**34**, R^1 = NHCOR^1, where CH_3, C_2H_5, CH_2Cl, C_6H_5, C_6H_4-*p*-F, C_6H_4-*o*-CH_3, C_6H_4-*m*-CH_3, C_6H_4-*p*-CH_3) had no substrate activity but behaved as linear competitive

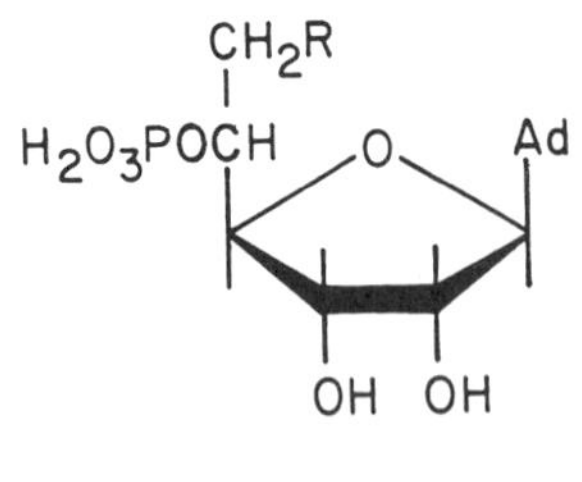

34 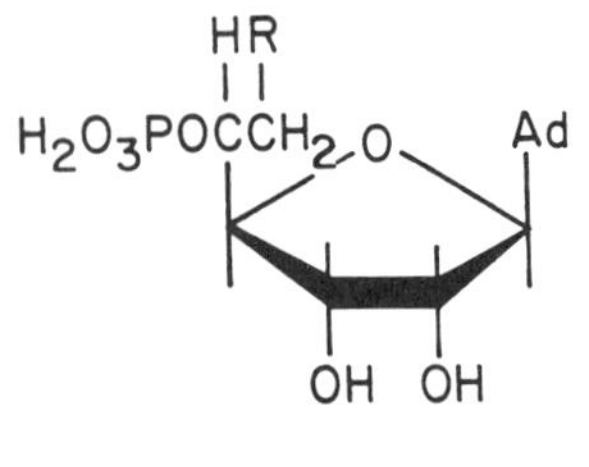   35

inhibitors of the rabbit enzyme with respect to AMP.[73] The inhibition constants indicated an interaction of the 5'-substituents of **34** with a lipophilic region of the enzyme. Thus, this region might be contiguous or identical to the region involved in nonpolar interactions with some of the substituents α to phosphorus in the phosphonate analogues of AMP described above. Compounds **34** were competitive inhibitors also of pig muscle AMP kinase.[73] The inhibition constants suggested that 5'-benzamidomethyl-AMP binds to the pig enzyme about one-third as strongly as 5'-acetamidomethyl-AMP. However, in the case of the rabbit enzyme, the former compound binds approximately twice as strongly as 5'-acetamidomethyl-AMP. This sixfold differential in affinity suggests that the pig enzyme possesses a structural feature (possibly a polar region near the benzene ring of enzyme-bound 5'-benzamido-AMP) which is absent or masked in the rabbit enzyme.

3.1.4. Binding of the Phosphoryl Group

Adenosine 5'-phosphorothioate, in which PO_2SH_2 is substituted for the phosphoryl group of AMP, is a substrate of AMP kinase. The ADP analogue formed by the enzyme contains a new chiral center at the α phosphorus, and only one diastereomer is produced, thus giving evidence regarding the stereospecificity of attack and confirming that the thiophosphoryl group, as expected from the inertness of adenosine as substrate or inhibitor, is enzyme bound and prevented from rotating during the enzyme-catalyzed reaction.[92] Several lines of evidence suggest that interaction of the phosphoryl moiety of AMP with the enzyme could play a major role, not only in the binding of AMP to its site, but also in the activation of the bond-forming atoms of AMP and ATP. Thus, with the rat muscle enzyme, adenosine itself is a weak inhibitor (K_i = 19 mM) noncompetitive with respect to AMP.[128] From this it can be calculated that, if adenosine possesses affinity for the same form of the site to which AMP binds, the dissociation constant K_a value) of the hypothetical complex would be in excess of 50 mM. In comparison, the K_a value of AMP with the rat muscle enzyme is approximately 0.65 mM, assuming that K_a and K_M are roughly equal in value, as they are in the case of the rabbit muscle enzyme.[76]

Replacement of the $CH_2OPO_3H_2$ system of AMP by $CH_2CH_2CO_2H$ (**36**) is reported to abolish substrate activity with pig and rabbit muscle AMP kinases and to produce competitive inhibition with respect to AMP with inhibition constants 6–7 times the K_M of AMP.[93] The lack of substrate activity of **36** is not ascribable to the absence of O-5' of AMP because the phosphonate isostere of

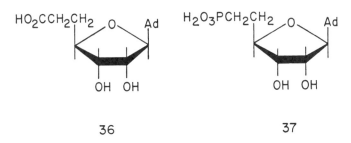

36   37

AMP (**37**), which also lacks O-5′, is a substrate of the pig and rabbit enzymes.[88] Also, a replacement of $CH_2OPO_3H_2$ in AMP by $CH_2CH_2AsO_3H$, to give the arsonomethyl analogue, allows substrate activity to persist with the rabbit enzyme.[94] Adenosine 5′-sulfate (**38**) is a competitive inhibitor ($K_i = 6.8$ mM and 5.7 mM, respectively) of the enzymes from rabbit muscle[95] and rat muscle and liver[78] with an affinity approximately 10% that of AMP. Compound **38** showed no substrate properties with the rat muscle or liver enzymes. Thus, the presence of one anionic charge in the O-5′ substituent of an AMP analogue apparently permits binding to occur at the AMP site, but no substrate activity is observed, presumably in part because of the higher thermodynamic barrier to enzyme-catalyzed formation of a mixed phosphoric–carboxylic or phosphoric–sulfuric anhydride.

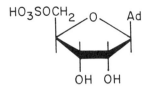

38

X-ray diffraction studies have shown that crystalline pig adenylate kinase can exist in either of two conformations depending upon the pH.[2] One of these, "A," is characterized by a closed adenosine pocket of restricted dimensions at the AMP site and a narrow cleft which can accommodate the phosphate residues of AMP and ATP. The other, "B," possesses an open AMP site and a wider phosphate cleft. It has been proposed[2] that "B" is related to the structure of the enzyme in solution prior to an AMP- and/or ATP-induced conformational change, whereas "A" is thought to correspond to the structure in solution subsequent to this change. The feeble affinity of adenosine for the AMP site, together with the absence of major conformational differences between adenosine and AMP in aqueous solution,[77] suggests that the phosphoryl group of AMP might play a direct play role in the initiation of the transition from "B" to "A." This possibility is consistent with the existence of crystalline form "A" when AMP and ATP are absent and when oligovanadate or sulfate anions are bound in the phosphate cleft.[2]

3.1.5. The Conformation of Enzyme-Bound AMP

The X-ray diffraction studies of crystalline pig muscle AMP kinase have indicated that the AMP pocket accommodates AMP in an *anti*-type conformation about the glycosidic bond,[2] a view supported by the inertness of the predominantly *syn* 8-bromo-AMP as a substrate or inhibitor (Section 3.1.1). The *trans* vinyl phosphonate analog of AMP (**39**) is a substrate of rabbit and pig muscle and rat liver AMP kinases with V_{max} values which are 13.4%, 6.6%, and 13%, respectively, those of AMP.[78,88] As discussed above, during catalysis the

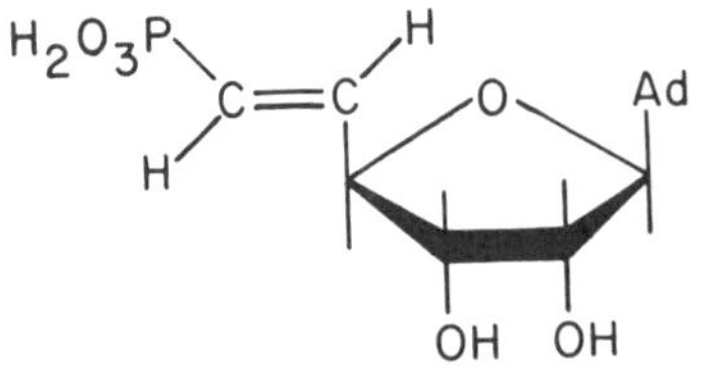

39

phosphate moiety of AMP becomes bound to the enzyme in a specific orientation required for the phosphoanhydride bond formation that produces ADP. A second factor involved in the action of AMP kinase appears to be an interaction of part of the pyrimidine ring of adenine with a specific enzymic site, because IMP is not a substrate. That **39** functions as a substrate indicates that both these factors are operative for this analogue during catalysis. Inspection of a molecular model of AMP shows that when the C(5')-O(5') torsional angle is varied, it is not possible, by rotation of other bonds, to maintain constant the spatial position and orientation of the O(5')-P bond in relation to the adenine ring. In light of this, the substrate activity of **39** favors the view that the C(5')-O(5') torsional angle of enzyme-bound AMP is similar to that of **39**—that is, that C-4' and P of enzyme-bound AMP are in a *trans*-type relationship.

3.2. Adenylate Aminohydrolase

Adenosine 5'-phosphate aminohydrolase (EC 3.5.4.6) catalyzes the conversion of AMP and IMP and is widely distributed among various species and tissues.[96] CMP and GMP are not substrates. The enzyme from mammalian tissues is inhibited by GTP and activated by ATP,[97–100] and it has been postulated that these regulatory effects are important in control of the interconversion of adenine and guanine nucleotides in mammalian systems.[99] In addition, the enzyme is thought to be involved in the control of glycolysis and gluconeogenesis[96] and in the regulation of intracellular levels of AMP, ADP, and AMP.[101] Most of the information on the substrate binding of AMP to AMP aminohydrolase pertains to preparations of the enzyme from rabbit muscle. Simple procedures have been described which readily furnish the enzyme from this source in crystalline form.[102] The findings to be discussed were all obtained in the presence of monovalent cations (usually K+), which activate the enzyme,[103] and in the absence of ATP and GTP.

3.2.1. Binding of the Adenine Moiety

1-Methyl-AMP is not a substrate and is a weak inhibitor of AMP aminohydrolase from rabbit muscle.[104] Small substituents are tolerated at N6. Thus, N^6-methyl-AMP is a substrate (Table IV), and the reaction was shown to

*Table IV. Substrate Constants of AMP Analogues and Derivatives
with AMP Aminohydrolase of Rabbit Muscle*

Compound	K_M (mM)	Rel. V_{max}	Reference
AMP	0.4	100	104
	0.6		108
N^6-Methyl-AMP	1.8	20.4	104
6-Chloroprine 5'-ribonucleotide		3.5	105
8-Bromo-AMP	0.33	0.03	108
2'-Deoxy-AMP	2.3	18.5	104
3'-Deoxy-AMP	0.2	0.22	87
2',3'-O-Isopropylidene-AMP	0.13	0.5	108
2',3'-Di-O-acetyl-AMP	0.67	0.07	108
allo-5'-Methyl-AMP	0.24	0.4	91
talo-5'-Methyl-AMP	1.35	38.0	91
Phosphonate analogue of AMP (**37**)	0.39[a]	21.8	88
Adenosine 5'-sulfate	3.2	10–16	104
Adenosine	20.0	1.1	104
5'-Carboxymethyl-5'-deoxyadenosine (**36**)	0.80	21.0	93
Adenosine 5'-phosphoramidate	49	73.4	104
ADP	0.8	0.2	104

[a] The K_M measured for AMP was 0.83 mM.

produce stoichiometric amounts of IMP and methylamine.[104] N^6-Ethyl-AMP is also a substrate of the rabbit muscle enzyme.[104] That 6-chloropurine ribonucleoside 5'-phosphate is a substrate of rat muscle AMP aminohydrolase[105] affords evidence that the catalytic mechanism probably involves a nucleophilic attack at C-6. In the case of the rat muscle enzyme, the presence of a chloro methoxy, methylthio, or ethylamino group at C-2 of AMP abolishes substrate and inhibitor properties and gives rise to stimulatory effects on the reaction velocity which are ascribable, from kinetic data, to binding of the derivatives to a regulatory site separate from the catalytic binding site for AMP.[106] 8-Bromo-AMP shows no substrate activity at normal assay levels of the rabbit muscle enzyme and does not inhibit the reaction at a level of 0.1 mM.[107] When a 1000-fold higher level of enzyme was employed, kinetic evidence for weak substrate activity of 8-bromo-AMP was found[108] (Table IV). This correlates with evidence for a predominantly *syn*-type glycosidic torsional angle in the molecule in aqueous solution (Section 3.1.1), which contrasts with the *anti*-type angle of ca. 70% of AMP in aqueous solution.[48,77] Replacement of N-7 of AMP by carbon appeared to abolish substrate and inhibitor activity. This effect appears to be unrelated to the torsional angle of the glycosidic bond, since the *anti* character of AMP is retained in 7-deaza-AMP.[77] In addition, transposition of C-8 and N-9 (giving formycin 5'-phosphate) permitted substrate activity.[104] These findings suggest that N-7 may play a direct role in binding and/or catalysis whereas N-9 does not.

The 5'-O-phosphoryl derivatives of 2'-deoxycoformycin (**9**) induces a powerful ($K_i = 10^{-9}$–10^{-10} M), time-dependent reversible inhibition of the enzyme that is counteracted by AMP,[109] suggesting that AMP aminohydrolase and adenosine aminohydrolase operate via the same general hydrolytic mechanism.

3.2.2. Binding of the Ribofuranose Ring

Replacement of the 2'-OH of AMP by H caused a 5-fold reduction in V_{max}, but replacement of the 3'-OH by H caused a 500-fold reduction (Table IV). This large effect is presumably associated with an interaction between the 3'-OH and the enzyme during catalysis rather than with possible variations in nucleotide conformations, because, in the cases of 5'-nucleotidase and AMP kinase, replacement of either the 2'- or 3'-OH of AMP by H led to reductions in V_{max} which were not higher than 10-fold.[87] An introduction of acetyl groups or an isopropylidene group at O-2' and O-3' did not significantly increase the K_M but strongly reduced the V_{max}. This would suggest a minimal interference with initial binding but inhibition at a later phase of catalysis, possibly because of limited bulk tolerance near the O-3' region of an AMP–enzyme complex.

3.2.3. The Ribose–Phosphate Bridge

Studies described below have established that certain groups can be attached at the C(5')—O(5') bridge area of AMP without preventing specific binding to the AMP site. Therefore, AMP derivatives substituted at the ribose–phosphate bridge are potentially useful tools for studies of catalytic mechanism or for the design of potent inhibitors. A number of AMP derivatives have been synthesized in which a substituent is attached to C-5' to produce either a 6'-deoxy-β-D-allofuranosyl configuration (**34**) or a 6'-deoxy-α-L-talofuranosyl configuration(**35**). Both epimeric forms of 5'-C-methyl-AMP (**34, 35**; R = H) possess substrate activity (Table IV) and behave as competitive inhibitors.[91] Two *allo* 5'-C-acylaminomethyl AMP derivatives (**34**, R = NHCOCH$_3$ and NHCOC$_2$H$_5$) and one *talo* derivative (**35**, R = NHCOCH$_3$) were weak substrates and competitive inhibitors which, as judged by their inhibition constants, bound to the enzyme between 2.5- and 5.9-fold more strongly than the corresponding 5'-C-methyl-AMP epimers This would suggest that the acylamino groups may interact with the enzyme.[108] The *talo* derivative **35** with R = NHCOC$_2$H$_5$ is a noncompetitive inhibitor and shows no substrate activity and hence appears to be hindered from binding to the AMP site by reason of its longer substituent. Replacement of the CH$_2$—O—P system of AMP by the phosphonate system CH$_2$—CH$_2$—P permits significant substrate activity to persist (Table IV). When acylaminomethyl groups CH$_2$NHCOR (R = Me, Et, OMe, OEt, OPr) are introduced α to the phosphorus of this phosphonate analogue of AMP, a series of competitive inhibitors are obtained that bind to the enzyme 3–4 times more strongly than the parent phosphonate.[108] This effect is similar to that observed when acylaminomethyl groups are attached to C-5'.

3.2.4. Binding of the Phosphoryl Group

The importance of the phosphoryl group of AMP in binding and catalysis is shown by the high K_M and low V_{max} values of adenosine relative to those of AMP (Table IV). The rate constant for heat inactivation of the enzyme is the same with adenosine as substrate as with AMP as substrate,[104] and hence adenosine is probably deaminated by AMP aminohydrolase itself rather than by another enzyme present in the preparation. That the phosphoryl binding site can accept a monoanionic residue is indicated by the finding that adenosine 5'-sulfate, 5'-carboxymethyl-5'-deoxyadenosine, and adenosine 5'-phosphoramidate are better substrates than adenosine (Table IV).

3.2.5. Conformation of Enzyme-Bound AMP

Studies have been carried out with conformationally restricted AMP analogues and have so far yielded little definitive information regarding the conformation of enzyme-bound AMP. The 5'(S)- and 5'(R)-epimers of 8,5'-cyclo-AMP (**40** and **41**, respectively) were not substrates of AMP aminohydrolase[51] whereas 8,2'-O-cyclo-AMP (**42**) and 8,3'-O-cyclo-AMP (**43**) were at best very weak substrates (rel. V_{max} 0.008% and <0.001% that of AMP).[87] Significantly higher substrate activity was shown by the arabinosyl analogue of AMP (**44**) and by the xylosyl analogue of AMP (**45**).[87] The substrate efficiency (V_{max}/K_M) of the *trans* vinyl phosphonate **39** is 40 times that of adenosine and twice that of 3'-deoxy-AMP, whereas V_{max}/K_M of the more flexible alkyl phosphonate **37** is 1760 times that of adenosine, amounting to 44% of that of AMP itself.[88] As described above, substrate specificity studies show that, during catalysis, important interactions occur between the enzyme and at least three segments of AMP, namely, the adenine ring, the 3'-OH, and the phosphoryl group. The substrate efficiencies of **37** and **39** suggest that the phosphonate group of **39** interacts in some manner with the phosphate binding site of AMP but that the additional rigidity of **39** due to its vinyl group may prevent the normal interactions between its phosphoryl group and/or 3'-OH group and the enzyme. Thus, when C-4' and P are in a fixed *trans* relationship, some substrate activity does occur, but it is not clear whether freedom of rotation about the C(5')—O(5') bond is a prerequisite for efficient catalysis or whether the C(5')—O(5') torsional angle of enzyme-bound AMP differs significantly from 180° (100% *trans* character). The substrate efficiency of the isosteric *trans* vinyl carboxylate **46** is similar to that of **39**, and the substrate efficiency of the corresponding saturated carboxylate **36** is similar to that of its phosphonate analogue **37**.[93]

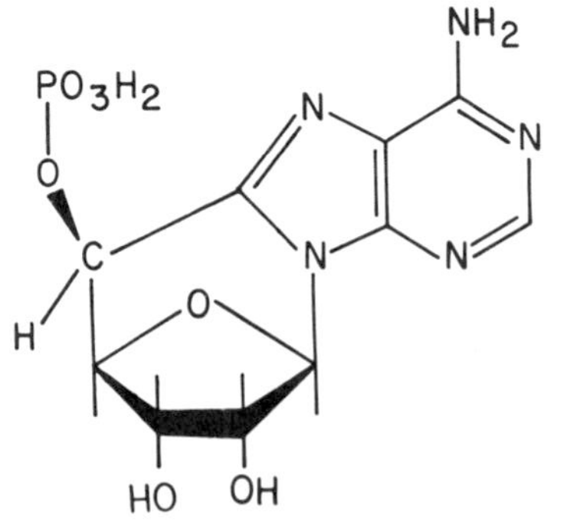

40

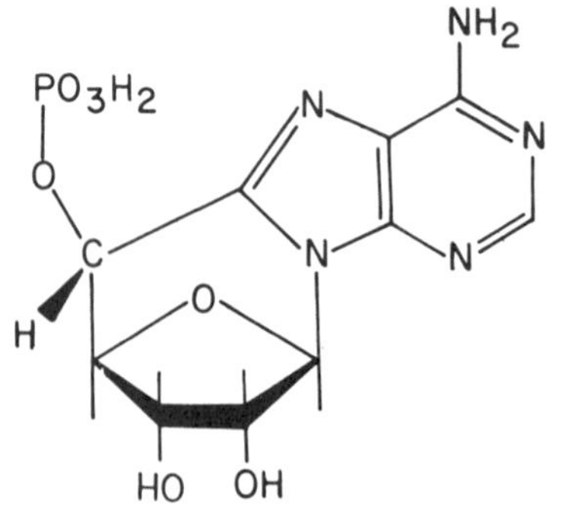

41

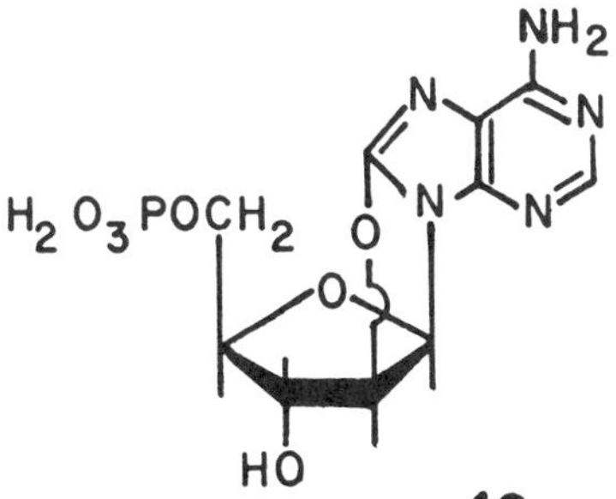

42

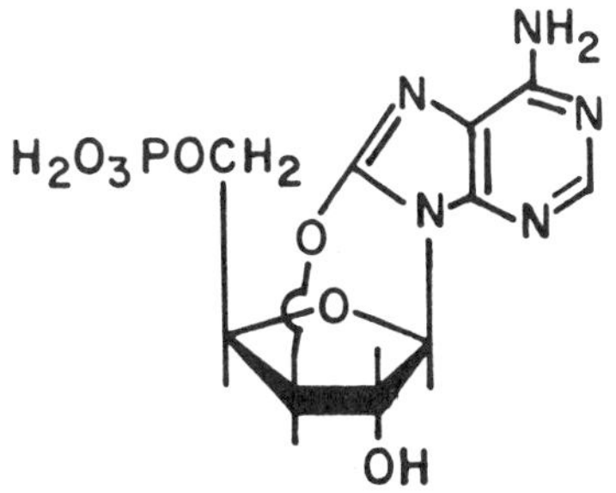

43

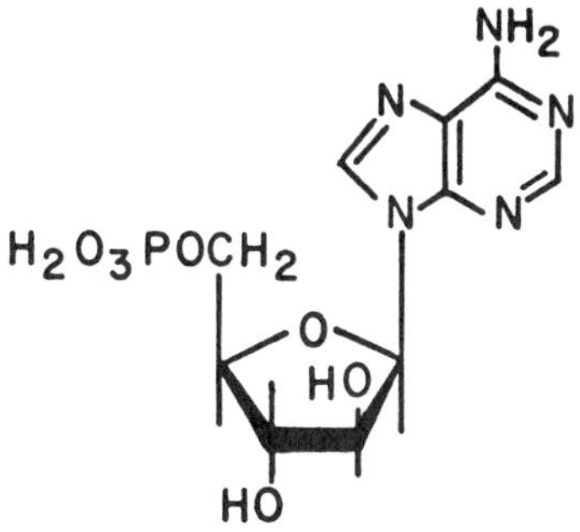

44

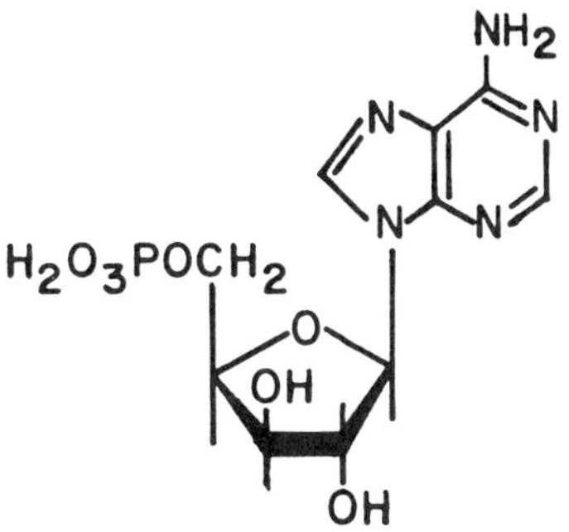

45

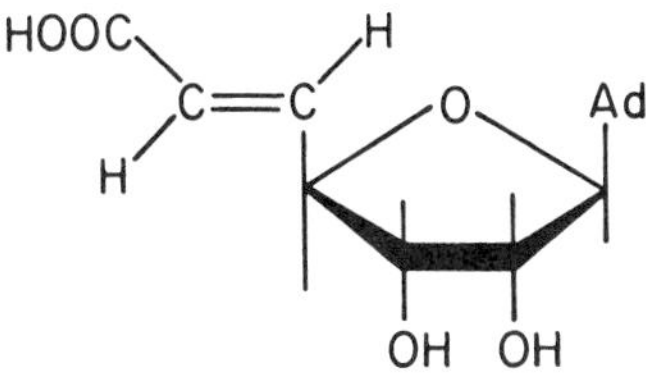

46

3.2.6. Active-Site-Directed Inactivation

AMP aminohydrolase is inactivated by a carboxylic–phosphoric mixed anhydride (**47**) in which the 5′-methylene group of AMP is replaced by a carbonyl group.[110] The effect is abolished by prior hydrolysis of the anhydride and is hence due to acylation or phosphorylation of the enzyme. Inactivation does not take place in the presence of a level of AMP equal to that of the anhydride, suggesting that the covalent reaction could occur either at the catalytic site or at the activating site proposed by Blakely and Vitols[111] and suggested by Maguire *et al.*[106] to bind 2-substituted derivatives of AMP.

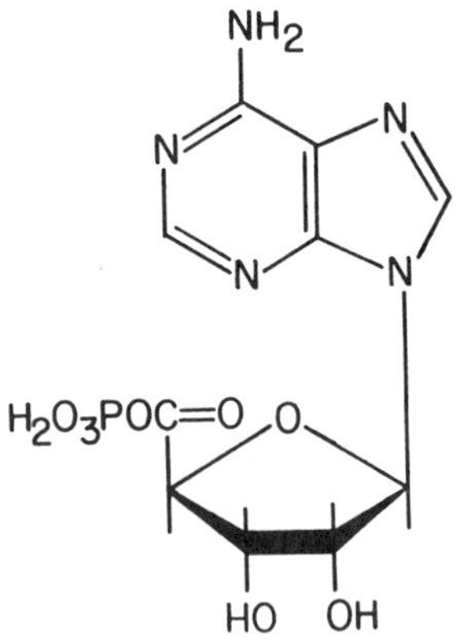

$$NH_2$$

47

3.3. Adenylosuccinate Lyase

Adenylosuccinate lyase (EC 4.3.2.2) catalyzes the final reaction involved in the *de novo* synthesis of AMP. This comprises a reversible cleavage of adenylosuccinic acid **(48)** to AMP and fumaric acid in which the equilibrium favors AMP formation. The enzyme was discovered in yeast by Carter and Cohen[112] and subsequently was found in a wide variety of other species.[113] Studies of initial-rate kinetics and of product inhibition have clearly indicated that, subsequent to cleavage of adenylosuccinate, the predominant order of release of products from the enzyme is fumarate followed by AMP.[114] The evidence is consistent with the view that AMP and adenylosuccinate both bind to the same form of the enzyme, from which it follows that modifications made in the AMP moiety can be

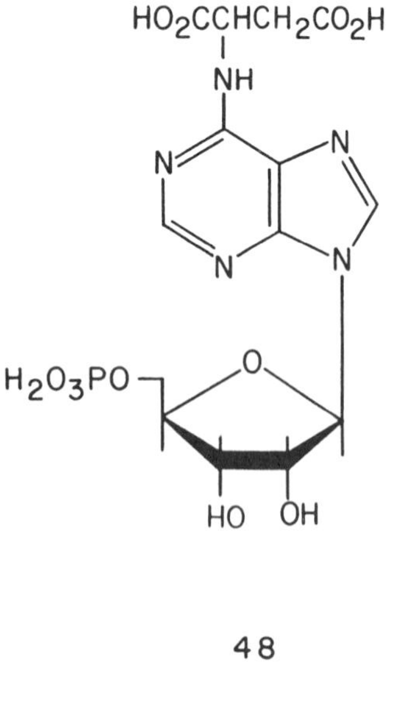

$$HO_2CCHCH_2CO_2H$$

48

expected to exert an effect on the binding of adenylosuccinate that parallels the effect on the binding of AMP itself. The rate of photooxidative inactivation of the enzyme is accelerated by AMP and unaffected by fumarate. On the basis of this observation, it was suggested that binding of AMP induces a conformational change in the enzyme.[115]

3.3.1. Binding of the Adenine Moiety

The pyrimidine ring of AMP is not a requirement for binding because the enzyme efficiently catalyzes the elimination of fumarate from the *de novo* purine nucleotide precursor 5-amino-4-*N*-succinocarboxamidoimidazole ribonucleotide (**49**). A number of lines of evidence support the view that the same enzyme and the same active site are utilized *in vivo* for the two fumarate elimination reactions in purine ribonucleotide biosynthesis that are performed, respectively, on adenylosuccinic acid and 5-amino-4-*N*-succinocarbox-amidoimidazole ribonucleotide.[113] That the 6-amino group of AMP assists binding is indicated by the finding, with the yeast enzyme, that unsubstituted purine ribonucleoside 5′-phosphate and IMP inhibit the cleavage reaction more weakly than AMP.[114] Additionally, with the mouse Ehrlich ascites enzyme, the inhibition constants of IMP, XMP, and GMP were in excess of 200 μM whereas the value for AMP was 19–35 μM.[116]

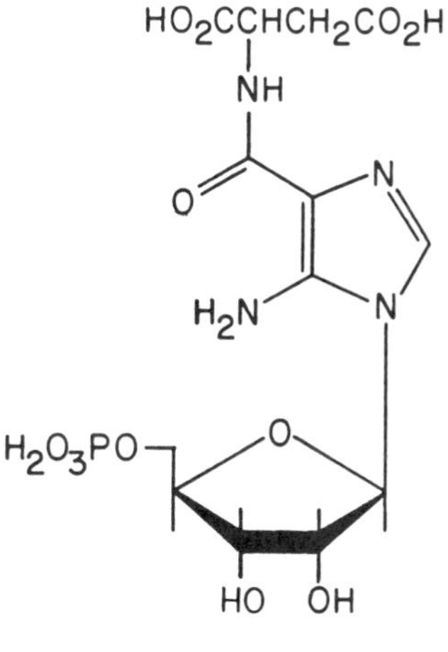

49

3.3.2. Binding of the Ribofuranose Ring

2′-Deoxy-AMP is a substrate of the yeast enzyme,[112] and 2′-deoxyadenylosuccinate is a substrate of the mouse Ehrlich ascites enzyme.[116] Inversion of the 2′-hydroxyl group of adenylosuccinate, producing the arabinofuranosyl analogue, also permits good substrate activity.[116] 2′-Deoxy-AMP and the arabinosyl analogue of AMP are competitive inhibitors of the mouse ascites enzyme with dissociation constants of 5 μM and 87 μM, respectively.[116]

3.3.3. Binding of the Phosphoryl Group

The involvement of the 5'-O-phosphoryl group in the binding of adenylosuccinate to the enzyme is indicated by the lack of substrate activity of
adenosine with the yeast enzyme[112] and by a virtual lack of inhibition of the
cleavage reaction when either adenosine or ATP is present at a level sixfold
higher than that of adenylosuccinate.[114] Adenosine 2'- and 3'-phosphates
likewise are not substrates of the yeast enzyme.[112]

The mixed carboxylic–phosphoric anhydride analogue of AMP (**47**) inactivates *E. coli* adenylosuccinate lyase.[117] Inhibition is prevented by prior hydrolysis of **47** or by prior addition to the enzyme of a near-saturating level of
adenylosuccinate. Relatively high levels of acetyl phosphate showed no tendency
to reversibly or irreversibly inhibit the enzyme. It is possible, therefore, that **47**
may acylate or phosphorylate the enzyme near the site at which the phosphoryl
group of AMP interacts.

4. Substrate Binding of Adenosine 5'-Triphosphate

ATP, in the form of its β,γ magnesium chelate complex, is utilized as a
substrate in numerous phosphoryl-transfer and adenyl-transfer reactions. Elucidation of all atomic interactions between ATP and an enzymatic ATP binding
site has not yet been achieved, although substantial progress toward this goal has
been made. In addition, definitive evidence is beginning to accumulate with
regard to the related question of the conformation of enzyme-bound MgATP
and other nucleoside triphosphates. Much of the evidence pertaining to enzyme–ATP complexes in solution has been derived by the measurement of NMR
relaxation rates of ^{31}P and ^1H nuclei in the presence of Mn^{2+} as a paramagnetic
probe and subsequent calculation of distances between the Mn of an MnATP
and the ^{31}P and ^1H nuclei.[118] More recently, measurement of nuclear Overhauser effects (NOEs) has been shown to be an effective method by which to
calculate interproton distances within enzyme-bound ATP and between ATP
and an enzyme.[119] Evidence regarding the identity of amino acid residues
located at ATP binding sites and the conformation of enzyme-bound ATP has
also been provided in several instances by X-ray diffraction studies of one or
more crystalline forms of an enzyme of known primary sequence with and
without absorbed ATP, ATP analogues, sections of the ATP molecule, or analogues of those sections. In some instances, the binding of ATP produces shifts in
the NMR signals of ATP and/or enzymic amino acid residues which help identify
residues located at the ATP site as well as residues distant from that site which
undergo a change of environment due to ATP-induced modification of enzyme
conformation. Other valuable lines of evidence regarding the conformation of
enzyme-bound ATP and the topography of the ATP binding area have been
obtained from studies of the substrate and inhibitor characteristics of derivatives
and analogues of ATP. An important example of this approach is the use of
Co(III)ATP and Cr(III)ATP complexes for assignment of the configuration of
the β-phosphorus of enzyme-bound MgATP (see Section 4.3).

No attempt is made here to present a comprehensive account of the extensive and often fragmentary information pertaining to substrate binding of ATP. Instead, a summary is given of recent knowledge of the mode of binding of ATP to two phosphotransferases that have been intensively studied. The findings furnish valuable insights into certain aspects of enzyme–ATP interactions and serve also to illustrate the scope of the approaches outlined above.

4.1. Adenylate Kinase

4.1.1. Assignment of the ATP Binding Site

Evidence that adenylate kinase possesses a MgATP–ATP (MgADP–ADP) binding site that is distinct from an AMP (ADP, ATP) binding site is summarized in Section 3.1. Upon addition of either AMP or MgATP, the pig muscle enzyme undergoes conformational changes which are clearly evidenced by shifts in the phenylalanine–tyrosine region of the ^{1}H-NMR spectrum and in the NMR peak assigned to H-2 peak His-36.[74] X-ray diffraction studies have been carried out with two conformationally different crystalline forms ("A" and "B") of the pig muscle enzyme. These forms appear to be related to the conformation of the enzyme in solution before ("B") and after ("A") the substrate-induced conformational change.[2] Binding of the adenosine moiety of ATP was concluded to occur in a pocket lined on one side by Val-67, Leu-69, Val-72, Leu-73, and Leu-76 and on the other by Ile-92, Tyr-95, Arg-97, and Gln-101. Measurements were limited to 0.6-nm resolution due to crystal instability, but model building suggested that Tyr-95 could be hydrogen bonded to N-1 of ATP and that the 2'- and 3'-hydroxyl groups of ATP could be hydrogen bonded to Gln-101 and to the amide carbonyl group of Val-67, respectively. [Other studies (Section 4.1.3.) have indicated that Tyr-95 is also near a phosphate-binding area of the enzyme.] ATP did not bind to the site under the conditions of the above X-ray diffraction studies. This is presumably due to competition with the $3\ M$ ammonium sulfate present inasmuch as the tripolyphosphate portion of ATP is responsible for most of the binding energy of ATP,[70] and the enzyme crystals contain two sulfate ions in the tripolyphosphate binding region.[2] The ATP site was, however, occupied under similar conditions by the powerful bisubstrate inhibitor P^1,P^5-di (adenosine-5')pentapolyphosphate[2] ($K_i = 0.03\ \mu M$ with the pig muscle enzyme [120]).

The above assignment of the binding site for the adenosine portion of ATP is in harmony with several noncrystallographic lines of evidence. For example, it has been observed that, upon binding of ATP, the ^{1}H-NMR signals of H-2 and H-8 undergo an appreciable shift attributable to the effect of ring currents from Tyr-95.[74] The AMP binding site, on the other hand, lacks an aromatic ring, and binding of AMP is not associated with an appreciable shift in the H-2 and H-8 signals of AMP. In addition, the postulated ATP binding site brings the γ phosphate of ATP and the phosphate of AMP close to His-36 in the active site, in agreement with data from ^{1}H-NMR studies that His-36 is in the active site and is located close to the Mn of β,γ-MnATP.[74] Furthermore, the pocket for the adenine portion of the proposed ATP site is relatively shallow and exposed to solvent in comparison with the almost completely enclosed pocket for the adenine ring at the AMP site. In accordance with this, the ATP site accommodates a benzoyl group at N[6], as evidenced by the good substrate activity of N[6]-benzoyl-ATP,[(72)] whereas the AMP site accommodates a methyl group at N[6] with difficulty or not at all (see Section 3.1.1). Finally, adenylate kinase shares a number of structural similarities with phosphoglycerate kinase, and the above assignment locates the ATP binding sites at topologically equivalent positions in both these kinases.[(2)] More recently ^{1}H-NMR studies of enzyme-bound CrATP and MgATP permitted the calculation of distances between the C-2 portion of His-36 and adenine H-2, on the one hand, and Cr^{3+}, on the other.[(121)] These estimates, together with intramolecular NOEs in enzyme-bound MgATP, were found to require an orientation or location of enzyme-bound metal–ATP complex in the pig enzyme that was different from that proposed from the above X-ray diffraction studies on enzyme crystals soaked in salicylate or in P^1,P^5-di(adenosine-5')pentaphosphate.[(2)]

4.1.2. Binding of the Adenine Moiety of ATP

The adenine moiety and, to some extent, the ribofuranose moiety of ATP exhibit less rigorous structural requirements for substrate activity than do the adenine and ribofuranose moieties of AMP. With rabbit muscle adenylate kinase, nucleoside triphosphates showed the decreasing order of reactivity ATP > CTP > GTP > UTP > ITP.[(69)] Magnetic resonance studies have shown that MnATP binds to the pig enzyme approximately 10-fold more tightly than do the Mn^{2+} salts of GTP or tripolyphosphate,[(70)] indicating that the major contributions to binding of MnATP, and hence presumably of MgATP, are made by the adenine and tripolydphosphate moieties. A study of nuclear Overhauser effects indicates that H-2 of ATP is within 0.5 nm of the C-2 proton of His-36 in the pig enzyme and that H-2 and the ribose H-1' are close to specific protons in Arg, Lys, or Leu and methyl protons of one or more Leu, Ile, or Val residues.[(121)] With the rabbit enzyme, intermolecular NOEs indicated that C-2, C-8, and H-1' of enzyme-bound ATP were within 0.4 nm of methyl or methylene groups of Leu, Ile, Lys, Arg, Val, and/or Thr residues.[(119,122)] Evidence for some degree of bulk tolerance at the N-1, C-2, and N[6] positions of ATP is provided by the substrate activities shown by 1,N[6]-etheno-ATP,[(71)] 2-aminopurine ribonucleoside 5'-triphosphate,[(123)] and N[6]-benzoyl-ATP,[(72)] the V_{max}/K_M values of which are 0.03, 0.63, and 0.07, respectively, compared to the V_{max}/K_M value of 1 for ATP. Maguire and co-workers[(80)] have shown that the enzyme converts 2-chloro-ADP to 2-chloro-ATP. The adenosine portion of ATP is apparently not mandatory for catalysis, since inorganic tripolyphosphate has been reported to be a weak substrate.[(124)]

4.1.3. Binding of the Tripolyphosphate Moiety of ATP

ATP-complexed Mg appears to make no contribution to ATP binding, inasmuch as free ATP ($K_i = 35 \mu M^{(70)}$ binds more effectively than MgATP ($K_i = 180 \mu M^{(72)}$ to the pig muscle enzyme. The X-ray diffraction studies with this enzyme have indicated that the phosphate residues of AMP and ATP are bound in a deep cleft lined with six arginine residues and a lysine residue.[2] The basic groups of some of these residues presumably form hydrogen bonds or salt linkages with the phosphate residues, processes which would tend to be aided by narrowing of the cleft in the catalytic cycle to exclude water. Consistent with this view, the crystal form "B" was found to possess a more open phosphate cleft than the crystal form "A."[2] The distance from Cr^{3+} of enzyme-bound tridendate CrATP to the C-2 proton of His-36 at the active site has been calculated from ^{1}H-NMR studies to be 1.29 ± 0.1 nm.[121] The ^{1}H-NMR studies with CrATP and MgATP and the X-ray structure of the enzyme yielded a metal–ATP binding site that accommodated all the NMR-derived distances; the assignment indicates that Lys-21 is near $P(\alpha)$ while Lys-27 is near $P(\beta)$ and $P(\gamma)$.[125]

The phosphate analogue potassium ferrate inactivates the pig enzyme by oxidation of Try-95, protection being afforded by substrates or competitive inhibitors.[126] These findings indicated that Try-95 is located near a phosphate-binding site and is essential for catalytic activity. It was inferred that Cys-25, which also became oxidized, may be near a phosphate-binding site but is not directly involved in catalytic events.[126] Additionally, 5'-O-(2,3-dibromosuc-cinyl)adenosine (**50**) is reported to inactivate pig heart adenylate kinase; the half-life is ca. 20 min. at 37°C with a 4 mM level of **50**.[127] Since AMP and ATP protect against inactivation, and since 5'-O-maleyladenosine is a competitive inhibitor ($K_i = 15$ mM) with respect to AMP,[127] it is possible that **50** may function as an ADP analogue which alkylates an amino acid residue near the phosphate cleft of the enzyme.

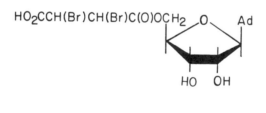

HO$_2$CCH(Br)CH(Br)C(O)OCH$_2$ O   Ad

HO   OH

50

4.1.4. Binding of the Ribofuranose Ring of ATP

The 2'-hydroxyl group of ATP apparently does not contribute to binding (despite the evidence that it could hydrogen bond to Gln-101[2], because the K_i values of Mn^{2+}dATP and free dATP with the pig enzyme are the same as those of Mn^{2+}ATP and free ATP, respectively.[70] 2',3'-O-Ethylidene-ATP (**51**) and 2',3'-O-isopropylidene-ATP (**52**) exhibit approximate 50% as much substrate efficiency (V_{max}/K_M) as does ATP with rabbit or pig muscle adenylate kinases.[73] This finding would suggest a relatively open character for the adenosine pocket at the ATP site[2] and agrees with the conclusion (Section 4.1.2) that the ribofuranose ring as a whole contributes little to the binding of ATP to these enzymes. Substrate and inhibition studies indicate that an acetylaminomethyl group attached in the *allo* configuration to C-5' of ATP (**53**) permits moderate

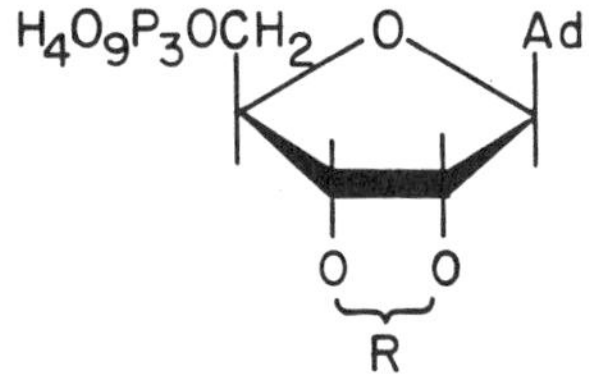

H₄O₉P₃OCH₂ ... O ... Ad

O   O
R

51, R = CHCH₃

52, R = C(CH₃)₂

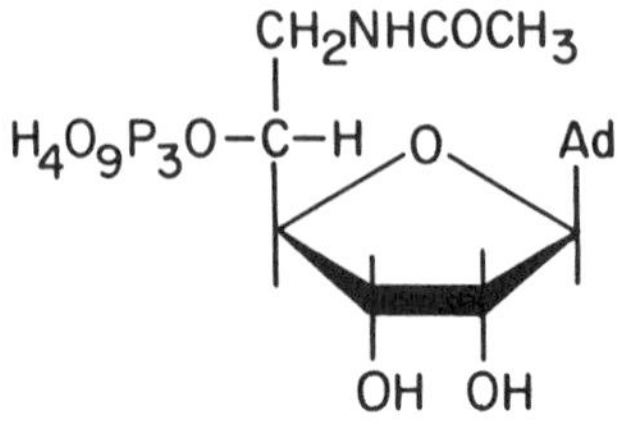

CH₂NHCOCH₃
H₄O₉P₃O–C–H ... O ... Ad

OH OH

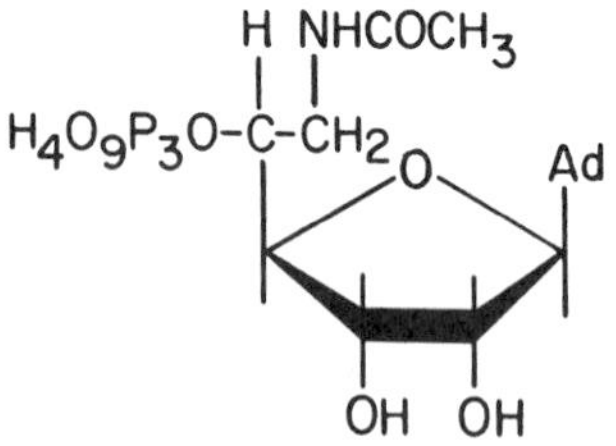

H NHCOCH₃
H₄O₉P₃O–C–CH₂ ... O ... Ad

OH OH

53   **54**

substrate activity (rel. V_{max}, 5%) whereas the same substituent in the *talo* configuration (**54**) permits little or no substrate activity and weakens binding.[73] With rat muscle adenylate kinase, *allo*-5'-C-methyl-ATP was a competitive inhibitor with respect to ATP [$K_M(ATP)/K_i$ = 0.32] whereas its *n*-propyl homologue was a noncompetitive inhibitor [$K_M(ATP)/K_i$ = <0.08]. In contrast, a methyl group in the *talo* configuration at C-5' of ATP produced a weak noncompetitive inhibition whereas an *n*-propyl group in that configuration gave moderately effective competitive inhibition [$K_M(ATP)/K_i$ = 1.9].[128] Replacement of 0-5' in ATP by a CH₂ group drastically reduced the substrate activity and affinity for the ATP site.[128]

4.1.5. Conformation of Enzyme-Bound ATP

8-Bromo-ATP, in which *syn* rotamers around the N(9)—C(1') bond are favored over *anti* rotamers in aqueous solution, is a poor substrate of the rabbit muscle enzyme (V_{max}/K_M 4% that of ATP).[83] This is consistent with an *anti*-type adenine–ribose torsional angle in enzyme-bound ATP. Likewise, larger groups (e.g., methylthio and propylthio) at C-8 of ATP produce weak noncompetitive inhibitors and permit only low substrate activity with rat muscle adenylate kinase. However, they produce good competitive inhibitors [$K_M(ATP)/K_i$ > 1] with good substrate activity (V_{max}/K_M ca. 25% that of ATP) toward the two predominant adenylate kinases of rat liver.[68,129] Nuclear Overhauser effects involving ribofuranose protons and H-8 of MgATP bound to porcine muscle adenylate kinase yielded interproton distances that indicated an *anti* conformation about the glycosidic bond.[121] Similar measurements with the rabbit muscle enzyme also indicated an *anti* glycosyl torsional angle ($\phi_{CN}^{(49)} = 65 \pm 10°$) and a 3' *endo* ribose pucker in enzyme-bound MgATP.[122]

4.2. Pyruvate Kinase

Pyruvate kinase catalyzes a reversible phosphoryl transfer reaction between phosphoenolpyruvate and ADP to give pyruvate and ATP. The muscle isozyme of the rabbit has been employed in the majority of studies of ADP–ATP binding. The reaction requires K^+ and Mg^{2+} or other divalent cations such as Mn^{2+} and Co^{2+}. Studies employing a variety of NMR and kinetic techniques have shown that, in the active site, one divalent cation binds to the enzyme while a second divalent cation forms a β,γ coordination complex with the phosphoryl groups of enzyme-bound ATP or an α,β complex with enzyme-bound ADP.[130]

4.2.1. Binding of the Adenine Moiety

There is little specificity in the interaction between the enzyme and the adenine ring of ADP, which is replaceable by guanine, hypoxanthine, uracil, or cytosine with little loss of substrate activity.[131,132] Likewise, the introduction into ADP of two methyl groups at N^6, replacement of N-7 by carbon, and the introduction of an amino group at C-2 had little effect on substrate activity.[132] N^6-Benzoyl-ADP had little substrate activity yet behaved as a competitive inhibitor with an affinity ($K_i = 630\ \mu M$)[72] comparable to that of ADP ($K_i = 800\ \mu M$).[133] With the rat muscle enzyme, 2-SMe-ADP and 2-NHMe-ADP were good substrates with possibly better affinity for the enzyme than ADP [$K_M(\text{ADP})/K_M > 2$].[134] The substrate properties of 8-substituted ADP derivatives are discussed in Section 4.2.3.

4.2.2. Binding of the Ribofuranose Ring

That the ribose moiety of ADP makes significant contributions to binding is shown by the observation that a replacement of either the 2'- or the 3'-hydroxyl group of ADP by hydrogen decreases the V_{max} by 80–90%.[132] Only minor conformational differences have been detected between AMP and dAMP in aqueous solution,[77] and, since ADP and dADP likewise are presumed to be conformationally similar to each other, the relatively poor substrate activity of dADP supports the notion that catalysis involves an interaction of the enzyme with the 2'-hydroxyl group of ADP. This interaction might occur via the oxygen of the 2'-hydroxyl group, insofar as, with the rat muscle enzyme, the substrate efficiency (V_{max}/K_M) of 2'-O-methyl-ADP is comparable to that of ADP itself.[134] In contrast, 3'-O-methyl-ADP had poor affinity and almost no substrate activity with the rat enzyme. This would suggest a possible role in catalysis of the 3'-hydroxylic hydrogen and/or insufficient space in the ADP–enzyme complex to accommodate the methyl group.[134] With the rabbit enzyme, inversion of either the 2'- or the 3'-hydroxyl group of ADP led to larger reductions in V_{max} than those seen with 2'- or 3'-dADP, and the 2',3'-O-isopropylidene derivative of

ADP showed no substrate activity and was not inhibitory at $7mM$.[132] These findings are also consistent with an interaction between the enzyme and the 2'- and 3' hydroxyl groups of ADP.

The 2',3'-dialdehydic product resulting from the periodate oxidation of ADP combines with and inactivates the rabbit enzyme irreversibly; protection is afforded by MgADP, and MgATP and, less effectively, by phosphoenolpyruvate or pyruvate. Kinetic analysis indicated that inactivation results from a reaction with one group per catalytic subunit, presumably a lysine residue, located near the MgADP binding site.[135] The rabbit enzyme is inactivated by a mixed anhydride that differs structurally from ATP only in that the 5'-methylene group has been replaced by a carbonyl group. This effect was prevented by MgADP, MgATP, or phosphoenolpyruvate and thus appears to be ATP site directed.[110] 5'-O-[p-Fluorosulfonyl)benzoyl]adenosine progressively inactivates the rabbit enzyme with the concomitant derivatization of tyrosine residues; both processes are slowed by MgATP or MgADP.[136] Modified tyrosines are found in three distinct peptides, possibly reflecting the low affinity of the 5' substituent for the MgDP portion of the MgADP binding site.

4.2.3. Conformation of Enzyme-Bound ATP

The conformation of ATP in the Mn^{2+}–enzyme–ATP complex of the rabbit muscle enzyme was partially elucidated by Mildvan and co-workers using NMR relaxation rate studies. These studies provided, to within ±10%, the six distances between enzyme-bound Mn^{2+} and H-2, H-8, H-1', and the three phosphorus atoms of enzyme-bound ATP.[137] The ribose–adenine conformation from these data is an *anti* type in which the plane of the imidazole ring forms an angle (ϕ_{CN})[49] of $-30°$ with the C(1')—O(4') bond. The lack of substrate specificity for the adenine ring (Section 4.2.1) suggests loose binding of the adenine ring to the enzyme and a variety of permissible torsion angles in the enzyme–ADP complex.

8-SMe-ADP, which, like 8-SMe-AMP,[84] probably exists in aqueous solution predominantly with a *syn*-type glycosyl conformation, is not a substrate of the muscle, liver, and kidney isozymes of rat pyruvate kinase and is a weak competitive inhibitor with respect to MgADP.[134] In contrast, 8-NHMe-ADP is an effective substrate (V_{max}/K_M ca. 50% that of ADP) for the three isozymes and appears to have a good affinity for them [K_M(ADP)/K_M = 1–2].[134] 8-NHMe-ADP, like 8-NHMe- and 8-NHEt-AMP,[138] is presumably in an *anti* form stabilized by an intramolecular hydrogen bond between the 8-NH hydrogen and 0-5'), and the above substrate properties of 8-NHMc-ADP hence support an *anti* glycosyl conformation in MgADP bound to the three rat isozymes. 8-NHEt-ADP showed barely detectable substrate activity with the rat isozymes with diminished affinity in two cases, whereas 8-NH(n-Pr)-ADP showed no activity,[134] suggesting steric interference with binding and/or subsequent events in the catalytic cycle by the additional methyl groups.

Phosphate residues in ATP, bound to the rabbit muscle enzyme, were concluded not to be directly coordinated to enzyme-bound Mn^{2+}.[137] Rather the Mn^{2+}–P distances (4.9–5.1 Å) indicated a predominantly second-sphere complex. In light of this, it was suggested that interaction between enzyme-bound

Mn^{2+} and ATP might be promoted by intervening water ligands or by ligands from the enzyme.[137] In addition to the above enzyme-bound divalent cation, a second, less strongly bound divalent cation is required for the activity of pyruvate kinase. The addition of an excess of Mg^{2+} to solutions of the Mg^{2+}–enzyme–ADP complex, or of the Mg^{2+}–enzyme–ATP–oxalate complex, produced changes in the α,β, and γ ^{31}P nuclear magnetic resonances and coupling constants which indicated a direct coordination of a second Mg^{2+} to the α- and β-phosphate residues of ADP and to the β- and γ- phosphate residues of ATP.[139]

The α,β,γ-tridentate coordination complex Cr(III)ATP can substitute for β,γ-Mg(II)ATP, to form a catalytically active Mg^{2+}–pyruvate kinase–Cr(III)ATP–pyruvate complex. This complex has been used as a paramagnetic probe to determine distances from Cr to the three carbons of enzyme-bound pyruvate.[140] Since other magnetic resonance studies had provided distances from the enzyme-bound Mn^{2+} to the carbons of pyruvate, H(2), H(8), H(1′), and the three phosphorus atoms of ATP, it was possible to obtain information regarding the conformations and relative positions of Cr(III)ATP, pyruvate, and the enzyme-bound divalent cation at the active site of pyruvate kinase.[141] X-ray diffraction studies, at 0.6-nm resolution, of crystalline cat muscle pyruvate kinase and its binary complexes with divalent metal cation or with substrates indicated, in agreement with the above magnetic resonance studies, that the divalent cation binds adjacent to ADP and phosphoenolpyruvate and that the Mn(II)ATP site partially overlaps the phosphoenolpyruvate site.[142]

4.3. Configuration of the β Phosphorus of Enzyme-Bound MgATP

Free and enzyme-bound ATP is present *in vivo* mainly as a β,γ coordination complex with Mg (dissociation constant, 14 μM). This complex exists in aqueous solution principally as a rapidly equilibrating mixture of two stereoisomers which differ in configuration at the β phosphorus. The isomers have been designated Δ (**55**) and Λ (**56**), respectively. Studies by Cleland and his associates[143] have shown that the β,γ coordination complexes of Cr(III) and Co(III) with ATP, in contrast to the MgATP complex, are sufficiently stable under nonbasic conditions to permit separation of the individual Δ and Λ isomers. In the case of Cr(III)ATP, these isomers, moreover, can serve as substrates of phosphokinases without undergoing cleavage of the chromium–oxygen bonds, thereby permitting an assignment in such cases of the configuration of the β phosphorus of enzyme-bound Cr(III)ATP, and thus of enzyme-bound MgATP.

The configurations of the Δ and Λ classes of isomers were established as

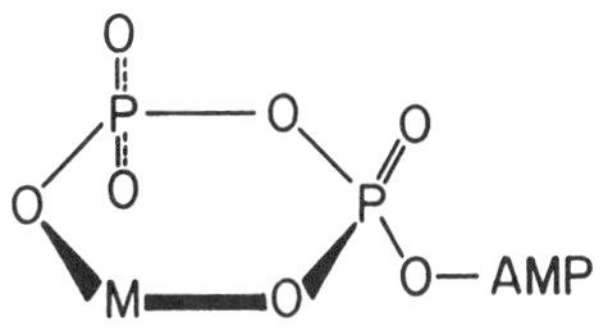

55 (Δ) 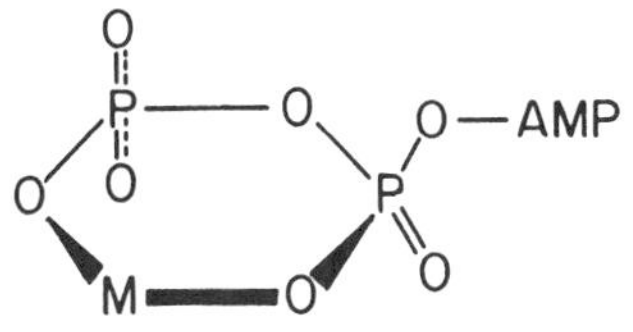   56 (Λ)

follows.[144] The Δ Co(III)(NH$_3$)$_4$ATP isomer was freed of its Λ isomer by selective conversion of the latter with hexokinase to Co(NH$_3$)$_4$ADP·glucose-6-P. The Δ isomer was then oxidized to the corresponding 2′,3′-dialdehyde which, upon treatment with aniline, eliminated Co(NH$_3$)$_4$PPP with a retention of configuration at the β phosphorus. The absolute configuration of Δ Co(NH$_3$)$_4$PPP was established by X-ray crystallography.[145] Identification of the Λ isomer of β,γ-bidentate Cr(III)ATP was made on the basis of its specific substrate activity with hexokinase. The broad usefulness of bidentate CrATP isomers in elucidating the configuration of the β phosphorus of enzyme-bound MgATP is illustrated by results[143] summarized in Table V. When CrATP functioned as a substrate, a complex composed of CrADP and the phosphorylated phosphate acceptor was slowly released from the enzyme.

The Cr(III)ATP complexes are valuable enzymatic probes in a second respect: as exemplified by studies with pyruvate kinase, outlined in Section 4.2, the paramagnetism of chromium enables a calculation of the distance from the chromium of enzyme-bound CrATP to various atoms (e.g., ^{13}C, ^{1}H, or ^{31}P) of a second substrate to be determined from the NMR spectra of those atoms. A number of studies of this type have now been performed by Mildvan and coworkers.[143]

5. General Aspects of Adenine Nucleoside and Nucleotide Binding Sites

5.1. Comparison of Adenosine Binding Sites

The effects of structural alterations in the purine ring of adenosine indicate that a common characteristic of adenosine aminohydrolase and adenosine kinase is that N-1 and N-3 play major roles in substrate recognition and/or transformation. In the case of adenosine aminohydrolase, a similar role is played

Table V. Specificities of Kinases for CrATP Isomers[a]

Kinase	CD assay[b]	^{14}C product assay[c]
Hexokinase	Λ	Λ
Glycerokinase	Λ	Λ
Pyruvate kinase	Δ	Δ
Phosphofructokinase	Δ	
Creatine kinase	—	Λ
Acetate kinase	—	
3-Phospho-glycerate kinase	—	
Myokinase	Δ	Δ
Arginine kinase	—	Λ

[a]From Ref. 143.
[b]Mixed bidentate CrATP isomers were incubated with enzyme and substrate. The specificity for the Λ or Δ isomer was inferred from the direction when a change in ellipticity at 575 nm was observed; —, no activity was observed.
[c]Carbon-14-labeled substrate was incubated with separated bidentate CrATP isomers, and CrADP-phosphorylated substrate complexes were isolated.

also by N-7. Simultaneous interaction of the aminohydrolase with N-1, N-3 and N-7 is presumably important for accurate positioning of the adenine ring for efficient catalysis. It is noteworthy that, in the case of AMP aminohydrolase, an interaction between the enzyme and N-1 and N-7 also appears to be involved in substrate recognition and/or transformation. It is possible that this enzyme, like adenosine aminohydrolase, may interact also with N-3, and this possibility is consistent with the lack of substrate activity of AMP derivatives that bear small substituents at C-2.

The ribose moiety of adenosine, in contrast to the adenine moiety, does not interact in a similar manner with the kinase and the aminohydrolase. With the kinase, the 2′-hydroxyl group of adenosine is essential for substrate activity, and a metabolic conversion of 2′-deoxyadenosine to 2′-deoxy-AMP is carried out by a separate kinase. With adenosine aminohydrolase, on the other hand, substrate activity does not require the presence of a 2′- hydroxyl group but does require an unsubstituted 5′-hydroxyl group.

As discussed in Section 2.2.3, the glycosidic torsional angle of adenosine, when bound to adenosine aminohydrolase, has not been precisely determined but appears to be *anti* in character. The question of the glycosidic torsional angle of adenosine bound to adenosine kinase is less clear. 8-Bromoadenosine is not a substrate of the aminohydrolase but is a substrate (V_{max} 31% that of adenosine) of the kinase.[10] 8-Methoxydenosine, however, is not a substrate of the kinase.[10] This could arise from limited bulk tolerance at C-8 of the kinase-bound adenosine, from steric interference by 8-substituents with attainment of an *anti* conformation, or from a combination of both effects.

5.2. Comparison of Adenosine 5′-Phosphate Binding Sites

The substrate properties of AMP analogues and derivatives indicate that the three major segments of AMP, namely, the adenine and ribofuranose rings and the phosphate group, all interact with AMP aminohydrolase and AMP kinase. The most striking difference in the interaction of AMP with the two enzymes resides in the ribofuranose ring region. Thus, in the case of AMP kinase the ribose hydroxyls each enhance V_{max} by a factor of only 2.5–5, whereas with AMP aminohydrolase the 2′-hydroxyl group enhances V_{max} 500-fold and the 3′-hydroxyl group enhances V_{max} 500-fold. The 3′-hydroxyl group presumably interacts directly with the aminohydrolase during catalysis.

The effect of structural alterations in the adenine ring of AMP show that, for both AMP aminohydrolase and AMP kinase, substrate binding and transformation are seriously hindered by the attachment of small groups of N-1 or C-2. The enzymes do, however, exhibit significant differences in their interactions with the exocyclic amino group and N-7 of AMP. Thus, N^6-methyl-AMP is not a substrate or inhibitor of the kinase, whereas N^6-ethyl-AMP is a substrate of the aminohydrolase. In addition, 7-deaza-AMP is not a substrate of AMP aminohydrolase but does serve as a substrate of AMP kinase. As discussed in Section 5.1, a function of N-7 may be to help establish the required orientation of the adenine ring at the active sites of adenosine aminohydrolase and AMP aminohydrolase.

5.3. Conformation of Free and Enzyme-Bound Adenine Nucleotides

5.3.1. The C(1')—N(9) Torsional Angle

The 5'-(R)-epimer of 8,5'-cyclo-AMP (**41**) is a substrate of snake venom 5'-nucleotidase,[51] suggesting an *anti* glycosidic character in enzyme-bound AMP. This view is consistent with the substrate inertness among AMP derivatives constrained in *syn* conformations by bulky C-8 substituents.[146,147] Proton and ^{13}C-NMR studies indicate that, in an aqueous solution, AMP is present predominantly as *anti* conformers.[48,77] X-ray analysis has shown that, in the crystalline state, simple purine nucleotides are *anti*, with ϕ_{CN} values (Section 2.2.3) in the range of $-43 \pm 20°$.[148]

The 1',9 torsional angle of free ATP in D_2O at pH 6 has been investigated using changes in the chemical shift in ^{1}H- and ^{31}P-NMR spectra induced by lanthanide cations, which bind predominantly to the β and γ phosphoryl groups.[149] In those conformations which best fit the NMR data, the 1',9 torsional angle, ϕ_{CN}, was ca. $-30°$. Studies of time dependence of nuclear Overhauser effects have indicated that, in an aqueous solution, 86% of P(β), P(γ)-bidentate $Co(NH_3)_4$ATP is present as two *anti* conformers with the 1',9 torsional angles centered at ϕ_{CN}-15° and $-55°$, respectively.[150] The 1',9 torsional angles of ATP bound to pyruvate kinase ($\phi_{CN} = -30°$; Section 4.2.3), adenylate kinase ($-65°$; Section 4.1.5), or phosphoribosylpyrophosphate ($-62°$) do not appear to differ significantly from that of free ATP, whereas the 1',9 torsional angles of ATP bound to protein kinase ($-84°$) and RNA polymerase ($-90°$) and of dATP bound to DNA polymerase (($-90°$) do appear to differ significantly from that of free ATP.[151] The 1',9 torsional angle of dATP bound to DNA polymerase is similar to the 1',9 torsional angles of deoxynucleotide residues in double-helical DNA. The 1',9 torsional angle of enzyme-bound dTTP is similar to that of dATP, and it has been suggested that the assumption of this angle by enzyme-bound deoxynucleoside triphosphates could serve to assist error-free duplication of the base sequences of template DNA.[118]

X-ray crystallographic analyses have shown that many enzymes that catalyze reactions involving ATP or NAD possess a common structural feature known as an NAD binding domain.[152] This comprises a polypeptide of approximately 150 residues arranged into four to six parallel strands connected by α-helical sections and is present in certain dehydrogenases[152] and phospho-kinases,[152,153] where it is involved in binding NAD and ATP, respectively. Available information regarding the conformation of NAD when bound to an NAD domain indicates a preference for an *anti* disposition about the adenosine 1',9 bond. Thus, magnetic resonance studies indicate an *anti* 1',9 conformation in solution for NAD bound to yeast or horse liver alcohol dehydrogenase.[154,155] In addition, X-ray analysis has shown that the *anti* conformation is present in ADP-ribose bound to the NAD site of crystalline liver alcohol dehydrogenase[156] and in NAD bound to the respective NAD sites of lactate dehydrogenase,[157] D-glyceraldehyde-3-phosphate dehydrogenase,[158] and malate dehydrogenase.[159] Analyses of NMR chemical shifts and of ORD data indicate that, in D_2O at pH 7, the purine and pyridine rings of NAD are in a stacked conformation in which the 1',9 bond of the adenosine moiety possesses an *anti* torsional angle.[160] It is of interest that propionyl-CoA, when bound to transcarboxylase in solution, is likewise *anti*, as indicated by magnetic resonance studies.[161]

5.3.2. The C(4')—C(5') Torsional Angle

Analysis of H(4')–H(5',5'') NMR coupling constants indicates that, in aqueous solution, the C(4')—C(5') rotamers of AMP resemble the classical 60° staggered type, 55% of AMP being present as the *gauche–gauche* (*gg*) conformer (Fig. 1, A), 25% as the *gauche–trans* (*gt*) conformer (Fig. 1, B), and 20% as the *trans–gauche* (*tg*) conformer (Fig. 1, C). That the preferred 4',5' conformation is *gg* has been confirmed by X-ray structural determinations and by theoretical potential energy calculations.[77] The (*R*)-epimer (**41**) of 8,5'-cyclo-AMP, which possesses a *gt* 4',5' conformation, is a substrate of snake venom 5'-nucleotidase, whereas the (*S*)-epimer (**40**), with a *tg* conformation, is not a substrate, indicating that the 4',5' conformation of AMP bound to 5'-nucleotidase is most likely of the *gt* type.[51]

Information on the 4',5' conformation of enzyme-bound ATP in solution can be expected to become increasingly available through the continued application and development of nuclear magnetic resonance techniques. Interatomic distances derived by these methods for pyruvate kinase-bound Cr(III)ATP appear to rule out *gg* or *tg* conformations about the 4',5' bond and are consistent with a *gt* type[116,162] which is similar to that of AMP bound to 5'-nucleotidase. On the other hand, the 4',5' conformation of the adenosine moiety of NAD bound to crystalline lactate dehydrogenase[157] or malate dehydrogenase[159] appears, from X-ray diffraction studies, to be *tg* rather than *gt*. Nuclear Overhauser enhancements have indicated, within the limits of the resolution achievable, that the adenosine of NAD is bound to yeast and horse liver alcohol dehydrogenase in solution with either a *gt* or *tg* conformation.[155] It is notable that the *gg* conformation is found in all crystal structure determinations of 5'-nucleotides[163] and appears to be the lowest energy conformer in solution but has apparently not yet been observed in an enzyme-bound 5'-nucleotide.

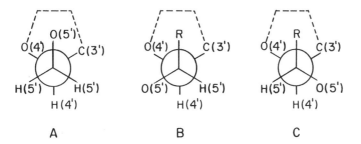

Figure 1. Conformations about the 4',5' nucleoside bond: (A) the *gauche–gauche* C(5')—C(4') nucleoside conformation; (B) the *gauche–trans* conformation of a nucleoside (R = H-5') and of the 5'(*R*)-epimer of an 8,5'-cyclonucleoside (R = C-8); (C) the *trans–gauche* conformation of a nucleoside (R = H-5') and of the (*S*)-epimer of an 8,5'-cyclonucleoside (R = C-8).

5.3.3. The C(5')—O(5') Torsional Angle

The substrate properties of the AMP analogue **39**, in which a *trans* vinyl group replaces the C(5')—O(5') system, indicate (Section 3.1.5) that phosphorus and C-4' might be in a *trans*-type relationship to each other when AMP is bound to AMP kinase.[88] A *trans* disposition about C(5')—O(5') has been concluded to be present also in 85% of free AMP at pH 8 on the basis of NMR studies and conformational energy calculations.[84,164,165]

Relatively little information is available as yet regarding the C(5')—O(5') torsion angle of ADP or ATP bound to enzymes in solution. The *trans* vinyl analogue of ADP, corresponding to the *trans* vinyl phosphonate analogue **39** of AMP, is a substrate of rabbit muscle pyruvate kinase.[88] This would suggest that a conversion of ADP to ATP by this enzyme can proceed with a transoidal form of the C(4')—C(5')-0(5')-P system and without rotation about the C(5')—O(5') bond. The *trans* vinyl ATP analogue corresponding to the AMP analog **39** and its saturated alkyl phosphonate counterpart are inhibitors with affinities for the ATP site of methionine adenosyltransferase that are similar to that of ATP itself, indicating that rotation about the C(5')—O(5') bond of ATP is not required for efficient binding of ATP to this enzyme.[166] The catalytic properties and substrate and inhibitor specificities of the enzyme suggest that it interacts simultaneously with the 2'-hydroxyl group and the triphosphate portion of ATP. On the basis of this view, the above findings were taken to favor a *trans* relationship of C-4' and P(α) about the C(5')—O(5') bond of enzyme-bound ATP.[166] The inhibition constants allow the *trans* character to be as much as 100% of maximal. X-ray diffraction data have indicated a *trans* C(5')—O(5') conformation in crystals of hydrated disodium ATP[167] and *trans* C(5')—O(5') conformations in the adenine nucleotide moiety in the range $\theta = 150-167°$ (*cis* is 0°) in NAD bound to lactate dehydrogenase,[157] malate dehydrogenase,[159] and glyceraldehyde-3-phosphate dehydrogenase[158] and in ADP ribose bound to alcohol dehydrogenase.[156]

5.4. Substrate-Induced Conformational Changes in Nucleotide-Utilizing Enzymes

5.4.1. Enzymes Acting on Purine 5'-Mononucleotides

Adenosine is a weak inhibitor of AMP kinase and produces noncompetitive kinetics with respect to AMP. This and several other lines of evidence (Section 3.1.4) indicate that the enzyme undergoes a conformational change upon binding AMP and that this change might be triggered, at least in part, by the phosphoryl group of AMP. It is of interest that the 5'-phosphoryl group appears to play a crucial role in substrate binding or transformation with a number of other enzymes which catalyze reactions of purine 5'-mononucleotides. Various lines of evidence indicate that many of these enzymes undergo a conformational change induced by the purine nucleotide. Thus, with IMP dehydrogenase, which converts IMP to XMP, and with adenylosuccinate synthetase, which converts IMP to adenylosuccinate, it has been found that inosine (in the presence or absence of inorganic phosphate), ribose 5-phosphate, and inosine 5'-phosphite are all either noninhibitory or weak noncompetitive inhibitors with respect to IMP. This would suggest an IMP-induced conformational change for which both of the two ionizable hydroxyl groups of the phosphate moiety are required.[168] That AMP induces a conformational change in adenylosuccinate lyase is indicated by the observation that the addition of AMP accelerates photooxidative inactivation of the enzyme.[115] In addition, kinetic analysis indicates that prior binding of IMP to IMP dehydrogenase and of AMP to adenylosuccinate lyase is necessary before the binding of second substrates can occur.[169,114] Ribose 5-phosphate is a competitive inhibitor of 5'-nucleotidase but is not a substrate, suggesting that the heterocyclic portion of the nucleotide is required to complete a conformational change in the enzyme.[170] This also appears to be the case with the two IMP-transforming enzymes discussed above. AMP aminohydrolase, which is a multisubunit enzyme, exhibits cooperative kinetics in the absence of activating monovalent cations,[103] an effect which appears to result from a binding of AMP to a regulatory site distinct from the catalytic site.[106] GMP reductase, which mediates the reductive deamination of GMP to IMP, does not utilize guanosine as a substrate, and XMP aminase, which converts XMP to GMP, likewise does not utilize xanthosine as a substrate. Furthermore, GMP reductase of *Aerobacter aerogenes* is subject to rapid GMP-site-directed inactivation by 6-chloropurine 5'-ribonucleotide but is unaffected by the corresponding nucleoside,[171] indicating that the 5'-phosphoryl group of GMP is essential for the binding of GMP to the enzyme. Initial-velocity kinetic analyses and product inhibition studies with this enzyme indicate a random order of substrate addition in which initial binding by either substrate (GMP or NADPH) increases the affinity for the second substrate by a factor of 12,[172] a property again suggestive of nucleotide-induced structural transitions.

5.4.2. Phosphoryl-Transfer Enzymes

In order to account for the remarkably low tendency of phosphoryl-transfer enzymes to promote abortive hydrolysis of phosphoanhydride bonds in ATP, Koshland[173] and Jencks[174] proposed that these enzymes undergo a substrate-induced conformational change which serves to restrict the access of water molecules to the catalytic center. A second effect of such a conformational change would be to enhance the nucleophilicity of the phosphoryl acceptor by removing it from the aqueous environment. Present knowledge of certain aspects of the overall tertiary structures of kinases and of the conformational changes exhibited by kinases has been reviewed by Steitz and co-workers.[175] Evidence from a variety of experimental approaches is providing support for the view that kinases undergo conformational changes in solution upon binding substrates. Notably, the crystal structures of adenylate kinase, hexokinase, phosphoglycerate kinase, pyruvate kinase, and phosphofructokinase have been elucidated. All five kinases are found to possess a deep cleft situated between two lobes, and in each instance the substrates bind in this cleft. In the case of adenylate kinase, AMP and ATP bind at opposite ends of the cleft with their phosphate residues occupying the central section. The cleft of porcine AMP kinase appears, from the crystallographic evidence, to narrow upon substrate binding (Section 4.1.1).[2] In addition, human adenylate kinase appears to undergo a large conformational change upon binding the bisubstrate inhibitor Ap_5A, as indicated by ^{1}H-NMR data.[176] For a second enzyme, yeast hexokinase, crystallographic evidence and data from X-ray scattering studies in solution indicate that the addition of a substrate (glucose) causes the two lobes to move toward one another, almost totally enclosing that substrate.[1] Additionally, NMR studies have indicated the existence of temperature-dependent forms of yeast hexokinase for which thermal transitions from one form to another are influenced by substrate binding.[177] The clefts of hexokinase, AMP kinase, and phosphoglycerate kinase are approximately equal in size and shape, as revealed by computer-generated drawings of space-filling models.[175] Findings in these areas prompted Steitz and co-workers[175,178] to propose that substrate-induced narrowing of a cleft between two lobes will prove to be a general, though by no means an exclusive, property of kinases.

6. References

1. W. S. Bennett and T. A. Steitz, Glucose-induced conformational change in yeast hexokinase, *Proc. Natl. Acad. Sci. U.S.A.* **75**, 4848–4852 (1978).
2. E. F. Pai, W. Sachsenheimer, R. H. Schirmer, and G. E. Schulz, Substrate positions and induced-fit in crystalline adenylate kinase, *J. Mol. Biol.* **114**, 37–45 (1977).
3. E. P. Anderson, in: *The Enzymes* (P. Boyer, ed.), Vol. 9, pp. 46–96, Academic Press, New York (1973).
4. C.-H. Chang, S. Cha, R. W. Brockman, and L. L. Bennett, Jr., Kinetic studies of adenosine kinase from L1210 cells: A model enzyme with a two-site ping-pong mechanism, *Biochemistry* **22**, 600–611 (1983).
5. A. Bloch, in: *Drug Design* (E. J. Ariens, ed.), Vol. 4, pp. 286–378, Academic Press, New York (1974).
6. R. J. Suhadolnik, *Nucleoside Antibiotics*, Wiley-Interscience, New York (1970).
7. T. D. Palella, C. M. Andres, and I. H. Fox, Human placental adenosine kinase. Kinetic mechanism and inhibition, *J. Biol. Chem.* **255**, 5264–5269 (1980).
8. H. P. Schnebli, D. L. Hill, and L. L. Bennett, Purification and properties of adenosine kinase from human tumor cells of type H. Ep. No. 2, *J. Biol. Chem.* **242**, 1997–2004 (1967).
9. B. Lindberg, H. Klenow, and K. Hansen, Some properties of partially purified mammalian adenosine kinase, *J. Biol. Chem.* **242**, 350–356 (1967).
10. R. L. Miller, D. L. Adamczyk, W. H. Miller, C. W. Koszalka, J. L. Rideout, L. M. Beacham, E. Y. Chao, J. J. Haggerty, T. A. Krenitsky, and G. B. Elion, Adenosine kinase from rabbit liver, *J. Biol. Chem.* **254**, 2346–2352 (1979).
11. A. Y. Divekar and M. T. Hakala, Adenosine kinase of sarcoma 180 cells. N^6-Substituted adenosines as substrates and inhibitors, *Mol Pharmacol.* **7**, 663–673 (1971).
12. A. Y. Divekar, M. T. Hakala, G. B. Chheda, and C. I. Hong, 6-Ureido derivatives of 9-[β-D-ribufuranosyl] purine as inhibitors of adenosine kinase, *Biochem. Pharmacol.* **22**, 545–548 (1973).
13. L. L. Bennett, Jr. and D. L. Hill, Structural requirements for activity of nucleosides as substrates for adenosine kinase. Orientation of substituents on the pentofuranosyl ring, *Mol. Pharmacol.* **11**, 803–808 (1975).
14. C.-H. Chang, R. W. Brockman, and L. L. Bennett, Jr., Adenosine kinase from L1210 cells, *J. Biol. Chem.* **255**, 2366–2371 (1980).
15. A. K. Drabikowska, L. Halec, and D. Shugar, Purification and properties of adenosine kinase from rat liver: Separation from deoxyadenosine kinase activity, *Z. Naturforsch* **40**, 34–41 (1985).

16. Y. Yamada, H. Goto, and N. Ogasawar, Purification and properties of adenosine kinase from rat brain, *Biochim. Biophys. Acta* **616**, 199–207 (1980).

17. V. Krygier and R. L. Momparler, Mammalian deoxynucleoside kinases, *J. Biol. Chem.* **246**, 2745–2751 (1971).

18. C. L. Zielke and C. H. Suelter, in: *The Enzymes* (P. D. Boyer, ed.), Vol. 4, pp. 47–78, Academic Press, New York (1971).

19. L. F. Thompson and J. E. Seegmiller, Adenosine deaminase deficiency and severe combined immunodeficiency disease, *Adv. Enzymol.* **51**, 167–210 (1980).

20. G. L. Tritsch (ed.), Adenosine deaminase in disorders of purine metabolism and in immune deficiency, *Ann. N.Y. Acad. Sci.* **451**, 1–345 (1985).

21. S. Frederiksen, Specificity of adenosine deaminase toward adenosine and 2′-deoxyadenosine analogues, *Arch. Biochem. Biophys.* **113**, 383–388 (1966).

22. J. L. York and G. A. LePage, A kinetic study of the deamination of some adenosine analogues, *Can. J. Biochem.* **44**, 331–337 (1966).

23. R. Wolfenden, J. Kaufman, and J. B. Macon, Ring-modified substrates of adenosine deaminases, *Biochemistry* **8**, 2412–2415 (1969).

24. G. Ronca and G. Zucchelli, Competitive inhibition of adenosine deaminase by purine and pyrimidine bases, *Biochim. Biophys. Acta* **159**, 203–205 (1968).

25. L. N. Simon, R. J. Bauer, R. L. Tolman, and R. K. Robins, Calf intestine adenosine deaminase. Substrate specificity, *Biochemistry* **9**, 573–577 (1970).

26. H. P. Baer, G. I. Drummond, and J. Gillis, Studies on the specificity and mechanism of action of adenosine deaminase, *Arch. Biochem. Biophys.* **123**, 172–178 (1968).

27. B. M. Chassy and R. J. Suhadolnik, Adenosine aminohydrolase. Binding and hydrolysis of 2- and 6- substituted purine ribonucleosides and 9-substituted adenine nucleosides, *J. Biol. Chem.* **242**, 3655–3658 (1967).

28. J. G. Cory and R. J. Suhadolnik, Structural requirements of nucleosides for binding by adenosine deaminase, *Biochemistry* **4**, 1729–1732 (1965).

29. M. Ikehara and T. Fukui, Deamination of adenosine analogs with calf intestine adenosine deaminase, *Biochim. Biophys. Acta* **338**, 512–519 (1974).

30. R. Wolfenden, D. F. Wentworth, and G. N. Mitchell, Influence of substituent ribose on transition state affinity in reactions catalyzed by adenosine deaminase, *Biochemistry* **16**, 5071–5077 (1977).

31. J. B. Macon and R. Wolfenden, 1-Methyladenosine. Dimroth rearrangement and reversible reduction, *Biochemistry* **7**, 3453–3458 (1968).

32. R. Wolfenden, On the rate-determining step in the action of adenosine deaminase, *Biochemistry* **8**, 2409–2412 (1969).

33. K. K. Ogilvie, L. Slotin, and P. Rheault, Novel substrate of adenosine deaminase, *Biochem. Biophys. Res. Commun.* **45**, 297–300 (1971).

34. M. Ikehara, S. Uesugi, and K. Yoshida, Studies on the conformation of purine nucleosides and their 5′-phosphates, *Biochemistry* **11**, 830–836 (1972).

35. M. Ikehara, S. Uesugi, and K. Yoshida, Conformation of purine nucleoside pyrophosphates as studied by circular dichroism, *Biochemistry* **11**, 836–842 (1972).

36. P. O. P. Ts'o, in: *Fine Structure of Proteins and Nucleic Acids* (G. D. Fasman and S. N. Timasheff, eds.), p. 49, Marcel Dekker, New York (1970).

37. R. E. Schirmer, J. P. Davis, J. H. Noggle, and P. A. Hart, Conformational analysis of nucleosides in solution by quantitative application of the nuclear Overhauser effect, *J. Am. Chem. Soc.* **94**, 2561–2577 (1972).

38. H. Follman and G. Gremels, Adenine nucleosides in solution: Stabilization of the *anti*-conformation by C-5′ substituents, *Eur. J. Biochem.* **47**, 187–197 (1974).

39. N. J. Leonard, M. A. Sprecker, and A. G. Morrice, Defined dimensional changes in enzyme substrates and cofactors. Synthesis of lin-benzo-adenosine and enzymic evaluation of derivatives of the benzopurines, *J. Am. Chem. Soc.* **98**, 3987–3994 (1976).

40. B. Evans and R. Wolfenden, Potential transition state analog for adenosine deaminase, *J. Am. Chem. Soc.* **92**, 4751–4752 (1970).

41. P. W. K. Woo, H. W. Dion, S. M. Lange, L. F. Dahl, and L. J. Durham, A novel adenosine and Ara-A deaminase inhibitor, (*R*)-3-(2-deoxy-β-D-*erythro*-pentofuranosyl)-3,6,7,8-tetrahydroimidazo[4,5-*d*][1,3]diazepin-8-ol, *J. Heterocycl. Chem.* **11**, 641–643 (1974).

42. A. Bloch, M. J. Robins, and J. R. McCarthy, The role of the 5′-hydroxyl group of adenosine in determining substrate specificity for adenosine deaminase, *J. Med. Chem.* **10**, 908–912 (1967).

43. A. J. Grant and L. M. Lerner, Unusual substrate for adenosine deaminase from calf intestinal mucosa, *Biochim. Biophys. Acta* **525**, 472–476 (1978).

44. R. H. Shah, H. J. Schaeffer, and D. H. Murray, Enzyme inhibitors IV. Syntheses of 6-substituted-9-(5-deoxy-β-D-xylofuranosyl)purines and their evaluation as inhibitors of adenosine deaminase, *J. Pharm. Sci.* **54**, 15–20 (1965).

45. H. J. Schaeffer, D. D. Godse, and G. Liu, Enzyme inhibitors. 3. Syntheses of cis-(6-substituted-9-purinyl)cycloalkylcarbinols as adenosine deaminase inhibitors, *J. Pharm. Sci.* **53**, 1510–1515 (1964).

46. J. Ott and F. Seela, *R*- and *S*-Alkylidene acetals of adenosine: Stereochemical probes for the active site of adenosine deaminase, *Bioorg. Chem.* **10**, 82–89 (1981).

47. H. J. Schaeffer, S. Gurwara, R. Vince, and S. Bittner, Novel substrate of adenosine deaminase, *J. Med. Chem.* **14**, 367–369 (1971).

48. R. Stolarski, C.-E. Hagberg, and D. Shugar, Studies on the dynamic *syn–anti* equilibrium in purine nucleosides and nucleotides with the aid of ¹H and ¹³C NMR spectroscopy, *Eur. J. Biochem.* **138**, 187–192 (1984).

49. J. Donohue and K. N. Trueblood, Base pairing in DNA, *J. MOl. Biol.* **2**, 363–371 (1960).

50. A. Hampton, P. J. Harper, and T. Sasaki, Substrate properties of cycloadenosines with adenosine aminohydrolase as evidence for the conformation of enzyme-bound adenosine, *Biochemistry* **11**, 4736–4739 (1972).

51. L. Dudycz and D. Shugar, Susceptibility to various enzymes of the carbon-bridged (*R*) and (*S*) diastereoisomers of 8,5′-cycloadenosine and their 5′-phosphates, *FEBS Lett.* **107** 363–365 (1979).

52. K. Tomita, T. Tanaka, M. Yoneda, T. Fujiwara, and M. Ikehara, Crystal and molecular structure of *S*-cycloadenosine derivatives, 8,3′-*S*-cyclo-adenosine and 8,2′-*S*-cycloadenosine-3′,5′-cyclic monophosphate, *Acta Crystallogr., Sect. A* **28**, S45 (1972).

53. J. Zemlicka, Formycin anhydronucleosides. Conformation of formycin and conformational specificity of adenosine deaminase, *J. Am. Chem. Soc.* **97**, 5896–5903 (1975).

54. H. J. Schaeffer, E. Odin, and S. Bittner, Enzyme inhibitors. XXIII. Syntheses of 9-(substituted araklyl)-6-substituted purines as inhibitors of adenosine deaminase, *J. Pharm. Sci.* **60**, 1184–1188 (1971).

55. H. J. Schaeffer, M. A. Schwartz, and E. Odin, Enzyme inhibitors. XVII. Kinetic studies on the irreversible inhibition of adenosine deaminase, *J. Med. Chem.* **10**, 688–689 (1967).

56. G. Ronca, M. F. Saettone, and A. Lucacchini, Alkylation of adenosine deaminase by benzylbromoacetate and 9-(*p*-bromoacetamindobenzyl)adenine, *Biochim. Biophys. Acta* **206**, 414–425 (1970).

57. A. Lucacchini, A. D. Bertolini, G. Ronca, D. Segnini, and C. A. Rossi, Affinity labeling of histidine and lysine residues in the adenosine deaminase substrate binding site, *Biochim. Biophys. Acta* **569**, 220–227 (1979).

58. H. J. Schaeffer and D. Vogel, Enzyme inhibitors. IX. Hydrophobic interactions of some 9-alkyladenines with adenosine deaminase. *J. Med. Chem.* **8**, 507–509 (1965).

59. H. J. Schaeffer, D. Vogel, and R. Vince, Enzyme inhibitors. VIII. Studies on the mode of binding of some 6-substituted 9-(hydroxyalkyl)purines to adenosine deaminase, *J. Med. Chem.* **5**, 502–506 (1965).

60. H. J. Schaeffer and R. Vince, Enzyme inhibitors. XVIII. Studies on the stereoselectivity of inhibition of adenosine deaminase by DL-, D-, and L-9-(2-hydroxylpropyl)adenine, *J. Med. Chem.* **10**, 689–691 (1967).

61. H. J. Schaeffer, R. N. Johnson, M. A. Schwartz, and C. F. Schwender, Enzyme inhibitors. 25. An equation to calculate the unknown K_i from two known values of K_i in an *R*, *S*, and *RS* series. Stereoselectivity of inhibition of adenosine deaminase by (*R*)-, (*S*)-, and (*RS*)-9-(1-hydroxy-2-alkyl and aralkyl)adenines, *J. Med. Chem.* **15**, 456–458 (1972).

62. H. J. Schaeffer and C. F. Schwender, Enzyme inhibitors. 26. Bridging hydrophobic and hydrophilic regions on adenosine deaminase with some 9-(2-hydroxy-3-alkyl)adenines, *J. Med. Chem.* **17**, 6–8 (1974).

63. P. W. K. Woo and D. C. Baker, Inhibitors of adenosine deaminase. Studies in combining high-

affinity enzyme-binding structural units. *erythro*-1,6-Dihydro-6-(hydroxymethyl)-9-(2-hydroxy-3-nonyl)purine and *erythro*-9-(2-hydroxy-3-nonyl)purine, *J. Med. Chem.* **25**, 603–605 (1982).

64. L. Noda, in: *The Enzymes* (P. Boyer, ed.), Vol. 8 3rd Edn., pp. 279–305, Academic Press, New York (1973).

65. D. E. Wilson, S. Povey, and H. Harris, Adenylate kinases in man: Evidence for a third locus, *Ann. Hum. Genet.* **39**, 305–313 (1976).

66. W. E. Criss, G. Litwack, H. P. Morris, and S. Weinhouse, Adenosine tri-phosphate: Adenosine monophosphate isozymes in rat liver and hepatomas, *Cancer Res.* **30**, 370–375 (1970).

67. T. K. Pradham and W. E. Criss, Three major forms of adenylate kinase from adult and fetal rat tissues, *Enzyme* **21**, 327–331 (1976).

68. A Hampton and D. Picker, Design of species- or isozyme-specific enzyme inhibitors. 3. Species and isozymic differences between mammalian and bacterial adenylate kinases in substituent tolerance in an enzyme–substrate complex, *J. Med. Chem.* **22**, 1529–1532 (1979).

69. W. J. O'Sullivan and L. Noda, Magnetic resonance and kinetic studies related to the manganese activation of the adenylate kinase reaction, *J. Biol. Chem.* **243**, 1424–1435 (1968).

70. N. C. Price, G. H. Reed, and M. Cohn, Magnetic resonance studies of substrate and inhibitor binding to porcine muscle adenylate kinase, *Biochemistry* **12**, 3322–3327 (1973).

71. J. A. Secrist III, J. R. Barrio, N. J. Leonard, and G. Weber, Fluorescent modification of adenosine-containing coenzymes. Biological activities and spectroscopic properties, *Biochemistry* **11**, 3499–3506 (1972).

72. A. Hampton and L. A. Slotin, Inactivation of rabbit, pig, and carp adenylate kinases by N^6-o- and p-fluorobenzoyladenosine 5′-triphosphates, *Biochemistry* **14**, 5438–5444 (1975).

73. A. Hampton, L. A. Slotin, F. Kappler, T. Sasaki, and F. Perini, Design of substrate-site-directed inhibitors of adenylate kinase and hexokinase. Effect of substrate substituents on affinity for the adenine nucleotide sites, *J. Med. Chem.* **19**, 1371–1377 (1976).

74. G. G. McDonald, M. Cohn, and L. Noda, Protein magnetic resonance spectra of porcine muscle adenylate kinase and substrate complexes, *J. Biol. Chem.* **250**, 6947–6954 (1975).

75. B. D. Rao, M. Cohn, and L. Noda, Differentiation of nucleotide binding sites and role of metal ion in the adenylate kinase reaction by ^{31}P NMR, *J. Biol. Chem.* **253**, 1149–1158 (1978).

76. L. Noda, in: *The Enzymes* (P. Boyer, ed.), Vol. 6 2nd Edn., pp. 139–149, Academic Press, New York (1962).

77. D. B. Davies, Conformations of nucleosides and nucleotides, *Prog. Nucl. Magn. Reson. Spectrosco.* **12**, 135–225 (1978).

78. T. T. Hai, D. Picker, M. Abo, and A. Hampton, Species or isozyme-specific inhibitors. 7. Selective effects in inhibitions of rat adenylate kinase isozymes by adenosine 5′-phosphate derivatives, *J. Med. Chem.* **25**, 806–812 (1982).

79. C. Chachaty, T. Zemb, G. Langlet, and T. D. Son, A proton-relaxation-time study of the conformation of some purine and pyrimidine 5′-nucleotides in aqueous solution, *Eur. J. Biochem.* **62**, 45–53 (1976).

80. G. Gough, M. H. Maguire, and F. Michal, 2-Chloroadenosine 5′-phosphate and 2-chloroadenosine 5′-diphosphate, pharmacologically active nucleotide analogs, *J. Med. Chem.* **12**, 494–498 (1969).

81. S. Takahashi, T. Miyazawa, and S. Higuchi, A conformational study of 5′-AMP and 2-substituted -5′-AMP's in aqueous solution, *Nucleic Acids Res. Spec. Publ.* No. 3, s61–s64 (1977).

82. A. Bloch, R. J. Leonard, and C. A. Nichol, On the mode of action of 7-deaza-adenosine (tubercidin), *Biochim. Biophys. Acta* **138**, 10–25 (1967).

83. I. Lascu, M. Kezdi, I. Goia, G. Jebeleanu, O. Barzu, A. Pansini, S. Papa, and H. H. Mantsch, Enzymatic properties of 8-bromoadenine nucleotides, *Biochemistry* **18**, 4818–4826 (1979).

84. R. H. Sarma, C.-H. Lee, F. E. Evans, N. Yathindra, and M. Sandaralingam, Probing the interrelation between the glycosyl torsion, sugar pucker, and the backbone conformation in C(8) substituted adenine nucleotides by ^{1}H and ^{1}H-{^{31}P} fast Fourier transform nuclear magnetic resonance methods and conformational energy calculations, *J. Am. Chem. Soc.* **96**, 7337–7348 (1974).

85. S. Uesugi and M. Ikehara, Carbon-13 magnetic resonance spectra of 8-substituted purine nucleosides. Characteristic shifts for the syn conformation, *J. Am. Chem. Soc.* **99**, 3250–3253 (1977).

86. R. Stolarski, L. Dudycz, and D. Shugar, NMR studies on the *syn–anti* dynamic equilibrium in purine nucleosides and nucleotides, *Eur. J. Biochem.* **108**, 111–121 (1980).

87. A. Hampton and T. Sasaki, Substrate properties of 8,2'- and 8,3'-*O*-cyclo derivatives of adenosine 5'-monophosphate with adenosine 5'-monophosphate utilizing enzymes, *Biochemistry* **12**, 2188–2191 (1973).

88. A. Hampton, F. Kappler, and F. Perini, Evidence for the conformation about the C(5')-O(5') bond of AMP complexed to AMP kinase: Substrate properties of a vinyl phosphonate analog of AMP, *Bioorgan. Chem.* **5**, 31–35 (1976).

89. A. Hampton, T. Sasaki, and B. Paul, Synthesis of 6'-cyano-6'-deoxy-homoadenosine-6'-phosphonic acid and its phosphoryl and pyrophosphoryl anhydrides and studies of their interactions with adenine nucleotide utilizing enzymes, *J. Am. Chem. Soc.* **95**, 4404–4414 (1973).

90. A. Hampton, F. Perini, and P. J. Harper, Synthesis of homoadenosine-6'-phosphonic acid and studies of its substrate and inhibitor properties with adenosine monophosphate utilizing enzymes, *Biochemistry* **12**, 1730–1736 (1973).

91. A. Hampton, P. Howgate, P. J. Harper, F. Perini, F. Kappler, and R. K. Preston, Interactions of the epimeric 5'-*C*-methyl and 5'-*C*-carbamyl derivatives of adenosine monophosphate with adenosine monophosphate utilizing enzymes, *Biochemistry* **12**, 3328–3332 (1973).

92. K.-F. Sheu and P. A. Frey, Enzymic and phosphorus-31 nuclear magnetic resonance study of adenylate kinase-catalyzed stereospecific phosphorylation of adenosine 5'-phosphorothioate, *J. Biol. Chem.* **252**, 4445–4458 (1977).

93. W. Meyer and H. Follmann, A study of the substrate and inhibitor specificities of AMP aminohydrolase, 5'-nucleotidase, and adenylate kinase with adenosine carboxylates of variable chain length, *Z. Naturforsch.* **35**, 273–278 (1980).

94. S. R. Adams, M. J. Sparks, and H. B. Dixon, The arsonomethyl analogue of adenosine 5'-phosphate. An uncoupler of adenylate kinase, *Biochem. J.* **221**, 829–836 (1984).

95. D. G. Rhoads and J. M. Lowenstein, Initial velocity and equilibrium kinetics of myokinase, *J. Biol. Chem.* **243**, 3963–3972 (1968).

96. C. L. Zielke and C. H. Suelter, in: *The Enzymes* (P. Boyer, ed.), Vol. **4** 3rd Edn., pp. 47–78 Academic Press, New York (1971).

97. B. Cunningham and J. M. Lowenstein, Regulation of adenylate deaminase by adenosine triphosphate, *Biochim. Biophys. Acta* **96**, 535–537 (1965).

98. B. Setlow and J. M. Lowenstein, Adenylate deaminase. II. Purification and some regulatory properties of the enzyme from calf brain, *J. Biol. Chem.* **242**, 607–615 (1967).

99. B. Setlow, R. Burger, and J. M. Lowenstein, Adenylate deaminase. I. The effects of adenosine and guanosine triphosphates on activity and the organ distribution of the regulated enzyme, *J. Biol. Chem.* **241**, 1244–1245 (1966).

100. R. Burger and J. M. Lowenstein, Adenylate deaminase. III. Regulation of deamination pathways in extracts of rat heart and lung, *J. Biol. Chem.* **242**, 5281–5288 (1967).

101. A. G. Chapman, A. L. Miller, and D. E. Atkinson, Role of the adenylate deaminase reaction in regulation of adenine nucleotide metabolism in Ehrlich ascites tumor cells, *Cancer Res.* **36**, 1144–1150 (1976).

102. K. L. Smiley, A. J. Berry, and C. H. Suelter, An improved purification, crystallization, and some properties of rabbit muscle 5'-adenylic acid deaminase, *J. Biol. Chem.* **242**, 2502–2506 (1967).

103. K. L. Smiley and C. H. Suelter, Univalent cations as allosteric activators of muscle adenosine 5'-phosphate deaminase, *J. Biol. Chem.* **242**, 1980–1981 (1967).

104. C. L. Zielke and C. H. Suelter, Substrate specificity and aspects of deamination catalyzed by rabbit muscle 5'-adenylic acid aminohydrolase, *J. Biol. Chem.* **246**, 1313–1317 (1971).

105. A. W. Murray and M. R. Atkinson, Adenosine 5'-phosphorothioate. A nucleotide analog that is a substrate, competitive inhibitor, or regulator of some enzymes that interacat with adenosine 5'-phosphate, *Biochemistry* **7**, 4023–4029 (1968).

106. M. H. Maguire, M. R. Atkinson, and P. G. Tonkes, Activation of rat skeletal muscle 5'-adenylate deaminase by non-substrate analogues of adenosine 5'-phosphate, *Eur. J. Biochem.* **34**, 527–534 (1973).

107. D. B. McCormick and G. E. Opar, Syntheses of 8-bromo-5'-adenylate-containing nucleotides, *J. Med. Chem.* **12**, 333–334 (1969).

108. A. Hampton, T. Sasaki, F. Perini, L. A. Slotin, and F. Kappler, Design of substrate-site-directed

irreversible inhibitors of adenosine 5'-phosphate aminohydrolase. Effect of substrate substituents on affinity for the substrate site, *J. Med. Chem.* **19**, 1029–1033 (1976).

109. C. Frieden, H. R. Gilbert, W. H. Miller, and R. L. Miller, Adenylate deaminase: Potent inhibition by 2'-deoxycoformycin 5'-phosphate, *Biochem. Biophys. Res. Commun.* **91**, 278–283 (1979).

110. A. Hampton, P. J. Harper, T. Sasaki, P. Howgate, and R. K. Preston, Carboxylic–phosphoric mixed anhydrides isosteric with AMP and ATP as reagents for enzymic AMP and ATP sites, *Biochem. Biophys. Res. Commun.* **65**, 945–950 (1975).

111. R. L. Blakely and E. Vitols, The control of nucleotide biosynthesis, *Annu. Rev. Biochem.* **37**, 201–224 (1968).

112. C. E. Carter and L. H. Cohen, The preparation and properties of adenylosuccinase and adenylosuccinic acid, *J. Biol. Chem.* **222**, 17–30 (1955).

113. S. Ratner, in: *The Enzymes* (P. Boyer, ed.), Vol. 7, 3rd Edn. pp. 167–197, Academic Press, New York (1972).

114. W. A. Bridger and L. H. Cohen, The kinetics of adenylosuccinate lyase, *J. Biol. Chem.* **243**, 644–650 (1968).

115. W. A. Bridger and L. H. Cohen, The kinetics of inactivation of adenylosuccinate lyase: Evidence for a substrate-induced conformational change, *Can. J. Biochem.* **47**, 665–672 (1969).

116. T. Spector, Mammalian adenylosuccinate lyase, *Biochim. Biophys. Acta* **481**, 741–745 (1977).

117. A. Hampton and P. J. Harper, The potential of mixed anhydrides as specific reagents for enzymic binding sites for alkyl phosphates. Inactivation of an enzyme by an anhydride isosteric with adenosine 5'-phosphate, *Arch. Biochem. Biophys.* **143**, 340–341 (1971).

118. A. S. Mildvan, Magnetic resonance studies of the conformations of enzyme-bound substrates, *Acc. Chem. Res.* **10**, 246–252 (1977).

119. D. C. Fry, S. A. Kuby, and A. S. Mildvan, Measurement of absolute interproton distances by frequency-dependent nuclear Overhauser effects: Conformation of bound MgATP on adenylate kinase and on a peptide fragment of the enzyme, *Biochemistry* **23**, 3357 (1984).

120. P. Feldhaus, T. Frohlich, R. S. Goody, M. Isakov, and R. H. Schirmer, Synthetic inhibitors of adenylate kinases in the assays for ATPases and phosphokinases, *Eur. J. Biochem.* **57**, 197–204 (1975).

121. G. M. Smith and A. S. Mildvan, Nuclear magnetic resonance studies of the nucleotide binding sites of porcine adenylate kinase, *Biochemistry* **21**, 6119–6123 (1982).

122. D. C. Fry, S. A. Kuby, and A. S. Mildvan, NMR studies of the MgATP binding site of adenylate kinase and of a 45 amino acid fragment of the enzyme, *Fed. Proc.* **43**, 1837 (1984).

123. W. R. McClure and K. H. Scheit, Enzyme kinetic parameters of the fluorescent ATP analog 2-aminopurine triphosphate, *FEBS Lett.* **32**, 267–269 (1973).

124. I. Lieberman, Inorganic triphosphate synthesis by muscle adenylate kinase, *J. Biol. Chem.* **219**, 307–317 (1956).

125. D. C. Fry, S. A. Kuby, and A. S. Mildvan, The ATP-binding site of adenylate kinase: Homology to F_1ATPase, RAS P21, myosin, and other nucleotide binding proteins, *Fed. Proc.* **44**, 672 (1985).

126. M. D. Crivellone, M. Hermodson, and B. Axelrod, Inactivation of muscle adenylate kinase by site-specific destruction of tyrosine 95 using potassium ferrate, *J. Biol. Chem.* **260**, 2657–2661 (1985).

127. J. Berghaeuser and A. Geller, An inactivator for the affinity labeling of adenine nucleotide dependent enzymes, *FEBS Lett.* **38**, 254–256 (1974).

128. F. Kappler, T. T. Hai, M. Abo, and A. Hampton, Species- or isozyme-specific enzyme inhibitors. 8. Synthesis of disubstituted two-substrate condensation products as inhibitors of rat adenylate kinases, *J. Med. Chem.* **25**, 1179–1184 (1982).

129. A. Hampton, F. Kappler, and D. Picker, Species- or isozyme-specific enzyme inhibitors. 4. Design of a two-site inhibitor of adenylate kinase with isozyme selectivity, *J. Med. Chem.* **25**, 638–644 (1982).

130. D. Dunaway-Mariano, J. L. Benovic, W. W. Cleland, R. K. Gupta, and A. S. Mildvan, Stereospecificity of the metal–adenosine 5'-triphosphate complex in reactions of muscle pyruvate kinase, *Biochemistry* **18**, 4347–4354 (1979).

131. K. M. Plowman and A. R. Krall, A kinetic study of nucleotide interactions with pyruvate kinase, *Biochemistry* **4**, 2809–2814 (1965).

132. D. C. Hohnadel and C. Cooper, The effect of structural alterations on the reactivity of the nucleotide substrate of rabbit muscle pyruvate kinase, *FEBS Lett.* **30**, 18–20 (1973).

133. S. Ainsworth and N. Macfarlane, A kinetic study of rabbit muscle pyruvate kinase, *Biochem. J.* **131**, 223–236 (1973).

134. T. T. Hai, M. Abo, and A. Hampton, Species- or isozyme-specific enzyme inhibitors. 9. Selective effects in inhibitions of rat pyruvate kinase isozymes by adenosine 5′-diphosphate derivatives, *J. Med. Chem.* **25**, 1184–1188 (1982).

135. M. V. Hinrichs and J. Eyzaguirre, Affinity labeling of rabbit muscle pyruvate kinase with dialdehyde-ADP, *Biochim. Biophys. Acta* **704**, 177–185 (1982).

136. D. L. DeCamp and R. F. Colman, Identification of critical tyrosyl peptides labeled by 5′-[*p*-(fluorosulfonyl)benzoyl]adenosine in the active site of pyruvate kinase, *Biochemistry* **24**, 3360 (1985).

137. D. L. Sloan and A. S. Mildvan, Nuclear magnetic relaxation studies of the conformation of adenosine 5′-triphosphate on pyruvate kinase from rabbit muscle, *J. Biol. Chem.* **251**, 2412–2420 (1976).

138. F. E. Evans and J. M. Wright, Proton and ^{31}P nuclear magnetic resonance study on the stabilization of the anti conformation about the glycosyl bond of 8-alkylamino adenyl nucleotides, *Biochemistry* **19**, 2113–2117 (1980).

139. R. K. Gupta and A. S. Mildvan, Structures of enzyme-bound metal·nucleotide complexes in the phosphoryl transfer reaction of muscle pyruvate kinase, *J. Biol. Chem.* **252**, 5967–5976 (1977).

140. R. K. Gupta, C. H. Fung, and A. S. Mildvan, Chromium (III)-adenosine triphosphate as a paramagnetic probe to determine intersubstrate distances on pyruvate kinase, *J. Biol. Chem.* **251**, 2421–2430 (1976).

141. A. S. Mildvan, D. L. Sloan, C. H. Fung, R. K. Gupta, and E. Melamud, Arrangement and conformations of substrates at the active site of pyruvate kinase from model building studies based on magnetic resonance data, *J. Biol. Chem.* **251**, 2431–2434 (1976).

142. D. K. Stammers and H. Muirhead, Three-dimensional structure of cat muscle pyruvate kinase at 6 Å resolution, *J. Mol. Biol.* **95**, 213–225 (1975).

143. W. W. Cleland and A. S. Mildvan, in: *Advances in Inorganic Biochemistry* (G. L. Eichorn and L. G. Marzilli, eds.), Vol. 1, pp. 163–191, Elsevier/North Holland, New York (1979).

144. R. D. Cornelius and W. W. Cleland, Substrate activity of (adenosine triphosphato)tetraaminecobalt(III) with yeast hexokinase and separation of diastereoisomers using the enzyme, *Biochemistry* **17**, 3279–3286 (1978).

145. E. A. Merritt, M. Sundaralingam, R. D. Cornelius, and W. W. Cleland, X-ray crystal and molecular structure and absolute configuration of (dihydrogen tripolyphosphate)tetraaminecobalt(III) monohydrate, $Co(NH_3)_4H_2P_3O_{10}\cdot H_2O$. A model for a metal–nucleoside polyphosate complex, *Biochemistry* **17**, 3274–3278 (1978).

146. G. I. Birnbaum and D. Shugar, A purine nucleoside unequivocally constrained in the *syn* form. Crystal structure and conformation of 8-(A-hydroxyisopropyl)adenosine, *Biochim. Biophys. Acta* **517**, 500–510 (1978).

147. R. Pless, L. Dudycz, R. Stolarski, and D. Shugar, Purine nucleosides unequivocally in the *syn* conformation: Guanosine and 5′-GMP with 8-*tert*-butyl and 8-(A-hydroxyisopropyl) substituents, *Z. Naturforsch., C: Biosci.* **33**, 902–907 (1978).

148. M. Sundaralingam, in: *Structure and Conformation of Nucleic Acids and Protein Nucleic Acid Interactions* (M. Sundaralingam and S. T. Rao, eds.), pp. 487–524, University Park Press, Baltimore (1975).

149. P. Tanswell, J. M. Thornton, A. V. Korda, and R. J. P. Williams, Quantitative determination of the conformations of ATP in aqueous solution using the lanthanide cations as nuclear magnetic resonance probes, *Eur. J. Biochem.* **57**, 135–145 (1975).

150. P. R. Rosevear, H. N. Bramson, C. O'Brian, E. T. Kaiser, and A. S. Mildvan, Nuclear Overhauser effect studies of the conformations of tetraamminecobalt(III)-adenosine 5′-triphosphate free and bound to bovine heart protein kinase, *Biochemistry* **22**, 3439–3447 (1983).

151. J. Granot, A. S. Mildvan, and E. T. Kaiser, Studies of the mechanism of action of regulation of cAMP-dependent protein kinase, *Arch. Biochem. Biophys.* **205**, 1–17 (1980).

152. M. G. Rossmann, A. Liljas, C. I. Branden, and L. J. Banaszak, in: *The Enzymes* (P. Boyer, ed.), Vol. **11**, 3rd Edn. pp. 61–102, Academic Press, New York (1975).

153. (a) T. N. Bryant, H. C. Watson, and P. L. Wendell, Structure of yeast phosphoglycerate kinase,

Nature (London) **247**, 14–17, (1974); (b) C. C. F. Blake and P. R. Evans, Structure of horse muscle phosphoglycerate kinase, *J. Mol. Biol.* **84**, 585–601 (1974).

154. D. L. Sloan and A. S. Mildvan, Magnetic resonance studies of the geometry of bound nicotin-amide adenine dinucleotide and isobutyramide on spin-labeled alcohol dehydrogenase, *Biochemistry* **13**, 1711–1718 (1974).

155. A. M. Gronenborn and G. M. Clove, Conformation of NAD$^+$ bound to yeast and horse liver alcohol dehydrogenase in solution, *J. Mol. Biol.* **157**, 155–160 (1982).

156. C.-I. Branden, H. Jornvall, H. Eklund, and B. Furugren, in: *The Enzymes* (P. Boyer, ed.), Vol. **11**, 3rd Edn., pp. 103–190, Academic Press, New York (1975).

157. K. Chandrasekhar, A. McPherson, M. J. Adams, and M. G. Rossman, Conformation of coenzyme fragments when bound to lactate dehydrogenase, *J. Mol. Biol.* **76**, 503–518 (1973).

158. M. G. Rossmann, in: *Structure and Conformation of Nucleic Acids and Protein–Nucleic Acid Interactions* (M. Sundaralingam and S. T. Rao, eds.), pp. 353–374, University Park Press, Baltimore (1975).

159. L. E. Webb, E. J. Hill, and L. J. Banaszak, Conformation of nicotin-amide adenine dinucleotide bound to cytoplasmic malic dehydrogenase, *Biochemistry* **12**, 5101–5109 (1973).

160. R. H. Sarma and N. O. Kaplan, 200 MHz proton nuclear magnetic resonance study of the geometric disposition of the base pairs in the oxidized and reduced pyridine nucleotides, *Biochem. Biophys. Res. Commun.* **36**, 780–788 (1969).

161. C. H. Fung, R. J. Feldmann, and A. S. Mildvan, ^{1}H and ^{31}P Fourier transform magnetic resonance studies of the conformation of enzyme-bound propionyl coenzyme on transcarboxylase, *Biochemistry* **15**, 75–84 (1976).

162. A. Mildvan, private communication.

163. M. Sundaralingam, in: *Symposia in Quantum Chemistry and Biochemistry* (E. Bergmann and B. Pullman, eds.), Vol. V, pp. 417–456, Israel Academy of Sciences and Humanities, Jerusalem (1973).

164. R. H. Sarma, R. J. Mynott, D. J. Wood, and F. E. Hruska, Determination of the preferred conformations constrained along the C(4')-C(5') and C(5')—O(5') bonds of β-5'-nucleotides in solution. Four-bond ^{31}P–^{1}H coupling, *J. Am. Chem. Soc.* **95**, 6457–6459 (1973).

165. F. E. Evans and R. H. Sarma, The intramolecular conformation of adenosine 5'-monophosphate in aqueous solution as studied by fast Fourier transform ^{1}H and ^{1}H–{^{31}P} nuclear magnetic resonance spectroscopy, *J. Biol. Chem.* **249**, 4754–4759 (1974).

166. F. Kappler, T. T. Hai, and A. Hampton, Use of a vinyl phosphonate analog of ATP as a rotationally constrained probe of the C5'—O5' torsion angle in ATP complexed to methionine adenosyl transferase, *Bioorg. Chem.* **13**, 289–295 (1985).

167. O. Kennard, N. W. Isaacs, J. C. Coppola, A. J. Kirby, S. Warren, W. D. S. Motherwell, D. G. Watson, D. L. Wampler, D. H. Chenery, A. C. Larson, K. A. Kerr, and L. R. Sanseverino, Three dimensional structure of adenosine triphosphate, *Nature (London)* **225**, 333–336 (1970).

168. A. W. Nichol, A. Nomura, and A. Hampton, Studies on phosphate binding sites if inosinic acid dehydrogenase and adenylosuccinate synthetase, *Biochemistry* **6**, 1008–1015 (1967).

169. A. Hampton, L. W. Brox, and M. Bayer, Analogs of inosine 5'-phosphate with phosphorus–nitrogen and phosphorus–sulfur bonds. Binding and kinetic studies with inosine 5'-phosphate dehydrogenase, *Biochemistry* **8**, 2303–2311 (1969).

170. D. E. Koshland, Jr., Mechanisms of transfer enzymes, in: *The Enzymes* (P. Boyer, ed.), Vol. 1, pp. 305–346, Academic Press, New York (1959).

171. L. W. Brox and A. Hampton, Inactivation of guanosine 5'-phosphate reductase by 6-chloro-, 6-mercapto, and 2-amino-6-mercapto-9-β-D-ribofuranosylpurine 5'-phosphates, *Biochemistry* **7**, 398–406 (1968).

172. L. W. Brox, Studies on IMP dehydrogenase and GMP reductase, Ph.D. Thesis, University of Alberta (1968).

173. D. E. Koshland, Jr., Application of a theory of enzyme specificity to protein synthesis, *Proc. Natl. Acad. Sci. U.S.A.* **44** 98–104 (1958).

174. W. P. Jencks, Binding energy, specificity, and enzyme catalysis: The Circe effect, *Adv. Enzymol.* **43**, 219–410 (1975).

175. C. M. Anderson, F. H. Zucker, and T. A. Steitz, Space-filling models of kinase clefts and conformation changes, *Science* **204**, 375–380 (1979).

176. H. R. Kalbitzer, R. Marquetant, P. Roesch, and R. H. Schirmer, The structural isomerization of human muscle adenylate kinase as studied by proton nuclear magnetic resonance, *Eur. J. Biochem.* **126**, 531–536 (1982).

177. B. Blicharska, H. Kolockzek, and Z. Wasylewski, A nuclear magnetic resonance study of conformational changes induced by substrate and temperature in bovine liver thiosulfate sulfurtransferase and yeast hexokinase, *Biochim. Biophys. Acta* **708**, 326–329 (1982).

178. T. A. Steitz, M. Shoham, and W. S. Bennett, Jr., Structural dynamics of yeast hexokinase during catalysis, *Phil. Trans. R. Soc. London* **B293**, 43–52 (1981).

Index

DATE DUE

APR 23 1993			